AF753910

MEDICINAL PLANTS OF SOUTH ASIA

MEDICINAL PLANTS OF SOUTH ASIA

NOVEL SOURCES FOR DRUG DISCOVERY

Edited by

MUHAMMAD ASIF HANIF
Department of Chemistry, University of Agriculture, Faisalabad, Pakistan

HAQ NAWAZ
Department of Chemistry, University of Agriculture, Faisalabad, Pakistan

MUHAMMAD MUMTAZ KHAN
Department of Crop Sciences, Sultan Qaboos University, Muscat, Oman

HUGH J. BYRNE
FOCAS Research Institute, Technological University of Dublin, Ireland

ELSEVIER

Elsevier
Radarweg 29, PO Box 211, 1000 AE Amsterdam, Netherlands
The Boulevard, Langford Lane, Kidlington, Oxford OX5 1GB, United Kingdom
50 Hampshire Street, 5th Floor, Cambridge, MA 02139, United States

Copyright © 2020 Elsevier Ltd. All rights reserved.

No part of this publication may be reproduced or transmitted in any form or by any means, electronic or mechanical, including photocopying, recording, or any information storage and retrieval system, without permission in writing from the publisher. Details on how to seek permission, further information about the Publisher's permissions policies and our arrangements with organizations such as the Copyright Clearance Center and the Copyright Licensing Agency, can be found at our website: www.elsevier.com/permissions.

This book and the individual contributions contained in it are protected under copyright by the Publisher (other than as may be noted herein).

Notices
Knowledge and best practice in this field are constantly changing. As new research and experience broaden our understanding, changes in research methods, professional practices, or medical treatment may become necessary.

Practitioners and researchers must always rely on their own experience and knowledge in evaluating and using any information, methods, compounds, or experiments described herein. In using such information or methods they should be mindful of their own safety and the safety of others, including parties for whom they have a professional responsibility.

To the fullest extent of the law, neither the Publisher nor the authors, contributors, or editors, assume any liability for any injury and/or damage to persons or property as a matter of products liability, negligence or otherwise, or from any use or operation of any methods, products, instructions, or ideas contained in the material herein.

Library of Congress Cataloging-in-Publication Data
A catalog record for this book is available from the Library of Congress

British Library Cataloguing-in-Publication Data
A catalogue record for this book is available from the British Library

ISBN: 978-0-08-102659-5

For information on all Elsevier publications visit our website at
https://www.elsevier.com/books-and-journals

Publisher: Susan Dennis
Acquisition Editor: Emily M. McCloskey
Editorial Project Manager: Redding Morse
Production Project Manager: Vignesh Tamil
Cover Designer: Miles Hitchen

Typeset by TNQ Technologies

Working together
to grow libraries in
developing countries
www.elsevier.com • www.bookaid.org

Dedication

The pioneering and dedicated contribution of Prof. Atta-ur-Rahman, FRS (N.I., H.I., S.I., T.I.) [UNESCO Science Laureate, Co-Chairman UN Committee on Science, Technology and Innovation (UNESCAP), President Network of Academies of Science of Islamic Countries (NASIC), Former Federal Minister of Science & Technology/Chairman HEC, Fellow of Royal Society (London), Academician (Foreign member) Chinese Academy of Sciences (CAS)] in the field of promoting higher education in developing countries and his innovative work in natural chemistry led us to dedicate our work to him.

Contents

4. Basil

Farwa Nadeem, Muhammad Asif Hanif, Ijaz Ahmad Bhatti, Muhammad Idrees Jilani, and Rashid Al-Yahyai

5. Bay Leaf

Saima Batool, Rasheed Ahmad Khera, Muhammad Asif Hanif, and Muhammad Adnan Ayub

6. Black Piper

Maryam Khan, Muhammad Asif Hanif, Rafia Rehman, and Ijaz Ahmad Bhatti

7. Caraway

Rafia Javed, Muhammad Asif Hanif, Rafia Rehman, Maryam Hanif, and Bui Thanh Tung

8. Chamomilla

Shaheera Rehmat, Rasheed Ahmad Khera, Muhammad Asif Hanif, Muhammad Adnan Ayub, and Muhammad Zubair

9. Chili Pepper

Saba Idrees, Muhammad Asif Hanif, Muhammad Adnan Ayub, Asma Hanif, and Tariq Mahmood Ansari

10. Chirayita

Asma Seher, Muhammad Asif Hanif, Maryam Hanif, and Asma Hanif

14. Curry Leaf

Saima Batool, Rasheed Ahmad Khera, Muhammad Asif Hanif, Muhammad Adnan Ayub, and Shahabuddin Memon

15. Cypress

Asma Shaheen, Muhammad Asif Hanif, Rafia Rehman, and Asma Hanif

16. Datura

Muhammad Saboor Nayyar, Muhammad Asif Hanif, Muhammad Irfan Mjaeed, Muhammad Adnan Ayub, and Rafia Rehman

17. Damask Rose

Fariha Shabbir, Muhammad Asif Hanif, Muhammad Adnan Ayub, Muhammad Idrees Jilani, and Shafiqur Rahman

28. Himalayan Birch

Muhammad Raffi Shehzad, Muhammad Asif Hanif, Rafia Rehman, Ijaz Ahmad Bhatti, and Khem Raj Bhatta

29. Hollyhock

Muhammad Raffi Shehzad, Muhammad Asif Hanif, Rafia Rehman, Ijaz Ahmad Bhatti, and Asma Hanif

30. Horseweed

Haq Nawaz, Muhammad Asif Hanif, Rafia Rehman, and Radosław Kowalski

31. Indian Globe Thistle

Naima Tariq, Anam Waheed, Muhammad Irfan Majeed, Muhammad Asif Hanif, Rafia Rehman, and Mohamed Eddouks

35. Lemon Balm

Kinza Waheed, Haq Nawaz, Muhammad Asif Hanif, Rafia Rehman, and Isiaka A. Ogunwande

36. Ma-Huang

Adan Iqbal, Rasheed Ahmad Khera, Muhammad Asif Hanif, Muhammad Adnan Ayub, and Muhammad Nadeem Zafar

37. Mango

Farooq Khalid, Haq Nawaz, Muhammad Asif Hanif, Rafia Rehman, and Abdullah Mohammed Al-Sadi

38. Moringa

Farwa Nadeem, Muhammad Asif Hanif, Ijaz Ahmad Bhatti, and Shahzad Maqsood Ahmed Basra

42. Puncture Vine

43. Saffron

44. Sesame

45. Sweet Lemon

46. Tomato

Kinza Waheed, Haq Nawaz, Muhammad Asif Hanif, and Rafia Rehman

47. Tree Turmeric

Ammara Shamim, Haq Nawaz, Muhammad Asif Hanif, Muhammad Idrees Jilani, and Ubirajara Lanza Júnior

48. Vanilla

Hafsa Ahmad, Rasheed Ahmad Khera, Muhammad Asif Hanif, Muhammad Adnan Ayub, and Muhammad Idrees Jilani

49. Walnut

Zarghouna Chudhary, Rasheed Ahmad Khera, Muhammad Asif Hanif, Muhammad Adnan Ayub, and Lamia Hamrouni

50. Yarrow

Fariha Ijaz, Haq Nawaz, Muhammad Asif Hanif, and Paulo Michel Pinheiro Ferreira

Contributors

Hassan Imran Afridi National Centre of Excellence in Analytical Chemistry, University of Sindh, Jamshoro, Pakistan

Christian Agyare Department of Pharmaceutics (Microbiology Section), Faculty of Pharmacy and Pharmaceutical Sciences, Kwame Nkrumah University of Science and Technology, Kumasi, Ghana

Hafsa Ahmad Department of Chemistry, University of Agriculture, Faisalabad, Pakistan

Shahzad Maqsood Ahmed Basra Department of Agronomy, University of Agriculture, Faisalabad, Pakistan

Abdullah Mohammed Al-Sadi Department of Crop Sciences, College of Agricultural and Marine Sciences, Sultan Qaboos University, Muscat, Oman

Rashid Al-Yahyai Department of Crop Sciences, Sultan Qaboos University, Muscat, Oman

Tariq Mahmood Ansari Institute of Chemical Sciences, Bahauddin Zakariya University, Multan, Pakistan

Muhammad Adnan Ayub Department of Chemistry, University of Okara, Okara, Pakistan

Muhammad Waqar Azeem Department of Chemistry, University of Agriculture, Faisalabad, Pakistan

Saima Batool Department of Chemistry, University of Agriculture, Faisalabad, Pakistan

Khem Raj Bhatta Department of Chemistry, St. Xavier's College Maitighar, Kathmandu, Nepal

Ijaz Ahmad Bhatti Department of Chemistry, University of Agriculture, Faisalabad, Pakistan

Yaw Duah Boakye Department of Pharmaceutics (Microbiology Section), Faculty of Pharmacy and Pharmaceutical Sciences, Kwame Nkrumah University of Science and Technology, Kumasi, Ghana

Zarghouna Chaudhry Department of Chemistry, University of Agriculture, Faisalabad, Pakistan

R.M. Dharmadasa Industrial Technology Institute, Colombo, Sri Lanka

Mohamed Eddouks Faculty of Sciences and Techniques Errachidia, Moulay Ismail University, Errachidia, Morocco

Paulo Michel Pinheiro Ferreira Department of Biophysics and Physiology, Laboratory of Experimental Cancerology, Federal University of Piauí, Teresina, Brazil

Lamia Hamrouni Laboratory of management and valorisation of forest resources, INRGREF- University of Carthage, Ariana, Tunisia

Muhammad Asif Hanif Department of Chemistry, University of Agriculture, Faisalabad, Pakistan

Maryam Hanif Department of Chemistry, University of Agriculture, Faisalabad, Pakistan

Asma Hanif Department of Chemistry, University of Agriculture, Faisalabad, Pakistan

Abdullah Ijaz Hussain Department of Chemistry, Government College University, Faisalabad, Pakistan

Saba Idrees Department of Chemistry, University of Agriculture, Faisalabad, Pakistan

Bazgha Ijaz Department of Chemistry, University of Agriculture, Faisalabad, Pakistan

Fariha Ijaz Department of Chemistry, University of Agriculture, Faisalabad, Pakistan

Adan Iqbal Department of Chemistry, University of Agriculture, Faisalabad, Pakistan

Zunaira Irshad Department of Chemistry, University of Agriculture, Faisalabad, Pakistan

Rafia Javed Department of Chemistry, University of Agriculture, Faisalabad, Pakistan

Muhammad Idrees Jilani Department of Chemistry, University of Sahiwal, Sahiwal, Pakistan

Ubirajara Lanza Júnior Federal University of Goiás – Special Academic Unit of Health Sciences-Regional Jataí, Course of Biomedicine, Campus Jatobá – University City-Brazil

Chandra Prakash Kala Ecosystem & Environment Management, Indian Institute of Forest Management, Bhopal, India

Farooq Khalid Department of Chemistry, University of Agriculture, Faisalabad, Pakistan

Ayesha Khalil Department of Chemistry, University of Agriculture, Faisalabad, Pakistan

Maryam Khan Department of Chemistry, University of Agriculture, Faisalabad, Pakistan

Muhammad Mumtaz Khan Department of Crop Sciences, Sultan Qaboos University, Muscat, Oman

Oli Khan Department of Botany, Bangabasi College, Kolkata, India

Sunil Khan Department of Botany, Haripal Vivekananda College, Hooghly, India

Selima Khatun Department of Botany, Government General Degree College, Simgur, Hooghly, India

Rasheed Ahmad Khera Department of Chemistry, University of Agriculture, Faisalabad, Pakistan

Radosław Kowalski Department of Analysis and Evaluation of Food Quality, University of Life Sciences in Lublin, Lublin, Poland

Mohamad Fawzi Mahomoodally Department of Health Sciences, Faculty of Science, University of Mauritius, Mauritius

Muhammad Irfan Majeed Department of Chemistry, University of Agriculture, Faisalabad, Pakistan

Najma Memon National Centre of Excellence in Analytical Chemistry, University of Sindh, Jamshoro, Sindh, Pakistan

Shahabuddin Memon National Centre of Excellence in Analytical Chemistry, University of Sindh, Jamshoro, Pakistan

Muhammad Mubeen Mohsin Department of Chemistry, University of Agriculture, Faisalabad, Pakistan

Ayesha Mushtaq Department of Chemistry, University of Agriculture, Faisalabad, Pakistan

Zahid Mushtaq Department of Biochemistry, University of Agriculture, Faisalabad, Pakistan

Farwa Nadeem Department of Chemistry, University of Agriculture, Faisalabad, Pakistan

Raziya Nadeem Department of Chemistry, University of Agriculture, Faisalabad, Pakistan

Haq Nawaz Department of Chemistry, University of Agriculture, Faisalabad, Pakistan

Muhammad Saboor Nayyar Department of Chemistry, University of Agriculture, Faisalabad, Pakistan

Huma Naz Department of Chemistry, University of Agriculture, Faisalabad, Pakistan

Shafaq Nisar Department of Chemistry, University of Agriculture, Faisalabad, Pakistan

Isiaka A. Ogunwande Natural Products Research Unit, Department of Chemistry, Faculty of Science, Lagos State University, Ojo, Lagos, Nigeria

Anca Racoti The National Institute for Research & Development in Chemistry and Petrochemistry – ICECHIM, Bucharest, Romania

Muhammad Raffi Shehzad Department of Chemistry, University of Agriculture, Faisalabad, Pakistan

Shafiqur Rahman Department of Pharmaceutical Sciences, South Dakota State University, Brookings, SD, United States

Rafia Rehman Department of Chemistry, University of Agriculture, Faisalabad, Pakistan

Shaheera Rehmat Department of Chemistry, University of Agriculture, Faisalabad, Pakistan

Muhammad Riaz Department of Chemistry, University of Sargodha, Sargodha, Pakistan

Mehrez Romdhane Laboratoire de Recherche: Energie, Eau, Environnement et Procédés, ENIG, University of Gabès, Tunisia

Shumaila Saif Department of Chemistry, University of Agriculture, Faisalabad, Pakistan

Sidra Sarwar Department of Chemistry, University of Agriculture, Faisalabad, Pakistan

Asma Seher Department of Chemistry, University of Agriculture, Faisalabad, Pakistan

Fariha Shabbir Department of Chemistry, University of Agriculture, Faisalabad, Pakistan

Asma Shaheen Department of Chemistry, University of Agriculture, Faisalabad, Pakistan

Shahzad Ali Shahid Chatha Department of Chemistry, Government College University, Faisalabad, Pakistan

Ammara Shamim Department of Chemistry, University of Agriculture, Faisalabad, Pakistan

Alexander Shikov St-Petersburg Institute of Pharmacy, Leningrad region, Vsevolozhsky district, Russia

Kiran Soomro National Centre of Excellence in Analytical Chemistry, University of Sindh, Jamshoro, Pakistan

Sajjad Hussain Sumrra Department of Chemistry, University of Gujrat, Pakistan

Naima Tariq Department of Chemistry, University of Agriculture, Faisalabad, Pakistan

Vahid Tavallali Department of Agriculture, Payame Noor University (PNU), Tehran, Iran

Bui Thanh Tung Department of Pharmacology and Clinical Pharmacy, School of Medicine and Pharmacy, Vietnam National University, Hanoi, Vietnam

Kinza Waheed Department of Chemistry, University of Agriculture, Faisalabad, Pakistan

Anam Waheed Department of Chemistry, University of Agriculture, Faisalabad, Pakistan

Shahzia Yousaf Department of Chemistry, University of Agriculture, Faisalabad, Pakistan

Muhammad Nadeem Zafar Department of Chemistry, University of Gujrat, Gujrat, Pakistan

Muhammad Zubair Department of Chemistry, University of Gujrat, Gujrat, Pakistan

About the Editors

Muhammad Hanif

Dr. Hanif is an Associate Professor and Group Leader at the Nano and Biomaterials lab within the Department of Chemistry at the University of Agriculture, Faisalabad, Pakistan. His work focuses on natural products and their analysis using advanced analytical chemistry, and he has published over 150 peer-reviewed journal articles, 10 books/manuals, and several book chapters. In addition, Dr. Hanif has supervised over 50 MPhil students and five PhD students. For outstanding contributions he has made to scientific development through the application of basic and applied scientific research, particularly in the field of chemical sciences (chemistry), and his unmatched services to the community as the benchmark of excellence, the Pakistan Academy of Sciences awarded him the prestigious "Gold Medal" in 2019. He holds 2 Patents on Natural Products from Portugal and Spain.

Affiliations and Expertise

Associate Professor and Group Leader, Nano and Biomaterials Lab, Department of Chemistry, University of Agriculture, Faisalabad, Pakistan

Haq Nawaz

Dr. Nawaz is an Assistant Professor with the Nano and Biomaterials Lab within the Department of Chemistry at the University of Agriculture, Faisalabad, Pakistan. His research focuses on natural products and their analysis using advanced organic chemistry and spectroscopy, and he has published over 35 peer-reviewed journal articles and several book chapters.

Affiliations and Expertise

Assistant Professor, Nano and Biomaterials Lab, Department of Chemistry, University of Agriculture, Faisalabad, Pakistan

Muhammad Mumtaz Khan

Muhammad Mumtaz Khan is an Associate Professor of Crop Production/Horticulture in the Department of Crop Sciences, College of Agriculture and Marine Sciences, Sultan Qaboos University, Oman. He has over 30 years' experience in research and teaching crop science with a particular focus on horticultural crops and nutraceutical horticultural plants. He has published more than 150 peer-reviewed journal and conference papers, edited one book on Lime and several book chapters. Since he started his academic career, he has taught a range of undergraduate/postgraduate courses and mentored a large number of MSc and PhD candidates. Due to his pioneer work and ability to express novel findings and his services to science/agriculture and the community, the Government of Pakistan awarded him the prestigious Presidential Award "Izazi Fazeelat" in 2007.

Affiliations and Expertise
Associate Professor of Crop Sciences/Horticulture, Department of Crop Sciences, College of Agriculture and Marine Sciences, Sultan Qaboos University, Muscat, Oman

Hugh J. Byrne
Prof. Byrne is Head of the FOCAS Research Institute at the Technological University of Dublin, Dublin, Ireland. He has over 30 years' experience in research science and has published over 300 peer-reviewed journal and conference papers. He has been responsible for over €25 million in funded projects and has supervised over 40 PhD students. His principal research interests are in applications of spectroscopy and the study of molecular and nanomaterials, biospectroscopy for diagnostics, cytological analysis, and nano–bio interactions.

Affiliations and Expertise
Head, FOCAS Research Institute, Technological University of Dublin, Ireland

Preface

Medicinal plants are an integral part of local heritage, with extraordinary global importance. Nowadays, our understanding of the synergism between herbs, drugs, and foodstuff is still in its infancy. Medicinal plants have been used traditionally in Desi, Unani, Chinese, Ayurveda, Siddha, Arabian, African, Korean, ancient Iranian, Latin American, and several other traditional medicinal systems. There are more than 35,000 plants species being used by various human cultures around the world for medicinal purposes. The World Health Organization reported that four billion people (80% of the world's population) rely mainly on traditional medicine, and a major part of the traditional therapies involve the use of plant extracts or their active constituents. Plants for which positive effects over consistent use have been found, tested on animals and birds. However, there is very little data relating to a standardized dosage available from traditional practitioners, which is problematic to chemists and pharmacists. Lack of information on dosage from traditional and orthodox medicine is considered an obstacle toward improvement of our understanding of the phytochemical components and their interactions. In recent years, an increased methodical interest in plant phytochemical (fruit, herb, spices, and vegetables) health benefits has been an important subject of plant-based research. Although the study of plant compounds is not new, scientists are only now starting the characterization of bioactive compounds to explore their impact on human health and disease. The uncertainty in the exact number of the species within the genus is largely attributed to the great variability in morphology, growth habit, flower color, leaves, and stem and chemical composition among the constituent species. Medicinal plants are available at many health foods stores, though the substantial scientific evidence for their usefulness in human health is inadequate. *Medicinal Plants of South Asia: Novel Sources for Drug Discovery* provides comprehensive and valuable information on 50 medicinal plants in easily understandable and similar style. It is a review of medicinal plants of this region, highlighting chemical components of high potential and applying the latest technology to reveal the underlying chemistry and active components of traditionally used medicinal plants. Drawing on the vast experience of its expert editors and authors, the book provides a contemporary guide source on these novel chemical structures, thus making it a useful resource for medicinal chemists, phytochemists, pharmaceutical scientists, and everyone involved in the use, sales, discovery, and development of drugs from natural sources. This book is also a step forward for obtaining substantial scientific evidence with the help of identification of active ingredients of medicinal plants.

Acknowledgments

This book would not have been possible without the support of our families who helped us to spare extra time to complete this tome.

CHAPTER

1

Alkanet

Asma Shaheen[1], *Muhammad Asif Hanif*[1], *Rafia Rehman*[1], *Muhammad Idrees Jilani*[2], *Alexander Shikov*[3]

[1] Department of Chemistry, University of Agriculture, Faisalabad, Pakistan; [2] Department of Chemistry, University of Sahiwal, Sahiwal, Pakistan; [3] St-Petersburg Institute of Pharmacy, Leningrad region, Vsevolozhsky district, Russia

OUTLINE

https://doi.org/10.1016/B978-0-08-102659-5.00001-X

Copyright © 2020 Elsevier Ltd. All rights reserved.

1. BOTANY

1.1 Introduction

Alkanet (*Anchusa officinalis L.*) is a plant belonging to the Boraginaceae family (Ozer et al., 2010). It is known with different common names such as alkanet, *Anchusa tinctoria*, orchanet in English, racine d'alcanna and racine d'orcanette in French, Dyers's Bugloss, radix anchusea (tinctoriae) in Latin, and rote ochsenzungenwurzel, alkannawurzel, and schminkwurzel in German. It is indigenous to the Mediterranean region. Alkanet and many other related plants have been used as henna and as a dye (Fig. 1.1). Ratanjot (Hindi, Urdu name) was used for the addition of natural, deep red color in various North Indian dishes. Now, synthetic red colors have replaced it in many places. In Tamil, it is called as vembalampattaiwhile. In English, it is known as alkanet. Alkanet roots are specifically used as a natural red dye, and it is grown extensively in Europe for this purpose. Alkanet flowers are bright blue in color. Roots of the plant are dark red, appearing blackish from the outside but blue-red inside with a whitish central part. A fine material of red color produced

FIGURE 1.1 Alkanet used in herbal medicine.

from the roots has been used as a dye since ancient times in the Mediterranean region. Root powder that is used as dyestuff is insoluble in water but soluble in organic solvents such as ether, alcohol, and oils. It also imparts color to varnishes, wines, vegetable oils, and alcoholic tinctures. When alkanet root powder is mixed with oil, it imparts a crimson color, but when applied on wood, the color of the wood changes to dark red-brown and accentuates the wood grain (Papageorgiou et al., 2008; Roeder, 1995). Alkanet root is also used for the treatment of various ailments like wounds, bites, pain, fever, and stings (Papageorgiou et al., 2008).

1.2 History/Origin

Its history goes back to at least 70 CE. It is called burlgoss, orachanet, *A. tinctoria*, or *Anchusa tinctoria*. In Latin, it means "used for dyeing or staining," and *Anchusa* comes from the Greek word anchousa, meaning "paint." It was reported by the Greek physician Hippocrates that alkanet root was used for treatment of skin ulcer from 460 to 370 BC. It was also used as dye and medicine by the botanist Theophrastus from 371 to 287 BC. Alkanet plant properties were also studied by Greek physician pharmacologist Dioscorides from 49–90 AD. Dioscorides described the properties of the red dyestuff in more detail (Papageorgiou et al., 2008).

1.3 Demography/Location

The genus *Alkanna* belongs to the Boraginacea family, which consists of 25 species widely distributed in the Mediterranean region and Asia. *A. tinctoria* has wide geographic distribution. It grows in arid maritime areas of southern Europe. The plant is a perennial herb with a prostrate bushy stem, blooming between March and May with small blue flowers. The propagation of the plant occurs from seeds, and percentage of seed germination is very small. Alkanet is indigenous to some parts of Turkey, southern Europe, and Hungry. It is also cultivated in other parts of Britain, Europe, and Northern Africa. This plant is also grown in and imported from Turkey, Albania, Egypt, and India (Roeder, 1995). In Pakistan, it is most commonly produced in Baluchistan, Landi Kotal, Chitral, Swat, and Kaghan.

1.4 Botany, Morphology, Ecology

The morphology of *A. tinctoria* (alkanet) was studied by Jeri and El-Gad. No information about the morphology of *A. tinctoria* was found in the literature except for some morphologic properties such as leaf shape, stem shape, corolla, anthers structure, bracts, and nutlet. Color and size of flower were used as taxonomic characters for the determination of specie (Gutierrez et al., 2004). It is a biennial or perennial herbaceous plant having a height of 0.3–0.6 m. It has blue to purple flowers of trumpet shape. Its dry, cylindrical, fissured rhizome contains brittle, exfoliating, and dark purple bark pointing toward the outside, and the remains of thick stem and leaf pieces are found close to the crown region (Bisset, 1994). The flower contains 4–5 mm calyxes, but in fruit, it is 5–6 mm. The corolla is blue and glabrous outside. The funnel is slightly long as compared to the calyx. There are five stamens, and anthers are fused with a corolla tube. The nutlets are 2 mm in diameter, irregularly reticulate, and tuberculate (Gutierrez et al., 2004). Alkanet is a short-bristled, perennial half-rosette shrub (Van Vleet, 2015). The stems are 10–20 cm, procumbent or ascending, and glandular. The basal leaves are 6–15 cm by 0.7–1.5 cm, linear-lanceolate, and lower ones are cauline, oblong-linear, and cordate at the base. The brackets are slightly longer than the calyx and oblong-lanceolate. The neck of the root is covered with remains of leaves and stems. The root is spindle-shaped, curved up to 25 cm long, and 1.5 cm thick, with purplish root bark. Alkanet plant requires fairly fertile, humus-rich, moist but well-drained soil. At room temperature, seeds germinate within 1–3 weeks. If sown in July, flowering will take place in spring. Plants require full sun and moist soil. Depending on conditions, common alkanet is a short-lived perennial or biennial, forming a rosette of leaves in the first year and flowering in the second year. The plant is 1–4 ft tall and is hardy down to −30°F or −34°C (Bisset, 1994).

2. CHEMISTRY

A red color pigment is present in alkanet root, especially in the bark, up to 5%–6%. The pigment contains fat-soluble naphthazarin (5, 8-dihydroxy-1, 4-naphthaquinone) constituents like alkannin and other related esters. These are soluble in fatty acids. Pyrrolizidine alkaloids are present in the plant root, but their level is still unknown (Papageorgiou and Digenis, 1980). Monounsaturated fatty acids (FAs) in seeds were found from 8.83% to 55.32% in oleic, from 0.22% to 6.21% in eicosenoic, from 0.04% to 8.94% in erucic, and from 0.08% to 2.71% for nervonic acid. While polyunsaturated FAs were found in between 1.41% and 68.44% for linoleic, 0.12% to 43.0% for a-linolenic, 0.04% to 24.03% for c-linolenic, and 0.02% to 14.59% for stearidonic acid. However, the total polyunsaturated FAs (13.91%–68.78%), monounsaturated (10.59%–73.28%), and saturated (9.3%–23.7%) varied significantly. Total unsaturated FAs ranged from 70.12% to 90.29% (Özcan, 2008).

Gas chromatography (GC) and GC-mass spectrometry (GC-MS) analysis showed the presence of 27 different compounds in alkanet essential oil, representing 93.32% of the total oil (Ozer et al., 2010). Major chemical compounds detected in the extracted essential oil of alkanet were α-terpinyl acetate, pulegone, isophytol, and 1, 8-cineole (Vuralb, 2004). *Alkanna* roots contain pyrrolizidine alkaloids that protect the plant against insects and herbivores. Other plant organs also have pyrrolizidine alkaloids, but it is more concentrated in the roots of these plants. It also contains phenolic compounds that contribute to quality as well as nutritional value in terms of modifying aroma color, flavor, and taste and in providing beneficial health effects. They are also responsible for plant defense mechanisms that counteract the reactive oxygen species to survive and provide protection from molecular damage and damage by herbivores, microorganisms, and insects (Chojkier, 2003). Important constituents of alkanet are shown in Fig. 1.2.

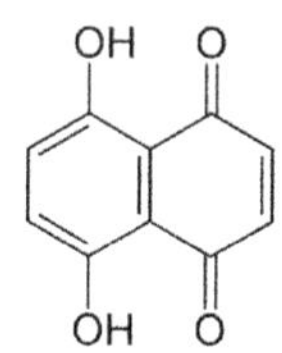

Naphthazarin

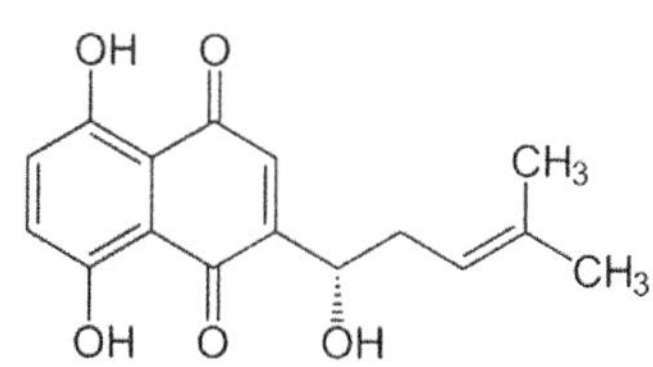

Alkannin

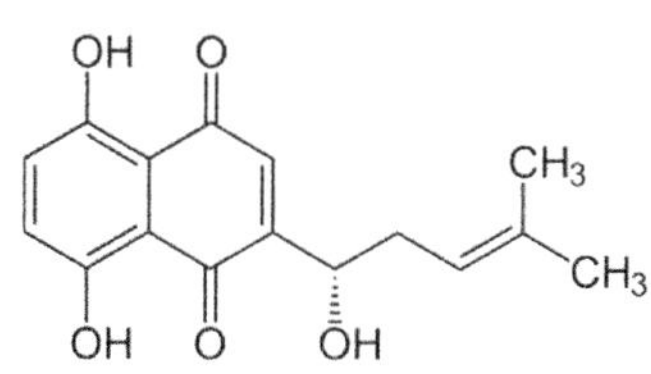

Shikonin

FIGURE 1.2 Important chemical compounds of alkanet.

3. POSTHARVESTING TECHNOLOGY

Roots, leaves, and flowers of alkanet are used by humans. The more appropriate time for harvesting is July and August (Barton and Castle, 1845). Alkanet root is used for various purposes. These roots are ground into fine powder, and oil is also extracted from alkanet roots. The whole process is done with due care from the initial to final stage. Safe and reliable packing is provided for alkanet root powder and other products. High hygienic levels are insured and human touch is avoided. Proper packing helps with easy and safe delivery of the product (Benli and Bahtiyari, 2015; Tung et al., 2013).

4. PROCESSING

Like other herbal plants, alkanet roots are used in a variety of ways for various purposes. In addition to the roots, its leaves are also used for different purposes. Different parts of alkanet plant are washed, dried, and then chopped into small pieces (Wendakoon et al., 2012). After slicing the plant is dried and mashed into powder. Oil can also be prepared from alkanet root. To make oil, the dried roots of alkanet are cut into small pieces and then crushed with the fingers. Coconut oil is poured over the crumbled roots of alkanet and kept under the sun. The oil color will change to red. After the color changes to deep red, the oil is preserved in a glass bottle (Gutierrez et al., 2004).

5. VALUE ADDITION

Natural dyes are extracted from alkanet for hair dyes, textiles, and the food industry. Alkanet root is mainly used as a natural dyeing agent, and it imparts a dark red color to natural wood, fibers, stone, wool, lip balm, soap, lipstick, lotion, ointments, vinegar, wine, tinctures, and varnishes. In the past, roots of alkanet were used for the improvement of appearance of low-quality wines. However, roots of alkanet are now primarily used as a dyeing agent and are not recommended for use internally. In soaps, roots of alkanet impart different shades of purple, pink, and blue depending on the amount, types of oil, and the soap alkalinity (Kourounakis et al., 2002). This plant also has a number of applications in various herbal health products because of the following features. Essential oil extracted from the alkanet root is helpful in reducing sleeping disorder such as chronic worries and insomnia. Roots of alkanet have proven to be very effective in increasing hair strength, as alkanet is a natural hair color dye that is very safe and has no effect on health. It has remarkable antibacterial as well as antiviral activities that are helpful in protecting essential body organs, especially skin. Roots of alkanet are quite beneficial for healing skin fungi. Roots of alkanet are beneficial in preventing nail cracking and also in reducing the inflammation that can happen to nails. These are also helpful in curing herpes. The natural cooling effect of alkanet root helps to cool a fever like traditional Ayurvedic healing does. The hypotensive effect of alkanet root is helpful in reducing cardiovascular system stress and also in reducing high blood pressure. Roots of alkanet are helpful in protecting the skin from any type of infection and recovering from inflammation. Because of the antiinflammation activity and the cooling effect of alkanet root, it is commonly used to heal burn scars. Hence, alkanet roots are useful as sunblock as well as sunburn relief (Jaradat et al., 2018; Papageorgiou and Assimopoulou, 2003).

6. USES

It contains a red pigment used in cosmetics. Its roots are used for the treatment of various skin diseases and wounds. Roots are also used for diarrhea and gastric ulcers. These also have radical scavenging activity and antiaging effects (Esfahani et al., 2012). In herbal medicine, alkanet is used as an expectorant (a substance that brings up phlegm). It helps in morphews, leprosy, jaundice, and spleen and kidney troubles. It stays fluxes of the belly, kills worms, helps calm fits, strengthens the back and eases its pain, and helps bruises and blemishes of the skin such as measles and small pox. It is also said to be a useful counter to poisons and venoms (Papageorgiou et al., 1999). Herbal medicines play an important role treating several diseases throughout the world. Therapeutic efficacy as well as less adverse effects of herbal products make them a good alternative for synthetic drugs. Isohexanol naphthazarin, a chemical constituent of alkanet plant, is responsible for the antitumor, antibacterial, radical scavenging, antigonadotropic, and antithrombosis properties of the alkanet plant (Kourounakis et al., 2002). Alkanet is used externally in the treatment of indolent ulcer, varicose veins, itching rashes, and bed sores. It is good for ulcers, hot inflammation, and burns. Alkanet is a healing herb used before antiseptics. A decoction of root and leaves of alkanet is helpful in curing inveterate coughs and all chest disorders. Alkanet tea is also specified for melancholy, to promote sweating, soothe the skin, break a fever, and as a blood purifier, diuretic, and astringent. The herb is also useful for the preparation of a homeopathic remedy frequently used in alleviating ulcers in the duodenum and stomach. The herb powder mixed with petroleum gel is applied as a balm for back pain or bruising. The root of alkanet has antibacterial properties and also has the property of relieving itching (Gunther, 1968).

7. PHARMACOLOGICAL USES

7.1 Wound Healing

Alkanet roots have been used for treatment of skin wounds and diseases. The root powder possesses antimicrobial properties (Sengul et al., 2009). The external layer of the root contains alkannin and shikonin, which are chiral pairs of naturally occurring isohexanol naphthazarins. Pharmaceutical formulations contain alkannin and shikonin as active ingredients that have wound healing properties. These compounds show activity against fungi, gram-negative bacteria, and gram-positive bacteria. Alkannin also shows bactericidal action against *Pseudomonas aeruginosa,* which forms biofilms against wound healing (Papageorgiou et al., 2008). A study was done on male rabbits with partial thickness, severe, and olive oil burns to see the effects of alkanet roots. When a 16% solution of alkanet root was applied on animal burn wounds, it was observed that it took 7–10 days to heal partial thickness burn wounds and 26 days in healing olive oil burn wounds. Though, severe burn wounds remained unresponsive (Ogurtan et al., 2002). A clinical study on 72 ulcus cruris patients showed that esteric pigments have outstanding antibiotic and wound healing properties (Papageorgiou, 1978).

7.2 Antiinflammatory and Analgesic Effects

Antiinflammatory characteristics of alkanet have also been considered for the preparation of traditional paste as well as ointment for wound bandages. Furthermore, the antinociceptive and analgesic affects were used in Iranian folk medicine for the treatment of inflammatory and pain-related diseases. Analgesic and antiinflammatory activities of ethanolic extracts of alkanet roots were evaluated by carrageenan-induced paw edema in rats and formalin testing in mice (Esfahani et al., 2012).

7.3 Antidiabetic Activity

Synthetic drugs like sulfonylurea and biguanides are used to lessen hyperglycemia in diabetes mellitus, but these drugs have many side effects such as hyperglycemic coma and hepatorenal disturbances and also are not safe for pregnancy. In a previous study, alkanet roots were used for antidiabetic activity. Alkanet roots were dried and then ground to a coarse powder and extracted with methanol using a percolator extraction method. The extract was concentrated by using reduced pressure. Antidiabetic effects of alkanet roots were tested on rats. Before inducing diabetes, weight and blood glucose level of animals of all groups were determined. It was observed that blood glucose level was reduced by 23.3% in normal fasted rats at single dose after 3 h administration of the plant extract (Kumar and Gupta, 2010).

7.4 Anticancer Activity

Naphthoquinone is the major bioactive compound isolated from the alkanet plant. It was found that some naphthoquinone compounds have anticancer activity, but information about the structural-functional relationship of naphthoquinone compound is limited (Kapadia et al., 1997). Anticolorectal cancer cell proliferation activities of alkannin and angelylalkannin (naphthoquinone compound) isolated from *A. tinctoria* were observed. Both compounds were found to have different anticancer potential, and the cancer cell inhibition of these two naphthoquinone compounds was related to the cell cycle arrest and apoptosis induction. This information encouraged the researchers to further isolate naphthoquinone compounds from this plant (Huu Tung et al., 2013). Eight naphthoquinone compounds were isolated and purified from *A. tinctoria*. Structure elucidation confirmed that three of them are novel chemical compounds (Tung et al., 2013). Besides low cost, novel natural chemical compounds were found effective against cancer, and their toxicity was found to be lower compare to commonly used chemotherapeutic drugs. There are several anticancer compounds present in *A. tinctoria* that may be used alone or as an adjunct to existing chemotherapy to improve efficacy and reduce drug-induced toxicity (Huu Tung et al., 2013).

7.5 Antioxidant Activity

Essential oil of alkanet has in vitro antioxidant activity potential. GC-MS analysis revealed that there are 27 different chemical compounds present in the oil, representing 93.32% of the total oil. These oil contents have antioxidant activity and contribute to disease prevention. A literature survey reported that root extracts of this plant shows antioxidant as well as some other biologic activities (Ozer et al., 2010).

7.6 Radical Scavenging Activity

Shikonin and alkannin are pharmaceutical substances present in alkanet roots with a wide spectrum of biologic properties. Radical scavenging activity is involved in anticancer, aging processes, wound healing, and antiinflammatory activities. Radical scavenging activity of shikonin and alkannin, both monomeric and oligomeric, and *A. tinctoria* root extract was studied and structure-activity relationship was determined. It was found that both oligomeric and monomeric alkannin and shikonin exhibit tremendously high radical scavenging activity (Assimopoulou and Papageorgiou, 2005). The presence of naphthoquinone components seems to be vital for that activity. Organic solvents and roots of *A. tinctoria* that contain active ingredients showed very good antiradical activity. Shikonin, alkannin, and their esters present in *A. tinctoria* root extract could be used auspiciously in cosmetic and pharmaceutical preparations because of their radical scavenging activity (Huu Tung et al., 2013).

7.7 Anti-MDR Bacterial Activity of *Alkanna tinctoria*

Leaves of *A. tinctoria* were used efficiently against multiple drug resistant (MDR) bacteria by diffusion method. In every plate of four extracts, i.e., hexane, aqueous, ethanol, and chloroform, *A. tintoria* leaves were poured (Esfahani et al., 2012). It was found that leaf extract of *A. tinctoria* proved effective against MDR isolates. Phytochemical analysis of *A. tinctoria* leaves revealed that these have versatile biochemical molecules that may kill or inhibit the drug-resistant human pathogenic bacteria (Khan et al., 2015).

7.8 Antiviral Action

Alkanet roots contain alkannin, a chemical compound that is very effective against viral action. Many pharmaceutical ointments like Helixderm and Histoplastin red have alkannin as the active ingredient. Alkannin shows antiviral action against the herpes simplex virus (Sengul et al., 2009).

7.9 Antiaging Effects

Alkanet roots have significant antiaging effects. Monomeric and oligomeric alkannin both exhibit antiaging activities. At room temperature an olive oil extract containing *A. tinctoria* possessed antiaging activity; however, when it was heated, this activity decreased. It was (Bisset, 1994; Ozer et al., 2010) concluded that *Alkanna* roots have many other effects in cosmetics beyond serving solely as a dye (Huu Tung et al., 2013).

8. SIDE EFFECTS AND TOXICITY

Roots of alkanet may cause pulmonary hypertension, acute liver failure, pneumonitis, cirrhosis, or heart failure (Mattocks, 1968; Roeder, 1995). These roots contain pyrrolizidine alkaloid. Hepatic metabolism of pyrrolizidine alkaloids releases toxic byproducts that are transported to the lungs and cause pulmonary toxicity. Veno-occlusive disease (sinusoidal obstruction syndrome) is a hepatic complication associated with transplantation of bone marrow that may occur in those patients who consume pyrrolizidine alkaloid–containing products. Pyrrolizidine alkaloids were found to be carcinogenic in animals, specifically associated with squamous cell, hepatocellular, liver angiosarcomas, and carcinomas. Pyrrolizidine alkaloids are substrates for the cytochrome P450 3A4 isoenzyme. Rifampin and phenobarbital are inducers of this enzyme that increase the pyrrolizidine alkaloid conversion to toxic metabolites (Huxtable, 1990). To study the toxicologic effect of pyrrolizidine alkaloid, an experiment was done on rats, and it was concluded that offspring with suckling young rats develop pyrrolizidine alkaloid–induced hepatotoxicity more than their mothers (Schoental, 1968).

References

Assimopoulou, A., Papageorgiou, V., 2005. Radical scavenging activity of *Alkanna tinctoria* root extracts and their main constituents, hydroxynaphthoquinones. Phytotherapy Research 19, 141–147.

Barton, B., Castle, T., 1845. The British Flora Medica; or, History of Themedicinal Plants of Great Britain. Oxford University, London.

Benli, H., Bahtiyari, M.İ., 2015. Combination of ozone and ultrasound in pretreatment of cotton fabrics prior to natural dyeing. Journal of Cleaner Production 89, 116–124.

Bisset, N.G., 1994. Herbal Drugs and Phytopharmaceuticals: A Handbook for Practice on a Scientific Basis. Medpharm Scientific Publishers, Stuttgart xvi, 566p. ISBN 3887630254 En Originally published in German (1984).(EBBD, 190000550).

Chojkier, M., 2003. Hepatic sinusoidal-obstruction syndrome: toxicity of pyrrolizidine alkaloids. Journal of Hepatology 39, 437–446.

Esfahani, H.M., Esfahani, Z.N., Dehaghi, N.K., Hosseini-Sharifabad, A., Tabrizian, K., Parsa, M., Ostad, S.N., 2012. Anti-inflammatory and anti-nociceptive effects of the ethanolic extracts of *Alkanna frigida* and *Alkanna orientalis*. Journal of Natural Medicines 66, 447–452.

Gunther, R., 1968. The Greek Herbal of Dioscorides Illustrated by a Byzantine A.D.

Gutierrez, S., Ang-Lee, M.K., Walker, D.J., Zacny, J.P., 2004. Assessing subjective and psychomotor effects of the herbal medication valerian in healthy volunteers. Pharmacology Biochemistry and Behavior 78, 57–64.

Huu Tung, N., Du, G.J., Wang, C.Z., Yuan, C.S., Shoyama, Y., 2013. Naphthoquinone components from *Alkanna tinctoria* (L.) Tausch show significant antiproliferative effects on human colorectal cancer cells. Phytotherapy Research 27, 66–70.

Huxtable, R.J., 1990. Activation and pulmonary toxicity of pyrrolizidine alkaloids. Pharmacology & Therapeutics 47, 371–389.

Jaradat, N.A., Zaid, A.N., Hussen, F., Issa, L., Altamimi, M., Fuqaha, B., Nawahda, A., Assadi, M., 2018. Phytoconstituents, antioxidant, sun protection and skin anti-wrinkles effects using four solvents fractions of the root bark of the traditional plant *Alkanna tinctoria* (L.). European Journal of Integrative Medicine 21, 88–93.

Kapadia, G.J., Balasubramanian, V., Tokuda, H., Konoshima, T., Takasaki, M., Koyama, J., Tagahaya, K., Nishino, H., 1997. Anti-tumor promoting effects of naphthoquinone derivatives on short term Epstein-Barr early antigen activation assay and in mouse skin carcinogenesis. Cancer Letters 113, 47–53.

Khan, U.A., Rahman, H., Qasim, M., Hussain, A., Azizllah, A., Murad, W., Khan, Z., Anees, M., Adnan, M., 2015. *Alkanna tinctoria* leaves extracts: a prospective remedy against multidrug resistant human pathogenic bacteria. BMC Complementary and Alternative Medicine 15, 127.

Kourounakis, A.P., Assimopoulou, A.N., Papageorgiou, V.P., Gavalas, A., Kourounakis, P.N., 2002. Alkannin and shikonin: effect on free radical processes and on inflammation-A preliminary pharmacochemical investigation. Archiv der Pharmazie 335, 262–266.

Kumar, N., Gupta, A.K., 2010. Wound-healing activity of *Onosma hispidum* (Ratanjot) in normal and diabetic rats. Journal of Herbs, Spices & Medicinal Plants 15, 342–351.

Mattocks, A., 1968. Toxicity of pyrrolizidine alkaloids. Nature 217, 723–728.

Ogurtan, Z., Hatipoglu, F., Ceylan, C., 2002. The effect of Alkanna tinctoria Tausch on burn wound healing in rabbits. Deutsche Tierärztliche Wochenschrift 109, 481–485.

Özcan, T., 2008. Analysis of the total oil and fatty acid composition of seeds of some *Boraginaceae taxa* from Turkey. Plant Systematics and Evolution 274, 143–153.

Ozer, M.S., Sarikurkcu, C., Tepe, B., Can, S., 2010. Essential oil composition and antioxidant activities of alkanet (*Alkanna tinctoria* subsp. tinctoria). Food Science and Biotechnology 19, 1177–1183.

Papageorgiou, V., 1978. Wound healing properties of naphthaquinone pigments from Alkanna tinctoria. Experientia 34, 1499–1501.

Papageorgiou, V., Assimopoulou, A., Ballis, A., 2008. Alkannins and shikonins: a new class of wound healing agents. Current Medicinal Chemistry 15, 3248–3267.

Papageorgiou, V., Digenis, G., 1980. Isolation of two new alkannin esters from *Alkanna tinctoria*. Planta Medica 39, 81–84.

Papageorgiou, V.P., Assimopoulou, A.N., 2003. Lipids of the hexane extract from the roots of medicinal *Boraginaceous* species. Phytochemical Analysis 14, 251–258.

Papageorgiou, V.P., Assimopoulou, A.N., Couladouros, E.A., Hepworth, D., Nicolaou, K., 1999. The chemistry and biology of alkannin, shikonin, and related naphthazarin natural products. Angewandte Chemie International Edition 38, 270–301.

Roeder, E., 1995. Medicinal plants in Europe containing pyrrolizidine alkaloids. Die Pharmazie 50, 83–98.

Schoental, R., 1968. Toxicology and carcinogenic action of pyrrolizidine alkaloids. Cancer Research 28, 2237–2246.

Sengul, M., Yildiz, H., Gungor, N., Cetin, B., Eser, Z., Ercisli, S., 2009. Total phenolic content, antioxidant and antimicrobial activities of some medicinal plants. Pakistan journal of pharmaceutical sciences 22, 102–106.

Tung, N.H., Du, G.-J., Yuan, C.-S., Shoyama, Y., Wang, C.-Z., 2013. Isolation and chemopreventive evaluation of novel naphthoquinone compounds from *Alkanna tinctoria*. Anti-Cancer Drugs 24.

Van Vleet, S.M., 2015. Invasive Weeds of Eastern Washington.

Vuralb, M., 2004. γ-Linolenic acid content and fatty acid composition of Boraginaceae seed oils. European Journal of Lipid Science and Technology 106, 160–164.

Wendakoon, C., Calderon, P., Gagnon, D., 2012. Evaluation of selected medicinal plants extracted in different ethanol concentrations for antibacterial activity against human pathogens. Journal of Medicinally Active Plants 1, 4.

CHAPTER

2

Asthma Weed

Bazgha Ijaz[1], *Muhammad Asif Hanif*[1], *Muhammad Adnan Ayub*[2], *Asma Hanif*[1], *Zahid Mushtaq*[3]

[1] Department of Chemistry, University of Agriculture, Faisalabad, Pakistan; [2] Department of Chemistry, University of Okara, Okara, Pakistan; [3] Department of Biochemistry, University of Agriculture, Faisalabad, Pakistan

OUTLINE

Medicinal Plants of South Asia
https://doi.org/10.1016/B978-0-08-102659-5.00002-1

Copyright © 2020 Elsevier Ltd. All rights reserved.

1. BOTANY

1.1 Introduction

Asthma weed (*Euphorbia hirta* L.) (Fig. 2.1) is annual plant that belongs to Euphorbiaceae family (Ping et al., 2013). Euphorbiaceae is one of the large families of angiosperms that comprises 300 genera and about 7500 species. Almost all members of this family are composed of herbs; some are trees and shrubs. Certain species of these genera are xerophytes. Euphorbiaceae family is widely dispersed in both hemispheres with a variety of morphologic arrangements, from trees to large, lush desert herbs (Cateni et al., 2003). The most varied genera in the plant kingdom are *Euphorbia*. Plants of *Euphorbia* may also be perennial herbs, trees, or woody shrubs with corrosive and poisonous latex. The roots of asthma weed are thick, and some plants contain fine and tuberous or fleshy roots. Numerous species are relatively thorny, succulent, or unarmed. In succulent species the leaves are frequently short-lived and small (Saeed-ul-Hassan et al., 2006). Various floras of this are of great economic value. Mostly, species cause sickness, poisoning, or even death if swallowed. Dermatitis, the skin disease, is also produced by several species of Euphorbiaceae, including asthma weed. Even rain water dripping from certain herbs is sufficient to cause skin diseases like

FIGURE 2.1 Asthma weed (*Euphorbia hirta*) plant commonly grows on road sides, gardens, and waste places.

dermatitis (Saeed-ul-Hassan et al., 2006). Asthma weed is a remedial, rhizomatous plant scattered in southwest Ghats of Asian country India and the northeast coastline of Tamil Nadu. Asthma weed is a recognized therapeutic herb with numerous pharmacological outlines. *E. hirta* is known by various names, which are given in different countries of the world according to their language. In English, it is typically called snake weed. In India and Pakistan, it is called dudhi or dudhani. Other vernacular names of *E. hirta* are sheer jiyah, dhudi Kalan (Unani), raktavinduchada (Sanskrit), asthma weed, milk weed, cat's hair (Australian), brokeruee, barakeru (Bengali), tawa-tawa (Kinaray), nanbala, bidarie (Telugu), nayeti, goverdhan, dudhali (Mah), and amumpatchaiyariss (Tami) (Saeed-ul-Hassan et al., 2006). *E. hirta* has many synonyms in different countries. In China, it is called feiyangcao, jiejiehua, dafeiyang, and daruzhicao. The following are its other synonyms: in Malaysia, kelusan and Ambin jantan; in Indonesia, daun biji kacang; in Papua New Guinea, kiki kana kuku, in the Philippines, botobotonis, (Tagalog), gatas-gatas; in Laos, ungl yang, mouk may; in Thailand, nam nomraatchasee (central) and yaa nam muek; in France, euphorbepilulifere and *Euphorbea fleusentete*; in Liberia, tuagbono; and in Norway, demba sindji

1.2 History/Origin

It is native to central America. A long time ago, it was introduced to southeast Asia and spread all over the world, to the subcontinent, subtropical, and tropical regions of Africa, Asia, and South America (Huang et al., 2012). For thousands of years, it has been used in traditional medicines in several areas throughout the world. Asthma weed plant is mostly known as asthma weed since it has properties for curing asthma and several lung-related respiratory diseases (Ogunlesi et al., 2009).

1.3 Demography/Location

Asthma weed is mostly cultivated in lowlands, gardens, paddy fields, waste places, and road sides (Rajeh et al., 2010). Dry environmental conditions are favorable for the better growth of asthma weed. Warm climate is the key parameter for the growth of asthma weed. Asthma weed prefers sunny and slightly shaded areas. This weed is considered an earlier colonizer on the field. The optimum temperature for the germination of this plant seed is 15–40°C, and it also requires sunlight for the germination. This is broadly dispersed in the Philippines, from sea level to an altitude of 500 m. This weed also arises in India, Pakistan, Indonesia, Yemen, Sri Lanka, Taiwan, Saudi Arabia, Nepal, Australia, Nigeria, Mali, Sudan, Kenya, Borneo, New Guinea, the United States, Mexico, and Brazil.

1.4 Botany, Morphology, Ecology

Several species of *Euphorbia* are succulent. The central stem of asthma weed and other side branches of this species are fleshy and thick, 6–36 inches (15–91 cm) tall. It is frequently branched, four angled, and covered with long, yellowish hairs. The short leaves are elliptical, opposite, oblong, or oblong-lanceolate, alternate, or are in whorls with a slightly toothed or rough margin, pale yellow on the lower side and dark green on the upper side (Basma et al., 2011). The base of the leaves of this plant is commonly unequal, rounded, or acute with three to four distinct main veins. The petioles are short in length, about 2–3 mm, with long stipules pectinate. Leaves are mostly short-lived and small in succulent species. The stipules of plants are generally small and somewhat morphed into glands or spines. Flowers are numerous, small, and crowed together in thick cymes having 1 cm diameter (Basma et al., 2011). Just like all members of Euphorbiaceae, flowers of asthma weed are unisexual. Male flowers of these plants are sessile, absent of perianth, along with fringed or linear bracteoles and one stamen. The female flower of the asthma weed has a superior ovary, short pedicel, rimmed perianth, is covered with tiny hairs, holds three styles, is three-celled, and the apex is 2-fid. The fruit of this plant is intensely three-lobed, covered in short hairs, base truncate, and three-seeded. The seeds of asthma weed are oblong, slightly wrinkled, pinkish-brown, four-sided prismatic, and a caruncle is absent. Duration of flowering of an individual herb is generally throughout a year.

2. CHEMISTRY

Asthma weed is an impressive, medicinally important plant. Due to unique properties of every part of this plant, it is of great significance. It has range of leaf colors from green to purple depending on environmental conditions. It possesses a volatile oil with several components including phytol, alcohol, fatty acids, and methanolic components depending upon the parts of the plant or weed. Asthma weed contains a white, milky latex that may cause skin irritation but is used in curing warts and snake bite in many countries. Asthma weed contains 0.3%–0.4% tannin, lycosidal substance, fatty acids, melissic acid, phorbic acid, eciphosterol, sterols, a small amount of alkaloids, essential oil, and sugars. Aerial parts of asthma weed contains small amounts of terpenoids. The study of leaves shows the presence of phytochemicals such as flavonoids, alkaloids, tannins, steroids, carbohydrates, and glycosides. But fats, saponins, and protein are absent in aerial parts (Alisi and Abanobi, 2012; Huang et al., 2012). Moreover, the roots and aerial parts possess terpenoids and ingenol-type and phorbol-type diterpenoid esters (Saeed-ul-Hassan et al., 2006). The mineral constituents of dried leaves of asthma weed sample are Ca 1.1%–2.0%, Fe 0.02%–0.04%, P 0.30%–0.4%, Mg 0.4%–0.6%, Mn 0.01%–0.02%, Cu 0.001%–0.003%, and Zn 0.01%–0.02%, while fresh leaves of asthma weed were found to have high levels of Cu (30–31 ppm), Mn (188–189 ppm), and Zn (152–153 ppm) (Saeed-ul-Hassan et al., 2006).

Asthma weed contains various classes of chemical compounds such as phenols, flavonoids, terpenoids, tannins, and phenolic acids (Fig. 2.2). Examples of polyphenols present in this weed are gallic acid, 3,4-di-O-galloylquinic acid, 2,4,6-tri-O-galloyl-D-glucose, 1,2,3,4,6-penta-O-galloyl-β-D-glucose, and myricitrin. Flavonoids present in the asthma weed plant are euphorbianin, quercitrin, leucocyanidol, and quercitol. Phytosterols and triterpenes found in this plant are β-amyrin, 24-methylenecycloartenol, and β-sitosterol. The aerial parts comprise triterpenes including taxaxerol, α-amyrin, friedlin, β-amyrin, and esters of it are 11α-oxidotaraxerol, 12α-oxidotaraxerol, cycloartenol, euphorbol hexacosoate, and 24-methylene-cycloartenol.

Tannins isolated from asthma weed plant are the euphorbin A, E, C, B, terchebin, and dimeric hydrolysable dehydroellagitannins, the monomeric

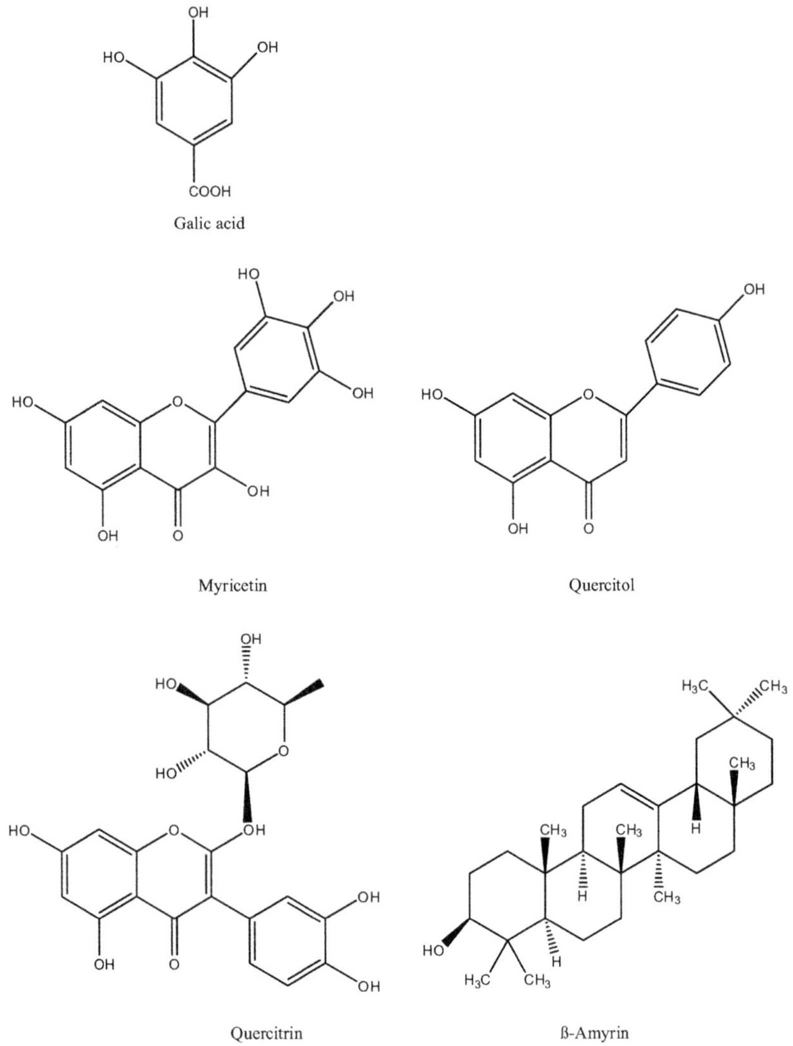

FIGURE 2.2 Some important chemical components of asthma weed.

hydrolyzable tannins geraniin, neochlorogenic acid, 2, 4, 6 tri-O-galloyl-β-D-glucose 1, 2, 3, 4, 6 penta-O-galloyl-β-D-glucose, and benzyl gallate. Isolated flavonoids include quercitrol, quercetin, and its derivatives comprising rhamnose, α-chlorophenolic acid, quercetin rhamoside, cyaniding 3, 5-diglucoside, rutin, myricitrin, leucocyanidin, camphol, pelargonium 3, 5-diglucoside, leucocyanidol, quercetrin, dihydroellagitannins. Another important compound flavonol glycoside xanthorhamnin was also sequestered. The latex also contains different components including inositol, friedelin, β-sitosterol, taraxerol, kaempferol, ellagic acid, quercitrin, and quercitol. Some other phenolic acids including gallic acid, ellagic acid, tannic acid, tartaric acid, and maleic acid were also extracted (Huang et al., 2012; Saeed-ul-Hassan et al., 2006). It also contains shikimic acid, quercitol derivatives, choline, tinyatoxin, camphol rhamnose, and chtolphenolic acid (Kumar et al., 2010b).

3. POSTHARVEST TECHNOLOGY

Plants are cut during flowering stage (or fruiting) and dried for use in infusions, liquid extracts, and tinctures. Fresh plant parts are used to obtain bioactive juice.

4. PROCESSING

Tender young leaves and shoots are cooked as a vegetable. It is used as a famine food when all else fails. The leaves can cause intestinal complaints.

5. VALUE ADDITION

The stem, taken internally, is famed as a treatment for asthma, bronchitis, and various other lung complaints. The herb relaxes the bronchioles but apparently depresses the heart and general respiration. It is usually used in combination with other antiasthma herbs such as *Grindelia camporum* and *Lobelia inflata*. It is also used to treat intestinal amebic dysentery. The leaves are mixed with those of *Datura metel* in preparing "asthma cigarettes."

6. USES

A number of constituents isolated from this plant showed antimicrobial activities against numerous pathogens. Some secondary metabolites and the phenolic compounds are most bioactive and act as membrane-permeable agents. Asthma weed plays a noteworthy role in the remedial field, where extracts from this herb reveal antiinflammatory, antioxidant, diastolic, diuretic, anthelminthic, and antibacterial effects. The role of herbs in the cure of disease is demonstrated by their occupation in all main systems of drugs regardless of the fundamental philosophical evidence. For instance, humans use Western medication with ancestries in Egypt and Mesopotamia, the Ayurvedic (Hindu), and Unani

(Islamic) systems focused in the Indian subcontinent and western Asia (Evans, 2002). Asthma weed has a positive effect on diverse female disorders and in the cure of respiratory illnesses including bronchitis and asthma (Rehman, 2009). Moreover, it has characteristics for cumulative milk flow because of its milky latex. It is more significant in curing respiratory infections, especially coryza, cough, and asthma bronchitis. In India, this plant is used for the treatment of worm infections in kids and also for gonorrhea, dysentery, jaundice, pimples, tumors, and digestive problems. On the warts and wounds, the fresh milky latex obtained from asthma weed plant is used. Roots are used in inflammation, sprains, miscarriage, and maggots in injuries, epilepsy, and irregular growth of teeth (Saeed-ul-Hassan et al., 2006).

In Africa, extracts of this herb are used in remediation of asthma (Ogbulie et al., 2007), malaria, respiratory infections, hypertension, gonorrhea, tumors, dysentery, and jaundice (Patel et al., 2009; Ragasa and Cornelio, 2013). The biologically active components that are found to be responsible for the aforementioned activities are saponins, sterols, flavonoids, and phenolic acids (Pióro-Jabrucka et al., 2011). It can also be used for chronic bronchitis, coughs, and other pulmonic disorders in Mauritius. This plant is also extensively used against colic diarrhea and especially amebic dysentery (Ogbulie et al., 2007). Extracts of the plant are used in the handling of ulcers, as ear drops, and for helping wound healing (Ogbulie et al., 2007). It also shows significant genotoxic and mitodepressive effects (Ping et al., 2013). Besides tawa-tawa, there are several further species of Euphorbial genera that play important roles in classical medications. When cracked, every specie of *Euphorbia* shows a milky liquid or juice that is used as a component in poisons. Asthma weed is a famous plant among physicians of classical medicine, and it is extensively used as an infusion or decoction to treat numerous ailments (Rajeh et al., 2010). In males, asthma weed has long been used as a sexual stimulant, both to increase libido and boost fertility. For males who want to boost their sex drive and improve their chances of starting a family, asthma weed can be a great solution. It can even help to prevent premature ejaculation. For women, asthma weed can stimulate the production of breast milk. However, it should not be given to pregnant women, as it can cause miscarriages.

7. PHARMACOLOGICAL USE

7.1 Prophylactic Agent

Asthma weed is broadly used as an infusion or decoction to treat numerous disorders including intestinal diarrhea, parasites, peptic ulcers, vomiting, heartburn, amebic, bronchitis, dysentery, asthma, laryngeal spasms, hay fever, coughs, emphysema, colds, menstrual problems, kidney stones, and venereal and sterility diseases. Furthermore, this plant can also be applied to treat skin infections and mucous membranes, like warts, tinea, scabies thrush, fungal afflictions, aphthae, measles, and Guinea-worm and as an antiseptic to treat wounds, conjunctivitis, and sores. Asthma weed plant has a character to treat severe toothache, headache, colic, rheumatism, and pains during pregnancy. It is used as pain relief and antidote of snakebites and scorpion stings (Rajeh et al., 2010).

7.2 Antidengue

Dengue fever is initiated by the arbovirus known as dengue virus, transferred by the mosquito *Aedes aegypti*. This fatal fever is treated by using traditional or herbal medicines. In the Philippines, asthma weed, natively known as "tawa-tawa," has applications in herbal medicine to treat dengue fever in rural areas (Kadir et al., 2013). The exact mechanism is yet unknown, but it has considerable significance for the treatment of dengue fever.

7.3 Antioxidant Activity

Commercial antioxidants may show unwanted side effects, so customer interest in natural antioxidants has increased considerably over recent years. The extract of asthma weed prepared in hot water was used for the determination of antioxidant potential. The crude extract exhibited noteworthy free radical scavenging action (Sharma et al., 2007). Flavonoids and phenolic components existing in asthma weed extract have the ability to scavenge the free radicals (Alisi et al., 2011; Alisi and Abanobi, 2012; Basma et al., 2011; Kumar et al., 2012).

7.4 Antimalarial Activity or Antiplasmodium Activity

Various extracts from different parts of asthma weed were used in traditional medication for the cure of malaria. Out of these plant species, asthma weed whole plant including stem, leaves, roots, and flower inhibited plasmodium growth more than 60%. In mice, *E. hirta* extracts showed noteworthy chemo suppression of parasitemia infected with *Plasmodium berghei*. It has been reported that antimalarial flavonol glycosides can also be produced from the tawa-tawa. The antiplasmodial action may be correlated to the presence of some components including steroids, terpenes, coumarins, lignans, flavonoids, phenolic acids, anthraquinones, and xanthones.

7.5 Antiinflammatory Activity

The main components isolated from aerial parts of asthma weed such as β-amyrin, triterpenes, β-sitosterol, and 24-methylenecycloartenol showed antiinflammatory effects. Both the triterpenes and the extract exerted major and dose-dependent antiinflammatory action in ear inflammation induced by phorbol acetate as a model in mice. The aqueous extracts also showed antiinflammatory activities.

7.6 Sedative and Anxiolytic Activity

Lyophilized extracts of asthma plant showed sedative and anxiolytic activity. Sedative characteristics can be established and verified with high doses by a decrease of interactive considerations measured in an unfamiliar or unknown environment, whereas anti-conflict response was observed at lower doses by an improvement in behavioral parameters by light/dark choice condition test. These lead the traditional use of asthma weed as a sedative and expose original anxiolytic characters.

7.7 Antidiarrheal Activity

Asthma weed plant extract by decoction was used to understand the antidiarrheal activity in mice. A flavonoid glycoside component, quercitrin isolated from asthma weed, showed antidiarrheal activity against castor oil–induced diarrhea in mice. It also seems to delay small intestinal transit. But, aglycone of quercitrin, in the occurrence of secretagogue composites, improved colonic liquid absorption, signifying that the antidiarrheal action of quercitrin component is only because of its aglycone, which is released by the glycoside present in the intestine.

7.8 Anticancer Activity

Extracts of asthma weed have seemed to demonstrate selective cytotoxicity action against numerous cancer cells. This plant is beneficial in effective cure of cancers, especially squamous cell carcinomas and malignant melanomas.

7.9 Diuretic Activity

The diuretic activity of asthma weed leaf was evaluated in rats using furosemide and acetazolamide as standard diuretic medications. Electrolyte excretion and urine output were considerably affected by ethanol and water extract of the plants. The study of asthma weed plant reveals that active constituents in the water extract have an analogous diuretic spectrum to acetazolamide (Perumal et al., 2012).

7.10 Antiamebic Activity and Antispasmodic Activity

Asthma weed whole plant showed inhibition in the growth of unicellular organisms, including *Entamoeba histolytica*, with the lowest active concentration of almost 10 pg/mL or less. The same extract also displayed more than 70% inhibition of induced KCl solution contractions and acetylcholine on isolated pig ileum at a concentration of about 80 μg/mL (Patel et al., 2009).

7.11 Molluscidal Activity

The aqueous extracts of leaf, stem and bark of herb asthma weed have strong molluscidal action. Sublethal doses of these extracts considerably change the levels of free amino acid, total protein, and nucleic acids. These also change the action mechanism of alkaline phosphatase, acid phosphatase, and enzyme protease in different tissues of the *Lymnaea acuminata*, a vector snail, by changing dose and time of asthma weed extracts (Patel et al., 2009).

7.12 Antifertility Activity

Asthma weed has exhibited significant antifertility activity. Asthma weed (50 mg/kg of body weight) reduces the density and sperm mobility or motility of cauda epididymal and leads to testis sperm interruption considerably, ultimately causing 100% infertility (Patel et al., 2009).

7.13 Antiplatelet Aggregation and Antiinflammatory

Aqueous extracts of asthma weed strongly reduced the release or freedom of prostaglandins I2, D2, and E2. Moreover, asthma weed extracts also exhibit an inhibitory activity on platelet accumulation and lower the development of carragenin in rats (Patil and Magdum, 2011).

7.14 Repellent and Antifeedant Effects

The alcoholic extracts of asthma weed exhibit considerable antifeedant and mosquito repellent effect (Wei et al., 2004) (Wei et al., 2005). The extracts of asthma weed contain quercitrin and polyphenols that might be responsible for the antifeedant effect and repellent activity (Panneerselvam et al., 2013).

7.15 Immunomodulatory Activity

Aqueous-alcoholic and aqueous extracts, possessing polyphenols, flavonoids, terpenes and sterols, revealed immunostimulant effect. The aqueous extract affects lymphoblast transformation, which is lectin-induced, and showed 45% immunomodulatory action (Ramesh and Padmavathi, 2010).

7.16 Antifungal Activity

The alcoholic extract revealed significant antifungal activity when verified against various plant pathogens including *Fusarium pallidoroseum, Colletotrichum capsici, Botryodiplodia theobromae, Penicillium citrinum, Alternaria alternata, Aspergillus niger,* and *Phomopsis caricae-papayae* through paper disc diffusion method (Gayathri and Ramesh, 2013; Patel et al., 2009). The antifungal activity of asthma weed was due to the cell membrane leakage of cellular proteins [34].

7.17 Larvicidal Activity

Asthma weed has exhibited significant larvicidal activity. Asthma weed was tested against the larvae of *Culex quinquefasciatus* (Say) and *Aedes aegypti* L. The larval mobility and mortality were detected after 24 hours of asthma weed extract exposure. Asthma weed extract showed larvicidal activity against *C. quinquefasciatus and A. aegypti* (Rahuman et al., 2008). Petroleum extract of asthma weed showed maximum larvicidal activity, such as 272.36 ppm (Rahuman et al., 2008).

7.18 Antibacterial Activity

Asthma weed demonstrates noteworthy action against bacteria such as *Klebsiella pneumoniae* (Singh and Kumar, 2012). This plant seems to have great antibacterial potential against various medically significant bacterial strains (Hussain et al., 2014) including *Staphylococcus epidermidis* (Shanmugaraju et al., 2007), *Bacillus subtilis, Pseudomonas pseudoalcaligenes, Salmonella typhimurium, Escherichia coli* (Shanmugaraju et al., 2007), and *Proteus vulgaris.* The antibacterial potential of methanol and aqueous extracts was examined by well diffusion and disk diffusion methods. The capability of asthma weed plant in methanol extracts toward antibacterial activity is more than other solvents. The methanol extract of asthma weed plant showed more activity against gram-positive stain. The bacteria that exhibit the most resistance is *P. vulgaris* (Parekh and Chanda, 2008). The presence of citronellal in asthma weed might be responsible for the potential of antibacterial activity (Saravanan et al., 2012; Shanmugaraju et al.,

2007). Gold nanoparticles manufactured using the leaves of asthma weed are also used for antibacterial activity (Elumalai et al., 2010).

7.19 Antidiabetic Activity

Antidiabetic activity is the one of the important activities of the asthma weed plant. Diabetes mellitus is such a metabolic ailment considering hyperglycemia-caused imperfections in insulin action, insulin secretion, or both. Chronic hyperglycemia is related with long-term impairment (Grover et al., 2002). Various studies have exposed that diabetes mellitus is concomitant with the decrease in antioxidant activity and increased creation of free radicals (Butterworth, 2005). So, free radicals have to be scavenged for its antidiabetic action. The antidiabetic action of asthma weed flowers extracts may be due to the presence of tannins, flavonoids, and other phenolic constituents, but the exact mechanism is still unknown (Kumar et al., 2010a). It has lowered the side effects related with synthetic medicines.

8. SIDE EFFECTS AND TOXICITY

The roots of asthma weed plant possess 12-deoxyphorbol-13-phenylacetate-20-acetate-ingenol triacetate, 12-deoxyphorbol-13-dodecanoate-20-acetate, as well as the highly toxic resiniferonol derivative. Besides these, other isolated terpenoids that are sterols, including cholesterol, β-sistosterol, stigmasterol, and campesterol, are also present in it (Saeed-ul-Hassan et al., 2006). Women should use this plant with extreme care during pregnancy.

References

Alisi, C., Ojiako, O., Osuagwu, C., Onyeze, G., 2011. Free radical scavenging and in-vitro antioxidant effects of ethanol extract of the medicinal herb chromolaena odorata linn. British Journal of Pharmaceutical Research 1, 141.

Alisi, C.S., Abanobi, S.E., 2012. Antimicrobial Properties of Euphorbia hyssopifolia and *Euphorbia Hirta* against Pathogens Complicit in Wound, Typhoid and Urinary Tract Infections.

Basma, A.A., Zuraini, Z., Sasidharan, S., 2011. A transmission electron microscopy study of the diversity of Candida albicans cells induced by *Euphorbia hirta* L. leaf extract in vitro. Asian Pacific journal of tropical biomedicine 1, 20–22.

Butterworth, L.M.a.P., 2005. Protective effects of moderate exercise with dietary vitamin C and E on blood antioxidative defense mechanism in rats with streptozotocin-induced diabetes. Canadian Journal of Applied Physiology 30, 172–185.

Cateni, F., Falsone, G., Zilic, J., 2003. Terpenoids and glycolipids from Euphorbiaceae. Mini Reviews in Medicinal Chemistry 3, 425–437.

Elumalai, E., Prasad, T., Hemachandran, J., Therasa, S.V., Thirumalai, T., David, E., 2010. Extracellular synthesis of silver nanoparticles using leaves of *Euphorbia hirta* and their antibacterial activities. Journal of Pharmaceutical Sciences and Research 2, 549–554.

Evans, W.C., 2002. Trease and Evan's Pharmacognosy, fifteenth ed., vol. 3. W. B. Suanders, Edinburgh, London.

Gayathri, A., Ramesh, K.V., 2013. Antifungal activity of *Euphorbia hirta* L. inflorescence extract against *Aspergillus flavus* A mode of action study. International Journal of Current Microbiology and Applied Sciences 2, 31–37.

Grover, J., Yadav, S., Vats, V., 2002. Medicinal plants of India with anti-diabetic potential. Journal of Ethnopharmacology 81, 81–100.

Huang, L., Chen, S., Yang, M., 2012. *Euphorbia hirta* (Feiyangcao): a review on its ethnopharmacology, phytochemistry and pharmacology. Journal of Medicinal Plants Research 6, 5176–5185.

Hussain, M., Farooq, U., Rashid, M., Bakhsh, H., Majeed, A., Khan, I.A., Rana, S.L., Rehman, M., Aziz, A., 2014. Antimicrobial activity of fresh latex, juice and extract of *Euphorbia hirta* and Euphorbia thymifolia: an in vitro comparative study. International Journal of Pharma Sciences 4, 546–553.

Kadir, S.L.A., Yaakob, H., Zulkifli, R.M., 2013. Potential anti-dengue medicinal plants: a review. Journal of Natural Medicines 67, 677–689.

Kumar, M.D., Prasad, M.A., Prabhudutta, P., 2012. Formulation and evaluation of topical dosage form of *Euphorbia hirta* L. and their wound healing activity. International Journal of Advances in Pharmaceutical Research 3, 1116–1121.

Kumar, S., Malhotra, R., Kumar, D., 2010a. Antidiabetic and free radicals scavenging potential of *Euphorbia hirta* flower extract. Indian Journal of Pharmaceutical Sciences 72, 533.

Kumar, S., Malhotra, R., Kumar, D., 2010b. *Euphorbia hirta*: its chemistry, traditional and medicinal uses, and pharmacological activities. Pharmacognosy Reviews 4, 58.

Ogbulie, J., Ogueke, C., Okoli, I.C., Anyanwu, B.N., 2007. Antibacterial activities and toxicological potentials of crude ethanolic extracts of *Euphorbia hirta*. African Journal of Biotechnology 6.

Ogunlesi, M., Okiei, W., Ofor, E., Osibote, A.E., 2009. Analysis of the essential oil from the dried leaves of *Euphorbia hirta* Linn (Euphorbiaceae), a potential medication for asthma. African Journal of Biotechnology 8.

Panneerselvam, C., Murugan, K., Kovendan, K., Kumar, P.M., Subramaniam, J., 2013. Mosquito larvicidal and pupicidal activity of *Euphorbia hirta* Linn.(Family: Euphorbiaceae) and Bacillus sphaericus against *Anopheles stephensi* Liston.(Diptera: Culicidae). Asian Pacific journal of tropical medicine 6, 102–109.

Parekh, J., Chanda, S., 2008. Antibacterial activities of aqueous and alcoholic extracts of 34 Indian medicinal plants against some Staphylococcus species. Turkish Journal of Biology 32, 63–71.

Patel, S.B., Naikwade, N.S., Magdum, C.S., 2009. Review on phytochemistry and pharmacological aspects of *Euphorbia hirta* linn. Asian Journal of Pharmaceutical Research and Health Care 1, 113–133.

Patil, S.B., Magdum, C.S., 2011. Determination of LC50 values of extracts of *Euphorbia hirta* linn and *Euphorbia nerifolia* linn using brine shrimp lethality assay. Asian Journal of Research in Pharmaceutical Science 1, 69–70.

Perumal, S., Pillai, S., Cai, L.W., Mahmud, R., Ramanathan, S., 2012. Determination of minimum inhibitory concentration of *Euphorbia hirta* (L.) extracts by tetrazolium microplate assay. Journal of Natural Products 5.

Ping, K.Y., Darah, I., Chen, Y., Sasidharan, S., 2013. Cytotoxicity and genotoxicity assessment of *Euphorbia hirta* in MCF-7 cell line model using comet assay. Asian Pacific journal of tropical biomedicine 3, 692–696.

Pióro-Jabrucka, E., Pawelczak, A., Przybyl, J., Baczek, K., Weglarz, Z., 2011. Accumulation of phenolic and sterol compounds in *Euphorbia hirta* (L.). Herba Polonica 57.

Ragasa, C.Y., Cornelio, K.B., 2013. Triterpenes from *Euphorbia hirta* and their cytotoxicity. Chinese Journal of Natural Medicines 11, 528–533.

Rahuman, A.A., Gopalakrishnan, G., Venkatesan, P., Geetha, K., 2008. Larvicidal activity of some Euphorbiaceae plant extracts against *Aedes aegypti* and *Culex quinquefasciatus* (Diptera: Culicidae). Parasitology Research 102, 867–873.

Rajeh, M.A.B., Zuraini, Z., Sasidharan, S., Latha, L.Y., Amutha, S., 2010. Assessment of *Euphorbia hirta* L. leaf, flower, stem and root extracts for their antibacterial and antifungal activity and brine shrimp lethality. Molecules 15, 6008–6018.

Ramesh, K.V., Padmavathi, K., 2010. Assessment of immunomodulatory activity of *Euphorbia hirta* L. Indian Journal of Pharmaceutical Sciences 72, 621.

Rehman, M.K.u., 2009. Dermatological Studies of Euphorbia Pilulifera L. M. Phil thesis, submitted to College of Pharmacy. University of the Punjab, Lahore (Pakistan).

Saeed-ul-Hassan, S., ur Rehman, M.K., Ansari, T., Bhatti, M.U., 2006. A review of *Euphorbia pilulifera* L. Pakistan Journal of Pharmacy 29.

Saravanan, R., Dhachinamoorthi, D., Senthilkumar, K., Srilakshmi, M., Sri, T.D., 2012. Antibacterial activity of *Euphorbia Hirta* extracts. International Journal of Research in Ayurveda and Pharmacy 3.

Shanmugaraju, V., Rajan, P.C., Abirami, N., Rajathi, K., 2007. Antibiogram and GC analysis of *Euphorbia hirta* leaf extract. Ancient Science of Life 26, 1.

Sharma, N.K., Dey, S., Prasad, R., 2007. In vitro antioxidant potential evaluation of *Euphorbia hirta* L. Pharmacologyonline 1, 91–98.

Singh, G., Kumar, P., 2012. Antibacterial potential of alkaloids of *Withania somnifera* L. and *Euphorbia hirta* L. International Journal of Pharmacy and Pharmaceutical Sciences 4, 78–81.

Further Reading

Ahmad, S.F., Khan, B., Bani, S., Kaul, A., Sultan, P., Ali, S.A., Satti, N., Bakheet, S.A., Attia, S.M., Zoheir, K.M., 2013. Immunosuppressive effects of *Euphorbia hirta* in experimental animals. Inflammopharmacology 21, 161–168.

Betancur-Galvis, L., Morales, G., Forero, J., Roldan, J., 2002. Cytotoxic and antiviral activities of Colombian medicinal plant extracts of the *Euphorbia* genus. Memórias do Instituto Oswaldo Cruz 97, 541–546.

Charles, C., Maribeth, L., Daniel, L., David, A., 2007. Floral Gigantism in Rafflesiaceae. Science Express, USA.

Joshi, B., 2011. The magical herb "*Euphorbia hirta* L." an important traditional therapeutic herb for wart disease among the vangujjars of forest near Kashipur, Uttarakhand. New York Sci. J 4 (2), 96–97.

Liu, Y., Murakami, N., Ji, H., Abreu, P., Zhang, S., 2007. Antimalarial flavonol glycosides from *Euphorbia hirta*. Pharmaceutical Biology 45, 278–281.

Perumal, S., Mahmud, R., Ramanathan, S., 2015. Anti-infective potential of caffeic acid and epicatechin 3-gallate isolated from methanol extract of *Euphorbia hirta* (L.) against *Pseudomonas aeruginosa*. Natural Product Research 29, 1766–1769.

Shih, M.-F., Cheng, Y.-D., Shen, C.-R., Cherng, J.-Y., 2010. A molecular pharmacology study into the anti-inflammatory actions of *Euphorbia hirta* L. on the LPS-induced RAW 264.7 cells through selective iNOS protein inhibition. Journal of Natural Medicines 64, 330–335.

Sivarajan, V., Balachandran, I., 1994. Ayurvedic Drugs and Their Plant Sources. Oxford and IBH Publishing.

Sudhakar, M., Rao, C.V., Rao, P., Raju, D., Venkateswarlu, Y., 2006. Antimicrobial activity of Caesalpinia pulcherrima, *Euphorbia hirta* and Asystasia gangeticum. Fitoterapia 77, 378–380.

CHAPTER

3

Bamboo

Muhammad Waqar Azeem[1], *Muhammad Asif Hanif*[1], *Muhammad Mumtaz Khan*[2]

[1] Department of Chemistry, University of Agriculture, Faisalabad, Pakistan; [2] Department of Crop Sciences, Sultan Qaboos University, Muscat, Oman

OUTLINE

https://doi.org/10.1016/B978-0-08-102659-5.00003-3

Copyright © 2020 Elsevier Ltd. All rights reserved.

1. BOTANY

1.1 Introduction

Bamboo (*Bambusa arundinacea*) (Fig. 3.1) is a woody, perennial, evergreen plant belonging to the Poaceae/Graminae, which is the fifth largest known ubiquitous family of monocotyledonous flowering plants containing all lower grasses along with some giant members (Sarangthem et al., 2010). Generally, bamboo is a name given to a number of fast-growing woody grasses constituting 75 genera and more than 1250 well-known species throughout the world except Europe (Singhal et al., 2013). They are usually found in diverse climatic conditions from hot tropical areas to colder mountainous regions. Bamboo is quite different from other members of family Poaceae due to the presence of internodes and numerous branches on each node, sometimes also called "bamboo tree" due to the woody appearance and structural integrity (Rathod Jaimik et al., 2011). This is the most versatile woody plant that is also known as "wonder plant" or "green gold" that can be cultivated through nodes, by stem layering, from vegetative portions, and by means of seeds that rapidly germinate and convert into seedlings within a week (Akinlabi et al., 2017; Nirala et al., 2017).

Bamboo has various common and regional names in different languages such as "baans" in Urdu, "vaans" in Punjabi, and "giant thorny bamboo" "bamboo manna" or simple "bamboo" in English. Region based names of bamboo are "vas-nu-mitha," "vanskapur" "wans" or "toncor" in Gujrati"banz" "vanoo" "banskapur" or "bans-lochana" in Hindi"bar-oowa bans" "baansh" "baans" or "bans-kapur" in Bengoli"venulavanam" or "vanshalochana" in Sanskrit"bansamitha" "bansa" or "baambii" in Marathi"tabashir" in Arabic"mungil" "mulmunkil" "mullumangila" or "munga-luppa" in Tamil"mullu-veduru," "mulkas veduru" or "vedur-uppu" in Telugu"vathega-kiyo," "vasan" "vd-chha," or "vathe gasu" in Burma"tavakshira" or "bidaruppu" in Kannad"tawashir" or "tabashir"

FIGURE 3.1 Bamboo tree, stem, and leaves.

in Unani; and "moleuppa" in Maliyalam (Rathod Jaimik et al., 2011). *B. arundinacea* is the most popular and wide spread specie. Some other important species of bamboo are *Phyllostachys bambusoides*, *Rhapsis excels*,

Phyllostachys pubescens, Coniogramme japonica, Lophatheri gracilis, Phyllostachys nigra, Bambusa vulgaris, Bambusa bambos, and *Acidosasa edulis* (Rastogi and Mehrotra, 1993; Soni et al., 2013). According to a recent report of International Network for Bamboo and Rattan trade database, annual bamboo export by the United States is 2.5 billion US dollars. Vietnam, Indonesia, and China are the biggest producers and exporters of different bamboo species all over the world. The total global export of raw bamboo was approximately 89 million US dollars in the year 2000, in which China contributed 25 million US dollars, which accounts for one-third of the total export. Similarly, some other countries like Indonesia, Vietnam, Singapore, and Hong Kong exported raw bamboo material worth 10.6, 7.7, 18.6, and 4.69 million US dollars with the percentage contribution of 12, 8.6, 20.9, and 5.3 of world total (Wu, 2014).

1.2 History/Origin

Fossil records indicate that humans have been using plants as medicines since Middle Paleolithic age approximately 60,000 years ago (Maridass and De Britto, 2008). Many parts of bamboo had extensively been used in ancient Ayurvedic system of medicines, but "tabashir" is of highest medicinal importance that originated from "adivasi" aboriginal tribes of India. It had been exported from India to Arabic traders for thousands of years during the medieval period. The town of Thane close to the Indian west coast has been known as a cleaning center for tabashir since the 12th century BCE (Ghosh, 2008).

1.3 Demography/Location

Bambusa bamboo, also known as the "Indian thorny bamboo" or "giant thorny bamboo" is a native species of Southeast Asia, characterized by tropical dense areas. Bamboo needs a humid environment for growth but can tolerate cold weather also. The best temperature for maximum growth usually ranges from 22°C to 30°C, but it can tolerate up to 36°C without significant damage. Some species of bamboo can even grow under extreme cold climatic conditions with temperature ranging from −28°C to 50°C. The preferable mean annual rainfall is 1200–2500 mm, but it can bear 700–4500 mm. The tolerable pH range is 4–7, while the best condition requires 4.5–6.5 pH values for maximum crop yield. However, numerous bamboo species are quite adaptive and can survive under harsh climatic conditions (Akinlabi et al., 2017; Khare, 2008). Bamboo is a plant of humid tropical lowland areas. Dappled shade is the best location for growth, but sunlight also promotes development if the soil is fully moist and adequately fertile. Sandy loam to loamy clay is more preferable in comparison with marshy locations and well-drained soil areas, but bamboo plant is unable to bear saline soil conditions. Larger bamboo species are found abundantly in subtropical and tropical areas, while smaller species favorably grow in temperate regions with high elevations. Bamboo is widely grown in Vietnam, Indonesia, China, India, Pakistan, Bangladesh, Myanmar, Peru, Malaya, Sri Lanka, Singapore, and Hong Kong (Tran, 2010).

1.4 Botany, Morphology, Ecology

Medicinally important, statistically emerging, economically favorable, and fast growing, the thorny plant of the earth consists of culms, branches, nodes, internodes, leaves, flowers, and dried fruit. Rapidly increasing or fast growing woody culms of bamboo have an average height ranging from 20 to 30 m and average diameter between 10 and 18 cm (Wang and Tsai, 2015). Internodes are thick walled with dark green coloration, while nodes are slightly swollen and also produces short aerial roots. Nodes of the plant constitute a dominant central branch with more than one lateral spiny branch in which lower ones are more wiry and long, usually bent toward the ground, whereas leafy branches on the upper end produce plumes bearing smaller spines. Leaves of the *B. arundinacea* have longer pointed tips, similar to the shape of a lens, having average length of about 15–30 cm and broadness of 8–15 mm. Leaves are linear, lanceolate, glabrous, with stiff tip, ciliate base, scabrous margins with narrow midrib and leaf sheath terminating on short bristly auricle and thick callus characterized by the pointed woody stem. The flowering process in different species of bamboo is quite prolonged and takes 30–35 years for reappearance, while a single clump is usually found loaded with 50–100 kg of seeds. Growth and development through germination of seeds is not an encouraging process; instead layering or cutting is widely used for continuation of species up to several generations. However, the flowering period determines the average life span of plant, which can be more than 100 years in the case of bamboo species (Banik, 2015; Maoyi and Baniak, 1995; Rao et al., 1989).

2. CHEMISTRY

Numerous chemical compounds found in the leaves of bamboo are therapeutically very important in a number of diseases and play a vital role as antioxidants (Figs. 3.2–3.4). Leaves mainly consists of benzoic acid, hydrocyanic acid, glutelin protein, flavonoids, proteolytic enzymes,

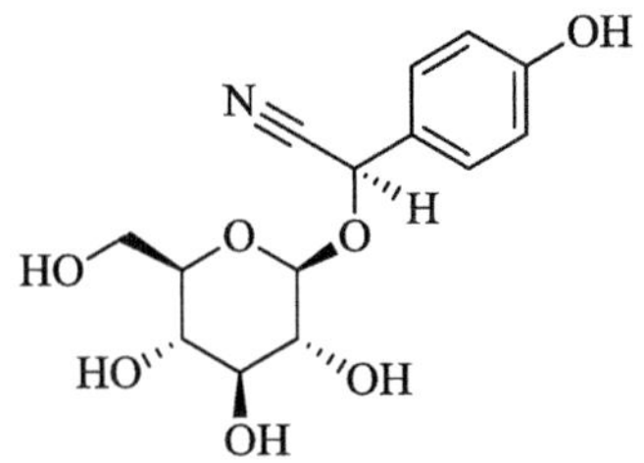

FIGURE 3.2 Structure of "taxiphyllin" present in bamboo shoots.

ureases, and nucleases (Menaria, 2016), while the shoot is known to have waxes and reducing sugars in abundance that do not have direct therapeutic effects, but when mixed with potentially active plant ingredients, they significantly enhance their medicinal effects as amylases, nucleases, and proteolytic enzymes, increases their splitting efficiency when found in combination with waxes and sugars. Shoot extract or exudate contains a high concentration of resin, germaclinum, and enzymes (silicone splitting, amygdalin splitting, amylase, proteolytic enzymes, deamidase, and

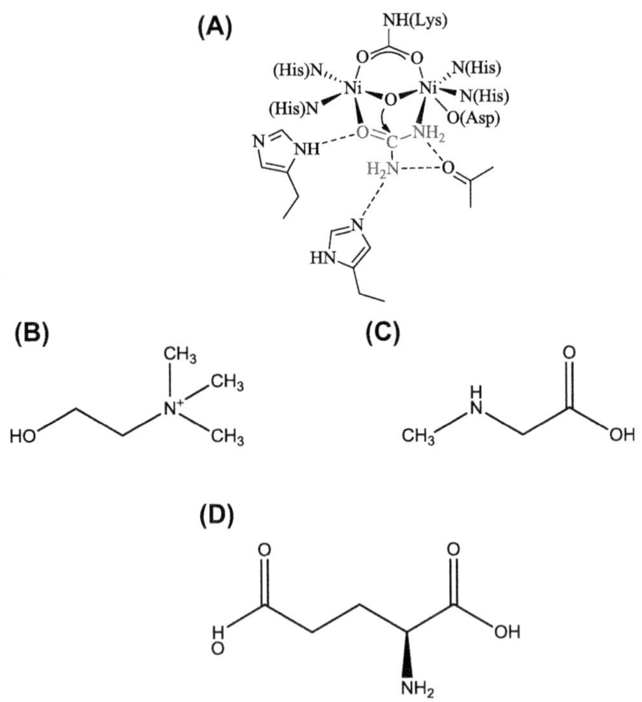

FIGURE 3.3 Active chemical components of bamboo leaves: (A) urease, (B) choline, (C) betaine, and (D) glutelin.

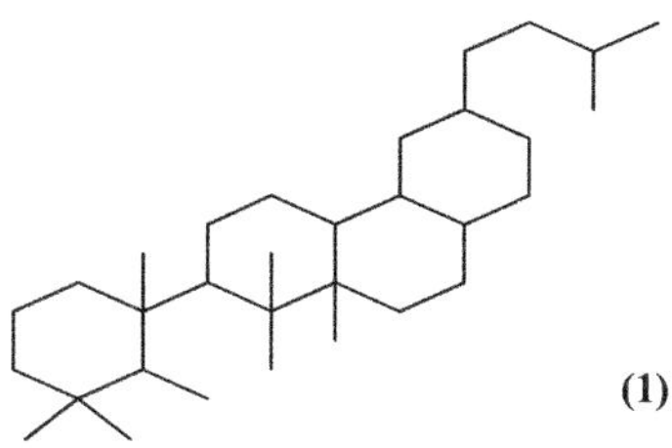

(1)

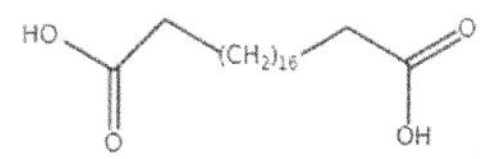

(2)

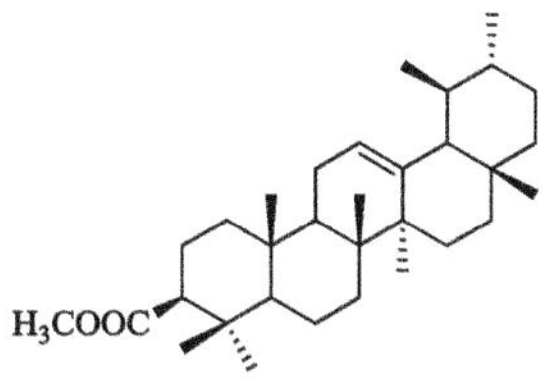

(3)

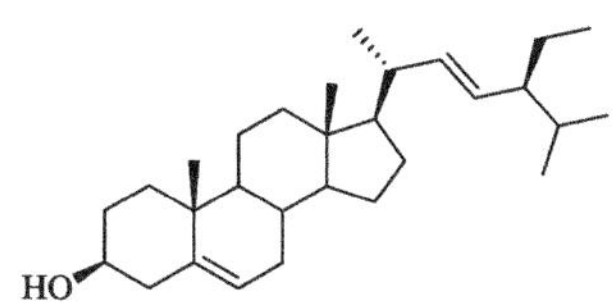

(4)

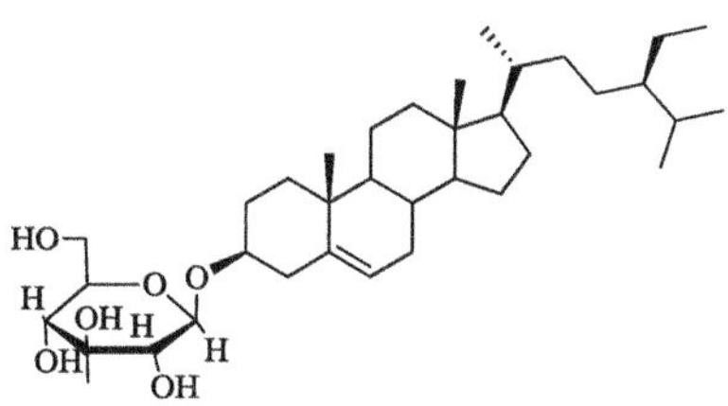

(5)

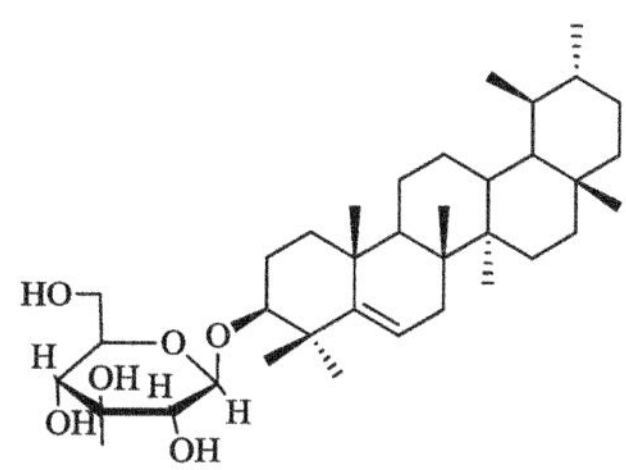

(6)

FIGURE 3.4 Structures of compounds obtained from ethanolic leaf extracts: (1) 17,20,20-tridemethyl-20-α-isoprenyl oleanane, (2) eicosanyl dicarboxylic acid, (3) α-amyrin acetate, (4) stigmast-5,22-dien-3β-ol, (5) stigmast-5-en-3β-ol-β-D glucopyranoside, and (6) urs-12-en-3β-ol-β-D-glucopyranoside.

nuclease) in addition to waxes and sugars. Chemical composition of the shoot with percentage contribution of each ingredient is listed in Table 3.1. Seeds of *Bambusa* are composed of histidine, cysteine, arginine, vitamins, amino acids, thiamine, riboflavin, tyrosine, phenylalanine, isoleucine, and leucine. Similarly, sprout juice is rich in hydrocyanic acid and benzoic acid.

Every part of the whole plant body consists of various chemical compounds, giving bamboo extreme medicinal importance as a siliceous, quite sticky, sweet, white crystalline camphor-like substance is found abundantly near joints insides the nodes. The shoot is composed of many potential and therapeutically active ingredients including (5,5′-di-(diferul-9,9′-dioyl)-[α-Larabinofuranosyl-(1 → 3)-O-β-D-xylopyranosyl-9(1 → 4)-D-xylopyranose] or taxiphyllin, benzoic acid, hydrogen cyanide, waxes, diferuloyl oligosaccharide, diferuloyl arabinoxylanhexasaccharide, resins, oxalic acid, and reducing sugars. Seeds of bamboo are rich in thiamine, riboflavin, niacin, phenylamine, threonine, tyrosine, valine, methionine, lysine, leucine, isoleucine, histidine, cysteine, or arginine. Leaves are known to have urease, nuclease, choline, betaine, methionine, lysine, and glutelin. Ethanolic extract obtained by the Soxhlet extraction of *B. arundinacea* contains the six most important ingredients, out of which two compounds, (1) 17,20,20-tridemethyl-20-α-isopranyl oleanane and (2) eicosanyl dicarboxylic acid, have natural origin, while other four compounds, (1) α-amyrin acetate, (2) Stigmast-5,22-dien-3β-ol, (3) Stigmast-5-en-3β-ol-β-D glucopyranoside, and (4) urs-12-en-3β-ol-β-D-glucopyranoside, are artificially synthesized (Umamaheswari, 2016).

3. POSTHARVEST TECHNOLOGY

Based on industrial utilization and nature of manufacturing products, different postharvesting techniques are adopted for preservation as 12% moisture content is required while using bamboo for constructional purposes. The air-drying of bamboo is the simplest, traditional, economical, and earliest known method for elimination of moisture content up to desired level. However, this is time consuming and the prolonged process requires more space for proper circulation of air around culm. Horizontal piles, vertical placements, strip arrangements, and bundle packing are used for drying of bamboo without additional heat supply. Therefore, a relatively efficient and quite expensive method of artificial heating through kiln-drying is highly appreciated these days to deal with larger bamboo masses. Nevertheless, bamboo drying is a relatively slow process, even while using kiln-drying, due to the cylindrical arrangement of thick-walled culms as compare to simple thin-walled wood cells. Moreover, fresh green culms requires more time for drying compared to older culms due to elevated moisture content (Tang et al., 2013). Discoloration, deep surface or node cracks, culm splits, cell collapse, and rupturing of the culm tissues are major problems associated with drying of bamboo that reduce its commercial utilization and market value. Therefore, slow drying at relatively lower temperature is favorable for maintenance of cell shape and entire structure of bamboo. Another major problem associated with bamboo culm is weakening: if younger culms are harvested to meet the increasing demands on industrial scale, due to high susceptibility and vulnerability toward termite, beetle, and fungal attacks owing to the availability of excess moisture, that can be overcome by using a number of chemical and nonchemical techniques. Nonchemical processes are used to prevent culms from infectious diseases and to enhance durability and resistance toward different types of microbial attacks without reducing sugar content, starches, or carbohydrates. Curing is used to prevent insecticidal attacks by cutting culms with leaves and branches to continue the normal process of respiration. Waterlogging is another important method of culm preservation in which younger culms are soaked into water for long period of time for elimination of carbohydrates and starches, which tends to avoid the attack of fungus and termites along with insects (Morrell, 2011). Plastering is commonly practiced, globally employed, and the oldest known method for protection of bamboo against insecticidal and fungal attacks. This method involves using whole bamboo mat for structure and covering it with mud through mixing with organic materials and lime for solidification. The main purpose of this layering is to get rid of rotting by fungi and various types of insects. Smoking is another cheaper method in which bamboo culms are piled with smaller empty spaces between at the top of a fire source for efficient release of moisture contents of freshly harvested culms. However, continuous contact with fire and the carbonaceous smoke changes and darken the color of culm, which reduces its applicability for ornamental uses. Heating of different bamboo species at temperature of about 120–150°C inside a kiln or heating ovens is meant to protect green culms against beetle attacks. During this process, various acidic chemical compounds are added to increase its usefulness by changing its chemical composition (Garcia et al., 1997).

Several chemical treatment processes such as butt treatment, old engine oil, steeping or sap displacement, and open tank treatments are also used and have been found effective for expanding life span and durability of bamboo culms. These methods are quite expansive in comparison with the nonchemical treatments but are essential to manufacture highly valued products of international export quality. Butt treatment method is employed for both the freshly harvested and previously dried bamboo culms in which a mixture of chemicals named Dursban and engine oil are mixed in specific proportion for immersion of bamboo culms. Sometimes, used engine oil is also coated on bamboo culms for preservation against microbial attacks but is not a scientifically proven method. In the steeping or sap displacement process, bamboo culms are dipped vertically in a mixture of chemical compounds to increase resistance against disease-causing and infectious entities. The open tank treatment method uses an aqueous mixture of chemical compounds for complete penetration of chemicals inside the cells and tissues of bamboo culms by dipping for several days and weeks. Penetration is made possible through diffusion process in all possible directions of culm.

Preservation through chemical methods is of two types, nonfixing and fixing, in which protective chemical compounds are applied on the outer side of bamboo culms along with inner cells and woody tissues more specifically in the dry season (Liese and Kumar, 2003; Liese and Schmitt, 2006). Organic acids (10% solution of acetic acid) are found to be more effective to avoid molding fungal attack on freshly harvested bamboo culms but for a short period of time due to high moisture content (Tang et al., 2012). Sometimes, pressure treatments are employed to ensure deep penetration and retention of chemicals inside the cells and tissues of bamboo culms by using vacuum apparatus (Tang, 2009). Roots and shoots of bamboo plant are of extreme medicinal importance and have extensively been used in Ayurvedic system of medicines that are dried under shade immediately after harvesting to be used as medicine. They are powdered and mixed with other medicinal compounds to be sold as tablets and pills for treatment of a number of diseases including respiratory and reproductive disorders. Leaves of the bamboo plant are used in stomach and joint problems along with various chest infections (Soni et al., 2013).

4. PROCESSING

Highly valuable siliceous secretions or exudation of components with extreme medicinal importance from the precious knot of stem of *B. arundinacea* to form true botanical quartz, which tends to restore elasticity or suppleness of tissue and is used for preparation of medicines, is generally known as "tabashir" or "tabasheer" This extracting material is also called "bamboo tears" "bamboo manna" or "bamboo silica" and is known to have aphrodisiac, antispasmodic, febrifuge, astringent, and stimulant effects when used for medicinal purposes. Tabashir is mainly constituted of silicic acid (96.9%) and organic matter (1%) and mixed with many other components used in skin care products, including antiwrinkle creams and face masks to maintain elasticity of collagen fibers, and acts as a catalyst for the calcium fixation in bones due to the high concentration of siliceous compounds. It slows down the aging process and a number of diseases associated with bones by strengthening tendons, cartilage, ligaments, and other connective tissues (Park and Zhao, 2012; Zhao et al., 2013). Process of preparation of medicines from tabashir is quite unique as it involves placement of sea salt at internodes of *Bambusa* culm that is covered by natural red clay and then baked at about 1000–15,000°C using pine as a source of combustible fuels for at least two to three times. However, best results are obtained by baking this loaded bamboo culm for at least nine times, which results in formation of "purple bamboo salt" It has been generally observed that all hazardous chemical constituents of bamboo are destroyed if salt is completely melted at much elevated temperature. Moreover, high temperature allows rapid absorption of essential elements like iron, potassium, zinc, copper, magnesium, manganese, phosphorous, and calcium that further enhance the therapeutic effects of all corresponding products obtained from bamboo plant (Nirmala and Bisht, 2015).

5. VALUE ADDITION

The *B. arundinacea* and all other known species of bamboo are extensively used in manufacturing of different products including medicines, cosmetics, skin care articles, food supplements, nutraceuticals, and edibles due to the presence of numerous bioactive compounds that make it an economically important crop to be properly vegetated and utilized on an industrial scale. Sitopaladi churna is an important medicine formed by mixing *Cinnamomum zylanicum*, *Elettaria cardamomum*, *Piper longum*, and *B. arundinacea* that is used to treat lung infections, cough, tuberculosis, sinus congestion, sore throat, and common cold. Similarly, many other Chinese medicines such as "Chenjin Wan" "Gualou Zhishi Tang" "Jupi Zhuru Tang" "Qinghuo Ditan Tang" "Qinggong Tang" "Qingluo Yin" "Xiaoer Qizhen Dan" " Zhuye Shigao Tang" and many other Ayurvedic preparations like "Vamsa Rochna" "Talisadi Churna" and "Chyawanprash" are formulated by adding tabashir as an essential chemical constituent (Nirmala and Bisht, 2015).

6. USES

Different parts of *B. arundinacea* can be used for different purposes in several capacities depending on chemical composition, nutritive values, and therapeutic effects because of high versatility and potential to act as a renewable source. Young shoots are cooked and used as food due to the presence of highly nutritive components and sugary sap from culm, or shoot is used to make drinks. The siliceous component from the internodes of the culms is used to make a number of "Ayurvedic" and "Chinese" medicines to treat debility, joint pain, skin infections, wrinkles, menstruation problems, stomach disorders, osteoporosis, osteoarthritis, and many other bone diseases. Some agroforestry uses of bamboo involve cultivation around farms and fields that act as a wind shield to protect other food crops and also to help to anchor soil to avoid soil erosion. The strong, woody stem can be utilized to make paper, furniture, rafts, rigging, ship sails, mats, chairs, screens, bedding, tables, household utensils, flutes, flasks, pitchers, pipes, baskets, hats, boxes, frames, umbrellas, ornaments, scaffoldings, packing and filling materials, along with many other objects (Umamaheswari, 2016).

Shoots of bamboo are considered to be highly nutritive and are used to make different dishes, but they contain an appreciable amount of a poisonous material "taxiphyllin" that immediately forms cyanide when it comes in contact with acids of the stomach lining. However, proper boiling and high temperature treatments are supposed to be appropriate for removal of all sorts of toxicants when using bamboo for edible purposes. The bamboo shoots are traditionally mixed with turmeric to prepare a Nepali dish "alu tama" through fermentation. Similarly, "khorisa" and "gulai rebung" are dishes prepared with bamboo in India and Indonesia respectively. In Tanzania, "ulanzi" is a very popular wine made with various bamboo species. Bamboo is one of the oldest known building materials extensively used for a number of constructional purposes and many industrial products including weapons, musical instruments, charcoal, boats, paper and pulp, flooring, furniture, toys, handicrafts, chopsticks, skewers, and food containers. Bamboo is also quite common to be used as scaffolding or supporting material during construction of houses and heavy bridges. Some high-profile applications of bamboo are to use it as a potential source of diesel fuel, airplane skin, phonograph needles, and scaffoldings. Moreover, bamboo ashes are used to manufacture electrical batteries and to polish jewels. It has also been used in light bulbs, filaments, cables, ropes, retaining walls, scales, windmills, dirigibles, and bicycles. In short, different parts of whole bamboo plant have plentiful, broad, and innumerable applications, beyond imagination (Hunter, 2003; Lobovikov, 2003; Xuhe, 2003).

Dried bamboo roots are found very effective when used as medicine for stimulation of blood circulation and to treat rheumatism, rabies, cancer, venereal diseases, restlessness, fever, sleeplessness, and anxiety due to diuretic, antipyretic, and astringent effects. The bamboo shoot is known to have antiinflammatory, antiviral, antifungal, antibacterial, anticancer, and antioxidant properties, so it is used for treatment of stomach diseases, hematuria, high blood pressure, gonorrhea, bleeding piles, and respiratory disorders and helps to induce labor during pregnancy, stimulate menstrual cycle, speed up the cycle measles, clean wounds, remove maggot-infested sores, and also is used as an appetizer. Leaves of bamboo are characterized by antipyretic and diuretic effects and aid in treatment of stomatitis, pharyngitis, common cold, chest infections, headache, arthritis, hemoptyses and febrifuge, leprosy, bleeding secretions, spasmodic disorders, diarrhea, asthma, and cough. Bamboo bark helps to stop bleeding, treats skin eruptions, and mostly is used as an antiemetic. Bamboo stems are used as splints in tinctures and to stimulate menstruation, stop nose bleeding, avoid vomiting, and prevent cerebral infections and bronchial diseases due to antitussive, sedative, and expectorant effects. Recently, another very valuable product of bamboo has been developed under the name "bamboo charcoal" that helps to adsorb unpleasant odors.

7. PHARMACOLOGICAL USES

7.1 Antidiabetic Activity

Seeds of the *B. arundinacea* were found to exhibit antidiabetic potential when compared with standard glibenclamide (Umamaheswari, 2016).

7.2 Antifertility Activity

Shoots of *B. arundinacea* are characterized by antifertility effects on both males and females when taken in appropriate amount for a specific period of time. If 1 kilogram of "Bambusa-arundinacea-tender-shoots" is taken just after parturition, the placenta will drop within the next 2 hours, which indicate that it has enough antifertility potential. Recently, an experiment was conducted by making an ethanolic extract of "Bambusa-arundinacea-tender-shoots" that was used in both male and female rats with an approximate dosage of 300 mg per kilogram per day continuously for 7 days, and it showed a 15% decrease in fertility in male rats after a whole week and 23% decline in fertility in the case of females just after treatment of 4 days. However, complete recovery was evident after 8 days of withdrawal of "Bambusa-arundinacea-tender-shoots." Similarly, a significant reduction in the amount of spermatozoa in cauda epididymis and caput was observed along with decrease in motility of spermatozoa obtained by cauda epididymides. The weight of prostate, vas deference, epididymides, and testis was also decreased; however, serum protein profile and pyruvic/oxaloacetic transaminase activity indicate that the extract is completely nontoxic in nature (Vanithakumari et al., 1989).

7.3 Antimicrobial Activity

The "water-phase-extract-of-bamboo-shavings" has enough potential to resist *Saccharomyces cerevisiae, Penicilliun citrinum, Aspergillus niger, Escherichia coli, Bacillus subtilis,* and *Staphylococcus aureus* depending on concentration. The minimum inhibitory concentration of "water-phase-extract-of-bamboo-shavings" against tested strains of bacteria was in the range of 4.9–32 mg per milliliter while using the two-fold dilution method (Zhang et al., 2010).

7.4 Antiinflammatory Effects

It is a known fact that almost all available antiinflammatory drugs are ulcerogenic. Therefore, a combination of phenylbutazone (nonsteroidal antiinflammatory drug) and bamboo methanol extract was examined and tested for antiinflammatory activity, which proved to be a potential antiinflammatory compound with minimum toxic effects and negligible ulcerogenic activities (Rathod Jaimik et al., 2011).

7.5 Antiulcer Activity

Leaves of the *B. arundinacea* are known to have antiulcer effects when used in the form of methanolic extract against carrageen-induced paw edema. During this experiment, rats were treated with this methanolic extract to observe the antiulcer activity in albinos, and all the corresponding results were compared with standard medicinal drug values. After conducting the whole experiment, it was deduced that methanolic extract of *B. arundinacea* when used in combination with nonsteroidal antiinflammatory drugs produces the best antiinflammatory drug for treatment of inflammatory conditions like rheumatoid arthritis along with peptic ulcer (Rathod Jaimik et al., 2011).

7.6 Protective Effects

Biologic activities of two pyrolyzates derived from bamboo were investigated to check their antiplasmin effects and protective effects using fibrinogen and fibrin degradation products assay and N-methyl-D-aspartate-induced cell death in primary cultured cortical neurons. Results of the investigation revealed that neuronal cell treatment with pyrolyzates of *P. bambusoides, P. nigra,* or *P. pubescens* resulted in restoration of viability of cell when compared with untreated N-methyl-D-aspartate-induced neuronal cell death assay. Moreover, cortical neurons treated with the *P. nigra* and *P. pubescens* showed significant reduction in apoptosis following exposure to N-methyl-D-aspartate through Hoechst 33,342 staining. Additionally, pyrolyzates of *P. nigra* also showed antiplasmin action in fibrinogen and fibrin degradation products assay. It is interesting to note that pyrolyzates also exhibit the activities of N-methyl-D-aspartate-receptor antifebrin and antagonist since combination of N-methyl-D-aspartate receptor, GABAergic drugs, heparin, and glucocorticosteroids are more useful to treat delayed postischemic injury (Rathod Jaimik et al., 2011).

7.7 Anthelmintic Activity

The ethanolic extract of *B. arundinacea* root was tested for anthelmintic effects against *Pheritima posthuma*, and investigations involve determination of death time of worms and paralysis time in varying concentration of extract doses like 10 mg/mL, 20 mg/mL, and 50 mg/mL. The extract exhibited very significant anthelmintic activity in a dose-dependent manner compared to control. The results of this activity were compared with two standards: Albendazole (10 mg/mL) and Piperazine citrate (15 mg/mL) (Kumar et al., 2012).

7.8 Insecticidal Activity

The shoots of bamboo are known to have a minute concentration of hydrogen cyanide that acts as a strong poison and has enough potential to induce death. Therefore, 8 g of fresh/raw or improperly cooked bamboo shoots when ingested results in sudden death of the individual due to the presence of 0.03% hydrogen cyanide. Hairs on bamboo species are the habitat of fungus and are responsible for dermatitis. Traces of cyanogenic glucosides and benzoic acid present in young shoots also have pronounced lethal effects on mosquito larvae and possess strong antiseptic and antiinsecticidal properties (Soni et al., 2013).

7.9 Antioxidant Activity

The existence of a double bond in conjunction with the 4-oxo functional group and 3/5-OH group imparts maximum antioxidant potential to leaves of *B. arundinacea* (Naz et al., 2012).

7.10 Aphrodisiac Effects

Numerous bamboo species are known to have strong aphrodisiac effects, so they are widely used in traditional, Chinese, and Ayurvedic systems of medicines. Shoots of whole bamboo plant when taken regularly increases sexual stamina and strengthen libido (Archana et al., 2010).

8. SIDE EFFECTS AND TOXICITY

Not enough is known about the use of bamboo during pregnancy and breastfeeding. Prolonged use of bamboo shoot might cause thyroid disorders (Chandra et al., 2004).

References

Akinlabi, E.T., Anane-Fenin, K., Akwada, D.R., 2017. Regeneration, Cultivation, and Sustenance of Bamboo, Bamboo. Springer, pp. 39–86.

Archana, A., Megha, A., Amit, R., 2010. Traditional remedy, kunch pak-a review. International Journal of Pharma Bio Sciences 1, 13.

Banik, R.L., 2015. Bamboo Silviculture, Bamboo. Springer, pp. 113–174.

Chandra, A.K., Ghosh, D., Mukhopadhyay, S., Tripathy, S., 2004. Effect of bamboo shoot, Bambusa arundinacea (Retz.) Willd. on thyroid status under conditions of varying iodine intake in rats. Indian Journal of Experimental Biology.

Garcia, C., Giron, M., Espiloy, Z., 1997. Protection of bamboo mat boards against powder-post beetle and fungal attack. FPRDI Journal 23, 53–61.

Ghosh, G.K., 2008. Bamboo: The Wonderful Grass. APH Publishing.

Hunter, I.R., 2003. Bamboo Resources, Uses and Trade: The Future?.

Khare, C.P., 2008. Indian Medicinal Plants: An Illustrated Dictionary. Springer Science & Business Media.

Kumar, H., Raju, M., Dinda, S., Sahu, S., 2012. Evaluation of anthelmintic activity of bambusa arundinacea. Asian Journal of Pharmacy and Technology 2, 62–63.

Liese, W., Kumar, S., 2003. Bamboo Preservation Compendium.

Liese, W., Schmitt, U., 2006. Development and structure of the terminal layer in bamboo culms. Wood Science and Technology 40, 4–15.

Lobovikov, M., 2003. Bamboo and rattan products and trade. Journal of Bamboo and Rattan 2, 397–406.

Maoyi, F., Baniak, R., 1995. Bamboo production systems and their management, propagation and management. In: Bamboo, People and the Environment: Proceedings of the Vth International Bamboo Workshop and the IV International Bamboo Congress Ubud, pp. 18–33.

Maridass, M., De Britto, A.J., 2008. Origins of plant derived medicines. Ethnobotanical Leaflets 44, 2008.

Menaria, J., 2016. Anti diabetic activity of leaves extract of bambusa arundinacea. The Pharmaceutical and Chemical Journal 3, 197–200.

Morrell, J.J., 2011. Resistance of selected wood-based materials to fungal and termite attack in non–soil contact exposures. Forest Products Journal 61, 685–687.

Naz, S.H., Zubair, M., Rizwan, K., Rasool, N., Jamil, M., Riaz, M., Imran, M., Abbas, M., 2012. Phytochemical, antioxidant and cytotoxicity studies of bambusa arundinacea leaves. International Journal of Phytomedicine 4, 220.

Nirala, D.P., Ambasta, N., Kumari, P., 2017. A review on uses of bamboo including ethno-botanical importance. International Journal of Pure Applied Bioscience 5, 515–523.

Nirmala, C., Bisht, M., 2015. Bamboo: a prospective ingredient for functional food and nutraceuticals. In: Proceedings of 10th World Bamboo Congress, September, pp. 17–22.

Park, K., Zhao, X., 2012. Increased in vitro anticancer activity in HepG2 human hepatoma cells and in vivo hepatitis protective effect of bamboo salt. Proceedings of the Nutrition Society 71.

Rao, I.R., Yusoff, A., Rao, A., Sastry, C., 1989. Propagation of Bamboo and Rattan through Tissue Culture.

Rastogi, R.P.M., Mehrotra, B., 1993. BN Compendium of Indian Medicinal Plants, vol. 4. Central DrugResearch Institute Lucknow & NISC, New Delhi, p. 1.

Rathod Jaimik, D., Pathak Nimish, L., Patel Ritesh, G., Bhatt Nayna, M., 2011. Phytopharmacological properties of Bambusa arundinacea as a potential medicinal tree: an overview. Journal of Applied Pharmaceutical Science 1, 27–31.

Sarangthem, K., Hoikhokim, T., Singh, N., Shantibala, G., 2010. Cyanogenic Glycosides in Bamboo Plants Grown in Manipur, India. In: Biology and Taxonomy, vol. 5, p. 2.

Singhal, P., Bal, L.M., Satya, S., Sudhakar, P., Naik, S., 2013. Bamboo shoots: a novel source of nutrition and medicine. Critical Reviews in Food Science and Nutrition 53, 517–534.

Soni, V., Jha, A.K., Dwivedi, J., Soni, P., 2013. Traditional uses, phytochemistry and pharmacological profile of Bambusa arudinacea Retz. Tang [Humanitas Medicine] 3, 20.21–20.26.

Tang, T., 2009. Bamboo Preservation in Vietnam. International Research Group on Wood Protection. IRG/W/40457.

Tang, T., Schmidt, O., Liese, W., 2012. Protection of bamboo against mould using environment-friendly chemicals. Journal of Tropical Forest Science 285–290.
Tang, T.K.H., Welling, J., Liese, W., 2013. Kiln drying for bamboo culm parts of the species Bambusa stenostachya, Dendrocalamus asper and Thyrsostachys siamensis. Journal of the Indian Academy of Wood Science 10, 26–31.
Tran, V.H., 2010. Growth and Quality of Indigenous Bamboo Species in the Mountainous Regions of Northern Vietnam. Citeseer.
Umamaheswari, S., 2016. Pharmacological Evaluation of Bambusa Arundinacea RETZ Roxb Seeds.
Vanithakumari, G., Manonayagi, S., Padma, S., Malini, T., 1989. Antifertility effect of Bambusa arundinacea shoot extracts in male rats. Journal of Ethnopharmacology 25, 173–180.
Wang, D.-H., Tsai, H., 2015. Bamboo Resources & Carbon Storage in Taiwan. 10th World Bamboo Congress, Korea.
Wu, J., 2014. International Trade of Bamboo and Rattan 2012. International Network for Bamboo and Rattan (INBAR), Beijing.
Xuhe, C., 2003. Promotion of bamboo for poverty alleviation and economic development. Journal of Bamboo and Rattan 2, 345–350.
Zhang, J., Gong, J., Ding, Y., Lu, B., Wu, X., Zhang, Y., 2010. Antibacterial activity of water-phase extracts from bamboo shavings against food spoilage microorganisms. African Journal of Biotechnology 9, 7710–7717.
Zhao, X., Song, J.-L., Kil, J.-H., Park, K.-Y., 2013. Bamboo salt attenuates CCl4-induced hepatic damage in Sprague-Dawley rats. Nutrition research and practice 7, 273–280.

Further Reading

Chacko, K., Jayaraman, M., 1988. Effect of container size on growth of bambusa arundinacea seedlings. Proceedings of International Bamboo Workshop 96–98.
Chandrashekara, U., 1996. Ecology of Bambusa arudinacea (Retz.) Willd. growing in teak plantations of Kerala, India. Forest Ecology and Management 87, 149–162.
Cheng, L., Adhikari, S., Wang, Z., Ding, Y., 2015. Characterization of bamboo species at different ages and bio-oil production. Journal of Analytical and Applied Pyrolysis 116, 215–222.
Macharla, S.P., Goli, V., Santhosha, D., Ravinder, N.A., 2012. Antdiabetic activity of Bambusa arundinaceae stem extracts on alloxan induced diabetic rats. Journal of Chemical, Biological and Physical Sciences 2, 832.
Mohmod, A.L., 1991. Effects of Age and Height on Selected Properties of Three Malaysian Bamboo Species. Univo Pertania Malaysia.
Thomas, T.P., 1988. Effect of N, P and K on growth of bambusa arundinacea seedlings in pots. In: Rao, I.V.R., Ganasharan, R., Sastry, C.B. (Eds.), Proceedings of International Bamboo Workshop, Bamboos Current Research, pp. 112–116. Cochin, India.

CHAPTER

4

Basil

Farwa Nadeem[1], *Muhammad Asif Hanif*[1], *Ijaz Ahmad Bhatti*[1], *Muhammad Idrees Jilani*[2], *Rashid Al-Yahyai*[3]

[1] Department of Chemistry, University of Agriculture, Faisalabad, Pakistan; [2] Department of Chemistry, University of Sahiwal, Sahiwal, Pakistan; [3] Department of Crop Sciences, Sultan Qaboos University, Muscat, Oman

OUTLINE

Medicinal Plants of South Asia
https://doi.org/10.1016/B978-0-08-102659-5.00004-5

Copyright © 2020 Elsevier Ltd. All rights reserved.

1. BOTANY

1.1 Introduction

Basil (*Ocimum basilicum* L.) (Fig. 4.1) is the most significant widely grown commercial ornamental crop and an annual aromatic herb belonging to Lamiaceae family with genus *Ocimum* constituting 50 to 150 species having maximum abundance in tropical regions of Asia, Africa, and Central and South America in spite of being worldwide in distribution (Nurzynska-Wierdak, 2007). Essential oil extracted from different basil varieties is of extreme economic importance, and the United States of America is known to have the largest export market in this regard, after the Netherlands, United Kingdom, France, Germany, and many European countries (Hiltunen and Holm, 1999). Global statistics of dried basil production is quite difficult to obtain, as large quantities of total world production specifically in California, India, and Mediterranean areas are locally consumed instead of internationally exported.

Basil

FIGURE 4.1 Basil plants and seeds.

However, even till today, the United States is the biggest dry basil consumer. Total import of dried basil leaves by the Netherlands is 80 tons/year, Germany 200 tons/year, United Kingdom 250 tons/year, and France 300–350 tons/year. Egypt is a major dry basil supplier (Davidson, 1961). Basil is currently produced in Argentina (0.2 tons), Hungary (0.3 tons), Albania (0.5 tons), Reunion (0.5 tons), Madagascar (1 ton), the United States (1 ton), Yugoslavia (1 ton), Israel (2 tons), Comoros (4.5 tons), Pakistan (4.5 tons), Egypt (5 tons), Bulgaria (7 tons), and India (15 tons) (Hiltunen and Holm, 1999).

1.2 History/Origin

O. basilicum is native to tropical Asia but more specifically found in Pakistan and India, where it has been cultivated for approximately 5000 years (Tucker and DeBaggio, 2000). Basil is thought to have been brought to England from India in the early 1500s and arrived in the United States in the 1600s (Lupton et al., 2016).

1.3 Demography and Location

Basil is worldwide in distribution, as all species of *Ocimum* can thrive in a variety of soils and are found to be highly resistant toward varying environmental factors, fluctuating climatic conditions, alternating temperature, and moderately changing soil pH (Pushpangadan and George, 2012). Basil grows well in soils having pH ranging from 4.3 to 8.2, but optimum conditions are obtained at a pH of 6.4. Basil has a medium to deep root system and requires high amounts of water (Treadwell et al., 2007).

1.4 Botany, Morphology and Ecology

O. basilicum is a branched herbaceous plant with an average height ranging from 0.6 to 0.9 m, having square, glabrous stems and branches, generally green to somewhat light purple in color. Leaves of basil are simple and oppositely arranged on stem with average length ranging from 2.5 to 5 cm or more. These longer leaves are ovate, having acute tip with lobed or toothed margins. Petiole is approximately 1.3–2.5 cm in length, and leaves are known to have several oil glands that strongly exude highly volatile scented oil. Inflorescence of basil is generally racemose, while terminal raceme is much longer than lateral ones. Bracts are stalked but shorter than acute, ovate, and calyx. Calyx is approximately 5 mm in length, enlarging inside the fruit that is known to have a shorter pedicel. The lower lip of the calyx is characterized by two central teeth and is longer than upper rounded lip. The corolla is approximately 8–13 mm in length with characteristic purplish, pink, or white color, slightly pubescent or glabrous. Nutlets are almost 2 mm in length, ellipsoid in shape, and black in color with slightly pitted texture. Floral sepals are five in number and found fused into two-lobed calyx. Furthermore, the ovary is superior and the fruit constitutes four achenes (Bilal et al., 2012).

2. CHEMISTRY

Sweet basil and other basil varieties are characterized by low fat content and known to have low caloric value. Basil is a rich source of minerals and vitamin A. Only a few fresh basil leaves having a weight of approximately 2.5 g contain 11.55 mg of potassium, 3.85 mg of calcium, 96.6 IU of vitamin A, and less than one calorie energy along with a small amount of vitamin C, fiber, protein, minerals, and many other chemical constituents (Hanif et al., 2011).

The characteristic odor of basil is mainly attributed to extensive existence of volatile essential oil, which is mainly confined to green basil leaves and known to have rich deposition of large quantities of aldehydes, terpenes, and phenols. The oil obtained from seeds is termed "fixed oil" and is largely composed of fatty acid contents. Some chemical compounds other than fixed and volatile oil include tannins, saponins, glycosides, alkaloids, carotenes, and ascorbic acid. Some major components of essential oil of *Ocimum* species are eugenol, δ-cadinene, eucalyptol, δ-gurjunene, linalool, α-bisabolol, α-terpineol, tau muralol, β-elemene, epibiciclosesquiphelandrene, α-bergamotene, β-farnesene, α-guaiene, methyl eugenol, cubenol, α-copaene, germacrene D, β-cadinene, tau-cadinol, elixen, camphor, α-cariophylene, bornyl acetate, and β-cariophylene (Hanif et al., 2011). The percentage contribution of different components is oil (18%–26%), triglycerides (94%–98%), palmitic acid (6.1%–11.0%), oleic acid (8.5%–13.3%), linoleic acid (17.8%–31.3%), and linolenic acid (43.8%–64.8%) (Angers et al., 1996). Structures of some chemical constituents is given in Fig. 4.2.

3. POSTHARVEST TECHNOLOGY

O. basilicum leaves are harvested early in the morning after evaporation of dew drops and just before intense sunshine to ensure maximum essential oil contents. During the drying process, leaves should not be shredded, as a huge amount of essential oil can be lost along with some major chemical constituents. Shade drying is mostly preferred instead of sun drying, as intense heating can cause loss of all aromatic compounds and volatile components (Pushpangadan and George, 2012).

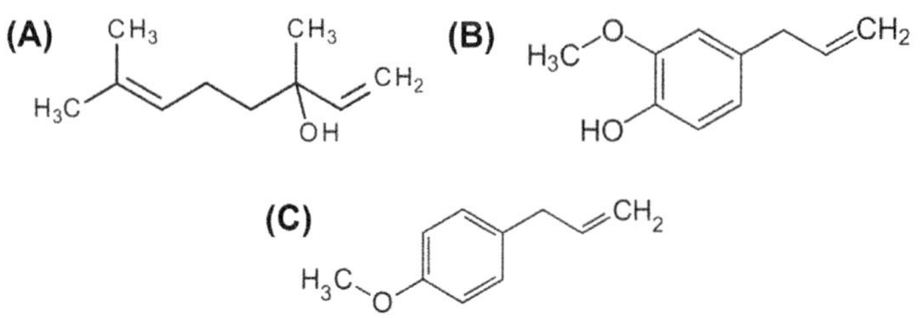

FIGURE 4.2 Major chemical constituents of essential oil in different basil varieties: (A) linalool, (B) eugenol, and (C) methyl chavicol.

4. PROCESSING

Conventionally, basil is dried by hanging the washed bundles in a shaded and dry place or spread thoroughly in between sheets of paper or towel to avoid discoloration and oxidation. However, hot air drying by artificial method is only used for industrial applications. Leaves of basil should be immediately dried after harvesting, as long-term exposure in open air results in darkening of leaves due to oxidation reaction. Drying is preferably done at temperature below 40°C to minimize the loss of highly volatile essential oil and all constituent bioactive compounds (Werker et al., 1993).

5. VALUE ADDITION

Leaves of basil can be used in combination with numerous other herbs such as juniper, thyme, garlic, rosemary, marjoram, sage, oregano, pepper, paprika, parsley, and mustard to be used as topping and a flavoring agent in soups, stew, stuffing, rice, meat, vegetables, roasted chicken, and fried fish. Basil is a major ingredient for manufacturing of beverages, liqueurs, vinegar, oils, drinks, teas, cheeses, and jams. Cinnamon and lemon basils are used to enhance taste in delicious deserts, while purple vinegar is used to glorify color and increase aroma. Larger basil leaves are chopped, torn, or minced before consumption or addition in pasta, rice, salads, and a number of vegetarian dishes. Branches and the soft woody stem of *O. basilicum* can be added as a flavoring agent in seafood, tender chicken, steamed meat, soups, and various drinks. Flowers of the basil are edible and can directly be added in dishes and salads for the purpose of garnishing at the end of cooking for maximum smell and prominent flavor. Uses of basil are plenteous and diverse, as it can directly be used in drinks, liqueurs, herbal teas, stews, sauces, dressings, fish, vegetables, and meat. It is extensively used for both industrial and domestic applications in preparation of pesto with varying combinations of nuts, oil, garlic, cheese, and basil. Basil is also used as a courtesy flavor to tomatoes (Simon et al., 1999).

6. USES

Uses and applications of basil are diverse, ranging from culinary to religious and religious to ritual, as a number of curious historical beliefs are associated with usage of basil in different forms. According to Hindu belief in India, basil leaves when buried with a dead body allow direct passport to heaven. In Portugal, basil is used as a gift to present to a loved one on religious holidays of Saint Antony and Saint John. Holy basil is known to have religious significance across a range of belief systems; the Serbian, Romanian, Bulgarian, and Greek Orthodox churches use basil for preparation of pots and holy water of basil beneath the church altars (Tilebeni, 2011). The essential oil of various basil species is known to have potential insecticidal properties when mixed with other solvents, as *Ocimum gratissimum* shows 100% repellency against *Musca domestic* (Nahak et al., 2011). Similarly, essential oil of *O. basilicum* showed strong repellency effects against *Tribolium castaneum*, commonly known as red flour beetle (Nahak et al., 2011). Basil has traditionally been used as an important herbal medicine for a wide range of symptoms ranging from analgesic to vermifuge and is reputed to treat numerous problems including fungal infections, headache, and acne (Adigüzel et al., 2005). *O. basilicum* has also been used in a number of traditional Chinese medicines for treatment of malaria, irregular menstrual cycle, anorexia, arthritis, earache, hemostyptic disorders, ulceration in gums, and kidney diseases (Deng et al., 2007). The use of sweet basil under trade name "Rihan" is most common in Arab countries, which is used for treatment of diarrhea, cold, and cataract in central and northern Oman (Ghazanfar and Al-Al-Sabahi, 1993). In Iran, "Reyhan" is used as medicine for treatment of influenza, colic ulcer, lung complaints, chest infections, urinary tract inflammations, and is known to be an excellent expectorant, appetizer, and tonic (Miller, 2014). The basil infusion in Jordan is used as antidiarrheal, antiemetic, and antihelminthic (Vieira and Simon, 2000). Basil flowers are considered to be digestive stimulant and highly antispasmodic (Bilal et al., 2012).

7. PHARMACOLOGICAL USES

7.1 Antioxidant Activity

O. basilicum L. shows strong antioxidant nature due to high flavonoid and phenolic contents such as anthocyanin, vicenin, eugenol, and orientin, which tends to vary significantly through variations in amount of these chemicals (Juliani and Simon, 2002).

7.2 Anticancer Activity

Basil is a natural anticancer agent at the cellular level due to a peculiar array of flavonoids including orientin and vicenin that are known to protect chromosomes and cellular structure from oxygen-based and radiation damage of white blood cells in the human body. Alcoholic extract of basil has modulatory impacts on several carcinogen metabolizing enzymes like glutathione transferase, cytochrome b_5, cytochrome P_{450}, and aryl hydrocarbon hydroxylase. Eugenol is an important flavonoid found in basil that inhibits early events of 7,12-dimethylbenz[a]anthracene (DMBA)-induced buccal pouch carcinogenesis, whereas basil leaf extract inhibits metabolic activation of carcinogens by the blockage and suppression of events mainly associated with carcinogenesis through inhibiting metabolic activation of carcinogen (Aggarwal and Shishodia, 2006).

7.3 Antiviral Activity

O. basilicum is a strong antiviral in nature, as it has enough potential against several RNA viruses including enterovirus 71 and coxsackievirus B1 and DNA viruses such as hepatitis B virus, adenoviruses, and herpes viruses due to the presence of ursolic acid, linalool, and apigenin (Chiang et al., 2005).

7.4 Antimicrobial Activity

Basil shows antibiotic activity that can be associated with high contents of linalool (Araújo Silva et al., 2016; Edris and Farrag, 2003).

7.5 Analgesic Activity

Methanolic basil extract was tested on mice for assessment of analgesic effects at various concentration levels. The results of the experiment demonstrated that analgesic effects were more pronounced at concentration of 200 mg per kilogram when compared with a standard aspirin drug (Mandoulakani et al., 2017).

7.6 Antiinflammatory Activity

To investigate the antiinflammatory effects of *O. basilicum* L., a tincture of it was induced with turpentine oil in male rats with severe inflammations. The results of whole research indicated that the tincture significantly reduced the total number of leukocytes, percentage of monocytes, and activation of circulating phagocytes but had poor inhibitory effect on synthesis of nitric oxide. Tincture of basil had very few inhibitory effects on several tested parameters when compared with a standard of diclofenac; however, it has strong antiinflammatory effects on bone marrow acute phase response and a reduced effect on synthesis of nitric oxide (Peana et al., 2002).

7.7 Antiulcer Activity

Fixed oil of basil was tested to investigate its antiulcer potential or activity against stress-, serotonin-, reserpine-, histamine-, alcohol-, indomethacin-, and aspirin-induced ulceration. The results indicated that significant inhibition was achieved in aspirin-induced gastric ulceration and gastric secretion in pylorus ligates rats. Numerous antisecretory, histamine antagonistic, and lipoxygenase inhibitory effects significantly contributed toward antiulcer activity. Hence, this fixed oil is considered to be a natural drug having both antiulcer and antiinflammatory effects on tested animals (Dharmani et al., 2004).

7.8 Antidiabetic Activity

The aqueous extract of *O. basilicum* was tested for hypoglycemic effects through in vitro investigations. Results of whole research indicated that aqueous extracts of basil are known to have strong antioxidant potentials and possesses the capability of inhibiting α-amylase and α-glucosidase, so it has proved to be useful in controlling diabetes of infected animals (El-Beshbishy and Bahashwan, 2012).

7.9 Antimalarial Activity

Essential oil of basil is known to have strong larvicidal properties, more specifically antimalarial activities, due to the presence of "eugenol" as a major chemical constituent, so it is used extensively in numerous mosquito repellant sprays. Leaves of basil are also effective against face marks, ringworm infections, facial wrinkles, cuts, ulcers, and wounds and helps to regain the elasticity of skin (Bansod and Rai, 2008).

7.10 Insecticidal Activity

Essential oil obtained from *O. gratissimum* and *O. basilicum* through steam distillation was tested against a number of insects, and results indicated that aromatic compounds have no significant influence on germination of seeds; however, 88% of treated seeds with aromatic powders were protected from insects as compare to other untreated seeds, which leads toward betterment in survival rate of the plant (Lopez et al., 2008).

7.11 Radio Protective Activity

The orientin and vicenin are two important flavonoids extracted from leaves of basil that are known to have superior radio protective effects as compared to numerous synthetic radio protectors by providing protection to lymphocytes of the human body against clastogenic radiation effects at nontoxic low concentrations. Extracts of three different plants such as *Plumobogo rosea, Ocimum sanctum,* and *Withania somnifera* were tested on bone marrow cells of mice that were radiated by gamma radiations of "2 gray" and results indicated that *O. sanctum* has maximum potential of protection from these radiations as measured by exogenous colony of spleen-forming unit assay (Uma Devi et al., 2000).

7.12 Immunomodulatory Activity

Scented oil of basil extracted by steam distillation presented modifications in humoral immune response in albino rats, which can be credited to release of hypersensitivity reaction mediator, production of antibodies, and response of tissues toward these mediators in target organs. Seed oil of basil seems to rapidly modulate both humoral and cell-mediated immune responsiveness along with GABAergic pathways liable for immune-modulatory effects. All the immune reactions of cells are improved by using basil through improvements in humoral and cellular immunity (Mukherjee et al., 2005).

7.13 Antistress Activity

Basil's immune-stimulant capacity accounts for its adaptogenic actions. Stress-induced ulcers and milk-induced leukocytosis are significantly reduced by basil. Stress is actually the nonspecific response of the body experienced by every individual because of an unsatisfied demand that can be either physical or psychologic. Extreme stress is very harmful for the body, so it requires immediate treatment. Many harmful diseases including ulcerative colitis, peptic ulcer, hypertension, male impotence, cognitive dysfunction, diabetes mellitus, endocrine disruption, anxiety, depression, psychiatric diseases, and immunosuppression are directly associated with stress. The excellent rejuvenating activity of basil proves to be very helpful in reducing stress and assisting the body in relaxation and improving memory. Basil's antihypnotic effect upturns survival time in extreme anoxic stress. Basil is known to reduce oxidative stress in the bodies of rabbits (Chattopadhyay et al., 1992).

7.14 Antipyretic Activity

The antipyretic activity of fixed oil of basil was tested against typhoid-paratyphoid A/B vaccine-induced pyrexia in the bodies of rats. Fixed oil of basil on intraperitoneal administration considerably decreased febrile response, showing its antipyretic potential. This fixed oil showed comparable antipyretic effects to aspirin at a dose of 3 mL per kilogram. Furthermore, fixed oil is known to have prostaglandin inhibitory effects that can comprehensively be explained by describing its antipyretic potentials (Singh et al., 2007).

7.15 Antiarthritic Activity

To evaluate the antiarthritic activity of fixed oil of basil, arthritis was produced intentionally in rats through incorporation of excessive formaldehyde. Fixed oil significantly reduced the diameter of inflamed paw. Striking improvements were observed in condition of arthritis in rats on intraperitoneal basil fixed oil administration continuously for 10 days. Antiarthritic effects of fixed oil of basil were observed by the dose of 3 milliliters per kilogram, which is analogous to 100 milligrams per kilogram of aspirin. Fixed oil of basil is known to inhibit mediators of inflammation such as bradykinin, PGE2, serotonin, and histamine along with carrageenan-induced inflammations. The end result shows that basil is a potential antiarthritic herbal medicine widely used in the treatment of inflammations (Singh and Majumdar, 1996).

7.16 Cardiovascular Effects

Basil has potential for lowering the excessive triglycerides and cholesterol level of the body, which helps in prevention of numerous cardiovascular disorders. The combination of high circulating triglycerides and low-density lipoprotein are major risks for atherosclerosis, strokes, and heart attack. Extracts of basil slowed thrombosis and aggregations of platelets, suggesting its potential for prevention of heart attack and strokes (Patel et al., 2012).

7.17 Antiosteoporotic Effects

Aqueous extracts of chicory, basil, and parsley were tested on rats severely affected by glucocorticoid-induced osteoporosis. Results of the whole research indicated that all three herbs have potential to resist osteoporosis, but chicory has more potential in comparison with basil and parsley due to rich deposition of inulin and flavonoids (Hozayen et al., 2016).

7.18 Anxiolytic and Sedative Effects

The sedative effects and anxiolytic potentials of essential oil and hydroalcoholic extracts of *O. basilicum* were investigated in mice, and results indicated that sedative effects and antianxiety effects of the essential oil are relatively higher than hydroalcoholic extracts using the same amount of medicine. The reason for this higher potential of scented oil is rich accumulation of phenolic contents that are not found in the case of aqueous extracts (Diantini and Subarnas, 2011).

7.19 Anticolic Effects

Scented or essential oil of basil has ameliorative effects on acetic acid–induced colitis in rats. Increased level of myeloperoxidase was decreased significantly after treatment with essential oil of basil with a treatment dose ranging from 200 to 400 microliter per kilogram. These results have shown that *O. basilicum* exhibits protective potential against acetic acid–induced colitis (Rashidian et al., 2016).

7.20 Cytotoxic Effects

Antioxidant potentials and anticancer activities of methanolic extracts of *O. basilicum* and *Mentha longifolia* were investigated, and results indicated that both have enough potential to act as strong antioxidants and provide protection against DNA damage because of the cytotoxic activities of these extracts against MCF-7 cell line (Shirazi et al., 2014).

7.21 Antihematotoxic Effects

Methanolic extract of the leaves of essential oil of *O. basilicum* L. has strong antihematotoxic effects against benzene-induced hematotoxicity in rats. Further investigations revealed that secondary metabolites of basil extract are mainly composed of monoterpene geraniol, which readily oxidizes into citral, which is known to have major modulatory effects in cell cycle's deregulation and hematological abnormalities induced by benzene in the bodies of mice (Miraj and Kiani, 2016).

7.22 Phytoremediatory Effects

Two species of basil *Ocimum minimum* L. and *O. basilicum* L. were tested for their phytoremediation potentials, and results indicated that both species can endure the pollution of endosulfan, but *O. basilicum* seems to be a more appropriate candidate in this regard (Ramírez-Sandoval et al., 2011).

7.23 Antihypertensive Effects

Recently, an experiment was conducted to study the antihypertensive effects of extracts of *O. basilicum* on rats affected by renovascular hypertension, and results indicated that basil has positive effects on embryo transference, cardiac hypertrophy, and blood pressure and are consistent with effects of embryo-transfer-converting enzyme, which needs further study and thorough investigation (Umar et al., 2010).

7.24 Vasorelaxant and Antiplatelet Effects

Several antiplatelet and vasorelaxant endothelium-dependent activities of the aqueous extract of *O. basilicum* were investigated, and corresponding results showed that hypercholesterolemic diet (HCD) statistically reduces vascular relaxation in Human chorionic gonadotropin (HCG) as compared to not-castrated group (NCG) ($P < .001$) and enhances vascular responses to phenylephrine ($P < .02$). Utilization of *O. basilicum* as a potential medicinal plant can be highly beneficial for numerous diseases associated with the cardiovascular system (Amrani et al., 2009).

7.25 Antithrombotic Effects

Aqueous extract of *O. basilicum* L. was investigated for experimental thrombosis and platelet aggregations. Several concluding results of this research indicated that basil has high potential to inhibit platelet aggregation through thrombin and adenosine diphosphate, which is dose dependent and takes 3 to 7 days for effective appearance. However, active plant ingredients responsible for such kind of behavior are still under investigation and need characterization (Özbek et al., 2007).

7.26 Synergistic Effects

Essential or scented oils extracted from different basil varieties through hydrodistillation were examined for antimicrobial activity against a wide range of foodborne molds, yeasts, and gram-positive and gram-negative bacteria by agar well diffusion method. Results of this current study indicated that there is a need for further investigation to completely understand the antimicrobial effects of essential oil of basil in the presence of several other food ingredients and preservation parameters (Lachowicz et al., 1998).

8. SIDE EFFECTS AND TOXICITY

Basil oils and extracts might slow blood clotting and increase bleeding. In theory, basil oils and extracts might make bleeding disorders worse. Basil extracts might lower blood pressure. Basil extracts might make blood pressure become too low in people with low blood pressure.

References

Aggarwal, B.B., Shishodia, S., 2006. Molecular targets of dietary agents for prevention and therapy of cancer. Biochemical Pharmacology 71, 1397–1421.

Adigüzel, A., Güllüce, M., şengül, M., Öğütcü, H., şahin, F., Karaman, I., 2005. Antimicrobial effects of *Ocimum basilicum* (Labiatae) extract. Turkish Journal of Biology 29, 155–160.

Amrani, S., Harnafi, H., Gadi, D., Mekhfi, H., Legssyer, A., Aziz, M., Martin-Nizard, F., Bosca, L., 2009. Vasorelaxant and anti-platelet aggregation effects of aqueous *Ocimum basilicum* extract. Journal of Ethnopharmacology 125, 157–162.

Angers, P., Morales, M.R., Simon, J.E., 1996. Fatty acid variation in seed oil among Ocimum species. Journal of the American Oil Chemists Society 73, 393–395.

Araújo Silva, V., Pereira da Sousa, J., de Luna Freire Pessôa, H., Fernanda Ramos de Freitas, A., Douglas Melo Coutinho, H., Beuttenmuller Nogueira Alves, L., Oliveira Lima, E., 2016. *Ocimum basilicum*: antibacterial activity and association study with antibiotics against bacteria of clinical importance. Pharmaceutical Biology 54, 863–867.

Bansod, S., Rai, M., 2008. Antifungal activity of essential oils from Indian medicinal plants against human pathogenic Aspergillus fumigatus and *A. niger*. World Journal of Medical Sciences 3, 81–88.

Bilal, A., Jahan, N., Ahmed, A., Bilal, S.N., Habib, S., Hajra, S., 2012. Phytochemical and pharmacological studies on *Ocimum basilicum* Linn-A review. International Journal of Current Research and Review 4.

Chattopadhyay, R., Sarkar, S., Ganguly, S., Medda, C., Basu, T., 1992. Hepatoprotective activity of Ocimum Sanctum leaf extract against Paracetamol included Hepatic damage in rats. Pharmacognosy Research.

Chiang, L.C., Ng, L.T., Cheng, P.W., Chiang, W., Lin, C.C., 2005. Antiviral activities of extracts and selected pure constituents of *Ocimum basilicum*. Clinical and Experimental Pharmacology and Physiology 32, 811–816.

Davidson, B., 1961. Black Mother; the Years of the African Slave Trade.

Deng, C., Liu, N., Gao, M., Zhang, X., 2007. Recent developments in sample preparation techniques for chromatography analysis of traditional Chinese medicines. Journal of Chromatography A 1153, 90–96.

Dharmani, P., Kuchibhotla, V.K., Maurya, R., Srivastava, S., Sharma, S., Palit, G., 2004. Evaluation of anti-ulcerogenic and ulcer-healing properties of *Ocimum sanctum* Linn. Journal of Ethnopharmacology 93, 197–206.

Diantini, A., Subarnas, A., 2011. Analysis of Indonesian spice essential oil compounds that inhibit locomotor activity in mice. Pharmaceuticals 4, 590–602.

Edris, A.E., Farrag, E.S., 2003. Antifungal activity of peppermint and sweet basil essential oils and their major aroma constituents on some plant pathogenic fungi from the vapor phase. Molecular Nutrition and Food Research 47, 117–121.

El-Beshbishy, H., Bahashwan, S., 2012. Hypoglycemic effect of basil (*Ocimum basilicum*) aqueous extract is mediated through inhibition of α-glucosidase and α-amylase activities: an in vitro study. Toxicology and Industrial Health 28, 42–50.

Ghazanfar, S.A., Al-Al-Sabahi, A.M., 1993. Medicinal plants of Northern and Central Oman (Arabia). Economic Botany 47, 89–98.

Hanif, M.A., Al-Maskari, M.Y., Al-Maskari, A., Al-Shukaili, A., Al-Maskari, A.Y., Al-Sabahi, J.N., 2011. Essential oil composition, antimicrobial and antioxidant activities of unexplored Omani basil. Journal of Medicinal Plants Research 5, 751–757.

Hiltunen, R., Holm, Y., 1999. Basil: The Genus Ocimum (Medicinal and Aromatic Plants-Iindustrial Profiles, vol. 10. CRC Press, pp. 1027–4502.

Hozayen, W.G., El-Desouky, M.A., Soliman, H.A., Ahmed, R.R., Khaliefa, A.K., 2016. Antiosteoporotic effect of *Petroselinum crispum, Ocimum basilicum* and *Cichorium intybus* L. in glucocorticoid-induced osteoporosis in rats. BMC Complementary and Alternative Medicine 16, 165.

Juliani, H., Simon, J., 2002. Antioxidant activity of basil. Trends in new crops and new uses 575.

Lachowicz, K., Jones, G., Briggs, D., Bienvenu, F., Wan, J., Wilcock, A., Coventry, M., 1998. The synergistic preservative effects of the essential oils of sweet basil (*Ocimum basilicum* L.) against acid-tolerant food microflora. Letters in Applied Microbiology 26, 209–214.

Lopez, M.D., Jordán, M.J., Pascual-Villalobos, M.J., 2008. Toxic compounds in essential oils of coriander, caraway and basil active against stored rice pests. Journal of Stored Products Research 44, 273–278.

Lupton, D., Khan, M.M., Al-Yahyai, R.A., Hanif, M.M., 2016. Basil, Leafy Medicinal Herbs: Botany, Chemistry. Postharvest Technology and Uses, pp. 27–41.

Mandoulakani, B.A., Eyvazpour, E., Ghadimzadeh, M., 2017. The effect of drought stress on the expression of key genes involved in the biosynthesis of phenylpropanoids and essential oil components in basil (*Ocimum basilicum* L.). Phytochemistry 139, 1–7.

Miller, N.F., 2014. Plants and humans in the near east and the caucasus: ancient and traditional uses of plants as food and medicine, a diachronic ethnobotanical review. Ethnobiology Letters 5, 22–23.

Miraj, S., Kiani, S., 2016. Study of pharmacological effect of *Ocimum basilicum*: a review. Der Pharmacia Lettre 8, 276–280.

Mukherjee, R., Dash, P., Ram, G., 2005. Immunotherapeutic potential of *Ocimum sanctum* (L) in bovine subclinical mastitis. Research in Veterinary Science 79, 37–43.

Nahak, G., Mishra, R., Sahu, R., 2011. Taxonomic distribution, medicinal properties and drug development potentiality of Ocimum (Tulsi). Drug Invention Today 3.

Nurzynska-Wierdak, R., 2007. Evaluation of morphological and developmental variability and essential oil composition of selected basil cultivars. Herba Polonica 53.

Özbek, H., Bahadır, Ö., Kaplanoğlu, V., Öntürk, H., 2007. Reyhan (*Ocimum basilicum* L.) uçucu yağının antienflamatuvar aktivitesinin araştırılması. Genel Tıp Derg 17, 201–204.

Patel, D., Prasad, S., Kumar, R., Hemalatha, S., 2012. An overview on antidiabetic medicinal plants having insulin mimetic property. Asian Pacific journal of tropical biomedicine 2, 320–330.

Peana, A.T., D'Aquila, P.S., Panin, F., Serra, G., Pippia, P., Moretti, M.D.L., 2002. Anti-inflammatory activity of linalool and linalyl acetate constituents of essential oils. Phytomedicine 9, 721–726.

Pushpangadan, P., George, V., 2012. Basil. In: Handbook of Herbs and Spices, second ed., vol. 1. Elsevier, pp. 55–72.

Ramírez-Sandoval, M., Melchor-Partida, G., Muñiz-Hernández, S., Giron-Perez, M., Rojas-García, A., Medina-Díaz, I., Robledo-Marenco, M., Velázquez-Fernández, J., 2011. Phytoremediatory effect and growth of two species of Ocimum in endosulfan polluted soil. Journal of Hazardous Materials 192, 388–392.

Rashidian, A., Roohi, P., Mehrzadi, S., Ghannadi, A.R., Minaiyan, M., 2016. Protective effect of *Ocimum basilicum* essential oil against acetic acid–induced colitis in rats. Journal of evidence-based complementary & alternative medicine 21, NP36–NP42.

Shirazi, M.T., Gholami, H., Kavoosi, G., Rowshan, V., Tafsiry, A., 2014. Chemical composition, antioxidant, antimicrobial and cytotoxic activities of *Tagetes minuta* and *Ocimum basilicum* essential oils. Food science & nutrition 2, 146–155.

Simon, J.E., Morales, M.R., Phippen, W.B., Vieira, R.F., Hao, Z., 1999. Basil: a source of aroma compounds and a popular culinary and ornamental herb. In: Perspectives on New Crops and New Uses, pp. 499–505.

Singh, S., Majumdar, D., 1996. Effect of fixed oil of *Ocimum sanctum* against experimentally induced arthritis and joint edema in laboratory animals. International Journal of Pharmacognosy 34, 218–222.

Singh, S., Taneja, M., Majumdar, D.K., 2007. Biological activities of *Ocimum sanctum* L. fixed oil—an overview. Indian Journal of Experimental Biology.

Tilebeni, H.G., 2011. Review to basil medicinal plant. International Journal of Agronomy and Plant Production 2, 5–9.

Treadwell, D.D., Hochmuth, G.J., Hochmuth, R.C., Simonne, E.H., Davis, L.L., Laughlin, W.L., Li, Y., Olczyk, T., Sprenkel, R.K., Osborne, L.S., 2007. Nutrient management in organic greenhouse herb production: where are we now? HortTechnology 17, 461–466.

Tucker, A.O., DeBaggio, T., 2000. Big Book of Herbs. Interweave Press.

Uma Devi, P., Ganasoundari, A., Vrinda, B., Srinivasan, K., Unnikrishnan, M., 2000. Radiation protection by the ocimum flavonoids orientin and vicenin: mechanisms of action. Radiation Research 154, 455–460.

Umar, A., Imam, G., Yimin, W., Kerim, P., Tohti, I., Berké, B., Moore, N., 2010. Antihypertensive effects of *Ocimum basilicum* L.(OBL) on blood pressure in renovascular hypertensive rats. Hypertension Research 33, 727–730.

Vieira, R.F., Simon, J.E., 2000. Chemical characterization of basil (*Ocimum* spp.) found in the markets and used in traditional medicine in Brazil. Economic Botany 54, 207–216.

Werker, E., Putievsky, E., Ravid, U., Dudai, N., Katzir, I., 1993. Glandular hairs and essential oil in developing leaves of *Ocimum basilicum* L.(Lamiaceae). Annals of Botany 71, 43–50.

CHAPTER

5

Bay Leaf

Saima Batool[1], *Rasheed Ahmad Khera*[1], *Muhammad Asif Hanif*[1], *Muhammad Adnan Ayub*[2]

[1] Department of Chemistry, University of Agriculture, Faisalabad, Pakistan; [2] Department of Chemistry, University of Okara, Okara, Pakistan

OUTLINE

https://doi.org/10.1016/B978-0-08-102659-5.00005-7

Copyright © 2020 Elsevier Ltd. All rights reserved.

1. BOTANY

1.1 Introduction

Bay leaf *(Laurus nobilis)* (Fig. 5.1) is an evergreen perennial shrub that belongs to the laurel family (Lauraceae). It has been used for 1000 years, and it is an essential ingredient in cooking and in many traditional practices (Parthasarathy et al., 2008). The genus *Laurus* has a range of 24,00 to 25,00 species, and their varieties are native to the Southern Mediterranean region, the subtropics and tropics of Eastern Asia, South and North America, the Balkans, and Asia Minor. The great variability among species is largely attributed to the uncertainty in the exact number of species. Due to the morphology, flower color, growth habitat, leaves, stems, and chemical composition, variability is found. Two laurel species are traditionally found: *Laurus azorica* and *L. nobilis*. There are number of plants outside the genus *Laurus* with the common name bay laurel, including bay rum tree, or simply bay *(Pimenta racemosa)* (Akgül et al., 1989). *L. nobilis* is known by different names. In Urdu, it is known as teejh pat. In English, it is typically called bay leaf or sweet bay. In Arabic, it is known as *waraq ghaar*. In German, it is known as *lorbeer*. In Greek, it is

FIGURE 5.1 Bay leaf.

called *dafni*. In India, specifically in Hindi, it is called *teejpatta*. In Meghalaya, bay leaf unit production ranges from 30 to 70 kg per tree per year, but in Nepal, the average range is 13 kg of the dry leaves. About 900 tons of bay leaf are produced in Udaipur district, and 2100 tons are exported by Nepal to India (Choudhary et al., 2014). Aegean and Eastern Mediterranean regions are the biggest collection areas of bay leaf for export (Nurbaş and Bal, 2005). Turkey exported 4869 tons of bay leaf to the United States in 2002 (Deniz, 2012).

1.2 History/Origin

The origin of bay leaf is most probably South Asia, from where it spread to Asia Minor and all over the world.

1.3 Demography/Location

Bay leaf is grown in different ecologic and climatic conditions. Wet, sandy soil that has a large quantity of water or some moist atmospheric conditions close to the ocean shore are optimum and the best conditions for rapid luxuriant growth (Patrakar et al., 2012). In warmer weather, leaves may burn; therefore partial sun shade, well-drained sandy soil that has some moisture, and a pH range of 4.5–8.2 are preferred. Bay bears black fruit and yellowish-white fluffy flowers in warmer areas. Temperatures below 28°F and extensive freezing will kill the bay (Kemp et al., 1983). Bay is widely growing in the following countries: India, Pakistan, other Southeast Asian countries, some Pacific islands, Australia, around the coast of the Mediterranean and Southern Europe, Greece, Portugal, France, Turkey, Spain, Algeria, Morocco, Belgium, Central America, Mexico, Southern United States, and the Canary Islands (Parthasarathy et al., 2008).

1.4 Botany, Morphology, Ecology

Bay leaf is native to South Europe (Patrakar et al., 2012). It is a multi-branched, deciduous shrub having height up to 6–8 m and diameter up to 15–40 cm with smooth, thin, and brown bark containing a shady crown (Patrakar et al., 2012). Leaves are alternate, lanceolate, and bipinnate compounds with smooth or sharp margins 29–30 cm long containing 24 leaflets that are lanceolate, 4.8–4.9 cm long, and 1.7–1.8 cm wide with 0.5 cm long petiole. Flowers are ebracteate, four-lobed, white, scented, and small, having eight to 12 male stamens and two to four female staminoids, and the fruit is 10–15 mm, in small clusters, ovoid, thin pericarp enclosing spinach-green seeds and black when ripe. Calyx is pubescent having five clefts and five petals along with glabrous glands, free and white.

2. CHEMISTRY

Bay leaf has a sharp and bitter taste. The difference in fragrance and aroma is due to the presence of essential oils in leaves and other parts of the plant. It has flavonoids, tannins, eugenol, citric acid, carbohydrate, steroids, alkaloids, triterpenoids, and essential oils. Antioxidant properties were discovered in the extract of bay leaf to have phenolic compounds. Each of these chemical constituents varies depending on the type of species. Tanine is a liquid glycoside derived from polypeptide and ester polymer that can be hydrolyzed by the secretion of bile (3, 4, 5–trinidrokside benzoic acid) and glucose (Sumono, 2008). Tanine or tanat acid isolated from some part of plants can be found in the market. It is a cream-colored powder, aromatic, with astringent taste (Sumono, 2008). Tanine is used as an astringent for the gastrointestinal tract or skin and can cause precipitation of the cell membrane protein. It also has a little penetration activity, so it can influence the permeability of the cell membrane.

Bay leaf has traces of fats; (that is, a low amount is present) so it has low caloric value. It is also known as a good and main source of vitamin A and many minerals. One ounce of bay leaf gives 54 calories, 1–1.2 g protein, 12–13 g carbohydrates, a trace of fat, 1–1.5 mg of iron (Fe), 51–53 mg of calcium (Ca), 2000–3000 IU of vitamin A, 14–15 mg of vitamin C, and a small amount of potassium. Bay seeds are rich in dietary fibers. In bay leaf, compounds like eugenol (11%–12%), methyl eugenol (9%–12%), and elemicin (1%–12%) are significant for the spicy aroma of bay leaves, and for determining effective quality of bay leaf, these are used as significant influencers (Biondi et al., 1993). The essential oils in leaves vary from 0.8% to 3% and dry bay fruits from 0.6% to 10%. Structures of some active compounds found in bay leaf are given in Fig. 5.2.

3. POSTHARVESTING TECHNOLOGY

Bay leaf can be harvested at any time of the year from a fully mature plant. Fresh bay leaves have a bitter and pungent taste; therefore before use, leaves should be dried. After picking the leaf, it should be left for 48–72 hours for drying. Better and deeper flavor is observed in freshly dried leaves. Harvesting should be avoided, when plant is wet.

4. PROCESSING

Bay is consumed in a variety of ways and for various purposes. In addition to its fresh leaves, other common processed forms of bay include whole dry leaves, frozen, powdered leaves, and extracted essential oils.

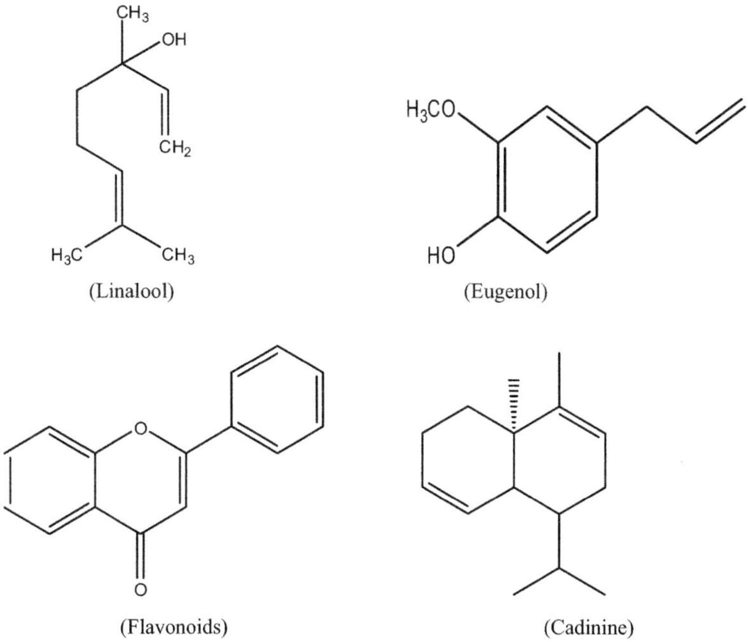

FIGURE 5.2 Structures of some active compounds found in bay leaf.

Leaves can be stored frozen for the sake of use for extended time beyond its fresh shelf life. For drying of bay leaf, different drying methods are available. Traditionally, it is dried in open air for 10–12 days. Sun drying has some disadvantages, like natural color loss and essential oil loss that result in low market value of bay leaf. Hot air drying at 60°C is the best method for producing bay leaves. Steam distillation is the best method for the recovery of essential oils from the bay leaf plant. Essential oil extracted from bay leaf is in two forms, fixed oil and volatile oil, that are collected from bay fruits (Bozan and Karakaplan, 2007).

5. VALUE ADDITION

Bay leaf can be combined with a variety of other herbs including cloves, thyme, tomato, mustard, parsley, paprika, sage, and pepper for use in soups, stews, as well as with fish, vegetables, and meat. Bay leaf with cloves and thyme is used to form tomato sherbet. Bay leaf with beef stock and large egg yolks forms Provencal bay tomato soup. Bay leaves with whole celery seeds, whole cloves, peppercorns, dried parsley, and thyme can be used in bouquet garni. Bay leaf pound cake can be made by using milk, sugar, butter, eggs, cake flour, and baking powder with bay leaves. The leaves of bay have a camphor-like volatile oil that can be used as a coolant, insecticide, germicide, and irritant. Roasting of bay seeds gives them a spicy, coffee-like flavor, and by removing pungency, they become crispy and brown. Small leaves of bay are used in salads, rice, and vegetarian dishes. Its woody branches can be used in steamed meat, drinks, and soups, while leaf bark is used as a condiment in many spices. Bay leaf has universal industrial importance as dried leaves and essential oils give courtesy flavor to foods as in meat products, canned soups, stews, baked goods, sausages, fish, cosmetics, and drugs. Spices and essential oils of bay leaf may extend storage life of foods, as they have antimicrobial and antioxidant activities (El et al., 2014). Chilling of bay leaf retains the taste of this shrub more effectively than drying.

6. USES

Many herbs and spices contribute significantly to health despite low amounts of consumption, as they are full of antioxidants and certain mineral compounds. It is not clear how much bay must be consumed to get its health benefits. Researchers do not have particular recommendations about the specific amount of use. Nevertheless, bay is full of antioxidants and is a good source of minerals and dietary fibers. It complements food flavor, and bay tea is used to treat stomachaches, clear up mucus in the lungs, colds, and sore throat. Poultice of bay leaves is used for the treatment of rheumatism and neuralgia (Goodrich et al., 1980). To treat headache, leaf of bay is kept in a nostril or under the headbands to relieve this pain. Traditionally, it has been used for the treatment of gastrointestinal problems such as impaired digestion, flatulence, eructation, and epigastric bloating and used as diuretic and has many analgesic effects (Elmastaş et al., 2006). Bay is great to add flavor and taste to food and many dishes with added health benefits.

Bay has many uses ranging from culinary to religious. There are number of curious beliefs associated with the historical use of bay leaf. The Temple of Delphi, dedicated to Apollo, used many bay leaves. The roof was made of bay leaves, and priestesses would have to eat bay before giving their oracles. This may have been aided by bay's slightly narcotic qualities. Thus bay leaves are said to aid with psychic powers, particularly prophetic dreams, clairvoyance, protection, healing, purification, strength, wishes, magic, exorcism, divination, visions, inspiration, wisdom, meditation, defense, and accessing the creative world. Israelite society consider the bay leaf as a symbol of victory over misfortune; they were very impressed by this tree. Ancient Mediterraneans said this tree radiates protective power and prevents them from misfortune, so it is planted near houses to keep lightning away. The Romans and Greeks used this as a head band mainly for their respected citizens, poets, heroes, and priests, and they consider sleeping with bay leaves to make a man a poet. Romans also believed that this tree protects from lightning, so Emperor Tiberius always kept a bay leaf hat because he had a fear of thunderstorms; and from witches and wizards. The French sometimes call bay the "berries of bay," and they crowned intelligent people with its berries and leaves, which are burned to increase the psychic powers and protect from evil and negativity. Chinese have a belief that to remove evil messes and crossed conditions, bay leaf with washed water can be used. Many people kept them in mojo bags to prevent unwanted interference from people.

Going beyond the ritualistic uses, bay has been used in cooking, and it is versatile as used in wide range of dishes, sauces, and condiments. It is an essential ingredient of many herbs and used in soups, stews, and stuffings, as well as fish, meats, vegetables, sauce, pickles, and sausages. It is easily blended with many other herbs such as garlic, mustard, pepper, parsley, rosemary, thyme, and oregano. Bay can also be an important ingredient in teas, oils, cheeses, and liquors, and its essential oil is used in the cosmetic industry for soaps, perfumes, prepared foods, beverages, and dental products. Bay has many traditional medical uses. Leaves are used for the treatment of skin rashes, earaches, and rheumatism. The leaves have aromatic fragrance, so they are kept in cloths and used to cover up bad mouth odor. The leaves of this plant, having a pepper odor and clove-like taste, are used in cooking. In addition to cooking, leaves and bark are used in treatment of rheumatism, nausea, vomiting, fever, anemia, body odor, diarrhea, and colic due to having astringent, aromatic, stimulant, and carminative qualities. Seeds mixed with honey or sugars are used in cough and dysentery in children.

Bay leaves having antidiarrheal, antiinflammatory, and antidiabetic activity are used for the improvement of the immune system. Antioxidants such as vitamin C, vitamin E, and carotenoids are used in many dietary sources and are used to lower blood cholesterol and uric acid level. Bay leaves have many sesquiterpene lactones that are responsible for inhibition of NO production, i.e., antiinflammatory, inhibition of alcohol absorption, and may improve liver glutathione S-transferase activity (Fang et al., 2005). Using bioassay-directed isolation study, different cytotoxic and apoptosis-induced compounds are identified in bay leaf. Many components of essential oil of bay leaf such as eugenol, methyl eugenol, and pinene have anticonvulsant activity, while eugenol, methyl eugenol, and cineole produce sedation and motor impairment (Sayyah et al., 2002). Essential oil of this leaf also has analgesic and many antiinflammatory activities (Barla et al., 2007). Many polar compounds such as flavones, flavonol, and phenols are present in the methanolic extract of bay leaf and show antioxidative activity.

Traditionally, it has been used as herbal medicine against number of diseases such as rheumatism, sprains, indigestion, earaches, and to enhance perspiration (Fang et al., 2005). It was reported by different researches that bay leaf can also be used to treat diabetes and migraine (Fang et al., 2005). It is used with warm water for drinking to treat internal ailments; as a result, excess water is removed by body by urination and acts as an emetic to induce vomiting. Fresh, mature leaves are used to treat blood dysentery, inflammation, and congestion of kidney. Bay leaf is also used to treat arthritis, headache, fungal diseases, anorexia, colds, cataracts, diarrhea, colic ulcer, appetizer, neuralgia, and digestive stimulant traditionally (Parthasarathy et al., 2008). Bay is found effective against many infections from fungi, viruses, bacteria, and protozoa. Bay is also helpful in inhibiting growth of carcinogenic cells. The leaves of bay are specific for many fevers, cough, flu, bronchitis, asthma, influenza, cough, cold, lowering blood cholesterol level, chicken pox, diarrhea, and antistress agents. Bay juice is an effective medication for sore eyes and night blindness, which is generally caused by deficit of vitamin A. Bay seeds are mucilaginous and relieve indigestion, sore throat, constipation, and diarrhea.

7. PHARMACOLOGICAL USES

7.1 Wound Healing Activity

The aqueous extract of *L. nobilis* were compared with the aqueous extract of *Allamanda* and found to have better wound healing activity. Many excision and incision wound healing models were used to estimate the wound healing activity. Many factors were studied to assess the wound healing activity such as tensile strength, weights of the granulation tissue, rate of wound closure, period of epithelialization, histopathology of the granulation tissue, and hydroxyproline content of the granulation tissue. Animals treated with bay leaf were found to have a reasonably high rate of wound contraction, hydroxyproline content, and weight of granulation tissue. Bay leaf–treated animals showed a higher number of inflammatory cells and less collagen compared with the animals that were treated with *Allamanda cathartica* (Nayak et al., 2006).

7.2 Antioxidant Activity

Ethanol extracts of *L. nobilis* showed powerful antioxidant activities. The antioxidant activity was determined by evaluating free radical scavenging, hydrogen peroxide scavenging, superoxide anion radical scavenging, reducing power, and metal chelating assays. Strong antioxidant activity of bay leaf was observed in linoleic acid emulsion at a concentration of 20, 40, and 60 μg/mL (94.2%, 97.7%, and 98.6% inhibition of lipid peroxidation, respectively). The antioxidant activity of ethanol extract may be due to phenolic compounds present in the extract (Elmastaş et al., 2006).

7.3 Anticonvulsant Activity

L. nobilis leaf essential oil showed anticonvulsant activity in mice. Essential oil components such as eugenol, pinene, and methyleugenol are responsible for this activity (Sayyah et al., 2002).

7.4 Analgesic and Antiinflammatory

L. nobilis essential oil showed analgesic and antiinflammatory activities in mice and rats (Sayyah et al., 2003). Ethanol extract obtained from the leaves and seeds of bay leaf also show the highest antiinflamatory activities by using a carrageenan-induced hind paw edema model (Kozan et al., 2006).

7.5 Antimutagenic Activity

Ethyl acetate extract of bay leaf has 3-kaempferyl *p*- coumarate antimutagen, which was identified experimentally and purified chromatographically. The antimutagenicity was due to a desmutagenic action that converted the Trp-P-2 metabolically activated form into its crucial carcinogenic form (Samejima et al., 1998).

7.6 Immunostimulant Activity

Immunostimulant effects of powder of bay leaf were shown on rainbow trout by giving them dietary constituents. Three groups of rainbow trout were fed with experimental diets. After 21 days, nonspecific immune parameters such as phagocytosis in blood leukocytes, extra- or intracellular respiratory burst activities, lysozymes, and protein levels were examined and showed immunostimulant activity (Bilen and Bulut, 2010).

7.7 Antiviral Activity

L. nobilis essential oil containing beta-ocimene, 1,8-cineol, alpha-pinene, and beta-pinene constituents were reported for inhibitory activity in vitro against SARS-CoV and HSV-1 replication. Essential oil has this activity with an IC_{50} value of 120 μg/mL and selectivity index of 4.16 (Bilen and Bulut, 2010).

7.8 Anticholinergic Activity

Essential oil, ethanolic extract, and decoction of *L. nobilis* were reported to have anticholinergic activity toward acetyl cholinesterase (AChE) enzyme and showed good anticholinergic activity. Ethanolic fraction of about 64% of bay leaf also shown this inhibitory activity (Ferreira et al., 2006).

7.9 Insect Repellent Activity

L. nobilis essential oils extracted from seeds were reported to have insect repellant activity against *Culex pipiens* (Erler et al., 2006).

7.10 Antimicrobial Activity

L. nobilis essential oil, methanolic extract of seed oil. and seed oil in vitro showed antibacterial activity. However, methanolic extract of seed oil has more effective antibacterial activity than essential oil and seed oil (Ozcan et al., 2010). Similarly, in another report the antibacterial activity of *L. nobilis* essential oil was determined against *Staphylococcus aureus, Bacillus subtilis,* and *Staphylococcus intermedius.* The *L. nobilis* essential oil showed good antibacterial activity with minimal inhibitory concentrations of 0.35 and 0.56 mg/mL, respectively. The major constituent of bay leaf, 1,8 cineol, might be responsible for its antibacterial activity (Derwich et al., 2009). Antifungal activity of *L. nobilis* was examined on seven strains of plant pathogenic fungi in vitro at different concentrations such as 50, 125, and 250 μg/mL. The greatest antifungal activity was obtained against the fungus *Botrytis cinerea* at a concentration of 250 μg/mL (Patrakar et al., 2012).

7.11 Acaricidal Activity

Acaricidal activity of bay leaf oils was observed against *Psoroptes cuniculi.* Acaricidal activity of bay oil led to a mortality rate of 73% at a concentration of 10% and at 5% average activity was considerably reduced to 51% (Macchioni et al., 2006).

8. SIDE EFFECTS AND TOXICITY

Bay leaf and bay leaf oil are likely safe for most people in food amounts. There is no choke possibility with ground bay leaf, as does exist with whole leaf. The whole leaf cannot be digested, so it remains intact while passing through the digestive system. There is not enough reliable information about the safety of taking bay leaf during pregnancy or breastfeeding. Bay leaf might interfere with blood sugar control and may not be safe to use during diabetes. Bay leaf might slow down the central nervous system (CNS). There is a concern that it might slow down the CNS too much when combined with anesthesia and other medications used during and after surgery. It is recommended to stop using bay leaf as a medicine at least 2 weeks before a scheduled surgery.

References

Akgül, A., Kivanc, M., Bayrak, A., 1989. Chemical composition and antimicrobial effect of Turkish laurel leaf oil. Journal of Essential Oil Research 1, 277–280.

Barla, A., Topçu, G., Öksüz, S., Tümen, G., Kingston, D.G., 2007. Identification of cytotoxic sesquiterpenes from *Laurus nobilis* L. Food Chemistry 104, 1478–1484.

Bilen, S., Bulut, M., 2010. Effects of laurel (*Laurus nobilis*) on the non-specific immune responses of rainbow trout (*Oncorhynchus mykiss*, Walbaum). Journal of Animal and Veterinary Advances 9, 1275–1279.

Biondi, D., Cianci, P., Geraci, C., Ruberto, G., Piattelli, M., 1993. Antimicrobial activity and chemical composition of essential oils from Sicilian aromatic plants. Flavour and Fragrance Journal 8, 331–337.

Bozan, B., Karakaplan, U., 2007. Antioxidants from laurel (*Laurus nobilis* L.) berries: influence of extraction procedure on yield and antioxidant activity of extracts. Acta Alimentaria 36, 321–328.

Choudhary, D., Kala, S., Todaria, N., Dasgupta, S., Kollmair, M., 2014. Effects of harvesting on productivity of bay leaf tree (Cinnamomum tamala Nees & Eberm): Case from Udayapur district of Nepal. Journal of Forestry Research 25, 163–170.

Deniz, H., 2012. Sustainable Collection of Laurel (*Laurus Nobilis* L.) Leaves in Antalya Province.

Derwich, E., Benziane, Z., Boukir, A., 2009. Chemical composition and antibacterial activity of leaves essential oil of *Laurus nobilis* from Morocco. Australian Journal of Basic and Applied Sciences 3, 3818–3824.

El, S.N., Karagozlu, N., Karakaya, S., Sahın, S., 2014. Antioxidant and antimicrobial activities of essential oils extracted from *Laurus nobilis* L. leaves by using solvent-free microwave and hydrodistillation. Food and Nutrition Sciences 5 (02), 97–106.

Elmastaş, M., Gülçin, I., Işildak, Ö., Küfrevioğlu, Ö., İbaoğlu, K., Aboul-Enein, H., 2006. Radical scavenging activity and antioxidant capacity of bay leaf extracts. Journal of the Iranian Chemical Society 3, 258–266.

Erler, F., Ulug, I., Yalcinkaya, B., 2006. Repellent activity of five essential oils against *Culex pipiens*. Fitoterapia 77, 491–494.

Fang, F., Sang, S., Chen, K.Y., Gosslau, A., Ho, C.-T., Rosen, R.T., 2005. Isolation and identification of cytotoxic compounds from Bay leaf (*Laurus nobilis*). Food Chemistry 93, 497–501.

Ferreira, A., Proença, C., Serralheiro, M., Araujo, M., 2006. The in vitro screening for acetylcholinesterase inhibition and antioxidant activity of medicinal plants from Portugal. Journal of Ethnopharmacology 108, 31–37.

Kemp, W.M., Twilley, R.R., Stevenson, J., Boynton, W., Means, J., 1983. The decline of submerged vascular plants in upper Chesapeake Bay: summary of results concerning possible causes. Marine Technology Society Journal 17, 78–89.

Kozan, E., Küpeli, E., Yesilada, E., 2006. Evaluation of some plants used in Turkish folk medicine against parasitic infections for their in vivo anthelmintic activity. Journal of Ethnopharmacology 108, 211–216.

Macchioni, F., Perrucci, S., Cioni, P., Morelli, I., Castilho, P., Cecchi, F., 2006. Composition and acaricidal activity of Laurus novocanariensis and *Laurus nobilis* essential oils against *Psoroptes cuniculi*. Journal of Essential Oil Research 18, 111–114.

Nayak, S., Nalabothu, P., Sandiford, S., Bhogadi, V., Adogwa, A., 2006. Evaluation of wound healing activity of Allamanda cathartica. L. and *Laurus nobilis*. L. extracts on rats. BMC Complementary and Alternative Medicine 6, 1.

Nurbaş, M., Bal, Y., 2005, Recovery of fixed and volatile oils from *Laurus nobilis* L. fruit and leaves by solvent extraction method. Journal of Engineering and Architectural Faculty of Eskişehir Osmangazi University.

Ozcan, B., Esen, M., Sangun, M.K., Coleri, A., Caliskan, M., 2010. Effective Antibacterial and Antioxidant Properties of Methanolic Extract of *Laurus Nobilis* Seed Oil.

Parthasarathy, V.A., Chempakam, B., Zachariah, T.J., 2008. Chemistry of Spices. Cabi.

Patrakar, R., Mansuriya, M., Patil, P., 2012. Phytochemical and pharmacological review on *Laurus nobilis*. International Journal of Pharmaceutical and Chemical Sciences 1, 595–602.

Samejima, K., Kanazawa, K., Ashida, H., Danno, G.-i., 1998. Bay laurel contains antimutagenic kaempferyl coumarate acting against the dietary carcinogen 3-amino-1-methyl-5 H-pyrido [4, 3-b] indole (Trp-P-2). Journal of Agricultural and Food Chemistry 46, 4864–4868.

Sayyah, M., Saroukhani, G., Peirovi, A., Kamalinejad, M., 2003. Analgesic and anti-inflammatory activity of the leaf essential oil of *Laurus nobilis* Linn. Phytotherapy Research 17, 733-736.

Sayyah, M., Valizadeh, J., Kamalinejad, M., 2002. Anticonvulsant activity of the leaf essential oil of *Laurus nobilis* against pentylenetetrazole-and maximal electroshock-induced seizures. Phytomedicine 9, 212–216.

Sumono, A., 2008. The use of bay leaf (Eugenia polyantha Wight) in dentistry. Dental Journal 41, 147–150.

CHAPTER

6

Black Piper

Maryam Khan, Muhammad Asif Hanif, Rafia Rehman, Ijaz Ahmad Bhatti
Department of Chemistry, University of Agriculture, Faisalabad, Pakistan

OUTLINE

https://doi.org/10.1016/B978-0-08-102659-5.00006-9

Copyright © 2020 Elsevier Ltd. All rights reserved.

1. BOTANY

1.1 Introduction

Black piper (*Piper nigrum* L.) (Fig. 6.1) belongs to Piperaceae family. The genus *Piper* contains more than 1000 species (Ahmad et al., 2012). Pepper is native to Southern India and Sri Lanka, although distributed in both hemispheres today. Both self- and cross-pollination can occur in *P. nigrum*, resulting in a wide range of species, but predominantly self-pollination occurs. Seeds have short viability and high sterility, so it is commercially cultivated through orthotropic stem cuttings with two to six nodes (Abbasi et al., 2010; Krishnamoorthy and Parthasarathy, 2011). *P. nigrum* has different common names according to its location. Its common English and Urdu names are pepper and kali mirch, respectively. Other common names include pipéri (in Greek), hujiao (in Chinese), poivre commun (in French), filfil (in Arabic), pfeffer (in German), and pimienta (in Spanish). World production of pepper had increased dramatically from 189,000 tons to 341,000 tons between 1997 and 2002, with the rate of over 12% per annum. In 2003, Brazil produced around 35,000 tons of pepper (32,000 tons black and 3000 tons white pepper), from around 41,000 ha cultivation. In the same year, Indonesia produced 57,000 tons, composed of 33,000 tons of black pepper and 24,000 tons of white pepper, while Malaysia produced 22,000 tons of pepper, comprising 19,800 tons black and 2200 tons white pepper. The total production in Sri Lanka and Vietnam was 12,750 tons and 88,000 tons of pepper, respectively, in 2003. Annual global export of *P. nigrum* was 291,125 metric tons in 2009, 290,234 metric tons in 2010, 279,225 metric tons in 2011, and 265,987 metric tons in 2012. The annual global import of *P. nigrum* was 266,859 metric tons in 2009, 306,276 metric tons in 2010, 280,831 metric tons in 2011, and 255,388 metric tons in 2012. Essential oil ranges from 1.2% to 3.5% in *P. nigrum*, which is rich in monoterpenes and sesquiterpenes. Black pepper has a long history of being used as a spice. The ripe and unripe fruits of *P. nigrum* are the source of white and black pepper, respectively. Black pepper has been used for the treatment of dyspepsia, cholera, gastric ailments, and arthritic disorders (Ravindran, 2003). Piperine is a major alkaloid constituent of *P. nigrum* L., responsible for the pungency of the black pepper. Piperine was reported to exhibit analgesic, antiinflammatory, antipyretic, and antidepressant activities. Insecticidal,

FIGURE 6.1 Black piper seeds and powder.

antiinflammatory, and antibacterial properties were also reported to be exhibited by the pepper plant (Lee et al., 2006, 2008; Zarai et al., 2013).

1.2 History/Origin

Pepper is native to Southern India, although it is widely distributed today and grown throughout the tropics. The name *Piper nigrum* is derived from Latin (*piper* meaning plant and *nigrum* meaning black). The history of black pepper is very old. Spices were in use in ancient Egypt from the age of the pyramids (2600–2100 BCE). The Assyrians and Babylonians traded pepper, cardamom, and cinnamon from the Malabar Coast of India from 3000 to 2000 BCE. The black pepper was grown extensively in the forest of Western Ghats and was grown in the southwestern region of India. The Malabar Coast was its center of origin. From India, pepper was taken to Indonesia, Malaysia, and then to other pepper-growing countries (Ravindran, 2003). Irritant smoke of burning peppers was used as a weapon against invaders by Americans (Barceloux, 2009; Cordell and Araujo, 1993).

1.3 Demography/Location

The *Piper* species are distributed in tropical and subtropical regions of the world (de Morais et al., 2007). *P. nigrum* requires high temperatures with heavy and frequent rainfalls and well-draining soil for optimum growth. These plants cannot tolerate frost. These conditions are usually met in the countries of India, Brazil, and Indonesia, so these are the greatest commercial exporters of peppercorns. Other important pepper-producing countries are Vietnam, China, the Philippines, Malaysia, and

Sri Lanka. Currently, it is grown in 26 countries (Barceloux, 2009; Ravindran, 2003).

1.4 Botany, Morphology, Ecology

It is a climbing, woody, perennial plant that usually grows to a height of 50–60 cm to 4 m (13 ft) (Bagheri et al., 2014; Barceloux, 2009; Tasleem et al., 2014). *P. nigrum* is a perennial climber, which adheres to the support tree by means of ivy-like roots. Old stems become thick and rough, and shoots arise from its base. Leaves are thick and of variable shape and size. The color of the upper surface of the leaf varies from dark to light green, and the lower surface is dull green. Under the surface of leaves and on young shoots, pearl glands are present. Spikes are whitish green or light purple when young, but mature ones turn green, light yellow, or light purple, and much variation was reported in spike length. The small, white flowers have two stamens, dithecous anthers, a single carpel, spherical ovary, and three to five lobed stigma; styles are absent. Fruit is a drupe, which turns red on ripening, and seeds are spherical and pungent. Wild forms are usually dioecious, while cultivated ones are bisexual (Barceloux, 2009; Ravindran, 2003). Humid tropics are suitable for pepper growth. Pepper plants require high temperatures with heavy and frequent rainfall and well-draining soil for optimum growth. Latitudes of 20 degrees north and 20 degrees south and altitude up to 2400 m from sea level are suitable for its growth. The plant can grow in a temperature range of 10–40°C with optimum growth between 25 and 40°C. Rainfall in the range of 1250–2000 mm is necessary for pepper production. The plant cannot bear frost.

2. CHEMISTRY

Black pepper has high nutritional value. It contains 47%–53% fiber, 11%–14% protein, and 10%–13.5% starch. It is also a source of iron, manganese, potassium, and minute amounts of vitamins K and C. Essential oil is present in a small portion of pepper plant material, which consists of terpenes, sesquiterpenes, and their derivatives, which are the reason for the aroma and flavor of pepper plant (Al-Jasass and Al-Jasser, 2012; Jayashree et al., 2009). Essential oils of black pepper are extracted from seeds and leaves. More than 250 volatiles have been reported to be present in this valuable spice. Some active components of black piper are shown in Fig.6.2. The main compounds detected are germacrene D, limonene, β-pinene, α-phellandrene, β-caryophyllene, α-pinene, and *cis*-β-ocimene.

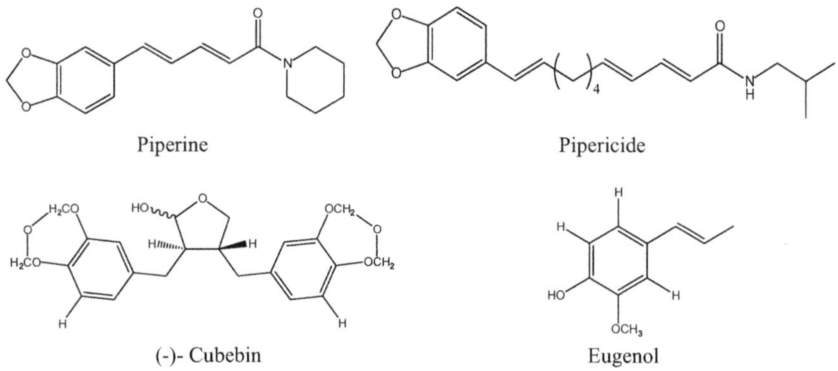

FIGURE 6.2 Active chemical constituents of black piper.

3. POSTHARVEST TECHNOLOGY

Flowering in pepper plants starts during May–June, and harvesting is usually done in December–January. Harvesting is usually done when one to two berries in a few spikes turn orange or red. Only the mature spikes should be harvested, as immature fruits shrivel up on drying, causing reduction in the quality of the products (Ravindran, 2003). After harvesting the fruit of black pepper is dried under sun for approximately 3 days until the moisture content becomes 8%–10%.

4. PROCESSING

The dried peppercorn is then cleaned in a process called blanching. During this process, peppercorns are passed through a conveyer belt, and boiling water is sprayed on them for 1 minute. After that, piper corn surface moisture is removed with the help of dried air. After drying the moisture content should not exceed 70%. It has a hygroscopic nature, so it can absorb moisture and is susceptible to fungus infections. So for long term storage, the moisture content should not be greater 10%–11%. Jute bags with polyethylene lining can be used for packing. To produced black peppercorn, ripened fruit is dried under the sun until it has 12% moisture content. While to produce white pepper, ripe berries are soaked in running water 7–10 days to soften skin. The skin is then removed with the help of hands or by treading on berries with the feet (Nair, 2011).

The essential oil of black pepper can be extracted by water or steam distillation. Its essential oil contains terpenic hydrocarbons along with their oxygenated compounds with a boiling point range of 80–200°C. The essential oil is susceptible to oxidative spoilage when exposed to sunlight and air, so it must be stored in airtight jars (Nair, 2011).

5. VALUE ADDITION

P. nigrum fruit is used by grinding green pepper corns and blending them with salt, vinegar, sugar, or other ingredients. BioPerine (standard extract of *P. nigrum,* containing 95% piperine) is used as a food additive for humans and animals. It can be coadministrated along with vitamins, antioxidants, minerals, amino acids, and herbal extracts. BioPerine along with beta-carotene was reported to cause a two-fold increase in the blood level of beta-carotene. It was also reported to enhance the bioavailability of drugs (Ahmad et al., 2012). Cosmoperine was also prepared from piperine, which enhances the natural power of skin to absorb nutrients. It is also reported to relieve pain and reduce skin reddening (Ahmad et al., 2012).

6. USES

Black pepper was thought to protect from evil. It is recommended to burn black pepper to get rid of bad vibrations at the home or office. It was thought to provide courage to face things. It is added to amulets as a protectant against the evil eye. It is mixed with salt and scattered on the property to dispel evil. Black pepper is mixed with other ingredients and believed to derive money, jobs, and health and to drive unwanted persons away. Black pepper and salt are sprinkled after the departure of an enemy to prevent him or her from returning. Black pepper brings positive changes and courage. It is especially helpful in confronting fears and provides strength, protection, and stamina.

Black pepper is used to cure fever, cold, and inflammation (Doucette et al., 2013; Parmar et al., 1997; Ravindran, 2003; Tasleem et al., 2014). It has also found uses in food preservation (Singh et al., 2004). It also exhibited antilarvicidal activities (Chaithong et al., 2006; Park et al., 2002). *P. nigrum* was reported to show antibacterial (Thakare, 2004), analgesic, and antiinflammatory activities (Tasleem et al., 2014), antiatherogenic activity (Agbor et al., 2012), anticancer (Deng et al., 2016), and antioxidant properties (Singh et al., 2004).

P. nigrum has been traditionally used to treat malaria in India and epilepsy in China (Ujam et al., 2017). Piperine, the major alkamide of *P. nigrum* was reported to possess a variety of biologic properties like analgesic, antipyretic, CNS stimulant, and antifeedant activities (Reddy et al., 2004). It has been also used to treat chronic indigestion, obesity, colon toxins, congestion, sinus, and cold extremities (Thakare, 2004).

7. PHARMACOLOGICAL USES

7.1 Analgesic and Antiinflammatory Activity

Piperine present in the black piper is known to show analgesic activity comparable to acetyl salicylic acid. It was reported that *P. nigrum* showed antiinflammatory activity by stimulating the production of cytokine and inhibiting the expansion of genes that encode nitric oxide synthase and cyclooxygenase-2 (Mueller et al., 2010). It was reported that various alkamides (piperrolein B, pellitorin, piperchabamide D, and dehydropipernonaline) isolated from *P. nigrum* inhibited the direct binding between lymphocyte function–associated antigen-1 (LFA-1) and its ligands, intercellular cell adhesion molecule-1 (ICAM-1), and hence exhibit antiinflammatory activity in an in vivo mouse model (Lee et al., 2008).

7.2 Antibacterial Activity

The petroleum ether extract of *P. nigrum* was reported to show strong antibacterial activity against both Gram +ve and Gram −ve bacterial strains. Five compounds were obtained on fractionation of petroleum ether extract of *P. nigrum,* namely pellitorine, 2E, 4E, 8Z-N-isobutyleicosatrienamide, pergumidiene, trachyone, and iso-piperolein B. All these compounds showed antibacterial activity, but 2E, 4E, 8Z-Nisobutyleicosatrienamide, trachyone, and pergumidiene were reported to possess high activity. Among all the compounds tested, 2E, 4E, 8Z-Nisobutyleicosatrienamide, trachyone, and pergumidiene showed major antibacterial activity against Gram +ve bacterial strains including *S. aureus, B. subtilis,* and *B. sphaericus.* Similarly, 2E, 4E, 8Z-N-isobutyleicosatrienamide, trachyone, and pergumidiene also showed antibacterial activity against Gram −ve bacterial strains, particularly against *K. aerogenes.* Hence, these compounds in the petroleum ether extract of *P. nigrum* were responsible for its antibacterial activity (Reddy et al., 2004). It was reported that ethanolic fruit extracts of *P. nigrum* showed the antibacterial activity against penicillin G–resistant strain of *S. aureus* (Thakare, 2004). Oil extract of *P. nigrum* was found effective against *Bacillus subtilis* and *P. aeruginosa* (Sasidharan and Menon, 2010). Aqueous, ethanol, methanol, and petroleum ether leaf extract of *P. nigrum* reported to show significant antibacterial activity against the *S. aureus, E. coli, S. typhimurium,* and *P. aeruginosa.* The alkaloids, flavonoids, anthraquinone, reducing sugars, tannins, saponins, and terpenoids in these extracts were reported to be responsible for it. Hence *P. nigrum* can be used as a potent antibacterial drug (Akthar et al., 2014).

7.3 Anticancer Activity

P. nigrum was reported to possess anticancer properties. The anticancer effect of piperine-free *P. nigrum* extract (PFPE) was evaluated by N-nitroso-N-methylurea (NMU)-induced mammary tumorigenesis rats and breast cancer cell lines MCF-7 and ZR-75-1. Protein levels induced by PFPE were studied by western blotting. PFPE was reported to increase p53 and decrease E-cadherin, matrix metalloproteinase 9, estrogen receptor, vascular endothelial growth factor, and matrix metalloproteinase 2, c-Myc levels in breast cancer rats. It also decreased protein levels of E-cadherin, vascular endothelial growth factor, and c-Myc in MCF-7 cells. Hence, PFPE enhances the response of breast cancer cells toward phytochemicals, inducing cell cycle arrest, and inhibits proliferation of cancer cells and decreases tumor size. PFPE also suppresses tumor cell invasion, angiogenesis, and migration. PFPE also prevents cancer by generating reactive oxygen species in higher cancer cells (Deng et al., 2016). Piperine was also reported to show anticancer property against lung cancer (Pradeep and Kuttan, 2002; Selvendiran et al., 2004). Pipernonaline was reported to inhibit the growth of prostate cancer cell lines PC3 and induce reactive oxygen species–related cell apoptosis (Lee et al., 2013). Purpurogallin was reported to suppress tumor cell metastasis and angiogenesis (Reed and Pellecchia, 2010).

7.4 Antioxidant Activity

Antioxidant potential of phenolic compounds from green pepper (*P. nigrum*) was evaluated. 1, 10-diphenyl-2-picrylhydrazyl (DPPH) radical scavenging ability, lipid peroxidation inhibition, and plasmid DNA damage protection upon exposure to gamma radiation were used to evaluate antioxidant potential. Major phenolic compounds of green pepper were reported to have high radical scavenging activity including 3, 4-dihydroxy-6-(N-ethylamino) benzamide, 3, 4-dihydroxyphenyl ethanol glucoside, and phenolic acid glycosides. It was reported that green pepper has high efficacy as an antioxidant due to the presence of phenolics that are lost during black pepper preparation. Hence, fresh spice and its extracts can be used as antioxidant agents (Chatterjee et al., 2007). The antioxidant activity of the solvent extracts of *P. nigrum*, purified piperine, and piperic acid was evaluated by DPPH (1, one diphenyl-2-picrylhydrazyl) free radical assay. Butylated hydroxytoluene was used as a control. It was reported that the ethanol extract, at a concentration of 50 mg/mL, showed a high antioxidant activity, which can be attributed to the highest amount of total phenolics in it. Water extract showed the lowest radical scavenging activity. It was reported that piperic acid presented a higher radical scavenger than piperine. In piperine, after transformation of piperine into piperic acid, appearance of the −COOH radical is reported that causes the increase in radical scavenging activity. The antioxidant activity of various solvent extracts, piperine, and piperic acid was also evaluated by reducing power assay. It was reported that reducing power of tested compounds increased with the increasing concentration of each extract. An aqueous emulsion of linoleic acid and β-carotene was used to evaluate the antioxidant activity of various solvent extracts, piperine, and piperic acid. The antioxidant reduced the extent of β-carotene destruction by reacting with the linoleic acid free radical or any other free radical formed within the system (Amarowicz et al., 2004; Jayaprakasha et al., 2001).

7.5 Antiatherogenic Activity

The antiatherosclerotic activities of three Piper species (*P. nigrum*, *Piper guineense*, and *Piper umbellatum*) were evaluated on atherogenic diet–fed hamsters. The hamsters were grouped into normal control (fed normal rodent chow), atherosclerotic control (fed the normal rodent chow supplemented with 0.2% cholesterol and 10% coconut oil), and six test groups (fed same diet as the atherosclerotic control group but with additional supplementation of two graded doses, i.e., 1 and 0.25 mg/kg body weight) of plant extracts for 12 weeks. The atherogenic diet caused a failure of the erythrocyte antioxidant defense system (decrease in superoxide dismutase, glutathione peroxidase, and catalase activities) and an increase in triglyceride, plasma total cholesterol, thiobarbituric acid reactive substances, and oxidation of low-density lipoprotein cholesterol, and accumulation of foam cells in the aorta was reported. Administration of the *Piper* species in six tested groups protected the antioxidant system and prevented low-density lipoprotein cholesterol oxidation. Hence, piper species can be used as an antiatherogenic agent (Agbor et al., 2012).

7.6 Antihypertensive Effects

Piperine was reported to possess a Ca^{2+} channel blockade effect that causes cardio-depressant and vasodilator activities, which in turn are the basis for the BP-lowering effect. Piperine was also reported to have the associated vasoconstrictor effects responsible for the decrease in blood pressure up to a certain limit and a small increase in the blood pressure after decline on administration of dose. Hence, piperine does not allow BP to decrease beyond a certain limit and with fewer side effects (Taqvi et al., 2008).

7.7 Antiasthmatic Effects

Antiasthmatic effect of piperine (the major alkaloid of *P. nigrum*) was reported by oral administration of piperine in different proportion to mice. It was reported that piperine suppressed the production of histamine, immunoglobulin E, interleukin-4, and interleukin-5, which in turn suppressed eosinophil infiltration, airway inflammation, and hyperresponsiveness. Moreover, polymerase chain reaction products for thymus and activation-regulated chemokine from lung cell RNA preparations were decreased and transforming growth factor-b products were increased in the piperine-treated group compared with control groups. Hence, piperine can be used in the treatment of asthma (Kim and Lee, 2009).

7.8 Antithyroid Activity

The antithyroid activity of *P. nigrum* was evaluated by injecting piperine isolated from dried fruits of *P. nigrum* in albino mice at a dose of 0.25 mg/kg/day for 15 consecutive days. A day after injection of the last dose, mice were sacrificed, serum was separated from their blood, and levels of both thyroid hormones (thyroxin and triiodothyronine) were evaluated. Serum level of both thyroid hormones was reported to be in lower concentrations than the control group. Hence, piperine can be used to inhibit thyroid function in euthyroid individuals (Panda and Kar, 2003).

8. SIDE EFFECTS AND TOXICITY

Taking large amounts of black and white pepper by mouth, which can accidently get into the lungs, has been reported to cause death. This is especially true in children.

References

Abbasi, B.H., Ahmad, N., Fazal, H., Mahmood, T., 2010. Conventional and modern propagation techniques in *Piper nigrum*. Journal of Medicinal Plants Research 4, 007–012.

Agbor, G.A., Vinson, J.A., Sortino, J., Johnson, R., 2012. Antioxidant and anti-atherogenic activities of three Piper species on atherogenic diet fed hamsters. Experimental & Toxicologic Pathology 64, 387–391.

Ahmad, N., Fazal, H., Abbasi, B.H., Farooq, S., Ali, M., Khan, M.A., 2012. Biological role of *Piper nigrum* L.(Black pepper): a review. Asian Pacific Journal of Tropical Biomedicine 2, S1945–S1953.

Akthar, M.S., Birhanu, G., Demisse, S., 2014. Antimicrobial activity of *Piper nigrum* L. and Cassia didymobotyra L. leaf extract on selected food borne pathogens. Asian Pacific Journal of Tropical Disease 4, S911–S919.

Al-Jasass, F.M., Al-Jasser, M.S., 2012. Chemical composition and fatty acid content of some spices and herbs under Saudi Arabia conditions. Science World Journal 2012 (859892), 1–5.

Amarowicz, R., Pegg, R., Rahimi-Moghaddam, P., Barl, B., Weil, J., 2004. Free-radical scavenging capacity and antioxidant activity of selected plant species from the Canadian prairies. Food Chemistry 84, 551–562.

Bagheri, H., Manap, M.Y.B.A., Solati, Z., 2014. Antioxidant activity of *Piper nigrum* L. essential oil extracted by supercritical CO_2 extraction and hydro-distillation. Talanta 121, 220–228.

Barceloux, D.G., 2009. Pepper and capsaicin (capsicum and piper species). Disease-a-Month 55, 380–390.

Chaithong, U., Choochote, W., Kamsuk, K., Jitpakdi, A., Tippawangkosol, P., Chaiyasit, D., Champakaew, D., Tuetun, B., Pitasawat, B., 2006. Larvicidal effect of pepper plants on *Aedes aegypti* (L.)(Diptera: Culicidae). Journal of Vector Ecology 31, 138–144.

Chatterjee, S., Niaz, Z., Gautam, S., Adhikari, S., Variyar, P.S., Sharma, A., 2007. Antioxidant activity of some phenolic constituents from green pepper (*Piper nigrum* L.) and fresh nutmeg mace (Myristica fragrans). Food Chemistry 101, 515–523.

Cordell, G.A., Araujo, O.E., 1993. Capsaicin: identification, nomenclature, and pharmacotherapy. The Annals of Pharmacotherapy 27, 330–336.

de Morais, S.M., Facundo, V.A., Bertini, L.M., Cavalcanti, E.S.B., dos Anjos Júnior, J.F., Ferreira, S.A., de Brito, E.S., de Souza Neto, M.A., 2007. Chemical composition and larvicidal activity of essential oils from Piper species. Biochemical Systematics and Ecology 35, 670–675.

Deng, Y., Sriwiriyajan, S., Tedasen, A., Hiransai, P., Graidist, P., 2016. Anti-cancer effects of *Piper nigrum* via inducing multiple molecular signaling in vivo and in vitro. Journal of Ethnopharmacology 188, 87–95.

Doucette, C.D., Hilchie, A.L., Liwski, R., Hoskin, D.W., 2013. Piperine, a dietary phytochemical, inhibits angiogenesis. Journal of Nutritional Biochemistry 24, 231–239.

Jayaprakasha, G., Singh, R., Sakariah, K., 2001. Antioxidant activity of grape seed (Vitis vinifera) extracts on peroxidation models in vitro. Food Chemistry 73, 285–290.

Jayashree, E., Zachariah, T., Gobinath, P., 2009. Physico-chemical properties of black pepper from selected varieties in relation to market grades. Journal of Food Science & Technology 46, 263–265.

Kim, S.H., Lee, Y.C., 2009. Piperine inhibits eosinophil infiltration and airway hyperresponsiveness by suppressing T cell activity and Th2 cytokine production in the ovalbumin-induced asthma model. Journal of Pharmacy and Pharmacology 61, 353–359.

Krishnamoorthy, B., Parthasarathy, V., 2011. Improvement of black pepper. Plant Sciences Reviews 37, 2010.

Lee, S.W., Kim, Y.K., Kim, K., Lee, H.S., Choi, J.H., Lee, W.S., Jun, C.-D., Park, J.H., Lee, J.M., Rho, M.-C., 2008. Alkamides from the fruits of Piper longum and *Piper nigrum* displaying potent cell adhesion inhibition. Bioorganic & Medicinal Chemistry Letters 18, 4544–4546.

Lee, S.W., Rho, M.-C., Park, H.R., Choi, J.-H., Kang, J.Y., Lee, J.W., Kim, K., Lee, H.S., Kim, Y.K., 2006. Inhibition of diacylglycerol acyltransferase by alkamides isolated from the fruits of Piper longum and *Piper nigrum*. Journal of Agricultural and Food Chemistry 54, 9759–9763.

Lee, W., Kim, K.-Y., Yu, S.-N., Kim, S.-H., Chun, S.-S., Ji, J.-H., Yu, H.-S., Ahn, S.-C., 2013. Pipernonaline from Piper longum Linn. induces ROS-mediated apoptosis in human prostate cancer PC-3 cells. Biochemical and Biophysical Research Communications 430, 406–412.

Mueller, M., Hobiger, S., Jungbauer, A., 2010. Anti-inflammatory activity of extracts from fruits, herbs and spices. Food Chemistry 122, 987–996.

Nair, K.P., 2011. Agronomy and Economy of Black Pepper and Cardamom: the" King" and" Queen" of Spices. Elsevier.

Panda, S., Kar, A., 2003. Piperine lowers the serum concentrations of thyroid hormones, glucose and hepatic 5′ D activity in adult male mice. Hormone and Metabolic Research 35, 523–526.

Park, I.-K., Lee, S.-G., Shin, S.-C., Park, J.-D., Ahn, Y.-J., 2002. Larvicidal activity of isobutylamides identified in *Piper nigrum* fruits against three mosquito species. Journal of Agricultural and Food Chemistry 50, 1866–1870.

Parmar, V.S., Jain, S.C., Bisht, K.S., Jain, R., Taneja, P., Jha, A., Tyagi, O.D., Prasad, A.K., Wengel, J., Olsen, C.E., 1997. Phytochemistry of the genus piper. Phytochemistry 46, 597–673.

Pradeep, C., Kuttan, G., 2002. Effect of piperine on the inhibition of lung metastasis induced B16F-10 melanoma cells in mice. Clinical & Experimental Metastasis 19, 703–708.

Ravindran, P., 2003. Black Pepper: *Piper Nigrum*. CRC Press.

Reddy, S.V., Srinivas, P.V., Praveen, B., Kishore, K.H., Raju, B.C., Murthy, U.S., Rao, J.M., 2004. Antibacterial constituents from the berries of *Piper nigrum*. Phytomedicine 11, 697–700.

Reed, J.C., Pellecchia, M., 2010. Methods and Compounds Useful to Induce Apoptosis in Cancer Cells. Google Patents.

Sasidharan, I., Menon, A.N., 2010. Comparative chemical composition and antimicrobial activity of berry and leaf essential oils of *piper nigrum* l. International Journal of Biological & Medical Research 1, 215–218.

Selvendiran, K., Banu, S.M., Sakthisekaran, D., 2004. Protective effect of piperine on benzo (a) pyrene-induced lung carcinogenesis in Swiss albino mice. Clinica Chimica Acta 350, 73–78.

Singh, G., Marimuthu, P., Catalan, C., Delampasona, M., 2004. Chemical, antioxidant and antifungal activities of volatile oil of black pepper and its acetone extract. Journal of the Science of Food and Agriculture 84, 1878–1884.

Taqvi, S.I.H., Shah, A.J., Gilani, A.H., 2008. Blood pressure lowering and vasomodulator effects of piperine. Journal of Cardiovascular Pharmacology 52, 452–458.

Tasleem, F., Azhar, I., Ali, S.N., Perveen, S., Mahmood, Z.A., 2014. Analgesic and anti-inflammatory activities of *Piper nigrum* L. Asian Pacific journal of tropical medicine 7, S461–S468.

Thakare, M.N., 2004. Pharmacological Screening of Some Medicinal Plants as Antimicrobial and Feed Additives.

Ujam, N., Eze, P., Umeokoli, B., Abbah, C., Okoye, F., Esimone, C., 2017. Evaluation of antiplasmodial and immunomodulatory activities of extracts of endophytic fungi isolated from four Nigerian medicinal plants. Planta Medica International Open 4. Tu-PO-80.

Zarai, Z., Boujelbene, E., Salem, N.B., Gargouri, Y., Sayari, A., 2013. Antioxidant and antimicrobial activities of various solvent extracts, piperine and piperic acid from *Piper nigrum*. Lwt-Food science and technology 50, 634–641.

CHAPTER

7

Caraway

Rafia Javed[1], *Muhammad Asif Hanif*[1], *Rafia Rehman*[1], *Maryam Hanif*[1], *Bui Thanh Tung*[2]

[1] Department of Chemistry, University of Agriculture, Faisalabad, Pakistan; [2] Department of Pharmacology and Clinical Pharmacy, School of Medicine and Pharmacy, Vietnam National University, Hanoi, Vietnam

OUTLINE

Medicinal Plants of South Asia
https://doi.org/10.1016/B978-0-08-102659-5.00007-0

Copyright © 2020 Elsevier Ltd. All rights reserved.

1. BOTANY

1.1 Introduction

Caraway (*Carum carvi* L.) (Fig. 7.1) is a biennial herb and belongs to the family Umbelliferae or Apiaceae (Furmanowa et al., 1991). The genus *Carum* has 20 species of flowering plants. Only the genus *Carum* has economic importance out of its all species. *C. carvi seeds* have been used for flavoring, as a spice, and oil of caraway is used for liquors, toothpaste, and as flavoring agent in different food products. It prefers cool temperate zones, meadows, and fields. It is cultivated and used in different countries like Pakistan, India, North America, and Northern Europe to Mediterranean regions, Iran, Russia, and Indonesia (Meshkatalsadat et al., 2012). In

FIGURE 7.1 Caraway seeds.

Pakistan, caraway is known as "shahi zeera" or "kaala zeera" Caraway has different names in different regions, as in English it is called caraway, and wild cumin, in German as kummel, in Norwegian, it is karve, in Russia, it is known as tamin, and in Swedish, it is known as kummin. Some other European common names are karwij in Nicaragua (NI) and carvi in France, Spain, and Italy (Houghton et al., 1995).

1.2 History/Origin

It is native to Western Asia and Europe and has been growing for many years. In Egypt, caraway is cultivated as one of the oldest spices. It is also present in Egyptian tombs. It is naturally found in Turkey, North Africa, Central Europe, Siberia, Pakistan, India, and Iran (Sadiq et al., 2010). In the first century, a Greek physician made a tonic that could restore color to the ladies' cheeks with pale complexions. The favorite bread of Emperor Julius Caesar was apparently made with caraway seeds, known as chara. It was favored for botanic medicines, known as pharmacopeia, in ancient Egypt, Persia, and Greece. Caraway has been used in the medication of digestive problems since early history (Fatemi et al., 2010). The use of caraway in England started in the 14th century when it was first written in a cook book compiled for King Richard II. The caraway was eaten at the end of the meal to cleanse the palate and freshen the breath in Elizabethan time. Caraway is particularly famous in German culture where it is an essential constituent in sauerkraut and some delicatessen meats. Black caraway seed has been used in Arabic tradition and used as herbal medicines for the medication of such diverse problems as rheumatism, gastrointestinal problems, hypertension, and diabetes. The English name caraway comes from Latin carum and Arabic al-karawya and from Caria. Caraway is particularly popular in Northern Europe, where the name caraway related to the Latin *cuminum*, e.g., Latvian kimenes, Danish kommen, Estonian koomen, Bulgarian kim, and Polish kminek. Some names originated from German names (Badura, 2003; Kazemipoor and Cordell, 2015).

1.3 Demography/Location

A well-drained soil rich in organic matter and a warm and sunny location is preferable for caraway plant. Caraway is mostly found in Pakistan, India, West Asia, and Europe. The yield of seeds of caraway is 1250 kg/ha in annual caraway and 900 kg/ha in biennial caraway. It grows rich in full sun and well-drained soil with pH ranges of 6.5–7.0. It has a minimum growing temperature of 4°C, and the optimum temperature is 13°C. It requires more fertile soil for better development, plant growth, and yield.

1.4 Botany, Morphology, Ecology

It relates to kingdom Plantae, Order Apiales, Family Umbelliferae, Genus *Carum*, specie *carvi*, and its binomial name is *Carum Carvi*. The length of the main flower stem is 40–60 cm (16–24 in). Flowers of caraway are in the form of a terminal or lateral cluster and small, pinkish, or white. Fruits of caraway have five pale ridges. Seeds of caraway have fine distinct tan, linear, ribs and are brown, oblong, and narrow. It has normally oblong, oval-shaped, and alternate leaves. Stem leaves mostly tend to drop but resemble carrots in shape. Caraway shoots are hollow, branching, erect, and slender. It has one or more shoots growing out from a single taproot, and mature plants are normally 1–3 feet tall. Roots are taproot. Fruits have a distinctive caraway odor.

2. CHEMISTRY

Caraway is one of the aromatic herbs that is known for excellence of its fragrant dried seeds. The essential oil yield of caraway fruit is 1–6% and it has characteristics odor. Its seeds retain 1.5% of waxes and small amount of tannins and resin, 5%–7% of ash, 13%–19% of crude fiber, 5%–10% of extractive nitrogen-free compounds, nitrogen compounds (25%–36%), fats (13%–21%), and water (9%–13%) (Sedláková et al., 2003). The fruits of caraway are distilled to obtain essential oil. There are several components present in caraway seeds. The main components of oil are limonene and carvone. Trace amounts of other compounds including camphene, thujone, pinene, carveol, furfural, acetaldehyde, etc., are also present in caraway seeds (Meshkatalsadat et al., 2012). Caraway seeds are a rich source of dietary fiber. One hundred grams of seeds provide 38 g of fiber, 100% of daily recommended intake of fiber. Soluble as well as insoluble dietary fiber helps prevent constipation by speeding up movement through the gut. The caraway seeds indeed are the storehouse for many vital vitamins. Vitamin A, vitamin E, vitamin C, as well as many B-complex vitamins like thiamin, pyridoxine, riboflavin, and niacin particularly concentrated in the seeds. Caraway spice is an excellent source of minerals like iron, copper, calcium, potassium, manganese, selenium, zinc, and magnesium (Laribi et al., 2013; Laribi et al., 2010; Riddhi and Yogesh; Seidler-Lozykowska et al., 2010).

The phenolic acids including ferulic acid, cinnamic acid, caffeic acid, catechuic acid, gallic acid, and flavonols such as kaempferol and quercetin are interestingly present in caraway. Some important components of caraway are shown in Fig. 7.2.

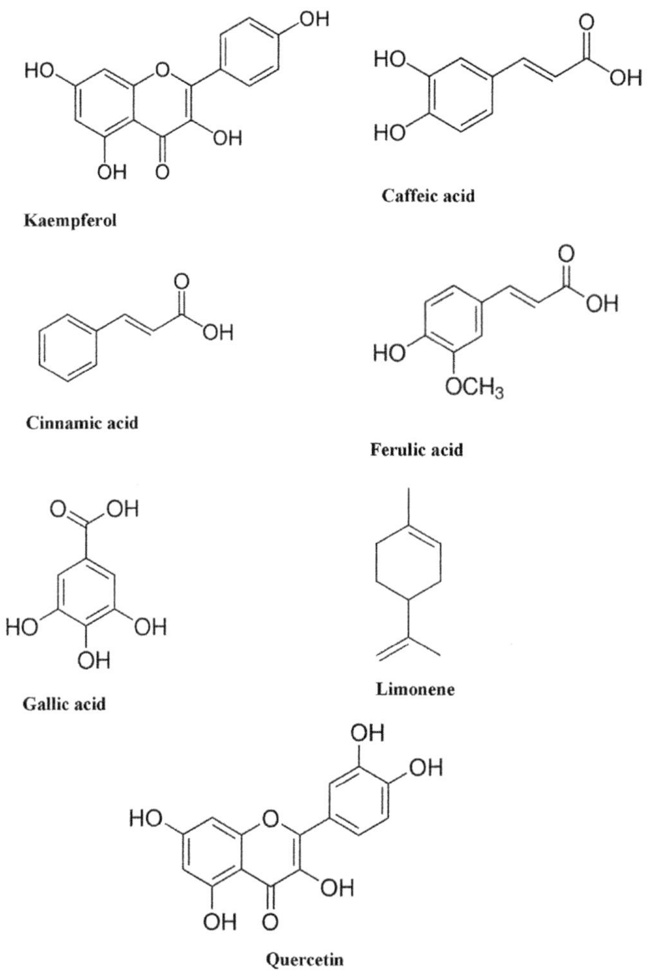

FIGURE 7.2 Some important chemical constituents of caraway.

3. POSTHARVEST TECHNOLOGY

Farmers harvest the caraway in two consecutive years. In the month of May, caraway flowers, and in the beginning of July, it is harvested. Harvesting is completed before the beginning of frost, and after that, flowering seeds are harvested in the second year. Adhering materials and dirt are removed from caraway by washing it. Then the seeds are allowed to turn brown and dry. Mostly sun drying is used for this purpose. Drying is completed by using a micropulverizer and powdering in disintegrators mills. After the seeds turn brown, they are placed in paper bags and then threshed when they are properly loose. Then seeds are dropped into a paper for separation of seed heads or collection. According to mode of marketing, the powder is tested and shifted. It is then packed for sales in the market. The annual form of caraway is spring caraway. It is not grown under a cover crop. It is very similar to biennial caraway from bolting to harvest. As compared to biennial caraway, this caraway flowers and harvests approximately 2.5 months later in the season.

4. PROCESSING

The concentration of essential oil is highest in achenes of caraway, so caraway is grown for its content of essential oil (Sedláková et al., 2003). Caraway leaves can be stored in the refrigerator in plastic bags, but fresh seeds are best. After drying, the seeds are stored for several months in an airtight containers and bags. Seeds of caraway can be stored for a long time if they are preserved properly. If bottled, caraway seeds are stored in dry and dark places, where they can be stored for 3 years. Caraway seeds in loose packs and airtight jars keep seeds fresh up to 6 months.

Steam and hydro-distillation are standard methods for distillation of caraway oil (Sedláková et al., 2003). The yield of essential oil per hectare depends directly on the seed yield per hectare and essential oil content in seeds. Average yield of caraway essential oil is around 20.2 kg/ha, but weather conditions and production area have a significant influence on this parameter, as well as its interaction (Aćimović et al., 2014).

5. VALUE ADDITION

The common use of caraway in different food products is due to its distinct flavor. Commonly, it is used in German peasantry, soup, cheese, and bread. It is mixed with butter, ginger, and salt to flavor it. In both Russia and Germany, it is also used in drinks. For fermented cabbage, caraway seed can be added to sauerkraut to have a distinct caraway flavor. Caraway seeds are added to baked products, dumplings, muffins, and scones. Caraway seeds are used in a coleslaw or carrot salad to bring new life to these family favorites. Tea of caraway seed can be used to improve digestion.

6. USES

It is used as food and in folk medicines in different countries. *C. carvi seeds* have been used for flavoring, as a spice, and oil of caraway is used for drinks, toothpaste, and as flavoring agent in different food products. It can also inhibit seed germination and sprouting of potato tubers. Caraway should be used in a small quantity just like other spices. Enhanced quantity in food can cause stomach ulcers and gastrointestinal irritation. It is helpful for people with eating problem because it restores a loss of appetite. Caraway has been significantly used for antiobesity and weight loss (Kazemipoor et al., 2013). In traditional European cooking, caraway is used in savory dishes as a chief spice. In Scandinavia, Holland, Hungry, Germany, and Austria the smoked and skimmed milk cheeses contain whole seed of caraway. The caraway-flavored cheese is still in use from the medieval recipes. Caraway seeds are used to treat diabetes, hypertension, and are used as diuretics in Moroccan traditional medicine (Lahlou et al., 2007). In Russia, for the treatment of pneumonia, caraway is used. In Poland, for the treatment of flatulence, indigestion, lack of appetite, caraway is recommended. In the United States and Great Britain, it is used as a carminative and stomachic. In Indonesia, to cure inflamed eczema, caraway is used. It is also used as a medicinal herb in Malay Peninsula (Johri, 2011).

The most satisfying use is caraway seed tea, but caraway seeds can be chewed for a digestion aid. One tablespoon of caraway seeds is poured in 12 ounces of water. Then seeds remain in a cup for 15 minutes. The seeds are removed by straining the tea into other bowl; honey is added to sweeten it. Caraway seeds are used to cook traditional rye bread. Rye flour is different and healthy in itself. Caraway seeds have a characteristic odor and aromatic, pleasant, warm, sharp taste. Caraway seed is used in many baked goods, breads, roasts, seafoods, and cabbage or potato soups (Sedlakova et al., 2001).

Caraway (*C. carvi* L.) possesses an insecticidal property for *Sitophilus zeamais* (Motsch), red flour beetle, *Tribolium Castaneum*, and maize weevil herbs. The caraway essential oil shows strong potential against *T. Castaneum* and *S. zeamais* adults with LC_{50} values of 2.53 mg/L and 3.37 mg/L, respectively. The essential oil of caraway shows strong fumigant activity against *T. castaneum* adults as compared to the others as, for example, essential oil of *Drimys winteri* ($LC_{50} = 9.0-10.5$ μL/L), *Perovskia abrotanoides* ($LC_{50} = 11.39$ μL/L), *Schinus terebenthifolius* ($LC_{50} = 20.50$ μL/L) (Arabi et al., 2008), *Citrus reticulata* ($LC_{50} = 19.47$ μL/L), and *Mentha microphylla* ($LC_{50} = 4.51$ μL/L) (Mohamed and Abdelgaleil, 2008). However, it shows lesser toxicity than the *Laurelia sempervirens* ($LC_{50} = 1.6-1.7$ μL/L).

Caraway essential oil has been researched for its biologic activities. A large number of medicinal herbs is used in traditional medicines. In different traditional medicines, the caraway seeds are used as an antispasmodic, eupeptic, carminative, and astringent. It is also used to cure bloating, dyspeptic headache, colic, morning sickness, flatulence, dyspepsia, diarrhea, and other digestive disorders. Caraway also assimilates the other herbs to improve liver function. Seeds of caraway (*C. carvi*) are traditionally used for weight loss. Caraway use has been widely spread in different ethnomedical systems from Mediterranean regions to Northern Europe, North America, Indonesia, Iran, and Russia, where it is used in their folk medicines. Caraway is used to cure scabies when mixed with castor oil and alcohol (Johri, 2011; Joshi, 2000; Schultes, 1998). Arabians have used black caraway as herbal medicine for the treatment of different problems such as hypertension, diabetes, rheumatism, and gastrointestinal problems.

7. PHARMACOLOGICAL USES

7.1 Prophylactic Activity

Carvone is a monoterpene present in caraway seeds, dill, and fennel fruits. These are mostly used as a folk remedy for diarrhea, acidity, and other gastric disorders. Carvone has a variety of pharmacological effects such as being an antioxidant, antinociceptive, insecticidal, anticancer, and having blood lipid lowering activity. It can be used as prophylaxis of different cardiovascular diseases since it can be included in our daily diet. Carvone is a calcium channel blocker (Yu et al., 2005).

7.2 Anticancer Activity

Different spices including caraway help in carcinogen detoxification by mechanism of induction of glutathione transferase (GST) by anticarcinogenic compounds that lower the risk of cancer. With treatment of 29 mg limonene and carvone, the performance of detoxifying enzyme GST in the liver was increased. Black caraway seeds have a major bioactive constituent known as thymoquinone (TQ). Experimental evidences revealed that TQ suppresses the growth of many types of cancer cells, modulates inflammatory responses, and acts as an antioxidant (Banerjee et al., 2010).

7.3 Fungicidal and Antimicrobial Properties

The caraway oil and seeds possess strong fungicidal and antimicrobial activity. The caraway extract exhibited the strongest inhibitory activity that was particularly expressed against *Emericella nidulans* and both *Penicillium* species (*P. commune* and *P. implicatum*), which did not grow at extract doses over 0.1%. At this level, the growth of *P. implicatum* was reduced (77.8%). A level over 0.5% is needed to completely inhibit *Aspergillus tamarii*. It has been commonly used as a laxative and antibacterial agent (Eddouks et al., 2004).

7.4 Antiinflammatory Effects

The disturbance in homeostasis of the human body due to any trauma or infection is termed inflammation. Currently, many antiinflammatory drugs are available, but their continuous administration leads to side effects. Epidemiological and experimental studies suggest that polyphenols possess antiinflammatory and antioxidant ability that may contribute, via the diet, to prevention of disorders like Alzheimer disease, inflammatory bowel disease, cancer, and cardiovascular disease. The phenolics of caraway extract show inhibitory activity on hyaluronidase, trypsin, and cellular enzymes such as 5-lipoxygenase, hyaluronidase, and trypsin, thus exerting an important antiinflammatory action. Triterpenes and flavonoids show inhibitory activity on the enzyme activity of hyaluronic acid–splitting enzymes (Thippeswamy et al.).

7.5 Antidiabetic Effects

The major cause of mortality and morbidity is diabetes in the world, in spite of all the advances. Caraway has a naturally antihyperglycemic ability. Aqueous extract of caraway has shown antihyperglycemic effect in streptozotocin-induced diabetic rats. The antiulcer genic agents are found in plant extract and volatile oil from *C. carvi* (Lemhadri et al., 2006). Insulin and various oral antidiabetic agents such as a-glucosidase inhibitors, glinides, and sulfonylureas are currently available therapies for diabetes.

7.6 Cardiac Health

Caraway essential oil is very helpful for cardiac health. Regular use of caraway essential oil reduces the risk of heart diseases and keeps the heart healthy for a long time. It helps to lower the cholesterol level and blood pressure, prevents hardening of the arteries and veins, strengthens the cardiac muscles, and maintain proper heart rate, thus helping the heart from all angles (Yu et al., 2005).

7.7 Antioxidant Properties

Many food and agricultural products, including oil seeds, vegetables, and grains retain antioxidants (Yu et al., 2005). Today, seed oils of cranberry, hemp, and caraway are commercially available, which are extracted by cold-pressed method. In vitro and in vivo antioxidant activity of caraway essential oil was tested. The effect of essential oil on lipid per oxidation, free radical scavenging capacity, and scavenging activity of essential oil on OH radicals and 2, 2-dphenyl-1-picrylhydrazyl (DPPH) was measured. The stable DPPH was reduced by essential oil. Lipid per oxidation in both systems of inductions was strongly inhibited by caraway essential oil (Samojlik et al., 2010).

7.8 Hepatoprotective Activity

Essential oils of *C. carvi* L. were analyzed for hepatoprotective effect and in vivo and in vitro antioxidant activity against carbon tetrachloride damage. Some liver biochemical parameters were determined in animals pretreated with essential oils and later intoxicated with carbon tetrachloride to assess in vivo hepatoprotective effect. Caraway essential oil strongly inhibited lipid per oxidation in both systems of induction (Samojlik et al., 2010).

7.9 Analgesic Uses

Herbology is a safer way of relieving pain. Analgesic herbs like *C. carvi* L. are most important for pain relief and popular remedies. Some uses for analgesic herbs are neuralgia, lower back pain, toothaches, headaches, arthritis pain, and sore muscles (Lima et al., 2017).

7.10 Antihistaminic Agent

Histamine is the major reason behind exhausting and disruptive coughs. People who suffer from seasonal allergies can continue to cough endlessly, the condition becoming so severe that the patient might collapse due to running out of breath in some cases. The caraway oil can miraculously cure these potentially dangerous coughs by neutralizing the effects of histamine and is also helpful in other ailments like allergies that are associated with histamine.

7.11 Diuretic Agent

Caraway essential oil is an ideal diuretic agent. It is literally a blessing for those who are suffering from high blood pressure, renal calculi, obstructed urination, and those who want to lose weight. An increase in the quantity and frequency of urination can help you in all of these situations. Caraway oil cleans the deposits from the kidney by urination and lowers blood pressure. Carum extract did not produce any renal toxicity or any other adverse effects during the study period. A lot of urination also frees the urinary tract from infections (Lahlou et al., 2007).

7.12 Stimulating Agent

Caraway oil is a stimulating agent. It is helpful in curing fatigue and depression. It stimulates the body's functioning, systems, and cycles within the body, including the excretory, nervous, endocrine, digestive, and circulatory systems. It also activates the brain and helps keep you awake and alert (Craig, 1999).

7.13 Estrogenic/Antiosteoporotic

Caraway seeds are reported to be estrogenic (Malini and Vanithakumari, 1987). Potential effects of caraway on reproductive parameters and hormones of female ovariectomized rats are demonstrated due to the presence of estrogenic isoflavonoids, apigenin luteolin. An aqueous and an ethanolic extract of caraway seeds produced significant antifertility effect via modulation of luteinizing hormone levels and follicle stimulating hormone, while the estrogen levels were increased (Johri, 2011).

8. SIDE EFFECTS AND TOXICITY

Caraway is safe for most people when taken by mouth in food and medicinal amounts for up to 3 months or when applied to the skin for up to 3 weeks. Caraway oil can cause belching, heartburn, and nausea when used with peppermint oil. It can cause skin rashes and itching in sensitive people when applied to the skin (Rodriguez-Fragoso et al., 2008). It is unsafe to take caraway in medicinal amounts during pregnancy. Caraway oil has been used to start menstruation, and this might cause a miscarriage. There is a concern that caraway might lower blood sugar. Caraway extract might increase the absorption of iron. Overuse of caraway extract with iron supplements or iron-containing food might increase iron levels in the body. This may be a problem for people who already have too much iron in the body (El-Shobaki et al., 1990).

References

Aćimović, M., Oljača, S., Tešević, V., Todosijevića, M., Djisalov, J., 2014. Evaluation of caraway essential oil from different production areas of Serbia. Horticultural Science 41, 122–130.

Arabi, F., Moharramipour, S., Sefidkon, F., 2008. Chemical composition and insecticidal activity of essential oil from Perovskia abrotanoides (Lamiaceae) against Sitophilus oryzae

(Coleoptera: Curculionidae) and *Tribolium castaneum* (Coleoptera: tenebrionidae). International Journal of Tropical Insect Science 28, 144–150.

Badura, M., 2003. Pimenta officinalis Lindl.(pimento, myrtle pepper) from early modern latrines in Gdańsk (northern Poland). Vegetation History and Archaeobotany 12, 249–252.

Banerjee, S., Padhye, S., Azmi, A., Wang, Z., Philip, P.A., Kucuk, O., Sarkar, F.H., Mohammad, R.M., 2010. Review on molecular and therapeutic potential of thymoquinone in cancer. Nutrition and Cancer 62, 938–946.

Craig, W.J., 1999. Health-promoting properties of common herbs. American Journal of Clinical Nutrition 70, 491s–499s.

Eddouks, M., Lemhadri, A., Michel, J.-B., 2004. Caraway and caper: potential antihyperglycaemic plants in diabetic rats. Journal of Ethnopharmacology 94, 143–148.

El-Shobaki, F., Saleh, Z., Saleh, N., 1990. The effect of some beverage extracts on intestinal iron absorption. Zeitschrift für Ernährungswissenschaft 29, 264–269.

Fatemi, F., Allameh, A., Khalafi, H., Ashrafihelan, J., 2010. Hepatoprotective effects of γ-irradiated caraway essential oils in experimental sepsis. Applied Radiation and Isotopes 68, 280–285.

Furmanowa, M., Sowińska, D., Pietrosiuk, A., 1991. Carum Carvi L.(Caraway): In Vitro Culture, Embryogenesis, and the Production of Aromatic Compounds, Medicinal and Aromatic Plants III. Springer, pp. 176–192.

Houghton, P.J., Zarka, R., de las Heras, B., Hoult, J., 1995. Fixed oil of Nigella sativa and derived thymoquinone inhibit eicosanoid generation in leukocytes and membrane lipid peroxidation. Planta Medica 61, 33–36.

Johri, R., 2011. Cuminum cyminum and *Carum carvi*: an update. Pharmacognosy Reviews 5, 63.

Joshi, S.G., 2000. Medicinal Plants. Oxford and IBH publishing.

Kazemipoor, M., Cordell, G.A., 2015. In: Mukherjee, P.K. (Ed.), Clinical Effects of Caraway, a Traditional Medicine for Weight Loss. Evidence-Based Validation of Herbal Medicine: Farm to Pharma. Elsevier Science, Amsterdam, Netherlands, pp. 339–362.

Kazemipoor, M., Hajifaraji, M., Haerian, B.S., Mosaddegh, M.H., Cordell, G.A., 2013. Anti-obesity effect of caraway extract on overweight and obese women: a randomized, triple-blind, placebo-controlled clinical trial. Evidence-based Complementary and Alternative Medicine 2013.

Lahlou, S., Tahraoui, A., Israili, Z., Lyoussi, B., 2007. Diuretic activity of the aqueous extracts of *Carum carvi* and Tanacetum vulgare in normal rats. Journal of Ethnopharmacology 110, 458–463.

Laribi, B., Kouki, K., Bettaieb, T., Mougou, A., Marzouk, B., 2013. Essential oils and fatty acids composition of Tunisian, German and Egyptian caraway (*Carum carvi* L.) seed ecotypes: a comparative study. Industrial Crops and Products 41, 312–318.

Laribi, B., Kouki, K., Mougou, A., Marzouk, B., 2010. Fatty acid and essential oil composition of three Tunisian caraway (*Carum carvi* L.) seed ecotypes. Journal of the Science of Food and Agriculture 90, 391–396.

Lemhadri, A., Hajji, L., Michel, J.-B., Eddouks, M., 2006. Cholesterol and triglycerides lowering activities of caraway fruits in normal and streptozotocin diabetic rats. Journal of Ethnopharmacology 106, 321–326.

Lima, T., da Nóbrega, F., de Brito, A., de Sousa, D., 2017. Analgesic-like activity of essential oil constituents: an update. International Journal of Molecular Sciences 18, 2392.

Malini, T., Vanithakumari, G., 1987. Estrogenic activity of Cuminum cyminum in rats. Indian Journal of Experimental Biology 25, 442–444.

Meshkatalsadat, M.H., Salahvarzi, S., Aminiradpoor, R., Abdollahi, A., 2012. Identification of essential oil constituents of caraway (*Carum carvi*) using ultrasonic assist with headspace

solid phase microextraction (UA-HS-SPME). Digest Journal of Nanomaterials and Biostructures 7, 637–640.
Mohamed, M.I., Abdelgaleil, S.A., 2008. Chemical composition and insecticidal potential of essential oils from Egyptian plants against Sitophilus oryzae (L.)(Coleoptera: Curculionidae) and *Tribolium castaneum* (Herbst)(Coleoptera: tenebrionidae). Applied Entomology and Zoology 43, 599–607.
Riddhi, M.P., Yogesh, T.J., Antioxidant Activity of Medicinal Spices and Aromatic Herbs.
Rodriguez-Fragoso, L., Reyes-Esparza, J., Burchiel, S.W., Herrera-Ruiz, D., Torres, E., 2008. Risks and benefits of commonly used herbal medicines in Mexico. Toxicology and Applied Pharmacology 227, 125–135.
Sadiq, S., Nagi, A.H., Shahzad, M., Zia, A., 2010. The reno-protective effect of aqueous extract of *Carum carvi* (black zeera) seeds in streptozotocin induced diabetic nephropathy in rodents. Saudi Journal of Kidney Diseases and Transplantation 21, 1058.
Samojlik, I., Lakic, N., Mimica-Dukic, N., Đakovic-Svajcer, K., Bozin, B., 2010. Antioxidant and hepatoprotective potential of essential oils of coriander (Coriandrum sativum L.) and caraway (*Carum carvi* L.)(Apiaceae). Journal of Agricultural and Food Chemistry 58, 8848–8853.
Schultes, R., 1998. History of Using Caraway as a Remedy. Caraway: the genus Carum, p. 186.
Sedlakova, J., Kocourkova, B., Kuban, V., 2001. Determination of essential oils content and composition in caraway (*Carum carvi* L.). Czech Journal of Food Sciences 19, 31–36.
Sedláková, J., Kocourková, B., Lojková, L., Kuban, V., 2003. Determination of essential oil content in caraway (*Carum carvi* L.) species by means of supercritical fluid extraction. Plant Soil and Environment 49, 277–282.
Seidler-Lozykowska, K., Baranska, M., Baranski, R., Krol, D., 2010. Raman analysis of caraway (*Carum carvi* L.) single fruits. Evaluation of essential oil content and its composition. Journal of Agricultural and Food Chemistry 58, 5271–5275.
Thippeswamy, N.B., Achur, R.N., 2014. Inhibitory effect of phenolic extract of carum carvi on inflammatory enzymes, hyaluronidase and trypsin. World Journal of Pharmaceutical Sciences 2 (4), 350–356.
Yu, L.L., Zhou, K.K., Parry, J., 2005. Antioxidant properties of cold-pressed black caraway, carrot, cranberry, and hemp seed oils. Food Chemistry 91, 723–729.

Further Reading

Al Yahya, M., 1986. Phytochemical studies of the plants used in traditional medicine of Saudi Arabia. Fitoterapia 57, 179–182.
Chu, S.S., Liu, Q.R., Liu, Z.L., 2010. Insecticidal activity and chemical composition of the essential oil of Artemisia vestita from China against Sitophilus zeamais. Biochemical Systematics and Ecology 38, 489–492.
Cronquist, A., 1981. An Integrated System of Classification of Flowering Plants. Columbia University Press.
Kameník, J., Pelán, J., Janoušková, V., 1996. Překlady. Triáda.
Khaled-Khodja, N., Boulekbache-Makhlouf, L., Madani, K., 2014. Phytochemical screening of antioxidant and antibacterial activities of methanolic extracts of some Lamiaceae. Industrial Crops and Products 61, 41–48.
Krkić, N., Šojić, B., Lazić, V., Petrović, L., Mandić, A., Sedej, I., Tomović, V., Džinić, N., 2013. Effect of chitosan–caraway coating on lipid oxidation of traditional dry fermented sausage. Food Control 32, 719–723.
Rutledge, C., McLendon, T., 2002. An Assessment of Exotic Plant Species of Rocky Mountain National Park: Summary Information for Remaining Exotic Plant Species. Department of Rangeland Ecosystem Science, Colorado State University.

Thippeswamy, N., Naidu, K.A., Achur, R.N., 2013. Antioxidant and antibacterial properties of phenolic extract from *Carum carvi* L. Journal of Pharmacy Research 7, 352–357.
Zheng, G., Kenney, P.M., Lam, L., 1992. Anethofuran, carvone, and limonene: potential cancer chemopreventive agents from dill weed oil and caraway oil. Planta Medica 58, 338–341.

CHAPTER

8

Chamomilla

Shaheera Rehmat[1], *Rasheed Ahmad Khera*[1], *Muhammad Asif Hanif*[1], *Muhammad Adnan Ayub*[2], *Muhammad Zubair*[3]

[1] Department of Chemistry, University of Agriculture, Faisalabad, Pakistan; [2] Department of Chemistry, University of Okara, Okara, Pakistan; [3] Department of Chemistry, University of Gujrat, Gujrat, Pakistan

OUTLINE

Medicinal Plants of South Asia
https://doi.org/10.1016/B978-0-08-102659-5.00008-2

Copyright © 2020 Elsevier Ltd. All rights reserved.

1. BOTANY

1.1 Introduction

Matricaria chamomilla L. (Fig. 8.1) commonly known as "chamomile" is a medicinal herb belonging to Compositae family. The Compositae family covers a wide variety of valuable medicinal genera like *Taraxacum, Silybum, Calendula, Tussilago, Artemisia, Achillea*, and *Matricaria*. In addition to these, the two most popular species of chamomile are chamomile and chamaemelum nobile, which are similar in many aspects of physical appearance, chemical properties, and in general applications. However, some minor variations have been reported in their size, type of flower, leaves and number of chemical compounds. In contrast to chamaemelum nobile, chamomile bears small fruits, is grassy, and has a sweet smell. Chamomile is called "babuna" in the Unani medicinal system and is used in sexual enhancement. Chamomile is a long-lasting drug derived from a

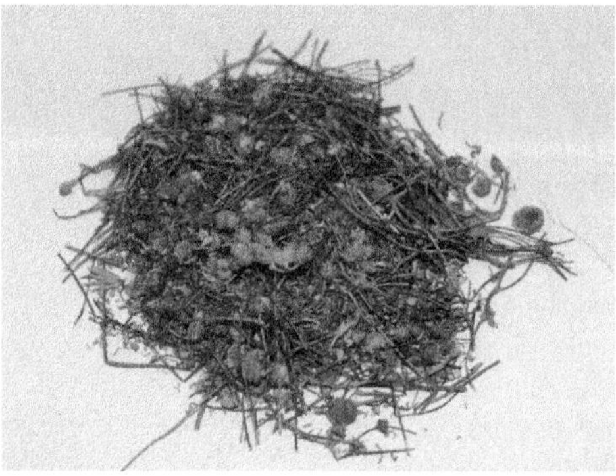

FIGURE 8.1 Dry chamomilla.

plant known by different names like scented mayweed, pinhead, sweet false chamomile, single chamomile, chamomilla flos chamomile, English chamomile, Roman chamomile, Hungarian chamomile, German chamomile, Babunj, Babuna camornile, Babuna and Baboonig. In Urdu, it is called Babuna (Franke and Schilcher, 2005).

1.2 History/Origin

"Chamomile" originates from two Greek words: "one" is "chamos," which means "ground," and the other is "milos," which means apple. Due to its growth close to the ground and smelling like apple, it is called chamomile. Chandra et al. (1968) first reported chamomile growth in alkaline soil of Lucknow. For about 200 years, chamomile had been cultivated in the Lucknow region of India. In Punjab, the plant was known during the Mughal's period about 300 years ago. It was first brought to North America by the Spanish colonists, probably in the early 16th century. Both German and Roman chamomile species are among the most widely used medicinal plants; they are traditionally employed by empirical herbalists in northern Mexico and the American southwest as a trivial tea to treat a variety of ailments, especially colic in small children (Davidow, 1999; Zadeh et al., 2014). As an old-age medicinal drug, dried chamomile flower came to be known in different countries like Rome, Greece, and earliest Egypt. Egyptians believed that chamomile concerted to the god of the sun, so it had a religious use. The flower is known as "manzanilla" in Spain (also meaning "little apple"). By the same name, it has long been employed to flavor dishes.

1.3 Demography/Location

Chamomile is widely distributed in Pakistan, India, Nepal, Sri Lanka, Bangladesh, and Afghanistan. It is also cultivated in German, Hungry, Finland, France, and South Africa (Galambosi et al., 1991). Mostly, German chamomile is cultivated in KwaZulu-Natal, Gauteng, Eastern Cape, Free State, Northwest and Mpumalanga provinces. The United States is the world's largest market of chamomile essential oil, followed by Japan and Europe. Soft drink companies are the main users of essential oils in the United States. About 10% from Japan accounts for the world demand. US perfume and flavoring industry dominates the Canadian market.

1.4 Botany, Morphology, Ecology

Chamomile is an annual plant. The thin, spindle-shaped roots flatly penetrate the soil. The stem grows to a height of 10–80 cm. The branched stem is heavily ramified and erect. Bi-to tri-pinnate leaves are long and narrow. The pedunculated and heterogamous flower heads are placed separately, having a diameter of 10–30 mm. The tubular florets are golden yellow, 1.5–2.5 cm long, ending always in a glandulous tube with five teeth. The concentrically arranged 11–27 white plant flowers are 6–11 mm long and 3.5 mm wide. In the beginning, 6–8 mm wide and flat receptacles present, which later become conical and cone shaped. It bears yellowish-brown achene fruit.

2. CHEMISTRY

Chamomile has a sweet, smoky odor that is largely compared with apples. A large group of sesquiterpene, flavonoid, polyacetylene, and coumarin compounds belong to chamomile. Chamomile infusions contain a large number of phenolic acids like hydroxyl-benzoic and hydroxycinnamic acid and their glycosides such as apigenin, luteolin, and chamazulene (Maciag et al., 2009). Umbelliferone, herniarin, and other minor ones are characterized coumarins in chamomile (Redaelli et al., 1981). Chamomile contains native compounds like glucose precursor of herniarin and (Z)-and (E)-2-b-D-glucopyranosyloxy-4-methoxy cinnamic acids (GMCA). Chamomile flowers contain 0.24% to 2.0% volatile oil that is usually blue in color. The two key constituents, (−)-alpha-bisabolol and chamazulene, account for 50%–65% of total volatile oil contents (Fig. 8.2). Flowers contain maximum concentration of essential oil when ray florets are in open mode; afterward, this concentration decreases. Approximately more than 100 chemical components were identified in the flower of chamomile, containing 28 terpenoids, 36 flavonoids (Singh et al., 2011), and 52 other compounds having prospective pharmacological activity. Flower buds contain the highest concentration of farnesene and bisabolol,

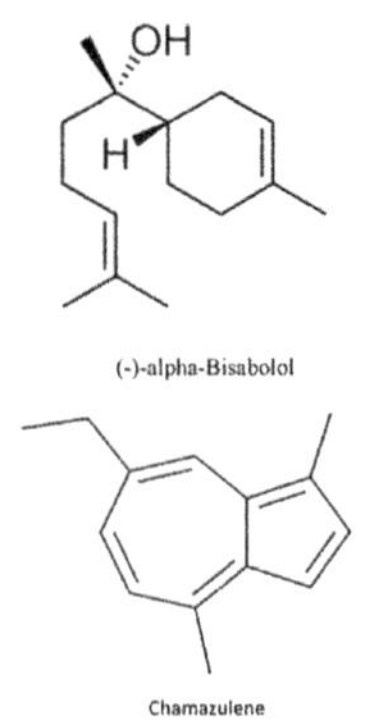

FIGURE 8.2 Key active components of chamomilla.

and the lowest concentration present in decaying flowers. The contents of bisabolol oxide and chamazulene increased from little buds to completely established flower buds. Depending upon the selection period and progress phase, diverse amounts and ratios of these compounds are present in different parts of inflorescence during the day. In flowers, the amount of α-bisabolol and α-bisabolol oxide A and B touched maximum level when blooming is at its highest point; after that, they drop (Arak et al., 1980). Due to its extensive pharmaceutical and pharmacological properties, this plant holds pronounced economic value and great demand in European countries. In another study, it was reported that chamazulene, α-bisabolol, A and B bisabolol oxides, and β-farnesene were the essential oil's main compounds, while α- and β-caryophyllene, caryophyllene-oxide, spathulenol, β-phellandrene, limonene, β-ocimene, and γ-terpinen were the minor compounds (Costescu et al., 2008). Similarly, essential oil isolated from Estonian chamomile contained bisabolol oxide A, bisabolone oxide A, (Z)-en-yne-dicycloether, bisabolol oxide B, α-bisabolol, and chamazulene as the main compounds (Singh et al., 2011). Root and shoot of chamomile also contain essential oil besides the capitula. Root oil of chamomile contains esters, ethers, and many oxides. Extraction of the chamomile also revealed bioactive phenolic compounds like naringenin (flavanone), quercetin and rutin (flavonols), luteolin and luteolin-7-O-glucoside (flavones), herniarin and umbelliferone (coumarin), chlorogenic acid, caffeic acid (phenylpropanoids), apigenin, and apigenin-7-O-glucoside. The tannin level in chamomile is less than 1%.

3. POSTHARVEST TECHNOLOGY

Dry leaves are used for the extraction of essential oils. The flowers of this plant can also be used for ornamental purposes. Fresh leaves are dried in the sun or sometimes in shaded areas. To prevent from oxidation and discoloration, leaves are dried by placing the whole spread leaves between two sheets of paper. Industrially chamomile leaves are dried by forced warm air-drying method. Before drying the chamomile leaves, make sure that these leaves are free from other plant species or herbs. After drying, the herb leaves are ground in a blender to convert into powder form. The dried form of the leaves can be stored for a very long time.

4. PROCESSING

Without affecting quality and oil yield, steam distillation could be possible on frozen, dried, as well as fresh flowers for the production of chamomile essential oil. The highest yield and quality of oil can be extracted by distillation for 12 h. On further distillation, oil of various compositions will be obtained. Steam distillation can also produce high-grade floral water and chamomile blue essential oil. Chamomile essential oil can be produced by solvent extraction using less complex and low-cost equipment. The price of oil extracted through solvent extraction decreases, as chamazulene is not formed due to absence of steam, and also because of this the oil will not be blue, although oil yields are higher. Dried flowers are prepared to be free of impurities and other external material for the tea market. Normally, oil packing requires 1-, 5-, and 10-kg aluminum flasks. However, in some cases, larger packs are also being used. In pharmaceutical and cosmetic purposes, bulk flowers are processed further for extraction in factories (Gates et al., 2007).

5. VALUE ADDITION

The oil has a broad range of uses in pharmaceutics and other fields like cosmetics, aromatherapy, perfumery, and the food industry. It was found that azulene is present in flower head essential oil. The applications of azulene are in toothpastes, skin lotions, hair preparations, perfumery, cosmetic creams, and in fine liquors (Subiza et al., 1990). The arid flowers of chamomile are greatly applied for baby massage oil, herbal tea, to cure cough and cold, and for accelerating the gastric flow secretion.

6. USES

Chamomile is used in a variety of ways for various purposes. In salads, fresh flower heads of chamomile plant are added. Fresh flowers are used in tea. Chamomile lemon cupcakes with honey buttercream frosting are made by using dried chamomile. The plant possesses more economic value due to its extensive pharmaceutical and pharmacological properties. Antiseptic, antispasmodic, mildly sudorific, and antiinflammatory properties are the major applications of the herb. Chamomile is very infrequently applied internally and more efficiently for painful menstruation, for irritation of the urinary tract, as a tisane in stomach disturbance allied with pain, diarrhea, nausea, and sluggish digestion. Externally, it is mainly used for hemorrhoids, soreness of mouth, the throat, eyes, also to injuries that are slower to heal, in skin eruption, and toxicities. Chamomile flower extraction produces tabulated products that are marketed in Europe and used for different diseases. Allergic conjunctivitis may be induced by chamomile tea eye washing. In these infusions, pollen of chamomile are allergens, so reactions are possible (Subiza et al., 1990).

7. PHARMACOLOGICAL USES

7.1 Antimicrobial Activity

Chamomile oil is well known to have a broad range of antiinflammatory activities. Against *Mycobacterium tuberculosis, Salmonella typhimurium,* and *Staphylococcus aureus,* chamomile has shown antibacterial activity.

7.2 Antioxidant Activity

Chamomile has shown prominent antioxidant activities due to the presence of considerable amounts of phenolic and flavonoid compounds (Al-Dabbagh et al., 2019).

7.3 Anticancer Activity

Apigenin has revealed chemo-preventive effects against cancer cells (Gardiner, 2007). Recent studies have shown that the evaluation of tumor growth inhibition can be carried out by apigenin. Recently, conducted studies have shown the promising growth inhibitory effects on the preclinical reproductions of skin, breast, prostate, and the ovarian cancers (Gates et al., 2007). Minimal growth inhibitory effects have been shown by studies on normal cells caused by chamomile extracts. However, in several human cancers cell lines, the chamomile extracts exhibited reduction of cell viability. Apoptosis is induced in cancer cells by chamomile exposure, yet not in ordinary cells at same amount. A test was conducted on the efficiency of innovative compound TBS-101. Both in vitro and in vivo, it is confirmed by different results that chamomile has a good safety profile against androgen-refractory prostate cancer cells with significant anticancer activities (Evans et al., 2009). A communal state with different types of etiologies is mouth ulcer (Gonsalves et al., 2008). Bolus 5-fluorouracil-based (5-FU) chemotherapy procedures are mainly involved in dose-limiting poisoning of stomatitis. A recent study was conducted as a placebo-controlled, double-blind clinical trial of 164 patients. At the time of their first life cycle of 5-FU-based chemotherapy, the patients were started on chamomile or placebo mouthwash. They were randomized three times daily for 14 days. After randomization of patients, no specific difference between patients was observed to either protocol arm. Toxicity differences were also not recommended in them. In similar conditions, another potential trial on chamomile provided the same results.

7.4 Antispasmodic Activity

Antispasmodic effects have been demonstrated by flavonoids and bisabolol in in vivo experiments. Many research studies revealed that colic is treated effectively in children by tea of chamomile in combination with other herbs. Considerable smooth muscle relaxants are present, as coumarins and flavonoids are extensively reported (Srivastava et al., 2010).

7.5 Antiphlogistic Activity

Alpha-bisabolol, alpha-bisabolol oxides component A and component B, and matricin are the components of volatile oils occupying 1%–2% of chamomile flower, which are converted into chamazulene and other flavonoids showing antiphlogistic and antiinflammatory properties (Peña et al., 2006). It was demonstrated in a study that essential oils and flavonoids of chamomile enter below the skin, apparent deeper into the skin's coatings (Merfort et al., 1994). Their use in topical antiphlogistic agents is of great demand.

7.6 Antiinflammatory Activity

Chamomile antiinflammatory activities are due to attenuation of cyclooxygenase enzyme activity without affecting the constitutive form and inhibition of LPS-induced prostaglandin E (2) release (Srivastava et al., 2009).

7.7 Protective Effects in Gastrointestinal Conditions

The course of diarrhea and relief of symptoms associated with conditions in children may decrease by using extract of apple pectin–chamomile. Two medical prosecutions have assessed the efficiency of chamomile for colic treatment in the children. The chamomile efficacy for the treatment of colic in children has been evaluated by two clinical trials. For administration, other herbs are mixed in chamomile tea. Herbal tea or the placebo received by colic had been reported in double-blind, prospective, placebo-controlled observation. Up to a 150-mL dose was given to each infant for three times per day. After 7 days of treatment, parents reported that the tea eliminated the colic in 57% of the infants, whereas placebo was helpful in only 26%. In either group relative to number of the nighttime awakenings, no confrontational special effects were observed (Gardiner, 2007). Effects of apple pectin preparation and chamomile extract were examined by another study in 79 children with noncomplicated, acute diarrhea, who were given either the pectin preparation/ chamomile or a placebo (Kell, 1997). Evidence offered by these results proves the effectiveness of chamomile use to cure infant colic disorder. Reasonably atopic eczema has been treated by chamomile (Hörmann and Korting, 1994). In association with eczema, Roman chamomile of the Manzana type may ease discomfort when used as a cream comprised of chamomile extract. Chamomile of Manzana type does not contain chamomile-related allergenic potential, but it exhibits a large number of active ingredients. To compare Kamillosan with 0.5% hydrocortisone, a half-side comparison was carried out. Kamillosan(R) cream was tested vs. 0.5% hydrocortisone cream and the vehicle cream as placebo in patients suffering from medium-degree atopic eczema (Patzelt-Wenczler and Ponce-Pöschl, 2000). After treatment of 2 weeks, a slight supremacy had been shown by Kamillosan over 0.5% hydrocortisone and minimal differences as related to placebo. To assess the applications of topical chamomile in handling eczema, further research is required. Traditionally, in several gastrointestinal conditions, chamomile has been used such for upset stomach, ulcers, gastrointestinal irritation, digestive disorders, and colic (Kroll and Cordes, 2006). Previously, the protective effect has been reported against the growth of gastric ulcers by using commercial preparations (STW5, Iberogast) containing the extracts of milk thistle fruit, greater celandine, angelica root, peppermint leaf, caraway fruit, liquor ice root, chamomile flower, lemon balm leaf, and bitter candy tuft (Khayyal et al., 2006). A dose-dependent antiulcer genic effect produced by the extraction of STW5 is linked with an increased mucin secretion, a reduced acid output, an increase in prostaglandin E (2) release, and reduction in leukotrienes. The observable results verified that STWF5 effectively lowered the gastric acidity as well as inhibiting the secondary hyperacidity efficiently. The results showed that hemorrhoids may be recovered by chamomile ointment. Sits bath format can also use the tinctures of chamomile. Inflammation associated with hemorrhoids may be decreased by using the tincture of Roman chamomile (Ramos-e-Silva et al., 2006). Many gastrointestinal disorders like diverticular disease, esophagus reflux, and inflammatory disease are linked with inflammation (Ramos-e-Silva et al., 2005). It is recommended in studies of preclinical models that chamomile can used to inhibit

Helicobacter pylori (Wu, 2006). Smooth muscle spasms linked with different gastrointestinal inflammatory disorders may be helpfully decreased by chamomile.

7.8 Skin Applications

It is often believed that chamomile is helpful to cure mild skin irritations like sunburn, sores, rashes, and smooth eye inflammations. However, evidence-based research has not been shown any value for the treatment of these conditions.

7.9 Protective Effects in Urinary Tract

In women of all ages, vaginal inflammation is common. Pain with urination, vaginal discharge, and itching are linked with vaginitis. A reduced level of estrogen causes atrophic vaginitis, and its occurrence is mostly associated with postmenopausal and menopausal women. The symptoms of vaginitis may improve by chamomile douche with few side effects (Benetti and Manganelli, 1985). In this condition, there is insufficient research data regarding probable potential benefits of chamomile to allow conclusions.

7.10 As Sleep Inducer and Mild Tranquillizer

Treatment of insomnia and induction of sedation have been traditionally carried out by chamomile preparations like tea and essential oil aromatherapy. Broad observations explained that chamomile has been used as a sleep inducer and mild tranquillizer. Benzodiazepine receptors in the brain linked with apigenin and flavonoids may show the sedative effects (Avallone et al., 1996). Studies in preclinical models have shown anticonvulsant and CNS depressant effects. Clinical trials are notable for their absence, although ten cardiac patients are reported to have immediately fallen into a deep sleep lasting for 90 minutes after drinking chamomile tea. Benzodiazepine-like hypnotic activity is demonstrated from chamomile extracts (Shinomiya et al., 2005).

8. SIDE EFFECTS AND TOXICITY

Chamomile overall and essential oil especially have many beneficial aspects of curing many harmful diseases. Even though chamomile is a good plant with versatile uses, it may produce harm when used as tea and essential oil. Results could be anaphylactic shock, contact dermatitis, and other severe allergic reactions. Persons who are allergic to asters, chrysanthemums, ragweed, and other members of the Asteraceae daisy family should avoid chamomile. It is also reported that chamomile essential oil is not recommended for use with pregnant women or children. Some essential oils may cause irritation or allergic reactions in people with sensitive skin, so it is wise to do a patch test before using regularly.

References

Al-Dabbagh, B., Elhaty, I.A., Elhaw, M., Murali, C., Al Mansoori, A., Awad, B., Amin, A., 2019. Antioxidant and anticancer activities of chamomile (Matricaria recutita L.). BMC Research Notes 12, 3.

Arak, E., Tammeorg, I., Myaeorg, U., 1980. Dynamics of some components of chamomile essential oil. Tartu Ulikooli Toimetised 19–32.

Avallone, R., Zanoli, P., Corsi, L., Cannazza, G., Baraldi, M., 1996. Benzodiazepine-like compounds and GABA in flower heads of *Matricaria chamomilla*. Phytotherapy Research 10, S177–S179.

Benetti, C., Manganelli, F., 1985. Clinical experiences in the pharmacological treatment of vaginitis with a camomile-extract vaginal douche. Minerva Ginecologica 37, 799–801.

Chandra, V., Singh, A., Kapoor, L.D., 1968. Experimental cultivation of some essential oil bearing plants in saline soils, *Matricaria chamomilla* L. Perfume and Essential Oil Review 59, 871.

Costescu, C., Hădărugă, N., Hădărugă, D., Riviş, A., Ardelean, A., Lupea, A.X., 2008. Bionanomaterials: synthesis, physico-chemical and multivariate analyses of the dicotyledonatae and pinatae essential oil/β-cyclodextrin nanoparticles. Revista de Chimie 59, 739–744.

Davidow, J., 1999. Infusions of Healing: A Treasury of Mexican-American Herbal Remedies. Simon & Schuster, New York.

Evans, S., Dizeyi, N., Abrahamsson, P.-A., Persson, J., 2009. The effect of a novel botanical agent TBS-101 on invasive prostate cancer in animal models. Anticancer Research 29, 3917–3924.

Franke, R., Schilcher, H., 2005. Chamomile: Industrial Profiles. CRC press.

Galambosi, B., Szebeni-Galambosi, Z., Repcak, M., Cernaj, P., 1991. Variation in the yield and essential oil of four chamomile varieties grown in Finland in 1985—1988. Agricultural and Food Science 63, 403–410.

Gardiner, P., 2007. Complementary, holistic, and integrative medicine: chamomile. Pediatrics in Review 28, e16.

Gates, M.A., Tworoger, S.S., Hecht, J.L., De Vivo, I., Rosner, B., Hankinson, S.E., 2007. A prospective study of dietary flavonoid intake and incidence of epithelial ovarian cancer. International Journal of Cancer 121, 2225–2232.

Gonsalves, W.C., Wrightson, A.S., Henry, R.G., 2008. Common oral conditions in older persons. American Family Physician 78, 845–852.

Hörmann, H., Korting, H., 1994. Evidence for the efficacy and safety of topical herbal drugs in dermatology: Part I: anti-inflammatory agents. Phytomedicine 1, 161–171.

Kell, T., 1997. More on infant colic. Birth Gazette 13, 3.

Khayyal, M., Seif-El-Nasr, M., El-Ghazaly, M., Okpanyi, S., Kelber, O., Weiser, D., 2006. Mechanisms involved in the gastro-protective effect of STW 5 (Iberogast®) and its components against ulcers and rebound acidity. Phytomedicine 13, 56–66.

Kroll, U., Cordes, C., 2006. Pharmaceutical prerequisites for a multi-target therapy. Phytomedicine 13, 12–19.

Maciag, P.C., Radulovic, S., Rothman, J., 2009. The first clinical use of a live-attenuated Listeria monocytogenes vaccine: a Phase I safety study of Lm-LLO-E7 in patients with advanced carcinoma of the cervix. Vaccine 27, 3975–3983.

Merfort, I., Heilmann, J., Hagedorn-Leweke, U., Lippold, B., 1994. In vivo skin penetration studies of camomile flavones. Die Pharmazie 49, 509–511.

Patzelt-Wenczler, R., Ponce-Pöschl, E., 2000. Proof of efficacy of Kamillosan (R) cream in atopic eczema. European Journal of Medical Research 5, 171–175.

Peña, D., Montes de Oca, N., Rojas, S., Parra, A., García, G., 2006. Anti-inflammatory and anti-diarrheic activity of Isocarpha cubana Blake. Pharmacologyonline 3, 744–749.

Ramos-e-Silva, M., Ferreira, A., Bibas, R., Carneiro, S., 2005. Clinical evaluation of fluid extract of Chamomilla recutita for oral aphthae. Journal of Drugs in Dermatology 5, 612–617.

Ramos-e-Silva, M., Ferreira, A., Bibas, R., Carneiro, S., 2006. Clinical evaluation of fluid extract of Chamomilla recutita for oral aphthae. Journal of Drugs in Dermatology 5, 612–617.

Redaelli, C., Formentini, L., Santaniello, E., 1981. HPLC determination of coumarins in *Matricaria chamomilla*. Planta Medica 43, 412–413.

Shinomiya, K., Inoue, T., Utsu, Y., Tokunaga, S., Masuoka, T., Ohmori, A., Kamei, C., 2005. Hypnotic activities of chamomile and passiflora extracts in sleep-disturbed rats. Biological and Pharmaceutical Bulletin 28, 808–810.

Singh, O., Khanam, Z., Misra, N., Srivastava, M.K., 2011. Chamomile (*Matricaria chamomilla* L.): an overview. Pharmacognosy Reviews 5, 82.

Srivastava, J.K., Pandey, M., Gupta, S., 2009. Chamomile, a novel and selective COX-2 inhibitor with anti-inflammatory activity. Life Sciences 85, 663–669.

Srivastava, J.K., Shankar, E., Gupta, S., 2010. Chamomile: a herbal medicine of the past with a bright future. Molecular Medicine Reports 3, 895–901.

Subiza, J., Subiza, J., Alonso, M., Hinojosa, M., Garcia, R., Jerez, M., Subiza, E., 1990. Allergic conjunctivitis to chamomile tea. Annals of Allergy 65, 127–132.

Wu, J., 2006. Treatment of rosacea with herbal ingredients. Journal of Drugs in Dermatology 5, 29–32.

Zadeh, J.B., Kor, N.M., Kor, Z.M., 2014. Chamomile (Matricaria recutita) As a Valuable Medicinal Plant. International journal of Advanced Biological and Biomedical Research 2 (3), 823–829.

CHAPTER

9

Chili Pepper

Saba Idrees[1], Muhammad Asif Hanif[1], Muhammad Adnan Ayub[2], Asma Hanif[1], Tariq Mahmood Ansari[3]

[1] Department of Chemistry, University of Agriculture, Faisalabad, Pakistan; [2] Department of Chemistry, University of Okara, Okara, Pakistan; [3] Institute of Chemical Sciences, Bahauddin Zakariya University, Multan, Pakistan

OUTLINE

Medicinal Plants of South Asia
https://doi.org/10.1016/B978-0-08-102659-5.00009-4

Copyright © 2020 Elsevier Ltd. All rights reserved.

1. BOTANY

1.1 Introduction

Chili Pepper (*Capsicum annuum* L.) (Fig. 9.1) is an annual herb that belongs to Solanaceae family (Greenleaf, 1986). It has been used for thousands of years and has become an important ingredient of cooking. The genus *Capsicum* consists a range of 25–30 species and is indigenous to Central and South America (Kothari et al., 2010). The uncertainty in the exact number of species within the genus is largely attributed to great variability among the constituent species. Fruits from different species of pepper differ in flavor, color, form, and size from very hot to mild or pleasantly pungent. Mostly, plants are self-pollinated; however about 17% cross-pollination through wind or insects generally occurs. *Capsicum* species are generally self-pollinated, and chili pepper is a moderately self-compatible crop; self-pollination can be enhanced by wind or related mechanical process (Raw, 2000).

Capsicum annuum is known by different names in the world depending upon types and place. In English is typically called chilies, long chilies, or red chilies. In Pakistan, in Urdu, it is called surkhmirch. In India, especially in Hindi, it called lalmirca. In Arabic, it is called filfil-e-ahmar. The most common *Capsicum* is *Capsicum annuum*. Only five species (*Capsicum frutescens, Capsicum pubescens, Capsicum annuum* L., *Capsicum chinense* Jacq., and *Capsicum baccatum* L.) are cultivated and domesticated

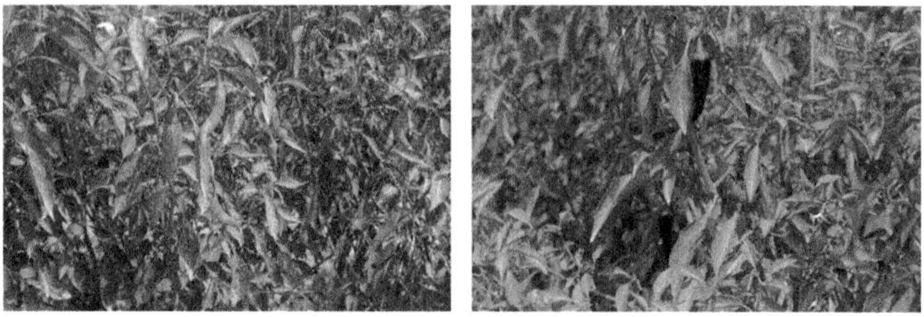

FIGURE 9.1 Chili pepper at various ripening stages.

(Costa et al., 2009; Orobiyi et al., 2013). Chili Peppers are perennial woody plants, grown as herbaceous annuals. The plant can vary in size from 2 to 4 feet tall depending on the species. Leaves are typically smooth, simple, or entire, glabrous, without hairs, flat, and differ in shape from ovate to elongate depending on the variety. The flowers are usually solitary, creamy white, and seeds are straw colored. All species of *Capsicum* are perennials when grown in favorable (semitropical or tropical) climates. Chili pepper is also cultivated ornamentally, particularly due to their bright, shiny fruits with a broad range of colors.

The essential oil content of chili pepper is equally variable between species and cultivar and is thought to be related to growing conditions, genetic factors, geographic origin, same chemo-types, and differences in the nutritional status of plants. The major chemicals of chili pepper essential oil are trans-b-ocimene, linalool, 2-methoxy-3-isobutylpyrazine, limonene, hex-cis-3-enol, and methyl salicylate. It is clear that chili pepper is morphologically and chemically highly variable. The origin, source, and growing conditions of chili pepper have an impact on the plant uses, flavors, aromas, and medicinal uses.

1.2 History/Origin

Capsicum annuum is native to South and Central America. It has been cultivated since 3500 BCE and has been used since 7000 BCE in Mexico. The generic name *Capsicum* comes from Latin word "capsa" meaning box or chest due to the fruit's shape that encloses seeds precisely, like in box. Chili has been known to the Western world since Christopher Columbus discovered American in 1493. *Capsicum* was brought toward Europe through Columbus in 1493 as a peppery spice. The ready appeal of *Capsicum* was such that within half a century it was distributed as far as Asia, and it was incorporated and continues to be diversified in cultures worldwide, as it had been originally in the Americas (Yamamoto and Nawata, 2004, 2005). Nowadays, chili pepper is present in several sizes, colors, and shapes all over the world. Mexican Indians probably used chili peppers before the birth of Christ (Govindarajan and Salzer, 1985). R. S. Macneish, an archaeologist,found seeds of pepper from about 7500 BCE in Mexico (DeWitt and Bosland, 1993). Chauca, a physician, on the next voyage of Columbus toward the New World noted medicinal and cooking use of chili peppers by Native Americans. Native Americans used irritant smoke, which was produced with burning of chili peppers, against invaders. In the early 19th century, chemical research on constituents present in chili peppers had been started. In 1846, Thresh crystallized the active constituent in chili peppers, and he named the component capsaicin. In the early 20th century, Dawson and Nelson found the chemical structure of capsaicin (Nelson, 1919; Nelson and Dawson, 1923).

1.3 Demography/Location

Chili pepper is grown in optimum conditions. It requires a warm and humid climate for growth and dry weather during maturity (Hussain and Abid, 2011). It is very susceptible to frost and grows poorly. Chili pepper is grown widely in Pakistan, India, China, Ethiopia, Myanmar, Mexico, Vietnam, Turkey, Peru, Ghana, Bangladesh, Japan, Africa, and America (Khan et al., 2012). The major producers of capsicum fruit are the United States, Italy, Pakistan, India, Mexico, Japan, and Brazil, where this crop holds economic value (Cruz et al., 2005). The world production is over 19 million tons fresh fruit, with 1.5 million hectares through Nigeria, which is the largest producer in Africa (Raji Abdul Ganiy et al., 2010). Pakistan is among the top five producers in the world. It is the biggest producer of red chilies,with an annual production at 85,000 tons.

1.4 Botany, Morphology, Ecology

C. annuum is an erect or prostrate annual herb and grows up to 0.75–1.8 m in cultivated varieties with several angular branches (Quresh et al., 2015). The leaves are simple, alternate, varying shapes, and oval to lanceolate, with smooth margins, generally wrinkled. The flowers are small, having a diameter of 1.5 or 1 in., and the color is white or violet, in clusters of two or more. The length of pedicel varies with cultivars, ranging from 3 to 8 cm. The color of petals is generally white with five to seven individual stamens that differ in color from pale blue to purple anthers (Berke, 2000). The fruits are berry-like and have several seeds that may be ovoid, elongated, cylindrical, obtuse, or oblong, but do not have sutures, and have a smooth, shiny surface and red color when ripe. The length of fruit is 12–25 cm and width is 7 mm. It has a characteristic aroma and pungent taste. Fruit colors are orange, green, black, yellow, and red to purple, white, and brown (Chaim et al., 2003). Chili peppers are warm-season and day-neutral plants. The best seed germination temperature ranges from 25 to 30°C. Optimum temperatures for productivity range from 18 to 30°C. Peppers are tolerant to a broad range of soil conditions. Though, well-drained soils and fertile, medium loams and a pH of 5.5–6.8 are generally considered most appropriate. If pH falls below 5.5, the result will be small growth of pepper plants and poor yields.

2. CHEMISTRY

The fruit of chili pepper possess capsaicin and numerous associated chemicals that have straight-chain alkyl vanillylamides and homologous series branched, together called capsaicinoids as their main chemical unit.

The main capsaicinoids present are capsaicin and 6,7-dihydrocapsaicin, and the minor capsaicinoids that are present are nordihydrocapsaicin, homodihydrocapsaicin, and homocapsaicin. Other parts of the plant have steroidal alkaloid and glycosides. Seeds contain the steroidal glycosides capsicoside A through D and allfurostanol. Chili pepper is rich in carotenoid pigments, involving cucurbitaxanthin A, carotene, zeaxanthin, capsorubin, lutein, zeaxanthin and capsanthin. Chili pepper contains no cholesterol, no fat content, and brings less caloric value (Bosland et al., 2012). It is a good source of several vitamins like vitamin E, vitamin C, vitamin A, and vitamin B complex and minerals such as thiamine, folate, molybdenum, manganese, potassium, calcium, iron, polyphenols (mainly luteolin), flavonoids, and quercetin (Chuah et al., 2008; Materska and Perucka, 2005). Peppers contain phenolics and flavonoids, carotenoids, and alkaloids (Materska and Perucka, 2005). Capsaicinoids are a group of alkaloids present in it. The major carotene pigments in chili peppers are lutein, β-carotene, and capsanthin, and they are mostly provitamin A (Howard and Wildman, 2006). Other phytochemicals present are alanine, scopoletin, chlorogenic acid, caffeic acid, linalool, amyrin, camphor, carvone, citric acid, linoleic acid, oleic, cinnamic, piperine, vitamins B1, C,B3, and E.Chili has seven times more vitamin C than oranges. Vitamins A and C and beta-carotenoids in chilies are powerful antioxidants that demolish free radicals (Simonne et al., 1997). Capsicum also has magnesium, sodium, phosphorus, sulfur, and selenium. The fruit of *Capsicum* species contain low volatile or essential oil contents that range from 0.1% to 2.6% in paprika. The pepper oil has major chemicals such as trans-b-ocimene, linalool, 2-methoxy-3-isobutylpyrazine, limonene, hex-cis-3-enol, and methyl salicylate. The other constituents recognized in higher amounts in oil separated at atmospheric pressure are nona-trans, non-1-en-4-one, nontrans-2-en-4-one, trans-2, 5-dien-4-one, benzaldehyde, and 2-entylfuran. The flavor component of chili pepper is 2-isobutyll-3-methoxy pyraxine. Components such as nona-trans, cis-2, 6-dienal and deca-trans, and trans-2, 4-dienal are responsible for aroma. Fixed oil obtained from seeds has triglycerides of about 60% in which linoleic acid and other unsaturation fatty acids are present. There are different fatty acids present such as palmitic acid, behenic acid, linolenic acid, stearic acid, arachidic acid, lignoceric acid, tricosanoic acid, linoleic acid and tricosanoic acid. Active components of red chilies are shown in Fig. 9.2.

3. POST HARVEST TECHNOLOGY

Chili peppers may be harvested in the green immature or red mature stage. Fruits can be harvested weekly. The best harvesting is in the cool hours of the day. For dry chili, it is essential to preserve red color of the

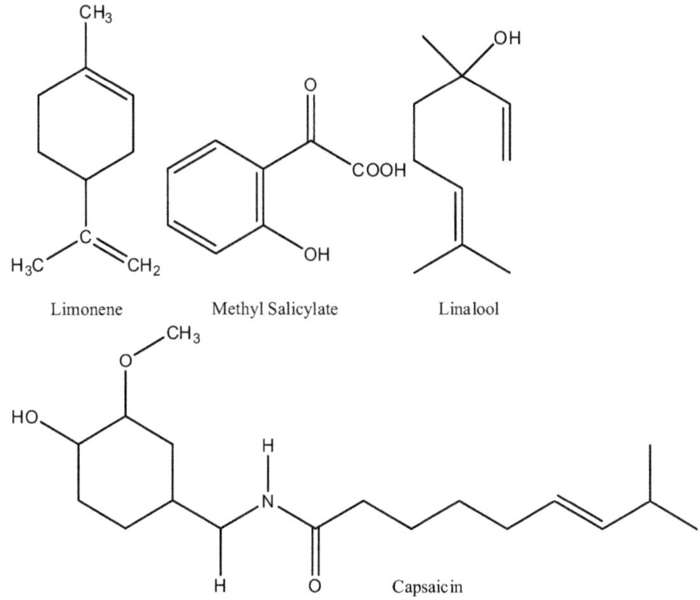

FIGURE 9.2 Active components of red chilies.

mature fruits. Drying of chili pepper in the sun is the common method, but this can bleach the fruits, and dew or rain support decay of fruit. Ordinarily, they have a short storage life of only 1–2 weeks. A wet, cool environment (45–50°F and 85%–90% comparative humidity) is the best condition for storing peppers. Another choice is to cut, wash, and freeze the peppers. The fruit of chilis are susceptible to chilling injury; therefore temperature management is important in maintaining quality. Mature green chili peppers hold best at temperatures between 10 and 12°C; if temperatures fall below 7°C, pepper injury will occur. Holding pepper fruit at the recommended temperatures and at 90–95% relative humidity allows peppers to be stored for up to 2 weeks.

4. PROCESSING

Chili pepper is an herbaceous plant, used in different ways for different purposes. Chili peppers can be used entire, chopped, or in several processed forms, for example, fresh, ground into powder, dried, or as an extract. They can be preserved in fresh form by refrigeration or dried through different drying methods, for example, electric drying, sun drying, solar drying, and oven drying. They can also be preserved by dehydration through freeze-drying or spray-drying. They can be processed by pickling, canning, or freezing and dehydrated to produce chili powder and paprika. Characteristic processing methods comprise freezing of peppers, in turn to store the fruit for a longer time. These processes may influence the content and bioaccessibility of the carotenoids contained in food (Pugliese et al., 2013). Chili pepper is traditionally dried in the sun (ONI, 2015). In the Southern United States, typically chili pepper is sun dried by scattering fruits on drying racks on the ground or on a roof. Blanching the fruits in warm water at 65°C for 3 min and removing calyx and pedicel can reduce drying time, enhance color retention, and reduce postharvest losses.

5. VALUE ADDITION

Dry chili is used widely as a spice in all kinds of curried dishes in India and abroad. Ground, roasted, dry chili combined with other condiments like turmeric, coriander, cumin, and farinaceous substances is used to make curry powder. It can be used in seasonings, eggs, meats, and fresh preparation. Chili pepper can be a chief ingredient in dairy foods, salad dressing, salsa, baked items, candies, mayonnaise, hot sauces, beverages, cosmetics, and pharmaceuticals (Bosland et al., 2012).

6. USES

Many herbs and spices have good health effect because they are full of antioxidants and mineral compounds. Chili pepper is a good source of vitamins and minerals; in addition to this, it is fat-free and full of antioxidants (Bosland et al., 2012). In security agencies, it is used in tear gas to control crowds. Chili pepper has many uses ranging from culinary to pharmaceutical. Chili is a vital spice used as a key ingredient in a vast variety of cuisines throughout the world. It is also used as a flavoring, colorant, and adds taste to the other bland foods. The species of *Capsicum* are used alone or entire and ground or in combination with other flavoring agents, mainly in stews, barbequed items, or pickles (Ravishankar et al., 2003).

In industries, they are used as flavoring and coloring agents in processed meats, alcoholic beverages, soups, lunches, sweets, and sauces (Kollmannsberger et al., 2011). They can be used dried, fresh, in conserves, or in the form of pepper sauces. Sweet peppers are a ingredient in pizza and pasta. The use of red paprika in tomato ketchup and sauces isalso encourage for improving color, so chili or paprika color may have a good demand as a natural plant colorant, as a substitute for synthetic color in the food industry. Besides the use in food processing industries, pharmaceutical and cosmetic industries use chili oleoresin of high pungency and low color. This oleoresin is used in pain balms, vapor rubs, and liniments since the pungent principle capsaicin serves as an effective counter irritant. They are also used in traditional medicine for their antimicrobial, sedative, and anticonvulsive properties (Cremer and Eichner, 2000). Capsaicin is used to relieve pain from migraines, cough, arthritis, or stuffy nose. It was also shown to have anticlotting activity.

7. PHARMACOLOGICAL USES

7.1 Antioxidant Activity

Chili pepper has maximum antioxidant action in the red mature stage, and antioxidants present in it arelycopene, vitamin C (ascorbic acid), p-coumaryl alcohol, ethoxyquin, and capsaicinoids. Phytochemicals, which are isolated from chilis, were reported efficient against Fe-induced lipid peroxidation (Oboh et al., 2007). In another study, methanolic extracts of chili pepper were reported to reduce H_2O_2-induced DNA and 4-hydroxy-2-nonenal-induced damage (Park et al., 2012). In another study, the antioxidant property of chili pepper fruits measured through DPPH radical scavenging activity augmented significantly with ripening in all cultivars. Strong positive correlations between antioxidant action and vitamin C, vitamin E, β-carotene, and total phenol during the entire ripening process were observed. These phytochemicals are good antioxidant compounds. Additionally, strong positive correlations among the phytochemicals have been observed, representing that accumulation of these substances was the major source of increase in antioxidant action in chili pepper fruits with progress of ripening (Aires et al., 2011).

7.2 Antiviral Property

Capsicum was found to be rich in those chemical components that are effective against viruses. Civamide (Cis-capsaicin) is active against HVS disease in guinea pigs and was also found to cure migraine headache pain. It is found to inhibit the viral replication cycle. It was found that capsaicin has a particular effect on sensory neurons involved in spreading and determination of herpes simplex virus disease (Bourne et al., 1999). Vanilloid capsaicin, isolated from chili pepper, was found active against pathogenesis of herpes simplex virus in models of animals, while chemicals separated from chili show antiviral action.

7.3 Anticancer Property

Capsaicin is efficient in vivo and in vitro against prostate cancer cell growth (Mori et al., 2006). Capsaicinoids, a class of compound in chilis, showed antitumour activity (Luo et al., 2011). Using Mexican plants in ethnomedicine, capsicum was found efficient against gastric cancer (Alonso-Castro et al., 2011). In another report, carotenoids, which are separated from red paprika, were noted for their cancer chemopreventive action (Maoka et al., 2001). Red pepper also has lycopene, which has anticancer activity (Simonne et al., 1997).

7.4 Antiinflammatory Activity

Red chili pepper possesses significant antiinflammatory activity. Capsaicinoids and capsianoside compounds have been found to reveal antiinflammatory actions and pain-reducing properties. In a lipopolysaccharide-stimulated macrophage model, maximum antiinflammatory activity of chili pepper was identified (Mueller et al., 2010). To reduce pain of rheumatoid arthritis, capsaicin was used in a previous study (Fraenkel et al., 2004).

7.5 Antiobesity Activity

Capsaicin, which is main chemical component of chili pepper, has antiobesity activity. Adipose tissue delivery between visceral and subcutaneous places is controlled by afferent nerves present in intestinal mucosa. Adipogenesis can be inhibited via capsaicin by activation of transient receptor potential vanilloid-1 channels. Sensitive sensory nerves' neurogenic mechanism affectsfat metabolism regulation by acting on transient receptor potential vanilloid-1, enabling selective activation of the network that regulates nerve action, causing a lipolytic effect in reaction toward gastrointestinal transient receptor potential channel stimulation. Expression of adiponectin and its receptor can be increased by nutritional capsaicin, thus reducing metabolic dysregulation of obese diabetic mice. The effects of capsaicin in liver and adipose tissue are due to its double action on transient receptor potential vanilloid-1activation and peroxisome proliferator-activated receptor alpha. Upon capsaicin action on white adipose tissue, proteins that are related to lipid metabolism and thermogenesis are changed. Capsaicin hinders adipogenesis in adipocytes and preadipocytes and induces apoptosis. Statistics of epidemiologic display that utilization of foods having capsaicin is related with a lesser incidence of obesity. Experimental data shows capsaicin acts as an antiobesity compound. Ingestion of capsaicin is related with an increase in energy expenses during the activation of brown adipose tissue, thus increasing fat oxidation and improving lipolysis (Leung, 2014).

7.6 Antifungal Activity

Peptides, which are separated from chili pepper seeds, repressed yeast growth of *Saccharomyces cerevisiae, Candida parapsilosis, Pichia membrane faciens, Candida tropicalis, Candida albicans, Kluyveromyces marxiannus,* and *Candida guilliermondii.* Peptides showed strong fungicidal action against *Schizosaccharomyces pombe, S. cerevisiae,* and *C. albicans* and also promoted numerous morphologic variations to *C. albicans*. It also reduced glucose-stimulated acidification of the medium mediated by H(+)-ATPase of *S. cerevisiae* cells in a dose-dependent mode and caused permeabilization of yeast plasma membrane toward dye SYTOX Green, as confirmed in confocal laser microscopy.

7.7 Antiplatelet Effects

Capsaicin has been found to be an effective inhibitor in aggregation of platelets and discharge reaction. It reduced hemolysis of red blood cells produced via hydrogen peroxide. Capsaicin has a membrane-stabilizing property through activation of phospholipase A2 interference (Wang et al., 1984).

8. SIDE EFFECTS AND TOXICITY

Eating chili peppers could cause heartburn, problems with ingestion, rectal pain, and skin problems.

References

Aires, A., Fernandes, C., Carvalho, R., Bennett, R.N., Saavedra, M.J., Rosa, E.A., 2011. Seasonal effects on bioactive compounds and antioxidant capacity of six economically important Brassica vegetables. Molecules 16, 6816–6832.

Alonso-Castro, A.J., Villarreal, M.L., Salazar-Olivo, L.A., Gomez-Sanchez, M., Dominguez, F., Garcia-Carranca, A., 2011. Mexican medicinal plants used for cancer treatment: pharmacological, phytochemical and ethnobotanical studies. Journal of Ethnopharmacology 133, 945–972.

Berke, T.G., 2000. Hybrid seed production in Capsicum. Journal of New Seeds 1, 49–67.

Bosland, P.W., Votava, E.J., Votava, E.M., 2012. Peppers: Vegetable and Spice Capsicums. Cabi.

Bourne, N., Bernstein, D., Stanberry, L., 1999. Civamide (cis-capsaicin) for treatment of primary or recurrent experimental genital herpes. Antimicrobial Agents and Chemotherapy 43, 2685–2688.

Chaim, A.B., Borovsky, Y., Rao, G., Tanyolac, B., Paran, I., 2003. fs3. 1: a major fruit shape QTL conserved in Capsicum. Genome 46, 1–9.

Chuah, A.M., Lee, Y.-C., Yamaguchi, T., Takamura, H., Yin, L.-J., Matoba, T., 2008. Effect of cooking on the antioxidant properties of coloured peppers. Food Chemistry 111, 20–28.

Costa, L.V., Lopes, R., Lopes, M.T.G., de Figueiredo, A.F., Barros, W.S., Alves, S.R.M., 2009. Cross compatibility of domesticated hot pepper and cultivated sweet pepper. Crop Breeding and Applied Biotechnology 9, 37–44.

Cremer, D.R., Eichner, K., 2000. Formation of volatile compounds during heating of spice paprika (*Capsicum annuum*) powder. Journal of Agricultural and Food Chemistry 48, 2454–2460.

Cruz, D.d.O., Freitas, B.M., Silva, L.A.d., Silva, E.M.S.d., Bomfim, I.G.A., 2005. Pollination efficiency of the stingless bee *Melipona subnitida* on greenhouse sweet pepper. Pesquisa agropecuária brasileira 40, 1197–1201.

DeWitt, D., Bosland, P.W., 1993. The Pepper Garden. Ten Speed Press, Berkeley, Calif.

Fett, D.D., 2003. Botanical briefs: capsicum peppers. Cutis 72, 21.

Fraenkel, L., Bogardus, S.T., Concato, J., Wittink, D.R., 2004. Treatment options in knee osteoarthritis: the patient's perspective. Archives of Internal Medicine 164, 1299–1304.

Govindarajan, V., Salzer, U.J., 1985. Capsicum-production, technology, chemistry, and quality part 1: history, botany, cultivation, and primary processing. Critical Reviews in Food Science & Nutrition 22, 109–176.

Greenleaf, W., 1986. Pepper breeding. Breeding Vegetable Crops 67–134.

Howard, L.R., Wildman, R.E., 2006. Antioxidant vitamin and phytochemical content of fresh and processed pepper fruit (*Capsicum annuum*). In: Handbook of Nutraceuticals and Functional Foods, second ed. CRC Press, pp. 165–191.

Hussain, F., Abid, M., 2011. Pest and diseases of chilli crop in Pakistan: a review. International Journal of Biology and Biotechnology 8, 325–332.

Khan, H.A., Ziaf, K., Amjad, M., Iqbal, Q., 2012. Exogenous application of polyamines improves germination and early seedling growth of hot pepper. Chilean Journal of Agricultural Research 72, 429.

Kollmannsberger, H., Rodríguez-Burruezo, A., Nitz, S., Nuez, F., 2011. Volatile and capsaicinoid composition of ají (*Capsicum baccatum*) and rocoto (*Capsicum pubescens*), two Andean species of Chile peppers. Journal of the Science of Food and Agriculture 91, 1598–1611.

Kothari, S., Joshi, A., Kachhwaha, S., Ochoa-Alejo, N., 2010. Chilli peppers—a review on tissue culture and transgenesis. Biotechnology Advances 28, 35–48.

Leung, F.W., 2014. Capsaicin as an Anti-obesity Drug, Capsaicin as a Therapeutic Molecule. Springer, pp. 171–179.

Luo, X.-J., Peng, J., Li, Y.-J., 2011. Recent advances in the study on capsaicinoids and capsinoids. European Journal of Pharmacology 650, 1–7.

Maoka, T., Mochida, K., Kozuka, M., Ito, Y., Fujiwara, Y., Hashimoto, K., Enjo, F., Ogata, M., Nobukuni, Y., Tokuda, H., 2001. Cancer chemopreventive activity of carotenoids in the fruits of red paprika *Capsicum annuum* L. Cancer Letters 172, 103–109.

Materska, M., Perucka, I., 2005. Antioxidant activity of the main phenolic compounds isolated from hot pepper fruit (*Capsicum annuum* L.). Journal of Agricultural and Food Chemistry 53, 1750–1756.

Mori, A., Lehmann, S., O'Kelly, J., Kumagai, T., Desmond, J.C., Pervan, M., McBride, W.H., Kizaki, M., Koeffler, H.P., 2006. Capsaicin, a component of red peppers, inhibits the growth of androgen-independent, p53 mutant prostate cancer cells. Cancer Research 66, 3222–3229.

Mueller, M., Hobiger, S., Jungbauer, A., 2010. Anti-inflammatory activity of extracts from fruits, herbs and spices. Food Chemistry 122, 987–996.

Nelson, E., 1919. The constitution of capsaicin, the pungent principle of capsicum. Journal of the American Chemical Society 41, 1115–1121.

Nelson, E., Dawson, L., 1923. The constitution of capsaicin, the pungent principle of Capsicum. III. Journal of the American Chemical Society 45, 2179–2181.

Oboh, G., Puntel, R., Rocha, J., 2007. Hot pepper (*Capsicum annuum*, Tepin and *Capsicum chinese*, Habanero) prevents Fe 2+-induced lipid peroxidation in brain–in vitro. Food Chemistry 102, 178–185.

Oni, S., 2015. Growth and fruit yield of pepper (*Capsicum frutescens* L.) as influenced by arbuscular mycorrhizal (AM) inoculation and fertilizers under greenhouse and field conditions. Applied Tropical Agriculture 15, 126–133.

Orobiyi, A., Dansi, M., Assogba, P., Loko, L., Vodouhe, R., Akouegninou, A., Sanni, A., 2013. Chili (*Capsicum annuum* L.) in Southern Benin: Production Constraints, Varietal Diversity, Preference Criteria and Participatory Evaluation.

Park, J.-H., Jeon, G.-I., Kim, J.-M., Park, E., 2012. Antioxidant activity and antiproliferative action of methanol extracts of 4 different colored bell peppers (*Capsicum annuum* L.). Food Science and Biotechnology 21, 543–550.

Pugliese, A., Loizzo, M.R., Tundis, R., O'Callaghan, Y., Galvin, K., Menichini, F., O'Brien, N., 2013. The effect of domestic processing on the content and bioaccessibility of carotenoids from chili peppers (Capsicum species). Food Chemistry 141, 2606–2613.

Quresh, W., Alam, M., Ullah, H., Jatoi, S.A., Khan, W.U., 2015. Evaluation and characterization of Chilli (*Capsicum annuum* L.) germplasm for some morphological and yield characters. Pure and Applied Biology 4, 628.

Raji Abdul Ganiy, O., Falade Kolawole, O., Abimbolu Fadeke, W., 2010. Effect of sucrose and binary solution on osmotic dehydration of bell pepper (chilli) (*Capsicum* spp.) varieties. Journal of Food Science and Technology 47, 305–309.

Ravishankar, G., Suresh, B., Giridhar, P., Rao, S.R., Johnson, T.S., 2003. Biotechnological studies on Capsicum for metabolite production and plant improvement. Capsicum: The Genus Capsicum 96–128.

Raw, A., 2000. Foraging behaviour of wild bees at hot pepper flowers (*Capsicum annuum*) and its possible influence on cross pollination. Annals of Botany 85, 487–492.

Simonne, A., Simonne, E., Eitenmiller, R., Mills, H., Green, N., 1997. Ascorbic acid and provitamin A contents in unusually colored bell peppers (*Capsicum annuum* L.). Journal of Food Composition and Analysis 10, 299–311.

Wang, J.-P., Hsu, M.-F., Teng, C.-M., 1984. Antiplatelet effect of capsaicin. Thrombosis Research 36, 497–507.

Yamamoto, S., Nawata, E., 2004. Morphological characters and numerical taxonomic study of *Capsicum frutescens* in Southeast and East Asia. Tropics 14, 111–121.

Yamamoto, S., Nawata, E., 2005. *Capsicum frutescens* L. in southeast and east Asia, and its dispersal routes into Japan. Economic Botany 59, 18–28.

CHAPTER

10

Chirayita

Asma Seher, Muhammad Asif Hanif, Maryam Hanif, Asma Hanif
Department of Chemistry, University of Agriculture, Faisalabad, Pakistan

OUTLINE

Medicinal Plants of South Asia
https://doi.org/10.1016/B978-0-08-102659-5.00010-0

Copyright © 2020 Elsevier Ltd. All rights reserved.

1. BOTANY

1.1 Introduction

Swertia, a genus in the family Gentianaceae, includes a large group of annual and perennial herbs, representing approximately 150 species growing in the mountains of Asia and Africa. *Swertia* species have been widely used as major ingredients in several herbal remedies in traditional, Ayurvedic, Unani, and Siddha medical systems since time immemorial. The reported species of Swertia are 11 in Pakistan (Flora of Pakistan), 5 in Afghanistan, 47 in Nepal, 75 in China, and 40 in India (Kumar and Van Staden, 2015). Chirayita (Fig. 10.1) is the commonly used term for different species of *Swertia* plant species. Out of all the species, *S. chirayita* is the most frequently used medicinal plant species and is native to Himalayan region.

1.2 History/Origin

"Chirayita" has traditionally been used since the 4th century BCE to control three key principles of energy related to human structure and body functions known as "Tridosha." Chirayita was first described by Roxburgh under the name of *Gentiana chyrayta* in 1814 (Scartezzini and Speroni, 2000). This annual plant is a native of North India, growing in the mountainous districts, and it has been held in considerable esteem as a

FIGURE 10.1 Dried chirata.

medicine by the Hindus. Chirayita goes all the way back to Sanskrit times and is mentioned in the Charaka Samhita, an ancient Ayurvedic healing text from India. This ancient herb is also sometimes known as the Nepali Neem because it is a common tree in the forests of Nepal. This plant was introduced to Europe in 1839 and has been used widely since (Aleem and Kabir, 2018).

1.3 Demography/Location

The plant inhabits temperate regions in the Himalayas. Loamy to sandy loam, friable, and well-drained soils are suitable for its cultivation. The soil should be enriched with FYM (farmyard manure), and if soils are clayey, addition of sand is recommended. The crop can be grown in areas having mild rainfall (100 cm) in the rainy season and in areas with a long, cold winter, receiving snowfall frequently.

1.4 Botany, Morphology, and Ecology

The chirayita flowers are used as well as embellishments in woodland backyards, comprising a sunny boundary, with limited shadow, within gloomy as well as in soggy lands. This plant generally develops to a height of 1–3 ft yearly. During the time period from September to October, these plants come into flower. Flowers of these plants are bisexual in nature and have green color along with purple tone. In other words, the flowers of chirayita consist of both female and male reproductive organs (Kumar et al., 2010b). Flowering in chirayita is in the appearance of many little, opposite, axillary, lax cymes set as tiny branches, and the entire inflorescence is 2.0 ft elongated. Flowers are tiny, green-yellow, stalked, purple in color, tetramerous, and rotate. The corolla is two times as long as the calyx and separated near the base into four ovate-lanceolate segments. A pair of nectarines covered by oval scales is present at the upper surface of the petal and end as brinks. Fruit of this plant is a tiny, one-celled case along with a translucent yellowish pericarp. It divides septicidally into two valves from the top. Seeds are frequent, small, angular, and multifaceted. The presence of nectarines and colorful corolla are the floral aspects that support cross-pollination in the species. Usually, the pollinators of chirayita are bees (Apoidea and Hymenoptera).

When the plant reaches flowering in July–September, it is harvested for the medicine industry (Joshi and Dhawan, 2005). Seed setting begins about October to November, and seeds develop immediately after dropping. Only a small number of dispersed reports in the literature recommend germination studies and nursery practices of chirayita. About 91% seed fertilization was found after 3°C freezing cure for 15 days, while an additional study revealed a maximum of 81% fertilization. The post-germination growth phase in chirayita is slow (Basnet, 2001; Raina et al., 1994). Low percentage of germination and feasibility of the seeds, extended development phase, and slight field handling are a number of factors that affect commercial farming of the plant (Joshi and Dhawan, 2005).

Chirayita flourishes and thrives in neutral, acidic, as well as alkaline or basic soils. However, chirayita prefers to grow in acidic soil condition with pH of 4.7–5.5 (Bhattarai and Acharya, 1996). The requirements for its growth are damp or humid soil, and this plant grows in good health in woodland or semishade conditions. In exact words, the plant blooms well in a moist and humus-rich soil in cloudy, light woodlands, by the side of the rivers, or in marshlands. This plant grows very well in regions where the summer season is chilly. Therefore, it is not surprising that the chirayita can flourish and thrive together in circumstances where there is semishade as well as full sunshine. These plants are capable of resisting temperatures as low as −15°C and still carry on growing up in good health. The chirayita herb is proliferated through its seeds. Generally, sowing is done when the temperature is less than 10°C during the spring season and in a condition when the soil is rich in humus. The seedlings are taken out separately when they have developed sufficiently to be handled and planted into separate containers or pots. In the early part of the summer season, the little plants are replanted outdoors. The plants are typically harvested immediately when the seeds start to set in and dried out in the sun for the next use (Kumar et al., 2010a).

2. CHEMISTRY

Chiratin and ophelic acid are the two bitter principles in this plant. The former has the larger fraction, and it yields, by steaming along with hydrochloric acid, ophelic acid and chiratogenin, but no sugar. Neither chiratin nor ophelic acid has been obtained in crystalline form. The chirayita's ash yields phosphates and carbonates of potassium, magnesium, and calcium. Tannin is almost completely absent. The wide-range biologic activities of chirayita are attributed to the presence of a diverse group of pharmacologically bioactive compounds belonging to different classes such as xanthones and their derivatives, lignans, alkaloids, flavonoids, terpenoids, iridoids, eocrinoids, and other compounds such as chiratin, ophelic acid, palmitic acid, oleic acid, and stearic acid (Kumar and Van Staden, 2015). The essential oil yield of chirayita is about 0.236 g/kg. This medicinal plant is mostly known for its bitter taste due to the existence of various chemical compounds such as swerchirin, swertiamarin, and amarogentin (Kumar and Van Staden, 2015). The percentages of microelements in the leaves of this plant are reported as

Zn (6.7%–6.9%), Cu (1.8%–2.0%), Mn (5.3%–5.5%), Fe (85.8%–86.0%), and Co (9.1%–9.3%). The following percentages of macroelements in the leaves of chirayita are reported as Na (28.8%–30.0%), K (92.4%–92.6%), Ca (20.3%–20.5%), and Li (3.8%–4.0%) (Negi et al., 2010).

From *S. delayayi* Franch the chemical components such as gentiopicroside, oleanolic acid, daucosterol, swertiamarin, isovitexin, isoorientin, and swertiadecoraxanthone-II have been separated. *S. cordata* consists of chemical constituents such as 1,7-dihydroxy-3,5,8-trihydroxyxanthone and 1-hrdoxy-3,5,7,8-tetramethoxyxanthone. Mangniferin, swertiamarin, oleanolic acid, 1,8-dihydroxy-3,7-dimethyl-oxyxanthone, and 1,5,8-trihydroxy-3-methoxyxanthone have been separated from *S. mussotti* Franch (Joshi et al., 2013). About 77 components of the essential oil of this plant belong to the classes of alcohol, acid, ester, aldehyde, and ketone, and miscellaneous hydrocarbons were recognized; among all these, the ketones were identified as a dominant class, with the maximum percentage of 28.57%. The major compounds that belonged to the ketones were camphor, 2-buten-2-one, 2-heptadecanone, 3-ethenylcyclohexenone, and (Z)-geranylacetone. The second major chemical group in this plant with the percentage of 27.45% was the alcoholic group. The most abundant alcoholic compound is cedrol, while β-eudesmol, patchoulol, p-cymen-3-ol, farnesol, and linalool were also detected as terpene alcohols. The volatile organic components of chirayita include ethyl formate, ethyl acetate, 2-butanone, 2-methylbutanal, 3-methylbutanal, ethanol, 3-buten-2-one, 2-ethyl furan, 3-methyl-2-butanone, pentanal, 2-butenal, methyl butanol, hexanal, β-pinene, 3-penten-2-one, 2-heptanone, heptanal, 2-hexanal, 2-pentylfuran, pentanol, octanal, 2,3-octanedione, 6-methyl-5-methylideneheptane-2-one, hexanol, 2,4-dimethylfuran, 4-methylhexanol, nonanal, (E)-2-octenal, limonene oxide, hexen-3-ol, furfural, 2-ethylhexanol, pyrrole, camphor, octanol, linalool, and 2,6-nonadienal (Gyawali et al., 2006). This plant has over 20 xanthones (Bhattacharya et al., 1976). The phytochemicals such as amarogentin (chirantin), amaroswerin, gentianine, swerchirin, mangiferin, lignan, triterpenoids, pentacyclic triterpenoids, and pentacyclic triterpenoids are also found in this plant (Tabassum et al., 2012). Structures of swertiamarin and swerchirin present in chirata are shown in Fig. 10.2.

3. POSTHARVESTING

November to December is the appropriate time for harvesting, and collection is done manually without using any equipment. The whole plant is pulled out, washed to remove the soil contents, and then sun dried for few days, or it could be oven dried at 80°C in a few hours (Bhatt et al., 2006).

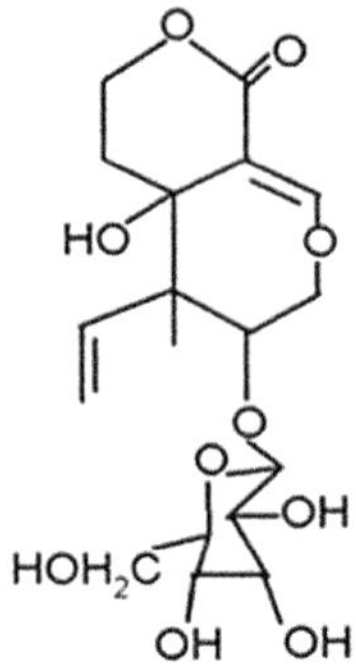

Swertiamarin

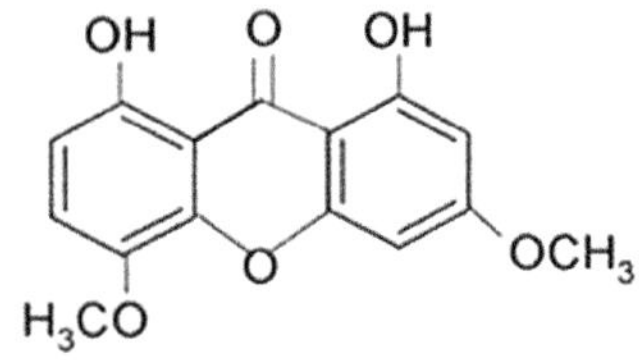

Swerchirin

FIGURE 10.2 Structures of swertiamarin and swerchirin present in chirata.

4. PROCESSING

Chirayita is a medicinally important plant and used in different traditional medicinal systems for various purposes. In addition to the fresh leaves, other common processed forms of chirayita are essential oil, freeze dried, sun dried, and air-dried leaves. To prevent discoloration and oxidation, the chirayita leaves are dried by placing them between two paper sheets. The dried leaves of chirayita are converted into a fine powder by pestle and mortar or grinding mill. The extract of powder is formed with 70% ethanol by cold maceration and then concentrated at 45°C and elevated pressure in the rotary evaporator. The extract of shade dried leaves is also formed with 70% ethanol by Soxhlet apparatus. To obtain crude solid extract, the ethanolic extract is concentrated in water bath and then lyophilized at the temperature of −55°C. The plant extract is preserved in the desiccators for future use. The reported yield of chirayita is 1.4% w/w (Nagalekshmi et al., 2011).

5. VALUE ADDITION

It is used in Ayurvedic industry as a constituent in an anticancer medication and in the skin tonic "Safi" as well as in skin soaps and cosmetic products. This medicinal plant is usually administered as a concentrated infusion, tincture, or as powder and fluid extract. It is also used in the cosmetic industry as an ingredient in facial creams, cleansers, scrubbers, and hair oil (Phoboo and Jha, 2010).

6. USES

Chirayita is well known for its pharmaceutical and therapeutic significance. This plant contains several flavonoids and alkaloids, and the majority of them have sufficient implementations. The roots of this plant have rising healing evidence and have ample analgesic and antipyretic effects. The chief phytochemicals of chirayita are swerchirin and amarogentin (Brahmachari et al., 2004). In the past, the whole plant was used in several traditional and indigenous systems of medicines, such as Ayurveda, Unani, and Siddha. In British and American pharmacopeias, it was utilized as tinctures and infusions. The roots of this plant served as a drug and an effective tonic for general weakness, fever, cough, joint pain, asthma, and the common cold. For headaches and blood pressure, the leaves and chopped stems are soaked overnight in water. A paste is prepared and filtered with one glass of water. The preparation is consumed once a day for 2–3 days. For tremor fever, whole chirayita plants are cut into small pieces and boiled in 1/2 L of water until the volume is reduced to less than half a glass. The filtered water is stored in a glass bottle, and a half spoon is given to children once a day for 2 days. For adults, the dosage is one spoon once a day for 2 days and varies to three times a day until cured. The decoction of boiled chirayita in water is taken orally to cure malaria, while paste of the chirayita plant is applied to treat skin diseases such as eczema and pimples. The plant is used in combination with other drugs in cases of scorpion bite (Kumar and Van Staden, 2015), utilized as anthelmintic, antipyretic, cathartic, antiperiodic, leucorrhoea, for asthma in Ayurveda, analeptic, to mitigate inflammation, stomachic, and soothing to nerve ending fevers and pregnant uterus (Tabassum et al., 2012). It is medication for gastrointestinal diseases, ulcers, skin problems, hiccup, cough, kidney and liver diseases, urinogenital tract diseases, and neurological diseases. Chirayita is also consumed as a disinfectant for breast milk and as a carminative and laxative. Gentianine as isoprene alkaloid isolated from chirayita has various pharmacological effects (Tabassum et al., 2012).

7. PHARMACOLOGICAL USES

7.1 Antidiabetic Activity

Ethanolic extracts of chirayita were found effective in blood sugar–lowering activity in rats. Chirayita is very effectual medication for stomach problems. It is utilized for the treatment of dyspepsia and diarrhea. Chirayita helps to stimulate the digestion process and maintains the normal sugar level in blood. This plant is beneficial for diabetics.

Chirayita lowers the high sugar level in the blood of animals and reduces the risk of hypoglycemia (Kumar et al., 2010a).

Alcoholic extract of chirayita with hexane fraction showed good hypoglycemic activity in albino rats. It is suggested that the hexane fraction of chirayita might not be able to reduce the glucose absorption in the intestine (Brahmachari et al., 2004).

7.2 Antioxidant Activity

The ethanolic extract of chirayita has shown in vitro and in vivo antioxidant effects (Chen et al., 2011).

7.3 Antihepatotoxic Activity

Chirayita has shown significant hepatoprotective activity in a previous study (Brahmachari et al., 2004).

7.4 Anthelmintic Activity

In vitro study of crude aqueous extract (CAE) and methanol extract (CME) of whole chirayita showed a significant decline in eggs per gram of feces when it was used as CAE, crude powder, and CME to the sheep that were naturally infected by gastrointestinal nematodes (Iqbal et al., 2006).

7.5 Antibacterial Activity

Both aqueous and ethanolic extracts of *S. chirayita* have antibacterial effect across the bacterial strains that are gram negative such as *P. vulgaris*, *E. coli* and *K. pneumoniae* (Rehman et al., 2011).

7.6 Antihepatotoxic Activity

The methanolic extract of entire plant of chirayita was estimated for its antihepatotoxic activity in liver necrosis, stimulated by CCl_4 in trail rats. This activity was more distinct for the fraction soluble in $CHCl_3$ in contrast to the fraction soluble in BuOH (Brahmachari et al., 2004).

7.7 Antimalarial Activity

It has been reported that the species of *Swertia* have medically important antimalarial activity. Swerchirin isolated from chirayita possessed antimalarial activity in a rodent test system affected by *Plasmodium berghei* (Brahmachari et al., 2004).

7.8 Antiinflammatory Activity

Benzene extract of chirayita exhibited major antiinflammatory activities in chronic, subacute, and acute models. Antiinflammatory activities of chirayita are due to prenylated xanthones, xanthone glycosides, mangostin, mangiferin, isomangostin, and mangostin triacetate. A bioactive compound, 1,5 dihydroxy-3,8-dimethoxyxanthone, isolated from chirayita was examined to have very effective antiinflammatory activity. The drug was orally used at a dosage of about 50 mg/kg for the inhibition of pedal edema induced by formalin and carrageenin by 57% and 56%, respectively (Brahmachari et al., 2004).

7.9 Insecticidal Activity

Chirayita was found to have good insecticidal activity under field and laboratory environments. The petroleum ether fraction of extract of different species of this plant (stem portion) revealed significant effects against the painted bug *Anguina gramini* (Brahmachari et al., 2004).

7.10 Antifeeding Activity

Chirayita extracts in MeOH, AcOEt, and benzene have shown antifeeding activities in *Anomis sabulifer*. A favorable antifeeding activity at the concentration of about 10% was shown by the AcOEt extract (Brahmachari et al., 2004).

7.11 Antiulcer and Antigastritis Activity

Efficacy of chirayita was investigated on experimentally induced stomach ulcers in rat models. Gastric mucosa was reduced by the ethanolic extract of chirayita (Brahmachari et al., 2004).

8. SIDE EFFECTS AND TOXICITY

Chirayita is likely to be safe in medicinal amounts when taken by mouth. However, reliable scientific data about chirayita use during pregnancy, breastfeeding, diabetes, and stomach ulcers is not available.

References

Aleem, A., Kabir, H., 2018. Review on *Swertia chirata* as traditional uses to its pyhtochemistry and phrmacological activity. Journal of Drug Delivery and Therapeutics 8, 73–78.

Basnet, D., 2001. Evolving nursery practices and method of cultivation of high value medicinal plant *Swertia chirata* Ham. Environment and Ecology 19, 935–938.

Bhatt, A., Rawal, R., Dhar, U., 2006. Ecological features of a critically rare medicinal plant, *Swertia chirayita*, in Himalaya. Plant Species Biology 21, 49–52.

Bhattacharya, S., Reddy, P., Ghosal, S., Singh, A., Sharma, P., 1976. Chemical constituents of gentianaceae XIX: CNS-depressant effects of swertiamarin. Journal of Pharmaceutical Sciences 65, 1547–1549.

Bhattarai, K., Acharya, N., 1996. Identification, Qualitative Assessment, Trade & Economic Significance of Chiraito (*Swertia* spp.) of Nepal. A report submitted to Asia Network for Sustainable Agriculture and Bioresources (ANSAB).

Brahmachari, G., Mondal, S., Gangopadhyay, A., Gorai, D., Mukhopadhyay, B., Saha, S., Brahmachari, A.K., 2004. Swertia (Gentianaceae): chemical and pharmacological aspects. Chemistry and Biodiversity 1, 1627–1651.

Chen, Y., Huang, B., He, J., Han, L., Zhan, Y., Wang, Y., 2011. In vitro and in vivo antioxidant effects of the ethanolic extract of *Swertia chirayita*. Journal of Ethnopharmacology 136, 309–315.

Gyawali, R., Ryu, K.-Y., Shim, S.-L., Kim, J.-H., Seo, H.-Y., Han, K.-J., Kim, K.-S., 2006. Essential oil constituents of *Swertia chirata* Buch.-Ham. Preventive Nutrition and Food Science 11, 232–236.

Iqbal, Z., Lateef, M., Khan, M.N., Jabbar, A., Akhtar, M.S., 2006. Anthelmintic activity of *Swertia chirata* against gastrointestinal nematodes of sheep. Fitoterapia 77, 463–465.

Joshi, P., Dhawan, V., 2005. *Swertia chirayita*-an overview. Current Science 89, 635. Bangalore.

Joshi, S., Mishra, D., Bisht, G., Khetwal, K.S., 2013. Mineral composition and antimicrobial activity of *Swertia cordata* (G. Don) Clarke aerial parts and roots. Indian Journal of Natural Products and Resources 4, 273–277.

Kumar, K.S., Bhowmik, D., Chandira, M., 2010a. *Swertia chirayita*: a traditional herb and its medicinal uses. Journal of Chemical and Pharmaceutical Research 2, 262–266.

Kumar, K.S., Bhowmik, D., Chiranjib, B., Chandira, M., 2010b. *Swertia chirata*: a traditional herb and its medicinal uses. Journal of Chemical and Pharmaceutical Research 2, 262–266.

Kumar, V., Van Staden, J., 2015. A review of *Swertia chirayita* (gentianaceae) as a traditional medicinal plant. Frontiers in Pharmacology 6.

Nagalekshmi, R., Menon, A., Chandrasekharan, D.K., Nair, C.K.K., 2011. Hepatoprotective activity of *Andrographis paniculata* and *Swertia chirayita*. Food and Chemical Toxicology 49, 3367–3373.

Negi, J.S., Singh, P., Rawat, M., nee Pant, G.J., 2010. Study on the trace elements in *Swertia chirayita* (Roxb.) H. Karsten. Biological Trace Element Research 133, 350–356.

Phoboo, S., Jha, P.K., 2010. Trade and sustainable conservation of *Swertia chirayita* (Roxb. ex Fleming) H. Karst in Nepal. Nepal Journal of Science and Technology 11, 125–132.

Raina, R., Johri, A., Srivastava, L., 1994. Seed germination studies in *Swertia chirata* L. Seed Research 22, 62–63.

Rehman, S., Latif, A., Ahmad, S., Khan, A., 2011. In vitro antibacterial screening of *Swertia chirayita* Linn. against some gram negative pathogenic strains. International Journal of Pharmaceutical Research and Development 4, 188–194.

Scartezzini, P., Speroni, E., 2000. Review on some plants of Indian traditional medicine with antioxidant activity. Journal of Ethnopharmacology 71, 23–43.

Tabassum, S., Mahmood, S., Hanif, J., Hina, M., Uzair, B., 2012. An overview of medicinal importance of *Swertia chirayita*. International Journal of Applied 2.

CHAPTER

11

Coleus

Shafaq Nisar[1], *Muhammad Asif Hanif*[1], *Kiran Soomro*[2], *Muhammad Idrees Jilani*[3], *Chandra Prakash Kala*[4]

[1] Department of Chemistry, University of Agriculture, Faisalabad, Pakistan; [2] National Centre of Excellence in Analytical Chemistry, University of Sindh, Jamshoro, Pakistan; [3] Department of Chemistry, University of Sahiwal, Sahiwal, Pakistan; [4] Ecosystem & Environment Management, Indian Institute of Forest Management, Bhopal, India

OUTLINE

Medicinal Plants of South Asia
https://doi.org/10.1016/B978-0-08-102659-5.00011-2

Copyright © 2020 Elsevier Ltd. All rights reserved.

1. BOTANY

1.1 Introduction

The coleus belongs to the Mint or Lamiaceae family. The main plant part that is used for the extraction of chemicals for medicinal purposes is the root. Coleus grows perennially over subtropical and tropical regions of Sri Lanka, Pakistan, India, Brazil, and East Africa at 1968–5905 ft high. It is cultivated in different states of India like Maharashtra, Gujarat, Tamil Nadu, Karnataka, and Rajasthan. Traditionally coleus roots were used in food additives, e.g., in preparation of pickles and in medicine by Ayurvedic schools of medicines. It has different common names such makandi (Urdu, Hindi), karmelo (Marathi), and gandhe jhar (Nepali) (Singh et al., 2011a).

1.2 History/Origin

Coleus is a native plant of tropical areas of Southeast Asia, India, Africa, and Australia, with the largest population being in Indonesia and Sri Lanka. Coleus found its way to Europe and, later, America through traders and botanists. Some traced coleus in Europe as early as the 17th century, and the Dutch botanist Karl Ludwig Blume is often credited for naming and introducing coleus to Europe.

1.3 Demography and Location

Coleus forskohlii is found in the subtropical Himalayan region of Bihar and the Deccan Plateau of southern India (Khatun et al., 2011; Valdes et al., 1987) as well as in Sri Lanka. Apparently, it is distributed to Arabia,

Egypt, Nepal, Thailand, Ethiopia, Myanmar, Brazil, and tropical East Africa (Shivaprasad et al., 2014). *C. forskohlii* grows well in sandy, red loam soils that have a pH in the range of 5.5–7.0, temperature in the range of 10–25°C, and humidity in the range of 83%–95%. An annual rainfall required for the proper growth of the crop is in the range of 100–160 cm, essentially between June to September (Shah and Kalakoti, 1996).

1.4 Botany, Morphology, and Ecology

Seeds and cutting stems both are used for the propagation of the plant. Plant propagation with the help of seeds is very slow and difficult, while the plant propagation with the help of terminal stem cutting is very easy and economical. A terminal stem of 4–4.7 inches is cut along with three to four pairs of leaves, and then this part is planted in the garden to induce rooting. The best period for planting is June–July or September–October. Regular care should be taken over weeding, plant protection, and watering (Rajamani and Vadivel, 2009).

C. forskohlii is a perennial herb that can grow to almost 0.45–0.60 m tall. Unlike other ornamental coleus, *C. forskohlii* lacks spectacularly showy, colored leaves. In fact, it has plain leaves of bright green color and flowers of blue color like lavender. It is an aromatic plant. Stems of the plant are branched and have four angles. Nodes are hairy in nature. Leaves are 3–5 inches in length and 1.2–2 inches in width, usually narrowed, pubescent into petioles. Inflorescence is of raceme type, 6–11.8 inches in length; flowers of the plant are stout, having size of 2–2.5 cm, generally perfect and the calyx hairy inside. Lower lobes of the calyx are elongated and have concave shape to enclose essential organs of the plants, while the upper lobe of the calyx is oval in shape. The ovary of the plant has four parts, and the stigma is two lobed. Cross-pollination occurs with the help of insects or wind (Lakshmanan et al., 2013). The roots of the plant are thick, fibrous, tuberous, and brown in color. Roots are strongly aromatic in nature. Root morphology differs with populations as they may be fibrous, tuberous, nontuberous, or semituberous in nature. Forskolin contents in roots within a population and between populations vary from 0.07% to 0.58%. Both tubers as well as leaves have somewhat different odors (Lakshmanan et al., 2013). Morphology of the different ecogeographic populations varies greatly. Growth habit of coleus plant is unusually variable: decumbent, procumbent, or erect. *C. forskohlii* can grow best in soil that contains fertile garden sand and loam. Mountain slopes in India are the native habitat of this species.

2. CHEMISTRY

The fragrance of *C. forskohlii* resembles the camphor plant. Chemical compounds of *forskohlii* are diterpenes in nature and usually belong to two different groups, i.e., 8, 13-epoxy-labd-14 en-11-one diterpenoids and abietane diterpenoids. The chemical compound that is present in coleus plant is forskolin, which was known as colenol and colforsin initially. But later on, it was changed to forskolin due to the identification of other diterpenoids and colenols (Lakshmanan et al., 2013). Chemically forskolin is a 7β-acetoxy-8, 13-epoxy-1α, 6β, nine α-trihydroxy-labd-14 en-11-one (Fig. 11.1).

C. Forskohlii is a rich source of different chemicals that are pharmacologically active compounds. These compounds are mostly diterpenes in nature and isolated from the different parts of the plant, but the most active compound is isolated from the root. Different diterpenoids that are isolated from *C. forskohlii* are categorized into different classes: (1) abietane diterpenoids; (2) 8,13-epoxy-labd-14-en-11-one-diterpenoids; (3)8,13-epoxy-labd-diterpenoids (Coleonone, Colenol, 3-hydroxy forskolin); (4) miscellaneous labdane diterpenoids (Coleonic acid, Colenolic acid, Forskoditerpene A); and (5) 8,13-epoxy-labd-14-en-11-one-diterpene glycosides. Furthermore, many compounds that are present in minor concentrations and isolated from the different parts of the plant are also reported by the researchers. Those include α-cedrol (Yao et al., 2002), coleoside (Ahmed and Vishwakarma, 1988), 4β,7 β,11-enantioeudesmantriol (Shan et al., 2007), coleonolic acid (Roy et al., 1990), α-amyrin (Xu et al., 2005), myrianthic acid (Shan and Kong, 2006), betulic acid (Xu et al., 2005), uvaol (Li et al., 2006), euscaphic acid (Shan and Kong, 2006), crocetin dialdehyde (Tandon et al., 1979), arjungenin (Shan and Kong, 2006), arjunic acid (Shan and Kong, 2006), â-sitosterol (Li et al., 2006), stigmasterol (Shah et al., 1989), genkwanin (Yao et al., 2002), colexanthone (Liu et al., 2007), caffeic acid (Ahmed and Vishwakarma, 1988), guaicol glycerin ether (Yao et al., 2002), coleoside B (Ahmed and

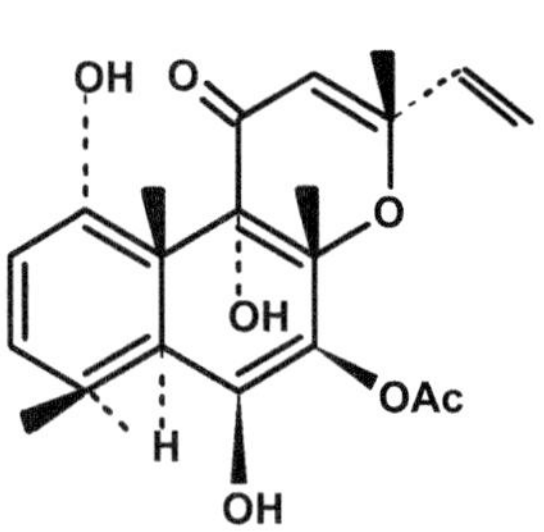

FIGURE 11.1 Structure of forskolin.

Merotra, 1991), sulfoquinovosyl diacylglycerol, digalactosyl diacylglycerol, monogalactosyl diacylglycerol, tetragalactosyl diacylglycerol, and trigalactosyl diacylglycerol (Mendes et al., 2006). The active compound "forskolin" is crystalline in nature, have molecular weight of 410.5 g/mole, melting point of 228–230°C, and maximum wavelength at 210 nm, 305 nm and is soluble in ethanol, dichloromethane, and methanol. It may also be dissolved in 2% v/v ethyl alcohol in water (Sapio et al., 2017).

Extracts of tuberous root of C. *Forskohlii* have minor diterpenoids, i.e., 9-deoxyforskolin, deactylforskolin, 1,9-dideoxy-7-deacetylforskolin, and 1,9-deoxyforskolin in addition to forskolin (Bhowal and Mehta, 2017). Forskolin occurred only in *C. forskohlii*, while all other six species of *Coleus*, i.e., *C. amboinicus, C. spicatus, C. parviflorus, C. malabaricus, C.canisus*, and *C.blumei*, do not show any positive test toward the presence of forskolin (Shah et al., 1980). In Japan, different studies carried out on hundreds of samples belonging to the species of *Orthosiphon, Plectranthus*, and *Coleus* revealed the absence of active compound "forskolin" in all samples. Second-generation derivatives of forskolin, i.e., 5-6-deoxy-7-deacetyl-7-methyl amino carbon forskolin (HIL568) and 6-(3-dimethylamino propionyl) forskolin hydrochloride (NKH477), showed antiglaucoma and cardiotonic activity, respectively (Lakshmanan and Manikandan, 2015). Six different compounds, demethyl crypto japonol, betulic acid, alpha-amyrin, beta-sitosterol, as 14-deoxycoleon U, and alpha-cedrol, were identified from tuberous roots of coleus plant that also showed good biologic activities (Lakshmanan and Manikandan, 2015). Two new diterpenes, forskolin I (1α, 6β-diacetoxy-7β, 9α-dihydroxy-8,13-epoxylabd-14-en-11-one) and J (1α, 9α-dihydroxy-6β,7β-diacetoxy-8,13-epoxylabd-14-en-11-one), were identified from a *C. forskohlii* plant sample that was collected from Yunnan Province (Shen and Xu, 2005). Further, two new compounds, forskoditerpenoside A, B and glycosides, were isolated from ethyl alcohol extract of the plant (Shan et al., 2007). In 2007, the existence of glycosides in the coleus plant was reported first to have relaxative effects. In 2008, a further three new diterpene glycosides, i.e., forskoditerpenoside C, D, and E, were identified in coleus plant that are present in minor quantity, and these compounds exhibited relaxative effects(Shan et al., 2008).

Major constituents of coleus essential oil are 13.15% β-sesquiphellandrene, 12.5% γ-eudesmol, 15% bornyl acetate, 7% 3-decanone, and 7.5% sesquiterpene hydrocarbon (Misra et al., 1994). Essential oil can also be extracted from flowers and stems of coleus plant. Essential oil extracted from the stem contains other important chemical constituent like β--caryophyllene, α-pinene, β-phellandrene, sabinene, α-humulene, limonene, α-copaene, and caryophyllene oxide (Kerntopf et al., 2002).

3. POSTHARVEST TECHNOLOGY

After 4.5–5 months planting, the crop of coleus gets ready for harvesting. The taproots of the coleus plant are harvested in autumn. It is thought that this is the best time for the extraction of the active compounds from the plant as the largest amount of active ingredients are present in roots during autumn. After harvesting, tuberous roots are cleaned and dried in sun.

4. PROCESSING

The fresh root tubers just after harvesting contain 75%–85% moisture level, which goes down to >12% due to drying. Root tubers can be dried by two methods: sun drying and mechanical drying. Sun drying requires more time to dry root tubers than that of the mechanical drying. Mechanical drying requires a temperature of 40°C to dry root tubers. After drying, root tubers having a slice thickness of 0.5 cm are packed in bag that is lined with polyethylene. Mechanical drying retains a high yield of forskolin compound as compared to sun drying. It was estimated that almost 800–1000 kg/ha yield of dry tubers are obtained, but the yield can be increased up to 2000–2200 kg/ha if the crop is cultivated properly (Rajamani and Vadivel, 2009).

5. VALUE ADDITION

Along with the medicinal applications, *C. forskohli* plant also has applications in different industries that increases its value in the world. Essential oil obtained from the coleus plant is very attractive for industries like the food industry and perfume industry due to its attractive fragrance and spicy nature. In the food industry, essential oil is used as flavoring (Kavitha et al., 2010). Forskolin can be used in combination with hydroxy citric acid, and this mixture can be used for decreasing body fat, so it maintains body shape. Root extract mixed with mustard oil is used to cure skin infections (Khan et al., 2012).

6. USES

Coleus plant was first used by Ayurvedic schools of medicines for medicinal purposes. In the beginning, the juice obtained from the roots of the plant was used to cure constipation in children, and later on the decoction of roots was used as tonic (Singh et al., 2011b). Now, it is commercially cultivated for making herbal medicines to cure diseases like skin infections, cardiac problems, cough, glaucoma, cancer, and eczema. The active compound "forskolin" is also an effective hypotensive and inotropic due to its vasodilatory characteristics (Seamon, 1984). Different medications are used to cure asthma, but forskolin is considered better as it activates adenylate cyclase used for the generation of cyclic adenosine monophosphate (AMP) in cells. This unique feature of forskolin makes the coleus plant important for medicinal purposes (Seamon et al., 1981). A standard extract of coleus roots "ForsLean" having 10%, 20%, and 40% forskolin is available on the market.

In India, different medicinal species of *Coleus* like *C. amboinicus, C. malabaricus, C. scutellaroides, C. forskohlii*, and *C. blumei* are used for the treatment of digestive disorders and dysentery. *C. forskohlii* is commonly used for the treatment of various ailments in different parts of the world. In Africa and Egypt, leaves are used as emmenagogue, diuretic, and expectorant, while in Brazil, these leaves are used for the treatment of intestinal disorders and as a stomach aid. In India, leaves are also used as a condiment. Tubers of the plant are prepared as a pickle, so this plant also has applications in the food industry. Worms can be treated with the help of root extract. A burning sensation can be lessened in festering boils with the help of root extract. The coleus plant is also important in the veterinary field (Kavitha et al., 2010).

Forskolin also has the ability to prevent the greying of hair and also restoring grey hair to its normal color (Ciotonea and Cernătescu, 2010). Though coleus plant gained importance due to its medicinal properties, it also has essential oil in flowers, tubers, and stems that has an attractive odor with a spicy note (Misra et al., 1994). Essential oil obtained from the coleus plant is used in the food flavoring industry. It is also used as an antimicrobial agent (Shivaprasad et al., 2014).

7. PHARMACOLOGICAL USES

7.1 Antiobesity Activity

There are arguments and counterarguments on the efficacy of forskolin for treatment of various diseases. One clinical study about the forskolin suggests that it plays an important role to increase lean mass and bone mass. Due to this research, pharmaceutical companies throughout the world started commercialization of forskolin as a bodybuilding supplement. But another research that was conducted in vitro and on animals demonstrated that forskolin stimulated the lipolysis in fat cells (Okuda et al., 1992) by activating the adenylate cyclase enzyme and increasing the level of cAMP (Majeed et al., 1998). It was suggested that *C. forskohlii* has no ability to promote weight loss in humans, but forskolin has ability to reduce weight in overweight human females without any side effects (Patel, 2010). The effects of forskolin as an antiobestic agent were studied in ovariectomized rats, and results are that forskolin may be a useful chemical compound for the treatment of obesity (Han et al., 2005).

7.2 Heart Disorder and Hypertension

Forskolin has positive inotropic action on cardiac tissues. Increased level of cAMP by forskolin helps to decrease elevated or normal blood pressure rate through vasodilation in different species of animals. *C. forskohlii* was traditionally used for the treatment of hypertension, angina, and congestive heart failure. The basic action of forskolin on the cardiovascular system is the lowering of blood pressure, but at the same time, it increases contractility of heart. This is considered to be due to the enhancement of cAMP level in cells that causes the relaxation of arteries, so the contraction force of the heart muscle increases (De Souza et al., 1983; Dubey et al., 1981).

7.3 Antiglaucoma Activity

Several studies on humans and animals showed that forskolin has the ability to lower high intraocular pressure, possibly by the activation of cAMP that reduces the aqueous flow (Caprioli et al., 1984).

7.4 Antiasthma Activity

Decreased levels of cAMP in smooth muscles of bronchi causes different allergic conditions and asthma in humans. Mast cells are degranulated due to the allergenic stimuli that causes the release of histamine. This in turn contracts the smooth muscles of bronchi. Forskolin activates generation of cAMP that inhibits degranulation of mast cells, which results in bronchodilation. Therefore, forskolin is considered to be a potential "bronchodilator agent" for the treatment of asthma (Ciotonea and Cernătescu, 2010). The blockage of bronchospasm is the main cause of the asthma that is mainly caused by the leukotriene C-4 and histamine (Marone et al., 1987). A study on the forskolin effects on asthmatic patients shows that powder formulations of forskolin cause bronchodilation in asthma patients (Bauer et al., 1993). Forskolin is regarded as a promising drug for the treatment of asthma if it is used in an appropriate dosage (Ciotonea and Cernătescu, 2010).

7.5 Anticancer Activity

Different studies carried out in mice reveal the ability of coleus plant to inhibit tumor colonization. Theoretically, it is considered possible to use coleus for the inhibition of tumor metastases in humans. Many tumor cells cause the aggregation of platelets both in vivo and in vitro. After the aggregation of platelets, different substances are released that promote tumor growth. Studies on forskolin as an anticancer agent reveal that aggregation of platelets is blocked in the presence of forskolin. In mice, it is observed that forskolin injection successfully reduces the tumor colonization in lungs by 70% (Agarwal and Parks, 1983).

7.6 Antithrombotic Effects

Forskolin has the ability to inhibit the platelet aggregation by the stimulation of adenylate cyclase enzyme that enhances the prostaglandins effect (Adnot et al., 1982; Siegl et al., 1982).

7.7 Antipsoriasis Effects

In psoriasis, cells started to divide at high rate, about 1000 times faster than that of normal cells. Coleus plant helps to slow down the rate of cell division by normalizing the cAMP to cGMP ratio. Like asthma, it is also caused by the decreased level of cAMP in skin in relation to other regulating substance, i.e., cyclic guanosine monophosphate (cGMP). It is reported that forskolin successfully cures psoriasis patients. The regulation of cAMP level in skin cells is the important feature of the forskolin that enables its use for the treatment of psoriasis (Dean, 2012).

7.8 Antidepressant Effects

Depression is considered to be associated with the imbalance between the neurotransmitters of brain, primarily dopamine and serotonin. When serotonin is in short supply, then other supplements may prove useful to perform the tasks of serotonin, like tryptophan, 5-HTP, Zoloft, or Prozac. If catecholamine neurotransmitters (norepinephrine, epinephrine) are in short supply, then amino acids (L-Tyrosine, L-Phenylalanine) or monoamine oxidase inhibitors (Gerovital, Deprenyl) prove to be helpful. An increase in the level of cAMP also increases the level of catecholamines. Since forskolin increases cAMP, it indirectly improves the function of neurotransmitters and hence relieves the depression. Clinical trials for the treatment of depression with coleus plant have not yet been done.

7.9 Increasing Lean Body Mass Effects

Forskolin drug also has health-promoting value as it increases lean body mass and stamina. The fatty tissues of the abdomen may cause cardiovascular disease, but with the use of forskolin drug, the level of cAMP increases, which increases the circulation of anabolic hormones that enhance utilization of fatty tissues. This, theoretically, leads to an increase in lean body mass. Forskolin also shows effect on numerous membrane transport proteins and also inhibits the transport of glucose in platelets, adipocytes, erythrocytes, and other cells (Mills et al., 1984).

7.10 Antihypothyroidism Activity

Forskolin also has ability to increase the production of thyroid hormone, and it also stimulates release of thyroid hormone by increasing the quantity of stimulatory guanine nucleotide binding proteins. Therefore, it is also used for the treatment of hypothyroidism (Patel, 2010).

7.11 Antidiabetic Activity

Forskolin is an important constituent for maintaining body composition (Kavitha et al., 2010). Forskolin has also proven to be effective for diabetic patients. Forskolin increases the level of cAMP in cells, due to which release of insulin also increases. cAMP actually activates two major signaling pathways of β cell. Though, different experiments have been done on the rats to check diabetic effect of forskolin, still further researches are needed to find the complete effect (advantages and disadvantages) of forskolin on diabetic patients as well as the exact dosage for the human population (Ríos-Silva et al., 2014).

8. SIDE EFFECTS AND TOXICITY

The safety of *C. forskohlii* plant and forskolin drug has not been completely evaluated. Forskolin should not be recommended for the patients that have ulcers because forskolin has the ability to increase the acid level in the stomach (Sapio et al., 2017).

References

Adnot, S., Desmier, M., Ferry, N., Hanoune, J., Sevenet, T., 1982. Forskolin (a powerful inhibitor of human platelet aggregation). Biochemical Pharmacology 31, 4071–4074.

Agarwal, K.C., Parks, R.E., 1983. Forskolin: a potential antimetastatic agent. International Journal of Cancer 32, 801–804.

Ahmed, B., Merotra, R., 1991. Celeoside-B: A new phenolic glycoside from Coleus Forskohlii. Die Pharmazie 46, 157–158.

Ahmed, B., Vishwakarma, R., 1988. Coleoside, a monoterpene glycoside from Coleus forskohlii. Phytochemistry 27, 3309–3310.

Bauer, K., Dietersdorfer, F., Sertl, K., Kaik, B., Kaik, G., 1993. Pharmacodynamic effects of inhaled dry powder formulations of fenoterol and colforsin in asthma. Clinical Pharmacology & Therapeutics 53, 76–83.

Bhowal, M., Mehta, D.M., 2017. Coleus forskholii: phytochemical and pharmacological profile. International Journal of Pharmaceutical Sciences and Research 8, 3599–3618.

Caprioli, J., Sears, M., Bausher, L., Gregory, D., Mead, A., 1984. Forskolin lowers intraocular pressure by reducing aqueous inflow. Investigative Ophthalmology & Visual Science 25, 268–277.

Ciotonea, C., Cernătescu, C., 2010. Biological active effects of Foskolin extract. Buletinul Institutului Politehnic DIN IASI 4, 95–106.

De Souza, N.J., Dohadwalla, A.N., Reden, Ü., 1983. Forskolin: a labdane diterpenoid with antihypertensive, positive inotropic, platelet aggregation inhibitory, and adenylate cyclase activating properties. Medicinal Research Reviews 3, 201–219.

Dean, W., 2012. Forskolin and cAMP. NaturoDoc. com. Available from: http://www.naturodoc.com/forskolin.htm.

Dubey, M., Srimal, R., Nityanand, S., Dhawan, B., 1981. Pharmacological studies on coleonol, a hypotensive diterpene from Coleus forskohlii. Journal of Ethnopharmacology 3, 1–13.

Han, L.-K., Morimoto, C., Yu, R.-H., Okuda, H., 2005. Effects of Coleus forskohlii on fat storage in ovariectomized rats. Yakugaku Zasshi: Journal of the Pharmaceutical Society of Japan 125, 449–453.

Kavitha, C., Rajamani, K., Vadivel, E., 2010. Coleus forskohlii A comprehensive review on morphology, phytochemistry and pharmacological aspects. Journal of Medicinal Plants Research 4, 278–285.

Kerntopf, M.R., de Albuquerque, R.L., Machado, M.I.L., Matos, F.J.A., Craveiro, A.A., 2002. Essential oils from leaves, stems and roots of Plectranthus barbatus Andr.(Labiatae) grown in Brazil. Journal of Essential Oil Research 14, 101–102.

Khan, B.A., Akhtar, N., Anwar, M., Mahmood, T., Khan, H., Hussain, I., Khan, K.A., 2012. Botanical description of coleus forskohlii: a review. Journal of Medicinal Plants Research 6, 4832–4835.

Khatun, S., Chatterjee, N.C., Cakilcioglu, U., 2011. The Strategies for Production of Forskolin vis-a-vis Protection Against Soil Borne Diseases of the Potential Herb Coleus forskohlii briq.

Lakshmanan, G.M.A., Manikandan, S., 2015. Review on pharmacological effects of Plectranthus forskohlii (Willd) briq. International Letters of Natural Sciences 1.

Lakshmanan, G.A., Manikandan, S., Panneerselvam, R., 2013. Plectranthus forskohlii (Wild) Briq.(Syn: coleus forskohlii)—a compendium on its botany and medicinal uses. International Journal of Research in Pharmacy and Science 3, 72–80.

Li, S., Yang, Q., Wang, X., Zou, G., Liu, Y., 2006. Chemical constituents of Coleus forskohlii replanted to Tongcheng. Zhong Cao Yao 37, 824–826.

Liu, Y., Wang, X., Wu, H., 2007. Main Components of Coleus forskohlii Extract and Relevant Extraction Method. Chinese Patent.

Majeed, M., Badmaey, V., Rajendran, R., 1998. Method of Preparing a Forskohlin Composition from Forskohlin Extract and Use of Forskohlin for Promoting Lean Body Mass and Treating Mood Disorders. Google Patents.

Marone, G., Columbo, M., Triggiani, M., Cirillo, R., Genovese, A., Formisano, S., 1987. Inhibition of IgE-mediated release of histamine and peptide leukotriene from human basophils and mast cells by forskolin. Biochemical Pharmacology 36, 13–20.

Mendes, B.G., Machado, M.J., Falkenberg, M., 2006. Screening of glycolipids in medicinal plants. Revista Brasileira de Farmacognosia 16, 568–575.
Mills, I., MORENO, F.J., FAIN, J.N., 1984. Forskolin inhibition of glucose metabolism in rat adipocytes independent of adenosine 3′, 5′-monophosphate accumulation and lipolysis. Endocrinology 115, 1066–1069.
Misra, L.N., Tyagi, B.R., Ahmad, A., Bahl, J.R., 1994. Variability in the chemical composition of the essential oil of Coleus forskohlii genotypes. Journal of Essential Oil Research 6, 243–247.
Okuda, H., Morimoto, C., Tsujita, T., 1992. Relationship between cyclic AMP production and lipolysis induced by forskolin in rat fat cells. Journal of Lipid Research 33, 225–231.
Patel, M., 2010. Forskolin: a successful therapeutic phytomolecule. East and Central African Journal of Pharmaceutical Sciences 13.
Rajamani, K., Vadivel, E., 2009. Marunthu Kurkan–Medicinal Coleus. Naveena Mulikai Sagupaddi Thozhil Nuttpangal. Tamil Nadu Agricultural University, Coimbatore, India, pp. 17–22.
Ríos-Silva, M., Trujillo, X., Trujillo-Hernández, B., Sánchez-Pastor, E., Urzúa, Z., Mancilla, E., Huerta, M., 2014. Effect of chronic administration of forskolin on glycemia and oxidative stress in rats with and without experimental diabetes. International Journal of Medical Sciences 11, 448.
Roy, R., Vishwakarma, R., Varma, N., Tandon, J., 1990. Coleonolic acid, a rearranged ursane triterpenoid from Coleus forskohlii. Tetrahedron Letters 31, 3467–3470.
Sapio, L., Gallo, M., Illiano, M., Chiosi, E., Naviglio, D., Spina, A., Naviglio, S., 2017. The natural cAMP elevating compound forskolin in cancer therapy: is it time? Journal of Cellular Physiology 232, 922–927.
Seamon, K.B., 1984. Forskolin and adenylate cyclase: new opportunities in drug design. Annual Reports in Medicinal Chemistry 19, 293–302.
Seamon, K.B., Padgett, W., Daly, J.W., 1981. Forskolin: unique diterpene activator of adenylate cyclase in membranes and in intact cells. Proceedings of the National Academy of Sciences 78, 3363–3367.
Shah, V., Kalakoti, B., 1996. Development of Coleus forskohlii as a medicinal crop. In: Domestication and Commercialization of Non-timber Forest Products, p. 212.
Shah, V., Bhat, S., Bajwa, B., Dornauer, H., De Souza, N., 1980. The occurrence of forskolin in the Labiatae. Planta Medica 39, 183–185.
Shah, V.C., D'sa, A.S., De Souza, N.J., 1989. Chonemorphine, stigmasterol, and ecdysterone: steroids isolated through bioassay-directed plant screening programs. Steroids 53, 559–565.
Shan, Y., Kong, L., 2006. Isolation and identification of terpenes from Coleus forskohlii. Chinese Journal of Natural Medicines 4, 271–274.
Shan, Y., Wang, X., Zhou, X., Kong, L., Niwa, M., 2007. Two minor diterpene glycosides and an eudesman sesquiterpene from Coleus forskohlii. Chemical and Pharmaceutical Bulletin 55, 376–381.
Shan, Y., Xu, L., Lu, Y., Wang, X., Zheng, Q., Kong, L., Niwa, M., 2008. Diterpenes from coleus forskohlii (WILLD.) BRIQ.(Labiatae). Chemical and Pharmaceutical Bulletin 56, 52–56.
shen, H., Xu, Y.,L., 2005. Two new diterpenoids from Coleus forskohlii. Journal of Asian Natural Products Research 7, 811–815.
Shivaprasad, H., Pandit, S., Bhanumathy, M., Manohar, D., Jain, V., Thandu, S.A., Su, X., 2014. Ethnopharmacological and phytomedical knowledge of Coleus forskohlii: an approach towards its safety and therapeutic value. Oriental Pharmacy and Experimental Medicine 14, 301–312.
Siegl, A., Daly, J., Smith, J., 1982. Inhibition of aggregation and stimulation of cyclic AMP generation in intact human platelets by the diterpene forskolin. Molecular Pharmacology 21, 680–687.

Singh, R., Gangwar, S.P., Singh, D., Singh, R., Pandey, R., Kalra, A., 2011a. Medicinal plant Coleus forskohlii Briq.: disease and management. Medicinal Plants 3, 1–7.
Singh, R., Gangwar, S.P., Singh, D., Singh, R., Pandey, R., Kalra, A., 2011b. Medicinal Plant Coleus Forskohlii Briq. Disease and management.
Tandon, J.S., Katti, S.B., Rüedi, P., Eugster, C.H., 1979. Crocetin-dialdehyde from coleus forskohlii BRIQ., Labiatae. Helvetica Chimica Acta 62, 2706–2707.
Valdes, L., Mislankar, S., Paul, A., 1987. Coleus barbatus (C. forskohlii)(Lamiaceae) and the potential new drug forskolin (Coleonol). Economic Botany 41, 474.
Xu, L., Lu, J., Li, W., Kong, L., 2005. Studies on the chemical constituents in root of Coleus forskohlii. Zhongguo Zhong Yao Za Zhi 30, 1753–1755.
Yao, C., Shen, Y., Xu, Y., 2002. The chemical constituents of Coleus forskohlii. Natural Product Research and Development 14, 1–6.

CHAPTER

12

Cubeb

Hafsa Ahmad[1], *Rasheed Ahmad Khera*[1], *Muhammad Asif Hanif*[1], *Muhammad Adnan Ayub*[2]

[1] Department of Chemistry, University of Agriculture, Faisalabad, Pakistan; [2] Department of Chemistry, University of Okara, Okara, Pakistan

OUTLINE

Medicinal Plants of South Asia
https://doi.org/10.1016/B978-0-08-102659-5.00012-4

Copyright © 2020 Elsevier Ltd. All rights reserved.

1. BOTANY

1.1 Introduction

Cubeb (*Piper cubeba*) (Fig. 12.1) belongs to Piperaceae (pepper family) and is a perennial herb. Piper species are extensively found in tropical and subtropical regions all over the world. They contain a large amount of lignans and neolignans, which are biologically active compounds (Parmar et al., 1997). It has been used for thousands of years and has become an essential ingredient in cooking as a spice. It is also widely used in drugs for various diseases treatments (Junqueira et al., 2007). The genus *Piper* contains more than 1000 species and is native to the tropical region of Asia. Cubebs have a bitter, pungent, and persistent taste, and its odor is

FIGURE 12.1 Cubeb.

aromatic. The manner of pollination in cubeb is wind or rain. Cubeb may also be pollinated by insects (Semple, 1974).

The fruits of *P. cubeba* are commonly known by different names depending upon where we are in the world. In English, is typically called tailed piper, Jawa peppercorn, or Jawanese pepper. In Pakistan, Bangladesh and India, specifically in Urdu, Hindi and Bangali, it is called Kabab-chini. In Brazil it called "pimenta de Java" ("Java pepper"). In Indonesia, it is called Cabé jawa and Kamukus. It is known as cubeba in Arabic (Chopra et al., 1956).

1.2 History/Origin

P. cubeba is native to Pakistan, India, Bangladesh, Nepal, Sri Lanka, and other tropical regions like Java and Sumatra. Through commerce with Arabs, cubeb arrives to Europe from India. The name cubeb is derived from the Arabic word kababah that means pepper, and the scientific name *Piper* is obtained from the Sanskrit term *pippali*, which indicates the long pepper (*P. longum*). Cubeb is found growing in forests without any cultivation. A small quantity of cubeb is also cultivated in the coffee plantations of Java. In 1640, John Parkinson tells in his *Theatrum Botanicum* that sale of cubeb was prevented by the king of Portugal to promote the black pepper (Gorman, 1969).

In ancient China, it was utilized as medicine. In the 17th century, from the South Pacific, British seafarer James Cook withstood cubeb during his trip. In Polynesia, cubeb grows up to 6 m long and is used for a drink called kawi (Gorman, 1970). In European markets, cubeb became famous in the 19th century. In the 20th century, cubeb was exported on a regular basis from Indonesia's islands to European countries and the United States (Khare, 2004). Cubeb is used as a spice and medicine for the treatment of various disorders.

1.3 Demography/Location

The ecologic requirements for the cultivation of cubeb are warm and wet tropical climate, humid conditions, and moisture. Cubeb pepper is grown widely in India, Indonesia, Malaysia, Sri Lanka, Pakistan, Reunion, Bangladesh, Vietnam, Kampuchea, Brazil, and Thailand (Cardeal et al., 2006).

1.4 Botany, Morphology, Ecology

The diameter of cubeb berry (small seed) is 3–4 mm. Cubeb berry may be brown, reddish brown, or dark brown in color. The seed's upper surface is hard, wrinkled, and oily. The stem of cubeb is grey in color. The height of cubeb plant is about 18–20 feet (Murthy and Bhattacharya, 1998). It is hollow and whitish from inside, reticulately wrinkled, has a strong odor, is spicy, blackish grey, tastes pungent, and is aromatic. Leaves are smooth, oval shaped, having a pointed tip, and are green in color. Flowers are small, arranged in spikes, and white in color. From these flowers, fruits develops that are stocked before ripe (Razafimandimby et al., 2017).

Cubeb plant grows well under shaded areas. As compared to black pepper, its cortical portion is less succulent and thinner, which contains seeds that are oily and whitish. Cubeb is commonly growing near bushes (its natural habitat) as it does not require too much sunlight. Stalked berries having a dry surface, and some are bigger than pepper corns. Cubeb grows well in the pH range 5.5–6.5. Fruits are harvested, washed with water, and dried under shade in the temperature range 35–40°C for a week.

2. CHEMISTRY

P. cubeba is an aromatic plant and is used as an herb and flavor. Aroma and fragrance are due to the presence of essential oil in the seeds. The main components in the essential oil include terpenes and their analogues, which are important for producing aroma, flavor, and odor (Govindarajan and Stahl, 1977).

The main components of the piper are flavonoids, lignans, dihydrochalcones, amides, neolignans, aristolactam, chalcones, long and short chain ester, terpenes, kawapyrones, piperolides, steroids, flavones, flavanones, propenylphenol, and alkaloids. Pepper is a flowering vine and extensively utilized in spices (Huang et al., 2014). The major constituent in cubeb is hinokinin, which is trypanosomicidal dibenzylbutyrolactone lignin. Due to the presence of piperine, cubeb has pungency. Piperine is a 2-trans, 4-trans piperidine amide of piperic acid. It has a fresh, pleasing, and warm aroma. On cooking, the flavor of cubeb mellows and combines with grains of other spice flavors. The first amide derivative that was obtained from *Piper* species was piperine.

Cubeb has 6%–8.5% resin, volatile oil 5%–20%, which is bluish green in color, fatty oil, starch, coloring substance, and gums. Cubebol, cubebin, and cubebic acid are the main components that are present in the resin. Minerals present in it are magnesium, iron, manganese, zinc, potassium, and calcium malates. Calcium and magnesium are present in abundant quantity. About 2.5% cubebin is found in cubeb. It has no odor or taste and is a crystallizable compound in alcoholic solution (Batterbee et al., 1969). Cubeb is a good source of vitamin A and vitamin C. Cubeb contains flavonoids like cryptoxanthin, zeaxanthin, carotene, and lycopene. Cubeb has vitamin B complex groups such as riboflavin, niacin, pyridoxine, and

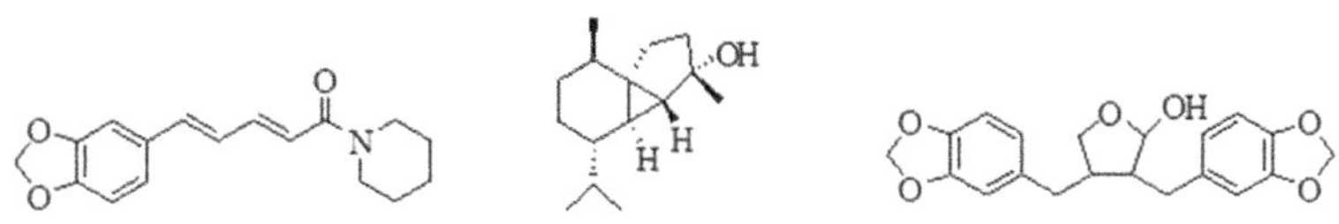

Piperine Cubebol Cubebin

FIGURE 12.2 Most active components in cubeb.

thiamin. Fatty acids are also found in cubeb abundantly. It contains glucose-6-phosphate dehydrogenase and glutathione peroxidase in large amounts, which plays an important role in scavenging activity. Piperine (alkaloids) is present abundantly in berries of cubeb. Cubeb has 47%–53% fiber, 11%–14% protein, and 10%–14% starch. The components that are isolated from the *Piper* genus are viridiflorol, aromadendrene, beta selinene, caryophyllene oxide, dillapiole, myristicin, and ascarcin. The main lignins that are present in cubeb are hinokinin, cubebin, isoyatein, and yatein (Zaveri et al., 2010).

Phytochemical studies of *P. cubeba* extracts indicated the presence of alkaloids, terpenes, lignans, benzoic acids, amide, chromenes, phenolics, and phenylpropanoids. The important fatty acid constituents are oleic, palmitic, linoleic, stearic, capric, myristic, and lauric acids (Tewtrakul et al., 2000).

Cubebene ($C_{15}H_{24}$) is the liquid part of cubeb that is a blue-yellow or pale-green, thick liquid having camphoraceous and warm woody odor. Cubebin is a crystalline substance, with formula $C_{20}H_{20}O_6$ also present in cubeb. Cubebin can be obtained by the distillation of oil or by cubebene. Tannins that are present in cubeb are cranesbill and proanthocyanidins (Nahak and Sahu, 2011). The most active components in cubeb are shown in Fig. 12.2.

3. POSTHARVEST TECHNOLOGY

Ripe fruits are harvested and then dried. Postharvest technology is necessary and plays an important role to improve the quality of cubeb. Usually, cubeb is harvested in October or November, and spikes are tossed in December or January. When spikes become mature, they are harvested by hand. From spikes, berries that are ripe are separated. After separation, they are immersed in water to remove pericarp and afterward dried in the sun for 3–4 days. Harvesting continues in major cubeb-producing countries for 2–3 months. Leaves are dried in an oven at 40°C for 4–8 days. After drying, the leaves are ground in a mechanical grinder and convert into powdered form (Kumoro et al., 2010). In Karnataka and Kerala States of India, blanching is a common precuring process that is adopted before drying. The reason is that blanched berries take less time to dry. Moreover, blanching prevents microbial growth, so a hygienic product is obtained. After blanching, cubeb berries are dried in sunlight. The drying time of cubeb berries also depends on climatic conditions. On a fully sunny day, they dry in 1–2 days, sometimes 4–5 days. Today, some farmers use a dryer to minimize the drying period. The most common artificial drying process are conduction drying, desiccated air drying, infrared drying, refrigerated air drying, and heated air drying. Solar dryers require far less time for drying than sun drying. Moreover, the berries that are drying through solar dryers give better quality products compared to sun drying (Vinay, 2012).

4. PROCESSING

Cubeb is an herbal plant that is consumed in a variety of ways and for various purposes. Cubeb berries are processed for product and for extraction of oil. After drying cubeb berries, they pass through the procedure of cleaning, grinding afterward, and finally packaging in the factory. From the peppercorns, to remove bits of stalk and twigs, dirt clods, dust, and some other impurities, gravity separators and blowers can be used (Barrett et al., 2004). Afterward, other treatments are also applied on dried and cleaned peppercorns to get rid of bacteria. Peppercorns are passed through a cold roll milling process that contains a series of rollers to grind the peppercorns. They are slightly crushed to cut peppercorns and release flavor (Indonesia, 1995). Additional grinding converts the crushed peppercorns into thick and then fine grinds of berries. Packaging may be done in cans, jars, and mills for home use and boxes, bags, or canisters for export or commercial sales. Blending of pepper with a variety of species is also included in packaging to prepare Cajun recipes, seafoods, sauces, and Italian foods, etc. (Eisai, 1995).

Essential oil of cubeb is extracted from dried cubeb berries via steam distillation method. Essential oil can be extracted from the leaves (0.35% –0.40%), flowers (0.5%–0.6%), and seeds (0.55%–0.58%).

5. VALUE ADDITION

Cubeb pepper is used to flavor many things including soaps, drinks, perfumes, cigarettes, sausages, and biscuits. Essential oil of cubeb is used as a cooling agent in toothpaste, chewing gum, and different confectionary items. It is also utilized in several herbal cough syrups. Cubeb oil is added to various perfumes, ornaments, cooking items, gourmet items, and medicine. It is used in restaurant dining, Cajun-style recipes, which depend on the spice, and in preparation of hygienic food to enhance the flavor and taste. Cubeb pepper is also placed at the table according to people's choice and taste. Cubeb has antioxidant activity. Due to this property, it is also utilized in conventional medicines and diets. Cubeb fruits are used as a natural antioxidant in pharmaceutical industries (Sudjarmoko et al., 2015).

6. USES

Cubeb is a very effective in treatment of aging and diseases. Cubeb used in the treatment of infection and inflammation. It provides relieve to muscle pain due to presence of a gentle, stimulating effect in the oil. It has been found most effective in cases of urinary tract and prostate gland infection (Lawless, 1995). It also helps in fertility. Cubeb has become popular all over the world and used in many traditional drugs for the treatments of several disorders (Magalhaes et al., 2012). Cubeb is effective for sore throat, chills, skin rashes, fever, flu, laryngitis, pain relief, colds, muscular aches, rheumatism, and throat problems.

In Pakistan and India, the powder form of cubeb peppercorn combined with salt and this mixture is used as an item in restaurants on the serving table. This mixture is also splashed on lemonades, fruit salad, vegetable salad, chaats, and soups. In the province of Punjab, a salt–spice mixture is used in Lassi to enhance the taste. It is mostly used in Indonesian curries. In India, cubeb has been used for several diseases, especially used for gonorrhea. In Europe, cubeb is used for the same purpose. Sushruta and Charaka utilized cubeb powder internally for dental and oral disorders. They advised cubeb for fevers, loss of voice, cough, and halitosis. Its paste is used as a mouthwash. Unani physicians externally apply the cubeb berries' paste on female and male genitals to increase the sexual pleasure, due to which cubeb was known as "Habb-ul-Uruus" (Khare, 2004). In China, it is utilized for the treatment of allergic problems. In Tibetan medicine, cubeb is the most effective drug that is allocated to spleen. It is used to produce asthma-alleviating cigarettes in the United States. In Malaysia, the powdered dried fruit is used to treat amebic dysentery. In Java, it is used as a popular aphrodisiac, often added by the wives to the husband's meal (Yu and Amri, 2016). In India and China, it is used in an anticancer medicine. In Europe, cubeb extensively is used as a natural medicinal plant. Cubeb's fruits have been used in therapy of asthma, dysentery, enteritis, diarrhea, gonorrhea, abdominal pain, and syphilis. It was also found that cubeb has inhibitory activity for hepatitis C virus. The extract of cubeb with methanol has been found most effective for analgesics and antiinflammatory activities. In Brazil and Mexico, it is utilized for antiinflammatory activity to reduce chest pains (Li and Brown, 2009).

Cubeb is utilized as appetizers, expectorants, stimulants, and stomachics in conventional medicines. Cubeb's fruit shows a significant effect in cases of enteritis, gastric pain, and diarrhea. The chemical composition of cubeb plays an important role in antioxidant activity (Dhar et al., 1968). It is used in curing of itching, constipation, and flatulence (Asprey and Thornton, 1954). Cubeb also shows antibacterial and gastroprotective activities. The main component of cubeb is piperine, which has importance in pharmacological application. It exhibited antipyretic, antileukemia, antidepressant, antimalarial, antiinflammatory, and analgesic activities. It has a bitter taste and astringent properties that make it a stimulant for the body's organs and improve the circulation of blood. Due to the presence of light properties, it helps to normalize the digestive tract (Kulshreshta et al., 1969). Cubeb ejects the extra mucus that is present in the tract and cleans the respiratory tract. It also acts as an aphrodisiac agent and helps in menstrual problems (Lala et al., 2004). It also acts as a diuretic agent. Cubeb is a medicinal plant that has an antioxidant property. Due to this property, plasma's antioxidant capacity increases, and chances of several disorders decrease. Cubeb contains flavonoids and phenolics acid, which are responsible for antioxidant activity. Cubeb displays antioxidant activity by preventing decay of hydroperoxide in free radicals or by inactivating lipid free radicals. Cubeb also shows antimutagenic properties (Duda-Chodak et al., 2011).

Cubeb extract has a significant role in the treatment of prostate cancer and breast cancer. Cubebin is the component of cubeb that has antimycobacterial, antiinflammatory, antiprotozoal, trypanocidal, vasorelaxant, and analgesic activities (de Souza et al., 2005). Butyrolactone lignan (hinokinin) is isolated from cubeb and shows a wide range of biologic application such as antimutagenic, trypanocidal, antiinflammatory, chemopreventive, analgesic, anticancer, and modulatory effects on human monoamine.

Cubeb oil also has antiinflammation activity. It is applied to wounds for early healing. It helps in the treatment of problems that are related to the mouth. In case of erectile dysfunction, it is applied on the penis. For the treatment of head-related problems, cubeb oil can be used as nasal drops. In case of pile, cubeb's powder is most impressive and helps to keep the digestive tract normal. It also gives strength to the heart. It is used to control asthmatic and cough conditions. It helps to regulate the menstrual cycle. In the urinary tract's toning up, cubeb pepper is most effective (Solomon, 1998). Cubeb also has hepatoprotective potential. Ethanolic extract of cubeb shows the effective hepatoprotective and antioxidant activity. It is a source of new amide and gastroprotective constituents (Morikawa et al., 2004). Cubeb encourages peristalsis, enhances flow of saliva, is a digestive tonic, stimulates appetite, and tones the colon muscles. The oleoresin of cubeb has fungistatic and bacteriostatic properties

(Cai et al., 2004). Cubeb is used as a diuretic for increasing urination. It is also is used for the treatment of gonorrhea, intestinal gas, cancer, and amebic dysentery (parasitic infection in the intestine) (Chande et al., 2006).

7. PHARMACOLOGICAL USES

7.1 Prophylactic Agent

Cubeb essential oil has excellent antimalarial activity. Juice from the leaves of cubeb is used to treat dysentery, asthma, syphilis, and enteritis. Piperine is the main component in cubeb that has bronchodilator and antitussive properties and is effective in several herbal cough syrups. Cubeb is useful in healing ulcers and wounds and in removing worms and parasites. Cubeb is an effective antioxidant. Cubeb contains flavonoids and phenolics acids that are responsible for the antioxidant activity. Antioxidants enhance the activity of superoxide dismutase and decrease the lipid per oxidation. Cubeb is a natural antioxidant that helps in reducing the chances of heart disorders and in maintaining good health (Pratt and Hudson, 1990). Cubeb berries provide a stimulating effect on the respiratory and urinary tracts. The essential oil of *P. cubeba* has excellent *Bacillus typhosus* and antiinfluenza properties (Ahmad et al., 2012). The fruits of cubeb show a wide range of biologic activities such as antimicrobial, bactericidal, antifungal, and antibacterial activities (Arruda et al., 2005). They are also used for cardiovascular diseases, liver diseases, and cancer (Perazzo et al., 2013).

7.2 Antioxidant Activity

Cubeb has health benefits due to the presence of antioxidant contents. Ethanolic extract of cubeb was evaluated by photochemical test through which the presence of flavonoids, glycosides, alkaloids, and tannins was confirmed. These constituents have great potential for antioxidant activity (Khalaf et al., 2008). Due to the presence of antioxidant activity, cubeb is extensively used in various medicines.

7.3 Hepatoprotective Activity

The hepatoprotective activity of cubeb extract was determined in animals. This activity was evaluated by using different doses of cubeb extract against CCl_4. It has been found that CCl_4-treated liver section causes the hepatocyte's fatty degeneration (liver fibrosis). The results show that ethanolic extract of cubeb fruits provides the effective protection in regeneration of hepatocyte against carbon tetrachloride's toxic effect. It inhibits liver fibrosis by decreasing both in vivo and in vitro lipid peroxidation (Pachpute and Deshmukh, 2013).

7.4 Insecticidal and Acaricidal Activity

Volatile oil of cubeb fruit has been found more effective for insect repellant and insecticidal activity (Park et al., 2002). Cubeb extract contains piperoctadecalidine and pipernonaline, which are piperidine alkaloids, and their toxicity was determined against arthropod pests. The results show that these alkaloids possess potent insecticidal activity (Park et al., 2002).

7.5 Antimicrobial Activity

The antimicrobial activity of several extracts of *P. cubeba* was investigated against bacterial pathogens including *Salmonella typhi, Scaphirhynchus albus, Bacillus megaterium, Escherichia coli, Pseudomonas aeruginosa,* and *Aspergillus niger* (Ghara Gheshlagh Alireza et al., 2018). The results also show that all extracts exhibited effective antimicrobial activity (Rios and Recio, 2005). Aqueous extract of cubeb did not exhibit antimicrobial activity. It has been found that n-hexane extract and constituents that were isolated from the cubeb possess the antimicrobial activity against all bacterial pathogens (Chitnis et al., 2007). *P. cubeba* exhibited the strong fungicidal activity (Siddiqui et al., 2007). Antifungal activity of cubeb was evaluated in phytopathogenic fungi including *Phytophthora infestans, Pyricularia oryzae, Botrytis cineria, Erysiphe graminis, Rhizoctonia solani,* and *Puccinia recondite* by using in vivo method.

7.6 Antiamebic Activity

The antiamebic activity of methanol extract of cubeb was studied against *Entamoeba histolytica* that affects the cecum in mice (Gayasuddin et al., 2011). It has been found that methanol extract reduces the cecal wall's ulceration severity (Sohni et al., 1995).

7.7 Antiasthmatic Activity

Antiasthmatic activity of cubeb was determined in rats. Cubeb fruit extract was prepared in milk. It was found that the extract decreased passive cutaneous anaphylaxis and has the ability to protect guinea pigs against antigen-induced bronchospasm. It can be concluded that cubeb fruit extract in milk is effective for antiasthmatic activity (Lim, 2012).

7.8 Antidiabetic Activity

The antilipidperoxidative and antihyperglycemic effects of cubeb fruit extract in ethanol were investigated in alloxan-induced diabetic mice. The level of glucose in blood, enzymes that metabolize carbohydrates, and grade of antioxidants and lipid peroxidation were attempted using colorimetric methods. Cubeb fruit ethanolic extract was administered to mice orally and showed potent antilipidperoxidative and antihyperglycemic effects in diabetic mice (Dhar et al., 1968).

7.9 Hypocholesterolemic Activity

Hypocholesterolemic activity of cubeb was evaluated by giving high cholesterol food to mice. It was found that methyl piperine reduced the cholesterol and maintained the cholesterol level in mice. It was also reported that the unsaponifiable fraction of cubeb's oil had the ability to reduce the hepatic and serum cholesterol in hypercholesterolemic mice.

7.10 Analgesic Activity

Analgesic activity of root of cubeb was evaluated by using acetic acid writhing method for Nonsteroidal anti-inflammatory drugs (**NSAIDs**)-type analgesia and rat tail flick method for opioid-type analgesia. Ibuprofen and pentazocine were used as drug controls. Water was added to powder of cubeb roots, and the suspension was administered orally to rats. The results showed that cubeb roots exhibited more powerful activity for NSAIDs-type analgesia as compared to the opioid-type analgesia (Vedhanayaki et al., 2003).

7.11 Immunomodulatory Activity

Immunomodulatory activity of alcoholic extract of cubeb fruit and its piperine component is reported. It was found that alcoholic extract showed cytotoxic activity, while the aqueous suspension of cubeb fruit powder exhibited giardicidal activity. They also provide protection against externally induced stress (Agarwal et al., 1994).

7.12 Antiinflammatory Activity

The antiinflammatory activity of cubeb was studied using plestimographic method against carrageenin-induced rat paw edema. Cubebin is the major component of cubeb and is effective against several inflammatory agents. In another report, it was found that hydroalcoholic extract of cubeb exhibited antiinflammatory activity. The results show that hydroalcoholic extract of cubeb possess antiinflammatory activity and could be used in many medicines for the treatment of several disorders that are associated with inflammation (Perazzo et al., 2013).

7.13 Anticancer Activity

Anticancer activity of *P. cubeba* was investigated against human cancer cells (Rajalekshmi et al., 2016). Cubebin contains amide and lactone groups that exhibited effective anticancer activity. It was found that the amide group showed more potency for anticancer activity compared to the lactone group (Wang et al., 2014). The analysis showed that these components act via apoptosis-mediated pathway of cell death. Cubebin could also be utilized in the synthesis of several anticancer agents (Rajalekshmi et al., 2016).

7.14 Antidepressant Activity

It has been reported that ethanol extract isolated from cubeb fruit produced piperine alkaloid, which was most effective for antidepressant activity. Piperine helps to reduce expression in cultured hippocampal neurons and releases stress. Therefore, it could be used as a therapeutic agent against depression (Li et al., 2007).

7.15 Antiulcer Activity

Antiulcer activity of cubeb was evaluated in rats against gastric ulcer. Piperine is an alkaloid present in *P. cubeba*. In mice, it stopped gastrointestinal transit, while in rats, it blocked gastric emptying of liquids or solids in time and in a dose-dependent mode. Gastric emptying inhibitory action of piperine is not dependent on pepsin and gastric acid secretion (Brinker, 2010).

7.16 Antinociceptive Activity

The antinociceptive activity of cubeb was demonstrated by writhing test in rat. Cubebin is the constituent of *P. cubeba* responsible for the antinociceptive activity. Cubebin (10 mg/kg) was administered orally to mice 30 minutes before intraperitoneal injection of prostacyclin and acetic acid. For 20 minutes, writhing number was counted. Antinociceptive activity was also reported by hot plate test in mice. The mice were placed on an aluminum plate that was adapted to a water bath. By watching the jumping movement and hind paw licking, reaction time was noted after the administration of cubebin (10 mg/kg) orally. The results showed that cubeb protects the tissue from damage and has the potential for antinociceptive activity (Wani et al., 2012).

7.17 Effects on Cardiovascular System

Dehydropipernonaline is an amide moiety in cubeb and can be used for coronary vasorelaxant activity. Methanolic extract of cubeb fruit was demonstrated for spasmolytic activity by using the rat ileum. It has been found that methanolic extract suppressed contraction in rat ileum. Moreover, ethanolic extract of cubeb is used to prevent platelet aggregation that is induced by thrombin. The results show that the constituents of cubeb were used as noncompetitive thromboxane A2 receptor antagonist to inhibit platelet aggregation (Shoji et al., 1986).

7.18 Effects on Respiratory System

Piperine is isolated from the *P. cubeba* and exhibits central stimulant action. It antagonized pentobarbitone or morphine-induced respiratory depression. Petroleum ether extract of cubeb in a small dose exerted respiratory stimulation, while it caused convulsion at a high dose due to presence of medullary stimulant factors. Cubeb's crude extract causes the ciliary movement suppression of the esophagus (Chaudhury et al., 2001).

8. SIDE EFFECTS AND TOXICITY

Cubeb seems to be safe for most people when taken by mouth, but the possible side effects are not known. Cubeb may irritate the gastrointestinal tract and kidneys (Nwaopara et al., 2008).

References

Agarwal, A., Singh, M., Gupta, N., Saxena, R., Puri, A., Verma, A., Saxena, R., Dubey, C., Saxena, K., 1994. Management of giardiasis by an immuno-modulatory herbal drug Pippali rasayana. Journal of Ethnopharmacology 44, 143–146.

Ahmad, Q.Z., Jahan, N., Ahmad, G., 2012. Nephroprotective effect of Kabab chini (*Piper cubeba*) in gentamycin-induced nephrotoxicity. Saudi Journal of Kidney Diseases and Transplantation 23, 773.

Arruda, D.C., D'Alexandri, F.L., Katzin, A.M., Uliana, S.R., 2005. Antileishmanial activity of the terpene nerolidol. Antimicrobial Agents and Chemotherapy 49, 1679–1687.

Asprey, G., Thornton, P., 1954. Medical plants of Jamaica. West Indian Medical Journal 3, 17–41.

Barrett, D.M., Somogyi, L., Ramaswamy, H.S., 2004. Processing Fruits: Science and Technology. CRC Press.

Batterbee, J., Burden, R., Crombie, L., Whiting, D., 1969. Chemistry and synthesis of the lignan (−)-cubebin. Journal of the Chemical Society C Organic (19), 2470–2477.

Brinker, F.J., 2010. Herbal Contraindications & Drug Interactions: Plus Herbal Adjuncts with Medicines. Eclectic Medical Publications.

Cai, Y., Luo, Q., Sun, M., Corke, H., 2004. Antioxidant activity and phenolic compounds of 112 traditional Chinese medicinal plants associated with anticancer. Life Sciences 74, 2157–2184.

Cardeal, Z.d.L., Gomes da Silva, M., Marriott, P., 2006. Comprehensive two-dimensional gas chromatography/mass spectrometric analysis of pepper volatiles. Rapid Communications in Mass Spectrometry 20, 2823–2836.

Chande, N., Laidlaw, M., Adams, P., Marotta, P., 2006. Yo Jyo Hen Shi Ko (YHK) improves transaminases in nonalcoholic steatohepatitis (NASH): a randomized pilot study. Digestive Diseases and Sciences 51, 1183–1189.

Chaudhury, M.R., Chandrasekaran, R., Mishra, S., 2001. Embryotoxicity and teratogenicity studies of an ayurvedic contraceptive—pippaliyadi vati. Journal of Ethnopharmacology 74, 189–193.

Chitnis, R., Abichandani, M., Nigam, P., Nahar, L., Sarker, S., 2007. Antioxidant and antibacterial activity of the extracts of *Piper cubeba* (Piperaceae). Ars Pharmaceutica 48, 343–350.

Chopra, R.N., Nayar, S.L., Chopra, I.C., 1956. Glossary of Indian Medicinal Plants. C SIR, New Delhi.

de Souza, V.A., da Silva, R., Pereira, A.C., Royo, V.d.A., Saraiva, J., Montanheiro, M., de Souza, G.H., da Silva Filho, A.A., Grando, M.D., Donate, P.M., 2005. Trypanocidal activity of (−)-cubebin derivatives against free amastigote forms of Trypanosoma cruzi. Bioorganic & Medicinal Chemistry Letters 15, 303–307.

Dhar, M., Dhar, M., Dhawan, B., Mehrotra, B., Ray, C., 1968. Screening of Indian plants for biological activity: I. Indian Journal of Experimental Biology 6, 232–247.

Duda-Chodak, A., Tarko, T., Rus, M., 2011. Antioxidant activity and total polyphenol content of selected herbal medicinal products used in Poland. Herba Polonica 1.

Eisai, P., 1995. Medicinal Herb Index in Indonesia. PT Eisai Indonesia, Jakarta, p. 91.

Gayasuddin, M., Shakil, S., Kavimani, S., 2011. Effect of ethanolic extract of *Piper cubeba* Linn. fruits on activity of pioglitazone. International Journal of Pharmacy & Industrial Research 1, 312–314.

Ghara Gheshlagh Alireza, G., Mehdi, R.R.S., Razzagh, M., Ata, K., Masoud, K., 2018. Antimicrobial effects of some herbal plants and spices on Staphylococcus epidermidis and *Pseudomonas aeruginosa*. International Journal of Food and Nutrition Science 9, 40–48.

Gorman, C., 1970. Hoabinhian: a pebble-tool complex with early plant associations in South-East Asia. Proceedings of the Prehistoric Society 35, 355–358.

Gorman, C.F., 1969. Hoabinhian: a pebble-tool complex with early plant associations in Southeast Asia. Science 163, 671–673.

Govindarajan, V., Stahl, W.H., 1977. Pepper—chemistry, technology, and quality evaluation. Critical Reviews in Food Science and Nutrition 9, 115–225.

Huang, S., Zhang, C.-P., Wang, K., Li, G., Hu, F.-L., 2014. Recent advances in the chemical composition of propolis. Molecules 19, 19610–19632.

Indonesia, P.E., 1995. Medicinal Herb Index in Indonesia. PT Eisai Indonesia, Jakarta, p. 275.

Junqueira, A.P.F., Perazzo, F.F., Souza, G.H.B., Maistro, E.L., 2007. Clastogenicity of Piper cubeba (Piperaceae) seed extract in an in vivo mammalian cell system. Genetics and Molecular Biology 30, 656–663.

Khalaf, N.A., Shakya, A.K., Al-Othman, A., El-Agbar, Z., Farah, H., 2008. Antioxidant activity of some common plants. Turkish Journal of Biology 32, 51–55.

Khare, C.P., 2004. Indian Herbal Remedies: Rational Western Therapy, Ayurvedic, and Other Traditional Usage. Springer science & business media, Botany.

Kulshreshta, V., Srivastava, R., Singh, N., Kohli, R., 1969. A study of central stimulant effect of Piper longum. Indian Journal of Pharmacology 1.

Kumoro, A.C., Hasan, M., Singh, H., 2010. Extraction of Sarawak black pepper essential oil using supercritical carbon dioxide. Arabian Journal for Science and Engineering 35, 7–16.

Lala, L., D'Mello, P., Naik, S., 2004. Pharmacokinetic and pharmacodynamic studies on interaction of "Trikatu" with diclofenac sodium. Journal of Ethnopharmacology 91, 277–280.

Lawless, J., 1995. Complete Essential Oils-A Guide to the Use of Oils in Aromatherapy and Herbalism. Element Books.

Li, S., Wang, C., Wang, M., Li, W., Matsumoto, K., Tang, Y., 2007. Antidepressant like effects of piperine in chronic mild stress treated mice and its possible mechanisms. Life Sciences 80, 1373–1381.

Li, X.-M., Brown, L., 2009. Efficacy and mechanisms of action of traditional Chinese medicines for treating asthma and allergy. The Journal of Allergy and Clinical Immunology 123, 297–306.

Lim, T., 2012. Piper Cubeba, Edible Medicinal and Non-medicinal Plants. Springer, pp. 311–321.

Magalhaes, L.G., de Souza, J.M., Wakabayashi, K.A., Laurentiz, R.d.S., Vinholis, A.H., Rezende, K.C., Simaro, G.V., Bastos, J.K., Rodrigues, V., Esperandim, V.R., 2012. In vitro efficacy of the essential oil of *Piper cubeba* L.(Piperaceae) against Schistosoma mansoni. Parasitology Research 110, 1747–1754.

Morikawa, T., Matsuda, H., Yamaguchi, I., Pongpiriyadacha, Y., Yoshikawa, M., 2004. New amides and gastroprotective constituents from the fruit of Piper chaba. Planta Medica 70, 152–159.

Murthy, C., Bhattacharya, S., 1998. Moisture dependant physical and uniaxial compression properties of black pepper. Journal of Food Engineering 37, 193–205.

Nahak, G., Sahu, R., 2011. Phytochemical evaluation and antioxidant activity of Piper cubeba and Piper nigrum. Journal of Applied Pharmaceutical Science 1, 153.

Nwaopara, A., Odike, M., Inegbenebor, U., Nwaopara, S., Ewere, G., 2008. A comparative study on the effects of excessive consumption of ginger, clove, red pepper and black pepper on the histology of the Kidney. Pakistan Journal of Nutrition 7, 287–291.

Pachpute, A.P., Deshmukh, T.A., 2013. Antioxidant and Hepatoprotective Activity of an Ethanol Extract of Piper Cubeba Fruits.

Park, B.-S., Lee, S.-E., Choi, W.-S., Jeong, C.-Y., Song, C., Cho, K.-Y., 2002. Insecticidal and acaricidal activity of pipernonaline and piperoctadecalidine derived from dried fruits of *Piper longum* L. Crop Protection 21, 249–251.

Parmar, V.S., Jain, S.C., Bisht, K.S., Jain, R., Taneja, P., Jha, A., Tyagi, O.D., Prasad, A.K., Wengel, J., Olsen, C.E., 1997. Phytochemistry of the genus piper. Phytochemistry 46, 597–673.

Perazzo, F., Rodrigues, I., Maistro, E.L., Souza, S., Nanaykkara, N., Bastos, J., Carvalho, J., de Souza, G., 2013. Anti-inflammatory and analgesic evaluation of hydroalcoholic extract and fractions from seeds of *Piper cubeba* L.(Piperaceae). Pharmacognosy Journal 5, 13–16.

Pratt, D.E., Hudson, B.J., 1990. Natural Antioxidants Not Exploited Commercially, Food Antioxidants. Springer, pp. 171–191.

Rajalekshmi, D.S., Kabeer, F.A., Madhusoodhanan, A.R., Bahulayan, A.K., Prathapan, R., Prakasan, N., Varughese, S., Nair, M.S., 2016. Anticancer activity studies of cubebin isolated from *Piper cubeba* and its synthetic derivatives. Bioorganic & Medicinal Chemistry Letters 26, 1767–1771.

Razafimandimby, H., Benard, A.-G., Andrianoelisoa, H., Leong Pock Tsy, J.-M., Touati, G., Levesque, A., Weil, M., Randrianaivo, R., Ramamonjisoa, L., Queste, J., 2017. Tsiperifery, The Wild Pepper from Madagascar, Emerging on the International Spice Market Whose Exploitation Is Unchecked: Current Knowledge and Future Prospects, vol. 72. Fruits.

Rios, J., Recio, M., 2005. Medicinal plants and antimicrobial activity. Journal of Ethnopharmacology 100, 80–84.

Semple, K.S., 1974. Pollination in piperaceae. Annals of the Missouri Botanical Garden 868–871.

Shoji, N., Umeyama, A., Saito, N., Takemoto, T., Kajiwara, A., Ohizumi, Y., 1986. Dehydropipernonaline, an amide possessing coronary vasodilating activity, isolated from *Piper iongum* L. Journal of Pharmaceutical Sciences 75, 1188–1189.
Siddiqui, Z.N., Khuwaja, G., Ahmad, J., 2007. Antifungal activity of cubebin from *Piper cubeba*. Journal of the Indian Chemical Society 84, 823–824.
Sohni, Y.R., Kaimal, P., Bhatt, R.M., 1995. The antiamoebic effect of a crude drug formulation of herbal extracts against *Entamoeba histolytica* in vitro and in vivo. Journal of Ethnopharmacology 45, 43–52.
Solomon, C., 1998. Cha Plu. Encyclopedia of Asian Food. HTML fulltext, Periplus Editions.
Sudjarmoko, B., Wahyudi, A., Ermiati, E., Hasibuan, A.M., 2015. Strategy for Developing Indonesian Pepper Export Based on Trade Performance Index and Analytic Hierarchy Process.
Tewtrakul, S., Hase, K., Kadota, S., Namba, T., Komatsu, K., Tanaka, K., 2000. Fruit oil composition of *Piper chaba* Hunt., *P. longum* L. and *P. nigrum* L. Journal of Essential Oil Research 12, 603–608.
Vedhanayaki, G., Shastri, G.V., Kuruvilla, A., 2003. Analgesic activity of *Piper longum* Linn. root. Indian Journal of Experimental Biology 41, 649–651.
Vinay, M., 2012. Market Dynamics of Pepper in Karnataka. University of Agricultural Sciences, Bangalore.
Wang, Y.H., Morris-Natschke, S.L., Yang, J., Niu, H.M., Long, C.L., Lee, K.H., 2014. Anticancer principles from medicinal Piper (胡椒 Hú Jiāo) plants. Journal of traditional and complementary medicine 4, 8–16.
Wani, T.A., Kumar, D., Prasad, R., Verma, P.K., Sardar, K.K., Tandan, S.K., Kumar, D., 2012. Analgesic activity of the ethanolic extract of Shorea robusta resin in experimental animals. Indian Journal of Pharmacology 44, 493.
Yu, E., Amri, H., 2016. China's other medical systems: recognizing Uyghur, Tibetan, and Mongolian traditional medicines. Global advances in health and medicine 5, 79–86.
Zaveri, M., Khandhar, A., Patel, S., Patel, A., 2010. Chemistry and pharmacology of Piper longum L. International Journal of Pharmaceutical Sciences Review and Research 5, 67–76.

CHAPTER

13

Cumin

Zarghouna Chaudhry[1], Rasheed Ahmad Khera[1], Muhammad Asif Hanif[1], Muhammad Adnan Ayub[2], Sajjad Hussain Sumrra[3]

[1] Department of Chemistry, University of Agriculture, Faisalabad, Pakistan;
[2] Department of Chemistry, University of Okara, Okara, Pakistan;
[3] Department of Chemistry, University of Gujrat, Pakistan

OUTLINE

https://doi.org/10.1016/B978-0-08-102659-5.00013-6

Copyright © 2020 Elsevier Ltd. All rights reserved.

1. BOTANY

1.1 Introduction

Aromatic plants of Apiaceae family have a hollow stem, and includes plants such as caraway, dill, parsley, parsnip, fennel, carrot, and a few highly toxic plants like hemlock. Apiaceae is the largest family, with about 300 genera and more than 3000 species.

Cumin (Fig. 13.1) is commonly known as "jeera" in India and called "zeera" in Pakistan and usually is used as a spice in household work.

FIGURE 13.1 Cumin.

Cumin seeds have various phytochemicals that are used as antiflatulent, carminative, and antioxidant, etc. In the cumin, the active and dynamic principles may increase the gastrointestinal tract motility, as well as enhancing the digestion power by increasing the secretions of gastrointestinal enzyme (Gohari and Saeidnia, 2011). Cumin is an essential spice, and its seeds are well used to add flavor to spicy dishes in almost all culinary preparations such as soups, cakes, breads, and cheeses (Raghavan, 2000). Cumin seeds and the essential oil components extracted from seeds are used in trade and in perfumery, food, beverage, and drug sectors of the industry (Beis et al., 2000).

1.2 History/Origin

Cumin has an extensive history that goes approximately 5000 years back to the ancient Egyptian civilization where cumin was used as a spice and a preservative in mummification. The Western world searched out cumin as a spice from Iran; cumin's name has its roots in the word *Kerman*, which is a city in Iran where it was extensively cultivated. The phrase "carrying cumin to *Kerman*" suitably articulates its relation to that city. *Kerman*, locally known as *kermun*, would have become *kumun*, and then finally called cumin in the European languages (Parthasarathy et al., 2008).

Throughout the middle ages, it was admired in the Europe, and a bride and groom carried a little quantity of cumin seeds in Germany at this time as a sign of commitment to each other. For 400 years beforehand, Portuguese and Spanish traders and colonists carried cumin seeds to the Americas and the New World after their journeys all the way through Africa. Approximately at the same time period, South East Asia and India finally got their hands on cumin seeds through the Ottoman Turks and their travels in the region. Right through history, cumin has played a vital function as a medicine and food; thus it had been a cultural symbol with varied features. Cumin is pointed out in the Bible not only as a seasoning for bread and soup but also was considered a currency used to pay tithes to the priests. It is not an actual revelation; it was used by the ancient Egyptians not only for cooking purposes but also for the ritual practices in their temples (Agrawal et al., 2001). Cumin is also distinguished for both its cosmetic and medicinal properties (Bahraminejad et al., 2012).

1.3 Demography/Location

For good quality growth and production of the cumin crop, a cool and dry climate is required, and after flowering of the crop, cloudy weather, high humidity, unseasonable rain, and higher dew are unfavorable factors. Sandy to sandy loam soil that has enough organic matter is suitable for the cultivation of the crop. This herb can also be developed under a preserved moisture environment in well-drained, medium black soil (Zohary et al., 2012). Color of leaf alters from green to purple at low temperature conditions. Early ripening of cumin can be induced at high temperature, and growth period is also reduced. Cumin is planted in Pakistan and India from October until the start of December, and the harvesting time begins in February. In Iran and Syria from November to mid-December, cumin is sown, and then it is harvested possibly in June or July (Divakara Sastry and Anandara, 2013). In India, China, Turkey, Pakistan, Morocco, Egypt, Syria, Mexico, Iran, and Chile, cumin has been widely grown (Benrejdal et al., 2012).

It is very difficult to obtain absolute figures about the production of cumin. In 2012, the estimated world production of cumin was about 300,000 tons, with 200,000 tons (about 70%) production from India. Iran, Turkey, and Syria are large producers of cumin as well, producing around 10,000, 10,000, and 20,000 tons of cumin, respectively. The production of cumin is estimated to be around 15,000–20,000 tons, which is 5%–7% of world production, in Afghanistan (Kokate et al., 2010). In Pakistan, cumin production is predicted to be about 5000 tons, which is 2% of the world production. While the largest producer of cumin is India, 90% of its total production is consumed by the country domestically, whereas Middle East producers (Iran, Turkey, Syria) export about 85%–90% of their production. The importing companies usually prefer 92%–96% purity (i.e., less than 5% stems, dust, dirt, etc.) of cumin seeds in the South Asian Association for Regional Cooperation region, while 100% purity for cumin (i.e., machine cleaned seeds) is preferred for the European market (Shivakumar et al., 2010). In 2012, the average global demand for cumin was around 90,000 tons, and its world exports are reported to about 105,737 tons. The countries that are major exporters of cumin are India 86%, Turkey 3.6%, and Egypt 2.1%, while Pakistan has a world export share of around 0.6%. For cumin seeds, the major import markets are the United States 15.8%, Egypt 13%, Brazil 6%, the United Kingdom 5.6%, and Spain 5.5%. The European Union countries imported a total quantity of 10,164 tons of cumin, and about 59% amount of this cumin arrived from India, Syria 26%, Turkey 10%, China 2.5%, and Pakistan (only 0.5%) (Chittora and Tiwari, 2013).

1.4 Botany, Morphology, Ecology

The cumin plant is harvested manually and is grown up to 30–50 cm (0.98–1.6 ft) in height. Cumin is an annual herbaceous plant, and it has a slender, branched stem that is 20–30 cm tall. It has a branched, glabrous stem that is dark green or gray in color, with 3–5 cm diameter. Cumin has alternate, compound, or simple leaves with sheathing leaf base below (Hussein and Batra, 1998). The flowers of cumin are pink in color, tiny, and naturally borne in umbels. A huge number of flowers having stalk in equal length from a common tip are spiraled, so they all bloom at the same level. Both male and female structures are present, and an incredible fruit known as cremocarp is developed from an inferior ovary. It is dry, capsuled, and at maturity habitually ruptures into two one-seeded fragments, with a grooved wall that has different longitudinal oil ducts, which provide the distinctive flavor, odor, and have incredible value toward the fruit itself. Grain-like fruits, though, are known as seeds, and actual seeds are present in them that come out through the breakdown of the fruit wall during the process of germination. Like the other members of the Apiaceae family (caraway, parsley, dill, etc.), these are ridged longitudinally and yellow-brown in color, resembling caraway seeds, and oblong in shape (Chandola et al., 1970).

For excellent production and growth, dry and cool (15–25°C) climate is required for cumin crops. Temperature below 3°C is essential for better germination. Cumin crops are very sensitive to rain because crop quality and yield is reduced during the time of harvesting due to rain. Due to diseases, the quality of a crop is negatively affected, and once the seed has turned black, it receives a lower price. Well-drained, sandy loamy soils are best for its cultivation, with a 6.8–8.3 pH range. The soils that are acidic or alkaline in nature decrease the crop yield unless the soil acidity is lowered to 7.51 pH (Baghizadeh et al., 2013).

2. CHEMISTRY

This spice is a good source of different minerals such as copper, iron, calcium, potassium, manganese, zinc, selenium, and magnesium. Moreover, it has an excellent quantity of B complex vitamins including vitamin B-6 (0.40–0.47 mg/kg), thiamin (0.60–0.68 mg/kg), riboflavin (0.30–0.35 mg/kg), and niacin (4.50–4.57 mg/kg) and some other important vitamins such as vitamin E (3.30–3.35 mg/kg) and vitamin C (7.67–7.72 mg/kg) that act as antioxidants. The seeds of this herb are a well-known source of numerous flavonoids and phenolics (Leung, 1980). Cumin seed has been acquired with different components such as cuminaldehyde (4-isopropyl-benzaldehyde), pyrazines i.e., 2-ethoxy-3-isopropyl-pyrazine, 2-methoxy-3-methyl-pyrazine, and 2-methoxy-3-sec-butylpyrazine. Additionally, cumin seed contains several phytochemicals that are well-known for their carminative, antioxidant, and antiflatulent activities (Takayanagi et al., 2003).

Cumin seeds include 5%–6% volatile oil that is composed of 60% aldehyde content. Additionally, cumin seeds give about 22% fat, various free amino acids, and have a mixture of flavonoid glycosides that are analogous to luteolin and apigenin (Bettaieb et al., 2010).

Due to its essential oil components, cumin has a strong, warm aroma and unique flavor. Cuminaldehyde content differs significantly, which is dependent on the source of oil (i.e., fresh vs. ground seeds). There is a loss of volatile oil up to 50% due to the fine grinding of cumin seeds, and within 1 hour of milling, maximum loss would have occurred. The minor constituents of cumin oil are sesquiterpenes, while hydrocarbons of monoterpenes are further major components. The chief and efficient elements of the distinctive aroma of unheated whole seeds are cumin alcohol and cumin aldehyde, in addition to other associated aldehydes (Derakhshan et al., 2010). Steam distilled essential oil of cumin seeds contains cumin aldehyde, γ-terpinene, *p*-mentha-1,3-dien-7-aL, β-pinene, *p*-mentha-1,4-dien-7-aL, *p*-cymene, in addition to camphene, limonene, myrcene, linalool, α-farnesene, α-phellandrene, α-terpinene, terpinolene, α-terpineol, β-farnesene, and safranal, respectively (Beis et al., 2000). Literature review showed that numerous bioactive compounds as well as essential secondary metabolites are present in cumin (Takayanagi et al., 2003). Toasted cumin seeds also contain some other substituted aroma compounds like cumin aldehyde (4-isopropyl-benzaldehyde) and pyrazines (2-ethoxy-3-isopropyl-pyrazine, 2-methoxy-3-methyl-pyrazine, 2-methoxy-3-sec-butylpyrazine) (Kitajima et al., 2003). The cumin leaves contain flavonoids (carotenes, zeaxanthin, and lutein), glycosides (quercetin, kaempferol), *p*-coumaric, rosmarinic, *trans*-2-dihydrocinnamic acids, and resorcinol (Dhaliwal et al., 2016). The roots of cumin plant include quercetin, and the stem contains *p*-coumaric, rosmarinic, *trans*-2-dihydrocinnamic acids, and resorcinol, and flowers have vanillic acid (Bettaieb et al., 2010). Some components present in the cumin volatile oil are shown in Fig. 13.2.

3. POSTHARVESTING TECHNOLOGY

About 4 months after planting, cumin seed is harvested when plants begin to shrink and seed color changes from dark green to brown-yellow. Cumin seed is small and boat-shaped, having nine ridges along the length. Cumin seeds are harvested by eliminating the entire plant from the ground (Tawfik and Noga, 2002). Cumin plants are dried in the partial sun or under sunlight. The seeds are beaten with the help of sticks by threshing the dried plants. Then seeds are dried further to 10% moisture level content by placing on trays or mats in the sun or with the help of a drier if too humid situation appears (Meena et al., 2013). Spices are more vulnerable to spoilage after grinding. The flavor and aroma of the compounds are not stable, and from the ground products, it will disappear quickly. Cumin seed is available as a ground powder as well as whole seeds (Morton, 2004).

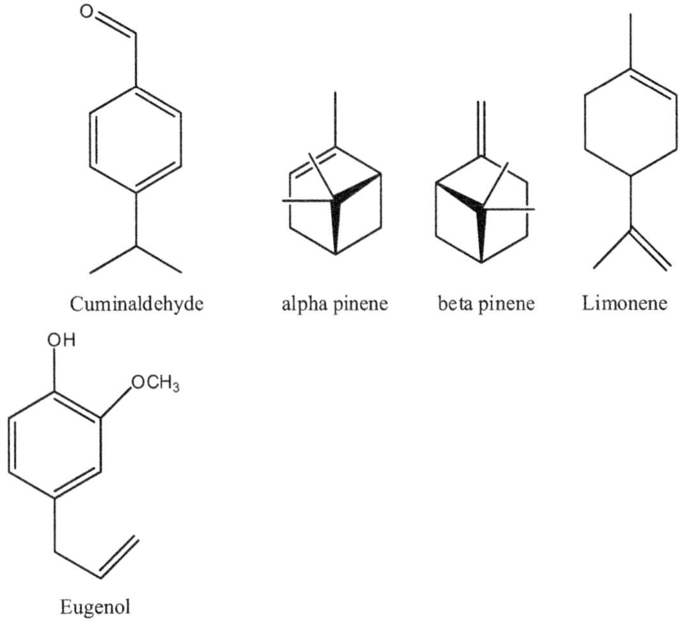

FIGURE 13.2 Components of cumin oil.

4. PROCESSING

Vacuum gravity separator, spiral gravity separator, and destoner are used on a commercial scale for cleaning cumin seeds. Dried seeds are frequently packed into sacks and also stored in a cool, dry room. Seeds should be stored at room temperature (25–28°C), where relative humidity and moisture level should not exceed than 81% and 13%–19%, respectively. Without the loss of volatile oil content, this spice can be stored in bags for 1 year. Storage in aluminum foil bags reduces quality of cumin powder. Insecticide should not be used directly on spices under any condition (Johri, 2011).

5. VALUE ADDITION

Cumin is a famous and significant ingredient in curry and chili powders. Cumin is usually found in garam masala, sofrito, adobos, as well as in achiote blends. Zeera pani is made from cumin and tamarind water, which is a pleasant, refreshing, and appetizing Indian drink (Kains, 1912). For many dishes, cumin can be used to add flavor, as it prolongs their natural sweetness. For curries, enchiladas, tacos, and some other Indian,

Cuban, Middle Eastern, and Mexican-style foods, it is used as a supplementary ingredient, and it is added to salsa to give it additional flavor. Cumin is also used on meat in addition to other familiar seasonings (Atta-Ur-Rahman et al., 2000). Also, it can used as an ingredient in some pickles and pastries (Herbst and Herbst, 2007).

6. USES

In traditional ways, cumin is mostly used for indigestion and diarrhea. Hot cumin water, made by boiling a teaspoon of roasted seeds into three cups of water, is considered an excellent remedy for fever and cold. Cumin is also tonic to the intestine, abortifacient, astringent, and carminative. It is believed that cumin is used to increase lactation and reduce nausea in pregnancy. Cumin stimulates the appetite, and its seeds are used in skin disorders, asthma, leprosy, bronchitis, cough, ulcer, etc. Cumin is still used in veterinary practices (Viuda-Martos et al., 2007).

Cumin is used as a stimulant as well as a relaxant. Some of the components of cumin essential oils are hypnotic in nature and have a tranquilizing effect (Kitajima et al., 2003). Cumin plant has some essential oils as well as caffeine, which act as a decongestant. Caffeine is a stimulating agent, and the abundantly aromatic essential oils present formulate the cumin as an anticongestive, useful for those who are suffering from respiratory disorders like asthma or bronchitis (Kumar et al., 2009).

Cumin is used for making hair shiny and glossy as well as to strengthen nails. Cumin may be effective for the treatment of carpal tunnel syndrome. Cumin makes the functions of the stomach stronger and arrests any bleeding. Being an excellent bactericide, cumin oil is very useful and effective for the treatment of cholera and diarrhea. The antiseptic property of cumin oil prevents wounds and cuts from becoming septic, and in fact, it also acts as a tonic for circulatory, excretory, and nervous systems. Oil extract of cumin is also useful for scalp treatment to get rid of dandruff, as well as for massage and aroma therapy. Ground seeds of cumin can help in preventing bleeding gums, while massaging the gums with seeds and inhaling cumin vapors help to relieve sinusitis (Willatgamuwa et al., 1998).

The seeds of cumin stimulate the production of the enzymes that help the body to break down fats, proteins, starch, and sugars and also help the liver to flush the toxins from the body. The antiseptic properties of this herb are found to help the immune system against flu and common cold. Cumin is used to prevent cough formation in the respiratory system. Cumin is a rich source of iron and has a significant amount of vitamin C (Thippeswamy and Naidu, 2005). Cumin has a good source of dietary fiber to help for the treatment of piles as well as constipation and acts as a natural laxative, which helps to heal up infections or wounds in digestive and excretory systems and seeps up digestion also (Singh et al., 2002). Cumin contains an incredible amount of calcium that is well accounted to secretions, and thymol activates the pancreatic secretions of acids, bile, and enzymes. The saliva helps in primary digestion, while thymol is responsible for the complete digestion of food in the stomach and intestine (Parashar et al., 2014).

7. PHARMACOLOGICAL USES

7.1 Antimicrobial Activity

Cumin has a fatty oil, mostly petroselinic acid and linoleic acid, that showed antimicrobial activity. The powder suspension of cumin has various inhibitory effects including mycelium growth inhibition and toxin or α-toxin production in *Aspergillus ochraceus*, *Caribena versicolor*, and *Cantharellus flavus*. Different investigations have exposed the antimicrobial activity of cumin oils as well as their solvent and aqueous extracts. The antibacterial action was evaluated against a number of useful and pathogenic gram-negative and gram-positive strains of bacteria (Iacobellis et al., 2005). The alcoholic extract and oil extracted from the seeds of cumin are found to inhibit the growth of *Klebsiella pneumoniae* and its clinical isolates. This property is attributed to the cuminaldehyde and linalool, while limonene, eugenol, pinenes, and other small components have contributed to the antimicrobial activity of cumin oil (Derakhshan et al., 2010). Cumin antifungal activity was recorded against food, soil, animal, and human pathogens including vibrio species, yeast, dermatophytes, mycotoxin, and aflatoxins producers (Hajlaoui et al., 2010).

7.2 Antidiabetic Activity

Cumin supplementations were established to be more effective than glibenclamide for diabetes mellitus treatment (Srinivasan, 2005). An active component of cumin oil was illustrated as cumin that inhibited the α-glycosidase and aldose reductase (Lee, 2005). Hyperlipidemia is an associated complication of diabetes mellitus. Oral administration of cumin reduced body weight, tissue and plasma cholesterol, free fatty acids, phospholipids, and triglycerides in alloxan diabetic rats. Cumin decreased the serum and liver cholesterol in rats when it was added to a hypercholesterolemic diet (Sambaiah and Srinivasan, 1991).

7.3 Anticancer Activity

Cumin may prevent the production of colon cancer cells and growth of breast cancer. Dietary supplementations of cumin were found to prevent the presence of rat colon cancer that was induced by a colon-specific carcinogen DMH (I,2-dimethylhydrazine). The animals that received cumin showed no colon tumors. Cumin was shown to protect the colon and to minimize the activity of β-glucuronidase and mucinase enzymes that are responsible for the liberation of toxin and also enhance the hydrolysis of protective mucins in the colon (Nalini et al., 2006).

7.4 Antioxidant Activity

In various test procedures, cumin oils as well as their solvent and aqueous extracts have shown important and significant antioxidant activities. These effects are known and documented as to their ability to significantly quench hydroxyl radicals, 2,2′-diphenyl-1-picryl hydrazyl (DPPH) radicals, and lipid peroxides. The other tests employed were ferric thiocyanate method in linoleic acid system, Fe^{2+} ascorbate-induced rat liver microsomal lipid peroxidation (LPO), soybean lipoxygenase dependent lipid peroxidation, and ferric reducing ability (Bettaieb et al., 2010). The cumin oil exhibited high antioxidant activity, which has been attributed largely to the presence of monoterpene alcohols, flavonoids, and other polyphenolic compounds (Martinez-Tome et al., 2001).

7.5 Antiosteoporotic Activity

Cumin seeds are considered estrogenic. Cumin showed significant antiosteoporotic effects because of the phytoestrogens present in it. Methanolic extract of cumin that was given to some animals showed a significant reduction in excretion and augmentation of urinary calcium and exhibited the mechanical strength in bones. Animals demonstrated better bone and ash densities and enhanced microarchitecture, with no severe effects such as body weight gain or weight of atrophic uterus (Shirke et al., 2008).

7.6 Immunomodulatory Activity

The activity of the immune system is enhanced due to a large quantity of vitamin A, vitamin C, iron, and essential oils of cumin. In some immune-suppressed animals, cumin's active compound countered the depleted T lymphocytes and decreased the size of adrenal glands and corticosterone levels, by which weight of thymus and spleen increased (Chauhan et al., 2010).

7.7 GastroIntestinal Disorders

Cumin is tremendously good for digestion problem. The aroma of cumin comes from the aromatic organic compound known as cumin aldehyde (a main component of essential oil) that stimulates salivary glands in the human mouth to smooth the progress of primary digestion of food (Milan et al., 2008). Thymol present in cumin exhibits stimulating properties that helps the secretion of enzymes and bile responsible for complete digestion of food in the stomach and intestines. Cumin is also carminative, and due to its magnesium and sodium contents as well as its essential oils, it promotes digestion (Dhandapani et al., 2002). Solvent and aqueous extracts of cumin increased the phytase, lipase, protease, and amylase activities (Vasudevan et al., 1999).

7.8 Effects on Central Nervous System (CNS)

Cumin exhibited antiepileptic acidity in garden snails. Essential oils (1%–3%) of cumin reduced epileptic activity dramatically by minimizing the firing rate of F1 neuronal cells (Sayyah et al., 2002).

7.9 Ophthalmic Effects

Cumin, especially the aqueous extract of cumin, postponed maturation and progression of cataracts including streptozotocin-induced cataracts in rats by avoiding the glycation of total soluble protein (Allahghadri et al., 2010).

8. SIDE EFFECTS AND TOXICITY

Cumin might slow blood clotting and lower blood sugar in medicinal doses.

References

Agrawal, S., Sastry, E.D., Sharma, R., 2001. Seed Spices: Production, Quality, Export. Pointer Publishers.

Allahghadri, T., Rasooli, I., Owlia, P., Nadooshan, M.J., Ghazanfari, T., Taghizadeh, M., Astaneh, S.D.A., 2010. Antimicrobial property, antioxidant capacity, and cytotoxicity of essential oil from cumin produced in Iran. Journal of Food Science 75, H54–H61.

Atta-Ur-Rahman, Choudhary, M.I., Farooq, A., Ahmed, A., Iqbal, M.Z., Demirci, B., Demirci, F., Baser, K.C., 2000. Antifungal activities and essential oil constituents of some spices from Pakistan. Journal of the Chemical Society of Pakistan 22, 60–65.

Baghizadeh, A., Karimi, M.S., Pourseyedi, S., 2013. Genetic diversity assessment of Iranian green cumin genotypes by RAPD molecular markers. International Journal of Agronomy and Plant Production 4, 472–479.

Bahraminejad, A., Mohammadi-Nejad, G., Kadir, M.A., Yusop, M.R.B., Samia, M.A., 2012. Molecular diversity of Cumin (*Cuminum cyminum* L.) using RAPD markers. Australian Journal of Crop Science 6, 194.

Beis, S., Azcan, N., Ozek, T., Kara, M., Baser, K., 2000. Production of essential oil from cumin seeds. Chemistry of Natural Compounds 36, 265–268.

Benrejdal, A., Dridi, F., Nabiev, M., 2012. Extraction and analysis of essential oil of cumin. Asian Journal of Chemistry 24, 1949.

Bettaieb, I., Bourgou, S., Wannes, W.A., Hamrouni, I., Limam, F., Marzouk, B., 2010. Essential oils, phenolics, and antioxidant activities of different parts of cumin (*Cuminum cyminum* L.). Journal of Agricultural and Food Chemistry 58, 10410–10418.

Chandola, R., Mathur, S., Anwer, M., 1970. A serious weed of cumin crop"zeeri"(Plantago pumila Willd.). Science and Culture 36, 168–169.

Chauhan, P.S., Satti, N.K., Suri, K.A., Amina, M., Bani, S., 2010. Stimulatory effects of Cuminum cyminum and flavonoid glycoside on Cyclosporine-A and restraint stress induced immune-suppression in Swiss albino mice. Chemico-Biological Interactions 185, 66–72.

Chittora, M., Tiwari, K., 2013. Biology and biotechnology of cumin. International Journal of Bioassays 2, 1066–1068.

Derakhshan, S., Sattari, M., Bigdeli, M., 2010. Effect of cumin (*Cuminum cyminum*) seed essential oil on biofilm formation and plasmid Integrity of *Klebsiella pneumoniae*. Pharmacognosy Magazine 6, 57.

Dhaliwal, H.K., Singh, R., Sidhu, J.K., Grewal, J.K., 2016. Phytopharmacological properties of *Cuminum cyminum* linn. as a potential medicinal seeds: an overview. World Journal of Pharmacy and Pharmaceutical Sciences 5, 478–489.

Dhandapani, S., Subramanian, V.R., Rajagopal, S., Namasivayam, N., 2002. Hypolipidemic effect of *Cuminum cyminum* L. on alloxan-induced diabetic rats. Pharmacological Research 46, 251–255.

Divakara Sastry, E., Anandara, J., 2013. Cumin, Fennel and Fenugreek. Soils, Plant Growth and Crop Production.

Gohari, A.R., Saeidnia, S., 2011. A review on phytochemistry of *Cuminum cyminum* seeds and its standards from field to market. Pharmacognosy Journal 3, 1–5.

Hajlaoui, H., Mighri, H., Noumi, E., Snoussi, M., Trabelsi, N., Ksouri, R., Bakhrouf, A., 2010. Chemical composition and biological activities of Tunisian *Cuminum cyminum* L. essential oil: a high effectiveness against Vibrio spp. strains. Food and Chemical Toxicology 48, 2186–2192.

Herbst, S.T., Herbst, R., 2007. New Food Lover's Companion. Barron's Educational Series, Inc.

Hussein, M., Batra, A., 1998. In vitro embryogenesis of cumin hypocotyl segments. Advances in Plant Sciences 11, 125–128.

Iacobellis, N.S., Lo Cantore, P., Capasso, F., Senatore, F., 2005. Antibacterial activity of *Cuminum cyminum* L. and *Carum carvi* L. essential oils. Journal of Agricultural and Food Chemistry 53, 57–61.

Johri, R., 2011. Cuminum cyminum and Carum carvi: an update. Pharmacognosy Reviews 5, 63.

Kains, M., 1912. In: Kains, M.G. (Ed.), Culinary Herbs; Their Cultivation, Harvesting, Curing and Uses.

Kitajima, J., Ishikawa, T., Fujimatu, E., Kondho, K., Takayanagi, T., 2003. Glycosides of 2-C-methyl-D-erythritol from the fruits of anise, coriander and cumin. Phytochemistry 62, 115–120.

Kokate, C., Purohit, A., Gokhale, S., 2010. Pharmacognosy, Vol. I & II. Nirali Prakashan, Pune. Appendix A1-A6.

Kumar, P.A., Reddy, P.Y., Srinivas, P., Reddy, G.B., 2009. Delay of diabetic cataract in rats by the antiglycating potential of cumin through modulation of α-crystallin chaperone activity. The Journal of Nutritional Biochemistry 20, 553–562.
Lee, H.-S., 2005. Cuminaldehyde: aldose reductase and α-glucosidase inhibitor derived from *Cuminum cyminum* L. seeds. Journal of Agricultural and Food Chemistry 53, 2446–2450.
Leung, A.Y., 1980. Encyclopedia of Common Natural Ingredients Used in Food, Drugs, and Cosmetics. Wiley.
Martinez-Tome, M., Jimenez, A.M., Ruggieri, S., Frega, N., Strabbioli, R., Murcia, M.A., 2001. Antioxidant properties of mediterranean spices compared with common food additives. Journal of Food Protection 64, 1412–1419.
Meena, S., Singh, B., Singh, D., Ranjan, J., Meena, R., 2013. Pre and post harvest factors effecting yield and quality of seed spices: a review. International Journal of Seed Spices 3, 1–11.
Milan, K.M., Dholakia, H., Tiku, P.K., Vishveshwaraiah, P., 2008. Enhancement of digestive enzymatic activity by cumin (*Cuminum cyminum* L.) and role of spent cumin as a bionutrient. Food Chemistry 110, 678–683.
Morton, M., 2004. Cupboard Love 2: A Dictionary of Culinary Curiosities. Insomniac Press.
Nalini, N., Manju, V., Menon, V., 2006. Effect of spices on lipid metabolism in 1, 2-dimethylhydrazine-induced rat colon carcinogenesis. Journal of Medicinal Food 9, 237–245.
Parashar, M., Jakhar, M., Malik, C., 2014. A review on biotechnology, genetic diversity in cumin (Cuminum cyminum). International Journal of Life Science and Pharma Research 4, L17–L34.
Parthasarathy, V.A., Chempakam, B., Zachariah, T.J., 2008. Chemistry of Spices. CABI.
Raghavan, S., 2000. Handbook of Spices, Seasonings and Flavorings. CRC press.
Sambaiah, K., Srinivasan, K., 1991. Effect of cumin, cinnamon, ginger, mustard and tamarind in induced hypercholesterolemic rats. Food/Nahrung 35, 47–51.
Sayyah, M., Peirovi, A., Kamalinejad, M., 2002. Anti-nociceptive effect of the fruit essential oil of *Cuminum cyminum* L. in rat. Iranian Biomedical Journal 6, 141–145.
Shirke, S.S., Jadhav, S.R., Jagtap, A.G., 2008. Methanolic extract of Cuminum cyminum inhibits ovariectomy-induced bone loss in rats. Experimental Biology and Medicine 233, 1403–1410.
Shivakumar, S., Shahapurkar, A., Kalmath, K., Shivakumar, B., 2010. Antiinflammatory activity of fruits of *Cuminum cyminum* Linn. Der Pharmacia Lettre 2, 22–24.
Singh, G., Kapoor, I., Pandey, S., Singh, U., Singh, R., 2002. Studies on essential oils: part 10; antibacterial activity of volatile oils of some spices. Phytotherapy Research 16, 680–682.
Srinivasan, K., 2005. Plant foods in the management of diabetes mellitus: spices as beneficial antidiabetic food adjuncts. International Journal of Food Sciences and Nutrition 56, 399–414.
Takayanagi, T., Ishikawa, T., Kitajima, J., 2003. Sesquiterpene lactone glucosides and alkyl glycosides from the fruit of cumin. Phytochemistry 63, 479–484.
Tawfik, A.A., Noga, G., 2002. Cumin regeneration from seedling derived embryogenic callus in response to amended kinetin. Plant Cell, Tissue and Organ Culture 69, 35–40.
Thippeswamy, N., Naidu, K.A., 2005. Antioxidant potency of cumin varieties—cumin, black cumin and bitter cumin—on antioxidant systems. European Food Research and Technology 220, 472–476.
Vasudevan, K., Vembar, S., Veeraraghavan, K., Haranath, P., 1999. Influence of intragastric perfusion of aqueous spice extracts on acid secretion in anesthetized albino rats. Indian Journal of Gastroenterology: Official Journal of the Indian Society of Gastroenterology 19, 53–56.

Viuda-Martos, M., Ruíz-Navajas, Y., Fernández-López, J., Pérez-Álvarez, J.A., 2007. Chemical composition of the essential oils obtained from some spices widely used in Mediterranean region. Acta Chimica Slovenica 54, 921.

Willatgamuwa, S., Platel, K., Saraswathi, G., Srinivasan, K., 1998. Antidiabetic influence of dietary cumin seeds (Cuminum cyminum) in streptozotocin induced diabetic rats. Nutrition Research 18, 131–142.

Zohary, D., Hopf, M., Weiss, E., 2012. Domestication of Plants in the Old World: The Origin and Spread of Domesticated Plants in Southwest Asia, Europe, and the Mediterranean Basin. Oxford University Press on Demand.

CHAPTER

14

Curry Leaf

Saima Batool[1], *Rasheed Ahmad Khera*[1], *Muhammad Asif Hanif*[1], *Muhammad Adnan Ayub*[2], *Shahabuddin Memon*[3]

[1] Department of Chemistry, University of Agriculture, Faisalabad, Pakistan; [2] Department of Chemistry, University of Okara, Okara, Pakistan; [3] National Centre of Excellence in Analytical Chemistry, University of Sindh, Jamshoro, Pakistan

OUTLINE

https://doi.org/10.1016/B978-0-08-102659-5.00014-8

Copyright © 2020 Elsevier Ltd. All rights reserved.

1. BOTANY

1.1 Introduction

This plant can grow at an altitude up to 1500 m and is abundantly found throughout Pakistan and India (Steinhaus, 2015). Moreover, it is native to Indonesia, Southern China, and tropical parts of the Asian subcontinent from Sri Lanka eastward to Myanmar and Hainan (China). Great variability among the constituent species is largely attributed due to uncertainty in the number of species within the genus. Variability is more common in growth habitat, morphology, chemical composition, stems, leaves, and flower color. Curry has a broad range of cultivars and varieties, varying in scents, flavors, and uses. Only two species of genus *Murraya* are found in India, namely *M. koenigii* (L.) Spreng and *M. Paniculata (L.)*. *M. koenigii* (L.) Spreng is more popular, having a large spectrum of therapeutic properties and also due to use of its leaves as a natural flavoring manager in a variety of food dishes and curries.

M. koenigii is recognized by different names. In Urdu it is called as kari patta. In English, it is commonly known as curry leaf. In Hindi, it is known as meetha neem. In Bangali, it is known as barsunga. In Gujarati, it is called as limdo or meetholimdo. In Kannada, it is known as karibue. In Marathi, it is specifically called kadhilimbu. In Telugu, it is called karivepku. In French, it is known as feuilles de cari, in German, curryblatter, in the Indonesian language, daun kari, in Italian, fogli de cari, and in Spanish, it is known as hoja.

1.2 History/Origin

The curry leaf plant is residential to Pakistan, India, Bangladesh, Sri Lanka, and other South Asian countries. Dating back to the 1st to 4th centuries AD, the use of curry leaf in vegetables as a flavoring agent is described in Tamil literature. Curry leaves are related with South India, as the term "curry" originates from the "kari" used for spices in Tamil, a place in India. Kari patta is an alternative name for curry leaf used throughout Pakistan. Nowadays, curry leaves are cultivated as a food flavoring agent in Australia, Pakistan, India, Africa, Sri Lanka, Southeast Asia, and the Pacific Islands. Except for the higher regions of the Himalayas, curry leaf plants are native in wastelands and forests all over the Indian subcontinent. From Pakistan's Ravi River, its range extends eastward from Assam in India and southward to Tamil Nadu in India and Chittagong in Bangladesh. This plant extended to South Africa, Malaysia, and Reunion Island with South Asian immigrants.

1.3 Demography/Location

Curry leaf growns in a variety of environmental and climatic conditions; the best conditions are full sun, slightly acidic and well-drained soil, and high temperatures (can sustain temperature above 40°C). Optimum temperature is 27°C with rising temperature of 27–47°C for germination. If curry leaf is container-grown, it requires an area with brilliant light and heat in the winter. It grows healthy in soil having a pH range from 5 to 7 and an optimum pH of 6.5. It is advisable to grow curry leaf in areas with no freezes. This plant is cultivated throughout India and distributed from Sikkim to Garhwal, Assam, Bengal, Western Ghats, and Travancore-Cochin, in other parts of the Asian region in moist forests of 500–1600 m height in Guangdong, Bhutan, Laos, Pakistan, Sri Lanka, Nepal, Malaysia, South Africa, Reunion Island, Thailand, and Vietnam. It is rarely found outside the Indian sphere of influence (Rana et al., 2004). India is one of the largest curry leaf producers, users, and exporters (Bhardwaj et al., 2011; Datta).

1.4 Botany, Morphology, Ecology

M. koenigii is a deciduous and perennial bush or small plant, getting up to 5–6 m in height. The plant case is 15–40 cm in diameter, which is short, smooth, and has a dense shady crown and brown and grayish bark (Mhaskar et al., 2000). The color of the chief stem is shady green to brownish, and the leaves are 15–30 cm long, having 11 to 25 leaflets rotate on rachis, 2.5–3.5 cm long ovate, lanceolate, oblique base and bipinnate. The leaf boundaries are unevenly serrate and 2–3 mm long petiole. Terminal cymes are in inflorescence, having 60 to 90 flowers, and each flower is white, complete, bisexual, stalked, ebracteate, sweetly scented funnel shaped, and regular, where a completely opened flower is 1.12 cm in average diameter. The calyx is pubescent and five clefts with deep lobes and five petals that are whitish, having glabrous glands. Fruits are 1.4–1.6 cm long, their diameter is 1–1.2 cm, and they are in close clusters, small, glandular, ovoid, and subglobose (Raghunathan and Mitra, 1982). Seeds are 10–11 mm long, diameter is 7–8 mm, enclosed in a thin pericarp, and their color is spinach green (Bonde et al., 2011).

2. CHEMISTRY

The strong distinguishing aroma of curry leaf is because of terpene hydrocarbons β-caryophyllene, β-phellandrene, β-gurjunene, and β-elemene [14]. Terpene constituents citral, linalyl acetate, menthone, menthol, and carvomenthone also contribute to its flavor, in addition to β-cryophyllene (Chowdhury et al., 2000). Curry leaves are also rich in calcium, carotene, nicotinic acid, vitamins A and C, fibers, and minerals. Higher antioxidant activity is shown by two carbazole alkaloids, named mahanimbine and koenigine, present in leaves. Plant root has murrayagetin, murrayanol, and girinimbine (Chakraborty et al., 1973). Mahanimbine, girinimbine, koenimbine, isomahanine, and mahanine were isolated from seeds (Reisch et al., 1994). Fruits of curry leaf plants have mahanimbine and koenimbine (Handral et al., 2012; Srivastava and Srivastava, 1993). The structures of some active compounds found in curry leaf are shown in Fig. 14.1.

3. POSTHARVEST TECHNOLOGY

Curry leaf is an ongoing plant, and there is no requirement to uproot it per annum. Harvest the leaves as the plant grows. Topping or pruning is also harvesting of leaves. If the plant is allowed to flower, then fruit/leaf quality will decrease. Harvesting of leaves should be done earlier than flowering. Similarly, leaves should be harvested before winter when they start shedding. For oil extraction, harvested leaves are graded, bundled, and sent to processing units. From the middle of April, flowering starts and often ends in the middle of May. The peak flowering season often observed is the last week of April, and fruiting season from the middle of July to August is observed. Leaf yield from 250 to 375 kg/ha is possible at the end of 1 year after transplanting. During the second and third years after planting the leaf yield is 1500–2000 kg/ha at 4-month intervals. The yield increases progressively to 2500 kg/ha at 3-month intervals for the fourth year and 3000–3500 kg/ha at 2.5-month intervals from the fifth year onward. Even after drying, leaves maintain their flavor. Therefore, curry leaves are sold in fresh and desiccated states. Fresh leaves are taken, washed, and drained in a colander and then dried under a fan by spreading on a towel. After this, leaves are removed from their stems and

FIGURE 14.1 Curry plant at various stages.

placed in a dry box covered with a paper at its bottom and top. Another method to store curry leaves is to make powder and keep in an airtight container. To make dry powder, stove drying at 50°C is best. High temperature may destroy powder quality. Polythene bags are also used to store, but darkness appears in leaves.

4. PROCESSING

Curry, like other herbal plants, is consumed in a variety of ways and for various purposes. In addition to its fresh leaves, other common processed forms of curry include whole dry leaves, frozen or powdered leaves, and extracted essential oils. Whole plant or chopped leaves can be frozen to be used for an extended time beyond their fresh shelf life. For cooking, fresh curry leaves are good. They can be used directly after harvesting, and they can be packed in plastic bags after harvesting and then refrigerated to keep them fresh up to 2 months. Curry leaves can be oven dried or air dried, and leaves produced as a result have longer shelf lif,e Desiccated leaves are accessible commercially. Pulverized curry trees are also available commercially, and if dried leaves are pulverized, then powder of actual curry leaves is obtained, which is used in various spices for cooking. Drying may be air drying, sun drying, shade drying, and tray drying. Curry leaves are also dried using a microwave, cross-flow tray, hot air oven, freeze drying, infrared, inert gas, and hot air frozen. Shade-dried leaves have a greater quantity of essential oil processed than microwave-dried leaves. Dehydration of curry leaves in a cabinet drier at 50°C for 3.5 hours was found to be more suitable than drying in shade, sun, and solar drying. The optimum pretreatment to retain the green color can be effectively achieved by dipping curry leaves in a cold solution containing magnesium oxide, potassium meta-bisulphite, and sodium bicarbonate.

(Vasudevan et al., 1997). Moreover, climatic conditions and geographic variations may also affect curry leaves. Fresh curry leaves yield 2.5% oil on steam distillation, which acts as a fixative in the soap industry.

5. VALUE ADDITION

Curry leaves can be combined with a variety of other herbs including garlic, pepper, thyme, red onion, chili, turmeric, mustard, sage, rosemary, parsley, ginger, and cloves and can be used in soups, stuffing, stews, and rice, as well as with fish, vegetables, chicken, and meats. Curry leaves and seeds can be used in many oils and cheeses. Famous curry leaf sambol (karapincha sambol) can be made by using curry leaves with fresh coconut, warm water or milk, green chili, red onion, cloves, garlic, ginger, ground black pepper, salt, and juice of lemon. Its oil can also be used in many soap and cosmetic industries (Rao et al., 2011).

6. USES

Curry has many uses ranging from culinary to religious. There are number of beliefs having relation with the historical use of curry leaf. Curry leaves are said to aid with psychic powers, particularly protection, healing, purification, strength, divination, purification, wishes, exorcism, inspiration, wisdom, meditation, and defense, etc. Many herbs and spices contribute significantly to health, despite the low amounts of consumption, as they are full of antioxidants and certain mineral compounds. Curry leaves are mostly used as a vegetable and as a spice in soups and meat dishes and many other medicinal foodstuffs (Matsuda et al., 2009). Many components are used for cosmetics and as depigmenting agents from its medicinal flowers. Curry leaf has tonic, stomachic, and carminative properties and has traditional use in Ayurvedic and Unani medicine. Curry leaves act as a tonic, antioxidant, antidiabetic, anticancer, antitumor, anticonvulsant, antiacne, anthelmintic, antipiles, analgesic, antiinflammatory, antihypertensive, and in the handling of bronchial respiratory diseases (Shah and Juvekar, 2006). Moreover, it is also effective against many diseases caused by fungi, viruses, bacteria, and many other microbes. The plant has specificity for many fevers, cough, flu, influenza, cold, lowering blood cholesterol level, diarrhea, bronchitis, and chicken pox.

Fresh juice of curry leaf root is used to relieve kidney pain. Root and mostly bark are used in the treatment of body eruptions and toxic bites (Yun-Cheung et al., 1986). Plant bark, leaves, and roots can be used as a stimulant and for stomach ache. Branches are used to clean the teeth as datun to gums and teeth stronger.

7. PHARMACOLOGICAL USES

7.1 Antimicrobial Activity

The essential oil from curry leaves has shown antibacterial effect against *Corynebacterium pyogenes, Bacillus subtilis, Proteus vulgaris, Staph aurous*, and *Pasteurella multocida*. The pure oil was found to be more active against these microorganisms. On fractionation, acetone extract of the leaves gives three bioactive carbazole alkaloids, mahanimbine, murrayanol, and mahanine, which showed mosquitocidal and antimicrobial activities (Narasimhan et al., 1975). The essential oil from tree of *M. koenigii* expresses antifungal activity against *Aspergillus niger, Microsporum gypseum, Candida tropicalis*, and *Candida albicans*. It was effective against *C. albicans* even after dilution (Vaijayanthimala et al., 2000). Ethanolic extract of leaves exhibited fungitoxicity against *Colletotrichum falcatum* and *Rhizoctonia solani* (Kishore et al., 1982). The ethanolic extract of the roots also displayed antifungal activity in a previous study (Kumar et al., 2012).

7.2 Hypoglycemic Activity

Curry leaf extract shows protective effect against cell damage and as an antioxidant defense system of pancreas and plasma by decreasing voxidative stress and pancreatic cell damage (Arulselvan and Subramanian, 2007). Curry leaves are considered a powerful antidiabetic diet

(Srinivasan, 2005). A curry leaf extract diet showed antihyperglycemic activity (drop in blood glucose level) against moderate diabetes induced in rats by alloxan (Yadav et al., 2004). Curry leaf extract was found to reduce blood glucose as well as blood cholesterol levels in diabetic mice (Ajay et al., 2011). In another study, ethanolic extract of curry leaves considerably decreased the levels of blood glucose, urea, creatinine, uric acid, and glycosylated hemoglobin in streptozotocin diabetic rats after 30 days (Arulselvan et al., 2006).

7.3 Antiprotozoal Activity

Ethanolic extracts of curry leaves showed good antiprotozoal activity against *Entamoeba histolytica,* antihypertensive action in cats/dogs, and antispasmodic action on guinea pig ileum (Bhakuni et al., 1969).

7.4 Hypolipidemic Activity

Albino rats fed with curry leaves for 90 days experienced a decrease of complete serum cholesterol low-density lipoprotein, an increase in the level of high-density lipoproteins, an increase in the lecithin cholesterol acyl transferase activity, and lower release of lipoprotein in the circulation (Adebajo et al., 2006).

7.5 Antilipid Peroxidative Activity

Curry leaves showed powerful antilipid peroxidative activity in a previous study (Khan et al., 1996).

7.6 Antihypertensive Activity

Curry leaves (*Murraya koenigii*) chutney supplementation showed promising antihypertensive activity in hypertensive subjects (Gaikwad et al., 2013).

7.7 Hepatoprotective and Antiulcer Activity

The carbazole isolated from the aqueous extract of curry leaves showed hepatoprotective activity (Mans et al., 2017; Sathaye et al., 2011). The dry bark powder of curry extracted by acetone showed important use in the defense of liver cells against ulcer (Pande et al., 2009). The extract has considerable antiulcer activity and reduced gastric volume and ulcerative lesion (Patidar, 2011).

7.8 Antiinflammatory Activity

Ethanolic extract of curry exhibited antihistaminic action and antiinflammatory activity (Gupta et al., 2011). The ethanolic extract of this plant has considerable antiinflammatory action on yeast-induced hyperpyrexia (Prasad et al., 2011).

7.9 Immunomodulatory Activity

Curry leaf extract exhibited good immunomodulation in experimental animals through antioxidant and immunosuppressant mechanisms that could be crucial in treatment of ethanolic liver injury wherein immune stimulation or autoimmunity is involved in its pathogenesis (Sathaye et al., 2011).

7.10 Anthelmintic Activity

Anthelmintic activity against *Pheretima posthuma* was observed by using ethanolic and aqueous extracts from curry leaves. Both extracts at concentration of 100 mg/mL have prominent anthelmintic activity (Kalola, 2007; Sharma et al., 2010).

7.11 Wound Healing Activity

Curry leaves possess significant wound healing activity (Gupta et al., 2009). Curry leaves speed up the wound healing mechanism by lessening the surface area of the wound (Patidar et al., 2010).

7.12 Anticancer Activity

The extract of curry showed in vitro anticancer activity. A major decrease in the cancer cell number and tumor weight was noted in tumor-bearing mice (Muthumani et al., 2009). Similarly, in another study, methanol extract of curry leaves collected from different regions of India showed good anticancer activity against breast cancer cell lines (Ghasemzadeh et al., 2014). MTT assay performed using methanol extract of curry leaf shown significant anticancer activity in a previous study (Nagappan et al., 2011).

8. SIDE EFFECTS AND TOXICITY

Curry leaves should be avoided if one is allergic to them, and a trained medical practitioner should be consulted before usage by pregnant or breastfeeding women and toddlers. Curry leaves can upset the stomach when used in large amounts.

References

Adebajo, A., Ayoola, O., Iwalewa, E., Akindahunsi, A., Omisore, N., Adewunmi, C., Adenowo, T., 2006. Anti-trichomonal, biochemical and toxicological activities of methanolic extract and some carbazole alkaloids isolated from the leaves of *Murraya koenigii* growing in Nigeria. Phytomedicine 13, 246–254.

Ajay, S., Rahul, S., Sumit, G., Paras, M., Mishra, A., Gaurav, A., 2011. Comprehensive review: *Murraya koenigii* Linn. Asian Journal of Pharmacy & Life Science. ISSN 2231, 4423.

Arulselvan, P., Senthilkumar, G., Sathish Kumar, D., Subramanian, S., 2006. Anti-diabetic effect of *Murraya koenigii* leaves on streptozotocin induced diabetic rats. Die Pharmazie-An International Journal of Pharmaceutical Sciences 61, 874–877.

Arulselvan, P., Subramanian, S.P., 2007. Beneficial effects of *Murraya koenigii* leaves on antioxidant defense system and ultra structural changes of pancreatic β-cells in experimental diabetes in rats. Chemico-Biological Interactions 165, 155–164.

Bhakuni, D.S., Dhar, M., Dhar, M., Dhawan, B., Mehrotra, B., 1969. Screening of Indian plants for biological activity: Part II. Indian Journal of Experimental Biology.

Bhardwaj, R.K., Sikka, B., Singh, A., Sharma, M., Singh, N., Arya, R., 2011. Challenges and constraints of marketing and export of Indian spices in India. In: Proc. International Conference on Technology and Business Management, pp. 28–30.

Bonde, S., Nemade, L., Patel, M., Patel, A., 2011. *Murraya koenigii* (Curry leaf): ethnobotany, phytochemistry and pharmacology—a review. International Journal of Pharmaceutical & Phytopharmacological Research 1, 23–27.

Chakraborty, D., Ganguly, S., Maji, P., Mitra, A., Das, K., Weinstein, B., 1973. Chemical taxonomy: XXXII, Murrayazolinine, a carbazole alkaloid from *Murraya koenigii*. Chemistry & Industry 7, 322–333.

Chowdhury, A., Kumar, S., Kukreja, A., Dwivedi, S., Singh, A., 2000. Essential oil from the leaves of *Murraya koenigii* (Linn.) spreng. Journal of Medicinal and Aromatic Plant Sciences 643–645. Central Institute of Medicinal and Aromatic Plants.

Datta, S., Prospects of Value Added Products and it's Future in Indian Market.

Gahlawat, D.K., Jakhar, S., Dahiya, P., 2014. *Murraya koenigii* (L.) Spreng: an ethnobotanical, phytochemical and pharmacological review. Journal of Pharmacognosy and Phytochemistry 3, 109–119.

Gaikwad, P., Khan, T.N., Nalwade, V., 2013 Apr. Impact of curry leaves (*Murraya koenigii*) chutney supplementation on hypertensive subjects. International Journal of Food Sciences and Nutrition 2, 68–72.

Ghasemzadeh, A., Jaafar, H.Z., Rahmat, A., Devarajan, T., 2014. Evaluation of bioactive compounds, pharmaceutical quality, and anticancer activity of curry leaf (*Murraya koenigii* L.). Evidence-based Complementary and Alternative Medicine 2014.

Gupta, P., Nahata, A., Dixit, V.K., 2011. An update on *Murraya koenigii* spreng: a multifunctional Ayurvedic herb. Zhong Xi Yi Jie He Xue Bao 9, 824–833.

Gupta, S., George, M., Singhal, M., Garg, V., 2009. Wound healing activity of methanolic extract of *Murraya koenigii* leaves. Pharmacologyonline 3, 915–923.

Handral, H.K., Pandith, A., Shruthi, S., 2012. A review on *Murraya koenigii*: multipotential medicinal plant. Asian Journal of Pharmaceutical and Clinical Research 5, 5–14.

Jain, V., Momin, M., Laddha, K., 2012. International Journal of Ayurvedic and Herbal Medicine 2 (4), 607–627, 2012.

Kalola, J., 2007. Pharmacognostical Phytochemical and Pharmacological Investigations on Inula Cappa.

Khan, B.A., Abraham, A., Leelamma, S., 1996. Role of *Murraya koenigii* (curry leaf) and *Brassica juncea* (Mustard) in lipid peroxidation. Indian Journal of Physiology and Pharmacology 40, 155–158.

Kishore, N., Dubey, N., Tripathi, R., Singh, S., 1982. Fungitoxic Activity of Leaves of Some Higher-Plants. National Academy Science Letters-India 5, 9–10.

Kumar, V., Suthar, S., Bandyopadhyay, A., Tekale, S., Dhawan, S., 2012. A Review on Traditional Indian Folk Medicinal Herb: *Murraya koenigii*. World Journal of Pharmacy and Pharmaceutical Sciences.

Mans, D., Grant, A., Pinas, N., 2017. Plant-based ethnopharmacological remedies for hypertension in Suriname—How efficacious are they? Herbal Medicine. IntechOpen.

Matsuda, H., Nakashima, S., Oda, Y., Nakamura, S., Yoshikawa, M., 2009. Melanogenesis inhibitors from the rhizomes of Alpinia officinarum in B16 melanoma cells. Bioorganic & Medicinal Chemistry 17, 6048–6053.

Mhaskar, K., Blatter, E., Caius, J., 2000. Kirtikar and Basu's Illustrated Indian Medicinal Plants: Their Usage in Ayurveda and Unani Medicines. Sri Satguru Publications.

Muthumani, P., Venkatraman, S., Ramseshu, K., Meera, R., Devi, P., Kameswari, B., Eswarapriya, B., 2009. Pharmacological studies of anticancer, anti inflammatory activities of *Murraya koenigii* (Linn) Spreng in experimental animals. Journal of Pharmaceutical Sciences and Research 17, 18.

Nagappan, T., Ramasamy, P., Wahid, M.E.A., Segaran, T.C., Vairappan, C.S., 2011. Biological activity of carbazole alkaloids and essential oil of *Murraya koenigii* against antibiotic resistant microbes and cancer cell lines. Molecules 16, 9651–9664.

Narasimhan, N., Paradkar, M., Chitguppi, V., Kelkar, S., 1975. Alkaloids of Murraya koenigii: structures of mahanimbine, koenimbine,(-)-mahanine, koenine, koenigine, koenidine & (+)-isomahanimbine. Indian Journal of Chemistry.

Pande, M., Gupta, S., Pathak, A., 2009. Hepatoprotective activity of *Murraya koenigii* Linn bark. Journal of Herbal Medicine and Toxicology 3, 69–71.

Patidar, D.K., 2011. Anti-ulcer activity of aqueous extract of *Murraya koenigii* in albino rats. International Journal of Pharma Bio Sciences 2, 524–529.

Patidar, D.K., Yadav, N., Nakra, V., Sharma, P., Bagherwal, A., 2010. Wound healing activity of Murraya koenigii leaf extract. International Journal of Comprehensive Pharmacy 4, 1–2.

Prasad, G., Dua, V., MATHUR, A., 2011. ANTI-Inflammatory activity of leaves extracts of *Murraya Koenigii* L. International Journal of Pharma Bio Sciences 2, 541–544.

Raghunathan, K., Mitra, R., 1982. Pharmacognosy of Indigenous Drugs, vol. 1. Central Council for Research in Ayurveda and Siddha, New Delhi, p. 433.

Rana, V., Juyal, J., Blazquez, M.A., 2004. Chemical constituents of the volatile oil of *Murraya koenigii* leaves. International Journal of Aromatherapy 14, 23–25.

Rao, B.R., Rajput, D., Mallavarapu, G., 2011. Chemotype categorization of curry leaf plants *Murraya koenigii* (L.) spreng. Journal of Essential Oil Bearing Plants 14, 1–10.

Reisch, J., Adebajo, A.C., Aladesanmi, A.J., Adesina, K.S., Bergenthal, D., Meve, U., 1994. Chemotypes of *Murraya koenigii* growing in Sri Lanka. Planta Medica 60, 295–296.

Sathaye, S., Amin, P., Mehta, V., Zala, V., Kulkarni, R., Kaur, H., Redkar, R., 2011. Immunomodulatory activity of aqueous extract of *Murraya Koenigii*, L in experimental animals. International Journal of Toxicological and Pharmacological Research 3 (4), 7–12.

Sathaye, S., Bagul, Y., Gupta, S., Kaur, H., Redkar, R., 2011. Hepatoprotective effects of aqueous leaf extract and crude isolates of *Murraya koenigii* against in vitro ethanol-induced hepatotoxicity model. Experimental & Toxicologic Pathology 63, 587–591.

Shah, K.J., Juvekar, A.R., 2006. Positive inotropic effect of *Murraya koenigii* (Linn.) Spreng extract on an isolated perfused frog heart. Indian Journal of Experimental Biology.

Sharma, U.S., Sharma, U., Singh, A., Sutar, N., Singh, P., 2010. In vitro anthelmintic activity of *Murraya koenigii* Linn. leaves extracts. International Journal of Pharma Bio Sciences 1.

Srinivasan, K., 2005. Plant foods in the management of diabetes mellitus: spices as beneficial antidiabetic food adjuncts. International Journal of Food Sciences and Nutrition 56, 399–414.

Srivastava, S., Srivastava, S., 1993. New Constituents and biological-activity of the roots of murraya-koenigii. In: Indian Chemical Soc 92 Acharya Prafulla Chandra Rd Attn: Dr Indrajit Kar/exec Sec, Calcutta 700009, India, pp. 655–659.

Steinhaus, M., 2015. Characterization of the major odor-active compounds in the leaves of the curry tree *Bergera koenigii* L. by aroma extract dilution analysis. Journal of Agricultural and Food Chemistry 63, 4060–4067.

Vaijayanthimala, J., Anandi, C., Udhaya, V., Pugalendi, K., 2000. Anticandidal activity of certain South Indian medicinal plants. Phytotherapy Research 14, 207–209.

Vasudevan, P., Tandon, M., Pathak, N., Nuennerich, P., Mueller, F., Mele, A., Lentz, H., 1997. Fluid CO_2 extraction and hydrodistillation of certain biocidal essential oils and their constituents. Journal of Scientific and Industrial Research 56, 662–672.

Yadav, S., Vats, V., Ammini, A., Grover, J., 2004. *Brassica juncea* (Rai) significantly prevented the development of insulin resistance in rats fed fructose-enriched diet. Journal of Ethnopharmacology 93, 113–116.

Yun-Cheung, K., Kam-Hung, N., Pui-Hay, B.P., Qian, L., Si-Xao, Y., Hong-Ta, Z., Kin-Fai, C., Doel, S.D., Woei-Song, K., Waterman, P.G., 1986. Sources of the anti-implantation alkaloid yuehchukene in the genus Murraya. Journal of Ethnopharmacology 15, 195–200.

CHAPTER

15

Cypress

Asma Shaheen, Muhammad Asif Hanif, Rafia Rehman, Asma Hanif

Department of Chemistry, University of Agriculture, Faisalabad, Pakistan

OUTLINE

Medicinal Plants of South Asia
https://doi.org/10.1016/B978-0-08-102659-5.00015-X

Copyright © 2020 Elsevier Ltd. All rights reserved.

1. BOTANY

1.1 Introduction

Trees of cypress (Fig. 15.1) are generally found native to the warm, temperate climate and are specifically located in Mediterranean regions, North America, and Asia. Most cypress species have similarities with various other genera such as *cupressocyparis*, *taxodium*, and *chamaecyparis*. These larger evergreen trees have tremendous variety with reference to color, size, and shape of plant. Furthermore, species of this plant are highly drought resistant. *Cupressus sempervirens* is an important specie of cypress that is native to North Africa, the Middle East, the Greek islands, and Turkey. This plant produces high-quality, pest-resistant, completely durable wood (Sahni, 1999). The wood of this plant can bear harsh and extreme weather conditions. However, medicinal and ornamental uses of this plant are of supreme importance, and soil conservation is at the top. *Cypress lawsoniana* is a widely cultivated plant of Europe, while *Cypress macrocarpa* and *Cypress arizonica* are grown in the United States, mainly for landscaping, wood, and ornamental values. Some cypress species are also found in Asia such as *Cypress cashmeriana*, which is the cypress of Kashmir and native to the Himalayas, thus grown for temples, avenues, and gardens for ornamental purposes. Cypress is the best source of timber, as *Cypress lusitanica*, also known as Mexican cypress, is prone to fire and is

FIGURE 15.1 Cypress tree.

grown all over the world, including the United States, Mexico, India, Nepal, and Pakistan. *Cypress torulosa* is also known as Bhutan cypress and is also a well-known source of high-quality timber. Formosan cypress or *Cypress formosensis* is abundantly found on the mountains of Taiwan and was introduced as an important specie in China to provide high-quality, lightweight timber for construction purposes. *Cypress japanica* is extensively grown in different regions of Japan. Cypresses are extremely beautiful trees made popular for landscaping. They can grow remarkably tall, slender, almost pencil shaped. Their heights can reach 80 ft but seldom spread more than 8 ft. The leaves are needle-like and dark green all year. Leaves and branches of cypresses are not in one plane. Cones develop in March through April.

Cypress common names are pencil pine, saru (Urdu, Marathi), churam (Tamil), churam, sooram (Malayalam), jeedakara, jeekaka, jeelakarra (Telugu), and surahba (Oriya).

1.2 History

Moving back to the fascinating history of the cypress tree, ancient Egypt had been using this high-strength durable wood to build mummy cases. Similarly, Greeks were big fans of this tree, as they use its wood to make burns to store the ashes of those who died in battles. The bark of this plant produces appreciable quantities of highly valuable essential oil. This tree was known to the local inhabitants since ancient times thus known to be everlasting tree. Some recent investigations on this tree have sparked research for its tremendous ability to resist relentless forest fire. This tree is known to have extreme religious significance in many different cultures all around the globe. This tree is used to make incense in Tibetan culture. Nevertheless, some people also use its cones to extract the essential oil for various useful purposes.

1.3 Demography and Location

According to an estimate, more than two dozen types of cypress trees are found in different regions all over the world. The "Leyland " is a relatively fast-growing specimen that can reach a height of about 50 feet. These evergreen plants are flat-branched, having soft, piney needles that can tolerate severe soil conditions. Such types of plants are most commonly found in coastal regions of the world and can survive for centuries. *C. torulosa* is also known as "Bhutan cypress" or "Himalayan cypress" is the specie of cypress most commonly found in South Asia.

1.4 Botany, Morphology, Ecology

Different types of cypress have different shapes. C. *torulosa* is a large tree, growing up to 45 m (150 ft) in height. Similarly, Monterey cypress can grow up to the height of about 70 feet with appreciable width and flat canopy that resembles an open umbrella. Furthermore, Arizona cypress is found similar to conventional Christmas trees that can be as long as 60 feet. The foliage of cypress is known to have different shades of green, starting from dark green to lighter bluish green in coloration depending upon the type of plants. The leaves of this plant range from fine needles to scaly textured, overlapping hairs that look like a braid attached to twigs (Ostfeld and Keesing, 2000). These plant species yield smaller cones, some of which are quite similar to nuts, and other woody cones that have the width of about 2 inches. Each tiny cone bears approximately 30 seeds. Cypress includes both deciduous and evergreen trees, whose branches differ from plant to plant, while the Leyland cypress sports the flat branches. Similarly, pond cypress has a number of spiny offshoots. All the cypress varieties have one thing in common: they prove to be heaven for all types of wildlife. Birds are especially fond of cypress trees due to their stronger branches and needle-like appearance, which makes it an excellent material for building of nests. Cypresses need full sun exposure. They also need moisture at the early stage of growth. Cramped spaces are avoided during cypress planting. All weeds are removed that grow near cypress trees. Weeds hold up needed nutrients from the roots of the trees and block sunlight and inhibit the growth of the young cypresses. When planted in rows, cypress trees can grow to be a formidable border able to withstand high winds and other inclement weather.

2. CHEMISTRY

Major components in essential oil of this plant are found to be carene and α-pinene for *Cypress atlantica*, α-cadinol and limonene for *Cypress cluclotrxiana*, sabinene and α-pinene for C. *cashmeriana*, sesquiterpene, terpinene, and α-pinene for *Cypress guadalupensis*, and α-cadinol and α-pinene for *Cypress macnabiana*. However, 15 major components were found to be the essential part of all cypress species, among which 11 are terpenoids (sabinene, α-pinene, β-pinene, mycene, bomyl acetate, carene, α-terpineol, p-cymene, terpinene, pociniene and terpinolene), one oxygenated sesquiterpene (α-cadinol), and two are diterpenoids (manoyl oxide and sandaracopimaradiene). Some previous investigations have revealed that fresh plant material of cypress is composed of 1.31% tannic agents, 1.67% free acid, 2.11% water-soluble minerals, 4.9% water-soluble polysaccharides, 2.07% reducing sugar, and 0.6% essential oil, while the essential oil of fresh plant material contains 5% sabines, 8% fenchone, and 2% α-pinene as a major monoterpene.

Some recent researches have revealed that essential oil of *Cypress chengiana* contains the highest contents of monoterpenes, while *Cypress duclouxiana* has the lowest concentration of the same class of compound. Sesquiterpenes are found to be abundant in *C. macnabiana*, where it contributes 47% of all the components. The maximum concentration of diterpene of about 16% is found in *C. guadalupensis*, while the major monoterpenes of *C. atlantica* are found to be carene (23.5%) and α-pinene (46.3%). Similarly, major components of *C. chengiana*, *C. cashmeriana*, and *C. ficnebris* are sabinene (32.1%, 17.4%, and 18.6%) and α–pinene (17.6%, 19.5% and 28.5%). Alcohols and ketones were found to be in maximum concentration in *C. duclouxiana* (5.9% and 2.9%), *C. chengiana* (11.3% and 4.5%), and *C. guadalupensis* (7.1% and 5.3%). Nevertheless, sesquiterpenes had a total contribution of almost less than 10% in actual composition of essential oil. Furthermore, essential oil of *C. duclouxiana* (32%) and *C. macnabiana* (47%) can significantly be differentiated by their sesquiterpenoid contents. Nonetheless, the major one in *C. duclouxiana* and *C. macnabiana* was α-cadinol with an approximate contribution of 8.2% and 24.1%, respectively. Various diterpenoids exhibited kaurane, abietane, isopimarane, and pimarane in actual skeleton that contributes less than 2% in whole composition of essential oil. The highest proportion of diterpenoids of about 15.9% was found in *C. guadalupensis*, whose major component was 8-P-hydroxysandaracopimaren, whereas *Cypress duclounrana* is the richest source of diterpenoids and sesquiterpenes (Pierre-Leandri et al., 2003).

Scented oil of various cypress species can be obtained from steam distillation of needles, stems, and younger twigs of plant. Cypress is a plant of deciduous and coniferous regions that is known to have a large number of needle-like structures. Its essential oil mainly consists of linalool, α-pinene, terpinolene, β-pinene, myrcene, α-terpinene, sabinene, bornyl acetate, cadinene, carene, cedrol, and camphene that are of high medicinal importance. Essential oil of this plant can be used to avoid a number of illnesses and devastating conditions (Smiroldo, 2006).

Some other monoterpenes like camphene, myrcene, and carvotanacetone, can also be present in its essential oil. Some recent investigations have revealed that these components possess excellent bioactive potentials. Similarly, high molecular weight polysaccharides or glycoproteins are also found in extracts of this plant, which is known to have

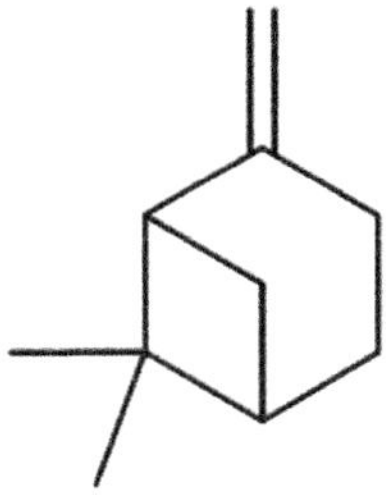

Beta-pinene

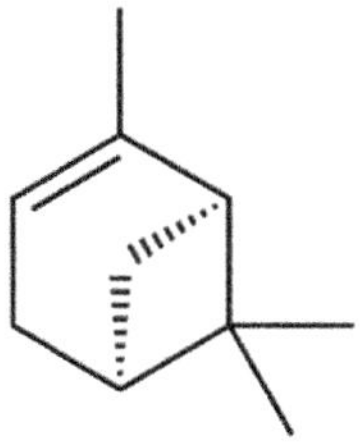

Alpha-pinene

FIGURE 15.2 Important aroma components of cypress.

astonishing antibiotic capacities. Some bioactive components of cypress are shown in Fig. 15.2.

3. POSTHARVEST TECHNOLOGY

After harvest, cypress can regenerate naturally by sprouting from the stump and by seed. Cypress leaves (needles) are shade dried to avoid losses of volatile components. Seeds are harvested when they attain complete maturity.

4. PROCESSING

The essential oil of cypress is steam distilled from the leaves (needles) and twigs, which are obtained by pruning the trees in autumn. However, it is an evergreen tree and can be used for processing of essential oil at any time of year. Good quality essential oil is obtained during cooler months of year.

5. VALUE ADDITION

Cypress essential oil also known as cupressus oil goes well with other woody oils. It is also blended with essential oils of lavender, citrus oils like bergamot, lime, lemon, orange, and grapefruit, bergamot, clary sage, frankincense, juniper, marjoram, pine, rosemary, and sandalwood. Essential oil is mostly produced by steam distillation and solvent extraction methods (Lardos et al., 2011).

6. USES

Cypress plant is known to have number of traditional uses, as this plant was utilized to make the cross of Jesus, and this tree is still supposed to be associated with death in some regions of Europe. The word "Cypress" is derived from a Greek word "Sempervirens" which is a botanical name meaning "lives forever" and this tree also gave its name to the "island of cypress" where it used to be worshiped by considering a holy plant. The world's most valued wood is produced by the cypress plant, as it is lighter in weight along with high durability that makes it an ideal building material for a number of constructional purposes. In addition to the extraordinary properties of the wood, it is also very popular because it does not generate sap nor undergo bleeding. Due to this unique feature, the wood of this tree takes well to sealers, paints, and stains. Wood of cypress is also very famous for oil and firewood. Wood of this plant burns clean, dries quickly, and splits very easily. Fatty and essential oil of this plant is abundantly used in hair shampoo and various beauty products.

Cypress is commonly used in manufacturing of tables, boats, sidings, bridges, barns, firewood, bed frames, roofing shingles, cabinets, porches, greenhouses, and boxes.

When cypress oil is topically applied, it reduces inflammation. It reduces the risk of septic in wounds by preventing bacterial and other infections. It also prevents spasms (Rosenberg et al., 2004). It possesses enough potential to shrink swollen blood vessels, as a result of which it helps to recover from edema and hemorrhoids. It is known to have significant pain-relieving effect. However, it is supposed to be a mild antioxidant when compared with other essential oils. It potentially reduces the chances of congestion in the lymphatic system, which has immense health benefits. It reduces the density of phlegm and even helps to remove mucus when used in combination with an expectorant. It strongly kills lice that infest humans. It also improves memory function (Joshi and Jain,

2014). Cypress essential oil is extracted from cypress trees belong to Cupressaceae family. These trees are rich in essential oil. Cypress essential oil is a stimulant for the circulatory system. It is used to heal several circulation-associated conditions like hemorrhoids and varicose veins. In aromatherapy, it is used to spiritually ground a person and impart a sense of emotional security. Essential oil of cypress has numerous health benefits, as it significantly contributes to a number of pharmaceutical products due to its antiseptic, astringent, diuretic, antispasmodic, hemostatic, sudorific, deodorant, styptic, and hepatic potentials and sedative properties, along with the capabilities to treat the severe respiratory disorders and conditions of vasoconstriction.

7. PHARMACOLOGICAL USES

7.1 Astringent Activity

Essential oil of this plant helps with stiff muscles and in strengthening the gums. However, contraction is the major function associated with astringency. Similarly, cypress essential oil helps in the contraction of hair follicles, muscles, skin, and gums, thereby preventing hair and teeth from falling out, along with tightening of loose muscles and skin.

7.2 Antiseptic Activity

Scented oil of cypress makes it an excellent choice for treating internal wounds and external injuries. It is also found to be the most common ingredient of medicinal creams, pharmaceutical formulations, and antiseptic lotions, and these kinds of properties are mainly attributed to the presence of camphene in appreciable concentration in essential oil of this plant (Cordell, 1995).

7.3 Diuretic Activity

Essential oil of cypress is known to have diuretic effects, as it enhances the quantity and frequency of urination, which appears to be highly beneficial and very important for the maintenance of good health. It forces the human kidney to excrete more urine, thereby properly detoxifying the entire body and whole blood circulatory system. When the volume of urine reaches up to 4%, it eliminates significant quantities of fats outside the human body. Therefore, more frequent urination helps in subsequent weight loss due to significant fat loss. Moreover, urination enhances proper digestion and prevents gas formation in the intestine, along with removal of excessive water and reducing swelling. One of the most pronounced effects of excessive urination is removal of hazardous toxicants. It cleans the kidneys and reduces blood pressure. It is interesting to note that a majority of pharmaceutical products or medicinal drugs for lowering blood pressure have high diuretic potential that enhances the process of urination (Richardson and Rejmánek, 2004).

7.4 Hemostatic and Styptic Properties

Styptic properties and hemostatic potentials are quite similar in meaning but differ slightly. The condition "hemostatic" indicates any chemical or biologic agent that can hinder the normal flow of blood and promote blood clotting, while the word "styptic" means caustic, which includes many other properties as astringent effects and vasoconstriction to avoid the excessive flow of blood through blood vessels. Both characteristics are highly valuable in terms of their areas of applications to a bleeding person that immediately needs the activation of hemostatic agents to promote the blood clotting to save individuals. So being styptic and astringent, essential oil of cypress promotes contraction of blood vessels, skin, muscles, hair follicles, and gums by stimulating the flow of blood through blood vessels. Both these properties are of high significance, as they possess enough potential to save life (Kuiate et al., 2006).

7.5 Respiratory Tonic Activity

Essential oil of cypress tones up the entire respiratory system of the human body and enhances the efficiency of lungs. It helps in reducing the accumulation of phlegm in lungs and the whole respiratory tract, thereby minimizing the harmful consequences of cough. It also clears blockages by making breathing easier when suffering from cold and cough (Cordell, 1995).

7.6 Sudorific Activity

Essential oil of cypress is considered to be a highly potential sudorific substance that can cause perspiration. Regular and periodic sweating filters the entire body and helps in quick removal of excessive moisture, salt residues, and harmful toxicants from body. Sudorific properties help in clearance and tightening of skin pores along with opening of sebaceous and sweat glands while keeping away a number of skin diseases including acne (Semwal et al., 2010).

7.7 Deodorant Activity

Essential oil of various cypress species possesses a masculine and fiery fragrance that can easily replace a number of synthetic deodorants having similar distinctive natural aroma (Semwal et al., 2010).

7.8 Hepatic Activity

Essential oil of cypress is found to be more effective for the liver as it ensures optimal health by regulating proper discharge of bile. It also protects the liver from contraction and a number of deleterious liver diseases (Samant et al., 1998).

7.9 Sedative Activity

This plant is known to have strong sedative effects on brain and body through proper relaxation by relieving anxiety and lowering stress. It stimulates feelings of happiness in case of sadness and at the time of anger. This property can prove to be helpful for pacifying mentally disturbed people that are suffering from major setbacks, trauma, and serious shocks in life (Samant and Dhar, 1997).

7.10 Vasoconstrictor Activity

Cypress essential oil is an effective vasoconstrictor, and it helps to narrow dilated blood vessels including varicose veins or broken capillaries.

7.11 Antibacterial Activity

Alcoholic extracts of cypress twigs are known for their strong antibacterial potentials against a number of gram-positive and gram-negative bacteria. Some recent investigations have revealed that cypress extracts possess strong potentials against *Pseudomonas aeruginosa*, *Staphylococcus aureus*, *Yersinia aldovae*, *Shigella flexenari*, *Citrobacter*, and *Escherichia coli*, which shows that cypress has a large number of bioactive compounds. The antifungal activities of these extracts have shown strong resistance against various fungal strains such as *Candida albicans*, *Saccharomyces cereviciae*, *Fusarium solani*, *Aspergillus parasiticus*, *Macrophomina*, and

Trichophyton rubrum. These alcoholic extracts showed significant results against various fungal strains (Shan et al., 2007).

7.12 Anticancer Activity

In vitro studies proved that cypress has anticancer potential. The crude ethanolic extract of this plant was used as homeopathic mother tincture to effectively treat a number of ailments, specifically tumors and moles. It was also used in numerous other traditional systems of medicines. Apoptosis and antiproliferative effects including in vitro studies showed more apoptotic, antiproliferative, and cytotoxic effects that caused a significant reduction in cell's viability, thereby inducing internucleosomal DNA fragmentation. It potentially collapses mitochondrial transmembrane and results in an increase in generation of Reactive oxygen species (ROS) and release of cytochrome *c* (Fernando and Rupasinghe, 2013).

7.13 Anti-HIV Activity

Some plants of *Cupressaceae* family have the ability to inhibit human immunodeficiency virus (HIV). Cell death was found dependent on final concentration (Naser et al., 2005).

7.14 Antispasmodic Activity

The antispasmodic activity of twigs of cypress was found and evaluated to have pronounced effects on isolated tissues (Rosenberg et al., 2004).

7.15 Antioxidant Activity

The activity of lipid peroxidation was carried out to evaluate its antioxidant potential on rats. Furthermore, activity of ethanolic fractions increases as the concentration of bioactive ingredients increases. The results obtained from this experimental investigation showed that this plant is a potential source of natural antioxidant compounds having excellent antioxidant activities. Aqueous and alcoholic extracts of this plant have also shown strong antioxidant and antiinflammatory potentials (Joshi and Jain, 2014).

7.16 Antidiabetic Activity

More recently, an experimental investigation was planned to evaluate the antidiabetic potentials of ethanolic extracts of various portions of cypress tree. Their antidiabetic potentials were assessed by intentionally inducing diabetes via serum biochemical analysis in alloxan, blood glutathione levels, and fasting blood sugar tests. These extracts significantly resisted diabatic activity even at the lower dose level of about 200 mg/kg. These also showed significant increase in glutathione level in blood due to strong antioxidant activity (Ahmed and Saeed, 2013).

7.17 Hepatoprotective Activity

Ethanolic extracts of various fractions of cypress are known to have excellent hepatoprotective potentials, mainly in rats, which were the center of focus in this experimental investigation. Various histopathological examinations showed that cypress plant is known to have excellent hepatoprotective effects. However, ethanolic extracts were more effective in comparison with other solvents used for extraction of bioactive compounds from cypress (Leather, 1996).

7.18 Insecticidal Activity

Insecticidal potentials of essential oil of cypress were studied against a number of adults and larvae of sycamore lace bug, *Corythucha ciliata*. This experimental investigation was performed at standard conditions and room temperature in a laboratory. All these activities were performed at three different concentrations in the study. Sycamore lace bug's larvae were preferably susceptible to tested products in comparison with adult organisms. Lower mortality rate was observed just 1 day after treatment (41.7%), while the highest mortality rate was observed 3 days after treatment (71.3%). In this study, both agents showed moderately satisfying activity in controlling the adults and larvae, but they also have obvious repellent activity that leads to better efficiency (Leather, 1996).

7.19 Neuropharmacological Activity

Aqueous extract of aerial parts was investigated for evaluation of neuropharmacological activity by using elevated plus-maze test (Tumen et al., 2012).

8. SIDE EFFECTS AND TOXICITY

Safety of cypress is not known. It causes kidney irritation, allergic reactions in people with sensitive skins, and problems in pregnant and breastfeeding mothers.

References

Ahmed, M., Saeed, F., 2013. Evaluation of insecticidal and antioxidant activity of selected medicinal plants. Journal of Pharmacognosy and Phytochemistry 2, 153–158.

Cordell, G.A., 1995. Changing strategies in natural products chemistry. Phytochemistry 40, 1585–1612.

Fernando, W., Rupasinghe, H.V., 2013. Anticancer properties of phytochemicals present in medicinal plants of North America. In: Kulka, M. (Ed.), Using Old Solutions to New Problems—Natural Drug Discovery in the 21st Century. InTech, Croatia.

Joshi, S.C., Jain, P.K., 2014. A Review on Hypolipidaemic and Antioxidant Potential of Some Medicinal Plants.

Kuiate, J., Bessière, J., Vilarem, G., Zollo, P., 2006. Chemical composition and antidermatophytic properties of the essential oils from leaves, flowers and fruits of *Cupressus lusitanica* Mill. from Cameroon. Flavour and Fragrance Journal 21, 693–697.

Lardos, A., Prieto-Garcia, J., Heinrich, M., 2011. Resins and gums in historical iatrosophia texts from Cyprus–a botanical and medico-pharmacological approach. Frontiers in Pharmacology 2.

Leather, S.R., 1996. Resistance to foliage-feeding insects in conifers: implications for pest management. Integrated Pest Management Reviews 1, 163–180.

Naser, B., Bodinet, C., Tegtmeier, M., Lindequist, U., 2005. Thuja occidentalis (Arbor vitae): a review of its pharmaceutical, pharmacological and clinical properties. Evidence-based Complementary and Alternative Medicine 2, 69–78.

Ostfeld, R.S., Keesing, F., 2000. Biodiversity and disease risk: the case of Lyme disease. Conservation Biology 14, 722–728.

Pierre-Leandri, C., Fernandez, X., Lizzani-Cuvelier, L., Loiseau, A.-M., Fellous, R., Garnero, J., oli, C.A., 2003. Chemical composition of cypress essential oils: volatile constituents of leaf oils from seven cultivated Cupressus species. Journal of Essential Oil Research 15, 242–247.

Richardson, D.M., Rejmánek, M., 2004. Conifers as invasive aliens: a global survey and predictive framework. Diversity and Distributions 10, 321–331.

Rosenberg, S.A., Yang, J.C., Restifo, N.P., 2004. Cancer immunotherapy: moving beyond current vaccines. Nature Medicine 10, 909–915.

Sahni, K., 1999. The Book of Indian Trees. Oxford University Press.

Samant, S., Dhar, U., 1997. Diversity, endemism and economic potential of wild edible plants of Indian Himalaya. The International Journal of Sustainable Development and World Ecology 4, 179–191.

Samant, S., Dhar, U., Palni, L., 1998. Medicinal Plants of Indian Himalaya: Diversity Distribution Potential Value. Gyanodaya Prakashan, Nainital, Uttaranchal, India, pp. 155–158.

Semwal, D., Saradhi, P.P., Kala, C., Sajwan, B., 2010. Medicinal plants used by local Vaidyas in Ukhimath block, Uttarakhand. Indian Journal of Traditional Knowledge 9, 480–485.

Shan, B., Cai, Y.-Z., Brooks, J.D., Corke, H., 2007. The in vitro antibacterial activity of dietary spice and medicinal herb extracts. International Journal of Food Microbiology 117, 112–119.

Smiroldo, T.L., 2006. Treatment review of chronic psoriasis. Australian Journal of Medical Herbalism 18, 103.

Tumen, I, Sezer Senol, F, Orhan, IE, 2012. Evaluation of possible in vitro neurobiological effects of two varieties of Cupressus sempervirens (mediterranean cypress) through their antioxidant and enzyme inhibition actions. Turk J Biochem 37, 5–13.

CHAPTER

16

Datura

Muhammad Saboor Nayyar[1], Muhammad Asif Hanif[1], Muhammad Irfan Mjaeed[1], Muhammad Adnan Ayub[2], Rafia Rehman[1]

[1] Department of Chemistry, University of Agriculture, Faisalabad, Pakistan;
[2] Department of Chemistry, University of Okara, Okara, Pakistan

OUTLINE

Medicinal Plants of South Asia
https://doi.org/10.1016/B978-0-08-102659-5.00016-1

Copyright © 2020 Elsevier Ltd. All rights reserved.

1. BOTANY

1.1 Introduction

Datura (Fig. 16.1) is a yearly weed of gardens, roadsides, and other waste places or developed land. It is broadly naturalized in warmer countries throughout the world, and it is very common in Pakistan, frequently showing up in open, aggravated spots, roadsides, pastures, livestock enclosures, agronomic and vegetable yield fields, plantations, vineyards, jettison banks, and bothered, unmanaged ranges. It has a place with Solanaceae, a family that incorporates the potato and tobacco, and numerous individuals from this family contain lethal substances. The plant's exact and normal appropriation is by all accounts all through the greater part of mild and hot districts of the world, inferable from its generic development and naturalization. There is a group of nine incredible species (*Datura stramonium, Datura ferox, Datura quercifolia, Datura pruinosa, Datura leichahhardtii, Datura inoxia, Datura discolor, Datura metel, Datura wrightii*) in the Datura genus, but the two famous species are *D. inoxia* and *D. strammonium* (Buchholz et al., 1935; Palazón et al., 2006). Both have been utilized as a part of a shamanic setting for spiritual drives on most landmasses since earlier documented history all through the old Americas, Europe, and South Asia.

FIGURE 16.1 Datura plant and seeds.

Datura is known by different names depending where you are in the world. In English, it is typically called datura; thorn apple; devils trumpet; Jamestown weed; mad-apple, and stink wort. In Spanish, it is called as belladona del pobre; cajon del diablo; chamico grande; chamisco; datura manzana; estramonio; manzana Espinosa, and peo de fraille. In French, it is belladone; conchombre diable; concombre a chein; datura stramonie; herbe des taupes; pomme epineuse; stramonie commune. In Arabic it is called datoora, tatoora, nafeer, and tuffah shoki (Bhatia et al., 2014; Sharawy and Alshammari, 2009). In Portuguese, it is estramanonio; figueira do inferno; figueire do inferno, and quinquilho. In Bhutan, it is dhaturo and nyangmo-throkchang. In Brazil, it is bem casado; estrasmónio; mamoninha brava; mata zombando; sia branca; trombeteira, and zabumba. In Cuba, it is campana and chamico. In Germany, it is known as stechapfel. In Indonesia, it is kecubung lutik and kecubung wulung. In Italy, it is indormia and stramonio comune. In Japan, it is called, shirobanachosenasagao (Konda and Shimizu, 2002). In Lebanon, it is daturah and nafir. In Netherlands, it is doornappel. In Norway, it is piggeple. In Poland, it is bielun dziedzierzawa. In South Africa, it is bloustinkolie; doringapple; gewone; iloqi; lechoe; lethsowe; makolieboom; makstinkblaar; makstinkolie; malpitte; olieblaar; olieblaarneut; olieneut; pietjielaporte; steekappel; stinkblaar; umhlavuthwa, and zaba-zaba. In Sweden, it is spikklubba. In Thailand, it is lampong. In Zimbabwe, it is chowa. Datura plant, called in Morocco "Chdek ejjmel," is used traditionally in medicine. In Nepali, its name is dhaturo. Madak and Rdardura are the names of datura in the Tibetan language (Gaire, 2008). In Hindko and Punjabi, it is called datura. In Pakistan, the Urdu name for this herb is dhutura (Shah and Khan, 2006).

1.2 History/Origin

For a long time, Datura's origin was thought to be from China, but the latest studies reveal that it is native to India (Gaire, 2008). Datura is widely naturalized in warmer countries throughout the world (Palazón et al., 2006; Shahid and Rao, 2014; Ullah et al., 2013; Hussain et al., 2011). The name datura originates from the early Sanskrit *dustura* or *dahatura* (Gaire, 2008). The genus name *Datura* is derived from dhatura, the Bengali name for the plant (Joy et al., 1998).

1.3 Demography/Location

Datura can be discovered effortlessly in all tropical nations (Vernay et al., 2008), and the principle fascination of this plant is the wonderful blossom in the form of a trumpet that reaches from white to pink shading. Datura is a yearly weed of nurseries, roadsides, and other waste places or developed land areas. It is usually grown in sunny circumstances (Palazón et al., 2006). Datura requires full sun and rich, fruitful earth that channels well. It spreads seeds outside into a readied bed in fall in hotter atmospheres and in early spring after all threat of ice has gone in cooler atmospheres. It thrives in most tolerably soil; however, it also develops best in calcareous rich soil or in decent sandy topsoil with pH of 6.5–8. Datura flourishes effortlessly under various climatic conditions. It has been found from deserts to cool mild timberlands. It is a heat-adoring plant and develops well over 25°C. Datura can endure temperatures down to −18°C (Hussain et al., 2011; Sharma et al., 2014). The entire plant has a disagreeable, foul, opiate scent, which decreases after drying (Sharma et al., 2014).

1.4 Botany, Morphology, Ecology

Datura is exclusively an herb; however, a 1-year-old plant develops a thick stem. The weed is yearly and can develop to statures of 1.5 m. On rich soil, it might achieve a height of even 6 feet. The plant is smooth, aside from a slight softness on the more youthful parts, which are secured with short, bended hairs, which tumble off as development continues. It exudes a rank, overwhelming, and to some degree, disgusting opiate smell. This foul smell emerges from the leaves, particularly when they are wounded; however the blooms are sweet-scented, which create a condition of trance if inhaled for a longer period of time (Priya et al., 2002). The root is long, thick, and whitish, emitting numerous strands. Root framework is composed of a taproot that is shallow for the span of the plant. The stem is green or purple and to a great extent is bald; albeit, youthful stems regularly have obvious hairs that are barrel shaped. The stem is heavy, erect, and verdant, smooth, a pale yellowish green in shading, fanning more than once in a forked way, and creating in the forks of the branches a leaf and a solitary erect blossom (Gaire, 2008). The plant blooms throughout late spring. The blooms are extensive and attractive, developing independently on tiny shoots springing from the axils of the leaves or at the forking of the branches. It has five sepals that are gamosepalous, tubular, five-toothed, sepaloid, bushy, persevering, valvate aestivation, and second rate. The calyx is long, tubular, and to some degree, swollen underneath, and strongly five-calculated, surmounted by five sharp teeth. The petals are five, gamopetalous, melded at the base, infundibuliform, surface shaggy, white, curved aestivation, and mediocre. The corolla, collapsed, and just half-opened is pipe molded of an unadulterated white with six noticeable ribs that reach out into a similar number of sharp-pointed sections. The blossoms open at night for the fascination of night-flying moths and radiate an intense aroma. The channel-shaped corolla of every bloom is up to 5″ long and 2″ across when completely open; its external edge has five shallow projections. Each of these projections shapes an intense point in the center. The corolla is white or pale violet all through, except at the throat of the blossom, where thick veins of dim violet occur. Stamens are five, polyandrous, alternipetalous, epipetalous, fiber long and smooth, anthers dithecous, basifixed, introse, and second rate. Carpels are two (bicarpellary), syncarpous, ovary back at a slant set to one side and front to one side, predominant, bilocular with numerous ovules on swollen placentae, axile placentation, style long, bilobed, and capitate. These containers are loaded with various cocoa to dark, kidney-shaped seeds of roughly 3 mm long. This herbaceous yearly weed is an erect plant, with dull green leaves, white blooms, and spiked containers that are loaded with various dark, kidney-shaped seeds (Jagatheeswari, 2014; Alfarhan et al., 2005; Ullah and Ullah, 2014). Whenever ready, this seed-vessel opens at the top, tossing back four valve-like structures, leaving a long, focal structure whereupon are various unpleasant, dull cocoa seeds (Hussain et al., 2011; Joy et al., 1998; Benvenuti et al., 1994; Reddy, 2008).

2. CHEMISTRY

The substance examinations of *Datura* species exhibited that leaves and seeds particularly were rich in alkaloids, including atropine, hyoscyamine, and scopolamine. These mixes are incorporated into numerous official medicines because of their anticholinergic properties. Datura also contains hyoscine, apohyoscine, and meteloidine. Poisonous stimulants are found in many parts of a datura plant and can bring about dazed conditions in animals and people. The alkaloids inside datura attack aggressively at muscarinic acetylcholine receptors (Joy et al., 1998; Abbasipour et al., 2011). Two active alkaloids of datura are shown in Fig. 16.2.

Datura contains tropane alkaloids together with hyoscyamine, hyoscine, littorine, acetoxytropine, valtropine, fastusine, and fastusinine (Monira and Munan, 2012). The leaves of the plant contain alkaloids. The developed plants contain higher levels of alkaloids. Seeds yield diploid I and tetraploid II other than alkaloids (Monira and Munan, 2012; Bo et al., 2003; Bellila et al., 2011).

3. POSTHARVEST TECHNOLOGY

At the point when the blossom of the datura plant blurs, a seed case begins to shape. The seed units are about the size of a walnut and can contain more than 100 seeds. The datura seed case turns from green to

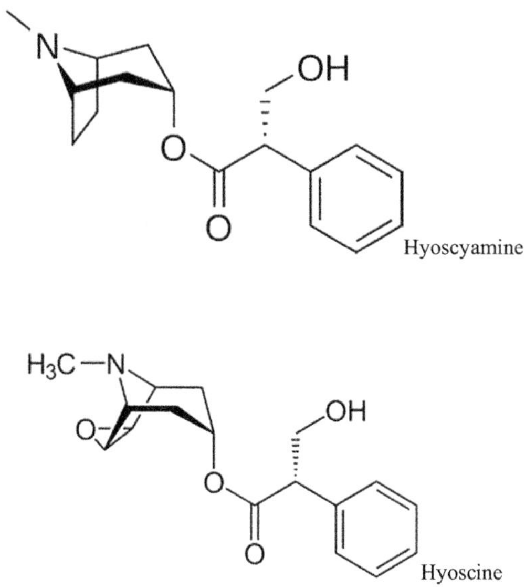

FIGURE 16.2 Two active alkaloids of datura.

chestnut as it ages and develops. The outside becomes distinctly fragile. Once the cases are ready, a split creates. This is the time when the unit is sought to be gathered from the plant. On the off chance that the unit is left on the vine, it will part into three or four areas and distribute the seeds in a wide region around the parent plant. It is hard to assemble the seeds once the case has opened. The aged seeds must be dried before putting away. Seeds are adjusted in case on three sides with an indent on the fourth, like the shape of a kidney (Shahid and Rao, 2014; Batanouny et al., 1999).

4. PROCESSING

Datura seed oil is extracted using cold press and solvent extraction. Datura oil has several industrial applications.

5. VALUE ADDITION

Datura seed oil is used in several formulations made to stop/regrow hairs and to overcome dandruff.

6. USES

The Chinese utilize the blooms of datura in homegrown arrangements. Middle Easterners in Africa used to smoke the dried leaves, blossoms, and seeds in hookahs as a solution for asthma and influenza. Today, individuals smoke the dried leaves and seeds due to their opiate impacts and to mitigate asthma. It is said that the inward breath of the smoke from such a blend unwinds the muscles and kills pain, in addition to calming irritation brought about by ailment and different illnesses. Datura is a metal-tolerant plant and is being utilized to consider Zn^{2+} accumulation. In datura plant, phytochelatins have been studied in the accumulation of Cd^{2+}. Datura has capacity for high biomass generation.

Types of datura have been utilized in different societies as source of atropine, hyoscamine, and hyoscine alkaloids for their pharmacological effects in human and veterinary medicines. Datura plants have been utilized for spiritual and religious purposes and as characteristic medication to treat asthma. Psychoactive impacts of datura attract the youngsters to it. Plants are devoured or smoked to accomplish psychedelic encounters. Oil produced from datura seeds is utilized to regrow hair, for treating gloom, and in India, people utilize it as an offering for master Shiva (Priya et al., 2002).

In Nepal, this plant is also viewed as holy to Shiva. Thistle apple blossoms and natural products are among the most critical offering endowments of the Newari tribe of Nepal. In Europe the plant was utilized for witchcraft and in balms or treatments. All through most European nations the seeds were utilized to blend drinks. In Mexico, different tribes have (i.e., Opata, Seri) utilized Toloache in religious customs. Datura was thought to cure those with deafness, to mitigate sleep, for deprived people, and to release the warmth of those with a fever (Kelly et al., 2002; Kagale et al., 2004; Hossain et al., 2014). In the Philippines, the Igorot, a Malayan community from Luzon, boil the green parts of the plant to make an intoxicating soup that is eaten mutually in a custom circle. In China, the white-bloomed variety of datura is viewed as holy, as it is trusted that flickering dew drops poured down from the sky onto its blossoms while Buddha was giving a sermon. The dried leaves and blossoms of datura might be smoked alone or with different herbs in a smoking mix to get relief from asthma. In Tunisia, datura ethereal parts are utilized as a part of people's medication for their antiasthmatic, antispasmodic, and antiparkinsonian properties. Datura is a standout among the most researched model plant species containing tropane alkaloids. These bioactive mixes are essential pharmaceuticals utilized as a part of an advanced drug as sleep inducing and spasmolytic cures. Datura has been utilized broadly in medication. It has been utilized as an analgesic for setting bones, treating wounds, skin ulcers, hemorrhoids, asthma, ailment, whooping cough, muscle fitness, sciatica, and menstrual cycle (Biswas et al., 2011; Ullah et al., 2014).

7. PHARMACOLOGICAL USES

7.1 Antimicrobial Activity

Datura has prominent antimicrobial activities against various organisms (Priya et al., 2002; Yousaf et al., 2008).

7.2 Antiinflammatory Effects

Datura is among the most vital therapeutic herbs utilized worldwide because of its calming property. In Pakistan, India, Indochina, and Africa, the controlled leaves or seeds are regularly blended with cannabis and smoked to diminish asthma and stiffness (Yang et al., 2014).

8. SIDE EFFECTS AND TOXICITY

Datura is unsafe when taken by mouth or inhaled. It is poisonous and can cause many toxic effects including dry mouth and extreme thirst, vision problems, nausea and vomiting, fast heart rate, hallucinations, high temperature, seizures, confusion, loss of consciousness, breathing problems, and death. The deadly dose for adults is 15–100 g of leaf or 15–25 g of the seeds.

References

Abbasipour, H., Mahmoudvand, M., Rastegar, F., Hosseinpour, M.H., 2011. Bioactivities of jimsonweed extract, *datura stramonium* L.(Solanaceae), against *Tribolium castaneum* (Coleoptera: tenebrionidae). Turkish Journal of Agriculture and Forestry 35, 623–629.

Alfarhan, A.H., Al-Turki, T.A., Basahy, A.Y., 2005. Flora of Jizan region. Final Report Supported by King Abdulaziz City for Science and Technology 1, 545.

Batanouny, K., Abou Tabl, S., Shabana, M., Soliman, F., 1999. Wild medicinal plants in Egypt. With contribution of. In: Aboutabl, E., Shabana, M., Soliman, F. (Eds.), With Support of the Swiss Development Co-operation (SDC). Academy of Scientific Research and Technology, Egypt. The World Conservation Union (IUCN), Switzerland, pp. 60–64.

Bellila, A., Tremblay, C., Pichette, A., Marzouk, B., Mshvildadze, V., Lavoie, S., Legault, J., 2011. Cytotoxic activity of withanolides isolated from Tunisian Datura metel L. Phytochemistry 72, 2031–2036.

Benvenuti, S., Macchia, M., Stefani, A., 1994. Effects of shade on reproduction and some morphological characteristics oi Abutilon theophrasti Medicos, *Datura stramonium* L. and Sorghum halepense L. Pers. Weed Research 34, 283–288.

Bhatia, H., Manhas, R., Kumar, K., Magotra, R., 2014. Traditional knowledge on poisonous plants of Udhampur district of Jammu and Kashmir, India. Journal of Ethnopharmacology 152, 207–216.

Biswas, K.R., Khan, T., Monalisa, M.N., Swarna, A., Ishika, T., Rahman, M., Rahmatullah, M., 2011. Medicinal plants used by folk medicinal practitioners of four adjoining villages of Narail and Jessore districts, Bangladesh. American-Eurasian Journal of Sustainable Agriculture 5, 23–33.
Bo, T., Li, K.A., Liu, H., 2003. Investigation of the effect of space environment on the contents of atropine and scopolamine in Datura metel by capillary zone electrophoresis. Journal of Pharmaceutical and Biomedical Analysis 31, 885–891.
Buchholz, J., Williams, L., Blakeslee, A., 1935. Pollen-tube growth of ten species of Datura in interspecific pollinations. Proceedings of the National Academy of Sciences 21, 651–656.
Gaire, B.P., 2008. Monographs on Datura Stramonium L.
Hossain, M.A., Al Kalbani, M.S.A., Al Farsi, S.A.J., Weli, A.M., Al-Riyami, Q., 2014. Comparative study of total phenolics, flavonoids contents and evaluation of antioxidant and antimicrobial activities of different polarities fruits crude extracts of Datura metel L. Asian Pacific Journal of Tropical Disease 4, 378–383.
Hussain, J., Khan, F.U., Ullah, R., Muhammad, Z., Rehman, N., Shinwari, Z.K., Khan, I., Zohaib, M., Din, I., Hussain, S.M., 2011. Nutrient evaluation and elemental analysis of four selected medicinal plants of Khyber Pakhtoonkhwa, Pakistan. Pakistan Journal of Botany 43, 427–434.
Jagatheeswari, D., 2014. Morphological studies on flowering plants (Solanaceae). International Letters of Natural Sciences 10.
Joy, P., Thomas, J., Mathew, S., Skaria, B.P., 1998. Medicinal plants. Tropical horticulture 2, 449–632.
Kagale, S., Marimuthu, T., Thayumanavan, B., Nandakumar, R., Samiyappan, R., 2004. Antimicrobial activity and induction of systemic resistance in rice by leaf extract of Datura metel against Rhizoctonia solani and Xanthomonas oryzae pv. oryzae. Physiological and Molecular Plant Pathology 65, 91–100.
Kelly, R.A., Andrews, J.C., DeWitt, J.G., 2002. An X-ray absorption spectroscopic investigation of the nature of the zinc complex accumulated in Datura innoxia plant tissue culture. Microchemical Journal 71, 231–245.
Konda, F., Shimizu, T., 2002. Naturalized plants of Mt. Fuji, central Japan. Mem. Natural Science and Museums 38, 95–107. Tokyo.
Monira, K.M., Munan, S.M., 2012. Review on Datura metel: a potential medicinal plant. Global Journal of Research on Medicinal Plants & Indigenous Medicine 1, 123.
Palazón, J., Moyano, E., Bonfill, M., Cusidó, R.M., Piñol, M.T., Teixeira da Silva, J., 2006. Tropane alkaloids in plants and genetic engineering of their biosynthesis. Floriculture, Ornamental and Plant Biotechnology 209–221.
Priya, K.S., Gnanamani, A., Radhakrishnan, N., Babu, M., 2002. Healing potential of Datura alba on burn wounds in albino rats. Journal of Ethnopharmacology 83, 193–199.
Reddy, C.S., 2008. Catalogue of invasive alien flora of India. Life Science Journal 5, 84–89.
Shah, G.M., Khan, M.A., 2006. Common Medicinal Folk Recipes of Siran Valley. Ethnobotanical leaflets, Mansehra, Pakistan, p. 5, 2006.
Shahid, M., Rao, N., 2014. Datura ferox and Oldenlandia corymbosa: new record to the UAE flora. Journal on New Biological Reports 3, 170–174.
Sharawy, S.M., Alshammari, A.M., 2009. Checklist of poisonous plants and animals in Aja mountain, Ha'il region, Saudi Arabia. Australian Journal of Basic and Applied Sciences 3, 2217–2225.
A. Sharma, R. IAAS, N. Chitwan, K.K. Pant, Aloe, (2014).
Ullah, R., Ullah, K., 2014. Summer weeds flora of district Dera Isamail Khan, Khyber Pakhtunkhwa, Pakistan. Pakistan Journal of Weed Science Research 20, 505–517.
Ullah, M., Khan, M.U., Mahmood, A., Malik, R.N., Hussain, M., Wazir, S.M., Daud, M., Shinwari, Z.K., 2013. An ethnobotanical survey of indigenous medicinal plants in

Wana district south Waziristan agency, Pakistan. Journal of Ethnopharmacology 150, 918–924.

Ullah, S., Khan, M.R., Shah, N.A., Shah, S.A., Majid, M., Farooq, M.A., 2014. Ethnomedicinal plant use value in the Lakki Marwat district of Pakistan. Journal of Ethnopharmacology 158, 412–422.

Vernay, P., Gauthier-Moussard, C., Jean, L., Bordas, F., Faure, O., Ledoigt, G., Hitmi, A., 2008. Effect of chromium species on phytochemical and physiological parameters in Datura innoxia. Chemosphere 72, 763–771.

Yang, B.-Y., Guo, R., Li, T., Wu, J.-J., Zhang, J., Liu, Y., Wang, Q.-H., Kuang, H.-X., 2014. New anti-inflammatory withanolides from the leaves of Datura metel L. Steroids 87, 26–34.

Yousaf, Z., Masood, S., Shinwari, Z.K., Khan, M.A., Rabani, A., 2008. Evaluation of taxonomic status of medicinal Species of the genus Hyoscyamous, Withania, Atropa and Datura based on poly acrylamide gel electrophoresis. Pakistan Journal of Botany 40, 229.239-2297.

CHAPTER

17

Damask Rose

Fariha Shabbir[1], *Muhammad Asif Hanif*[1], *Muhammad Adnan Ayub*[2], *Muhammad Idrees Jilani*[3], *Shafiqur Rahman*[4]

[1]Department of Chemistry, University of Agriculture, Faisalabad, Pakistan; [2] Department of Chemistry, University of Okara, Okara, Pakistan; [3] Department of Chemistry, University of Sahiwal, Sahiwal, Pakistan; [4]Department of Pharmaceutical Sciences, South Dakota State University, Brookings, SD, United States

OUTLINE

Medicinal Plants of South Asia
https://doi.org/10.1016/B978-0-08-102659-5.00017-3

Copyright © 2020 Elsevier Ltd. All rights reserved.

1. BOTANY

1.1 Introduction

Damask rose (*Rosa damascena* Mil.) (Fig. 17.1), a deciduous shrub, is the most significant of aromatic, medicinal, and ornamental plants. It is the most valuable *Rosa* species cultivated for the production of rose water and rose oil, largely utilized in the perfume industry and as a flavoring agent in food products (Mirzaei et al., 2016). Damask is actually a hybrid rose of two different categories; the autumn damasks and the summer damasks. The autumn damasks have Portland roses, which are shorter, highly compact, having the ability to repeat flowers in the red range in the autumn. While the flowers of summer damasks appear once only, having thorns that are quite open, along with growing shrubs that have intensely pink to white colors (Huxley, 1992; Ruba et al.). It is a member of Rosaceae family. The genus *Rosa* contains more than 200 species and 18,000 cultivars around the world. Damask rose as the king of flowers has been the symbol of love, purity, faith, and beauty since ancient times (Mahboubi, 2016). There is considerable disagreement about the exact number of rose species. They make a set of plants that could be of stiff shrubs, climbing or trailing with the stems, usually provided with sharp prickles. Rose flowers usually vary in form, and size is typically large and attractive, found in a range of colors starting from white through yellow and red. Most of the species are native to Asia, with a few numbers being native to North America, Europe, and Northwestern Africa. Roses have acquired cultural significance in different societies. Rose plants are found in different size ranges starting from compact, small roses, to the climbers

FIGURE 17.1 Fresh and dry petals of damask rose.

that can reach about 7 m tall. There is a great development in a wide selection of garden roses due the ability of different species to hybridize easily (Chandraju et al., 2012).

Due to its utilization as an ingredient in various multiherbal formulation such as "*Safoof-e-Muhazzil*" for obesity treatment, it has become famous as "*Gul-e-Surkh*" in the Unani system of medicine (Naquvi et al., 2014; Giray, 2012, p. 2). As an ornamental plant in Iran, it is also known as "Gole Mohammadi," i.e., the flower of Prophet Mohammad (Mahboubi, 2016).

1.2 History/Origin

Traditionally, its origin has been believed to be in the Middle East. Although from recent genetic analysis, its more probable origin is in the home of its pollen parents, i.e., the foothills of central Asia, because it is a hybrid of *Rosa fedtschenkoana* crossed with the pollens of *Rosa moschata* x *Rosa gallica* (Iwata et al., 2000, p. 3). The credit to bring damask rose to Europe from Syria between 1254 and 1276 is given to Crusader Robert de Brie. The fragrance of damask rose has been produced historically in Afghanistan (Kabul Province). UN Development program helped Afghanistan to develop rose oil industry on modern basis, a business that could give poppies a run for their money. For centuries, it has been considered a symbol of love and beauty. In the Middle East, from biblical times, an ancient method has been used to preserve the fragrance of damask rose in rose water form, later done in the Indian subcontinent. In the early 11th century, Iranian physician Avicenna (Ibn Sīnā) discovered a method of extraction of rose water from the rose petals. He invented steam distillation and used it to produce essential oils such as rose essence, forming the foundation of what later became aromatherapy. During the reign of Henry VIII, damask roses were introduced into England.

1.3 Demography/Location

During the flowering season, damask rose needs humid air and moderate temperatures for oil-rich contents. The temperate climates found between 300 and 1800 m altitude are suitable for the growth of damask rose (Loghmani-Khouzani et al., 2007). It is cultivated mainly in Turkey, Bulgaria, Pakistan, Italy, France, Russia, India, and Morocco for the large-scale production of rose oil. In Pakistan the main areas of rose cultivation are Chakwal, Kallar Kahar, Shahiwal (near Sargodha), Pattoki, Choha Syedan Shah, Hyderabad, Islamabad, and Faisalabad (Farooq et al., 2011). A minute amount of essential oil is present in the petals of fresh damask roses. From 3000 kg rose petals, about 1 kg rose oil is extracted (Verma et al., 2011). The largest production of rose oil (about 10 tons oil annually) is being done together in Morocco, Turkey, and Bulgaria.

1.4 Botany, Morphology, Ecology

The damask rose is a tall, about 2.2 m (7 ft 3 in), growing deciduous shrub, having densely packed stems with stouts along with stiff bristles and curved prickles. It has pinnate leaves with five and rarely seven leaflets. The color of roses is in the range of light to moderate pink and light red. It has comparatively small flowers growing in group form. The flowers have 17–25 petals in bloom form (Harkness, 2003). A single flower's weight varies between 2.09 and 3.44 g, the number of flowers petals is between 22 and 28, and a bushes' diameter varies between 53 and 118 cm (Kovatcheva et al., 2011).

2. CHEMISTRY

According to the report of international standard of rose oil, it requires 15%–22% geraniol, 8%–15% nonadecane, and 20%–34% citronellol as the major components of the essential oil (Mirzaei et al., 2016). Further rose oil is characterized by the typical ratio of citronellol/geraniol (C/G) in range

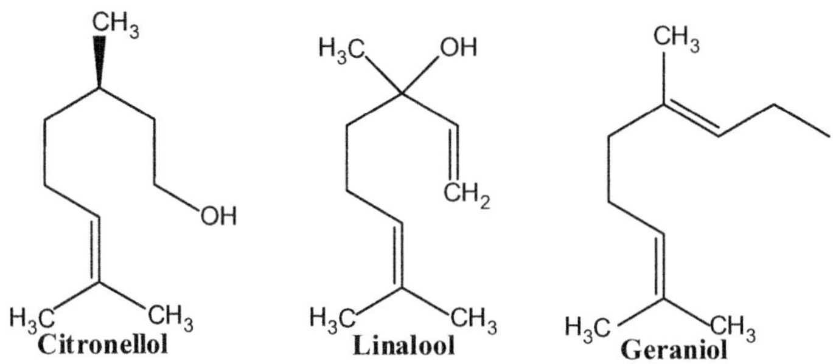

FIGURE 17.2 Important chemical components of damask rose.

of 1.25–1.30, preferably, by the perfumery industries (Baser, 1992; Mirzaei et al., 2016). The damask rose seed oil is of great importance due to its richness in ω-3 fatty acid, i.e., α-linolenic acid, while in other plants, this fatty acid is either absent or present in a very minute amount. The damask seed oil usually contains palmitic acid (C16:0), stearic acid (C18:0), oleic acid (ω-9) (C18:1), linoleic acid (ω-6) (C18:2), and α-linolenic acid (ω-3) (C18:3). It also contains ascorbic acid, α-tocopherol, and β-carotene (Kazaz et al., 2009). The hip of damask rose is also rich in vitamins A, B_3, C, D, and E (Popescu et al., 2015). One kilogram damask rose contains about 1224 mg phosphorus in the fruits and 67–1459 mg in the seeds and fruit flesh. About 2243–12,454 mg potassium, 3885–11,162 mg calcium, 441–1501 mg magnesium, 98–163 mg sodium, 4 mg copper present in fruits and fruits parts, 11–118 mg iron contents in the seeds, fruit flesh, and fruits, 24–73 mg manganese, 7–14 mg zinc in the seeds, fruit flesh, and fruits, and 1 mg boron in the seeds (Kazaz et al., 2009). Flavonoids glycoside and kaempferol-3-O-β-D-glucopyranosyl(1 → 4)-β-D-xylopyranoside, named roxyloside A, are separated from damask rose's buds along with four known compounds, afzelin, isoquercitrin, quercetin gentiobioside, and cyaniding-3-O-β-glucoside, which demonstrate that damask rose and its flavonoids may be active in improving the cardiovascular system (Kwon et al., 2009). Some important chemical components of damask rose are shown in Fig. 17.2.

3. POSTHARVEST TECHNOLOGY

Damask rose plant is a onetime flowering plant per year, having 35–45 days flowering period starting from the second half of March to the end of April (sometimes to the mid of May, depending upon weather conditions). During the rose flowering period, rose flowers are handpicked on a daily basis during the early morning from 05:00 to 10:00 a.m. By nipping just below the calyx, rose flowers are plucked. On average, manually about 2–3 kg rose flowers are plucked per hour and are transported to the rose oil factories for the process of distillation (Harkness, 2003).

4. PROCESSING

The storage of damask rose flowers for 20 days at −20°C has maintained the essential oil content and quality of rose oil per international standards. In fact, it is better to use the fresh flowers for the production of essential oil. But, when flowers were collected in large quantities, it becomes impossible to distill them together in the distillation unit due to flower glut or technical fault in the distillation unit, as some parts of the flowers undergo varying degrees of fermentation until distillation (Sharma and Kumar, 2016). There is a remarkable reduction in steep decline of rose oil contents with the cold storage in different packaging materials of the damask rose flowers (Kazaz et al., 2010). To decrease the rate of respiration of freshly prepared products by manipulation of O_2 and CO_2 inside the packets is achieved by the utilization of proper temperature along with the management of relative humidity level, and modified atmosphere technology is used to maintain the quality and prolong the postharvest life of the freshly prepared products of damask rose. Providing the proper circumstances of rose oil postharvest handling can interrupt the sharp degradation of geraniol, although the citronellol content rises in the oil.

5. VALUE ADDITION

Rose water, or colorless liquid with the common name Urq e Golab (in Pakistan) due to its calming and relaxing properties, is used in religious ceremonies like washing the God House in Mecca (Saudi Arabia) and shrines in Pakistan and India. It is also for flavoring foods in Pakistan, India, and Iran. The dried buds and petals of rose are sold in groceries as flavor and laxative agents. Its fragrant flowers are used in foods as rose water, marmalade, and pastry, while its precious essential oil has a worldwide growing demand. Dried rose petals, in addition to rose oil, are also utilized for various purposes. Medically, to cure digestive problems, its has great importance as a food supplement. Furthermore, rose petals are utilized in Gul-e-Roghan. In Persian cuisine, chicken with rose is most famous. In "zuhurat" herbal tea, rose petals or the whole flower is used. Most popularly, it is used as a flavoring ingredient in many desserts including yogurt, jam, ice cream, rice pudding, Turkish delight, etc.

6. USES

In cooking, damask rose is utilized as a flavoring ingredient. In religious ceremonies, rose water is most widely utilized. The decoction of flowers is used for treatment of chest and abdominal pains, menstrual bleeding, and digestive ailments as a gentle laxative for constipation. It is famous as cardiotonic agent for heart strengthening (Mahboubi, 2016; Shahriari et al., 2007). Rose water traditionally has been used as an antiseptic agent for eye washing (Gochev et al., 2008) and mouth disinfecting (Akhmadieva et al., 1992; Mahboubi, 2016) and as antispasmodic agent for alleviating abdominal pains and bronchial and chest congestions. The decoctions of dried rose water are used as diuretic and are recommended for relieving fever, breast pain, and menstrual problems (Foster and Duke, 1990; Mahboubi, 2016). Rose essential oil has been used traditionally for treatment of cardiac diseases via massage on the skin. Nowadays, it has been shown that damask rose aqueous extract increases heart rate and contractility in guinea pig via stimulatory effect on b-adrenoceptor (Shafei et al., 2011) and suppresses the activity of ACE (angiotensin-I-converting) enzyme (Kwon et al., 2009). The efficacy of damask rose extract on primary dysmenorrheal syndrome (PMS) was confirmed in a double-blind cross-over clinical trial on 92 single girls. Damask rose extract decreased the average of pain density in PMS such as mefenamic acid without any side effects (Bani et al., 2014).

7. PHARMACOLOGICAL USES

7.1 Anticancer Activity

Damask rose shows antitumor, anticarcinogenic, and cytotoxic activities (Zamiri-Akhlaghi et al., 2011) against cancer cells. Geraniol as the main compound of damask rose acts via different mechanisms. It induces the apoptosis in cancer cells and increases the expression of apoptotic protein Bak (Burke et al., 1997; Mahboubi, 2016), arrests the G0/G1 phase of cell cycle and reduces cdk2 activity(Wiseman et al., 2007), and inhibits the 3-hydroxy-3-methylglutaryl-CoA (HMG-CoA) reductase (Elson, 1995; Mahboubi, 2016) and ornithine decarboxylase activity (Carnesecchi et al., 2001) that finally causes the death of cancerous cells.

7.2 Antimicrobial Activity

Rose essential oil showed considerable antibacterial activity against *Xanthomonas axonopodis* spp. *Vesicatoria, Chromobacterium violaceum,* and *Erwinia carotovora* strains (Basim and Basim, 2003; Mahboubi, 2016;

Ulusoy et al., 2009, p. 1), *Staphylococcus aureus* (Jirovetz et al., 2006; Mahboubi, 2016), *Bacillus cereus, Staphylococcus epidermidis, Pseudomonas fluorescens* (Gochev et al., 2008), *Pseudomonas aeruginosa* (Jirovetz et al., 2006; Mahboubi, 2016), *Escherichia coli, Proteus vulgaris, Klebsiella pneumoniae, Candida albicans,* and *Enterococcus faecalis* (Mahboubi, 2016).

7.3 Antiinflammatory Effects

The antiinflammatory effects of damask rose ethanol, chloroform extracts were studied in animal models, but rose essential oil failed to produce antiinflammatory effects (Hajhashemi et al., 2010; Rakhshandeh et al., 2008). Indeed, the components that have analgesic effects in ethanol extract are not found in rose essential oil. It has been shown that rose hip powder (10 g) for 1 month has no antiinflammatory or antioxidant effects in rheumatoid arthritis patients (Kirkeskov et al., 2011), while others have identified rose hips as an antiinflammatory agent (Willich et al., 2010). Unsaturated fatty acids, triterpenoic acids, or unidentified compounds and their synergistic effects exhibit antiinflammatory effect (Larsen et al., 2003) via inhibiting cyclooxygenase 1 and 2 (Jäger et al., 2007; Mahboubi, 2016). Additional studies are required to confirm these mixed and conflicting results.

7.4 Adaptogenic/Antistress Activity

Damask rose shows relaxant activity by inhibiting the histamine H1 receptors and blocking the calcium channels of tracheal chain (Shafei et al., 2010), inhibiting the KCl-related contraction and electrical field stimulation (Sadraei et al., 2012). Damask rose aqueous and ethanol extracts shown bronchodilatory and antitussive effects likely by inhibition of tachykinin and decrease of citric acid that induced coughs (Shafei et al., 2010).

7.5 Antidiabetic Activity

The extract of damask rose has noncompetitive inhibitory effect for α-glucosidase enzyme. In aldose-dependent manner, the oral intake of (100–1000 mg/kg body wt.) of damask plant's extract was shown to decrease the glucose level in the blood after the insertion of maltose in diabetic and normal rats. From this experiment, the postprandial glucose level was found to be decreased due to the reduction in the absorption of carbohydrates from the intestine, suggesting damask rose acts as an antidiabetic agent (Gholamhoseinian and Fallah, 2009).

7.6 Cardiovascular Effects

There is little reported work on the effect of damask rose in the cure of cardiovascular disorders. Enhanced contractility and improved heart rate are observed in the heart of guinea pig by injecting damask rose aqueous ethanolic extract. Mechanism of this effect is still unknown. But in the isolated heart of guinea pig, the plant extract showed probable stimulatory effect on β-adrenoceptor. At present, cyanidin-3-O-β-glucoside, a new compound, is separated from the damask rose "buds." This compound could considerably suppress the activity of ACE. As ACE is a key vital enzyme in the production of angiotensin II, damask rose could be effective in enhancing cardiovascular function (Kwon et al., 2009).

7.7 Anti-HIV Effects

In vitro, the influence of methanol and water extracts of damask rose for the infection of HIV was reported (Boskabady et al., 2011). In this research, nine compounds along with a new compound, 2-phenylethanol-O-(6-Ogalloyl)-β-D-glucopyranoside, refined from the methanolic extract were assessed on HIV-1MN-infected C8166 human T lymphoblastoid cells and for chronically HIV−1IIIB-infected H9 human T-cell lymphoma cells for the test of anti-HIV activities. On C8166 cells, Kaempferol 1 and its 3-O-β-D-glucopyranosides 3 and 6 showed major action against the infection of HIV, whereas kaempferol-7-O-β-D-glucopyranoside did not exhibit any major effect.

7.8 Hypnotic Effects

Hypnotic effect is one of the major activities of damask rose on the central nervous system. The aqueous, chloroformic, and ethanolic extracts of damask rose were studied for the estimation of hypnotic effect in the mice. The 500- and 1000-mg/kg dose of the aqueous and ethanolic extracts of damask rose predominantly enhanced the sleeping time of induced pentobarbital in the mice, which is comparable to diazepam. However, the hypnotic effect has not been observed in chloroformic extract of damask rose (Rakhshandah et al., 2010a).

7.9 The Analgesic Effects

The chloroformic, aqueous, and ethanolic extracts of damask rose were evaluated in mice on tail flick and hot plate for the study of analgesic effects, but only ethanolic extract exhibited the analgesic effects (Rakhshandeh et al., 2008). The essential oil of damask rose and hydroalcoholic extract tested for analgesic activity in mice by tail flick, acetic acid, and formalin tests showed no analgesic effects. Moreover, potent analgesic effects in formalin and acetic acid tests of hydroalcoholic extract were observed (Hajhashemi et al., 2010). Based on the analgesic effects of ethanolic and hydroalcoholic extracts of damask rose, it is suggested that water insoluble ingredients could be responsible for the analgesic effect. Water insoluble kaempferol and quercetin are thought to be responsible for observed analgesic effects (O'Neil et al., 2001).

7.10 Neuroprotective Effects

Several studies have reported the neuroprotective effects of damask rose, e.g., its use in the cure of dementia disease. For example, damask rose extract was found to increase neurite outgrowth activity while amyloid β (Aβ) was reduced (Awale et al., 2011). Aβ is a well-known major pathologic marker of Alzheimer disease. From the chloroform extract of the damask rose, an active constituent was isolated with a very long polyunsaturated fatty acid, which may be neuroprotective and could be responsible for strong neurite outgrowth activity. Therefore, damask rose may have valuable effects in patients with dementia.

7.11 Anticonvulsant Effects

The damask rose essential oil was found to delay the initiation of epileptic seizures in acute pentylenetetrazole (PTZ)-induced seizure in rats and shown to minimize tonic-clonic seizure (stage 4) duration (Kheirabadi et al., 2008). Furthermore, this plant was reported to prolong the lateral periods in PTZ-induced seizure in a chronic model before the generalization of tonic-clonic seizures (Kheirabadi et al., 2008). Moreover, damask rose's essential oil was reported as an adjunct treatment for refractory seizures in children. It is noteworthy to mention that flavonoids were suggested to be involved in this activity by acting on brain GABAergic system, and benzodiazepine's effect on the GABA receptors was found to be enhanced by the flavonoids (Kheirabadi et al., 2008). Some of the components of damask rose such as eugenol and geraniol were reported to show the antiepileptic effect (Aggarwal et al., 2011). The exact mechanisms of these pharmacological effects are still unknown, so future studies are required to investigate further.

7.12 Antioxidant Activity

Decoctions (Cho et al., 2003), aqueous extract (Cho et al., 2003), essential oil (Köse et al., 2012), absolute (Ulusoy et al., 2009), methanol (Baydar and Baydar, 2013), and ethanol extracts (Shahriari et al., 2007) of rose petals have shown antioxidant activity in different systems. The antioxidant activity of rose absolute with higher concentration of carotene, α, β, γ-tocopherols is higher than that of rose essential oil and rose water (Ulusoy et al., 2009). The antioxidant activity of damask rose is not related to anthocyanin level (VanderJagt et al., 2002), but it is correlated to total phenolic, flavonol contents of damask rose. Leaf methanol extract of rose with high concentration of (+)-catechin and (−)-epicatechin as phenolic compound has shown antioxidant activity higher than that of BHT and trolox (Baydar and Baydar, 2013). The beneficial effects of rose essential oil against formaldehyde inhalation on the reproductive system are related to the antioxidant activity of rose essential oil. Pretreatment with rose essential oil was found to decrease abnormal sperm and increased the sperm counts in rats (Köse et al., 2012). Rose hips as an herbal tea is consumed as a strong antioxidant beverage (Halvorsen et al., 2002). Therefore, the medicinal effects of damask rose in scavenging of free radicals are recognized, as it as good beverage for health. Consistent with this notion, in Iranian cultures, rose water was added to cold beverages as a refreshing agent (Mahboubi, 2016).

8. SIDE EFFECTS AND TOXICITY

Rose water may cause burning, stinging, redness, or irritation in some persons.

References

Akhmadieva, A., Zaichkina, S., Ruzieva, R., Ganassi, E., 1992. The protective action of a natural preparation of anthocyan (pelargonidin-3, 5-diglucoside). Radiobiologiia 33, 433–435.

Awale, S., Tohda, C., Tezuka, Y., Miyazaki, M., Kadota, S., 2011. Protective effects of *Rosa damascena* and its active constituent on Aβ (25–35)-induced neuritic atrophy. Evidence-based Complementary and Alternative Medicine 2011.

Aggarwal, B., Prasad, S., Reuter, S., Kannappan, R., R Yadav, V., Park, B., Hye Kim, J., C Gupta, S., Phromnoi, K., Sundaram, C., 2011. Identification of novel anti-inflammatory agents from Ayurvedic medicine for prevention of chronic diseases:"reverse pharmacology" and "bedside to bench" approach. Current Drug Targets 12, 1595–1653.

Bani, S., Hasanpour, S., Mousavi, Z., Garehbaghi, P.M., Gojazadeh, M., 2014. The effect of *Rosa damascena* extract on primary dysmenorrhea: a double-blind cross-over clinical trial. Iranian Red Crescent Medical Journal 16.

Baser, K., 1992. Turkish Rose OiI.

Basim, E., Basim, H., 2003. Antibacterial activity of *Rosa damascena* essential oil. Fitoterapia 74, 394–396.

Baydar, N.G., Baydar, H., 2013. Phenolic compounds, antiradical activity and antioxidant capacity of oil-bearing rose (*Rosa damascena* Mill.) extracts. Industrial Crops and Products 41, 375–380.

Boskabady, M.H., Shafei, M.N., Saberi, Z., Amini, S., 2011. Pharmacological effects of Rosa damascena. Iranian Journal of Basic Medical Sciences 14 (4), 295.

Burke, Y.D., Stark, M.J., Roach, S.L., Sen, S.E., Crowell, P.L., 1997. Inhibition of pancreatic cancer growth by the dietary isoprenoids farnesol and geraniol. Lipids 32, 151–156.

Carnesecchi, S., Schneider, Y., Ceraline, J., Duranton, B., Gosse, F., Seiler, N., Raul, F., 2001. Geraniol, a component of plant essential oils, inhibits growth and polyamine biosynthesis in human colon cancer cells. Journal of Pharmacology and Experimental Therapeutics 298, 197–200.

Chandraju, S., Thejovathi, C., Kumar, C.C., 2012. Distillery spentwash as an effective liquid fertilizer and alternative irrigation medium in floriculture. Research in Plant Biology 2.

Cho, E., Yokozawa, T., Rhyu, D., Kim, S., Shibahara, N., Park, J., 2003. Study on the inhibitory effects of Korean medicinal plants and their main compounds on the 1, 1-diphenyl-2-picrylhydrazyl radical. Phytomedicine 10, 544–551.

Elson, C.E., 1995. Suppression of mevalonate pathway activities by dietary isoprenoids: protective roles in cancer and cardiovascular disease. Journal of Nutrition 125, 1666S–1672S.

Farooq, A., Khan, M.A., Ali, A., Riaz, A., 2011. Diversity of morphology and oil content of *Rosa damascena* landraces and related Rosa species from Pakistan. Pakistan Journal of Agricultural Sciences 48, 177–183.

Foster, S., Duke, J., 1990. Rosa Rugosa Thunb. Medicinal Plants. Houghton Mifflin Co, New York, NY, p. 234.

Gholamhoseinian, A., Fallah, H., 2009. Inhibitory effect of methanol extract of *Rosa damascena* Mill. flowers on α-glucosidase activity and postprandial hyperglycemia in normal and diabetic rats. Phytomedicine 16, 935–941.

Gochev, V., Wlcek, K., Buchbauer, G., Stoyanova, A., Dobreva, A., Schmidt, E., Jirovetz, L., 2008. Comparative evaluation of antimicrobial activity and composition of rose oils from various geographic origins, in particular Bulgarian rose oil. Natural Product Communications 3, 1063–1068.

Hajhashemi, V., Ghannadi, A., Hajiloo, M., 2010. Analgesic and anti-inflammatory effects of *Rosa damascena* hydroalcoholic extract and its essential oil in animal models. Iranian Journal of Pharmaceutical Research 163–168.

Halvorsen, B.L., Holte, K., Myhrstad, M.C., Barikmo, I., Hvattum, E., Remberg, S.F., Wold, A.-B., Haffner, K., Baugerød, H., Andersen, L.F., 2002. A systematic screening of total antioxidants in dietary plants. Journal of Nutrition 132, 461–471.

Harkness, P., 2003. The Rose: An Illustrated History. Firefly books.

Huxley, A. (Ed.), 1992. New RHS Dictionary of Gardening. Macmillan, ISBN 0-333-47494-5.

Iwata, H., Kato, T., Ohno, S., 2000. Triparental origin of Damask roses. Gene 259, 53–59.

Jäger, A.K., Eldeen, I.M., van Staden, J., 2007. COX-1 and-2 activity of rose hip. Phytotherapy Research 21, 1251–1252.

Jirovetz, L., Buchbauer, G., Denkova, Z., Slavchev, A., Stoyanova, A., Schmidt, E., 2006. Chemical composition, antimicrobial activities and odor descriptions of various Salvia sp. and Thuja sp. essential oils. Nutrition-Vienna 30, 152.

Kazaz, S., BaydaR, H., ERBaS, S., 2009. Variations in chemical compositions. Czech Journal of Food Sciences 27, 178–184.

Kazaz, S., Erbas, S., Baydar, H., Dilmacunal, T., Koyuncu, M.A., 2010. Cold storage of oil rose (*Rosa damascena* Mill.) flowers. Scientia Horticulturae 126, 284–290.

Kheirabadi, M., Moghimi, A., Rakhshande, H., Rassouli, M.B., 2008. Evaluation of the anticonvulsant activities of *Rosa damascena* on the PTZ induced seizures in wistar rats. Journal of Biological Sciences 8, 426–430.

Kirkeskov, B., Christensen, R., Bügel, S., Bliddal, H., Danneskiold-Samsøe, B., Christensen, L.P., Andersen, J.R., 2011. The effects of rose hip (Rosa canina) on plasma antioxidative activity and C-reactive protein in patients with rheumatoid arthritis and normal controls: a prospective cohort study. Phytomedicine 18, 953–958.

Köse, E., Sarsılmaz, M., Taş, U., Kavaklı, A., Türk, G., Özlem Dabak, D., Sapmaz, H., Ögetürk, M., 2012. Rose oil inhalation protects against formaldehyde-induced testicular damage in rats. Andrologia 44, 342–348.
Kovatcheva, N., Zheljazkov, V.D., Astatkie, T., 2011. Productivity, oil content, composition, and bioactivity of oil-bearing rose accessions. HortScience 46, 710–714.
Kwon, E.-K., Lee, D.-Y., Lee, H., Kim, D.-O., Baek, N.-I., Kim, Y.-E., Kim, H.-Y., 2009. Flavonoids from the buds of *Rosa damascena* inhibit the activity of 3-hydroxy-3-methylglutaryl-coenzyme a reductase and angiotensin I-converting enzyme. Journal of Agricultural and Food Chemistry 58, 882–886.
Larsen, E., Kharazmi, A., Christensen, L.P., Christensen, S.B., 2003. An antiinflammatory galactolipid from rose hip (*Rosa c anina*) that inhibits chemotaxis of human peripheral blood neutrophils in vitro. Journal of Natural Products 66, 994–995.
Loghmani-Khouzani, H., Sabzi Fini, O., Safari, J., 2007. Essential oil composition of *Rosa damascena* mill cultivated in central Iran. Scientia Iranica 14, 316–319.
Mahboubi, M., 2016. *Rosa damascena* as holy ancient herb with novel applications. Journal of traditional and complementary medicine 6, 10–16.
Mirzaei, M., Sefidkon, F., Ahmadi, N., Shojaeiyan, A., Hosseini, H., 2016. Damask rose (*Rosa damascena* Mill.) essential oil is affected by short-and long-term handling. Industrial Crops and Products 79, 219–224.
Naquvi, K.J., Ansari, S., Ali, M., Najmi, A., 2014. Volatile oil composition of *Rosa damascena* mill.(Rosaceae). Journal of Pharmacognosy and Phytochemistry 2.
O'Neil, M.J., Smith, A., Heckelman, P., 2001. The Merck Index, vol. 309. Merck & Co. Inc, Whitehouse Station, NJ, p. 405.
Popescu, A., Matei, N., Roncea, F., Miresan, H., Pavalache, G., 2015. Determination of caftaric acid in tincture and rose water obtained from Rosae damascenae flores. Ovidius University Annals of Chemistry 26, 12–19.
Rakhshandah, H., Hosseini, M., Dolati, K., 2010a. Hypnotic effect of *Rosa damascena* in mice. Iranian Journal of Pharmaceutical Research 181–185.
Rakhshandeh, H., Vahdati-Mashhadian, N., Dolati, K., Hosseini, M., 2008. Antinociceptive effect of *Rosa damascena* in mice. Journal of Biological Sciences 8, 176–180.
Ruba, P.H., Maheshwari, M., Gupta, A., Therapeutic Values of Rose.
Sadraei, H., Asghari, G., Emami, S., 2012. Inhibitory effect of *Rosa damascena* Mill flower essential oil, geraniol and citronellol on rat ileum contraction. Research in Pharmaceutical Sciences 8, 17–23.
Shafei, M.N., Rakhshandah, H., Boskabady, M.H., 2010. Antitussive effect of *Rosa damascena* in Guinea pigs. Iranian Journal of Pharmaceutical Research 231–234.
Shafei, M.N., Saberi, Z., Amini, S., 2011. Pharmacological effects of *Rosa damascena*. Iranian Journal of Basic Medical Sciences 14, 295–307.
Shahriari, S., Yasa, N., Mohammadirad, A., Khorasani, R., Abdollahi, M., 2007. In Vivo Antioxidant Potentials of Rosa Damascene Petal Extract from Guilan, Iran, Comparable to Alpha-Tocopherol.
Sharma, S., Kumar, R., 2016. Effect of temperature and storage duration of flowers on essential oil content and composition of damask rose (Rosa× damascena Mill.) under Western Himalayas. Journal of Applied Research on Medicinal and Aromatic Plants 3, 10–17.
Ulusoy, S., Boşgelmez-Tınaz, G., Seçilmi°-Canbay, H., 2009. Tocopherol, carotene, phenolic contents and antibacterial properties of rose essential oil, hydrosol and absolute. Current Microbiology 59, 554–558.
VanderJagt, T., Ghattas, R., VanderJagt, D., Crossey, M., Glew, R., 2002. Comparison of the total antioxidant content of 30 widely used medicinal plants of New Mexico. Life Sciences 70, 1035–1040.

Verma, R.S., Padalia, R.C., Chauhan, A., 2011. Chemical investigation of the volatile components of shade-dried petals of damask rose (*Rosa damascena* Mill.). Archives of Biological Sciences 63, 1111–1115.

Willich, S., Rossnagel, K., Roll, S., Wagner, A., Mune, O., Erlendson, J., Kharazmi, A., Sörensen, H., Winther, K., 2010. Rose hip herbal remedy in patients with rheumatoid arthritis—a randomised controlled trial. Phytomedicine 17, 87–93.

Wiseman, D.A., Werner, S.R., Crowell, P.L., 2007. Cell cycle arrest by the isoprenoids perillyl alcohol, geraniol, and farnesol is mediated by p21Cip1 and p27Kip1 in human pancreatic adenocarcinoma cells. Journal of Pharmacology and Experimental Therapeutics 320, 1163–1170.

Zamiri-Akhlaghi, A., Rakhshandeh, H., Tayarani-Najaran, Z., Mousavi, S.H., 2011. Study of cytotoxic properties of *Rosa damascena* extract in human cervix carcinoma cell line. Avicenna Journal of Phytomedicine 1, 74–77.

Further Reading

Karami, A., Parviz, Z., Khosh-Khui, M., Salehi, H., Saharkhiz, M.J., 2012. Analysis of essential oil from nine distinct genotypes of Iranian damask rose (*Rosa damascena Mill*). Journal of Medicinal Plants Research 6, 5495–5498.

Kovacheva, N., Rusanov, K., Atanassov, I., 2010. Industrial cultivation of oil bearing rose and rose oil production in Bulgaria during 21st century, directions and challenges. Biotechnology & Biotechnological Equipment 24, 1793–1798.

Maurice, T., Lockhart, B.P., Privat, A., 1996. Amnesia induced in mice by centrally administered β-amyloid peptides involves cholinergic dysfunction. Brain Research 706, 181–193.

Nikbakht, A., Kafi, M., Mirmasoudi, M., Babalar, M., 2005. Micropropagation of damask rose (*Rosa damascena mill.*) cvs azaran and ghamsar. International Journal of Agriculture and Biology 7, 535–538.

Nyeem, M., Alam, M., Awal, M., Mostofa, M., Uddin, S., Islam, N., Rouf, R., 2007. CNS Depressant Effect of the Crude Ethanolic Extract of the Flowering Tops of *Rosa Damascena*.

Rakhshandah, H., Shakeri, M.T., Ghasemzadeh, M.R., 2010b. Comparative hypnotic effect of *Rosa damascena* fractions and Diazepam in Mice. Iranian Journal of Pharmaceutical Research 193–197.

CHAPTER

18

Dill

Muhammad Mubeen Mohsin[1], *Muhammad Asif Hanif*[1], *Muhammad Adnan Ayub*[2], *Ijaz Ahmad Bhatti*[1], *Muhammad Idrees Jilani*[3]

[1] Department of Chemistry, University of Agriculture, Faisalabad, Pakistan;
[2] Department of Chemistry, University of Okara, Okara, Pakistan;
[3] Department of Chemistry, University of Sahiwal, Sahiwal, Pakistan

OUTLINE

Medicinal Plants of South Asia
https://doi.org/10.1016/B978-0-08-102659-5.00018-5

Copyright © 2020 Elsevier Ltd. All rights reserved.

1. BOTANY

1.1 Introduction

Anethum graveolens L. is an industrially important herb, member of Apiaceae family, and is popular with the common name "dill." From ancient times, this plant has been used as a flavoring agent in food recipes (Hemphill, 2000). Dill (Fig. 18.1) is the sole member of *Anethum* genus. The name *Anethum* is a Greek word that means strong smell. One of the common varieties of *Anethum graveoeloens* L. is Sowa, also known as Indian dill, and it grows in cold climatic conditions of Japan, the Indian subcontinent, and Malaysia (Jana and Shekhawat, 2010).

A. graveolens is known by different names depending where you are in the world. The most frequently known and popular common name of *A. graveolens* is dill, and it is a plant native to Oriental and Mediterranean countries. It is also found growing wild in various parts of Africa, Asia, Europe, and South Russia. In English, *A. graveolens* is typically called dill,

FIGURE 18.1 Dill plants, flowers and fruits (seeds).

and in French it is known as dilly, aneton, and garden dill. It is also known as indische dille, gurkenkraut in German, and the *A. graveolens* is known as aneto in Italian. *A. graveolens* is also cultivated in India and Pakistan, and it is commonly known as surva or sowa. It is cultivated in England, Germany, Hungary, and the United States of America, and the cultivated variety is often known as '"garden dill." *A. graveolens* is grown in China, Japan, Arab countries, and Greece and is commonly known as shih lo, diru, shibith, and anithos, respectively.

1.2 History/Origin

The beginning of *A. graveolens* is believed to be in the Mediterranean region. The dill plant was previously used as a culinary and medicinal herb in so many countries. From 5000 years ago, the dill plant has been known as a medicinal herb, and it is referred as a "demulcent" in Egypt (Hegazi et al.). Before that, Babylonians in the ancient Akkadian had been reported to grow this plant in their fields and gardens (Hobhouse, 2004). In the Greek culture, the dill plant is a well-known and widely used plant for herbal medicine. The people of Greece obtained essential oils from the dill plants at their homes and also used this oil to make drinks for their uses (Heilmeyer, 2007). Pedanius Dioscorides was the first Greek doctor as well as a surgeon who used dill seeds for wound healing (Castleman and Hendler, 1995). In previous history, the Gladiators used dill plant in their food stuff. They thought dill plant was the source of motivation in fight and provided courage. The seeds of the dill plant are known as "meeting house seeds." They used to be crushed during long religious ceremonies to keep followers wakeful or children quiet. Dill plant seeds were also used for the purpose of keeping the breath fresh and stomach healthy. Dill (*A. graveolens*) is an industrially and economically valuable plant. During the period of King Edward I, a tax was applied on dill plant import and export to get money for the London bridge restoration (Duke and Duke, 1983). It was the most popular kitchen garden and medicinal plant in Europe during the 17th century (Halberstein, 2005).

1.3 Demography/Location

It is a cold weather condiment crop, found in the northern plains of Pakistan, India, and Bangladesh. It is also grown in Southeast Asia and Japan for its aromatic leaves and fruit (Saleh-e-In et al., 2010). It is grown widely in Africa, South Russia, France, Italy, Germany, China, Hungary, Arab countries, and the United States of America (Radulescu et al., 2010). Dill needs enough water to survive during drought period, and the soil of the dill plant crop must be kept moist all the time (Small and Canada, 1997).

1.4 Botany, Morphology, Ecology

Dill can be grown as a biennial or annual depending upon the variety. Dill height is 1–4 ft. in ideal conditions, and the leaf size is 3–7 inches. Dill's inflorescences are arranged in umbels, in which the flower stalks develop from a common point. The dry fruit of dill is called a schizocarp, which contains dill seeds. The shape of the seeds is oval and dark brown in color. Leaves are aromatic, soft, needle-like, and lacy in appearance (Wright, 2010). Full sun is best for good production of the dill plant. But very warm climate can cause early flowering and reduce leaf growth. The dill plant is grown in the cold environment in the temperature range from 5 to 25°C. The production of the dill plant is affected by high winds, as dill stems bend easily. Dill crop is also affected by hail and low moisture content in the soil (Small and Canada, 1997). The medium-textured soil, which can reserve moisture and has satisfactory drainage, is very suitable to produce a dill crop. The soil pH range of 5.0–8.2, with an average of 6.6, is considered good for producing a dill crop (Tucker and DeBaggio, 2000).

2. CHEMISTRY

Dill has different classes of compounds such as volatile oils, fatty oils, inorganic elements (Ca, K, Mg, P, Cu, Mn, Fe and Na), fiber, proteins, carbohydrates, phenolic and flavonoid compounds, vitamins, and carotenoid compounds (Isopencu and Ferdes, 2012; Hedges and Lister, 2007). The existence of glycosides, steroids, tannins, saponins, glycosides, and reducing sugars in dill essential oil has also been previously found (Dahiya and Purkayastha, 2012). The essential oil yield of dill plant is almost 0.3%–0.4% (Saleh-e-In et al., 2010). Volatile oil is present in different parts of dill plant such as leaves, flowers, and fruits. The chemical composition of dill volatile oil varies depending on the plant parts. The major chemical components of volatile oil are phellandrene, limonene, and carvone. The flavor and antiflatulent property of dill and its volatile oil are attributed to carvone. Some other compounds such as monoterpene hydrocarbons and oxygenated monoterpenes are also present in the dill volatile oil (Saleh-e-In et al., 2010). These compounds include α-pinene, α-thujene, n-heptacosane, sabinene, β-myrcene, p-cymene, n-heneicosane, p-menth-3-en-2-one, cis-dihydrocarvone, *trans*-dihydrocarvone, n-docosane, α-copaene, cadinol, γ-muurolene, n-nonadecane, p-menth-1,8-dien-6-ol, and neophytadiene. Oil obtained from dill weed is used in the food industry for flavoring. Seed oil is used to produce soaps and perfumes (Wright, 2010). Some important components of dill are shown in Fig. 18.2.

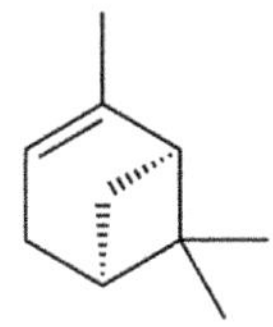

Alpha-pinene

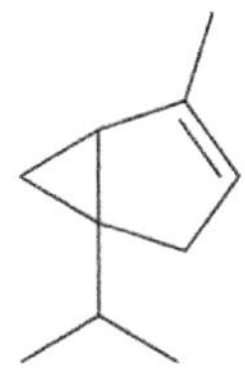

α-Thujene

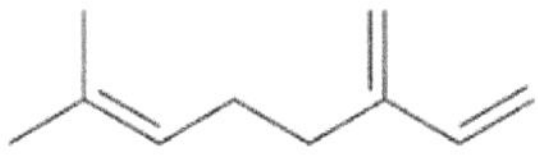

β-Myrcene

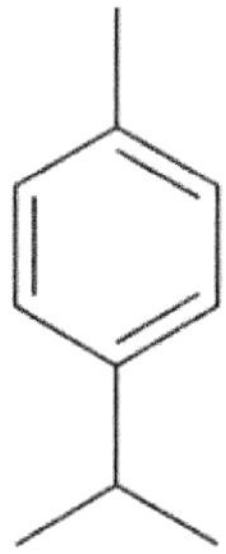

p-Cymene

FIGURE 18.2 Some important components of dill.

3. POSTHARVEST TECHNOLOGY

The best time for harvesting of dill is early in the morning, before the plant is fully grown and flower buds are fully open. At this time, the plant has its highest moisture content and gives its best aroma, as well as essential oil yield, and has minimum chance of seed shattering. The harvested plant can be stored for 2 to 3 days under cooled conditions, usually in a refrigerator. The plant can be freeze dried for longer preservation and to maintain the aroma and texture. Another way to keep the leaves fresh for the period of a day is to put the stem in a cup of water. Mostly, chefs prefer dry dill leaves to fresh as they give better flavor. Several methods have been reported for the preservation of dill weed. One of the most common and conventional method is the air-drying method. In this method, plant material is layered on filter paper and placed in a dark war room with proper air circulation (Wright, 2010).

4. PROCESSING

Whole plant, leaves, and seeds, depending on your interest, can be preserved by the drying method. Cut the stem or leafy area of the plant by a sharp knife or scissors. Wash the cut leaves and stem with clean water to remove dust particles and insects. Dry the cut leaves in open air in a dark room and remove the stem when the seeds are fully ripen and turn a brown color. Fresh dill stems give superior aroma and grassy flavor compared to air dried, which is the way it is added at the end of the cooking process in food dishes. With the passage of time, fresh leaves lost aromatic compounds, so dry leaves have less flavor than fresh leaves. On the other hand, dill seeds have a stronger aroma than leaves, and their flavor is enhanced by drying and with the passage of time. Seeds have been used in those dishes where we need a strong aroma (Wright, 2010).

The essential oil obtained from the aboveground parts of dill plants, including unripe fruits, is known as "dill herb oil." It is obtained from fresh, tender herbs by steam distillation for 2–4 h. Mature herbs usually are steam distilled for 8–10 h for the complete extraction of oil. The oil obtained from the mature herbs is usually rich in carvone content and resembles dill fruit oil. Herb oil containing less than 20% carvone is preferred for the flavoring purposes. The typical flavor of herb oil is due to the presence of phellandrene. The Europeans dill herbs yield about 0.29%–1.5% oil, containing caraphene, carvone, dillapiole, isomyristicin, limonene, myristicin, n-octyl alcohol, α-phellandrene, α-pinene, terpinene, waxes, etc.

5. VALUE ADDITION

Dill seed and leaves are an important ingredient of Asian food dishes. Mostly, they are added in dips, rice, chicken, breads, salads, vinegar, potato chips, vegetables, sauces, cheese, bakery products, creams, yogurts, fishes, meat, food snacks, and condiments as a flavoring agent.

6. USES

Dill has been used in food, soap, flavorings, cosmetics, mouthwash, fragrances, and the medicinal industry. In the Middle Ages, this plant had been used against black magic and witchcraft. In traditional medicinal systems, it has been used for the treatment of digestion issues, liver problems, cold, spasms, menstruation cramps, loss of appetite, throat swelling, gallbladder complaints, kidney diseases, nerve pain, bronchitis, genital ulcer, sleeping disorders, and hemorrhoids. (Wright, 2010).

The volatile oil isolated from dill leaves and seeds has been used in the food industry as a flavoring agent for drinks, meat products, baked goods, cheese, yogurts, snacks, and condiments (Wright, 2010).

Dill (*A. graveolens)* is used as for antiinflammatory, analgesic, antimicrobial, antisecretory, gastric mucosal protective, and smooth muscle relaxant effects, increased progesterone concentration, hyperlipidemic, and many other medicinal effects (Al-Snafi, 2014).

7. PHARMACOLOGICAL USES

7.1 Antibacterial Activity

By using a micro-broth dilution assay evaluated process, dill showed antibacterial activity against 11 microorganisms. It has also shown antibacterial potential against five bacteria strains with average MIC values of 10 mg/mL (Ruangamnart et al., 2015).

7.2 Antifungal Activity

Wild growing dill showed strong antifungal activity against seven tested fungi. It was observed that dill inhibited the microbial growth as low as 1/500 v/v MICs against all strains. The highest antifungal potential was observed against *Alternaria alternate*, where the observed MIC value was 1/6500 v/v (Meddah et al., 2015).

7.3 Antioxidant Activity

Ethyl acetate, ethanol, n-hexane, and dichloromethane extracts of dill (*A. graveolens* L.) plants were grown under organic as well as conventional conditions. Their inhibitory effect against butyrylcholinesterase, tyrosinase, and acetylcholinesterase was tested. Different antioxidant assays were used to evaluate the antioxidant properties of the dill essential oil. It was observed that dill essential oil showed good antioxidant potential against different antioxidant assays (Orhan et al., 2013).

8. SIDE EFFECTS AND TOXICITY

Dill is likely to be safe in food and medicinal application. However, dill essential oil causes skin irritation. A similar effect was observed with the application of fresh dill juice to skin.

References

Meddah, B., Khaldi, Achraf, Moussaouia, Abdellah, Sonnet, Pascal, Akermy, Moulay, M.h., 2015. Chemical composition and antifungal activity of essential oil of *Anethum graveolens* L. from South-Western Algeria (Bechar). Journal of Chemical and Pharmaceutical Research 7, 615–620.

Al-Snafi, A., 2014. The pharmacological importance of Anethum graveolens—A review. International Journal of Pharmacy and Pharmaceutical Sciences 6, 11–13.

Castleman, M., Hendler, S.S., 1995. The Healing Herbs: The Ultimate Guide to the Curative Power of Nature's Medicines. Bantam.

Dahiya, P., Purkayastha, S., 2012. Phytochemical analysis and antibacterial efficacy of dill seed oil against multi-drug resistant clinical isolates. Asian Journal of Pharmaceutical and Clinical Research 5, 62–64.

Duke, J.A., Duke, P.-A.K., 1983. Medicinal Plants of the Bible. Trado-medic books.

Halberstein, R.A., 2005. Medicinal plants: historical and cross-cultural usage patterns. Annals of Epidemiology 15, 686–699.

Hedges, L., Lister, C., 2007. Nutritional Attributes of Herbs, Crop and Food Research Confidential Report.

M. Hegazi, M. Metwaly, E. Belal, Influence of Plant Growth-Promoting Bacteria (PGPB) on Coriander (*Coriandrum Sativum*, L.) and DILL (*Anethum Graveolens*, L.) Plants.

Heilmeyer, M., 2007. Ancient Herbs. Getty Publications.

Hemphill, I.R., 2000. Spice Notes: A Cook's Compendium of Herbs and Spices. Pan Macmillan.

Hobhouse, P., 2004. Plants in Garden History. Pavilion.

Isopencu, G., Ferdes, M., 2012. The effect of *Anethum graveolens* upon the growth of *E. Coli*. U.P.B. Scientific Bulletin, Series B 74.

Jana, S., Shekhawat, G., 2010. Anethum graveolens: an Indian traditional medicinal herb and spice. Pharmacognosy Reviews 4, 179.

Orhan, I.E., Senol, F.S., Ozturk, N., Celik, S.A., Pulur, A., Kan, Y., 2013. Phytochemical contents and enzyme inhibitory and antioxidant properties of *Anethum graveolens* L.(dill) samples cultivated under organic and conventional agricultural conditions. Food and Chemical Toxicology 59, 96–103.

Radulescu, V., Popescu, M.L., Ilies, D.-C., 2010. Chemical composition of the volatile oil from different plant parts of *Anethum graveolens* L.(Umbelliferae) cultivated in Romania. Farmacia 58, 594–600.

Ruangamnart, A., Buranaphalin, S., Temsiririrkkul, R., Chuakul, W., Pratuangdejkul, J., 2015. Chemical compositions and antibacterial activity of essential oil from dill fruits (*Anethum graveolens* L.) cultivated in Thailand. Mahidol Univeristy Journal of Pharmaceutical Sciences 42, 135–143.

Saleh-e-In, M.M., Sultana, A., Husain, M., Roy, S.K., 2010. Chemical constituents of essential oil from *Anethum sowa* L. Herb (leaf and stem) growing in Bangladesh. Bangladesh Journal of Scientific & Industrial Research 45, 173–176.

Small, E., Canada, N.R.C., 1997. Culinary Herbs. NRC Research Press.
Tucker, A.O., DeBaggio, T., 2000. Big Book of Herbs. Interweave Press.
Wright, J., 2010. The Herb Society of Americas Essential Guide to Dill. American press.

CHAPTER

19

Fennel

Rafia Javed[1], *Muhammad Asif Hanif*[1], *Muhammad Adnan Ayub*[2], *Rafia Rehman*[1]

[1] Department of Chemistry, University of Agriculture, Faisalabad, Pakistan;
[2] Department of Chemistry, University of Okara, Okara, Pakistan

OUTLINE

Medicinal Plants of South Asia
https://doi.org/10.1016/B978-0-08-102659-5.00019-7

Copyright © 2020 Elsevier Ltd. All rights reserved.

1. BOTANY

1.1 Introduction

Fennel (*Foneiculum vulgare*) (Fig. 19.1) belongs to the family Umbelliferae and is commonly known as "saunf." This perennial herb has different medicinal uses and is grown almost all over the world. Different parts of fennel like stalks, leaves, and fruits have been used in cooking for meat dishes, pickles, confectioneries, pastries, sauces, and soups (Khan and Musharaf, 2014). Subtropical and temperate regions are suitable for its cultivation. Ideal growth of fennel herb can be obtained with well-drained, rich, and moist soil. The major fennel growing countries are Pakistan, India, Italy, Russia, the United States, Germany, and France. It is mainly grown in Pakistan and India. English recognizes it as fennel, Punjabi as saunf, Hindi as bari saunf, the Sanskrit name is madhurika, Unani name is baadiyaan, and the Ayurvedic name of fennel is mishreyaa (Kaur and Arora, 2010).

1.2 History/Origin

Fennel is cultivated in temperate areas, but it is native to the Mediterranean region and Southern Europe (He and Huang, 2011). The

FIGURE 19.1 Fennel plants, flowers, and fruits (seeds).

first settlers on Madeira were given the name Funchal due to abundance of wild fennel. This word derived from the word *"funcho"* (*fennel*) and the suffix *"-al"* (Frutuoso, 1873). Spanish missionaries cultivated fennel in North America in their medical gardens. Cultivation of fennel spread from mission gardens to California, where it was grown as wild anise. In New England colonies and kitchen gardens, this herb was brought by English settlers (Malhotra, 2012). Fennel was given the name "marathron" by the ancient Greeks. A battle was fought in a field of marathron, so it was named after this plant as the Battle of Marathron (Maheshwari et al., 2014). Due to its edible shoots and aromatic fruits, ancient Romans cultivated it.

Pliny considered its remedial and medicinal properties and observed that serpents use juice of this plant by rubbing it to sharpen their sight. Before Norman Conquest, it was mentioned in Anglo-Saxon medical and cookery recipes. It was not cultivated in northern Europe, although Romans used its shoots as vegetable. In Spanish agriculture dating CE 961 fennel seeds, fennel water and fennel shoots were used. Charlemagne diffused the plant in Central Europe where it was cultivated in imperial farms. In ancient China, fennel was considered a snake bite remedy, and Egyptians also used it as medicine and food (Blumenthal, 1998; Bown, 2001).

1.3 Demography/Location

Fennel is grown in moderate climates in early spring and the winter season. It is a Mediterranean perennial herb. It is mostly found near the sea in dry, stony, calcareous soils. Butterflies and beneficial insects are attracted toward gardens due to this plant. A pH of approximately 5.5–6.5, well-drained, rich soil, and full sun light are required for its proper growth. The main international suppliers of fennel are Egypt, China, and India. Rabi season is suitable for fennel growth in Pakistan and India. The annual production of fennel is 650 tons in Pakistan (lMa Yousef, 2008).

1.4 Botany, Morphology, Ecology

Fennel is a branched, perennial herb that has hairy, feathery, and soft foliage and grows up to 2 m tall. Due to its anise-flavored seeds and foliage, it is grown in herb and vegetable gardens and harvested for its usage in cooking. Leaves of fennel are branched and cut into the fine segments, while the stem is smooth and seems polished, bright green, cylindrical, and erect. Fennel leaves have filiform (thread-like) segments, which are finely dissected, and grow up to 40 cm and 0.5 mm wide. The flowers bloom in July and August, and 13–20 rays of bright golden flowers grow in large and flat terminal umbels. On short pedicels of each umbel section, 20–50 tiny yellow flowers are present, and these flowers produce terminal compound umbels that are 5–15 cm wide. Fruits are 1.5–2.0 mm broad, 0.12–0.2 inches (3 to 5 mm) long, ridged, and oblong to ovoid in shape. Strong ribs are present on elongated fruits. Dry seeds are fruits of fennel that are grooved. The fennel seeds are obtuse at the ends and have slightly curved and elliptical five lines in length. Fennel is of two types. Butterflies are attracted toward yellow flowers that use fennel as a host plant for their eggs and caterpillars and for food. Florence fennel (*Foeniculum vulgare* var. azoricum) is grow up to 2–3 feet in height.

The fennel term is derived from hay because its color is greenish-yellow. Wild fruits are less flavored than sweet fennel and have dark color blunt at their ends. The seeds, leaves, stalk, and bulb all are edible. Just before the seeds ripen, the seed heads are harvested. Along with all the healing properties the fennel seeds work as fabulous flavor enhancers and are sweetish in taste. The seeds ripening period is September to October. Fennel reproduce easily from seeds but can also reproduce from crown or root (Badgujar et al., 2014).

2. CHEMISTRY

Volatile oil is present in fennel seeds. *Trans*-anethole and fenchone are major components. Alpha-pinene, camphene, and limonene are also present in essential oil. The extent of each of these chemical constituents varies depending on cultivation conditions such as soil type, weather, irrigation, and other horticulture practices. Different parts of fennel like roots, leaves, and seeds are safe and edible. But even a small amount of its essential oil is toxic (Afify et al., 2011).

Potassium, fiber, and calcium are present in considerable amount, but vitamin C is present in a high amount in fennel bulb. The mineral and vitamin profile of fennel seeds include nitrates weight ranging from 650 to 3767 mg/kg, pseudo bulb weight from 199 to 383 g, dry matter 61–75.8 g/kg, Na 77–512 mg/kg, Magnesium 82–389 mg/kg, potassium 4241 to 5851 mg/kg, calcium 56–363 mg/kg, dietary fiber 5.75–7.59 g/kg, and vitamin C 87–347 mg/kg (Koudela and Petrikova, 2008). Inflorescence contains the lowest moisture content (71.31 g/100 g), while stems and leaves exhibit the highest moisture content (77.46 and 76.36 g/100, respectively). The less abundant macronutrients are fats, reducing sugars, and proteins. The amount of protein present in inflorescence is 1.37 g/100 g, while in the stem, it is 1.08 g/100 g. The richest macronutrients in all parts are carbohydrates, ranging from 18.44 to 22.82 g/100 g. From the fresh portion of 100 g of these parts, approximately 94 Kcal of energy can be obtained. Stem and leaves give the lowest energy, while inflorescence give highest values of energy. In fennel seeds, the total phenolic contents are 7.55 mg GAE/g and 1.1% of essential oil (Badgujar et al., 2014).

Flavonoids aglycones, flavonoid glycosides, and hydroxyl cinnamic acid derivatives have been reported in fennel (Parejo et al., 2004). Fruit of this plant contains apigenin, quercetin, rosmarinic acid, cinnamic acid, hesperidin, 1,5 di-caffeoylquinic acid, ferulic acid, quercetin-7-*o*-glucoside, ferulic acid-7-*o*-glucoside, *p*-coumaric acid, caffeic acid, chlorogenic acid, gallic acid, and neochlorogenic acid (Gutiérrez-Grijalva et al., 2018). While, chlorogenic acids, rosmarinic acid, 1, 5-*O*-di-caffeoylquinic acid, 1, 4-*O*-di-caffeoylquinic acid, 1, 3-*O*-di-caffeoylquinic, 5-*O*-caffeoylquinic

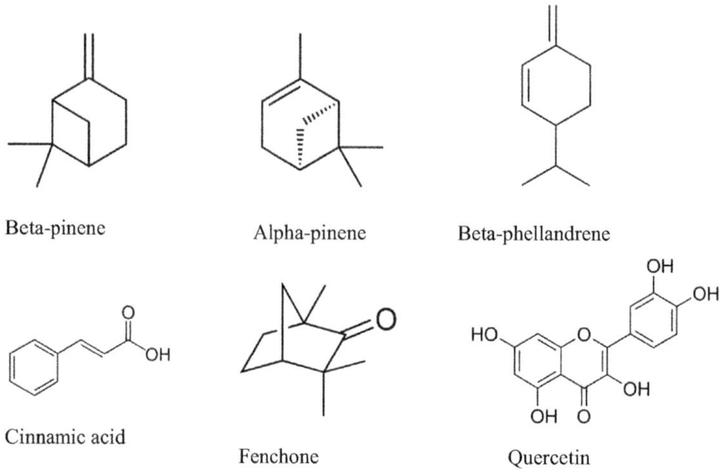

FIGURE 19.2 Structure of some important chemicals present in fennel.

acid, 4-*O*-caffeoylquinic acid and 3-*O*-caffeoylquinic acid, apigenin, and quercetin are the major phenolics and flavonoids present in fennel seeds. Moreover, the concentration of flavonoid compounds is less than phenolic compounds (Faudale et al., 2008). Recently, two novel compounds 3′,8′-binaringenin and 3,4-dihydroxyphenethylalchohol-6-*O*-caffeoyl-*β*-D-glucopyranoside were separated from wild fennel (Ghanem et al., 2012). The most prevalent flavonoids in fennel are isorhamnetinglucoside, kaempferol-3-arabinoside, kaempferol-3-glucuronide, quercetin-3-arabinoside, isoquercitrin, and quercetin-3-glucuronide (Kunzemann and Herrmann, 1977). Aqueous extract of fennel revealed kaempferol-3-*O*-glucoside, kaempferol-3-*O*-rutinoside, and quercetin-3-*O*-galactoside. Structures of important chemical constituents of fennel are shown in Fig. 19.2.

3. POSTHARVEST TECHNOLOGY

Fresh leaves give optimum flavor of fennel. So, at the time of blooming, it is harvested. The proper harvest time of fennel seeds is when the color of green seeds turns brown on the stalk. In late summer, mostly the color changing takes place. To avoid the spreading of seeds, these are collected carefully. The fennel bulb is harvested when it reaches the size of a small tennis ball. A sharp knife is used for the cutting the bulb from the base line. Seed heads are placed in paper bags and shaken to isolate the seeds. After separation of seeds from stem, these seeds are placed on a warm and ventilated screen. Airtight containers of fennel seeds are placed in a cool, dark place. In this way, seeds will remain flavored for months. Fennel bulbs are stored in the fridge and used within few days to get the best flavor. Well-dried seeds can be stored for more than 6 months (Telci et al., 2009).

4. PROCESSING

For preservation of fennel for home usage, drying is the most suitable way. On drying, fennel leaves lose quite a bit of flavor. When seed heads are on the plant, place small paper bags over them. To dry the seed heads, hang the seed head bags upside down in a dry and airy place. Within a few weeks the seeds should be separate from heads. The vacuum-sealed bags or airtight jars can be stored in a dark and dry place, even in the freezer. Fennel seeds are used in stuffing for fatty meats and fish, cakes, breads, salad dressings, stews, soups, and sauces. For the cooking of meat, the Mediterranean use dried stems. Fennel can be dried by gas stove by spreading it on cookie sheets or in the oven. The leaves will dry quickly because the fronds are so feathery (Sidhu et al., 2007).

Fennel leaves remain suitable for cooked dishes, but they lose their crisp texture during freezing. Leaves can be frozen as whole or chopped form after stripping leaves off the stalk. Leaves are dried before freezing (Green, 2006). Leaves can be placed in an oven at 120°F for 4–6 hours for drying purpose. Freeze drying is an alternative method for long-term storage of fennel leaves. Airtight containers and plastic bags are used for storage. The plastic bags or containers should be vapor and moisture resistant so they will not crack at low temperature. These plastic bags should not absorb odors or flavors. Dry fennel leaves and seeds can be stored for 6 months, but frozen seeds and leaves are stored for 1 year at 0°F (McCormack, 2004).

5. VALUE ADDITION

Fennel is used daily in baked, grilled, boiled, and stewed dishes, snacks, and salads in the raw form. The light-green dye is used as a food colorant for coloring textiles or wooden items. For wool fabrics, the natural dye obtained from fennel leaves gives a light brown hue and the flower produces a yellow tint. Moreover, fennel is used as a food preservative. Different parts of fennel like seeds, leaves, fruit stem, and whole plant are used in different traditional food dishes in Bangladesh, India, Pakistan, Spain, Italy, and Portugal. In Pakistan and India, uncoated and sugar-coated seeds are used as *mukhwas* (mouth freshener), which are helpful as a digestive aid (Badgujar et al., 2014).

6. USES

The roots, leaves, and seeds of fennel are edible and used in stuffing various foods. Grip water is made by mixing its water with sodium bicarbonate and a syrup like dill water or anise water. It is used to ease flatulence in children. In sundry desserts and sweet meats, the dry leaves and seeds are employed (Barros et al., 2010). For a long time, fennel has been used for remedial and culinary purposes. For enhancement of appetite, fennel has also been used in Italian cuisine. In some natural toothpastes, fennel is also used as a flavoring and antibacterial agent. In salads, the fresh roots are added sometimes. Fennel is used as a remedy for flatulence, digestive aid, and acts as a carminative. Dried seeds speed up the digestion of fatty foods, ease stomach pains, and are antispasmodic. A mild infusion of the leaves is safe and remedial for colic in infants. It is also beneficial for colds and cough. Problem with bronchial tissues can be cured by fennel. The phlegm loosening in lungs can be improved by alpha-pinene that acts as an expectorant. Eye strain and irritation reduction and improvement in eyesight can be obtained by crushed seeds of fennel. Fennel has also been used as a galactagogue. In nursing mothers, seeds boiled in barley water enhance the flow of milk. Intestinal bacteria and hook worm can be expelled by using fennel seeds and leaf tea (McIntyre, 1997; Ody, 1993). Fennel is used for protection purpose. It is employed as a talismanic herb because it is known to protect the individual from negativity and harm (Lepe-Camacho).

Botanical mosquito repellents, which cause little risk to the environment or human health, may be feasible alternatives to synthetic chemical repellents such as DEET (N, N-diethyl-meta-toluamide). Thus, many people prefer to use natural repellents extracted from plants, such as citronella oil from *Cymbopogon nardus*, p-menthane-3,8-diol obtained from *Eucalyptus maculate citriodora*, and fennel oil (Yoon et al., 2015). Fennel essential oil has insecticidal properties against *Tribolium confusum*, and the mortality rate was found to be dose dependent (Li et al., 2011). The fruit-derived phytochemical constituents ((+)-fenchone, estragole, and phenylpropenes (E)-anethole etc.) of *Foeniculum vulgare* exhibited prominent insecticidal activities against *Lasioderma serricorne, Callosobruchus chinensis,* and *Sitophilus oryzae*. From ancient times, fennel has been used to treat different diseases like stomachache, mouth ulcer, liver pain, leucorrhoea, laxative, kidney ailments, irritable colon, insomnia, gastritis, gastralgia, flatulence, fever, emmenagogue, dieresis, diarrhea, depurative, constipation, conjunctivitis, colic in children, cancer, arthritis, aperitif, antiemetic, and abdominal pains. Fennel seeds have been used as an ingredient for removing any foul smell of the mouth (Badgujar et al., 2014).

7. PHARMACOLOGICAL ACTIVITIES

7.1 Anticancer Activity

Anethole is the main active component of fennel seeds oil and exhibits anticancer activity. In Swiss albino mice, the anticancer activity of anethole was studied through Ehrlich ascites carcinoma induced in a tumor model. The study revealed that anethole reduces the volume and weight of tumor and also increases survival time. Significant cytotoxic effect was also produced in the Ehrlich ascites tumor cells in the paw. It also decreased the levels of nucleic acids and malondialdehyde, and it increased glutathione concentrations (Mohamad et al., 2011). Anticancer activity of methanol extract of fennel was evaluated by different anticancer assays against B16F10 melanoma cell line. The results showed that fennel extract revealed significant anticancer activity with 200 μg/mL MIC value in Trypan blue exclusion assay. In micronucleus assay, standard drug doxorubicin showed 0.018% micronucleus, while fennel methanol extract exhibited 0.006% micronucleus (Pradhan et al., 2008).

7.2 Antimicrobial and Antiviral Activities

Fennel has ethnic remedial property against several infectionous disorders of myco-bacterial, viral, fungal and bacteria. Several studies have been conducted to validate antimicrobial, antimycobacterial, and antiviral potential of fennel (Miraj and Kiani, 2016).

7.3 Antiinflammatory Activity

Fennel fruit methanol extract inhibited 69% paw edema and 70% ear edema in mice. Moreover, this concentration significantly increased serum transaminase, aspartate aminotransferase, and alanine aminotransferase levels.

7.4 Antioxidant Activities

Fennel seed extracts exhibited excellent antioxidant activity determined through different assays such as reducing power, total antioxidant, hydrogen peroxide scavenging, DPPH free radical scavenging, superoxide anion radical scavenging, and metal chelating activities and reducing power (Anwar et al., 2009; Chang et al., 2013; Oktay et al., 2003; Roby et al., 2013).

7.5 Antiallergic Activity

After oral intake of 200 mg/kg once a day for 7 days the methanolic extract of fennel fruit showed significant inhibitory effect on DNFB-(2,4-dinitrofluorobenzene)-induced delayed-type hypersensitivity. The possible immunosuppressive properties of fennel are expected to be due to an inhibitory effect on immunologically induced swelling (Choi and Hwang, 2004).

7.6 HepatoProtective Activity

Essential oil of fennel seeds revealed a potent hepatoprotective effect against acute hepatotoxicity produced by carbon tetrachloride in rats. Oral administration of essential oil decreased the levels of serum alanine amino transferase, alkaline phosphatase, bilirubin, alkaline phosphatase, and aspartate aminotransferase, as compared to the control group. Compounds like D-limonene and β-myrcene present in essential oil may responsible for protection of the liver from CCl_4 toxicity (Özbek et al., 2003).

7.7 Anxiolytic Activity

The ethanolic extract of fennel fruit was tested for its anxiolytic activity (Kishore et al., 2012). The 100- and 200-mg dose of extract per kg of body weight of animal showed significant results as compared to 1 mg/kg diazepam (reference anxiolytic drug) (Badgujar et al., 2014).

7.8 Antistress Activity

Bioactive compounds isolated from natural resources play a key role in health and treatment of stress-related disorders (Padma and Khosa, 2002). Fennel extract showed a significant antistress activity against the stress induced by vigorous swimming of test animals. To evaluate antistress activity, different parameters like urinary levels of vanillylmandelic acid and ascorbic acid in rats were studied. Fennel extract at a concentration of 200, 100, and 50 mg/kg body weight exhibited an improvement in urinary levels of vanillylmandelic acid ($P < .001$) and excretion level of ascorbic acid ($P < .001$). In conclusion, the extract of fennel acted as an antistress agent (Koppula and Kumar, 2013).

7.9 Memory-Enhancing Property

A previous study show that fennel extract possesses memory-enhancing properties (Koppula and Kumar, 2013). Scopolamine-induced amnesia in the extract-treated groups took 3–5 days in recovery compared to the control group (normal group), which took 6 days to recover.

7.10 Nootropic Activity

Methanolic extract of fennel was administered for 8 days, successively, which resulted in deficits of scopolamine and memory-induced aging in mice. This suggested that fennel may be used in the treatment of dementia and Alzheimer diseases (Joshi and Parle, 2006).

7.11 Antihirsutism Activity

The ethanolic extract of fennel seed was mixed in a cream that was used against idiopathic hirsutism. The effectiveness of the cream was more in the case of 2% fennel extract than 1% fennel. By using the cream, 0%, 2%, and 1%, the hair diameter reduced 0.5%, 18.3%, and 7.8% (Javidnia et al., 2003).

7.12 Estrogenic Properties

The administration of acetone extract (50, 150, and 250 μg/100 g body wt) of fennel seeds resulted in an increase of total concentrations of nucleic acids and protein as well as an increase in the weight of mammary glands and oviducts (Devi et al., 1985). Fennel oil significantly decreases uterine contraction, which is induced by prostaglandin E2. Thus, the extracts of fennel have strong estrogenic activity (Ostad et al., 2001).

7.13 Expectorant Activity

Fennel seeds enhance the external transport of extraneous corpuscles by stimulating the ciliary motility of the respiratory apparatus. This showed that fennel is used in particularly polluted environments and in treating bronchial and bronchopulmonary afflictions. The smooth muscle contraction of the trachea can be stimulated by inhalation of fennel volatile oil. This results in ease of the expectoration of bacteria, mucus, and other corpuscles extraneous to the respiratory tracts (Reiter and Brandt, 1984).

7.14 Anticolitic Activity

Essential oil of fennel reduces intestinal gas and regulates the motility of smooth muscles of the intestine (Chakŭrski et al., 1980).

7.15 Antinociceptive Activity

Antinociceptive activity was observed in fennel methylene chloride extract and *n*-hexane extract, being less than a peripheral antinociceptive reference drug (acetyl salicylic acid) (Nassar et al., 2010).

7.16 Diuretic Activity

The ethanolic extract of *Foeniculum vulgare* fruit revealed excellent diuretic activity (Bardai et al., 2001).

7.17 Cardiovascular Activity

An aqueous extract of *Foeniculum vulgare* leaves possesses potential cardiovascular action. A significant dose-related reduction was obtained in arterial blood pressure, without affecting the respiratory rate and heart rate, by an intravenous administration of lyophilized boiled water extract of fennel leaves. Very little hypotensive activity was observed by nonboiled aqueous extract (Abdul-Ghani and Amin, 1988).

7.18 Antimutagenic Effects

Foeniculum vulgare inhibits the oxidative stress induced by cyclophosphamide (Tripathi et al., 2013).

7.19 Gastrointestinal Effects

The aqueous extract of *Foeniculum vulgare* showed remarkable antiulcerogenic effect. Fennel fruit had clearly a protective effect against ethanol-induced gastric mucosal lesion in rats (Birdane et al., 2007).

7.20 Antipyretic Activity

Ethanolic extract of fennel fruit showed a moderate antipyretic activity against hyperpyrexia (Badgujar et al., 2014).

7.21 Hypoglycemic Activity

The essential oil of fennel exhibits potential hypoglycemic and antioxidant activity against streptozotocin-induced diabetes in rats (Badgujar et al., 2014).

7.22 Antispasmodic Activity

An alcoholic extract of the fruits of fennel possesses antispasmodic activity, which inhibited the acetylcholine and histamine-induced guinea pig ileal contractions in vitro. An essential oil extracted from the fruits of fennel inhibited oxytocin and prostaglandin (Forster et al., 1980).

7.23 Human Liver Cytochrome P450 3A4 Inhibitory Activity

Human liver cytochrome P450 3A4 inhibitory activity was observed due to 13 compounds isolated from the methanolic extract of fennel. Among these compounds, 5-methoxypsoralen (5-MoP) gave the highest inhibition with an IC50 value of 18.3 μm (Zaidi et al., 2007).

7.24 Antiaging Effects

A formulation containing 4% concentrated seed extract of fennel showed notable antiaging effect with supporting experimental data related to transepidermal water loss and skin moisture (Rasul et al., 2012).

7.25 Bronchodilatory Effects

Ethanol extract and essential oil extracted from fennel exhibited bronchodilatory activity. Moreover, anethole bears a striking resemblance to the catecholamines epinephrine, norepinephrine, and dopamine. This structural similarity appears to be responsible for the various sympathomimetic activities of fennel including bronchodilatory effect (Albert-Puleo, 1980).

8. SIDE EFFECTS AND TOXICITY

In nursing mothers, the fennel seeds boiled in barley water enhance the flow of milk. However, fennel seems unsafe during breastfeeding, as it could possibly damage the nervous system of infants. Any solid conclusions cannot be drawn without further studies. Fennel might slow blood clotting. Fennel might act like estrogen in hormone-sensitive conditions such as uterine cancer, ovarian cancer, breast cancer, endometriosis, or uterine fibroids (Romm, 2017).

References

Abdul-Ghani, A.-S., Amin, R., 1988. The vascular action of aqueous extracts of *Foeniculum vulgare* leaves. Journal of Ethnopharmacology 24, 213–218.

Afify, A.E.-M.M.R., El-Beltagi, H.S., Hammama, A.A.E.-A., Sidky, M.M., Mostafa, O.F.A., 2011. Distribution of trans-anethole and estragole in fennel (*Foeniculum vulgare* Mill) of callus induced from different seedling parts and fruits. Notulae Scientia Biologicae 3, 79–86.

Albert-Puleo, M., 1980. Fennel and anise as estrogenic agents. Journal of Ethnopharmacology 2, 337–344.

Anwar, F., Ali, M., Hussain, A.I., Shahid, M., 2009. Antioxidant and antimicrobial activities of essential oil and extracts of fennel (*Foeniculum vulgare* Mill.) seeds from Pakistan. Flavour and Fragrance Journal 24, 170–176.

Badgujar, S.B., Patel, V.V., Bandivdekar, A.H., 2014. *Foeniculum vulgare* Mill: a review of its botany, phytochemistry, pharmacology, contemporary application, and toxicology. BioMed Research International 2014.

Bardai, S.E., Lyoussi, B., Wibo, M., Morel, N., 2001. Pharmacological evidence of hypotensive activity of Marrubium vulgare and *Foeniculum vulgare* in spontaneously hypertensive rat. Clinical and Experimental Hypertension 23, 329–343.

Barros, L., Carvalho, A.M., Ferreira, I.C., 2010. The nutritional composition of fennel (*Foeniculum vulgare*): shoots, leaves, stems and inflorescences. LWT-Food Science and Technology 43, 814–818.

Birdane, F.M., Cemek, M., Birdane, Y.O., Gülçin, İ., Büyükokuroğlu, M.E., 2007. Beneficial effects of *Foeniculum vulgare* on ethanol-induced acute gastric mucosal injury in rats. World Journal of Gastroenterology: WJG 13, 607.

Blumenthal, M., 1998. Therapeutic Guide to Herbal Medicines.

Bown, D., 2001. New Encyclopedia of Herbs & Their Uses. DK Pub.

Chakŭrski, I., Matev, M., Koĭchev, A., Angelova, I., Stefanov, G., 1980. Treatment of chronic colitis with an herbal combination of *Taraxacum officinale*, Hipericum perforatum, Melissa officinaliss, Calendula officinalis and *Foeniculum vulgare*. Vutreshni bolesti 20, 51–54.

Chang, S., Bassiri, A., Jalali, H., 2013. Evaluation of antioxidant activity of fennel (*Foeniculum vulgare*) seed extract on oxidative stability of olive oil. Journal of Chemical health risks 3.

Choi, E.-M., Hwang, J.-K., 2004. Antiinflammatory, analgesic and antioxidant activities of the fruit of *Foeniculum vulgare*. Fitoterapia 75, 557–565.

Devi, K., Vanithakumari, G., Anusya, S., Mekala, N., Malini, T., Elango, V., 1985. Effect of *Foeniculum vulgare* seed extract on mammary glands and oviducts of ovariectomised rats. Ancient Science of Life 5, 129.

Faudale, M., Viladomat, F., Bastida, J., Poli, F., Codina, C., 2008. Antioxidant activity and phenolic composition of wild, edible, and medicinal fennel from different Mediterranean countries. Journal of Agricultural and Food Chemistry 56, 1912–1920.

Forster, H., Niklas, H., Lutz, S., 1980. Antispasmodic effects of some medicinal plants. Planta Medica 40, 309–319.

Frutuoso, G., 1873. As Saudades da terra, pelo doutor Gaspar Fructuoso: Historia das ilhas do Porto-Sancto, Madeira, Desertas e Selvagens. Typ. funchalense.

Ghanem, M.T., Radwan, H.M., Mahdy, E.-S.M., Elkholy, Y.M., Hassanein, H.D., Shahat, A.A., 2012. Phenolic compounds from *Foeniculum vulgare* (Subsp. Piperitum)(Apiaceae) herb and evaluation of hepatoprotective antioxidant activity. Pharmacognosy Research 4, 104.

Green, A., 2006. Field Guide to Herbs & Spices: How to Identify, Select, and Use Virtually Every Seasoning at the Market. Quirk books.

Gutiérrez-Grijalva, E., Picos-Salas, M., Leyva-López, N., Criollo-Mendoza, M., Vazquez-Olivo, G., Heredia, J., 2018. Flavonoids and phenolic acids from oregano: occurrence, biological activity and health benefits. Plants 7, 2.

He, W., Huang, B., 2011. A review of chemistry and bioactivities of a medicinal spice: *Foeniculum vulgare*. Journal of Medicinal Plants Research 5, 3.

Javidnia, K., Dastgheib, L., Samani, S.M., Nasiri, A., 2003. Antihirsutism activity of fennel (fruits of *Foeniculum vulgare*) extract—a double-blind placebo controlled study. Phytomedicine 10, 455—458.

Joshi, H., Parle, M., 2006. Cholinergic basis of memory-strengthening effect of *Foeniculum vulgare* Linn. Journal of Medicinal Food 9, 413—417.

Kaur, G.J., Arora, D.S., 2010. Bioactive potential of Anethum graveolens, *Foeniculum vulgare* and Trachyspermum ammi belonging to the family Umbelliferae-Current status. Journal of Medicinal Plants Research 4, 087—094.

Khan, M., Musharaf, S., 2014. *Foeniculum vulgare* Mill. A medicinal herb. Medicinal Plant Research 4.

Kishore, R.N., Anjaneyulu, N., Ganesh, M.N., Sravya, N., 2012. Evaluation of anxiolytic activity of ethanolic extract of *Foeniculum vulgare* in mice model. International Journal of Pharmacy and Pharmaceutical Sciences 4, 584—586.

Koppula, S., Kumar, H., 2013. *Foeniculum vulgare* Mill (Umbelliferae) attenuates stress and improves memory in wister rats. Tropical Journal of Pharmaceutical Research 12, 553—558.

Koudela, M., Petrikova, K., 2008. Nutritional compositions and yield of sweet fennel cultivars-*Foeniculum vulgare* Mill. ssp. vulgare var. azoricum (Mill.). Thell. Hort. Sci 35, 1—6.

Kunzemann, J., Herrmann, K., 1977. Isolation and identification of flavon (ol)-O-glycosides in caraway (Carum carvi L.), fennel (*Foeniculum vulgare* Mill.), anise (Pimpinella anisum L.), and coriander (Coriandrum sativum L.), and of flavon-C-glycosides in anise. I. Phenolics of spices (author's transl). Zeitschrift fur Lebensmittel-untersuchung und-Forschung 164, 194—200.

Lepe-Camacho, A., Natural Wellness Essentials Solutions.

Li, X.H., Ge, L.Y., Wang, J., 2011. Repellent Effects of the Insecticide Based on Porous Starch and Fennel Essential Oil against *Tribolium confusum*, Advanced Materials Research. Trans Tech Publ, pp. 476—479.

lMa Yousef, I., 2008. Accessions from different regions of Pakistan. vulgare Mill. Journal of the Chemical Society of Pakistan 30.

Maheshwari, R.K., Chauhan, A., Mohan, L., Maheshwari, M., 2014. Spice up for scrumptious tang, cologne & wellbeing. Journal of Global Biosciences 3, 304—313.

Malhotra, S., 2012. Fennel and Fennel Seed In: Handbook of Herbs and Spices, second ed., vol. 2. Elsevier, pp. 275—302.

McCormack, J., 2004. Seed Processing and Storage: Principles and Practices of Seed Harvesting, Processing, and Storage. An Organic Seed Production Manual for Seed Growers in the Mid-Atlantic and Southern US. McCormack.

McIntyre, A., 1997. The Medicinal Garden: How to Grow and Use Your Own Medicinal Herbs. Henry Holt.

Miraj, S., Kiani, S., 2016. Study of antibacterial, antimycobacterial, antifungal, and antioxidant activities of *Foeniculum vulgare*: a review. Der Pharmacia Lettre 8, 200—205.

Mohamad, R.H., El-Bastawesy, A.M., Abdel-Monem, M.G., Noor, A.M., Al-Mehdar, H.A.R., Sharawy, S.M., El-Merzabani, M.M., 2011. Antioxidant and anticarcinogenic effects of methanolic extract and volatile oil of fennel seeds (*Foeniculum vulgare*). Journal of Medicinal Food 14, 986—1001.

Nassar, M.I., Aboutabl, E.-s.A., Makled, Y.A., El-Khrisy, E.-D., Osman, A.F., 2010. Secondary metabolites and pharmacology of *Foeniculum vulgare* Mill. Subsp. Piperitum. Revista latinoamericana de química 38, 103—112.

Ody, P., 1993. The Complete Medicinal Herbal, 192pp. Dorling Kindersley, London, ISBN 156458187X. En 120.

Oktay, M., Gülçin, İ., Küfrevioğlu, Ö.İ., 2003. Determination of in vitro antioxidant activity of fennel (*Foeniculum vulgare*) seed extracts. LWT-Food Science and Technology 36, 263–271.

Ostad, S., Soodi, M., Shariffzadeh, M., Khorshidi, N., Marzban, H., 2001. The effect of fennel essential oil on uterine contraction as a model for dysmenorrhea, pharmacology and toxicology study. Journal of Ethnopharmacology 76, 299–304.

Özbek, H., Uğraş, S., Dülger, H., Bayram, I., Tuncer, I., Öztürk, G., Öztürk, A., 2003. Hepatoprotective effect of *Foeniculum vulgare* essential oil. Fitoterapia 74, 317–319.

Padma, P., Khosa, R., 2002. Anti-stress agents from natural origin. Journal of Natural Remedies 2 (1), 21–27.

Parejo, I., Jauregui, O., Sánchez-Rabaneda, F., Viladomat, F., Bastida, J., Codina, C., 2004. Separation and characterization of phenolic compounds in fennel (*Foeniculum vulgare*) using liquid chromatography-negative electrospray ionization tandem mass spectrometry. Journal of Agricultural and Food Chemistry 52, 3679–3687.

Pradhan, M., Sribhuwaneswari, S., Karthikeyan, D., Minz, S., Sure, P., Chandu, A.N., Mishra, U., Kamalakannan, K., Saravanankumar, A., Sivakumar, T., 2008. In-vitro cytoprotection activity of *Foeniculum vulgare* and Helicteres isora in cultured human blood lymphocytes and antitumour activity against B16F10 melanoma cell line. Research Journal of Pharmacy and Technology 1, 450–452.

Rasul, A., Akhtar, N., Khan, B., Mahmood, T., Zaman, S.U., Khan, H., 2012. Formulation development of a cream containing fennel extract: in vivo evaluation for anti-aging effects. Die Pharmazie-An International Journal of Pharmaceutical Sciences 67, 54–58.

Reiter, M., Brandt, W., 1984. Relaxant effects on tracheal and ileal smooth muscles of the Guinea pig. Arzneimittel Forschung 35, 408–414.

Roby, M.H.H., Sarhan, M.A., Selim, K.A.-H., Khalel, K.I., 2013. Antioxidant and antimicrobial activities of essential oil and extracts of fennel (*Foeniculum vulgare* L.) and chamomile (Matricaria chamomilla L.). Industrial Crops and Products 44, 437–445.

Romm, A., 2017. Botanical Medicine for Women's Health E-Book. Elsevier Health Sciences.

Sidhu, K., Kaur, J., Kaur, G., Pannu, K., 2007. Prevention and cure of digestive disorders through the use of medicinal plants. Journal of Human Ecology 21, 113–116.

Telci, I., Demirtas, I., Sahin, A., 2009. Variation in plant properties and essential oil composition of sweet fennel (*Foeniculum vulgare* Mill.) fruits during stages of maturity. Industrial Crops and Products 30, 126–130.

Tripathi, P., Tripathi, R., Patel, R.K., Pancholi, S.S., 2013. Investigation of antimutagenic potential of *Foeniculum vulgare* essential oil on cyclophosphamide induced genotoxicity and oxidative stress in mice. Drug and Chemical Toxicology 36, 35–41.

Yoon, J.K., Kim, K.-C., Cho, Y., Gwon, Y.-D., Cho, H.S., Heo, Y., Park, K., Lee, Y.-W., Kim, M., Oh, Y.-K., 2015. Comparison of repellency effect of mosquito repellents for DEET, citronella, and fennel oil. Journal of parasitology research 2015.

Zaidi, S.F., Kadota, S., Tezuka, Y., 2007. Inhibition on human liver cytochrome P450 3A4 by constituents of fennel (*Foeniculum vulgare*): identification and characterization of a mechanism-based inactivator. Journal of Agricultural and Food Chemistry 55, 10162–10167.

CHAPTER

20

Fenugreek

Sidra Sarwar[1], Muhammad Asif Hanif[1], Muhammad Adnan Ayub[2], Yaw Duah Boakye[3], Christian Agyare[3]

[1] Department of Chemistry, University of Agriculture, Faisalabad, Pakistan; [2] Department of Chemistry, University of Okara, Okara, Pakistan; [3] Department of Pharmaceutics (Microbiology Section), Faculty of Pharmacy and Pharmaceutical Sciences, Kwame Nkrumah University of Science and Technology, Kumasi, Ghana

OUTLINE

Medicinal Plants of South Asia
https://doi.org/10.1016/B978-0-08-102659-5.00020-3

Copyright © 2020 Elsevier Ltd. All rights reserved.

1. BOTANY

1.1 Introduction

Trigonella foenum-groecum L. (fenugreek) (Fig. 20.1) is an annual dicotyledonous plant belonging to the subfamily Papilionaceae, family Fabaceae or Leguminosae (Fig. 20.1). This plant is native to Southern Europe, Asia, and the Mediterranean region (Snehlata and Payal, 2012). The seeds and leaves of the plant are generally used as an ingredient in traditional medicine and as a spice in food preparation in the Indo-Pak subcontinent and other Oriental countries (Syeda et al., 2008). Almost 260 species of this plant may exist, and currently, 18 species have been recognized (Acharya et al., 2006). The uncertainty in the exact number of the species within the genus is largely attributed to the great variability in morphology, growth habit, flower color, leaves, and stem and chemical composition among the constituent species (Svecova and Neugebauerova, 2010). This plant is a highly self-pollinated crop in nature because of its flower structure. Emasculation and manual pollination have been used effectively for crossing different lines of fenugreek. Due to the adaptation of fenugreek for self-pollination, artificial crossing is not easy in this crop. To improve *T. foenum-groecum* genetically, mutation breeding has been recommended. Moreover, hybridization has also been effectively used in this crop for development of specific traits (Snehlata and Payal, 2012).

T. foenum-groecum is known by different names depending upon where you are in the world. In English, it is typically called heyseed, hulba in Arabic, fenegriek in Dutch, sambelil in Farsi, bukkehonrkler in Norwegian, alholva in Spanish, and dari in Persian. In Hindi, Urdu, Punjabi, and Marathi it is called methi, and uluva in Malayalam. Kasuri methi is a type

FIGURE 20.1 Feenugreek plants, seeds and sold forms.

of fenugreek with superior aroma known worldwide being cultivated in Kasur (Punjab, Pakistan) (Ahmad et al., 2016).

1.2 History/Origin

T. foenum-groecum is found growing in Iran and other neighboring countries up to northern India (Al-Asadi, 2014). The species "*foenum-graecum*" means "Greek hay," which indicates its use as a forage crop in the past. One of the oldest medicinal plants is fenugreek (*Trigonella foenum-graecum* L.), belonging to the family Fabaceae, which originated in central Asia approximately 4000 BCE. Seeds of fenugreek were used in trade in 2000——1700 BCE and recovered from India (Punjab). Benefits and the description of fenugreek had been reported earlier in 1500 BCE in the Ebers

Papyrus in Egypt (Ahmad et al., 2016). Fenugreek plant was introduced to Central Europe by Benedictine monks, and it is promoted in the 9th century by Charlemagne. It was grown extensively in the imperial gardens of Charlemagne (Popova, 2017). Fenugreek was grown in the 16th century in England.

1.3 Demography/Location

Fenugreek is an annual, semiarid crop that is grown from seeds by the line sowing method. This crop is quite tolerant to extremely low temperature and frost because it is a cold season crop. This crop can also be cultivated on black cotton soils and is grown widely in various countries including Pakistan, India, Bangladesh, Nepal, Afghanistan, Austria, France, Germany, Greece, Switzerland, Portugal, Spain, Turkey, Russia, the United Kingdom, Egypt, Ethiopia, Kenya, Morocco, Sudan, Tanzania, Tunisia, China, Iran, Israel, Japan, Lebanon, Argentina, Canada, the United States, and Australia. India is the main producer of fenugreek in the world (Edison, 1995; Poynter, 1971). Annual production of the crop ranges from 4352 to 4981 kg ha^{-1} and the maximum yield of the seeds ranges from 1724 to 1886 kg ha^{-1} (Pavlista and Santra, 2016).

1.4 Botany, Morphology, Ecology

This plant is an erect or prostrate, straight or profusely branched plant, almost 20–130 cm in length. The stem of fenugreek is circular to slightly quadrangular in structure, greenish in color, which is often characterized by pinkish color due to the accumulation of anthocyanin under field conditions. The diameter of the stem is 0.5–1 cm, and the leaves are trifoliate, simple, stipulate, and distinctly petiolate with orbicular or oval leaf lamina with an entire margin. Leaf lamina and petioles vary from greenish to pinkish color in the field. The petiole is pubescent, pale green in color, often anthocyanin tinged, and very small in size, ranging 0.5–1.1 mm. The flowers of this plant are yellow in color when young, but on maturity, it turns to a white color and is 1.6–2.2 cm in size. Midsummer is generally the flowering season for the herb fenugreek. Calyx is campanulate, pale green, pubescent, and 6–8 mm in size. An individual sepal is pubescent, pale green in color, and 13–19 mm in size. Corolla is papilionaceous, white, papery, and 1.5–1.9 cm in size. Filament is hyaline and tubular. Ovary is deep green, glaucous, and 1.8–2.5 mm in size. Stigma is pale green in color, glaucous, and 1.5–2.1 mm in size. Style is pale green/hyaline glaucous and 0.2–0.5 mm in size. Pollen grains are oval to circular, ellipsoidal, orbicular, hyaline and when pollen grains are treated with 0.5% acetocarmine, these become stained red or pink. Pollen grains are 70%–90% orbicular and 10%–30% are ellipsoidal in shape. The shape of the seeds is rectangular to oval, and there are deep grooves between the radical and cotyledon. The color of seeds varies from pale brown to golden yellow. Fenugreek requires 5–10 days for germination, while the first trifoliate leaf appears 5–8 days after germination. It is a fast-growing plant, which may grow on dry grasslands, cultivated or uncultivated lands, hillsides, and plains, as well as field edges, but it requires a fair amount of sunlight. Fenugreek needs 4–7 months to reach maturity (Mehrafarin et al., 2011; Montgomery, 2009; Petropoulos, 2003). The flowering period is mid-summer (June to August), and the ripening period of seeds is during late summer (August to September). It is a drought-tolerant plant and grows well in a tropical climate with mild winter and cool summer. However, the development of its leaf and flower is temperature dependent (Chayut et al., 2014). Fenugreek tolerates pH ranges from 5.3 to 8.2. Fenugreek requires watering in dry conditions and also requires full sunlight (MIN, 2011).

2. CHEMISTRY

The composition of fenugreek includes many chemical components. They include alkaloids (trimethylamine, trigonelline, choline, neurin, gentianine, carpaine, and betain), amino acids (isoleucine, 4-hydroxyisoleucine, histidine, leucine, lysine, L-tryptophan, arginine), saponins (graecunins, fenugreekine, trigofoenosides A–G, fenugrin B), steroidal (yamogenin, smilagenin, diosgenin), flavonoids (rutin, quercetin, vitexin, isovitexin), fibers (neutral detergent fiber, gum), coumarin, vitamins, minerals, lipids, mucilage, and proteins (Sowmya and Rajyalakshmi, 1999; Yadav and Kaushik, 2011). Fenugreek contains lipids such as triacylglycerols, diacylglycerols, monoacylglycerols, phosphatidylcholine, phosphatidylethanolamine, phosphatidylinositol, and free fatty acids (Wani and Kumar, 2018).

Fenugreek seed contains high contents of protein and carbohydrates, minerals, vitamins, and phytonutrients and low contents of fiber, ash, and moisture. It is also known as a good source of vitamin A and minerals like calcium, magnesium, iron, zinc, phosphorus, and potassium. Fresh leaves of fenugreek contain about 220.97 mg per 100 g of leaves of ascorbic acid and about 19 mg per 100 g of carotene. About 11 g seeds of fenugreek have 35.5 calories, 6.6 IU of vitamin A, 84.7 mg of potassium, 32.6 mg of phosphorus, 19.4 mg of calcium, and small amounts of vitamin C and other vitamins, minerals, protein, and fibers. Fenugreek seeds are rich in dietary fibers. Dietary fibers comprise 6.28%–9.3% of total weight of fenugreek seed (El Nasri and El Tinay, 2007). The oil of fenugreek is an essential oil, and it should be ingested or applied on skin after dilution.

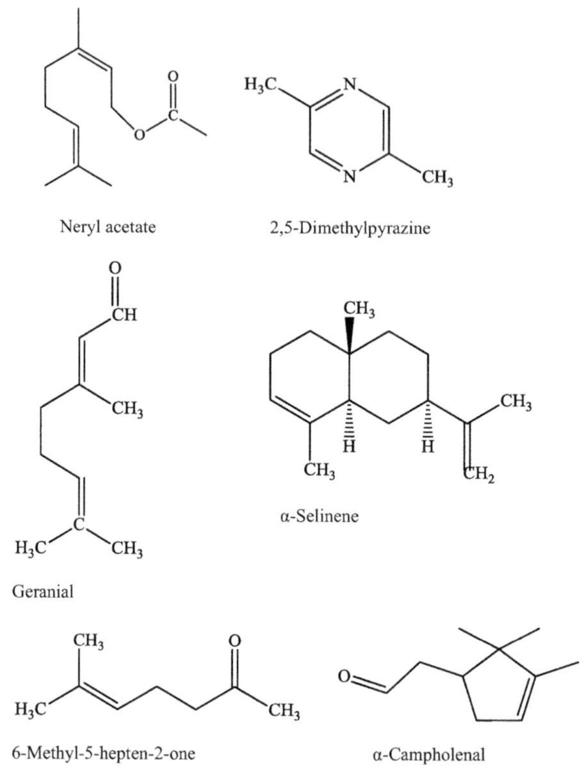

FIGURE 20.2 Fenugreek important bioactive isolates.

Aroma of fenugreek seeds is due to the presence of volatile oils. The stem contains mucilage, bitter fixed oil, protein, and alkaloids. The leaves contain seven saponins called graecunins, which are glycosides of diosgenin. During roasting a large number of alkaloids are degraded to related pyridines and nicotinic acid. These products of degradation are responsible for the aroma and taste of seeds. Fenugreek oil contains aromatic components such as neryl acetate, camphor, β-pinene, β-caryophyllene, 2,5-dimethylpyrazine, geranial, 3-octen-2-one, α-selinene, 6-methyl-5-hepten-2-one, α-terpineol, α-pinene, α-campholenal, and γ-terpinene (Hamden et al., 2011). Fenugreek's important bioactive isolates are shown in Fig. 20.2.

3. POSTHARVEST TECHNOLOGY

It is very important to harvest the crop in time, from late August to early October, as the grains remain small and immature in early harvest and seed losses are due to pod bursting in late harvesting. The best time to harvest the crop is early morning. As the harvesting process is completed, the crop is bound in "Judi" and packed in netted bags or cloth or in bamboo baskets. The crop should be dehydrated in the threshing yard and threshed by the method of trampling under the feet of bullocks. The seeds must be separated and cleaned by the method of winnowing. It was reported that when fenugreek leaves were sun dried, 84.94% ascorbic acid was reduced, and 83.79% ascorbic acid was reduced when fenugreek leaves were oven dried. Fresh leaves of fenugreek are used as vegetables in diets. It was found that there was a better retention of nutrients in the leaves of fenugreek. The leaves of fenugreek must be stored either in refrigeration conditions, dried in an oven, or blanched for some time (about 5 minutes) and should be cooked in a pressure cooker (Yadav and Sehgal, 1997).

4. PROCESSING

Fenugreek like other herbal plants is consumed in a variety of ways and for various purposes. In addition to its fresh leaves, other common processed forms of fenugreek include whole dry leaves, frozen, powdered leaves, and extracted essential oils. First, the plant is harvested, followed by the removal of the inedible parts of the plant and removal of the leaves from branches. The leaves are spread on trays and cooled under 2 to 5°C for 30 minutes. After this, leaves are put into in plastic baskets and overwrapped with cling film to store the leaves (Gomez et al., 2003).

The essential oil is isolated from the seeds of fenugreek by employing steam distillation method. Another method of extraction of fenugreek oil from leaves, stem, and seeds is supercritical carbon dioxide fluid. The temperature of 40°C and pressure of 25 MPa is found optimum for better yields and aroma. The separation temperature is 60°C, and separation pressure is 1 atmospheric pressure (1 atm). The highest extraction ratio is 8.95 mg per g of dry seeds of fenugreek by adding 50% (V/W) absolute alcohol as modifier (Ren and Zhu, 2011).

5. VALUE ADDITION

The flavor of fenugreek herb and spice is the same, but it can be used in both ways. In the form of fresh, dried, and frozen, leaves of fenugreek are available. Fresh, green leaves are folded into fry breads and used in curries. Fresh leaves from the plant (often sold as methi) can be added to salads and cooked dishes. When leaves are dried, they retain most of their flavor and make excellent last-minute additions to curries, sauces, and soup. Fenugreek has some flavor affinities. Fenugreek is the perfect bittersweet counterpart. One of the many components in heavily spiced dishes is fenugreek, and it works well with strong flavors like cumin, coriander, and paprika.

Fenugreek blooms white flowers during the summer and has very aromatic seeds. Seeds of fenugreek can be ground and roasted. The ground seeds are often used in curry powder, spice blends, and dry rubs. A pinch can also be sprinkled over yogurt, cooked greens, or sauce. Fresh leaves can be added to salads and cooked dishes. The tea extracted from the seeds of this plant can be used in the treatment of fever. While both (leaves and seeds) smell like caramel or maple syrup when heated, their taste is rather bitter, like burnt sugar.

Addition of fenugreek fiber to the refined flours helps to fortify them with a balance of soluble and insoluble fiber. Flour fortified with 8%–10% fenugreek fiber has been used to prepare bakery foods such as bread, muffins, pizza, and cakes with suitable sensory properties (Srinivasan, 2005). Fenugreek flour has been integrated up to a 10% level in the formation of biscuits without affecting their quality. The sensory, nutritional, and physical characteristics generally revealed that biscuits that contain 10% germinated fenugreek flour were the best among all the composite fenugreek flour biscuits. Hence the development and utilization of such functional foods will not only improve the nutritional status of the general population but also help those people who suffer from degenerative diseases (Hooda and Jood, 2005).

6. USES

Fenugreek has many uses ranging from culinary to religious; its uses are often steeped in ritual. There are several curious beliefs associated with the historical use of fenugreek. In some countries, fenugreek is considered a seed of the sun and used for all manner of luck and success spells. In some countries, fenugreek is used to call up and control the spirits of the dead (Ahmad et al., 2016).

Ancient Egyptians used fenugreek as a vegetable, and its seeds were used to make incense and to preserve mummies. They used the roasted fenugreek seeds as a coffee and for medicinal purposes. It had less importance for cooking, but later, it was cultivated on a comparatively large scale as a medicinal agent. When the leaves are dried, they give a sweet hay scent due to the presence of coumarin in the leaves. The plant has been added to bad hay to make it more attractive fodder, so cows enjoy its taste. Scent of the hay passes into the milk of cows. Seeds were once thought to be used for the treatment of baldness. The seeds were roasted and gobbled by women in harems, who hoped to make their body hair more handsome (Manniche and Museum, 1989).

In India, fresh leaves and stems of fenugreek are generally cooked as a winter vegetable, and the seeds are used year-round as a flavoring agent for a variety of dishes. The fenugreek seeds are also eaten raw as sprouts and used medicinally. In Ethiopia and Egypt, it is used in baking bread as methi, and the Swiss use it for flavoring cheese. In the United States, it is primarily used to make spice blends for stews and soups (Passano, 1995). A traditional Egyptian use of the seeds was to slow roast and use as a coffee substitute. Fenugreek is used in the Jewish sweet dish halva (Shipard, 2004). The herb of fenugreek has been used as a cooking spice in European countries for centuries, and it remains a popular component in pickles, curry powders, and mixtures of spice in Pakistan, India, Bangladesh, and other Asian countries (Madar and Stark, 2002). The aroma of the seed is spicy, the flavor slightly pungent and bitter. Lightly roasting fenugreek seed releases a nutty and sweet maple syrup-like flavor. It has been used as a major flavoring of commercial maple syrup, vanilla essence, butterscotch, and caramel production. The fragrance of maple and fenugreek flavor have led to its use in artificial maple syrup (Shipard, 2004).

Fenugreek has been used in folk medicines for treatment of cellulitis, boils, and tuberculosis. Fenugreek remained a key component in a 19th-century patent medicine for postmenopausal and dysmenorrheal symptoms. It has also been suggested for lactation promotion. The seeds of fenugreek have been used orally as a substitute for insulin in reducing blood glucose, and the extracts from the seed have been reported to lower glucose levels of blood (Madar and Stark, 2002). The seeds of fenugreek are used to treat cervical cancer as a pessary in China. The aerial parts of fenugreek like stem and leaves are used as a folk medication for abdominal cramps, which are associated with gastroenteritis and menstrual pain, and it also used to relieve labor pains in the Middle East (Didarshetaban et al., 2013). Fenugreek has a favorable action on the purification of blood, and it is believed to be able to detox the body mainly via sweat. Fenugreek is also known as a lymphatic cleansing herb. Fenugreek is a valuable herb for all mucus conditions of the body, mainly the lungs, by helping to clear congestion. Fenugreek is a strong antioxidant, which acts as a throat cleanser and mucus solvent. Accumulated masses of cellular debris dissolve by drinking water in which seeds of fenugreek have been soaked. Fenugreek can be used for the treatment of head colds, catarrh, influenza, constipation, bronchial complaints, emphysema, asthma, pneumonia, tuberculosis, hay fever, sore throat, pleurisy, laryngitis, and sinusitis (Didarshetaban et al., 2013; Wani and Kumar, 2016). The seeds of fenugreek are a rich source of protein and a natural emollient, which is called lecithin; therefore, the seeds are used in the treatment of baldness and hair loss. It also helps in moisturization and strengthening the hair and keeping away dandruff and lice (Didarshetaban et al., 2013). Fenugreek oil is available at many health food stores, though the substantial scientific evidence for its usefulness in human health is inadequate.

7. PHARMACOLOGICAL USES

7.1 As Lactation Aid

Fenugreek is considered to be the most popular and utilized herbal galactagogue in the world (Lawrence and Lawrence, 2010). It is not clearly known how fenugreek increases milk flow, but it likely has to do with the phytoestrogens and diosgenin constituents. Breasts are modified sweat glands, and fenugreek has been found to stimulate sweat production, as it contains a hormone precursor to increase milk formation. Some scientists reported that fenugreek can enhance milk supply of nursing mother within 24–72 hours after first taking the herb (Snehlata and Payal, 2012).

7.2 In Diabetes Management

There are a significant number of studies that have been carried out to show the capability and strength of fiber, particularly the efficiency of dietary fiber of fenugreek, which is soluble on insulin production and glucose management of blood and serum. Blood glucose level decreased among diabetes type II patients by 25% when 100 g of fenugreek powder, which contains 50% dietary fiber, was administered for 10 days. It has been reported that postprandial elevation reduced by soluble fiber fraction in the blood glucose level by hindering the digestion of sucrose of diabetic type II rats. When the soluble fiber of fenugreek was orally administered twice a day for 28 days at a dose of 0.5g per kg, then the level of fructosamine in serum reduced with no considerable change in insulin level, in contrast with the control. However, the conclusion is that fenugreek soluble fiber had a valuable effect on dyslipidemia, and it could inhibit the platelet in model diabetic type II rats (Khorshidian et al., 2016).

7.3 Immunological Activity

Three doses of aqueous extract of fenugreek 50 mg per kg, 100 mg per kg, and 200 mg per kg of body weight were given to Swiss albino mice to study their immune system for 10 days; the immunomodulatory effect was revealed (as from relative thymus weight, body weight, delayed type of hypersensitivity response, cellularity of lymphoid organs, hemagglutination titer, phagocytosis, lymph proliferation, quantitative hemolysis assay, and major increase in phagocytic capacity of macrophages and phagocytic index) (Raju and Bird, 2006).

7.4 Hypoglycemic Effects

The hypoglycemic effect of fenugreek has been especially documented in humans and animals with type 1 and type 2 diabetes mellitus (Roberts, 2011). The result suggested that the hypoglycemic effect may be mediated through stimulating insulin synthesis and/or secretion from the beta pancreatic cells. Upon prolonged administration of the same dose of the active principle for 30 days to the severely diabetic rabbits, fasting blood glucose lowered significantly, but it could elevate the fasting serum insulin level to a much lower extent, which suggests an extrapancreatic mode of action for the active principle. The effect may due to increasing the sensitivity of tissues to available insulin. The hypoglycemic effect was observed to be slow but sustained, without any risk of developing severe hypoglycemia (Puri et al., 2002). A high-fiber fenugreek diet is useful in the management of diabetes (Wani and Kumar, 2016). It may be concluded that fenugreek extract can lower kidney/body weight ratio and blood glucose and also improves hemorheological properties in experimental diabetic rats following repeated treatment for 6 weeks (Xue et al., 2007).

7.5 Hypocholesterolemic Effects

Hypocholesterolemic problems result when the level of cholesterol in the blood decreases abnormally, but using aqueous extracts and methanolic extracts of fenugreek seeds at a dose of 1 gram per kilogram body weight in mice resulted in a hypoglycemic effect (Zia et al., 2001).

7.6 Antioxidant Activity

Fenugreek seed extracts with methanol, acetone, dichloromethane, ethanol, hexane, and ethyl acetate have a significant radical scavenging activity (Dash, 2011). Fenugreek has a protective effect on lipid peroxidation and on enzymatic antioxidants (Bhatia et al., 2006). The highest protein and saponin contents are present in fenugreek husk, cotyledons, and seeds, and a higher proportion of total polyphenols is present in husk.

Fenugreek seeds exhibited 72% antioxidant activity by free radical scavenging activity. The extract of endosperm and husk from fenugreek shows 56% and 64% antioxidant activities by free radical scavenging activity, respectively. The differentiation of fenugreek seeds into endosperm and husk could have an edge on process viability with respect to prior selective fractionation of bioactive components for their effective separation (Naidu et al., 2011). In the treatment of calcic urolithiasis patients, fenugreek can be used (Laroubi et al., 2007).

7.7 Antitumor Activity

The extract of fenugreek seeds considerably restrained 7, 12-dimethyl benz(a) anthracene-induced mammary hyperplasia and reduced its prevalence in rats. Protective effects of fenugreek to counteract breast cancer may be due to increased apoptosis (Amin et al., 2005). Furthermore, alcoholic extracts of fenugreek exhibited in vitro cytotoxicity opposed to various cancer cell lines of humans like HT29 (a cancer cell line), neuroblastoma cell line, or IMR-32 (Verma et al., 2010). A selective cytotoxic effect of fenugreek extract in vitro to a panel of cancer cell lines has been observed, including T-cell lymphoma (Alsemari et al., 2014).

7.8 Antibacterial Activity

Fenugreek seed oil and aqueous extract have a potent antibacterial activity against *Salmonella typhi*, *Escherichia coli*, and *Staphylococcus aureus*. Aqueous extract is prepared by boiling of fenugreek seed in water (Verma et al., 2015).

7.9 Anthelmintic Activity

Seeds of fenugreek showed effective anthelmintic activity. Alcoholic extract of fenugreek had shown a favorable result regarding anthelmintic activity, and water extract has also shown activity to a lesser degree (Buchineni and Kondaveti, 2016).

7.10 Antiulcer Activity

A gel fraction and the aqueous extract, which is extracted from seeds of fenugreek, showed considerable ulcer protective effects. It has relaxing effect on gastritis and gastric ulcer (Srinivasan, 2006).

7.11 Effects on Body Weight and Obesity

It has been indicated in some studies that fenugreek seed extract supplementation is effective in reducing body and adipose tissue weights. The probable mechanism may be due to flushing out carbohydrates from the body before entering the blood stream, resulting in weight loss and a high content of soluble fiber in fenugreek that forms a gelatinous structure; this in turn has effects on slowing the digestion and absorption of food from the intestine and creates a sense of satiety (Khorshidian et al., 2016).

8. SIDE EFFECTS AND TOXICITY

Side effects include bloating, gas, diarrhea, stomach upset, dizziness, headache, and a "maple syrup" odor in urine. Fenugreek can cause nasal congestion, wheezing, facial swelling, coughing, and severe allergic reactions in hypersensitive people. Fenugreek might lower blood sugar (Losso et al., 2009).

References

Acharya, S., Thomas, J., Basu, S., 2006. Fenugreek: an "old world" crop for the "new world". Biodiversity 7, 27–30.

Ahmad, A., Alghamdi, S.S., Mahmood, K., Afzal, M., 2016. Fenugreek a multipurpose crop: potentialities and improvements. Saudi Journal of Biological Sciences 23, 300–310.

Al-Asadi, J.N., 2014. Therapeutic uses of fenugreek (*Trigonella foenum-graecum* L.). American Journal of Social Issues and Humanities.

Alsemari, A., Alkhodairy, F., Aldakan, A., Al-Mohanna, M., Bahoush, E., Shinwari, Z., Alaiya, A., 2014. The selective cytotoxic anti-cancer properties and proteomic analysis of *Trigonella Foenum-Graecum*. BMC Complementary and Alternative Medicine 14, 1.

Amin, A., Alkaabi, A., Al-Falasi, S., Daoud, S.A., 2005. Chemopreventive activities of *Trigonella foenum graecum* (Fenugreek) against breast cancer. Cell Biology International 29, 687–694.

Bhatia, K., Kaur, M., Atif, F., Ali, M., Rehman, H., Rahman, S., Raisuddin, S., 2006. Aqueous extract of *Trigonella foenum-graecum* L. ameliorates additive urotoxicity of buthionine sulfoximine and cyclophosphamide in mice. Food and Chemical Toxicology 44, 1744–1750.

Buchineni, M., Kondaveti, S., 2016. In-vitro anthelmintic activity of fenugreek leaves (aqueous extract) in Indian earthworms. The Pharma Innovation 5, 70.

Chayut, N., Sobol, S., Nave, N., Samach, A., 2014. Shielding flowers developing under stress: translating theory to field application. Plants 3, 304–323.

Dash, B., 2011. Antibacterial activities of methanol and acetone extracts of fenugreek (*Trigonella foenum*) and coriander (Coriandrum sativum). Life Sciences and Medicine Research.

Didarshetaban, M.B., Pour, S., Reza, H., 2013. Fenugreek (Trigonella foenum-graecum L.) as a valuable medicinal plant. International Journal of Advanced Biological and Biomedical Research 1, 922–931.

Edison, S., 1995. Spices-Research Support to Productivity. The Hindu Survey of Indian Agriculture, Kasturi and Sons Ltd., National Press, Madras, pp. 101–105.

El Nasri, N.A., El Tinay, A., 2007. Functional properties of fenugreek (*Trigonella foenum graecum*) protein concentrate. Food Chemistry 103, 582–589.

Gomez, S., Roy, S.K., Pal, R., 2003. Primary processing of fenugreek (*Trigonella foenum graecum* L.)—An eco-friendly approach for convenience and quality. Plant Foods for Human Nutrition 58, 1–10.

Hamden, K., Keskes, H., Belhaj, S., Mnafgui, K., Allouche, N., 2011. Inhibitory potential of omega-3 fatty and fenugreek essential oil on key enzymes of carbohydrate-digestion and hypertension in diabetes rats. Lipids in Health and Disease 10, 1.

Hooda, S., Jood, S., 2005. Organoleptic and nutritional evaluation of wheat biscuits supplemented with untreated and treated fenugreek flour. Food Chemistry 90, 427–435.

Khorshidian, N., Yousefi Asli, M., Arab, M., Adeli Mirzaie, A., Mortazavian, A.M., 2016. Fenugreek: potential applications as a functional food and nutraceutical. Nutrition and Food Sciences Research 3, 5–16.

Laroubi, A., Touhami, M., Farouk, L., Zrara, I., Aboufatima, R., Benharref, A., Chait, A., 2007. Prophylaxis effect of *Trigonella foenum graecum* L. seeds on renal stone formation in rats. Phytotherapy Research 21, 921–925.

Lawrence, R.A., Lawrence, R.M., 2010. Breastfeeding: A Guide for the Medical Professional. Elsevier Health Sciences.

Losso, J.N., Holliday, D.L., Finley, J.W., Martin, R.J., Rood, J.C., Yu, Y., Greenway, F.L., 2009. Fenugreek bread: a treatment for diabetes mellitus. Journal of Medicinal Food 12, 1046–1049.

Madar, Z., Stark, A.H., 2002. New legume sources as therapeutic agents. British Journal of Nutrition 88, 287–292.

Manniche, L., Museum, L.B., 1989. An Ancient Egyptian Herbal. University of Texas Press, Austin.

Mehrafarin, A., Rezazadeh, S., Naghdi Badi, H., Noormohammadi, G., Zand, E., Qaderi, A., 2011. A review on biology, cultivation and biotechnology of fenugreek (*Trigonella foenum-graecum* L.) as a valuable medicinal plant and multipurpose. فصلنامه علمی پژوهشی گیاهان دارویی 1, 6–24.

MIN, O.Y., 2011. Effect of Fenugreek Seeds on Short Term Memory and Morphology of Cornu Ammonis in Female Mice. Universiti Tunku Abdul Rahman.

Montgomery, J., 2009. The Potential of Fenugreek (*Trigonella Foenum-Graecum*) as a Forage for Dairy Herds in Central Alberta. University of Alberta.

Naidu, M.M., Shyamala, B., Naik, J.P., Sulochanamma, G., Srinivas, P., 2011. Chemical composition and antioxidant activity of the husk and endosperm of fenugreek seeds. LWT-Food Science and technology 44, 451–456.

Passano, P., 1995. The many uses of methi. Manushi 31–34. November.

Pavlista, A.D., Santra, D.K., 2016. Planting and harvest dates, and irrigation on fenugreek in the semi-arid high plains of the USA. Industrial Crops and Products 94, 65–71.

Petropoulos, G.A., 2003. Fenugreek: The Genus Trigonella. CRC Press.

Popova, T., 2017. New archaeobotanical evidence for *Trigonella foenum-graecum* L. from the 4th century Serdica. Quaternary International 460, 157–166.

Poynter, F., 1971. The Book of Spices. by Frederic Rosengarten Jr. Livingston Publishing Company, Wynnewood, Pennsylvania, p. 489, 1969, 330 illus.(73 in colour), $20.00. Medical history 15, 108-108.

Puri, D., Prabhu, K., Murthy, P., 2002. Mechanism of action of a hypoglycemic principle isolated from fenugreek seeds. Indian Journal of Physiology and Pharmacology 46, 457–462.

Raju, J., Bird, R., 2006. Alleviation of hepatic steatosis accompanied by modulation of plasma and liver TNF-α levels by Trigonella foenum graecum (fenugreek) seeds in Zucker obese (fa/fa) rats. International Journal of Obesity 30, 1298–1307.

Ren, X.F., Zhu, W.J., 2011. Optimal Conditions for Extraction of Oil from Fenugreek (*Trigonella Foenum-Graecum* L.) by Supercritical CO2 Fluids (SFE-CO2), Advanced Materials Research. Trans Tech Publ, pp. 2980–2983.

Roberts, K.T., 2011. The potential of fenugreek (*Trigonella foenum-graecum*) as a functional food and nutraceutical and its effects on glycemia and lipidemia. Journal of Medicinal Food 14, 1485–1489.

Shipard, I., 2004. How Can I Use Herbs in My Daily Life. David Stewart.

Snehlata, H.S., Payal, D.R., 2012. Fenugreek (*Trigonella foenum-graecum* L.): an overview. International Journal of Current Pharmaceutical Research 2, 169–187.

Sowmya, P., Rajyalakshmi, P., 1999. Hypocholesterolemic effect of germinated fenugreek seeds in human subjects. Plant Foods for Human Nutrition 53, 359–365.

Srinivasan, K., 2005. Role of spices beyond food flavoring: nutraceuticals with multiple health effects. Food Reviews International 21, 167–188.

Srinivasan, K., 2006. Fenugreek (*Trigonella foenum-graecum*): a review of health beneficial physiological effects. Food Reviews International 22, 203–224.

Svecova, E., Neugebauerova, J., 2010. A study of 34 cultivars of basil (Ocimum L.) and their morphological, economic and biochemical characteristics, using standardized descriptors. Acta Universitatis Sapientiae, Alimentaria 3, 118–135.

Syeda, B., Muhammad, I., Shahabuddin, M., 2008. Antioxidant activity from the extract of fenugreek seeds. Pakistan Journal of Analytical & Environmental Chemistry 9, 78–83.

Verma, S., Yadav, S., Singh, A., 2015. In vitro antibacterial activity and phytochemical analysis of mangifera indica L flower. Extracts against pathogenic. Microorganisms. Journal of Pharmacology & Clinical Toxicology 3, 1053.

Verma, S.K., Singh, S.K., Mathur, A., 2010. Journal of chemical and pharmaceutical research. Journal of Chemistry 2, 861–865.

Wani, S.A., Kumar, P., 2016. Fenugreek: a review on its nutraceutical properties and utilization in various food products. Journal of the Saudi Society of Agricultural Sciences.

Wani, S.A., Kumar, P., 2018. Fenugreek: a review on its nutraceutical properties and utilization in various food products. Journal of the Saudi Society of Agricultural Sciences 17, 97–106.

Xue, W.-L., Li, X.-S., Zhang, J., Liu, Y.-H., Wang, Z.-L., Zhang, R.-J., 2007. Effect of Trigonella foenum-graecum (fenugreek) extract on blood glucose, blood lipid and hemorheological properties in streptozotocin-induced diabetic rats. Asia Pacific Journal of Clinical Nutrition 16, 422–426.

Yadav, R., Kaushik, R., 2011. A study of phytochemical constituents and pharmacological actions of Trigonella foenum-graecum: a review. International Journal of Pharmacy and Technology 3, 1022–1028.

Yadav, S.K., Sehgal, S., 1997. Effect of home processing and storage on ascorbic acid and β-carotene content of bathua (*Chenopodium album*) and fenugreek (*Trigonella foenum graecum*) leaves. Plant Foods for Human Nutrition 50, 239–247.

Zia, T., Hasnain, S.N., Hasan, S., 2001. Evaluation of the oral hypoglycaemic effect of *Trigonella foenum-graecum* L.(methi) in normal mice. Journal of Ethnopharmacology 75, 191–195.

CHAPTER

21

Figs

Shumaila Saif[1], *Muhammad Asif Hanif*[1], *Rafia Rehman*[1], *Maryam Hanif*[1], *Oli Khan*[2], *Sunil Khan*[3]

[1] Department of Chemistry, University of Agriculture, Faisalabad, Pakistan; [2] Department of Botany, Bangabasi College, Kolkata, India; [3] Department of Botany, Haripal Vivekananda College, Hooghly, India

OUTLINE

Medicinal Plants of South Asia
https://doi.org/10.1016/B978-0-08-102659-5.00021-5

Copyright © 2020 Elsevier Ltd. All rights reserved.

1. BOTANY

1.1 Introduction

The common fig (*Ficus carica* L.) (Fig. 21.1) is a deciduous tree that belongs to the Moraceae family (*Mulberries*). In ancient times, it had been used as an ornamental plant and a fruit. The genus *Ficus* contains

FIGURE 21.1 Fig tree and fruits.

approximately 2000 subtropical and tropical trees, shrubs, and vine species native to the hot parts of world. The uncertainty in the exact number of species within the genus is largely attributed to a huge variability among the constituent species. Variability is prevalent in morphology, growth habit, flower color, leaves, stems, and chemical composition. Fig self-pollinates very easily, and it does not require pollination by a wasp of another tree, but it can be pollinated by fig wasp to produce seeds (Stover et al., 2007).

F. carica is known by different names in different regions of world. In English, it is typically known as "fig." In Pakistan, specifically in Urdu, it is called "anjir" or "angeer." It is also known as "higo" or "brevo" in Spanish, "fieguier" in French, "feige" in German, "kerma" in Arabic, and "figo" in Italian and Portuguese (Flaishman et al., 2008; Joseph and Raj, 2011a). Probably the most familiar fig is the common fig (*F. carica*); however, it has a wide range of varieties and cultivars. The cultivars of fig have various unique features such as compressed to spreading growth habits, fruit color, taste, shape, size, and plant hardiness (Himelrick, 1999).

1.2 History/Origin

F. carica is native to the Mediterranean area from Afghanistan to Portugal and southwest Asia, where it was generally cultivated for its fruits in ancient times. The term fig originates from a Latin word "*ficus*." In the southern parts of the Arabian Peninsula, it was first brought into cultivation by at least 3000 BC. After its cultivation, it circulated into Iran, Turkey, Syria, and all the Mediterranean countries. The fig was taken to most subtropical areas of the Western Hemisphere during the era of exploration following the discovery of America by Columbus. Dushevskii and Kazas (1985) described two forms of *F. carica* from the vicinity of Mangup Kale, a settled area on Mt. Baba Dag in Crimea abandoned in 1783. They are thought to be descendants of figs cultivated on the plateau from the 12th century (Hiwale, 2015).

In the 1500s, fig trees were familiarized to England and Mexico, and in 1669 to the Eastern United States, and to California in 1881. Common figs were successfully cultivated all over California and the Gulf states, while Smyrna fig, which was not native to California, did not fruit until the cross-pollination process in plants by a tiny wasp. Then in 1899 the wasp (*Blastophaga psenes*) was introduced that cross-pollinates the common fig (*F. carica*) and closely related *Ficus palmate*. The Greek, Roman, and Egyptian civilizations were fond of the fruit. Egyptians made pastries from figs, and the Greeks forbade their export due to their high value. Romans considered the fig tree to be symbolic of the civilization's prosperity (Condit, 1947).

The most cited fruit in the Bible and in the Garden of Eden is a fig. It was also mentioned in the book of Genesis that Adam and Eve, after eating the "prohibited fruit" from Tree of Knowledge of Good and Evil, covered themselves with fig leaves. Moreover, illustrations of leaves of fig were used to cover the genitals of naked figures in sculpture and painting. The use of the fig leaf as a defender of humility has entered the language (Flaishman et al., 2008). Different countries have strong cultural connections with figs, and the fruit has many roles in political and religious histories. For example, the Hadith quotes Hazrat Muhammad (PBUH) claiming that figs surely "descended from paradise."

1.3 Location/Demography

Although fig is grown in a variety of climatic and environmental conditions, the optimum conditions are found in countries with dry and sunny areas. Warmth, light, and moisture are the key ecologic requirements for fig cultivation. Fig is less resistant to frost, and these spend the summer outdoors and are overwintered in a frost-free, cool place. However, it requires basic soil for cultivation. Fig cultivation is widely observed in different countries: Pakistan, Afghanistan, and some regions of Eastern and South India. The top five producers of figs are Egypt, Algeria, Morocco, Turkey, and Iran. The United States placed at the sixth position in world production (Crisosto et al., 2011).

1.4 Botany, Morphology, Ecology

Fig is a monoecious, deciduous tree or large shrub. It is generally 15–20 ft tall, with spreading branches, and the diameter of the trunk is hardly beyond 7 ft. Leaves are single, large, alternate (equal to 1 ft in length), and bright green in color. Flowers rise from the axils of old leaves. It exhibits a structure of receptacles. Female flowers occupy the upper part of the receptacle, and the lower part has male flowers. Syconium, the ripened receptacle, comprises a huge amount of small whitish seeds, which can be small, large, medium, and range in number per fruit from 30 to 1600. Fruit is a syconium, is pyriform fleshy, solitary, 5–8 cm long, very sweet, and tasty (Mawa et al., 2013). The bark is flat. Outer bark is ash-colored, exfoliated through asymmetrical, rounded flakes. The sections of middle bark look light reddish brown. The layers of light yellowish or orange-brown-colored granular tissue are present in the inner bark (Badgujar et al., 2014).

Fig requires warm, temperate, or Mediterranean conditions. The areas having an arid or semiarid environment, plenty of sunshine, high summer temperature, and moderate winter are suitable for the plant development. Although plants can survive at temperature as high as 45°C, beyond 39°C the quality of fruit deteriorates. Fig requires calcareous, well-drained, medium to heavy and deep (about 1 m) soil. While for better root establishment, light sandy, shallow, and deep soils are suitable. It grows well in soils with a pH ranging from 7 to 8. In general, the limiting factor for its cultivation is climate rather than soil.

2. CHEMISTRY

The fruit of the fig tree has a fleshy and succulent pulp and is sweetened slightly. It is appreciated for dessert. A large number of fig ecotypes have been described on the basis of their taste, flavor, and other phenotypic characters. Fig color varies from dark purple to green. Cedrol, γ-muurolene, manoyl oxide, α-terpinyl acetate, abietatriene, α-pinene, and pentadecanal are the main components of leaves, while cedrol, α-terpinyl acetate, manoyl oxide, α-pinene, and abietadiene are main components of the fruits. However, the aroma of leaves and fruits is the outcome of a complex mixture of aldehydes, esters, alcohols, terpenoids compounds, and others, at low concentrations that reach the olfactory epithelium, intensely contributing to the taste of foods. Most of recognized mixtures certainly add to the pleasant taste and aroma of fig fruits to different extents (Soltana et al., 2017).

F. carica contains a large amount of minerals, carbohydrates, vitamins, sugars, dietary fiber, organic acids, and phenolic compounds. Fig is an extremely nourishing fruit. It has an excellent amount of calories, proteins, iron, calcium (higher than milk), and maximum fiber content. Figs are fat and cholesterol free and contain a high concentration of amino acids. Sugars and organic acids are also found in fig that influence their quality, like other fruit species. It also contain phenolic substances, which contribute significantly to their value, because it was proved that their consumption shown a positive result on health of humans (Veberic et al., 2008). The dried and fresh figs contain a huge amount of fiber and polyphenols. Figs are a rich source of phenolic compounds like proanthocyanidins. Fig is a very poor source of vitamin C but very rich in sugar, next to dates (Mawa et al., 2013). Total sugar content of fresh fig is 16% and of dried is 52%. Besides essential or fixed oils, the plant also includes coumarins, sterols, flavonoids, anthocyanins, and triterpenoids, etc., in different parts of the plant. The compounds found in *F. carica* leaf extracts are quercetin, rutin, carotene, luteolin, psoralen, and bergapten. The latex contains resin, rennin, albumin, cerin, caoutchouc, sugar and malic acid, proteolytic enzymes, lipase, diastase, esterase, catalase and peroxidase, 6-O-linoleyl-β-D-glucosyl-β-sitosterol, 6-O-oleyl-β-D-glucosyl-β-sitosterol, and 6-O-palmitoyl-β-D-glucosyl-β-sitosterol. Fruits have cyanidin-3-

1

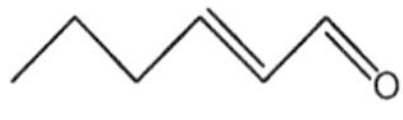

(E)-2-Hexenal

2

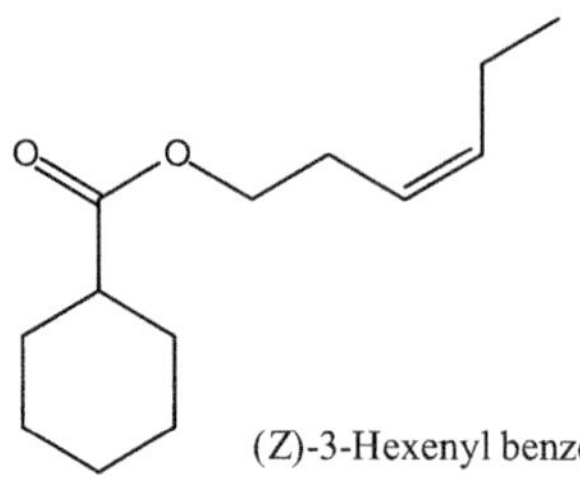

(Z)-3-Hexenyl benzoate

3

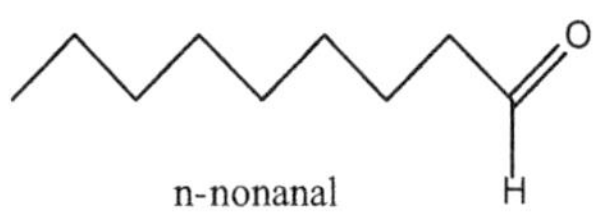

n-nonanal

4

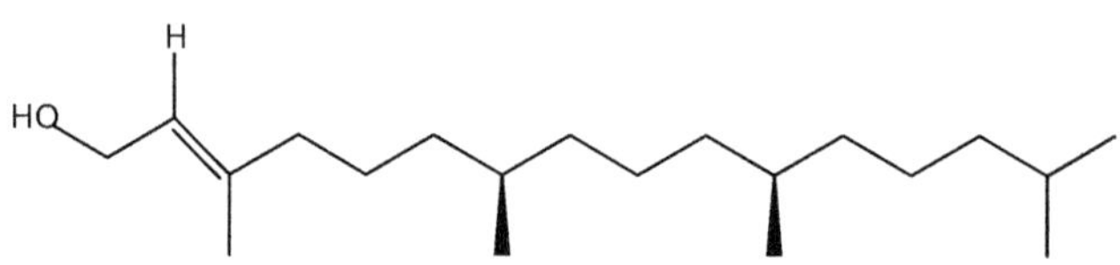

Phyton

FIGURE 21.2 Potent chemical components of Fig.

orhamnoglucoside, cyanidin-3-O-glucoside, cholesterol, insoluble sugars, protein, saturated fat, vitamin A, vitamin C, sodium, iron, and calcium. Roots contain bergapten and psoralen (Chawla, 2012; Joseph and Raj, 2011b; Patil and Patil, 2011b). The fig essential oil contains (Z)-3-hexeny benzoate, n-nonanal, n-tetracosane, (E)-2-hexenal n-hexadecanoic acid, and n-docosane (Ayoub et al., 2010). Some potent chemical components of Fig are shown in Fig. 21.2.

3. POSTHARVEST TECHNOLOGY

Conventionally, the best harvesting times of figs are, first, in early summer (late June) and, second, in late summer or early fall (August or September). The exact harvesting timing of the main crop depends on conditions and climate. For example, cultivators in cooler coastal areas generally harvest their figs during October and November. The typical harvest time is between June and September for warmer and inland climates. Figs must be allowed to ripen completely on the tree before they are picked. They will not ripen if picked when immature. A ripe fruit will be slightly soft and starting to bend at the neck. Harvest the fruit lightly to prevent yellowing. Fresh figs do not keep well and can be stored in the refrigerator for only 2–3 days. Some fig cultivars are delicious when dried. They take 4–5 days to dry in the sun and 10–12 hours in a dehydrator. Dried figs can be stored for 6–8 months (Yemiş et al., 2012).

In some tropical locations, fig trees may bear certain fruit all year, with increased production in early summer and midwinter. Figs are broadly used up fresh, either peeled or not. Fresh fruits obviously have a short, postharvest life of 7–10 days, but with a combination of cooler conditions and a CO_2-enriched atmosphere, the fruit can be stored for up to 2–4 weeks. Figs are also very widespread as dried fruit, since drying prolongs their storability (Veberic et al., 2008). It is best to use, eat, dry, or freeze figs as soon as possible after harvest. If the figs are dried either in the sun or using a dehydrator, they will last for up to 3 years in the freezer. The figs must be washed and dried and placed on a baking sheet and frozen until hard. Once the fruit is hard, it can be transferred to a container and stored in the freezer for up to 3 years. Fresh figs will keep in the refrigerator when placed in a single layer on a tray. The tray should be placed in the coldest part of refrigerator, generally the crisper. However, the figs must not be placed close to fresh vegetables, as they can cause the veggies to rot quickly.

4. PROCESSING

Fig is consumed in diverse ways for various purposes. Fresh figs are subjected to two different drying processes, i.e., oven drying and sun drying. The fig is completely ready when it drops to ground. To remove soil and rubbish, the fig must be washed. Trim off damaged and cracked parts. Before proceeding, perfectly dry them with a dishcloth or paper towel. A sharp peeling knife must be used to cut the figs in half from stem to tip. They will dry more rapidly by cutting in half. For this purpose, a drying rack must be used, or purchase screening with adequate ventilation holes. A cover of cheesecloth must be used to line the rack before settings the figs on top. The figs must not be dried on a solid sheet. For proper drying, they need airflow from below and above. If a rack with very large holes is used, then a double layer of cheesecloth must be used. As they dry, they will be protected from insects by using this double layer of cheesecloth. The cheesecloth must be strongly placed around the drying rack, locking it with tape if necessary, to guaranteed it will not come loose. This method works best when it is very dry and hot outside.

The figs must not be placed in the shade because they will not dry as rapidly, and then they can spoil before they are preserved properly. Every evening, take them inside if temperature does not fall more than 20° at night. Turn the figs over, so they dry uniformly from all sides in the mornings. When the outside seems rubbery and no juice may be seen on the inside on squeezing, then the figs are ready. An oven can be used to finish if the figs remain a little tacky. To make them last even longer, the dried figs must be stored in a dry and cool place or freeze them. The temperature of the oven must be adjusted to 140°F (60°C) because it is necessary to dry the figs at a low temperature. A food dehydrator can also be used to dry figs. Wash figs carefully with water. Any broken parts are carefully trimmed away and perfectly dry them with a paper towel or dishcloth. Cut the figs in half, and by using a sharp trimming knife, they can be sliced from stem to tip, lengthwise. A rack with ventilation holes must be used to place them cut side up so the figs dry from below and above. The figs that do not dry evenly must be dried by using a regular baking pan. After placing those in the oven, prop the door of oven somewhat open to permit moisture to escape and prevent the figs from getting too hot and cooking instead of drying. During the drying process the figs must be turned occasionally. Figs must be allowed to stay in oven for up to 8–24 hours. When the outsides are rubbery and no juice may be seen on the inside on splitting one open, then the figs are dried. The dried figs can be stored in airtight vessels or freezer bags and must be kept in a refrigerator or freezer for approximately 18–24 months when appropriately stored. Essential oil can be extracted from figs through hydro distillation. Essential oil is concentrated in the leaves. Essential oil has rich amount of oxygenated compounds (Ayoub et al., 2010; Zito et al., 2013).

5. VALUE ADDITION

The fresh or dried fig can be eaten but is mainly used in making of jam and pickling. The fig paste, fig powder, fig concentrate, fig nuggets, and diced and sliced figs are considered food products including figs. In fig jam, preserves, and paste the natural flavor of figs can be preserved. For making fig concentrate, which replaces corn syrup and sucrose, the water is extracted from figs. Diced, chopped, and sliced figs are fused into food products. Figs are added to bars, cookies, and snacks when dried. Moreover, high-quality figs are used for fresh consumption, so few figs are canned. The oil is present in dried seeds of figs and is 30% fatty acids. The beneficial ingredients for health and beauty products such as soap, moisturizers, and fragrance are formed from the natural humectants of figs. In India, after the harvest of fruit, the leaves of figs are plucked and used for fodder. The leaves of figs are used as a source of perfume material because the leaves create a woody, mossy scent in southern France. Since the beginning of civilization, fresh and dry fig have mostly been used as food. The syrup of fig is used as a medicine for minor constipation. For domestic animals, the leaves of figs are traditionally consumed as a fodder. For the production of tremendously familiar milk products such as cheese through different indigenous communities, the latex of the plant is used as a clotting agent. Moreover, the wood is used for jewels, crowns, and for decoration purposes (Badgujar et al., 2014).

6. USES

Many herbs and spices contribute significantly to health despite the low amounts of consumption, as they are full of antioxidants and certain mineral compounds. In addition to this, they are a good source of certain minerals and dietary fiber. Fig is a tree, and its fruit is usually eaten. To make medicine, the fruit and leaves are used. Fig fruit is used as a cathartic to get rid of constipation. Fig fruits contain a high content of dietary fiber, vitamins, minerals, and amino acids, and fig is fat and cholesterol free, so this fruit is ideal to control hypertension, diabetics, and colon cancer. Fig leaf is used for diabetes, high cholesterol, and skin disorders such as psoriasis, eczema, and vitiligo. Some people apply milky sap from the tree directly to skin for treatment of skin tumors and warts (Mawa et al., 2013).

Fig has been used for many years as a culinary vegetable and for medicinal and spiritual purposes. Green, unripe fig is used as a vegetable for preparing soup. Fresh or dried fruits of *F. carica* can be eaten and are also used as jam. In warm and humid climates, fresh and raw figs are usually eaten without peeling, and they are frequently obliged with cream and sugar. Pies, cakes, puddings, and other bakery products such as jam, jellies, and preserves are made from processed figs. Dried figs can be used in appetizers and desserts such as cakes, porridge, muesli bars, muffins, oatmeal, or breakfast cereals. Fresh figs are usually used in healthy salads. Peeled or unpeeled, the fruits can be cooked by different methods, such as in cakes, pastries, bread, desserts, or bakery goods, or added to ice cream mixture. Home owners can preserve complete fruits in sugar syrup or cook them as jam, paste, or marmalade. Fig paste forms the filling for bakery products. Other current uses are as icepack, lotion, gargle, and drink (Ahmad et al., 2013; Chawla, 2012; Chawla et al., 2017).

Fig has an extensive list of traditional medical uses. It is considered a good nutritional support for diabetics. For controlling menorrhagia, through its sharp action, these compounds bring a styptic effect. Fruits of figs are used in lithotripter, for nose bleeding, as aphrodisiacs, for leprosy, hair health, and as antipyretics, demulcents, and cathartics. Different inflammations such as paralysis, chest pain, liver diseases, and piles are also treated by fig fruits. The fruit's juice mixed with honey is used for hemorrhage. However, fruits are used as a minor laxative, medicine, and diuretic in Indian medicine. It is used as an aid in spleen and liver diseases. The dry fruit of *F. carica* is a supplement food for diabetics. It is commercialized in the market as sweet due to its high level of sugars. To get rid of pain, fruit paste is applied to swellings, tumors, and inflammations (Mawa et al., 2013). For treatment of leukoderma, stimulant, and ringworm contamination, roots are used. Latex is used as anthelmintic, diuretic, antianemic, and expectorant. Leaves are used as antidiabetic, vermifuge, and for dermatitis in humans or phototoxicity in animals. The seeds are used as cooked oil, grease, etc. (Ahmad et al., 2013; Chawla, 2012; Joseph and Raj, 2011b). Physical and mental exertion is removed by fig, and it provides the body with improved potency and strength. Fig is an excellent stimulant for those who suffer from cracks in tongue, lips, and mouth (Hiwale, 2015).

Figs are valued as sacred, denoting a symbol of peace, fertility, or prosperity by many civilizations from ancient times. Figs have been in existence since the beginning of the world according to the Old Testament. In the Holy Bible, it was also revealed when Eve picked the forbidden fruit that it was not an apple tree, but some scholars believe it was a fig. Plato accepted that Greek athletes at Olympia were fed a diet of figs to increase their total strength and running speed. Angeer as a dry fruit is also considered a good nutritional support.

7. PHARMACOLOGICAL USES

7.1 Antibacterial Activity

Antibacterial activity of *F. carica* extracts has been widely studied against several bacterial strains. It has been shown that *F. carica* leaf methanol extract inhibits the growth of clinical isolates of *Staphylococcus aureus* resistant to penicillin. In addition, this extract had an additive effect when tested in synergy with antibiotics. Its action was mainly linked to the cell loss of viability as a property of phenolic compounds to cross the bacterial membrane. The methanol and ethanol extracts of the fruit of *F. carica* were tested against *Escherichia coli, Pseudomonas aeruginosa, Streptococcus* sp., *Enterobacter* sp., *Klebsiella pneumonia, Salmonella typhi,* and *Salmonella paratyphi* (Jasmine et al., 2014). The antibacterial activity of the extracts of chloroform, hexane, ethyl acetate, and alcohol has also been investigated. Ethyl acetate extract showed a significant zone of inhibition against *S. aureus*. While in another study, the ethanolic leaf extracts were tested against several bacterial strains, and the results showed an inhibitory activity against *Streptococcus anginosus* at a concentration of MIC = 0.156 mg/mL (Bouyahya et al., 2016).

7.2 Antiviral Activity

A small contagious agent that replicates only inside the living cells of other organisms is a virus. Methanol, chloroformic, ethyl acetate, hexanic, and hexane-ethyl acetate extracts of *F. carica* have been confirmed for antiviral activity. In vitro antiviral potential action has been considered via observing cytopathic effect toward echovirus type 11 (ECV-11), herpes simplex type 1 (HSV-1), and adenovirus (ADV). It has been concluded that hexane-ethyl acetate and hexanic extracts prohibited development of viruses at concentrations of 78 mg/mL. For herbal medicines, these two extracts are the most effective candidates. Viral contagious diseases like adenovirus, herpes virus, and echovirus are being treated by the treatment of these extracts (Badgujar et al., 2014).

7.3 Antioxidant Activity

It is well known that *F. carica* products from latex and fruit are rich in phenolic compounds (polyphenols, flavonoids, tannins, etc.) with an antioxidant power. Using different systems, several organic extracts from *F. carica* proved capable of reducing free radicals. A study was carried out on polyphenols and flavonoids that extracted from *F. carica* latex and used the radical scavenging activity in vivo assay system via the determination of the action of superoxide dismutase and glutathione reductase. It showed a significant reduction in the rate of these two enzymes in liver cells. An in vitro antioxidant action of *F. carica* leaf methanolic extracts was also evaluated using the scanning technique of the radical DPPH, revealing its antioxidant capacity (IC50) to be 0.0903 mg/mL. It also found an inhibition of 10.222 DPPH radical at a concentration of 250 mg/mL (Bouyahya et al., 2016).

7.4 Antiinflammatory Activity

In traditional medicine, leaves of *F. carica* are used to get rid of different inflammatory ailments such as hemorrhoids, insect bites, and stings. The antiinflammatory effect in carrageenan-induced rat paw edema and cotton pellet granuloma method was studied (Patil and Patil, 2011a). The ethanolic extract at 600 mg/kg/day of body weight showed best antiinflammatory activity of 75.90% in acute swelling, and in long-lasting inflammation, there was a 71.66% reduction in granuloma weight. All extracts showed a greater antiinflammatory effect than that of indomethacin, which is a standard drug. This antiinflammatory activity might be correlated to antiradical activity of extracts and, by extension, to their chemical composition (Ali et al., 2012).

7.5 Antipyretic Activity

An antipyretic is used to lower body temperature when a fever is present. Several studies have revealed the antipyretic activity of figs. Indeed, the ethanolic extracts of leaves were tested and showed a significant reduction in body temperature; their effect is comparable to that of paracetamol. This effect may be related to the inhibitory action of these extracts on heat shock proteins or due to their effect on the thermoregulatory center (Bouyahya et al., 2016).

7.6 Antiangiogenic Activity

The development of new blood vessels is called angiogenesis. If we can stop cancers from increasing blood vessels, we can slow the evolution of the cancer, or occasionally shrink it. Antiangiogenic drugs are treatments that stop tumors from growing their own blood vessels. The antiangiogenic and antiproliferative possibilities of *F. carica* latex extract was examined by using human umbilical vein endothelial cells. It has been concluded that latex extracts of *F. carica* contain solid antiangiogenic and antiproliferative activities. Latex extract may be a suitable candidate for inhibition of angiogenesis in cancer and other long-lasting ailments, and as a potential agent (Badgujar et al., 2014).

7.7 Hypoglycemic and Hypocholesterolemic Activity

Hypocholesterolemic activity has also been observed in fig leaves. The aqueous decoction of fig leaves is used to prepare the chloroform extract. It causes a decrease in the total cholesterol/HDL cholesterol ratio and deterioration in the levels of total cholesterol, together with a reduction of hyperglycemia. Additionally, in hepatocellular carcinoma cell line (HepG2), the cell content of cholesterol appreciates the decrease of blood cholesterol level in streptozotocin-induced diabetic rats (Canal et al., 2000).

8. SIDE EFFECTS AND TOXICITY

Eating too many figs can be heavy on the stomach and can even cause a stomachache, as too much fiber is bad for the stomach. It causes gas and bloating, increases skin sensitivity to sunlight, is harmful for the liver and intestines, increases the risk of calcium deficiency, causes retinal, rectal, and vaginal bleeding, increases the risk of hypoglycemia, causes allergic reactions, and is harmful for individuals suffering kidney and gallbladder problems. Eating figs can lower the blood sugar level in body and is harmful for the individuals who have to undergo surgery. Consult with a doctor before eating figs during pregnancy or during the breastfeeding stage.

References

Ahmad, S., Bhatti, F.R., Khaliq, F.H., Irshad, S., Madni, A., 2013. A review on the prosperous phytochemical and pharmacological effects of *Ficus carica*. International Journal of Bioassays 2, 843–849.

Ali, B., Mujeeb, M., Aeri, V., Mir, S.R., Faiyazuddin, M., Shakeel, F., 2012a. Anti-inflammatory and antioxidant activity of Ficus carica Linn. Natural Product Research 26 (5), 460–465. https://doi.org/10.1080/14786419.2010.488236. Epub 2011 Jun 12.

Ayoub, N., Singab, A.N., Mostafa, N., Schultze, W., 2010. Volatile constituents of leaves of *Ficus carica* Linn. grown in Egypt. Journal of Essential Oil Bearing Plants 13, 316–321.

Badgujar, S.B., Patel, V.V., Bandivdekar, A.H., Mahajan, R.T., 2014. Traditional uses, phytochemistry and pharmacology of *Ficus carica*: a review. Pharmaceutical Biology 52, 1487–1503.

Bouyahya, A., Bensaid, M., Bakri, Y., Dakka, N., 2016. Phytochemistry and ethnopharmacology of *Ficus carica*. International Journal of Biochemistry Research & Review 14, 1–12.

Canal, J., Torres, M.D., Romero, A., Pérez, C., 2000. A chloroform extract obtained from a decoction of *Ficus carica* leaves improves the cholesterolaemic status of rats with streptozotocin-induced diabetes. Acta Physiologica Hungarica 87, 71–76.

Chawla, A., 2012. *Ficus carica* Linn.: a review on its pharmacognostic, phytochemical and pharmacological aspects. International Journal of Pharmaceutical & Phytopharmacological Research 1, 215–232.

Chawla, A., Kaur, R., Sharma, A.K., 2017. *Ficus carica* Linn.: a review on its pharmacognostic, phytochemical and pharmacological aspects. International Journal of Pharmaceutical & Phytopharmacological Research 1, 215–232.

Condit, I.J., 1947. The Fig. Chronica Botanica Co., USA.

Crisosto, H., Ferguson, L., Bremer, V., Stover, E., Colelli, G., 2011. Fig (*Ficus Carica* L.), Postharvest Biology and Technology of Tropical and Subtropical Fruits: Cocona to Mango. Elsevier, pp. 134–160e.

Dushevskii, V.P., Kazas, A.N., 1985. Byulletin Gosudartvennogo Nikitskogo Botanicheskogo Sada 58, 50–53.

Flaishman, M.A., Rodov, V., Stover, E., 2008. The fig: botany, horticulture, and breeding. Horticultural Reviews-Westport Then New York 34, 113.

Himelrick, D., 1999. Fig Production Guide. Alabama A & M Auburn Universities, Alabama Cooperative Extension System. ANR-1145 1.

Hiwale, S., 2015. Fig (*Ficus carica*), Sustainable Horticulture in Semiarid Dry Lands. Springer, pp. 159–175.

Jasmine, R., Manikandan, K., Niveditha, B., Thirupathi, K., Manikandan, G., 2014. Evaluation the efficiency of Ficus carica fruits against a few drug resistant bacterial pathogens. World Journal of Pharmacy and Pharmaceutical Sciences 3, 1394–1400.

Joseph, B., Raj, S.J., 2011a. A comparative study on various properties of five medicinally important plants. International Journal of Pharmacology 7, 206–211.

Joseph, B., Raj, S.J., 2011b. Pharmacognostic and phytochemical properties of *Ficus carica* Linn–An overview. International Journal of Pharmtech Research 3, 8–12.

Mawa, S., Husain, K., Jantan, I., 2013. *Ficus carica* L.(Moraceae): phytochemistry, traditional uses and biological activities. Evidence-based Complementary and Alternative Medicine 2013.

Patil, V.V., Patil, V.R., 2011a. Evaluation of Anti-inflammatory Activity of Ficus Carica Linn. Leaves.

Patil, V.V., Patil, V.R., 2011b. *Ficus carica* Linn. An overview. Research Journal of Medicinal Plant 5, 246–253.

Soltana, H., Flamini, G., Hammami, M., 2017. Volatile compounds from six varieties of *Ficus carica* from Tunisia. Records of Natural Products 11, 6.

Stover, E., Aradhya, M., Ferguson, L., Crisosto, C.H., 2007. The fig: overview of an ancient fruit. HortScience 42, 1083–1087.

Veberic, R., Colaric, M., Stampar, F., 2008. Phenolic acids and flavonoids of fig fruit (*Ficus carica* L.) in the Northern Mediterranean region. Food Chemistry 106, 153–157.

Yemiş, O., Bakkalbaşı, E., Artık, N., 2012. Changes in pigment profile and surface colour of fig (*Ficus carica* L.) during drying. International Journal of Food Science and Technology 47, 1710–1719.

Zito, P., Sajeva, M., Bruno, M., Rosselli, S., Maggio, A., Senatore, F., 2013. Essential oils composition of two Sicilian cultivars of *Opuntia ficus-indica* (L.) Mill.(Cactaceae) fruits (prickly pear). Natural Product Research 27, 1305–1314.

Further Reading

Balázs, A., Ficsor, E., Győry, H., 2010. The history of the fig tree (*Ficus carica* L.) and its use in phytotherapy. Orvosi Hetilap 152, 72–75.

Bercu, R., Popoviciu, D., 2014. Anatomical study of *Ficus carica* L. leaf. Annals of the Romanian Society for Cell Biology 19, 33.

Chandrasekar, S., Bhanumathy, M., Pawar, A., Somasundaram, T., 2010. Phytopharmacology of *Ficus religiosa*. Pharmacognosy Reviews 4, 195.

Robinson, J.P., Nithya, K., Ramya, R., Karthikbalan, B., Kripa, K., 2014. Effect of vesicular arbuscular mycorrhiza Glomus fasciculatum on the growth and physiological response in *Sesamum indicum* L. International Letters of Natural Sciences 18.

CHAPTER

22

Frangipani

Saba Idrees[1], *Muhammad Asif Hanif*[1], *Muhammad Adnan Ayub*[2], *Muhammad Idrees Jilani*[3], *Najma Memon*[4]

[1] Department of Chemistry, University of Agriculture, Faisalabad, Pakistan; [2] Department of Chemistry, University of Okara, Okara, Pakistan; [3] Department of Chemistry, University of Sahiwal, Sahiwal, Pakistan; [4] National Centre of Excellence in Analytical Chemistry, University of Sindh, Jamshoro, Sindh, Pakistan

OUTLINE

Medicinal Plants of South Asia
https://doi.org/10.1016/B978-0-08-102659-5.00022-7

Copyright © 2020 Elsevier Ltd. All rights reserved.

1. BOTANY

1.1 Introduction

Frangipani (*Plumeria rubra* L.) (Fig. 22.1) is a perennial shrub or laticiferous tree belonging to family Apocynaceae. It has been used for thousands of years, and the fruit is also eaten as food in many countries (Zaheer et al., 2010b). The genus *Plumeria* consists of eight species (Ye et al., 2009). The uncertainty in the exact number of species within the genus is largely attributed to the great variability among the constituent species. Variability is prevalent in morphology, growth habit, flower color, leaves, and stems

FIGURE 22.1 Frangipani plant and flowers.

and chemical composition. Bees and butterflies help the plants to pollinate. *Plumeria* reproduces fast and easy. The resulting diversity and variation have led some authors to reclassify a portion of the genus.

P. rubra is known by different names depending where you are in the world. In English, it is generally called frangipani. In Hindi, it is known as lal champa (Sultana et al., 2015), while in Malaysia, it is mainly known as "kemboja," but numerous other names such as "pokok kubur" and "bunga kubur" were used to indicate various hybrids and species of *Plumeria* (Devprakash et al., 2012). In Urdu, it is known as gul e yas. Probably the most familiar frangipani is *P. rubra*. The plant grows as a small tree or spreading shrub with the height of 2–8 m and a similar width. The large leaves can reach 30–50 cm long and are organized alternately and flocked at the end of branches. The flowers are very prominent, terminal, and strongly fragrant and appear on the ends of branches. The colors range from ordinary pink to white with a yellow color in the center of the flower (Gilman and Watson, 1994a). Frangipani are generally grown as ornamental plants in gardens, graveyards, and parks due to their beautiful fragrant flowers of several colors and sizes (Tohar et al., 2006a). The essential oil components of frangipani are equally variable between species and cultivars and are thought to be related to growing conditions, geographic origins, genetic factors, different chemo-types, and differences in the nutritional status of plants. The essential oil from orange flowers of frangipani contained both nonterpene esters such as 2-phenylethyl benzoate, benzyl salicylate, alkanoic acids, and benzyl benzoate in considerable amounts. The orange-flowered cultivar has the highest concentration of geraniol (4.1%) and (E)-nerolidol (14.4%) among the species studied (Tohar et al., 2006a). It is reported that essential oil yield is 0.39% in frangipani (orange flowered), 0.06% in frangipani (pink flowered), and 0.03% in frangipani (yellow flowered). Some variations have been observed between studied oil samples. (*E*)-Non-2-en-1-ol, the major chemical of oil of flower was not detected in leaves, whereas (*E*)- β-farnesene and α-patchoulene, present in the oil of leaves, have not been recognized in flower oil. The majority of essential oil is concentrated in the flowers and leaves (Lawal et al., 2015). It is obvious that frangipani is morphologically and chemically highly variable. The origin, source, and growing conditions of frangipani have an impact on the plant uses, flavors, aromas, and medicinal uses.

1.2 History/Origin

Charles Plumier, who was a French botanist of the 17th century, used the name *Plumeria*, but a spanish priest, Francisco de Mendoza, was the first who gave the name in 1522 (Choudhary et al., 2014). The natural perfume of the flowers of *Plumeria* reminded people of fragranced gloves, so these flowers have been named frangipani. Species of *Plumeria* were already widely being used throughout the world for their ornamental and medicinal applications, but in the period from the early 19th century, there was particular interest in the research about the structure elucidation, extraction, isolation, and establishment of pharmacological action of chemical components of plants of *Plumeria* species. Frangipani tree originates from South America, and it was introduced in Tahiti in 1852. Phytochemical studies on *Plumeria* genus started as far back as 1870 as separation of iridoid glycoside plumieride was done from frangipani stem bark (Choudhary et al., 2014).

1.3 Demography/Location

Although frangipani is grown in a variety of climatic and environmental conditions, the optimum conditions are found in countries with a warm climate. Warmth, light, and moisture are the key ecologic requirements for frangipani cultivation. Frangipani is susceptible to frost, so outdoor cultivation is restricted in frost-free regions of the world. They favor and grow best in a hot, dry climate. They are very fire hardy and drought resistant. Frangipani is grown widely in Pakistan, China, India, Brazil, Australia, Italy, the United States, North America, West Indies, Malay Archipelago, Jamaica, and Guiana (Aziz et al., 2013; Gilman and Watson, 1994b).

1.4 Botany, Morphology, Ecology

The plant is an upright, laticiferous tree or spreading shrub that grows to a height of 7–8 m (20–25 ft). Leaves are lanceolate to obovate, scattered, nerves several, horizontal, and 12.5–20 cm long. Flowers are very fragrant, generally red, purple, or pink and yellow from the center. They are large in terminal, deciduous, two to three dichotomous cymes, bracts several and wide. Calyx is small, five fids, glandular within; obtuse, lobes broad. Corolla salver shaped, throat necked, stamens near the base of the tube, anthers obtuse, cell rounded at the base. Carpels are two, distinct; style short; stigma, two lobed; ovules several seriate in each cell. Flowers of most cultivars are highly fragrant. The hybrids vary in their abundance of blooms, with some producing more than 200 flowers per cluster and other just 50 to 60 flowers. Follicles are ellipsoid, linear, or oblong. Seeds are winged, oblong, or lanceolate, plano convex, albumen fleshy, thin; ovate cordate, oblong, or cotyledons (Zaheer et al., 2010a).

Frangipani requires warm temperate or Mediterranean conditions; the optimum temperature for growth is 26–32°C. The plants are found to develop better in long day, full sunny conditions and require well-drained soils or potting soils. They can tolerate drought; however, they grow better in moist but not wet soil, with the optimum pH of the soil between 6.0 to 6.7, tolerating a range of pH from slightly acidic to slightly alkaline (Gilman and Watson, 1994b).

2. CHEMISTRY

Frangipani is an aromatic plant used as a shrub and is sweet smelling (Tohar et al., 2006a). A large number of frangipani ecotypes have been described based on their taste, flavor, and other phenotypic characters (Gilman and Watson, 1994a). Pink flowers of *Plumeria* are due to a phenolic compound (Wong et al., 2011). Bitter glycosides, lupeol, plumieride, plumeric acid, β-sitosterol, plumieride, plumieride, fulvoplumierin and amyrin are found in frangipani. Plumericin, Plumieride, isoplumericin, 4-hydroxy acetophenone, coumaroyl plumieride and protoplumericine are present in the bark of the plant. Flowers have essential oil. Roots have fulvoplumierin, plumericin, and three new compounds: β-dihydroplumericinic acid, isoplumericin, and β-dihydroplumericin. The flowers have resin, quercetin, and traces of kamempferol and a cyanidin diglycoside. It has seven volatile components: nanodecane, β-phenylethyl alcohol, heneicosane, 2-methylbutan-1-ol, benzyl salicylate, tetradecanoic acid, and phenylacetald; among them, 2-methylbutan methylbutan-1-ol could be considered the chemical marker in characterizing its essential oil (Omata et al., 1992).

It is good source of minerals such as calcium, magnesium, potassium, sodium, and copper. In a previous study, the yields of the volatile oils were 0.12% and 0.23% (v/w), respectively, for the leaves and flowers (Lawal et al., 2013). Frangipani is also known for flavonoids and antioxidant properties. Frangipani was found to possess compounds of phenolics, tannins, alkaloids, glycosides, saponins, and steroids (Sharma et al., 2017). The chemical classes present in the oil of flowers have been monoterpene hydrocarbons, aliphatic compounds, sesquiterpene hydrocarbons, and fatty acids. The major compounds in the oil of the flower were *n*-tetradecanal, phenyl acetaldehyde, (*E*)-non-2-en-1-ol, and limonene with considerable quantities of *cis*-9-tricosene, γ-elemene, (*E*, *E*)-α-farnesene, octadecanal, *n*-nonanal, α-copaene, and octadecanal. Sesquiterpene hydrocarbons and monoterpene hydrocarbons were the main classes of compounds identified in the leaf oil. The major components of the leaf oil were *E*-β-farnesene, (Z)-β-farnesene, α-patchoulene, and limonene. Phytol and α-copaene also found in significant amounts. The minor compounds comprise phenyl acetaldehyde, β-bisabolene, ledol, and *n*-nonanal (Lawal et al., 2013). Methanol extract of true frangipani has reducing sugar, tannins, flavonoids, terpenoids, and alkaloids

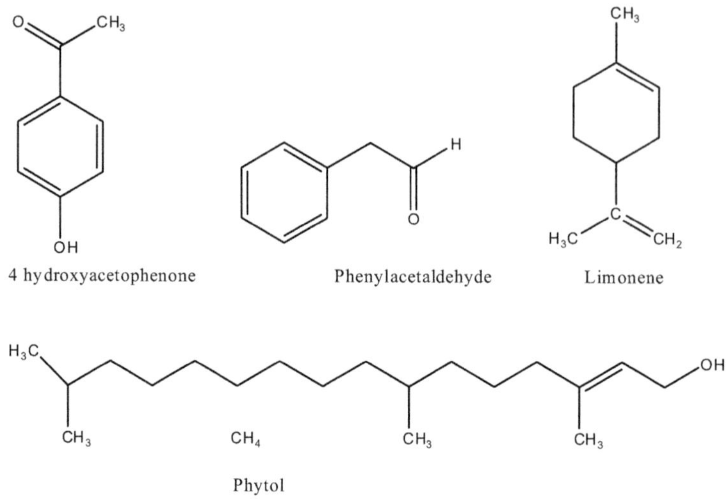

FIGURE 22.2 Important bioactive components of frangipani.

in flower and leaf and other tannins, carbonyl, saponins, steroids in leaf only (Zaheer et al., 2010a). There are different kinds of fatty acids present in it, such as myristic acid, lauric acid, palmitic acid, and linoleic acid. Some active components of frangipani are shown in Fig. 22.2.

3. POSTHARVEST TECHNOLOGY

Conventionally, the best harvesting time of frangipani is after ripening. The fresh flowers are consumed fresh and show flavor complexity and intensity that is greatly lost in dried flowers. However, many contrary findings reported that there is no difference between fresh and dried frangipani in relation to flavor contents. It is difficult to store in succulent conditions for longer term storage purposes, so it is advised to dry flowers appropriately for long-term storage. While drying, flowers should not be broken or shredded because broken flowers show reduced flavor due to loss of essential oil content. Flowers can be kept for numerous days in a plastic bag in 48–55°F temperatures. Further, all parts of the plant give out a milky sap when damaged. The sap may irritate skin and eyes (Criley, 2005), so precautions need to be taken during harvesting.

4. PROCESSING

Frangipani, like other shrubs, is consumed in a variety of ways and for various purposes. In addition to its fresh leaves, other common processed forms of frangipani include whole dry leaves, frozen, powdered leaves, and extracted essential oils. Whole plants or chopped leaves can be stored frozen, with or without oils, to be used for an extended time beyond its fresh shelf life. Alternative methods for preserving frangipani flower include storage in river sand and silica gel in the form of oil concentrate (Karunananda and Peiris, 2007). Essential oil can be extracted from different methods such as hydro distillation, steam distillation, and supercritical fluid extraction. Although the flowers of various *Plumeria* species are fragrant in nature, in general, percentage yields of the volatile constituents are very low. The essential oil yield is reported as 0.03% in frangipani (yellow flowered), 0.06% in frangipani (pink flowered), and 0.39% in frangipani (orange flowered) grown in peninsular Malaysia (Tohar et al., 2006a). In another report, the flower essential oil yield was reported 0.03%–0.12% in frangipani of Malaysian origin (Tohar et al., 2006b). In India, there is 0.04%–0.07% yield of frangipani (Kumari et al., 2012).

5. VALUE ADDITION

Frangipani is used in several ways. The latex mixed with coconut oil is used for itching. The juice with camphor is used for itching and as rubefacient in rheumatic pains. A poultice of heated leaves is useful for swellings. Decoction of leaves is used for eruptions and cracks of the soles of the feet. Frangipani flower tea brewed with or without black/green tea is believed to have a beneficial cooling effect and to be good for digestion. Regular consumption of frangipani flower tea is useful in maintaining good health. Fresh frangipani flowers are cooked as vegetables, providing a complementary savory flavor, giving therapeutic effects and health benefits.

6. USES

Many shrub and spices contribute significantly to heath despite a low amount of consumption, as they are full of antioxidants and mineral compounds. It is not clear how much frangipani should be consumed to gain its health benefits. Researchers do not have recommendations about the precise amount of uses; nevertheless, frangipani is full of antioxidants; in additions to this, it is also a good source of minerals. Frangipani complements food flavor. Frangipani oil and tea are available at many health food stores, though the substantial scientific evidence for its usefulness in human health is inadequate. Frangipani has many uses ranging from culinary to religious. In much of southeastern Asia, *Plumeria* is a symbolic plant that is religiously treated. In India, it is called "temple tree" because it is used in religious practices. In India, local folk believe that it provides shelter to ghosts and demons. In Malay, the aroma of the *Plumeria* has an association with vampires. It is a very valued plant in Hawaii where they use it for making leis, a type of necklace.

It is the national flower of Nicaragua, and its appears on several of their bank notes. It is the national tree of Laos, where it is known as dok jampa. It is noticed as a holy tree in Laos and all Buddhist temples and planted in courtyards. The frangipani is the flower of the city of Palermo in Sicily, Italy. These trees previously were treated as taboo in Thai homes due to superstitious relations with the plant's Thai name, lantom, which is related to ratom, the Thai word for sadness. Consequently, frangipanis have been considered to giving unhappiness. Today, though, the blossoms are presented as fragrant contributions, as Thai and Buddhist people wear them on unique event days such as Songkran (Thai New Year).

According to Vietnamese myth, ghosts live in trees with white and scented flowers, including the frangipani. In Vietnam and China, the color white is related to death and funerals. In Hindu culture, the flower means loyalty. Hindu women place the flower in their hair for their wedding ceremony to display their loyalty toward their husbands. The frangipani flower is found in numerous cosmetic products with soothing properties, in addition to in perfumery for its exhilarating and unusual aroma. In Asia, a flower infusion can be applied on the body following the bath. This infusion will tone up skin while perfuming it with a fine and delicate aroma. They are known as excellent ornamental plants and frequently seen in the graveyards (Misra et al.). It is a plant that is famous for its scented flowers and attractiveness.

The flowers of this plant are bechic and aromatic, broadly used in pectoral syrups. Flower essential oils are used for aromatherapy purposes and perfumery. Decoction of frangipani flowers have been reported to be used in Mexico for control of diabetes mellitus (Zaheer et al., 2010c). The leaves of frangipani are used in rheumatism, inflammations, ulcers, leprosy, and as rubefacient (Dabhadkar et al., 2012). The milky sap of leaves and stem was applied to skin ailments like scabies and herpes (Shinde et al., 2014). The bark of the root is bitter, carminative pungent, heating, laxative, and conventionally used to cure constipation, asthma, leprosy, and ulcers (Gopi et al., 2011). In Indonesia, frangipani bark is being used to treat gonorrhea, whereas in the Philippines, bark is used as a febrifuge, emmenagogue, and purgative (Sharma and Kumar, 2012). The fruit is noted to be eaten in West Indies. In India, it is used like an abortifacient (Patil and Bairagi). *P. rubra* L. cv. *acutifolia*, two species of this genus occurring in China, is a little tree whose flowers are used to cure fever and cold, infective hepatitis, whooping cough, diarrhea, tracheitis, mastitis, and calculus of urethra. Extracts of the plant are identified to have some biologic applications like anthelmintic, antimicrobial, antioxidant, antiinflammatory, anticancer, analgesic, antitumor, abortifacient, analgesic, antiulcer, antifertility, hypolipidemic, and antipyretic (Misra et al., 2012; Shinde et al., 2014).

Frangipani is known to have strong antioxidant properties. Research has shown the plant possesses significant antimicrobial, antiviral, and anticancer properties. Leaf extract revealed considerable antibacterial action against *Salmonella typhe* and moderate action against *Streptococcus saprophyticus, Streptococcus agalactin, and Enterococcus coli* (Egwaikhide et al., 2009). There is extensive diversity in the phytochemical constituents of frangipani; these constituents differ significantly with time, cultivation processes, and storage The nutritional and pharmacological properties of whole shrub in natural form, as it has been traditionally used, result from the interaction of many different active phytochemicals; consequently, the overall benefits of frangipani cannot be completely duplicated using single, isolated constituents. There is very little data relating to a standardized dosage available from traditional practitioners, which is problematic to chemists and pharmacists. Lack of information on dosage from traditional and orthodox medicine is considered an obstacle toward improvement of our understanding of the phytochemical components and their interactions. In recent years an increased methodical interest in plant phytochemical (fruit, herb, spices, and vegetables) health benefits has been an important subject of plant-based nutritional research. Although the study of plant compounds is not new, scientists are only now starting the characterization of bioactive compounds to explore their impact on human health and disease.

7. PHARMACOLOGICAL USES

7.1 Antimicrobial Activity

The small secondary metabolites of stem bark of frangipani show antimicrobial action. Four new iridoids, viz., epiplumeridoid C and plumeridoids A, B, and C, have been separated from frangipani stem bark. Compounds showed antifungal, antibacterial, and antialgal, actions (Kuigoua et al., 2010). The in vitro antibacterial action of ethyl acetate, ethanol, chloroform, and aqueous extracts of frangipani leaves was performed by disc diffusion method against bacterial strains. The relative study of extract by ciprofloxacin, particular standard, showed considerable antibacterial action (Baghel et al., 2010). The crude frangipani ethanolic extract displayed antioxidant action, and it may contain reasonable antimicrobial action. The antimicrobial action of frangipani has been tested via the disc diffusion method. The antimicrobial action has been evaluated against six pathogenic bacterial strains (both gram negative and positive). Leaf extract revealed considerable antibacterial action against *Salmonella typhe* and moderate action against *Streptococcus saprophyticus, Streptococcus agalactin*, and *Enterococcus coli*. Antimicrobial action and phytochemical substances of methanol extract of frangipani leaf and flower has been investigated. Phytochemical screening of crude extract exhibited presence of tannins, flavonoids, phlobatannins, and steroids, reducing sugar, saponins, terpenoids, and cardiac glycosides. Phlobatanins have been found to be absent in methanolic extract of frangipani (flower). All the crude extract exhibited higher inhibitory effects at tested concentration apart from *Bacillus anthracis* and *Corynebacterium pyogenes* of frangipani leaf (Egwaikhide et al., 2009).

7.2 Antioxidant Activity

Flavone glycoside separated from frangipani displays hypolipidemic and antioxidant action. In the TLC-based qualitative antioxidant determination using DPPH assay, frangipani displayed free radical scavenging properties. Frangipani ethanolic extract showed a considerable dose-dependent inhibition of DPPH action (Ramproshad et al., 2012).

7.3 Antiplasmodial Activity

Alcoholic and aqueous extracts of fresh frangipani shoot displayed considerable antiplasmodial action against a drug-sensitive strain, whereas dry alcoholic extract displayed considerable antiplasmodial action, too. A study was held to find out effect of frangipani extracts in controlling proliferation of *Plasmodium falciparum* within red blood cells. The infected red blood cells in culture manifested an elevated level of lipid peroxidation. Infected blood cell cultures treated with hydro-methanolic extracts displayed considerable reduction in lipid peroxidation. Scientist revealed considerable reduction in lipid peroxidation in heart microsome in vitro through ursolic acid, flavones, and glycosides, which are present in large quantities in frangipani. The compounds of frangipani may be considered strong inhibitors of *Plasmodium* growth and the life cycle of erythrocytic stages (Hung and Yen, 2002; Mathew et al.).

7.4 Antihelmintic Activity

Helminths infections are the most extensive infections in domestic animals and humans, disturbing a large number of world populations. The majority of these infections, due to worms, are usually restricted mostly to tropical areas, and the occurrence is accelerated due to unhealthy lifestyle and results in the development of pneumonia, anemia, and eosinophilia (Bundy, 1994). Saponin extract of frangipani leaves displayed considerable anthelmintic effect in concentration of 25 mg/mL comparable with standard piperazine citrate (Rastogi et al., 2009).

7.5 Anticancer Activity

Anticancer action of ethanolic extract of frangipani leaves against Ehrlich ascites carcinoma in Swiss albino mice was studied. The extract was administered orally, which enhanced the life span of Ehrlich ascites carcinoma–treated mice and restored hematological parameters compared with Ehrlich ascites carcinoma–bearing mice (Rekha and Jayakar, 2011).

7.6 Analgesic Activity

Analgesic action of frangipani was tested through acetic acid–induced writhing model in mice. The extract produced a considerable writhing inhibition. Ethanolic extracts may contain centrally and peripherally mediated analgesic activities. The peripheral analgesic effect of the plant's extract may be mediated via inhibition of cyclooxygenases and/or lipoxygenases or other inflammatory mediators, although central analgesic action of the extract may be mediated during inhibition of central pain receptors. The crude extract of experimental plant exhibited analgesic action (Rastogi et al., 2009).

7.7 Abortifacient Activity

The chloroform, ethanol ethyl acetate, and aqueous extract of frangipani pods displayed abortifacient action in female albino rats. The extracts considerably reduced the number of live fetuses throughout postimplantation phase, while resorption index and postimplantation losses augmented considerably. Saponins, alkaloids, phenolic, and steroids present in pod extract, which act alone or in combination, perhaps were moderately responsible for observed pregnancy-terminating effects (Dabhadkar and Zade, 2012). It suggested a reduction of ovarian steroidogenesis, which may be probable mechanism of action of this plant in reducing fertility (Dhanapal et al., 2012).

7.8 Antidiabetic Activity

P. rubra extract was subjected to antidiabetic study in an alloxan-induced diabetic model and hypoglycemic activity at three-dose levels

100, 200, and 400 mg/kg, respectively. Diabetes was induced by alloxan monohydrate (150 mg/kg, i.p.). *P. rubra* extract and standard drug glibenclamide (10 mg/kg, p.o.) were administered to animals for 28 days. A significant reduction ($P < .001$) in fasting blood glucose levels in normal and alloxan-induced diabetic mice was recorded for *P. rubra* extracts. The results of the present study provide support to the traditional usage of the plant in diabetes (Yadav et al., 2016).

8. SIDE EFFECTS AND TOXICITY

Scientific studies about safe use of *P. rubra* are lacking. The milky sap can be a skin irritant in sensitive individuals, causing rashes and blistering. Ingestion of the sap or bark can cause vomiting and diarrhea.

References

Aziz, A., Khan, I.A., Munawar, S.H., Sadr-ul-Shaheed, S.-u.-S., 2013. Antipyretic Study of Methanolic Bark Extract of *Plumeria rubra*, Linn. In Various Pyrexia Induced Models.

Baghel, A.S., Mishra, C.K., Rani, A., Sasmal, D., Nema, R.K., 2010. Antibacterial activity of *Plumeria rubra* Linn. plant extract. Journal of Chemical and Pharmaceutical Research 2, 435–440.

Bundy, D., 1994. 1. The global burden of intestinal nematode disease. Transactions of the Royal Society of Tropical Medicine and Hygiene 88, 259–261.

Choudhary, M., Kumar, V., Singh, S., 2014. Phytochemical and pharmacological activity of genus plumeria: an updated review. International Journal of Biomedical and Advance Research 5, 266–271.

Criley, R.A., 2005. Plumeria in Hawai 'i.

Dabhadkar, D., Zade, V., 2012. Abortifacient Activity of *Plumeria Rubra* (Linn) Pod Extract in Female Albino Rats.

Dabhadkar, D., Zade, V., Rohankar, P., Pare, S., Wikhe, M., 2012. Estrogenic and anti-estrogenic potentials of ethanolic pod extract of Plumeria rubra in female albino rats. Global Journal of Pharmacology 6, 142–147.

Devprakash, T.R., Gurav, S., P SKG, M.T., 2012. An review of phytochemical constituents & pharmacological activity of Plumeria species. International Journal of Current Pharmaceutical Research 4, 1–6.

Dhanapal, R., Ratna, J.V., Gupta, M., Sarathchandiran, I., 2012. Ovarian antisteroidogenic effect of three ethnomedicinal plants in prepubertal female mice. International Journal of Biological & Pharmaceutical Research 3, 30–36.

Egwaikhide, P., Okeniyi, S., Gimba, C., 2009. Screening for anti-microbial activity and phytochemical constituents of some Nigerian medicinal plants. Journal of Medicinal Plants Research 3, 1088–1091.

Gilman, E.F., Watson, D.G., 1994a. Plumeria Rubra-Frangipani. Fact Sheet ST-491. Florida Cooperative Extension Service. Institute of Food and Agricultural Sciences, University of Florida, pp. 1–4.

Gilman, E.F., Watson, D.G., 1994b. Plumeria Rubra Frangipani. Fact Sheet ST-490. Environmental Horticulture Department, Florida Cooperative Extension Service, Institute of Food and Agriculture Sciences, University of Florida.

Gopi, J., Khatri, P., Singh, N., Gaud, H., Patel, R., 2011. Phytochemical and pharmacological potential of *Plumeria rubra* Linn.(Apocynaceae): a review. International Journal of Pharmaceutical Science 3, 1162–1168.

Hung, C.-Y., Yen, G.-C., 2002. Antioxidant activity of phenolic compounds isolated from *Mesona procumbens* Hemsl. Journal of Agricultural and Food Chemistry 50, 2993–2997.

Karunananda, D., Peiris, S.E., 2007. Preservation of *Plumeria rubra* L.(Rathu Araliya) for dry flower arrangements. Tropical Agricultural Research 19, 160–169.

Kuigoua, G.M., Kouam, S.F., Ngadjui, B.T., Schulz, B., Green, I.R., Choudhary, M.I., Krohn, K., 2010. Minor secondary metabolic products from the stem bark of *Plumeria rubra* Linn. displaying antimicrobial activities. Planta Medica 76, 620–625.

Kumari, S., Mazumder, A., Bhattacharya, S., 2012. In-vitro antifungal activity of the essential oil of flowers of *Plumeria alba* Linn.(Apocynaceae). International Journal of Pharmacy and Technology 4, 208–212.

Lawal, O.A., Ogunwande, I.A., Opoku, A.R., 2013. Chemical Composition of Essential Oils of *Plumeria Rubra* L. Grown in Nigeria. Methodology.

Lawal, O.A., Opoku, A.R., Ogunwande, I.A., 2015. Phytoconstituents and insecticidal activity of different solvent leaf extracts of *Chromolaena odorata* L., against Sitophilus zeamais (Coleoptera: Curculionidae). European Journal of Medicinal Plants 5, 237–247.

Mathew, S., Jani, D., George, L., In-Vitro Evidence of Effective Anti-plasmodium Activity by *Plumeria Rubra* (L) Extracts.

Misra, V., Uddin, S.M., Srivastava, V., Sharma, U., 2012. Antipyretic activity of the *Plumeria rubra* leaves extract. International Journal of Pharmacy 2, 330–332.

Omata, A., Nakamura, S., Hashimoto, S., Furukawa, K., 1992. Volatile components of plumeria flowers. Part 2.1 Plumeria rubra L. cv.'Irma Bryan'. Flavour and Fragrance Journal 7, 33–35.

Patil, P., Bairagi, V., Phytopharmacological Review of Plumeria Species.

Ramproshad, S., Afroz, T., Mondal, B., Khan, R., Ahmed, S., 2012. Screening of phytochemical and pharmacological activities of leaves of medicinal plant *Plumeria rubra*. International Journal of Research in Pharmacy and Chemistry 2, 1001–1007.

Rastogi, S., Rastogi, H., Singh, V., 2009. Anti-inflammatory and anthelmintic activities of methanolic extract of Plumeria rubra leaves. Indian Journal of Natural Products 25, 15–18.

Rekha, J.B., Jayakar, B., 2011. Anti cancer activity of ethanolic extract of leaves of *Plumeria rubra* (Linn). Current Pharma Research 1, 175–179.

Sharma, A., del Carmen Flores-Vallejo, R., Cardoso-Taketa, A., Villarreal, M.L., 2017 Aug 17. Antibacterial activities of medicinal plants used in Mexican traditional medicine. Journal of ethnopharmacology 208, 264–329.

Sharma, S.K., Kumar, N., 2012. Antimicrobial potential of *Plumeria rubra* Syn Plumeria acutifolia bark. Der Pharma Chemica 4, 1591–1593.

Shinde, P., Patil, P., Bairagi, V., 2014. Phytopharmacological review of Plumeria species. Scholars Academic Journal of Pharmacy 3, 217–222.

Sultana, S., Asif, H.M., Akhtar, N., Ahmad, K., 2015. Medicinal plants with potential antipyretic activity: a review. Asian Pacific Journal of Tropical Disease 5, S202–S208.

Tohar, N., Awang, K., Mohd, M.A., Jantan, I., 2006a. Chemical composition of the essential oils of four Plumeria species grown on Peninsular Malaysia. Journal of Essential Oil Research 18, 613–617.

Tohar, N., Mohd, M.A., Jantan, I., Awang, K., 2006b. A comparative study of the essential oils of the genus *Plumeria* Linn. from Malaysia. Flavour and Fragrance Journal 21, 859–863.

Wong, S.K., Lim, Y.Y., Abdullah, N.R., Nordin, F.J., 2011. Antiproliferative and phytochemical analyses of leaf extracts of ten Apocynaceae species. Pharmacognosy Research 3, 100.

Yadav, A.V., Undale, V.R., Bhosale, A.V., 2016. Antidiabetic activity of *Plumeria rubra* L. in normal and alloxan induced diabetic mice. International Journal of Basic & Clinical Pharmacology 5, 884–889.

Ye, G., Li, Z.X., Xia, G.X., Peng, H., Sun, Z.L., Huang, C.G., 2009. A new iridoid alkaloid from the flowers of *Plumeria rubra* L. Cv. acutifolia. Helvetica Chimica Acta 92, 2790–2794.

Zaheer, Z., Konale, A., Patel, K., Subur, K., Farooqui, M., 2010a. *Plumeria rubra* Linn.: an Indian medicinal plant. International Journal of Pharmacy and Therapeutics 1, 116–119.

Zaheer, Z., Konale, A.G., Patel, K.A., Khan, S., Ahmed, R.Z., 2010b. Comparative phytochemical screening of flowers of *Plumeria alba* and *Plumeria rubra*. Asian Journal of Pharmaceutical and Clinical Research 3, 88–89.

Zaheer, Z., Konale, A.G., Patel, K.A., Khan, S., Ahmed, R.Z., 2010c. Comparative phytochemical screening of flowers of *Plumeria alba* and *Plumeria rubra*. Asian Journal of Pharmaceutical and Clinical Research 3.

CHAPTER

23

Garlic

Shumaila Saif[1], *Muhammad Asif Hanif*[1], *Rafia Rehman*[1], *Muhammad Riaz*[2]

[1] Department of Chemistry, University of Agriculture, Faisalabad, Pakistan;
[2] Department of Chemistry, University of Sargodha, Sargodha, Pakistan

OUTLINE

Medicinal Plants of South Asia
https://doi.org/10.1016/B978-0-08-102659-5.00023-9

Copyright © 2020 Elsevier Ltd. All rights reserved.

1. BOTANY

1.1 Introduction

Garlic (*Allium sativum* L.) (Fig. 23.1) is a bulbous herb belonging to Alliaceae family. It has been known as a valuable seasoning plant for thousands of years and has been used as a common medicine for different diseases and physiologic conditions (Singh and Singh, 2008). The genus *Allium* contains approximately 600 species (300 varieties are native to South and Central Asia) present over the whole northern hemisphere, and it is cultivated all over the world mainly in dry and hot places. There is a great variability among the constituent species, which is prevalent in bulb weight for every plant, number of bulbils per umbel, bulbil size, clove weight, and total sugars. The variability of phenotypic factors is higher in magnitude than genotypic factors (Sandhu et al., 2015). Garlic self-pollinates very easily (Mann, 1952). There are a number of plants with the common name garlic, e.g., British wild garlic and American wild garlic, but these are not true garlic.

A. sativum L. is known by different names in the world. In English, it is typically called garlic. In Urdu, it is called as lahsun. *Lashuna* is the Sanskrit name for garlic. In Bengali, it is called *Rasun*, in Gujarati, it is called *Lasan*, and in Kannada, its name is *Belluli* (Belemkar et al., 2013). Among the species of *Allium* the most familiar is garlic (*Allium sativum* L.). Another popular example includes *Allium ophioscorodon*. Several varieties exist, but the most common include Indian garlic, Greek garlic, and Tunisian garlic. Garlic is distinguished from other family members due to its clove-like bulbs and flat leaves. The plants can be an herbaceous, biennial vegetable, perennial bulbous plant, and may be treated as annuals and usually have a typical leek odor. Most common varieties of garlic are treated as annuals, but in some climates, some are permanent that may grow 2 feet high or more. A wide range of forms, colors, and sizes has elevated garlic's ornamental importance in recent years, increasing the plant's economic value globally (Devi et al., 2014).

Garlic contains 2% (max.) essential oil on a dry basis with extensive variation of chemical composition as a result of genetic diversity, environmental conditions, and agronomic behavior of culture (Lawrence and

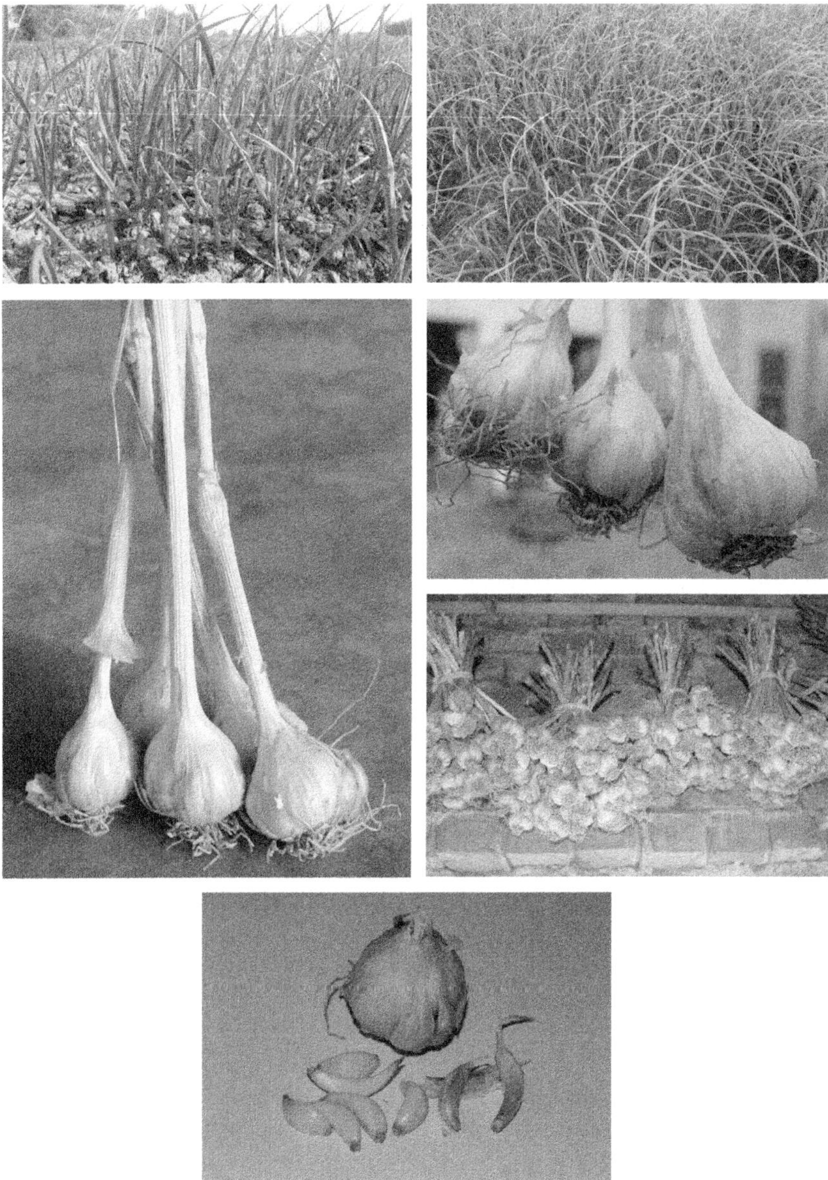

FIGURE 23.1 Garlic.

Lawrence, 2011). The main components of garlic essential oil are allicin, diallyl disulfide, and allyl propyl disulfide. Garlic is morphologically and chemically highly variable; these variations appear to be strongly influenced by environment. The origin, source, and growing conditions of garlic have an impact on the plant uses, flavors, aromas, and medical uses. The variability of garlic is reflected in the wide range of uses, which will be explored in more detail in the chapter.

1.2 History/Origin

Garlic is native to Central Asia (Kyrgyzstan, Turkmenistan, Kazakhstan, Tadzhikistan, and Uzbekistan) and northeastern Iran, where it has been cultivated for many years. The generic name *Allium* comes from the Celtic word "all" meaning pungent (Singh and Singh, 2008). There are number of suggested origins for the word garlic. The word "garlic" originated from the Anglo-Saxon *gar-leac* or spear plant: *gar* means "spear" (because the shape of a clove resembles the head of a spear) and *leac* means "leek." It was considered a medicinal plant for the treatment of various varieties of ailments and has been used as an important part of Mediterranean, Asian, and European diets as a food item (Goldy, 2000). About 5000 years ago, Sanskrits used garlic for medicinal purposes, and Chinese also used it in medicine for at least 3000 years. The Egyptians, Greeks, Babylonians, and Romans used garlic for medicinal purposes. Garlic was cultivated in Egypt as far back as 3200 BC. Pasteur was the person who distinguished garlic's antiseptic activity in 1858; then it was used as an antibacterial to stop gangrene during World War I and World War II (Tattelman, 2005).

First of all the healing qualities of garlic were utilized by Sumerians (2600–2100 BC), and it was believed that they brought garlic to China, from where garlic was spread to Korea and Japan. Garlic was used as a form of currency by Ancient Egyptians.

The cloves of garlic were buried in King Tut's tomb. The Greek physicians Galen and Hippocrates and, during the Middle Ages, Hildegard von Bingen used garlic for different purposes. Garlic was used to keep away the evil eye, witches, and vampires in the Middle Ages. Garlic has been also used as an aphrodisiac. The Chinese consider garlic to be a forbidden food for Buddhist monks because of its reputation as a sexual stimulant (Kemper, 2000; Bradley, 1992). The Italian mafia believed garlic oil was poisonous and could cause death. Garlic was valued as an offering fit for the gods and hated as an ingredient appropriate only to be fed to hogs (Rosen et al., 2008). In Russia, during the preparation of military projects and for piloting, garlic was used, and it was renamed Russian penicillin (Petrovska and Cekovska, 2010).

1.3 Demography/Locations

Garlic is grown widely in a variety of climatic and environmental conditions. Optimum conditions are found in countries with a warm climate. Garlic requires well-drained, fertile, and sandy soils, and good amount of organic material is required for optimum bulb growth. It grows well in soils with a pH range of 5.5–6.8. Garlic is often grown as a winter/spring crop also. Temperature, light, and moisture are the ecologic requirements for garlic cultivation (Adam, 2006). Garlic is grown widely in the countries including Pakistan, China, Republic of Korea, Indonesia, India, the United States, and Thailand. It is grown throughout Pakistan including Punjab, Sindh, Khyber Pakhtunkhwa, Kashmir, Gilgit-Baltistan, and Baluchistan (Hallam, 2004).

Absolute figures for garlic oil production are difficult to acquire. The major producer of garlic oil in the world is Indonesia. World garlic oil export is largely designed to the European Union and the North America. However, the pattern of the product's import is indicated in the US import statistics. During the era 2000–05, garlic oil imported to the United States averaged 73 tones, with an annual average growth rate of 17%. The US demand for garlic oil is assumed to be one-third of the total global demand. The total global demand for garlic is estimated to be 219 tons per annum (Kerckhoffs et al., 2002).

The leading producer of garlic is China, with an annual production of about 20 million tons, which accounts for 81% of world production. India with 4.6%, South Korea with 1.4%, Egypt with 1.2%, and the United States with 0.8% follow for annual world production. A large quantity of garlic produced in the United States is centered in Gilroy, California (Boriss and Kreith, 2006).

1.4 Botany, Morphology, and Ecology

Garlic is an annual herb native to South and Central Asia. *A. sativum* L. is an important herb, about 40 cm tall when fully grown. The leaves of garlic are very flat, slender, and 1–2 feet long. All the leaves arise from the swollen stem that takes the form of a bulb. The bulb of garlic is of a compound nature, consisting of several bulbets, called cloves, of different size, overall enclosed by layers of white scaly leaves. The length of bulb is 12–18 inches (30–45 cm) and its diameter is 9–12 in. (22.5–30 cm). The ovoid cloves are three to four sided with an acute summit, contracted like a fiber, and have a truncate base. Each clove is individually surrounded by white scales and enclosed in a pinkish-white skin. From the central clove, the plant shoots a quill-like, round, hollow, and unbranched stalk, which is covered at the bottom by long, flat, and narrow, grass-like leaves. The white flowers are located at the end of a stalk, rising directly from the bulb, and assembled together in a globular head. The flowers produce egg-shaped bulbils, which have an important function in the propagation of the plant (Lardo & Kreuter).

Garlic requires a cooler environment during the early stages for vegetative growth. For germination, optimum temperature is not available, and it grows from cloves with growing temperature of 13–24°C. Long days favor bulb development. The garlic shoots can bear temperatures as low as −6°C without damage, while temperatures below −12°C can kill shoots and cause poor bulb development. Garlic has medium, white, hairless, deep, and quite strong roots (Rosen et al., 2008).

2. CHEMISTRY

Garlic is an impressively important spice known as "stinking rose," and it may be recognized for its scent and flavor, but actually garlic is fragrance-free until the cells of garlic are ruptured by crushing or cutting (Simon, 1984). Garlic has a strong, spicy flavor that sweetens considerably with cooking. Sulfur compounds are responsible for garlic's signature scent. Whole garlic bulb gives very little aroma, but when clove is cut the allicin is released, which gives a pungent, spicy, and mellow smell (Tucker and DeBaggio, 2000). Besides the scent, allicin is the cause of many of garlic's health benefits, including its antimicrobial, antioxidant, cholesterol-lowering, and blood-thinning properties, and it also plays an important role in garlic's anticancer effects (Koch and Lawson, 1996). Some major components of garlic are shown in Fig. 23.2.

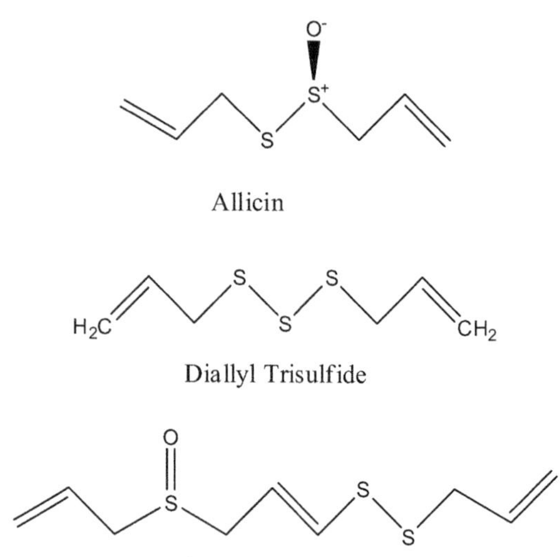

FIGURE 23.2 Some major components of garlic.

Fresh garlic contains numerous vitamins, minerals, and trace elements. It is also a rich source of carbohydrates, proteins, manganese, and vitamin B_6, vitamin C, copper, selenium, vitamin B_1, phosphorus, and calcium. The fresh peeled cloves of garlic have 6.30% protein, 1.0% mineral matter, 29.0% carbohydrates, 60%–65% (wb) moisture, 0.10% fat, 0.80% fiber, 0.03% calcium, 0.31% phosphorous, 0.001% iron, 13 mg/100g vitamin C, and 0.40 mg/100g nicotinic acid. Garlic has higher sulfur content than any member of the *Allium* genus. Garlic contains about 0.5% volatile oil, which is comprised of sulfur-containing compounds, vitamin B, and flavonoids (Mikail, 2010).

A. sativum has a strong characteristic odor and taste, due to which the bulb of garlic is used as a flavoring agent. The scented garlic oil is obtained from the crushed bulbs of garlic. The bulbs are comprised mainly of 0.5%–2% of volatile oil/essential oil, which have sulfur-containing compounds including diallyl sulphide, diallyl disulphide, allyl methyl sulphide, allyl methyl trisulphide, dimethyl trisulphide, 3-vinyl-1,2-dithiin, 2-vinyl-1,3-dithiin, diallyl trisulfide, and allyl propyl disulphide, alliin, allicin, ajoene enzymes, minerals, vitamin B, and flavonoids (Chekki et al., 2014). The information available about the chemical properties of garlic is centered on the numerous studies that have been conducted in various parts of the world. Garlic has a large number of fatty acids; the major components of Tunisian garlic are lauric acid (49.3%) and linoleic acid (20.4%), while Indian and Greek garlic have low lauric acid (0.5%, 0%) and high linoleic acid (64.8% and 53.6%), respectively.

3. POSTHARVEST TECHNOLOGY

The best harvesting time of garlic ranges from early July to mid-September, although harvesting can occur slightly earlier or much later. The garlic is harvested when the leaves are half to three-quarters brown. The plant can be collected out of the ground by using a pitchfork while digging the soil. The plants planted in the spring can be planted later in the year than the plants planted in the fall (Hannan and Sorensen, 2001). Garlic and its dried products must be stored at low-humidy conditions, because at intermediate temperature, sprouting can occur. Storage life is affected by the variety of garlic. Approximately 1 to 2 months at ambient temperature 20–30°C under low relative humidity (<75%) is required for storing garlic in a perfect condition. Bulbs will ultimately become spongy, withered, and soft under these situations after this storage time. In case of long-term storage, −1 to 0°C temperature with low relative humidity of 60%–70% will be suitable. Perfect air flow is very important to stop moisture accumulation. The growth of bulbs indicates that garlic loses dormancy at the temperature of 5–18°C. High humidity will favor mold growth. Garlic must be stored separately to avoid the transmission of garlic odor to other goods. A ventilation system should be designed for bulk storage of garlic, and for this purpose, air can be provided to the store room from the bottom. Container rows must be arranged parallel toward the direction of air flow and must be placed 6–7 inch apart (Kaur et al.).

4. PROCESSING

Garlic like other plants is consumed in a variety of ways and for various purposes. The most frequently used method for dehydration of garlic is conventional air drying. First, fresh, firm cloves of garlic are dried without cutting. The cloves are separated and peeled for preparing. The cloves are cut lengthwise. Additional predrying treatment is not necessary. The cloves are then dried for 2 hours at 60°C, and then the temperature is reduced to 55°C till the cloves are fully dried. Garlic salt can also be prepared from dried garlic if required. A food processor or blender is used for making dried garlic powder. Then, four parts of salt are added to one part of garlic powder, and it is shaken for 1 to 2 seconds. Very fine salt will form if blended longer. Propane flaming can be used to remove the tops after drying them. Care should be taken so that garlic bulbs are not exposed to flaming damage. Complete top removal is done by flailing. For a short curing period of 1–2 days, garlic can be dug with a fine potato digger and windrowed on the soil surface; after that, it is placed into bins or sacks for complete drying up to 10–14 days in the fields. Particularly during high temperature (over 32°C) and bright sunlight, garlic must be protected from sun blister. Greenery occurs due to extreme contact by sun. Garlic powder is a seasoning/spice, while complete garlic is a vegetable, not a flavor (Medina and Garcia, 2007).

5. VALUE ADDITION

Garlic has carminative, antirheumatic, and digestive properties. It is used in Ayurvedic formulations. Since ancient times, it has been used in Ayurvedic formulations for curing giddiness, muscular pain, heating intestinal ulcers and lungs, etc. Garlic is used as an ingredient to flavor the different vegetarian and nonvegetarian dishes and pickles, and it is consumed green as well as dried in spice form. Chutneys, curry powders, and good tasty pickles are prepared from garlic cloves. The smell and flavor of salted meat and fish is removed by using garlic. In many countries, instead of using fresh bulbs, dehydrated garlic is used in powdered or granulated form. The main advantages of preparation of value-added products obtained from garlic are to increase the storage life and reduce the transportation cost, as well as to earn more profit. It is sold either in the form of dehydrated powder or dehydrated flakes. The dehydrated garlic attains uniform importance in the world these days, and the dehydrated products are more uniform in flavor than the stored bulbs. Large amounts of dehydrated products including garlic oil, dehydrated garlic, and powder flakes are exported from South Asia to the United Kingdom, Turkey, Germany, Italy, France, and Japan. Other garlic products are crushed or chopped garlic, peeled processed garlic, garlic granules, and fried/roasted garlic. Fresh garlic bulbs are used to prepare fried garlic granules, and they have good demand in the foreign market. However, the cost of fresh garlic changes widely changing over the year. For preparing quality products from garlic with a long shelf life, during the season for their use in the off-season, a simple process technology needs to be made available. Moreover, due to short available time for kitchen work and increase in number of working women in urban areas the demand for a ready-to-use product is increasing (Bensky et al., 2004). Capsules of garlic oil that are available commercially contain vegetable oil and a minor quantity of garlic essential oil because of bitter odors (Yang et al., 1995).

6. USES

Garlic (*A. sativum*) has been used for many years for culinary, medicinal, and spiritual purposes. The medicinal effects of garlic on different diseases and particular systems have been studied (Valente et al., 2014). Today, it is frequently used in many cultures for flavoring. It is also used as a seasoning and food preservative. Garlic was regarded as a force for both good and evil in myths. Garlic has been used for white magic or protection according to different cultures of Europe. According to Central European traditional beliefs, garlic could be worn, rubbed on chimneys and keyholes, and hung in windows to ward off vampires. In Islam, it is strictly prohibited to eat garlic while going to the mosque, as its odor could disturb other Muslims during prayers (Hodges and Bennett, 2006).

Going beyond ritualistic uses, garlic has also been used for culinary purposes. It is a valuable culinary spice used in a wide range of meat dishes, condiments, and seasoning for vegetables. Garlic has been used as a flavor and food component of cooking over the world, as it enhances the smell and flavor of foods when it combines with an enormous number of foods. The oil of garlic is also used in food flavoring, and it is also pickled. Flower stems of garlic are edible and used in cooking. These are used like chives but have a minor taste. Fresh garlic, powder of garlic, and oil of garlic are used in food and beverages to add flavor (Biloba, 1999).

Garlic has also been used to inhibit and treat cardiovascular disease by dropping cholesterol and blood pressure, as an antimicrobial, and a preventive agent for cancer (Tattelman, 2005). Garlic is one of those plants that was used for thousands of years to treat infectious diseases. Historically, garlic has been used to treat severe diarrhea, leprosy, earaches, deafness, constipation, parasitic diseases, and to fight infections, lower fever, and relieve stomach aches.

In Europe and India, colds, hay fever, coughs, and asthma were treated by garlic medications. Garlic oil has also been used as ear drop to heal pain of ear infection by modern herbalists and doctors (Kemper, 2000). Diarrhea, abdominal discomfort, otitis media, and respiratory tract infections are treated by garlic in Africa, particularly in Nigeria. Garlic was used externally for the treatment of ringworm, milled with honey for use against rheumatism, boiled with sugar and vinegar for treatment of asthma, made as a drink for the treatment of epilepsy, and mixed with milk for use as a vermifuge. In Europe and Asia, garlic is particularly used in healing wounds. Garlic has several properties for treating skin problems. Garlic contains several powerful antioxidants that help people to stay young by regenerating skin and tissues. It is an extremely effective remedy for acne treatment cases and pimples.

7. PHARMACOLOGICAL USES

7.1 Antibacterial Activity

The antimicrobial activity of garlic is due to the major compounds like allicin and other sulfur-containing compounds. It has been observed that ethanolic, chloroform, and aqueous extracts of garlic strongly inhibited the growth of pathogenic bacteria with a fluctuating degree of susceptibility. The *Staphylococcus aureus* gram-positive bacteria growth was more affected by garlic compared to gram-negative bacteria. On the other hand, the garlic aqueous extracts may be used as conventional antibiotics to fight mediators of nosocomial toxicities that are predominant in hospitals. The effects of ethanolic and aqueous extracts of garlic against bacteria like *Salmonella typhi* and *Escherichia coli* were studied in an in vitro study. It was observed that ethanolic extract had a greater inhibitory effect, while the aqueous extract had minute inhibitory effect. Pure form of allicin has been found to show antibacterial action against multidrug-resistant enterotoxigenic strains of *Escherichia coli*. In another report, the aqueous extract showed antibacterial action against gram-negative (*Klebsiella pneumoniae*) and gram-positive (*Bacillus subtilis*) strains, and methanolic extracts exhibited antimicrobial action against all tested microorganisms, while ethanolic extracts of garlic exhibited the highest activity against

Bacillus subtilis. In China, an exclusive garlic derivative, allitrida, was successfully used to treat all bacterial infections (Mikaili et al., 2013). The antibacterial activity of certain concentrations (50, 100, 200, 300, and 500 mL/L) of essential oil extracts of garlic against two bacteria, *Salmonella enteritidis*, and *Staphylococcus aureus*, was studied. Due to the highest inhibition effect of garlic, it was observed that the essential oil extracts of *Allium* plants (garlic) showed good antibacterial activity. Moreover, high inhibitory activity of garlic extract was observed for all tested concentrations. Essential oil of garlic showed less sensitivity toward *Staphylococcus aureus*. While the growth rate of *Salmomella enteritidis* was highly inhibited by garlic essential oil extracts. Finally, essential oil extracts of garlic can be used as natural antibacterial additives by combining in different food products for flavoring (Benkeblia, 2004).

7.2 Antiparasitic Activity

Garlic has been found to have antiparasitic activity against *Trypanosoma brucei* and *Plasmodium falciparum* due to allicin. Many intestinal protozoan parasites such as *Giardia lamblia* and *Entamoeba histolytica* are also affected by allicin. DTS is the chemically stable final transformation product of allicin. In an in vitro study, the action of diallyl trisulfide was studied against different protozoan parasites. It was concluded that diallyl trisulfide has the potency to treat different parasitic infections of humans and animals like *Entamoeba histolytica*, *Giardia lamblia*, and *Trypanosoma* species. Ajoene isolated from *Allium sativum* L. is an inhibitor of human Trypa cruzi trypanothione reductase and glutathione reductase. Cytostatic and the antiparasitic property of ajoene may be because of its several effects on main enzymes of antioxidant thiol metabolism. *Echinacea purpurea*, *cocoa*, and *A. sativum* L. extracts are different substances that are used for preparing alchinal. It has been proven that the quantity of adult types and muscular larvae of *Trichinella spiralis* considerably decreases due to this preparation. Many microorganisms such as *Trypanosoma* spp, *Giardia* spp, *Plasmodium* spp, *Leishmania* spp, and *Cochlospermum planchonii* are affected by garlic oil. While the aqueous extract of garlic has also been found to be effective against giardiasis and hymenolepiasis (Mikaili et al., 2013).

7.3 Antiviral Activity

Garlic has been shown to have antiviral activity. It was reported that antiviral activity of garlic was attributed to the following components: ajoene > allicin > allyl methyl > thiosulfinate > methyl allyl thiosulfinate. The major component, allicin, of *A. sativum* L. showed antimicrobial actions in both in vivo and in vitro studies. The human cytomegalovirus (HCMV), herpes simplex virus type 1, vaccinia virus, influenza B virus, herpes simplex virus type 2, vesicular stomatitis virus, parainfluenza virus type 3, and human rhinovirus type 2 are sensitive to the extracts of garlic. It has been showed that attacks of common cold virus are prevented by allicin-containing complements. The major antimicrobial action of allicin is the result of chemical reaction of allicin with thiol groups of different enzymes, i.e., alcohol dehydrogenase. It has been observed in an in vivo study that mice models are protected by administrating garlic against intranasal immunization with influenza viruses, which increases the production of neutralizing antibodies when a vaccine is given. Ajoene, separated from garlic extracts, can prevent adhesive interaction and synthesis of leukocytes. In another study, allitridin's (DTS) effect on the copying of HCMV and appearance of viral immediate-early genes showed that this element has anti-HCMV efficiency (Mikaili et al., 2013).

7.4 Anticancer Activity

The anticancer activities were shown by compounds derived from garlic including diallyl sulfide, diallyl trisulfide (DATS), and diallyl disulfide (DADS). The cytotoxicity produced by diallyl trisulfide is stimulated by ROS and successive initiation of the ROS-dependent caspase path in U937 leukemia cells. In an in vitro study, it was shown that apoptosis in different human cancer cell lines was induced by DATS, and in an in vivo study, it gave significant protection against animal tumor models such as colorectal cancer. In another study, it was suggested that tumor cell motility and invasion is inhibited by DADS treatment, and it acts as a dietetic source to reduce the danger of cancer metastasis. In recent times, a strong compound, S-allylcysteine (SAC), derived from garlic showed an in vitro chemopreventive property. It can be used for treatment of cancer. The famous biologically active component allicin (diallyl thiosulfinate) of freshly crushed extract of garlic is very effective on cell proliferation of colon cancer cells (Mikaili et al., 2013).

7.5 Antiinflammatory Activity

The antiinflammatory activity has been observed in garlic extracts. It is reported that *Eimeria papillata* toxicities that cause swelling and harm to the liver are significantly reduced by garlic treatment. It has been observed that garlic oil mostly showed the antiinflammatory action by obstructing the assembly–disassembly routes of the cytoskeleton. For the development of antiinflammatory drugs with minute side effects, a lead compound derived from allicin is considered a good starting point (Mikaili et al., 2013).

7.6 Immunomodulatory Activity

It was observed that allicin showed an inhibitory immunomodulatory influence on intestinal epithelial cells, and it has ability to reduce abdominal infection. Allicin also exhibited an in vitro immunomodulatory activity toward different functions of peripheral blood cells (Mikaili et al., 2013).

7.7 Cardioprotective Activity

The protection of rat heart from ischemic reperfusion injury was studied in in vitro experiments (Banerjee et al., 2003). The major thiosulfinate compound such as allicin in homogenized garlic extract and different sulfur compounds in garlic oil showed antioxidant and antihypertensive properties in a variety of doses. The cholesterol-lowering effect of garlic is more obvious than other biologic effects. It has been suggested that sulfur compounds in garlic can be the foundation of inhibition of synthesis of cholesterol. It was suggested that a large number of medicinal properties of garlic are related with its antioxidant activities (Petchdee, 2012).

8. SIDE EFFECTS AND TOXICITY

Garlic is safe when used in food amounts. However, care should be taken in using garlic in medicinal amounts during pregnancy and breastfeeding, as no reliable data exist about garlic effects. Garlic is unsafe in bleeding disorder, diabetes, and for people who are suffering from gastrointestinal tract problems.

References

Adam, K.L., 2006. Community Supported Agriculture. ATTRA-National Sustainable Agriculture Information Service Butte, MT.

Banerjee, S., Mukherjee, P.K., Maulik, S., 2003. Garlic as an antioxidant: the good, the bad and the ugly. Phytotherapy Research 17, 97–106.

Belemkar, S., Dhameliya, K., Pata, M.K., 2013. Comparative study of garlic species (*Allium sativum* and *Allium porrum*) on glucose uptake in diabetic rats. Journal of Taibah University Medical Sciences 8, 80–85.

Benkeblia, N., 2004. Antimicrobial activity of essential oil extracts of various onions (*Allium cepa*) and garlic (*Allium sativum*). LWT-Food Science and Technology 37, 263–268.

Bensky, D., Gamble, A., Kaptchuk, T.J., 2004. Chinese Herbal Medicine: Materia Medica. Eastland Press, Seattle.

Biloba, G., 1999. Herbal remedies: adverse effects and drug interactions. American Family Physician 59, 1239–1244.

Boriss, H., Kreith, M., 2006. Commodity Profile: Peanuts, Agricultural Issues Center. University of California.

Bradley, P., 1992. British Herbal Compendium: A Handbook of Scientific Information on Widely Used Plant Drugs/published by the British Herbal Medicine Association and Produced by its Scientific Committee. The Association, Bournemouth, Dorset.

Chekki, R.Z., Snoussi, A., Hamrouni, I., Bouzouita, N., 2014. Chemical composition, antibacterial and antioxidant activities of Tunisian garlic (*Allium sativum*) essential oil and ethanol extract. Mediterranean Journal of Chemistry 3, 947–956.

Devi, A., Rakshit, K., Sarania, B., 2014. Ethnobotanical Notes on Allium Species of Arunachal Pradesh, India.

Goldy, R.G., 2000. S.D.V. Agent, Producing Garlic in Michigan. Michigan State University Extension.

Hallam, D., Tropical, H.P.S.R.M, 2004. The Market for Non-traditional Agricultural Exports. FAO Rome.

Hannan, R., Sorensen, E., 2001. Crop Profile for Garlic in Washington. Washington State University, Pullman.

Hodges, S., Bennett, B.C., 2006. The ethnobotany of Pluchea carolinensis (Jacq.) G. Don (Asteraceae) in the Botánicas of Miami, Florida 1. Economic Botany 60, 75–84.

M. Kaur, P. Kaur, G. Kaur, Post Harvest Management of Garlic.

Kemper, K.J., 2000. Garlic (*Allium Sativum*), the Longwood Herbal Task Force and the Center for Holistic Pediatric Education and Research, pp. 1–49.

Kerckhoffs, D.A., Brouns, F., Hornstra, G., Mensink, R.P., 2002. Effects on the human serum lipoprotein profile of β-glucan, soy protein and isoflavones, plant sterols and stanols, garlic and tocotrienols. Journal of Nutrition 132, 2494–2505.

Koch, H.P., Lawson, L.D., 1996. Garlic: The Science and Therapeutic Application of *Allium Sativum* L. and Related Species. Williams & Wilkins xv, Baltimore, Maryland, p. 329.

A. Lardo, M. Kreuter, Historical Aspects *Allium sativum* Is Supposed to Originate From Central Asia, From Where Its Cultivation has Spread.

Lawrence, R., Lawrence, K., 2011. Antioxidant activity of garlic essential oil (*Allium sativum*) grown in North Indian plains. Asian Pacific Journal of Tropical Biomedicine 1, S51–S54.

Mann, L., 1952. Anatomy of the garlic bulb and factors affecting bulb development. California Agriculture 21, 195–251.

Medina, J.D.L.C., Garcia, H., 2007. Garlic: Post-harvest Operations. INPhO Post-harvest Compendium. Food and Agriculture Organization of the United Nations.

Mikail, H., 2010. Phytochemical screening, elemental analysis and acute toxicity of aqueous extract of *Allium sativum* L. bulbs in experimental rabbits. Journal of Medicinal Plants Research 4, 322–326.

Mikaili, P., Maadirad, S., Moloudizargari, M., Aghajanshakeri, S., Sarahroodi, S., 2013. Therapeutic uses and pharmacological properties of garlic, shallot, and their biologically active compounds. Iranian Journal of Basic Medical Sciences 16, 1031–1048.

Petchdee, S., 2012. Cardioprotective effects of garlic. KKU Veterinary Journal 22, 242–254.

Petrovska, B.B., Cekovska, S., 2010. Extracts from the history and medical properties of garlic. Pharmacognosy Reviews 4, 106.

Rosen, C.J., Becker, R., Fritz, V., Hutchison, B., Percich, J., Tong, C., Wright, J., 2008. Growing Garlic in Minnesota. University of Minnesota Extension.

Sandhu, S., Brar, P., Dhall, R., 2015. Variability of agronomic and quality characteristics of garlic (*Allium sativum* L.) ecotypes. SABRAO J. Breed. Genet 47, 133–142.

Simon, J.E.C., 1984. Herbs an Indexed Bibliography 1971–1980: The Scientific Literature on Selected Herbs and Aromatic and Medicinal Plants of the Temperature Zone.

Singh, V.K., Singh, D.K., 2008. Pharmacological effects of garlic (*Allium sativum* L.). Annual Review of Biomedical Science 10, 6–26.

Tattelman, E., 2005. Health effects of garlic. American Family Physician 72, 103–106.

Tucker, A., DeBaggio, T., 2000. The Big Book of Herbs: A Comprehensive Illustrated Reference to Herbs of Flavor and Fragrance. Interweave Press, Loveland, CO.

Valente, C., Aboua, G., Du Plessis, S.S., 2014. Garlic and Its Effects on Health With Special Reference to the Reproductive System. Antioxidant-Antidiabetic Agents and Human Health.
Yang, E., Zha, J., Jockel, J., Boise, L.H., Thompson, C.B., Korsmeyer, S.J., 1995. Bad, a heterodimeric partner for Bcl-x L and Bcl-2, displaces Bax and promotes cell death. Cell 80, 285–291.

Further Reading

Hickey, M., 2012. Growing Garlic in NSW. Department of Primary Industries, New South Wales.
M.S. Nimbarte, K. Zakiuddin, Garlic Peeling Machine–A Past Review.

CHAPTER

24

Ginkgo biloba

Ayesha Khalil, Haq Nawaz, Muhammad Asif Hanif, Rafia Rehman

Department of Chemistry, University of Agriculture, Faisalabad, Pakistan

OUTLINE

Medicinal Plants of South Asia
https://doi.org/10.1016/B978-0-08-102659-5.00024-0

Copyright © 2020 Elsevier Ltd. All rights reserved.

1. BOTANY

1.1 Introduction

Ginkgo (the maidenhair tree) is among the oldest living species of trees on earth, and that is why it is called the "living fossil" Ginkgo belongs to the family Ginkgoaceae that flourished in large forests over 150 million years and is currently grown around the world for ornamental and medicinal purposes (Fig. 24.1) (Huh and Staba, 1992). The genus *Ginkgo* contains only one specie, *Ginkgo biloba*, which has survived; hence, it is called "living fossil." There is no variability in the morphology, stem habit, or chemical composition, so it is classified in a separate division, the Ginkgophyta (Zhang et al., 2016). In ancient times, Ginkgo itself and its fossil cousins were widespread, and leaves of ginkgo fossils were well-known from each continent, being native to Asia. Now, Ginkgo grows in North America and also in Europe (Crane et al., 2013).

G. biloba is known by different names in different places. In English, it is typically called "*maidenhair tree,*" and in Urdu, it is named "*pankha plant*" (Traditional and Berry Seeds). It called "*balkuwari*" in Hindi. In Japan, it is called "*ginkyo,*" and its German, French, and Italian name is "*ginkgo.*" It is called "*noyer du Japon*" and "*l'arbre aux quarante ecus*" in France (Jacobs and Browner, 2000).

1.2 History/Origin

G. biloba is native to China and spreads up to 1100 m in broad-leaved mixed mesophytic forest. It is to be found in the hill countries and on the border of the Yangtze River valley. Although, China, Korea, and Japan are its natural habitat, and the place of its origin is believed to be in the mountain valleys of the eastern China province Zhejiang (Wang, 1961). The generic name, ginkgo, comes from the Chinese word yin-kuo meaning "silver apricot," and biloba comes from the leaf of the maidenhair tree, referring to its fan-shaped, two-lobed leaves. It is called maidenhair tree due to its resemblance to "maidenhair fern" (*Adiantum*) foliage (DeFeudis, 1991). A German physician and botanist named Engelbert Kaempfer used the Ginkgo name first. He was the first Westerner that

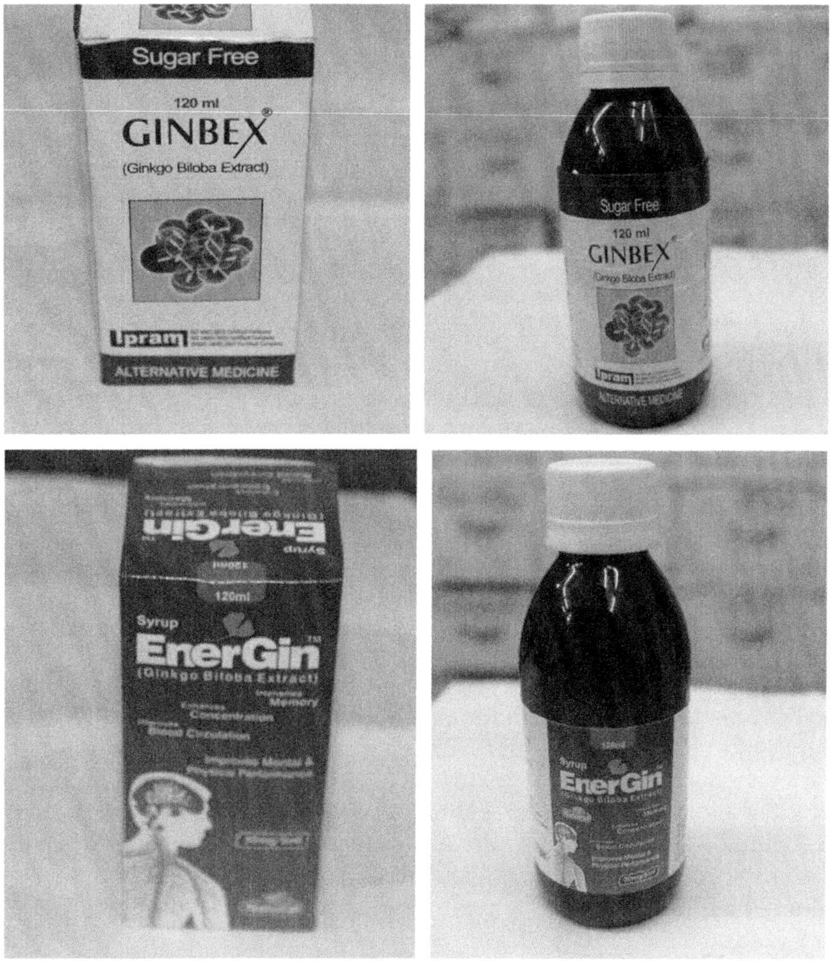

FIGURE 24.1 Commonly sold medicinal formulations of *Ginkgo biloba*.

described the plant when he was stationed at an outpost of the Dutch East India Company (VOC) from 1690 to 1692 in Nagasaki, Japan. In 1712, this name was published in his *Amoenitatum Exoticarum* and was contained in its fifth fascicle, which is a flora of Japan. In 1771, Linnaeus introduced the Latin name *Ginkgo biloba* in his *Mantissa Plantarum Altera*. He added the specific name "*biloba*" to the genus name "*Ginkgo*" that was adopted in 1712 by Kaempfer (Nagata et al., 2015).

From 1730 onward, maidenhair trees were abundantly planted all over Europe in Geetbets (Belgium), Anduze (France), Padova (Italy), Vienna (Austria), Slavkov (Czech Republic), Daruvar (Croatia), Kew (United Kingdom), Montpellier (France), and Harbke (Germany). In North

America, the first maidenhair tree was planted in Philadelphia in 1784 (Zhao et al., 2010).

Goethe wrote a poem as a tribute to the beauty of this tree in Heidelberg in 1815 (Unseld, 2010). About 350 years ago, the knowledge of this plant was limited only to China. In China, more than 100 types of *G. biloba* plant are still grown around the temples, and have been for thousands of years, because its nuts are used in worship (Hori et al., 2012). This plant is also found widely in South Asian countries.

1.3 Demography/Location

Ginkgo is mostly grown on rocky slopes, along stream beds, on the edge of soil beds, and on degraded sites. Ginkgo is grown in every climate. Maidenhair tree is tolerant of smoke, poor soil, extreme cold, shortage of water, and automobile exhaust (Huxtable, 1992).

1.4 Botany, Morphology, Ecology

G. biloba is a deciduous, ancient, tough, and high plant that has irregular lobes and fan-shaped disposed leaves. The maidenhair trees are extremely long-living, more than 1000 years, and can reach 20–30 m in height (Lorenzi, 2003). The female and male trees are separate. Female ovules are more rounded, and male pollen is borne among the leaves on catkin-like cones (Tupper, 1911). The wood parenchyma and crystal cells are present in the roots of maidenhair tree in series or rows that run longitudinally in the radial planes all over the roots. Its ovules are 2–3 mm long and are produced at the end of stalk in the form of a pair, and the stalk is 1.5–2.0 cm long. The seed coat consists of three layers, where the inner layer is thin membranous, the middle layer is hard and stony, and the outer layer, called a sarcotesta, is soft and fleshy (Friedman, 1987). The fleshy sarcotesta of seed is generally called a Ginkgo nut. The ovules of ginkgo from development up to maturation are green in color, and they turn to a yellow color as that of leaves in cold temperature. The seeds fall down from trees a month after fertilization (Hori, 1996). The maidenhair tree needs well-drained soil and full sunlight with a soil temperature of 15–27°C. Ginkgo tree may be capable of producing two or three secondary trunks just below the ground level when harsh growing conditions such as stress and higher accumulation of food material occur. Ginkgos can grow in a wide variety of seasonal climate ranges from Mediterranean to cold temperate; in winter, temperature minimum can reach −22°C. The soil in which a Ginkgo plant grows can be of any type because this plant can exist in extreme pH soil (Ziegler et al., 1996).

2. CHEMISTRY

Maidenhair tree has been well investigated for different classes of constituents. Ginkgo seeds have a foul odor only after full maturity. The color of *G. biloba* leaves turns to a beautiful golden hue when they are ready to fall on the ground. The female tree contains a hard, edible seed with foul smelling fruit. The seed of maidenhair tree is variously referred to as malevolent, disgusting, abominable, odious, unpleasant, and repulsive. It is often compared to the odor of vomit. This odor is due to butyric acid, a malodorous chemical compound, present in sour butter, while giving it a distinguishing smell, which is found in the integument of the seed (Parliment, 1995). *G. biloba* has a rotten smell that is due to two volatile compounds present in the hexanoic acids and sarcotesta-butanoic (Parliment, 1995).

G. biloba was chemically characterized in bioactive and nutritional components named sugar, tocopherols, fatty acids, flavonoids, phenolics, and organic acids. The leaves of this plant contain a high content of flavonoid compounds, while their nuts are very nutritious. They contain virtually zero fat and about 50 calories per ounce. When the nuts dry, they contain 13% proteins, 68% starch, 3% fats, and 6% sucrose (Jacobs and Browner, 2000). The main fatty acids that are present in ginkgo are oleic, palmitic, and α-linolenic acid, the most abundant sugar is fructose, the most abundant vitamin is α-tocopherol, and the main organic acid is quinic acid (Pereira et al., 2013). *G. biloba* seeds contain two proteins: ginkgo seed albumin protein (GAP) and ginkgo seed globulin protein. The contents of GAP, ginkgo seed protein isolate (GPI), and ginkgo seed globulin proteins are 87.8%, 91%, and 93.4%, respectively (Deng et al., 2011).

The maidenhair tree leaf contains many active components including diterpenes lactones, sesquiterpenes, flavonol, ginkgolides, flavones glycoside, catechin, ascorbic acid, and hydroxybenzoic acid (Pereira et al., 2013). Oleic (C18:1n9), α-linolenic (C18:3n3), and palmitic (C16:0) are the abundantly found fatty acids (11.18%, 18.03%, 35.9%, respectively). Shikimic, oxalic, malic, and quinic acids are also present (Pereira et al., 2013). Some structures of *G. bilboa* bioactive compounds are shown in Fig. 24.2.

3. POSTHARVEST TECHNOLOGY

The quality of the ginkgo leaves depends on the time of harvesting. *G. biloba* is harvested in October. Maidenhair trees start to change their color in the fall season and become yellow-gold. At this stage, this tree has

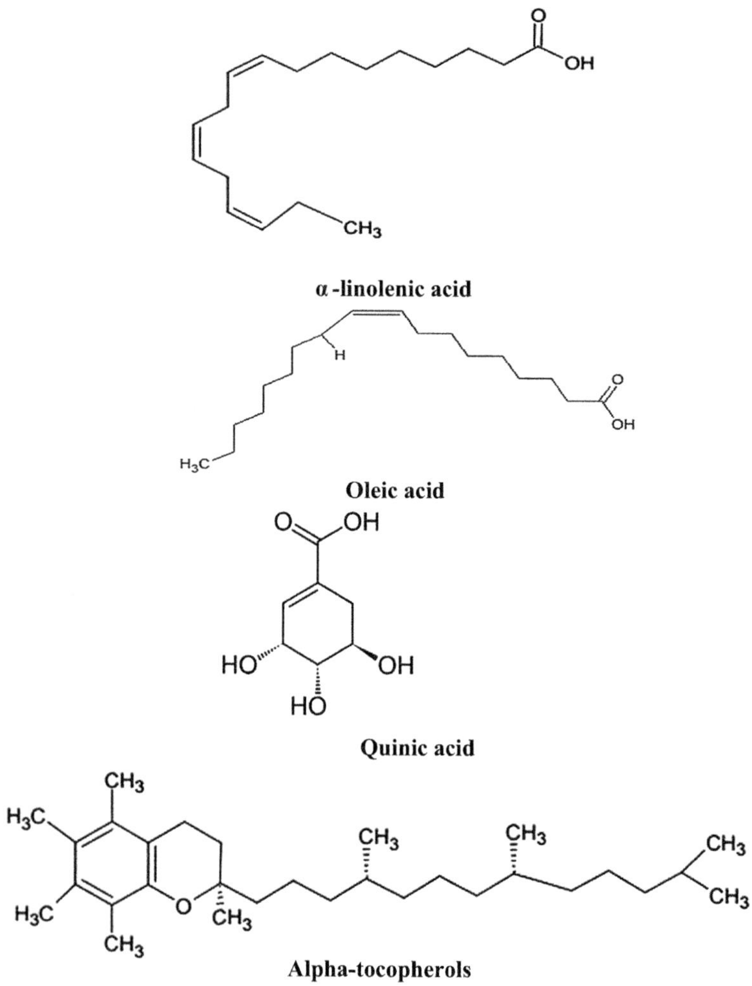

FIGURE 24.2 Structures of *Ginkgo biloba* bioactive compounds.

more potent medicinal activities. It can also be harvested in summers and in the months when they are green in color. The yellow color of the leaves indicates a high quantity of secondary metabolites. The best time for harvesting leaves with maximum amount of ginkgolides and bilobalide is at the end of summer when the leaves start to change color. The branches of ginkgo are flexible and can be easily harvested by pulling down and snapping with a backward hand movement (Hobbs and Miovic, 1991).

G. biloba seeds can be stored at 4 or 25°C and die within 6 months when they are stored at 25°C. The seeds tissues can be preserved for 1 year when storing at cold temperatures. However, seed capability to germinate cannot be preserved for more than 6 months. Glutathione and ascorbate are the major two antioxidants in the two main parts of ginkgo seeds (Tommasi et al., 2006).

4. PROCESSING

Ginkgo leaves can also be used as teas. Dried seeds will not germinate. The seeds that mature on the tree and fall on the ground should be washed to remove fruit pulp and then put in water. The best seeds will sink to the bottom of the water. These seeds are mixed with wet river sand in plastic bags to keep them buried underground. The temperature should be 1–5°C. The seeds of ginkgo are dried and canned to use in stew and soups. Seeds can also be eaten as a snack (Ottariano, 1998).

The active components of ginkgo plants are present mostly in the leaves. One standard dose of plant extract can be yielded from processing of approximately 50 fresh leaves of ginkgo. The purified extract of *G. biloba* leaves is produced by solvent extraction with different solvents such as toluene and n-butanol. Soxhlet ethanol extraction method showed the best performance on total yield and the amount of terpene, lactones, and flavonoids (Diamond et al., 2000). Dry leaf extracts were found to have a potency of 6% terpenes and 24% flavone glycoside (Duke, 1997).

5. VALUE ADDITION

G. biloba has positive effects on memory and learning in humans (deLuna, 2000). *G. biloba* is mostly used with other herbs, especially brahmi (bacopa), tulsi, ginseng, rhodiola and gotu kola. *G. biloba* is mostly used with gotu kola due to its ability to enhance the blood flow rate of the entire circulatory system and brain. This plant also has the ability to increase energy and helps to improve the ability to focus and concentrate when using with tulsi, rhodiola, or ginseng. Today, *G. biloba* is used with coriolus mushroom for overcoming dementia (Fang et al., 2015).

6. USES

G. biloba has international importance, and multistrategic efforts are required for conservation of this tree species, involving all stakeholders, such as local communities, scientists, foresters, and NGOs for its micro- and macropropagation and subsequent forestation programs (Chakravarty et al., 2016). Ginkgo tree leaves and seeds have been used for medicinal purposes since 1509 AC. In traditional Chinese medicine, the extract of ginkgo leaves has been used to treat circulatory disorders, cognitive problems, asthma, tinnitus, and vertigo (Mostafa et al., 2016). Ginkgo is also called the "brain herb," and it has been studied for the treatment of cerebral atherosclerosis, cerebral insufficiencies, and depression.

Traditional medicinal companies in China have been using *G. biloba* in many medicines for the last 5000 years (Oken et al., 1998). The herbal extract of maidenhair tree has been used for cosmetic and medicinal value. This herb is extensively used after transformed into capsules, teas, and tablets. It is a natural sunscreen. The mask of *G. biloba* protects our skin from the harmful effects of the sun (Pykett et al., 2007). It can also reduce hair loss and thinning. It also proved to be useful for various types of male baldness. It can also be used in combination with many other herbal oils to reduce the hair problems. *G. biloba* extract is used to prevent skin problems such as stretch marks, scars, pimples, acne, and patchiness due to availability of vitamin E in it. It keeps skin fresh and healthy and also replenishes the skin (Begoun, 2004). It is also used to enhance the flow of blood in our body, especially in the cerebrum. It stimulates tone in the venous system and is used as a circulatory system tonic. This herb proved to be a very helpful and effective mixture for various diseases that are caused by constrained flow of blood. Extract of this plant is ingested to treat leg ulcers, while large doses are used in treatment of varicose veins. In Europe, it is generally recommended in the treatment of stroke. The blood capillaries can be strengthened in the whole body by the dried leaf extract of ginkgo to avoid hemorrhagic stroke. The standardized extract of plant has been usually sold in the United States as dietary enhancements and also in Europe as phytomedicine (Strømgaard et al., 2005).

7. PHARMACOLOGICAL USES

7.1 Anticancer Activity

Both in vivo and in vitro experimental studies that were conducted for humans and animals have shown that *G. biloba* extracts contains some constituents that have anticancer activities and have been used in the treatment of cancer. Its leaf extract is widely used as a drug in medicine preparations.

G. biloba extracts and its ginkgolide constituents have been found to be effective in inhibiting peripheral-type benzodiazepine receptor–enriched breast cancer cell line (Papadopoulos et al., 1999).

7.2 Antiangiogenic Effects

G. biloba extracts show antiangiogenic activities by inhibition of the activation of human neutrophils to avoid "oxidative (respiratory) burst." *G. biloba* extract's antiangiogenesis property is also due to significant decrease in content of vitreous tissues transforming growth factor-β2 in retinal, and platelet-derived growth factor (due to which oxygen-induced ischemic retinopathy occur). It also inhibits angiogenesis in the internal retinal membrane (Juarez et al., 2000; Ottariano, 1998).

7.3 Antimicrobial Activity

Gingko leaf extract in ethanol, hexane, methanol, and chloroform solvents shows the antibacterial activity against animals and plant pathogenic strains (*Bacillus subtilis, Erwinia chrysanthemi, Agro bacterium tumefaciens, Xanthomonas phaseoli, and Escherichia coli*) by employing broth-dilution and disc-diffusion assays (Sati and Joshi, 2011). *G. biloba* is effective against insects due to the presence of bilobalide (Atzori et al., 1993).

7.4 Antistress Effects

Circulating concentrations of corticosterone, nor-epinephrine, and epinephrine have been found to increase the stress condition (Rapin et al., 1994). *G. biloba* extract decreases the concentration of nor-epinephrine, adrenocorticotrophic hormones, epinephrine, and corticosterone (Oliver et al., 1994). *G. biloba* extract showed effects on the central nervous system to decrease arginine vasopressin in the hypothalamus, adrenocorticotropic hormone in pituitary, and corticotrophin-releasing hormones of the anterior pituitary gland (Marcilhac et al., 1998). The leaf extract of *G. biloba* inhibits monoamine oxidase, which regulates biogenic amines present in the brain (White et al., 1996).

7.5 Antioxidant Effects

The proanthocyanidins and flavonol may play an important role in the inhibition of the process of atherosclerosis due to free radical scavenging activities (Joyeux et al., 1995; Packer et al., 1998). *G. biloba* extract may affect the neurosensory systems due to its free radical scavenging and antioxidant properties (Zhao et al., 2018). Except ginkgolides A, all terpenes present in *G. biloba* extract are superoxide scavengers (Scholtyssek et al., 1997). Flavonoids present in *G. biloba* extract are hydroxyl and superoxide radicals scavengers (Emerit et al., 1995). *G. biloba* extract releases lipoperoxide to reduce and even to inhibit the morphologic and functional retina impairments (Droy-Lefaix et al., 1995). It also scavenges free radicals and stimulates glucose fertilization that protects the pancreatic cells against alloxan toxicity effects. *G. biloba* extracts have both anti-lipoperoxidative and radical scavenging effects that inhibit the degradation of prostaglandin 12 and nitric oxide (Ellison and DeLuca, 1998).

7.6 Antiinflammatory Effects

Antiinflammatory activity of *G. biloba* extracts is due to the presence of ginkgolides B that inhibits the platelet activating factors. This extract plays an important role in the pathogenesis of asthma. Ginkgolide B effectively inhibits the increase of T-helper 2 cytokines (which are interleukin IL-13 and IL-5), in bronchovascular lavage fluid by decreasing eosinophils count. It also inhibited the mucus hypersecretion by goblet cell and ovalbumin-induced eosinophils in lung tissue in the airway. It was concluded in a previous study that ginkgolides B is the most effective component in asthma treatments (Chu et al., 2011).

7.7 Antitumor Activity

Radiosensitivity of the tumor cells can be increased after decreasing hypoxic fraction in the tumor cells and by improving the blood flow. *G. biloba* extract delayed the tumor growth and inhibits the radiation damage of the normal tissues (Ha et al., 1996). Flavonoids in the *G. biloba* extract have anticarcinogenic activity and show chemopreventive effect against the BP-induced gastric carcinogenesis (Agha et al., 2001).

8. SIDE EFFECTS AND TOXICITY

G. biloba leaf extract is likely safe for most people when taken by mouth in recommended doses. Ginkgo acid present in *G. biloba* extracts causes allergic dermatitis in persons with sensitive skin on contact (Kochibe, 1997). It can cause some minor side effects such as headache, stomach upset, dizziness, forceful heartbeat, constipation, and allergic skin reactions. The fresh seed of *G. biloba* is dangerous and is likely unsafe, as eating it could cause seizures and death. *G. biloba* is also unsafe for pregnant women and people suffering from diabetes, seizure, infertility, and surgery.

References

Agha, A., El-Fattah, A., Al-Zuhair, H., Al-Rikabi, A., 2001. Chemopreventive effect of Ginkgo biloba extract against benzo (a) pyrene-induced forestomach carcinogenesis in mice: amelioration of doxorubicin cardiotoxicity. Journal of Experimental & Clinical Cancer Research: Climate Research 20, 39–50.

Atzori, C., Bruno, A., Chichino, G., Bombardelli, E., Scaglia, M., Ghione, M., 1993. Activity of bilobalide, a sesquiterpene from *Ginkgo biloba*, on Pneumocystis carinii. Antimicrobial Agents and Chemotherapy 37, 1492–1496.

Begoun, P., 2004. The Complete Beauty Bible: The Ultimate Guide to Smart Beauty. Rodale.

Chakravarty, S., Rai, P., Puri, A., Shukla, G., Pala, N.A., 2016. The plant that survived atomic explosion, can it survive human threat? Indian Forester 142, 264–276.

Chu, X., Ci, X., He, J., Wei, M., Yang, X., Cao, Q., Li, H., Guan, S., Deng, Y., Pang, D., 2011. Article A Novel Anti-inflammatory Role for Ginkgolide B in Asthma via Inhibition of the ERK/MAPK Signaling Pathway.

Crane, P.R., Nagata, T., Murata, J., Ohi-Toma, T., DuVal, A., Nesbitt, M., Jarvis, C., 2013. 773. *GINKGO BILOBA*–Connections with people and art across a thousand years. Curtis's Botanical Magazine 30, 239–260.

DeFeudis, F.V., 1991. *Ginkgo biloba* Extract (EGb 761): Pharmacological Activities and Clinical Applications. Elsevier.

deLuna, A., 2000. The effects of *Gingko biloba* on learning and memory. Nutrition Noteworthy 3.

Deng, Q., Wang, L., Wei, F., Xie, B., Huang, F., Huang, W., Shi, J., Huang, Q., Tian, B., Xue, S., 2011. Functional properties of protein isolates, globulin and albumin extracted from *Ginkgo biloba* seeds. Food Chemistry 124, 1458–1465.

Diamond, B.J., Shiflett, S.C., Feiwel, N., Matheis, R.J., Noskin, O., Richards, J.A., Schoenberger, N.E., 2000. *Ginkgo biloba* extract: mechanisms and clinical indications. Archives of Physical Medicine and Rehabilitation 81, 668–678.

Droy-Lefaix, M., Cluzel, J., Menerath, J., Bonhomme, B., Doly, M., 1995. Antioxidant effect of a *Ginkgo biloba* extract (EGb 761) on the retina. International Journal of Tissue Reactions 17, 93–100.

Duke, J.A., 1997. The Green Pharmacy: New Discoveries in Herbal Remedies for Common Diseases and Conditions From the World's Foremost Authority on Healing Herbs. Rodale, p. 230.

Ellison, J.M., DeLuca, P., 1998. Fluoxetine-induced genital anesthesia relieved by *Ginkgo biloba* extract. Journal of Clinical Psychiatry 59, 199–200.

Emerit, I., Oganesian, N., Sarkisian, T., Arutyunyan, R., Pogosian, A., Asrian, K., Levy, A., Cernjavski, L., 1995. Clastogenic factors in the plasma of Chernobyl accident recovery workers: anticlastogenic effect of *Ginkgo biloba* extract. Radiation Research 144, 198–205.

Fang, X., Jiang, Y., Ji, H., Zhao, L., Xiao, W., Wang, Z., Ding, G., 2015. The synergistic beneficial effects of *Ginkgo* flavonoid and Coriolus versicolor polysaccharide for memory improvements in a mouse model of dementia. Evidence-based Complementary and Alternative Medicine 2015.

Friedman, W.E., 1987. Growth and development of the male gametophyte of *Ginkgo biloba* within the ovule (in vivo). American Journal of Botany 1797–1815.

Ha, S.W., Chun, J.Y., Cho, C.K., Cho, M.J., Shin, K.H., Park, C.-I., 1996. Enhancement of radiation effect by *Ginkgo biloba* extract in C3H mouse fibrosarcoma. Radiotherapy & Oncology 41, 163–167.

Hobbs, C., Miovic, M., 1991. *Ginkgo*, Elixir of Youth: Modern Medicine from an Ancient Tree. Botanica Press.

Hori, T., 1996. Ginkgo to the Japanese people. Microscopia 13, 184–185.

Hori, T., Ridge, R.W., Tulecke, W., Del Tredici, P., Trémouillaux-Guiller, J., Tobe, H., 2012. *Ginkgo Biloba* A Global Treasure: From Biology to Medicine. Springer Science & Business Media.

Huh, H., Staba, E.J., 1992. The botany and chemistry of *Ginkgo biloba* L. Journal of Herbs, Spices, & Medicinal Plants 1, 91–124.

Huxtable, R.J., 1992. The pharmacology of extinction. Journal of Ethnopharmacology 37, 1–11.

Jacobs, B.P., Browner, W.S., 2000. *Ginkgo biloba*: a living fossil. The American Journal of Medicine 108, 341–342.

Joyeux, M., Lobstein, A., Anton, R., Mortier, F., 1995. Comparative antilipoperoxidant, antinecrotic and scavenging properties of terpenes and biflavones from Ginkgo and some flavonoids. Planta Medica 61, 126–129.

Juarez, C., Muino, J., Guglielmone, H., Sambuelli, R., Echenique, J., Hernandez, M., Luna, J., 2000. Experimental retinopathy of prematurity: angiostatic inhibition by nimodipine, *ginkgo-biloba*, and dipyridamole, and response to different growth factors. European Journal of Ophthalmology 10, 51–59.

Kochibe, N., 1997. Allergic Substances of *Ginkgo biloba*, Ginkgo Biloba A Global Treasure. Springer, pp. 301–307.

Lorenzi, H., 2003. Árvores exóticas no Brasil: madeireiras, ornamentais e aromáticas. Instituto Plantarum de Estudos da Flora.

Marcilhac, A., Dakine, N., Bourhim, N., Guillaume, V., Grino, M., Drieu, K., Oliver, C., 1998. Effect of chronic administration of *Ginkgo biloba* extract or Ginkgolide on the hypothalamic-pituitary-adrenal axis in the rat. Life Sciences 62, 2329–2340.

Mostafa, R.E., Ibrahim, B.M., Jaleel, G.A.A., 2016. Neuro-Protective Effects of *Ginkgo Biloba* Leaves Extract on Cerebral Ischemia-Reperfusion Injury Induced Experimentally in Ovariectomized Rats.

Nagata, T., DuVal, A., Crane, P.R., 2015. Engelbert Kaempfer, Genemon Imamura and the origin of the name *ginkgo*. Taxon 64, 131–136.

Oken, B.S., Storzbach, D.M., Kaye, J.A., 1998. The efficacy of *Ginkgo biloba* on cognitive function in Alzheimer disease. Archives of Neurology 55, 1409–1415.

Oliver, C., Guillaume, V., Hery, F., Bourhim, N., Boiteau, K., Drieu, K., 1994. Effect of *Ginkgo biloba* extract on the hypothalamo-pituitary-adrenal axis and plasma catecholamine levels in stress. European Journal of Endocrinology 130, 207.

Ottariano, S.G., 1998. Medicinal Herbal Therapy: A Pharmacist's Viewpoint. Nicolin Fields Pub.

Packer, L., Saliou, C., Droy-Lefaix, M.-T., Christen, Y., 1998. *Ginkgo biloba* Extract EGb 761: Biological Actions, Antioxidant Activity, and Regulation of Nitric Oxide Synthase.

Papadopoulos, V., Kapsis, A., Li, H., Amri, H., Hardwick, M., Culty, M., Kasprzyk, P.G., Carlson, M., Moreau, J.-P., Drieu, K., 1999. Drug-induced inhibition of the peripheral-type benzodiazepine receptor expression and cell proliferation in human breast cancer cells. Anticancer Research 20, 2835–2847.

Parliment, T.H., 1995. Characterization of the Putrid Aroma Compounds of *Ginkgo biloba* Fruits. ACS Publications.

Pereira, E., Barros, L., Ferreira, I.C., 2013. Chemical characterization of *Ginkgo biloba* L. and antioxidant properties of its extracts and dietary supplements. Industrial Crops and Products 51, 244–248.

Pykett, M.A., Craig, A.H., Galley, E., Smith, C., Long, S.P., 2007. Skincare Composition Against Free Radicals. Google Patents.

Rapin, J.R., Lamproglou, I., Drieu, K., Defeudis, F.V., 1994. Demonstration of the "anti-stress" activity of an extract of *Ginkgo biloba* (EGb 761) using a discrimination learning task. General Pharmacology: The Vascular System 25, 1009–1016.

Sati, S., Joshi, S., 2011. Antibacterial activities of *Ginkgo biloba* L. leaf extracts. Science World Journal 11, 2237–2242.

Scholtyssek, H., Damerau, W., Wessel, R., Schimke, I., 1997. Antioxidative activity of ginkgolides against superoxide in an aprotic environment. Chemico-Biological Interactions 106, 183–190.

Strømgaard, K., Vogensen, S.B., Nakanishi, K., 2005. Ginkgo Biloba. Marcel Dekker, New York.

Tommasi, F., Paciolla, C., de Pinto, M.C., De Gara, L., 2006. Effects of storage temperature on viability, germination and antioxidant metabolism in *Ginkgo biloba L.* seeds. Plant Physiology and Biochemistry 44, 359–368.

Traditional, M.U., Berry Seeds, B.D.G., Ginkgo (*Ginkgo Biloba* L.) in Hindi-(India).

Tupper, W.W., 1911. Notes on *Ginkgo biloba*. Botanical Gazette 51, 374–377.

Unseld, S., 2010. Goethe and the *Ginkgo*: A Tree and a Poem. University of Chicago Press.

Wang, C.-W., 1961. The Forests of China: Maria Moors Cabot Foundation. Harvard University, Cambridge, Massachusetts.

White, H.L., Scates, P.W., Cooper, B.R., 1996. Extracts of *Ginkgo biloba* leaves inhibit monoamine oxidase. Life Sciences 58, 1315–1321.

Zhang, Z., Mei, N., Chen, S., Guo, L., Guo, X., 2016. Assessment of Genotoxic Effects of Selected Herbal Dietary Supplements, Nutraceuticals. Elsevier, pp. 883–892.

Zhao, L.-j., Liu, W., Xiong, S.-h., Tang, J., Lou, Z.-h., Xie, M.-x., Xia, B.-h., Lin, L.-m., Liao, D.-f., 2018. Determination of total flavonoids contents and antioxidant activity of *Ginkgo biloba* leaf by near-infrared reflectance method. International journal of analytical chemistry 2018.

Zhao, Y., Paule, J., Fu, C., Koch, M.A., 2010. Out of China: distribution history of *Ginkgo biloba L.* Taxon 59, 495–504.

ZIEGLER, A.M., REES, P.M., ROWLEY, D.B., Bekker, A., Qing, L., HULVER, M.L., 1996. 17 Mesozoic Assembly of Asia: Constraints From Fossil Floras, Tectonics, and Paleomagnetism.

CHAPTER

25

Ginseng

Muhammad Mubeen Mohsin[1], Muhammad Asif Hanif[1], Muhammad Adnan Ayub[2], R.M. Dharmadasa[3]

[1] Department of Chemistry, University of Agriculture, Faisalabad, Pakistan; [2] Department of Chemistry, University of Okara, Okara, Pakistan; [3] Industrial Technology Institute, Colombo, Sri Lanka

OUTLINE

Medicinal Plants of South Asia
https://doi.org/10.1016/B978-0-08-102659-5.00025-2

Copyright © 2020 Elsevier Ltd. All rights reserved.

1. BOTANY

1.1 Introduction

Panax ginseng (ginseng) (Fig. 25.1) is the slow-growing perennial plant that belongs to the family Araliaceae. The genus *Panax* contains 10 to 15 species that are distributed in tropical regions of North America and in Asia. Ginseng is a smooth perennial plant with green leaves, small white flowers, red berries, and yellowish-brown roots. All *Panax* species contain ginsenosides, which are responsible for the pharmaceutical activity of these plants (Lee et al., 2010). Ginsenosides, based on a broad group of chemicals such as triterpene and saponin (Hobbs, 2002).

P. ginseng has the common name Chinese ginseng due to its origin and *Aralia quinquefolia* (L.) Decne & Planch. is a specie of *Panax* (American ginseng), found in Southern Canada and the United States of America. *Panax japonicus* (Japanese ginseng) is found in Japan, and Sanchi ginseng is also present in China's Yunnan region. *Panax trifolius* L. (diminutive person ginseng) originated in Nova Scotia to Wisconsin and further south, and *Panax real Ting, Panax omeiensis, Panax pseudoginseng* Wallich are present in Nepal and the eastern Himalayas. The common name of *Panax ginseng* is ginseng or Korean ginseng (Vo et al., 2015).

FIGURE 25.1 Ginseng roots.

1.2 History/Origin of Ginseng

The origin of ginseng dates to prehistory. In China, ginseng is used as a home-grown drug that began around 5500 years back (Zheng, 1985; Wang, 1987; Yun, 2001). From the beginning, the ginseng plants were considered a food, and after that, it was used for the strength-giving and restoring of power of the human. In the 3rd century AD, ginseng was used for trade between China and Korea to exchange Chinese silk and medicine in exchange for wild ginseng. In the 1900s, the claim for ginseng exceeded the available wild supply and Korea started the profitable cultivation of ginseng (Craker et al., 2003). Between 196 and 200 AD, ginseng was used as medicines of Shanghan Lun (treatise on fevers) (Yun, 2001). Ginseng naturally grows on mountains in Nepal, India, and Pakistan. Since ginseng possesses an array of beneficial impacts, it is widely used for South Asian herbal preparations.

1.3 Demography/Distribution

Ginseng grows on moist, shaded mountainsides in China, Nepal, India, Pakistan, Korea, and Russia. The most favorable conditions for optimum growth and development of the ginseng plant are a forest area with weakly acidic soil that is well drained and rich in mold (Khrolenko et al., 2012). Ginseng is also grown widely in the countries of Russia, China, Korea, America, and Vietnam.

1.4 Botany, Morphology, Ecology

P. ginseng is a smooth perennial herb, with a large, fleshy, slow-growing root, 2–3 in. in length (occasionally twice this size) and from 0.5 to 1 in. in thickness. Its main portion is spindle-shaped and heavily annulated (ringed growth), with a roundish summit, often with a slight terminal, projecting point. At the lower end of this straight portion, there is a narrower continuation, turned obliquely outward in the opposite direction, and a very small branch is occasionally borne in the fork between the two. Some small rootlets exist upon the lower portion. The color of the roots varies from pale yellow to brown. It has a mucilaginous sweetness, approaching that of licorice, accompanied with some degree of bitterness and slight aromatic warmth, with little or no smell. The stem is simple and erect, about a foot high, bearing three leaves, each divided into five finely toothed leaflets, and a single terminal umbel with a few small, yellowish flowers. The fruit is a cluster of bright red berries (Grieve, 1971).

Ginseng grows profusely under conditions that simulate its natural habitat. It requires 70%–90% natural or artificial shade. Ginseng thrives in a climate with 40–50 in. of annual precipitation and an average temperature of 50°F. It requires several weeks of cold temperatures for removing dormancy. Ginseng generally prefers a loamy, deep (12-in.), well-drained soil with a high organic content and a pH near 5.5. Ginseng is typically a shade-preferring plant whose normal growth and development is ensured exclusively under forest shade without durable impact of direct sunshine. Although, the plant prefers shade, heavy shade conditions may also impede the growth of ginseng, which periodically enters a state of dormancy. Extremely sandy soil tends to produce long, slender roots of inferior quality. Most ginseng crops are started from seeds, rather than roots or seedlings. This is the least expensive way to start a plantation and may help to prevent the introduction of soil-borne disease to new plantations. Ginseng requires 3–5 years to produce a marketable crop from seeds (Harrison et al., 2000).

As mentioned earlier, wild ginseng develops under trees in the profound mountains and favors to a great degree a chilly atmosphere. The ideal temperature is between 10 and 20°C during the leafing stage and between 21 and 25°C during the blossoming and fruiting stages. High temperature harms ginseng by bringing on a suspension of photosynthesis, drying of leaves, and early defoliation. Leaf spot malady, anthracnose, and root decay are likewise brought about by high temperature. The rate of leaf spot infection significantly increases at the point when the temperature surpasses 21°C. At temperatures above 30°C, the photosynthesis rate diminishes, which builds the breathing rate and debases the development and improvement of ginseng (Ryu et al., 2012).

2. CHEMISTRY

Chemical constituents and other properties of ginseng are varied among different species of *Panax*. The chemical composition of ginseng is dependent on several factors including age, variety, species of the plant, location, growing conditions, and the time of harvesting. Level of bioactive compounds increases with the age of plant, and the best chemicals components are found in 4- to 6-year-old plants. In earlier studies, aromatic oils from Chinese ginseng root, *P. ginseng*, were separated, and different sesquiterpene hydrocarbons and its oxygenated derivatives were identified in it. Recently, three novel sesquiterpene hydrocarbons were identified from its plant extract including panaxene, panaginsene, and ginsinsen (Smigielski et al., 2006). Ginseng also contains other classes of compounds such as unsaturated fatty acids, polysaccharides, phenolic compounds, saponins, polyacetylenes, peptidoglycans, and starches (Tang and Eisenbrand, 1992). Many saponins that were isolated from the roots of Japanese, Himalayan, American, San-ch'i, and Siberian ginseng have similar properties, but the overall components in these *Panax* species were quite different (Hou, 1977). The essential oil yield of ginseng varies from 0.07% to 0.09% depending on the species. The main constituent of *P. ginseng* is falcarinol, and according to recent literature, this compound reveals cytotoxic and antiproliferative action and can stop development of tumor cells (Smigielski et al., 2006). *P. ginseng* has essential oils mostly confined to the roots and thus has a specific aroma. Ginseng root's essential oil chiefly comprises compounds of sesquiterpenes, aliphatic hydrocarbons, and very small amounts of monoterpenes. The main components of ginseng essential oil are spathulenol, 2-epi-(E)-β-caryophyllene, and falcarinol (Fig. 25.2). The other components present in *P. ginseng* essential oil are calarene, bicyclogermacrene, α-neoclovene, α-humulene, allo-aromadendrene, n-docosane, β-elemene, pacifigoria-1(6)-10-diene, humulene epoxide II, α-isocomene, cyperene, n-tetracosane, eudesma-4(14)-7(11)-diene, β-selinene, globulol, ginsenol, n-heneicosane, 2-pentylfuran, (E,E)-deca-2,4-dienal, β-caryophyllene, aromadendrene, n-pentacosane, germacrene A, γ-cadinene, δ-cadinene, caryophyllene oxide, n-tricosane, epi-α-muurolol, n-nonadecane, ethyl

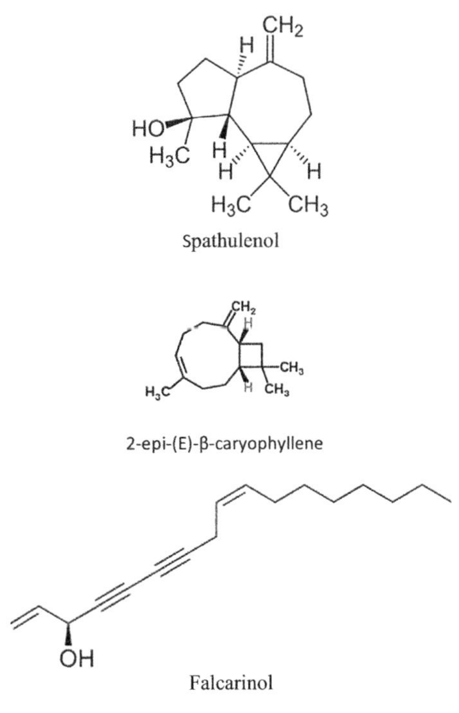

FIGURE 25.2 Major components of ginseng essential oil.

hexadecanoate, and methyl hexadecanoate (Smigielski et al., 2006). The root of ginseng comprises a sugar, mucilage, resin, a volatile oil, starch, several steroids, and a saponin complex (Hou, 1977).

3. POSTHARVEST TECHNOLOGY

Ginseng plant takes a minimum of 7 years to gain optimum growth. The best harvesting time for the roots is September to December, when plant drops the seeds. After harvesting, the roots are washed properly with water and dried in a dark room with appropriate temperature conditions (32–35°C), air circulation, and humidity level. Roots can also dry under sunlight, but the quality of the air drying method is much better than sun drying (Carroll and Apsley, 2004).

4. PROCESSING

The dried ginseng is called white ginseng or sun-dried white ginseng. The white ginseng can be processed to obtain red ginseng. Red ginseng has a red color that might be attributed to a change in chemical composition during the additional steaming and drying process (Chuang and Sheu, 1994).

5. VALUE ADDITION

Ginseng is a medicinal plant and has been used in Unani, Chinese, Ayurveda, and African traditional medicinal systems. It has been used as a tonic and body stress resister. Ginsenosides are the major active components of ginseng and responsible for their biologic activities. Recent scientific studies have shown that this can be used against various pathologic conditions. Ginseng can be sold in the form of various value-added products in addition to raw form. Ginseng products include extracts, teas, creams, and drinks (Braz et al., 2009).

6. USES

Ginseng is a powerful plant that helps the body to be more resistant, stronger, and more resilient, while promoting greater alertness and improved concentration and memory, even if it is only consumed occasionally. Ginseng is effective in fortifying and protecting the body and is also one of the best plants for nervous balance. Ginseng, the root of *Panax* species, is a well-known folk medicine. The efficacy of ginseng as a therapeutic option for the treatment of male erectile dysfunction and the improvement in the survival of patients with advanced stomach cancer during postoperative chemotherapy have been demonstrated in double-blind clinical trials (Braz et al., 2009).

7. PHARMACOLOGICAL USE

7.1 Antioxidant Activity/Capacity

Antioxidant activity of ginseng plant extracts and isolated compounds have been extensively studied in the last few decades. It has been reported that methanol extract rich in saponins significantly inhibited the lipid peroxidation in rats (Keum et al., 2000). Moreover, methanol extract has the ability to scavenge the superoxide free radicals. In another study, water extract of ginseng completely inhibited the DPPH free radicals at the concentration of 2 mg/mL, 40% hydroxyl free radical scavenged at concentration of 0.1 mg/mL, and 80% carbon-centered free radical scavenging at the concentration of 0.5 mg/mL (Kim et al., 2002). *In vivo* antioxidant activity of ginseng leaf extract was also studied in diabetic rats. It was reported that ginseng leaf extract significantly reduced the lipid peroxide production in diabetic rats (Jung et al., 2005).

7.2 Antisterility Activity

In an *in vivo* study, ginseng extracts significantly improved the motility as well as sperm count in male rats (Lakshmi et al., 2011).

7.3 Antiproliferative Activity

Antiproliferative activity of American ginseng extracts treated with high-temperature steaming and unsteamed conditions were studied against human colorectal cancer cells. The higher temperature steaming ginseng extracts showed larger antiproliferative activity than unsteamed extract. Moreover, it was reported that the higher concentration of ginsenoside Rg3 in steamed extract might be responsible for higher antiproliferative activity (Wang et al., 2006). Similarly, in another scientific report, steam-dried Asian and American ginseng roots showed higher anticancer activity than air-dried white Asian and American ginseng roots (Sun et al., 2011). In another scientific report, anticancer activity of polar solvent (water) and nonpolar solvent (n-hexane) extracts were compared against HegG2 and MCF-7 cells. The reported experimental results showed that n-hexane extract revealed higher anticancer activity than water (Lee et al., 2009).

7.4 Adaptogenic Activity

Ginseng is an effective plant that protects the body in extreme stress conditions. In tradition medical systems, this plant has been used under extreme stress conditions, and the recent scientific studies in animals prove that it is a stress-releasing plant. Ginsenosides are the major bioactive component of ginseng roots that reduce aging and stress and enhance sexual performance (Nocerino et al., 2000). In another study, ginseng extract significantly reduced the stress level in rats by maintaining the plasma glucose and cholesterol levels (Rai et al., 2003).

7.5 Antidiabetic Activity

Ginseng is a useful herbal plant to maintain the blood glucose level. Antidiabetic potential of ginseng was studied in diabetes type 2 patients. It was reported that the daily dose of 100–200 mg of ginseng extract for 8 weeks significantly reduced the blood glucose level and enhanced physical performance (Lakshmi et al., 2011). In another study, 150 mg/kg of ginseng berry extract injection for a period of 12 days improved the blood glucose level in obese diabetic mice (Attele et al., 2002). Malonyl ginsenosides (MGR) are ginsenosides found in dried as well as fresh ginseng plant. The effect of MRG on diabetic rats was studied by different antidiabetic tests. It was reported that 50–100 mg/kg dose per day significantly reduced the blood glucose level and improved the body weight of diabetic rats (Liu et al., 2013).

7.6 Antiinflammatory Activity

A recent paper proposed an antiinflammatory role of *P. ginseng* in the sequence of progression to promotion in a model of carcinogenesis. *P. ginseng* affects multiple points within the inflammatory cascade, including inhibition of cyclooxygenase2, inducible nitric oxide synthase, and nuclear factor kappa B. *P. ginseng* has a radioprotective effect associated with antioxidant and immune modulation properties (Lakshmi et al., 2011).

8. SIDE EFFECTS AND TOXICITY

The short-term application of ginseng to skin has no side effect, but the long-term intake of ginseng is not safe, as it causes hormonal imbalance and insomnia. Similarly, it may cause vaginal bleeding, allergic reactions, high blood pressure, diarrhea, breast pain, mood changes, menstrual disturbance, liver damages, rash, and headache.

References

Attele, A.S., Zhou, Y.-P., Xie, J.-T., Wu, J.A., Zhang, L., Dey, L., Pugh, W., Rue, P.A., Polonsky, K.S., Yuan, C.-S., 2002. Antidiabetic effects of Panax ginseng berry extract and the identification of an effective component. Diabetes 51, 1851–1858.

Braz, A.d.S., Diniz, M., de ALMEIDA, R.N., 2009. Recent advances in the use of Panax ginseng as an analgesic: a systematic review. Boletin Latinoamericano y del Caribe de Plantas Medicinales y Aromaticas 8, 188–194.

Carroll, C., Apsley, D., 2004. Growing American Ginseng in Ohio: An Introduction. Ohio State University extension fact sheet F-56-04.

Chuang, W.-C., Sheu, S.-J., 1994. Determination of ginsenosides in ginseng crude extracts by high-performance liquid chromatography. Journal of Chromatography A 685, 243–251.

Craker, L.E., Gardner, Z., Etter, S.C., 2003. Herbs in American fields: a horticultural perspective of herb and medicinal plant production in the United States, 1903 to 2003. HortScience 38, 977–983.

Grieve, M., 1971. A Modern Herbal: The Medicinal, Culinary, Cosmetic and Economic Properties, Cultivation and Folk-Lore of Herbs, Grasses, Fungi, Shrubs, & Trees with All Their Modern Scientific Uses. Courier Corporation.

Harrison, H., Parke, J., Oelke, E., Kaminski, A., Hudelson, B., Martin, L., Kelling, K., 2000. Alternative Field Crops Manual: Ginseng. Purdue Univ., Dept. of Hortic. and Landscape Architecture, Extension/Outreach, New Crops Center. Online publication.

Hobbs, C., 2002. Medicinal Mushrooms: An Exploration of Tradition, Healing, and Culture. Book Publishing Company.

Hou, J.P., 1977. The chemical constituents of ginseng plants. The American Journal of Chinese Medicine 5, 123–145.

Jung, C.-H., Seog, H.-M., Choi, I.-W., Choi, H.-D., Cho, H.-Y., 2005. Effects of wild ginseng (Panax ginseng CA Meyer) leaves on lipid peroxidation levels and antioxidant enzyme activities in streptozotocin diabetic rats. Journal of Ethnopharmacology 98, 245–250.

Keum, Y.-S., Park, K.-K., Lee, J.-M., Chun, K.-S., Park, J.H., Lee, S.K., Kwon, H., Surh, Y.-J., 2000. Antioxidant and anti-tumor promoting activities of the methanol extract of heat-processed ginseng. Cancer Letters 150, 41–48.

Khrolenko, Y., Burundukova, O., Burkovskaya, E., Zhuravlev, Y., 2012. Mesophyll structure and chloroplast density in Panax ginseng leaves from the Sikhote-Alin Mts. Acta Biologica Cracoviensia Series Botanica 54, 54–60.

Kim, Y.K., Guo, Q., Packer, L., 2002. Free radical scavenging activity of red ginseng aqueous extracts. Toxicology 172, 149–156.

Lakshmi, T., Roy, A., Geetha, R., 2011. Panax ginseng-A universal panacea in the herbal medicine with diverse pharmacological spectrum–a review. Asian Journal of Pharmaceutical and Clinical Research 4, 14–18.

Lee, S.D., Yoo, G., Chae, H.J., In, M.J., Oh, N.S., Hwang, Y.K., Hwang, W.I., Kim, D.C., 2009. Lipid-soluble extracts as the main source of anticancer activity in ginseng and Ginseng Marc. Journal of the American Oil Chemists Society 86, 1065.

Lee, S.Y., Kim, Y.K., Park, N.I., Kim, C.S., Lee, C.Y., Park, S.U., 2010. Chemical constituents and biological activities of the berry of Panax ginseng. Journal of Medicinal Plants Research 4, 349–353.

Liu, Z., Li, W., Li, X., Zhang, M., Chen, L., Zheng, Y.-n., Sun, G.-z., Ruan, C.-c., 2013. Antidiabetic effects of malonyl ginsenosides from Panax ginseng on type 2 diabetic rats induced by high-fat diet and streptozotocin. Journal of Ethnopharmacology 145, 233–240.

Nocerino, E., Amato, M., Izzo, A.A., 2000. The aphrodisiac and adaptogenic properties of ginseng. Fitoterapia 71, S1–S5.

Rai, D., Bhatia, G., Sen, T., Palit, G., 2003. Anti-stress effects of Ginkgo biloba and Panax ginseng: a comparative study. Journal of Pharmacological Sciences 93, 458–464.

Ryu, K.-R., Yeom, M.-H., Kwon, S.-S., Rho, H.-S., Kim, D.-H., Kim, H.-K., Yun, K.W., 2012. Influence of air temperature on the histological characteristics of ginseng ('Panax ginseng'CA Meyer) in six regions of Korea. Australian Journal of Crop Science 6, 1637.

Smigielski, K., Dolot, M., Raj, A., 2006. Composition of the essential oils of ginseng roots of *Panax quinquefolium* L. and Panax ginseng CA Meyer. Journal of Essential Oil Bearing Plants 9, 261–266.

Sun, S., Qi, L.-W., Du, G.-J., Mehendale, S.R., Wang, C.-Z., Yuan, C.-S., 2011. Red notoginseng: higher ginsenoside content and stronger anticancer potential than Asian and American ginseng. Food Chemistry 125, 1299–1305.

Tang, W., Eisenbrand, G., 1992. Panax Ginseng CA Mey. Springer.

Vo, H.T., Ghimeray, A.K., Vu, N.T., Jeong, Y.-H., 2015. Quantitative estimation of ginsenosides in different ages of Panax vietnamensis and their anti-proliferation effects in hela cells. African Journal of Traditional, Complementary and Alternative Medicines 12, 79–83.

Wang, C.-Z., Zhang, B., Song, W.-X., Wang, A., Ni, M., Luo, X., Aung, H.H., Xie, J.-T., Tong, R., He, T.-C., 2006. Steamed American ginseng berry: ginsenoside analyses and anticancer activities. Journal of Agricultural and Food Chemistry 54, 9936–9942.

Wang, J., 1987. The earliest officially published pharmacopoeia in the world. Journal of traditional Chinese medicine Chung i tsa chih ying wen pan/Sponsored by All-China Association of Traditional Chinese Medicine, Academy of Traditional Chinese Medicine 7, 155–156.

Yun, T.K., 2001. Brief introduction of Panax ginseng CA Meyer. Journal of Korean Medical Science 16, S3.

Zheng, B., 1985. Shennong's herbal–one of the world's earliest pharmacopoeia. Journal of traditional Chinese medicine Chung i tsa chih ying wen pan/Sponsored by All-China Association of Traditional Chinese Medicine, Academy of Traditional Chinese Medicine 5, 236, 236.

CHAPTER

26

Guava

Zunaira Irshad[1], *Muhammad Asif Hanif*[1], *Muhammad Adnan Ayub*[2], *Muhammad Idrees Jilani*[3], *Vahid Tavallali*[4]

[1] Department of Chemistry, University of Agriculture, Faisalabad, Pakistan;
[2] Department of Chemistry, University of Okara, Okara, Pakistan;
[3] Department of Chemistry, University of Sahiwal, Sahiwal, Pakistan;
[4] Department of Agriculture, Payame Noor University (PNU), Tehran, Iran

OUTLINE

Medicinal Plants of South Asia
https://doi.org/10.1016/B978-0-08-102659-5.00026-4

Copyright © 2020 Elsevier Ltd. All rights reserved.

1. BOTANY

1.1 Introduction

Guava (*Psidium guajava* L.) (Fig. 26.1) is a momentous fruit grown in many subtropical and tropical regions all around the world (Rai et al., 2009). It belongs to family Myrtaceae. This family is further divided in to two subfamilies, including Leptospermoideae, which comprises dehiscent capsulated fruits, and Myrtoideae, comprising thickset fruits (Wilson et al., 2001). *Eugenia, Myrcianthes, Campomanesia,* and *Psidium* genera also belong to this family. The *Psidium* genus contains more than 3800 species of shrubs (Chalannavar et al., 2013). Guava is an annual plant and the toughest among fruiting trees of tropical areas, and it has high production rate and is most adaptable to any sort of environment (Pino et al., 2004). It provides food to millions of people around the world. The tree grows fast and starts fruiting within 2 to 4 years. Guava has many medicinal uses, that is why it is commonly called the common man's apple (Joseph and Priya, 2011).

The name of the guava in different languages of the world is different like in Bengali (goaachhi, piyara, peyara); Arabic (juafa, juava, guwâfah); Filipino (bayabas, guyabas); English (common guava, guava); Hawaiian (kuawa); Dutch (goejaba); French (goyava, goyavier); German (guavenbaum, guava); Indonesian (jambu biji); Japanese (banjiro); Hindi (goaachhi, jamba, amrud, amarood, sapari, safed safari); Mandinka (biabo); Sanskrit (mansala); Spanish (araza-puita, gauyaba blanca, perulera, guaiaba dulce, guayaba, guayaba agria, guayaba común, guayabillo, agria); Tamil (koyya); Tigrigna (zeitun); and Urdu (amrud or amrood).

FIGURE 26.1 Guava plants, leaves and fruits.

1.2 History/Origin

The guava is believed to have originated in America (Mexico and Peru). Then it spread into many regions of the world during the 19th century, from Mpumalanga to Mozambique and from Western Cape to Madeira. It is harvested in the southern United States, subtropical and tropical Asia, and tropical Africa. Guavas are grown in many of the countries around the world. Different varieties of the guava are grown commercially, in which the apple guava is most important. The adult trees of guava can survive temperatures down to −4°C, but the younger plants are vulnerable to freezing in the condition of low temperature. Guavas were first grown in Florida in the 19th century and now are grown in many of the north areas such as Sarasota, Chipley, Waldo, and Fort Pierce. Guava can be sown in pots and even fruiting starts in pots. Usually the guava plants start fruiting in 2–3 years and continue to give fruits for 40–60 years. It is found in many regions of the Americas, the Bahamas, Bermuda, and southern Florida, where it was reportedly introduced in 1847 and was common in over half of the State by 1886 (Morton, 1987). Almost 300 years ago, most varieties were grown in the United States, but today, this plant is also grown in India, Pakistan, China, and many other countries.

1.3 Demography/Location

Guava can be grown in all types of soil (Morton, 1987). Despite its origin in tropical America, it can be grown in tropical and subtropical countries around the world (Richardson and Rejmánek, 2011). It is being commercially cultivated in many lands, including Pakistan, India, Bangladesh, Thailand, Brazil, Cuba, South Africa, the Philippines, New Zealand, California, Vietnam, Venezuela, Haiti, Florida, Thailand, and West Indies. Pakistan is the second largest producer of guava fruit, while India ranks first in the world. The combined global production of guava fruit is about 40 million tons.

1.4 Morphology, Botany, Ecology

Guava is an evergreen shrub that grows up to 6–25 ft. in height with bending spread branches. Its leaves are long and oppositely arranged and contain branched veins. The flowers are white in color, containing four to five petals, 2 cm in length, having brush-like stamens with aroma. The fruits are yellow, round, 3–10 cm in diameter, having a weight of 100–400 g, with the four to five petals remaining on the fruits. The fruits turned reddish-yellow when developed. Guava fruit shape is round, ovoid, or pear-like (Mitra, 1997). The guava is made up of a thick mesocarp of variable breadth and a fragile endocarp with many tiny, stiff, yellowish seeds inserted all over it (Malo and Campbell, 1994). Guava flesh is comprised of two kinds of cell wall tissues: parenchyma and stone cells. Stone cells are an extremely hard substance, accountable for a typical grimy or coarse feeling when the fruit is eaten; because of their character, they are tough enough that cannot be degraded by the enzymes

(Marcelin et al., 1993). External peel color varies from different tones of green to yellow when developed; its flesh might remain white or turned to yellow, pink, or light red. Young fruit of guava are hard, a little bit dry, and sour. Once it ripens, the fruit becomes very soft, sweet, nonacidic, and its skin becomes thin and edible (Malo and Campbell, 1994). Numerous variations of guava exist today. On the other hand, classification can be done based on color, i.e., pink or white. Seedless varieties are becoming more popular in many countries, as they possess more nutrition (Yadava, 1996).

2. CHEMISTRY

Guava leaves, fruit, and seeds contain a significant amount of essential oil phenols, tannins, vitamins, lectins, and vitamins. Guava contains an appreciable amount of vitamin C as well as vitamin A and pectins. A lot of flavonoids are present in the guava leaves, especially quercetin. Guava is considered a superfruit because it contains dietary fiber, dietary minerals, potassium, manganese, copper, vitamins A and C, and folic acid. It contains four times more vitamin C compared to oranges. Mostly low-calorie nutrients are present in guava (Kumari et al., 2013). Guava is rich in antioxidant compounds and contains a high level of ascorbic acid, myricetin, and apigenin acid in its fresh fruit. The chemical composition of guava varies significantly with variety, stage of maturity, and season (Lim et al., 2006). The concentration of each of the chemical components differs depending on the type of species or cultivar as well as cultivation conditions such as soil type, weather, irrigation, pruning, and other horticultural practices. Guava is rich in dietary fiber (Mamede et al.). Guava fruit exhibits moisture (77%–86%), crude fiber (2.8%–5.5%), protein (0.9%–1.0%), fat (0.1%–0.5%), ash (0.43%–0.7%), carbohydrates (9.5%–10%), minerals, and vitamins. The powdered guava seed contains chemical compounds like lesser protein (5%–10%) and greater fiber (65%–70%) content (Mandal et al., 2009). The polysaccharides were found to contain 2-O-methyl-L-arabinose, 2-O-acetyl-D-galactose, and D-methyl galacturonate in a molar ratio of approximately 1:1:1 (Mandal et al., 2009). Many minerals including calcium, magnesium, sulfur, iron, manganese, zinc, sodium, potassium, and phosphorus are present in guava. The proteins present in guava seeds are albumin (1%–3%), globulin (3%–7%), prolamin (1%–3%), and glutelin (42%–46%). The percentages of the insoluble residue were appropriately 85%–90%. Glutelin is the major protein fraction from guava seed. The crude oil extracted from guava seeds showed high levels of unsaturated fatty acids (85%–90%), mainly linoleic acid (75%–80%). The amount of tocopherol and total phenolic contents in the oil are 25–30 and 90–95 mg/100 g,

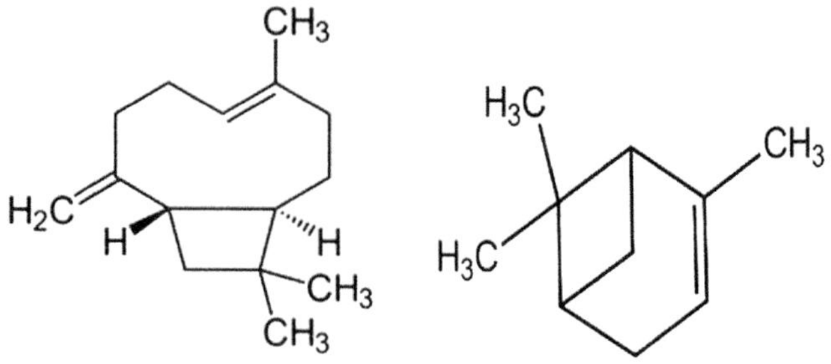

Caryophyllene Pinene

FIGURE 26.2 Structures of important chemical constituents of guava.

respectively (Malacrida and Jorge, 2013). Guava leaves contain several phenolic and flavonoid compounds (Shao et al., 2012; Gutiérrez et al., 2008; Paniandy et al., 2000). Structures of important chemical constituents of guava are shown in Fig. 26.2.

3. POSTHARVEST TECHNOLOGY

The picked fruits are placed in a cool place away from the sun. The fruits are put in an area with proper ventilation if overnight storage is required. Green fruits are stored for a long time and then matured by ethephon. The ethylene synthesis regulates the process of ripening. Guava fruits have a rapid rate of ripening, so they have a relatively short shelf life ranging from 3 to 8 days depending on variety, harvest time, and environmental conditions. Ethylene production and respiration (CO_2 production) increases after the first day of harvest. Guava reaches its climacteric peak between day 4 and 5 postharvest (mature, green harvested fruits) and then declines (Bashir and Abu-Goukh, 2003).

4. PROCESSING

Guava fruit has been used in the food industry for the production of jams, jellies, and marmalades. The processing of guava fruit is mostly done by hand, and hard fruits are picked from the plant. Guavas can be stored for 8–12 days at 6–14°C. In the global market, the demand of the pink guava is greater compared to white guava. Pink guava is used in sauces to decrease the acidity. The guava pulp can be easily stored for a long time by treatment of heat or chemicals, or by dipping in sugar syrup. The pulp from the fruit is extracted by mixing it with water and then seeds are separated.

The pulp can be stored by heating at 70–80°C. Juices can be sequestered from the guava pulp by simple hydraulic pressing or can be diluted in water to adjust the consistency of the juices (Kanwal et al., 2017).

5. VALUE ADDITION

Value-added products of guava are in great demand in the global market. The use of the juice and other products made from guava is becoming popular due to its nutritious value. In different fruit juices like mango, apple, pear, etc., guava juice is added to enhance taste, flavor, and vitamin C content. Papaya and aonla juices are blended with the guava juice to prepare the nectar. Guava jellies and cheese are prepared by cutting the fruit into pieces, blending it with water, and then boiling. After boiling the liquid part is further used in jellies and pulp is used in cheese preparation. Canned fruits are also an important product. The fruits are first dipped in a brine solution and then canned in a dilute citric acid solution (Sinha and Sinha, 2017).

6. USES

The fruit is mostly used fresh (Morton, 1987). Guava has been used in different traditional medicine systems to cure various diseases (Medina and Pagano, 2003). Guava has a very rich taste and fragrance (Thaipong and Boonprakob, 2005). The fruit contains very high mineral and vitamin content. Due to the high nutritional value, the guava is used in diverse ways. The guava agua fresca drug is popular in Mexico. The entire fruit or the juice can be used in sauces (hot or cold) and in candies, fruit bars, dried snacks, desserts, or dipped in chamoy. In the Philippines, ripe guava is used in cooking sinigang. Guava is used as a snack in the hot season. In East Asia, the guava is used along with the sour and sweet plum powder. It can also be used as a juice or salad. The wood of the guava tree is yellow or reddish brown, and it is used for many purposes, including furniture. Guatemalans make spinning tops of this wood, and combs are also made from it. It is considered an important source of wood and charcoal. The bark of this plant is darker and used to make hides. Its leaves and bark are used for dying silk and in some countries for cotton. All parts of this plant are used for the treatment of gastrointestinal problems. Crushed leaves are used for injuries and for tooth pain; the leaves are chewed. The guava leaf extracts are used to treat throat and chest problems, coughs, gargled to relieve oral ulcers, and also used to cure leucorrhea. The extracts are used in the treatment of epilepsy and convulsions. The placenta is expelled by using the bark and leaf extract orally by the mother. In

Nigeria, the twigs of the guava are used as a tooth cleaning product, and they also stop the development of plaque in the teeth.

7. PHARMACOLOGICAL USES

7.1 Antioxidant Activities

The antioxidant potential of guava fruit extracts was assessed by means of different in vitro antioxidant assays (Martínez et al., 2012). The extracts of branch and leaf showed relatively higher antioxidant properties than fruits and seeds. The guava seed oil exhibited a great DPPH scavenging activity and antiradical efficiency (Malacrida and Jorge, 2013). Guava leaf essential oil has been proven to be a potent source of antioxidant compounds (Lee et al., 2012). Different phenolic compounds are present in the guava extracts, and studies have proved that there is a linear relationship between phenolic compounds and the radical scavenging ability (Chen and Yen, 2007; Chen et al., 2007), and antihyperglycemic effect is also linked to the antioxidant potential (Huang et al., 2011). Pink guava (fruit pulp is pink colored) proved to possess increased the antioxidant enzyme activity (Nor and Yatim, 2011).

7.2 Antibacterial Activity

Four antibacterial flavonoids, namely, morin-3-O-lyxoside, morin-3-O-arabinoside, quercetin, and quercetin-3-O-arabinoside, were isolated from leaves of guava (Rattanachaikunsopon and Phumkhachorn, 2010). The antibacterial activity of guava extracts was analyzed against different strains of *Listeria monocytogenes, Escherichia coli, Staphylococcus aureus, Vibrio parahaemolyticus, Salmonella Enteritidis, Bacillus cereus, Pseudomonas aeruginosa, Aeromonas hydrophila, Pseudomonas putida*, and *Alcaligenes faecalis* (Mahfuzul Hoque et al., 2007). In addition, the guava extract possesses greater antimicrobial potential for killing gram-positive bacterial and fungal strains (Nair and Chanda, 2007).

7.3 Antidiarrhoeal Activity

Diarrhea is an infection in the bowels that is usually caused by bacteria, viruses, or parasites. The guava leaf extracts have been tested against diarrhea-causing bacteria: *S. aureus, Salmonella* spp., and *E. coli*. The methanol extract showed the highest bacterial growth inhibition. *S. aureus* strains were most inhibited by the extracts and essential oil of guava (Gonçalves et al., 2008). *Vibrio cholera* is inhibited by the use of the guava bark extracts (Rahim et al., 2010). Guava fruit products were also found effective to reduce the abdominal pain during diarrhea.

7.4 Antiviral Activity

The antiviral activity of guava extracts was determined against growth of A/Narita/1/2009 (amantadine-resistant pandemic 2009 strain) at an IC_{50} of 0.05% and the growth of A/Yamaguchi/20/06 (sensitive strain) and A/Kitakyushu/10/06 (oseltamivir-resistant strain). The growth of these strains was inhibited strongly by the guava extracts. Guava tea is effective for influenza virus, and it has also been proven to develop viral resistance in the body (Sriwilaijaroen et al., 2012).

7.5 Antidiabetic Activity

Diabetes is the disease in which the body's insulin production decreases or the body stops responding to insulin, or both may happen. Guava leaves are a potential antidiabetic agent, as these reduce blood glucose and improve plasma insulin (Subramanian et al., 2009; Soman et al., 2010). Guava decreased the damage, lipid oxidation, and DNA breakage (Huang et al., 2011). The aqueous extracts were reported to improve the glucose uptake by the cells, and the phenolic compounds present in these extracts may responsible for antidiabetic activity (Cheng et al., 2009). Guava peels are also reported to reduce diabetes (Rai et al., 2007). In another report, it was mentioned that long-term use of guava peels led to decrease of blood glucose level and improved plasma insulin (Shen et al., 2008). In a comparison study, it was reported that guava leaves extracts showed a greater decrease in blood glucose level than its peels (Wu et al., 2009).

7.6 Inotropic Activity

Inotropic compounds are responsible for muscle movement. Guava extracts prepared in hexane, water, and methanol have been proven to decrease smooth muscle contractile force. They also reduce acetylcholine release in neuromuscular junctions due to interaction with calcium channels of presynaptic membranes (Conde Garcia et al., 2003).

7.7 Immunomodulatory Activity

Immunomodulators are the chemical compounds that change the immune response of the human immune system. There are various natural products that are being used as immunomodulators. Extracts made from guava have revealed immunomodulatory activities (Kaileh et al., 2007).

7.8 Antiinflammatory Activity

Inflammation and swellings are the major cause of pains and can be avoided by the use of antiinflammatory substances. Guava leaf extract has proven to be a beneficial antiinflammatory substance, as it is used for the treatment of acne (Qa'dan et al., 2005). NF-κ B and STAT1 costimulated with TNF-α and INF-γ activation is stopped by the ethyl acetate extract of the guava leaves. Aqueous extract of guava is important for skin infections (Choi et al., 2012). The antiinflammatory activity is also possessed by the essential oil of guava, and it was proven to be due to presence of the pinene and caryophyllene compounds (Siani et al., 2013). Guava extract in water has proven to be effective on nociceptive pain in rats in a dose-dependent manner (Ojewole, 2005).

7.9 Antiparasitic Activity

Antiparasitic compounds are used to cure the parasitic diseases that are induced by ectoparasites, protozoa, parasitic fungi, ameba, and helminths, etc. In an in vitro antiparasitic assay, as a host for *Toxoplasma gondii,* guava leaf essential oil showed significant results. The potential therapeutic activity of guava leaf essential oil may have contributed to the *in vitro* inhibition of free radicals associated with toxoplasmosis pathology (Lee et al., 2013).

7.10 Anticancer Activity

The guava leaf extract contains flavonoids that possess anticancer activity. They cause apoptosis and induction in cells (Bontempo et al., 2012). The guava extracts were effective to treat epidermal lesions in oral cancer (Fathilah, 2011). It can also be used as an important chemoprotective agent in different cancers (Peng et al., 2011). Guava leaf extracts decrease Tr cells, so they can reduce the chance of tumor (Seo et al., 2005). Water extract can be used to treat prostate cancer.

7.11 Hepatoprotective Activity

Hepatoprotection represents the protection of the liver from damage caused by hepatotoxins. Ethyl acetate and methanolic extracts of guava leaf lessened the amount of hepatotoxic substances. Psiguadials A and B, sesquiterpenoid-diphenyl-methane, and meroterpenoids were separated from the leaves of guava, which showed strong protective influences on the development of hepatoma cells in human body.

7.12 Gastroprotective Activity

The extracts of guava were tested for various types of ulcers in rats. The guava extract showed equal effect as omeprazole. Secretory volume, increased gastric pH, and acid secretion in the stomach were reduced after use of guava extract (Livingston and Sundar, 2012).

8. SIDE EFFECTS AND TOXICITY

Pregnant or breastfeeding women should stick with food amounts until more is known about guava fruits and should avoid using leaves and bark extracts without proper advice.

References

Bontempo, P., Doto, A., Miceli, M., Mita, L., Benedetti, R., Nebbioso, A., Veglione, M., Rigano, D., Cioffi, M., Sica, V., 2012. *Psidium guajava* L. anti-neoplastic effects: induction of apoptosis and cell differentiation. Cell Proliferation 45, 22–31.

Bashir, H.A., Abu-Goukh, A.-B.A., 2003. Compositional changes during guava fruit ripening. Food Chemistry 80, 557–563.

Chalannavar, R.K., Narayanaswamy, V.K., Baijnath, H., Odhav, B., 2013. Chemical constituents of the essential oil from leaves of *Psidium cattleianum* var. cattleianum. Journal of Medicinal Plants Research 7, 783–789.

Chen, H.-Y., Yen, G.-C., 2007. Antioxidant activity and free radical-scavenging capacity of extracts from guava (*Psidium guajava* L.) leaves. Food Chemistry 101, 686–694.

Chen, H.-C., Sheu, M.-J., Lin, L.-Y., Wu, C.-M., 2007. Chemical composition of the leaf essential oil of *Psidium guajava* L. from Taiwan. Journal of Essential Oil Research 19, 345–347.

Cheng, F.C., Shen, S.C., Wu, J.S.B., 2009. Effect of guava (*Psidium guajava* L.) leaf extract on glucose uptake in rat hepatocytes. Journal of Food Science 74, H132–H138.

Choi, J.H., Park, B.H., Kim, H.G., Hwang, Y.P., Han, E.H., Jin, S.W., Seo, J.K., Chung, Y.C., Jeong, H.G., 2012. Inhibitory effect of Psidium guajava water extract in the development of 2, 4-dinitrochlorobenzene-induced atopic dermatitis in NC/Nga mice. Food and Chemical Toxicology 50, 2923–2929.

Conde Garcia, E., Nascimento, V., Santiago Santos, A., 2003. Inotropic effects of extracts of *Psidium guajava* L.(guava) leaves on the Guinea pig atrium. Brazilian Journal of Medical and Biological Research 36, 661–668.

Fathilah, A., 2011. *Piper betle* L. and *Psidium guajava* L. in oral health maintenance. Journal of Medicinal Plants Research 5, 156–163.

Gonçalves, F.A., Andrade Neto, M., Bezerra, J.N., Macrae, A., Sousa, O.V.d., Fonteles-Filho, A.A., Vieira, R.H., 2008. Antibacterial activity of GUAVA, *Psidium guajava* Linnaeus, leaf extracts on diarrhea-causing enteric bacteria isolated from Seabob shrimp, Xiphopenaeus kroyeri (Heller). Revista do Instituto de Medicina Tropical de São Paulo 50, 11–15.

Gutiérrez, R.M.P., Mitchell, S., Solis, R.V., 2008. Psidium guajava: a review of its traditional uses, phytochemistry and pharmacology. Journal of Ethnopharmacology 117, 1–27.

Huang, C.-S., Yin, M.-C., Chiu, L.-C., 2011. Antihyperglycemic and antioxidative potential of *Psidium guajava* fruit in streptozotocin-induced diabetic rats. Food and Chemical Toxicology 49, 2189–2195.

Joseph, B., Priya, M., 2011. Review on nutritional, medicinal and pharmacological properties of guava (*Psidium guajava* Linn.). International Journal of Pharma and Bio Sciences 2, 53–69.

Kaileh, M., Berghe, W.V., Boone, E., Essawi, T., Haegeman, G., 2007. Screening of indigenous Palestinian medicinal plants for potential anti-inflammatory and cytotoxic activity. Journal of Ethnopharmacology 113, 510–516.

Kanwal, N., Randhawa, M., Iqbal, Z., 2017. Influence of processing methods and storage on physico-chemical and antioxidant properties of guava jam. International Food Research Journal 24.

Kumari, N., Gautam, S., Ashutosh, C., 2013. Psidium guajava a fruit or medicine-an overview. The Pharma Innovation 2, 63.

Lee, W.C., Mahmud, R., Pillai, S., Perumal, S., Ismail, S., 2012. Antioxidant activities of essential oil of *Psidium guajava* L. leaves. APCBEE Procedia 2, 86–91.

Lee, W.C., Mahmud, R., Noordin, R., Pillai Piaru, S., Perumal, S., Ismail, S., 2013. Free radicals scavenging activity, cytotoxicity and anti-parasitic activity of essential oil of *Psidium guajava* L. Leaves against *Toxoplasma gondii*. Journal of Essential Oil Bearing Plants 16, 32–38.

Lim, Y.Y., Lim, T.T., Tee, J.J., 2006. Antioxidant properties of guava fruit: comparison with some local fruits. Sunway Academic Journal 3, 9–20.

Livingston, R.N., Sundar, K., 2012. *Psidium guajava* Linn confers gastro protective effects on rats. European Review for Medical and Pharmacological Sciences 16, 151–156.

Mahfuzul Hoque, M., Bari, M., Inatsu, Y., Juneja, V.K., Kawamoto, S., 2007. Antibacterial activity of guava (*Psidium guajava* L.) and neem (*Azadirachta indica* A. Juss.) extracts against foodborne pathogens and spoilage bacteria. Foodborne pathogens and disease 4, 481–488.

Malacrida, C., Jorge, N., 2013. Fatty acids and some antioxidant compounds of *Psidium guajava* seed oil. Acta Alimentaria 42, 371–378.

Malo, S.E., Campbell, C., 1994. The Guava. University of Florida Cooperative Extension Service, Institute of Food and Agriculture Sciences, EDIS.

Mamede, A.M.G.N., Barboza, H.T.G., Soares, A.G., Neves Jr., A.C.V., de Oliveira Fonseca, M.J., Postharvest physiology and technology for fresh guavas, From Cultivation to Consumption and Health Benefits, 91.

Mandal, S., Sarkar, R., Patra, P., Nandan, C.K., Das, D., Bhanja, S.K., Islam, S.S., 2009. Structural studies of a heteropolysaccharide (PS-I) isolated from hot water extract of fruits of *Psidium guajava* (Guava). Carbohydrate Research 344, 1365–1370.

Marcelin, O., Saulnier, L., Williams, P., Brillouet, J.-M., 1993. Reexamination of composition and physico-chemical characteristics of water-soluble pectic substances from guava (*Psidium guajava* L.). Carbohydrate Research 242, 315–321.

Martínez, R., Torres, P., Meneses, M.A., Figueroa, J.G., Pérez-Álvarez, J.A., Viuda-Martos, M., 2012. Chemical, technological and in vitro antioxidant properties of mango, guava, pineapple and passion fruit dietary fibre concentrate. Food Chemistry 135, 1520–1526.

Medina, M., Pagano, F., 2003. Characterization of Guava Pulp (Psidium Guajava L.)"Criolla Roja", vol. 20. Revista de la Facultad de Agronomia, Universidad del Zulia, pp. 72–86.

Mitra, S.K., 1997. Postharvest Physiology and Storage of Tropical and Subtropical Fruits. Cab International.

Morton, J.F., 1987. Fruits of Warm Climates. JF Morton.

Nair, R., Chanda, S., 2007. Antibacterial activities of some medicinal plants of the western region of India. Turkish Journal of Biology 31, 231–236.

Nor, N.M., Yatim, A.M., 2011. Effects of pink guava (*Psidium guajava*) puree supplementation on antioxidant enzyme activities and organ function of spontaneous hypertensive rat. Sains Malaysiana 40, 369–372.

Ojewole, J., 2005. Hypoglycemic and hypotensive effects of *Psidium guajava* Linn.(Myrtaceae) leaf aqueous extract. Methods and Findings in Experimental and Clinical Pharmacology 27, 689–696.

Paniandy, J.-C., Chane-Ming, J., Pieribattesti, J.-C., 2000. Chemical composition of the essential oil and headspace solid-phase microextraction of the guava fruit (*Psidium guajava* L.). Journal of Essential Oil Research 12, 153–158.

Peng, C.-C., Peng, C.-H., Chen, K.-C., Hsieh, C.-L., Peng, R.Y., 2011. The aqueous soluble polyphenolic fraction of *Psidium guajava* leaves exhibits potent anti-angiogenesis and anti-migration actions on DU145 cells. Evidence-based Complementary and Alternative Medicine 2011.

Pino, J.A., Bello, A., Urquiola, A., Marbot, R., Martí, M.P., 2004. Leaf oils of *Psidium parvifolium* Griseb. and *Psidium cattleianum* Sabine from Cuba. Journal of Essential Oil Research 16, 370–371.

Qa'dan, F., Thewaini, A.-J., Ali, D.A., Afifi, R., Elkhawad, A., Matalka, K.Z., 2005. The antimicrobial activities of *Psidium guajava* and *Juglans regia* leaf extracts to acne-developing organisms. The American Journal of Chinese Medicine 33, 197–204.

Rahim, N., Gomes, D.J., Watanabe, H., Rahman, S.R., Chomvarin, C., Endtz, H.P., Alam, M., 2010. Antibacterial activity of Psidium guajava leaf and bark against multidrug-resistant *Vibrio cholerae*: implication for cholera control. Japanese Journal of Infectious Diseases 63, 271–274.

Rai, P.K., Rai, N.K., Rai, A., Watal, G., 2007. Role of LIBS in elemental analysis of Psidium guajava responsible for glycemic potential. Instrumentation Science & Technology 35, 507–522.

Rai, M.K., Jaiswal, V.S., Jaiswal, U., 2009. Shoot multiplication and plant regeneration of guava (*Psidium guajava* L.) from nodal explants of in vitro raised plantlets. Journal of Fruit and Ornamental Plant Research 17, 29–38.

Rattanachaikunsopon, P., Phumkhachorn, P., 2010. Contents and antibacterial activity of flavonoids extracted from leaves of *Psidium guajava*. Journal of Medicinal Plants Research 4, 393–396.

Richardson, D.M., Rejmánek, M., 2011. Trees and shrubs as invasive alien species—a global review. Diversity and Distributions 17, 788–809.

Seo, N., Ito, T., Wang, N., Yao, X., Tokura, Y., Furukawa, F., Takigawa, M., Kitanaka, S., 2005. Anti-allergic *Psidium guajava* extracts exert an antitumor effect by inhibition of T regulatory cells and resultant augmentation of Th1 cells. Anticancer Research 25, 3763–3770.

Shao, M., Wang, Y., Huang, X.-J., Fan, C.-L., Zhang, Q.-W., Zhang, X.-Q., Ye, W.-C., 2012. Four new triterpenoids from the leaves of *Psidium guajava*. Journal of Asian Natural Products Research 14, 348–354.

Shen, S.C., Cheng, F.C., Wu, N.J., 2008. Effect of guava (*Psidium guajava* Linn.) leaf soluble solids on glucose metabolism in type 2 diabetic rats. Phytotherapy Research 22, 1458–1464.

Siani, A.C., Souza, M.C., Henriques, M.G., Ramos, M.F., 2013. Anti-inflammatory activity of essential oils from Syzygium cumini and *Psidium guajava*. Pharmaceutical Biology 51, 881–887.

Sinha, M., Sinha, A.M.P., 2017. Value addition of guava cheese cv. Allahabad safeda by medicinal herbs. Journal of Pharmacognosy and Phytochemistry 6, 856–859.

Soman, S., Rauf, A.A., Indira, M., Rajamanickam, C., 2010. Antioxidant and antiglycative potential of ethyl acetate fraction of *Psidium guajava* leaf extract in streptozotocin-induced diabetic rats. Plant Foods for Human Nutrition 65, 386–391.

Sriwilaijaroen, N., Fukumoto, S., Kumagai, K., Hiramatsu, H., Odagiri, T., Tashiro, M., Suzuki, Y., 2012. Antiviral effects of *Psidium guajava* Linn.(guava) tea on the growth of clinical isolated H1N1 viruses: its role in viral hemagglutination and neuraminidase inhibition. Antiviral Research 94, 139–146.

Subramanian, S., Banu, H.H., Ramya Bai, R.M., Shanmugavalli, R., 2009. Biochemical evaluation of antihyperglycemic and antioxidant nature of *Psidium guajava* leaves extract in streptozotocin-induced experimental diabetes in rats. Pharmaceutical Biology 47, 298–303.

Thaipong, K., Boonprakob, U., 2005. Genetic and environmental variance components in guava fruit qualities. Scientia Horticulturae 104, 37–47.

Wilson, P.G., O'Brien, M.M., Gadek, P.A., Quinn, C.J., 2001. Myrtaceae revisited: a reassessment of infrafamilial groups. American Journal of Botany 88, 2013–2025.

Wu, J.-W., Hsieh, C.-L., Wang, H.-Y., Chen, H.-Y., 2009. Inhibitory effects of guava (*Psidium guajava* L.) leaf extracts and its active compounds on the glycation process of protein. Food Chemistry 113, 78–84.

Yadava, U.L., 1996. Guava Production in Georgia Under Cold-Protection Structure, Progress in New Crops. ASHS Press, Arlington, VA, pp. 451–457.

CHAPTER

27

Henna

Shaheera Rehmat[1], Rasheed Ahmad Khera[1], Muhammad Asif Hanif[1], Muhammad Adnan Ayub[2], Abdullah Ijaz Hussain[3]

[1] Department of Chemistry, University of Agriculture, Faisalabad, Pakistan;
[2] Department of Chemistry, University of Okara, Okara, Pakistan;
[3] Department of Chemistry, Government College University, Faisalabad, Pakistan

OUTLINE

https://doi.org/10.1016/B978-0-08-102659-5.00027-6

Copyright © 2020 Elsevier Ltd. All rights reserved.

1. BOTANY

1.1 Introduction

Henna (*Lawsonia inermis* L.) (Fig. 27.1) is a perennial shrub belonging to the Lythraceae family (Saadabi, 2007). The genus *Lawsonia* contains a broad range of varieties from areas across North and East Africa, the Middle East, South Asia, and the Arabian Peninsula (Cartwright-Jones, 2006). Henna flowers contain both types of reproductive organs and are able to accomplish self-pollination (Phirke and Saha, 2013). Depending upon the region of the world, *L. inermis* is known by different names. In English, it is typically called camphire, cypress shrub, Egyptian privet, henna plant, Jamaica mignonette, mindie, and tree mignonette. In Pakistan and India, it is called hena or mehndi. It is known as alhenna, henna, henneh, hinna, and yoranna in Arabic. Kypros is the commonly used name in Greek (Trivedi, 2006). On the basis of their color, fragrance, flavor, and other phenotypic characters, a broad range of henna ecotypes has been evaluated (Roy, 2004). Henna has different leaf colors, from dark green to reddish brown, and plants may grow up to 6 m in height, depending on the species. *L. inermis* exhibits a broad range of varieties and cultivars, differing in scent, flavors, and uses. However, it is considered that small-leaved cultivars are more effective and of better quality as compared to large-leaved cultivars.

1.2 History/Origin

Henna has been used for thousands of years and has become an important constituent for decorating hands and feet for cooling purposes (Chaudhary et al., 2010). The English name "henna" (*L. inermis,* also

FIGURE 27.1 Henna plants, dried leaves and powder.

known as hina, the mignonette tree, and the henna tree) comes from the Arabic word *hinna* (Bailey et al., 1976). For the word henna, there are a number of suggested origins. It is known as being worn in ancient Egypt (Wurstbauer et al., 2001). According to Watt, Arabic poet Imaru-e-al-Qais first referred to the word *Kafur* in his poetry. Watt has written that in this poem the word *Kafur* might have meant hinna. The Quranic name *Kafur* has many important traditions (Muhammad, 2014). The history of henna is steeped in legend. Henna has been used in Pakistan, India, Africa, and the Middle East for over 5000 years for decoration and medicine (Basipogu and Syed, 2015). Moreover, henna has also been used traditionally in other parts of the world like the Horn of Africa, the Arabian Peninsula, North Africa, South Asia, and the Near East for centuries (Cartwright-Jones, 2006). The use and commercialization of henna in Europe in the late 19th century is often attributed to Adelina Patti, an Opera singer (Singh et al., 2015). Henna was popular among the Pre-Raphaelite artists of England in the 1800s and women connected to the aesthetic movement (Sherrow, 2006).

1.3 Demography/Location

Even though henna grows in different environmental and climatic situations, for henna cultivation the preferable optimum conditions are found in countries that have warm climates and long droughts. As a small desert plant, henna can also grow in the form of a houseplant. In the outdoors, it can be grown if the temperature never drops below 11°C (Cartwright-Jones, 2004). India is the largest producer of henna in the market (Chen et al., 2016). For growth, germination, and development, henna requires high temperature and Mediterranean conditions. Optimum temperature for germination is 25°C with growing temperatures of 11–30°C. The plant can tolerate extreme heat, long drought, and low air humidity as well as poor, stony, and sandy soils. Henna is well adapted to fully drained, heavy, fertile clay soils. It grows better in soils where annual precipitation is 0.2–4.2 m and soil pH is 4.3–8.0 (Kidanemariam et al., 2013). The total export of the henna is over 10,000 t/year. Pakistan, Iran, Sudan, India, and Egypt are the major exporters, while the main importers are the Middle East and North Africa, Western Europe, and North America. The total henna herb export from Sudan is about 1000 t/year. Saudi Arabia is the biggest importer (3000 t/year), followed by the United States (500–600 t/year), France (250 t/year), and Great Britain (100 t/year).

1.4 Botany, Morphology, Ecology

L. inermis is a tall, multibranched shrub, 1.8–7.6 m high with square, glabrous stem and oppositely arranged branches, usually green in color. The leaves are simple, subsessile, and elliptical, and grow opposite each other on the stem. They are 1.5–5 cm or more long, acuminate with tapering at the long point, the margins are entire, and they have depressed veins on the dorsal surface (Danzarami et al.). The petiole is 1.5–3.5 cm long and 0.5–1.3 cm broad. *Lawsone* is a red orange dye present in the henna leaves, and petioles of the smaller, young leaves contain its highest concentration. The inflorescence is usually pyramidal, many flowered and large, with terminal panicle up to 25 cm long. The bracts are leafy, minute, caduceus, linear, and 0.4–0.7 mm long. Calyx is a 1–1.7 mm long tube and ovate-triangular. The fruit has 2–3.5 mm long pedicel. Petals are 1.5–4 mm long, usually whitish but sometimes reddish in color. Stamens are eight in number and are inserted in pairs on the rim of the calyx tube. Ovary is superior, and fruit is in the form of a globose capsule 4–8 mm in diameter, purplish green, and opens irregularly into four splits. Seeds are 2–3 mm long, 4-angular, with thick coat (Kumar et al., 2005).

2. CHEMISTRY

Henna is an aromatic plant with an earthy and grassy smell (Tadesse and Mesfin, 2010). Noble sweet henna's taste is bright, bitter, warm, and more pungent, offering unique flavor. The presence of essential oils in leaves and other parts of the henna plant differentiates the fragrance and aroma in many henna cultivars. The fragranced essential oil is mainly composed of terpenoids, xanthones, phenols, and aldehyde. Depending upon the type of species and cultivar, the extent of each of these chemical constituents varies. For example, β-ionone is largely responsible for the pungent odor extracted from flowers, lawsone produces a sweet, medium-bodied flavor, and tannin's aroma is leather- and oak-like with a bitter flavor. Other aroma-containing compounds are citronellol, limonene, linalool, eugenol, and α-terpineol (Ogunbinu et al., 2007).

Henna contains a low amount of fats present in the seeds and carries low caloric value. The oil isolated from seeds of henna is called fatty oil and is comprised of arachidic acid, stearic acid, oleic acid, and behenic acid (Uddin et al., 2013). In addition to fatty oil, plants also contain resin, glycosides, steroids, saponins, tannins, amino acids, lipids, and mannitol compounds. Small amounts of henna tannic acid and gallic acid are also present in leaves (Agarwal et al., 2014). It is also well known as a good source of multivitamins, minerals, and other trace elements. In leaves, the Mg content is less than 0.2%, while Cu, Zn, and Fe are present in greater quantity. Small amounts of vitamin C, carbohydrates, proteins, gums, minerals, and glycosides are also present in the leaves. Henna seeds are rich in carbohydrates, fibers, proteins, and phytosterols (Zumrutdal and Ozaslan, 2012). Henna is also popular for flavonoids and antiinflammatory properties (Alia et al., 1995). On the USDA's GRAS (Generally Recognized as Safe) register, the dried leaves and petiole of henna are listed to be used safely (Kirkland and Marzin, 2003). Structures of some important chemical constituents of henna are shown in Fig.27.2.

3. POSTHARVEST TECHNOLOGY

Conventionally, in the Indo-Pak region the best harvesting time for henna is at the end of the dry seasons just before the two rainy seasons begin. It has been observed that the lawsone content is substantially higher in hot, dry climates. The crop following the cool season has a lower amount of lawsone. A henna crop harvested after passing the rainy season has greatly reduced the amount of lawsone content (Chowdhury et al., 2010).

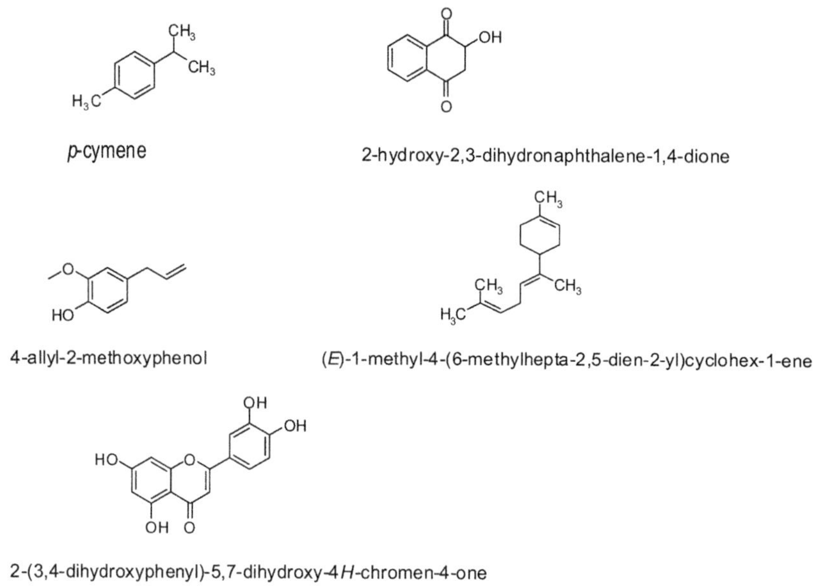

FIGURE 27.2 Structures of some important chemical constituents of henna.

Henna can be stored for many weeks when it is wrapped in different layers of aluminum foil and placed in well-sealed and well-filled airtight containers and stored in heavy plastic or glass jars (Sghaier et al., 2017).

4. PROCESSING

It is problematic to store henna in moist environments for longer term storage purposes. Consequently, it is preferable to dry leaves properly for long-term storage. Care should be taken while drying the leaves. Leaves should not be broken or tattered because broken leaves will show reduced amount of essential oil content, resulting in loss of flavor. Therefore, leaves should be dried in a dry, dark place, out of the sunlight, otherwise they will lose fragrance due to the volatility of the essential oil. By keeping dried henna in closed glass jars and heavy plastic containers away from the heat and light, it can be stored for many months. Freezing the henna could be another preferred way for long-term handling. However, freezing can cause the blackening of leaves. To avoid this, leaves are chopped and tightly packed in aluminum sheets, then thawed and used just as fresh (Sghaier et al., 2017). The henna can also be stored by forming in ice cube trays sealed in a freezer bag. In some cases, layering in salt alone can also preserve the henna leaves (Sghaier et al., 2017). As *L. inermis* acts as an antibacterial agent, there are very rare chances of bacterial growth in the stored form. Moreover, strictly following the food safety instructions may also reduce the development of bacterial growth (Babu and Subhasree, 2009).

5. VALUE ADDITION

Henna paste is used to decorate feet, hands, nails, and hair for its cooling purposes (Chaudhary et al., 2010) and to create month-long destruction and deterrence of fungi (Tripathi et al., 1978), as well as to prevent skin cracking and moisture loss (Mutluoğlu and Uzun, 2009). Henna powder is applied principally as aqueous extracts for skin and hair dyeing. However, women also used henna on their hair to protect it from the UV rays that damage and dry the hair and ultimately clear the scalp of lice and dandruff. In medicine, henna is used for its properties as a skin healer and cleanser. As part of marriage rituals, henna powder is mixed with water for decorating the bride (Al-Suwaidi and Ahmed, 2010). It is used on hair as a colorant and natural conditioner (Bartuska and Silverman, 1980).

6. USES

L. inermis is a medicinal herb. In Ayurvedic and Unani medicines, the seeds and bark of henna are used (Chaudhary et al., 2010). From gastronomic to religious, henna has been used for many purposes, and its uses are frequently immersed in ceremonial practices (Laouer et al., 2003; Pasricha et al., 1980). There is a wide range of probing beliefs linked with the historical applications of henna. It was considered good luck, especially for women, because it protected the wearer against the evil eye (Sharaby, 2006). In Europe, henna was considered La Lune Rousse by Parisian courtesan Cora Pearl for dying her hair red. Henna is considered a Sunnah by Muslims, a praiseworthy convention of the Prophet Muhammad (PBUH) (Hasan et al.). Moreover, Muslim women colored their hands and nails with henna to assert femininity and differentiate them from men (Hasan et al.).

Religious uses aside, as a medicinal pant, henna is helpful for a variety of diseases ranging from skin disorders to leprosy. The roots of henna plant have been used in premature greying of the hair as well as to cure amenorrhea, skin diseases, leprosy, and dysmenorrhea. Henna leaves have been effectively used in jaundice, hair loss, hemorrhages, boils, leukoderma, dysentery, burning sensation, ulcers, wound, liver tonic, expectorant, greyness of the hair, amenorrhoea, anemia, scabies, leprosy, diarrhea, bronchitis, cough, and for hematinic, antiinflammatory, and diuretic purposes. The oil of henna flower has been applied to cure muscular pains, amentia, fever, burning sensation, cardiopathy, and insomnia, whereas its seeds are used for constipation, insanity, gastropathy, eliminating odors, intermittent fevers, promoting intellect, and regulating menstruation (Jain et al., 2010).

7. PHARMACOLOGICAL USES

7.1 Antioxidant Activity

In a positive control comparison, polar extracts of the henna plant were established to have effective properties in different antioxidant evaluations. The antioxidant properties of a variety of the pure compounds such as lawsone demonstrated the role of the phenolic components (Hosein and Zinab, 2007). The methanol extract from henna leaves was used to screen the free radical scavenging activity using the 2,2-diphenyl-1-picrylhydrazyl (DPPH) assay as well as to examine the antioxidant activity by using ferric thiocyanate (FTC) and thiobarbituric acid (TBA) methods. When compared with positive control, the 0.02% methanol extract showed a good activity in TBA and FTC assays, whereas when the same extract was used in DPPH assay, it showed 67.7% decolorization of DPPH (Endrini et al., 2002; Hsouna et al., 2011). The oxidation of linoleic acid was prevented by using an n-butanol henna leaf extract; strong antioxidant activities have been displayed by the isolated extracts of the phenolic glycosides against DPPH (Hsouna et al., 2011). In addition to α-tocopherol, the essential oil of the leaf also exhibited the same antioxidant properties at 0.02% concentration in the TBA and FTC assays (Endrini et al., 2002). Furthermore, when compared with ascorbic acid, strong antioxidant activities have been displayed by the compounds isolated from the leaves of henna, including apigenin, cosmosiin, lawsone, p-coumaric acid, apiin, and naphthaquinone against the 2,2′-azino-bis (3-ethylbenzothiazoline-6-sulphonic acid) (ABTS) (Mikhaeil et al., 2004).

7.2 Antiinflammatory, Analgesic, Antipyretic, and Antiarthritic Activities

The antiinflammatory activities of the henna plant have been linked with the polar extracts of phenolic compounds. Ethanol extract of the henna leaf has been associated with analgesic, antiinflammatory, and antipyretic activities displayed in rats at doses ranging from 250 to 2000 mg/kg (Alia et al., 1995). In contrast to the phenylbutazone (100 mg/kg) and positive control, another component, lawsone, extracted from chloroform eluents was demonstrated to have modest antiinflammatory activity at 500 mg/kg. Moreover, the chloroform and the n-butanol components isolated from the leafy extracts of henna displayed good activities when taken orally at 500 mg/kg (Alia et al., 1995). Another study investigated the components isolated from the roots and stem of the henna plant, such as lawsaritol and isoplumbagin, demonstrating anti-inflammatory activities. In rats, inflammation of the carrageenan-induced hind paw edema was reduced by aqueous leaf extract of henna at a dose of 250 mg/kg (Al Saif, 2016).

7.3 Anticancer Activity

It has been revealed that the aqueous leaf extracts of henna exhibit very good antitumor activity in mice at a dose 1000 mg/kg. This activity was shown in B16F10 melanoma tumor model as well as in DMBA-induced 2-stage skin carcinogenesis (Raja et al., 2009). In animal models, the aqueous extract of leaves of henna plant was used at a dose of 10 mg/kg for decreasing the number of cancer cells effectively and elicited a significant action against the Ehrlich ascites carcinoma cells. The aqueous leaf extract of *L. inermis* significantly reduced the tumor effects and prolonged the average as well as the mean survival rates in mice (Ozaslan et al., 2009).

The ethanol extract from the leaves of henna has been used effectively for anticarcinogenic activities against the benzopyrene-induced forestomach at a dose 200 mg/kg, decreasing the tumor incidence and the tumor burden in mice (Dasgupta et al., 2003). Ethanol extract of henna root has been applied to elicit antitumor activity against Dalton lymphoma ascites-induced mice at a dose of 180 mg/kg, through reducing hemoglobin content, the number of red blood cells and monocytes, as well as by means of reducing the proliferation of white blood cells, lymphocytes, and platelets (Priya et al., 2011). Moreover, essential oil extracted from the henna leaves displayed efficient cytotoxic activity against the liver cell lines (Rahmat et al., 2006). It was reported that a compound extracted from the stem bark of the henna may play an important role as a cytotoxic agent against melanoma and colon cancer cell lines and toward several nonsmall cell lung, central nervous system, colon, and renal cell lines (Wagini et al.).

7.4 Antiulcer and Antitubercular Activities

Ethanol, chloroform, and aqueous leaf extracts of *L. inermis* were compared with a ranitidine positive control to determine the antiulcer activity and in pylorus ligation as well as reduce the total acidity, gastric acid secretion, and ulcer index in aspirin-induced rats (Goswami et al., 2011). In comparison with the negative control gum acacia the same extracts displayed effective properties to cure the chronic and acute gastric ulcers at doses of the 200 and 400 mg/kg in two rat prototypes. Many other researchers also elaborated on the antiulcer properties of extract of the henna leaf (Goswami et al., 2011). Sucralfate, in the positive control, was demonstrated to exhibit antiulcer activity at a dose of 250 mg/kg. Both in vitro and in vivo, strong antitubular activity was shown by the aqueous extracts of henna leaves through preventing growth of the *Tubercle bacillus* from *Mycobacterium tuberculosis* and the septum at a dose of 6 μg/mL. It has also been displayed to counter *M. tuberculosis* infections in mice and guinea pig models at a concentration of 5 mg/kg (Sharma, 1990).

7.5 Antimicrobial Activity

Antimicrobial activities of the henna extract as well as its active and effective compounds against a variety of the human pathogens have been determined (Rahmoun et al., 2013). Different parts of henna herb have been investigated to determine their antimicrobial activities, and it was reported that root extracts of henna showed antimicrobial activity. The increasing number of microorganisms that have established resistance to antibiotics has brought about an increased interest in plant-derived antimicrobial therapy (El-Hag et al., 2007).

7.6 Antifertility, Ovicidal, and Abortifacient Activities

There are some reports that indicate that exposure of organisms to henna extracts may affect their capability to propagate and cause spontaneous abortion and reduced fertility. However, in spite of all these teratogenic effects, no abortions were documented in Sprague Dawley rats at a dose of 40–1000 mg/kg of henna per day. The pregnancy rate in rats was seen to decrease by up to 40%–60% after daily administration of henna leaf powder from plant, at a concentration of 300 mg/kg. Henna has been recognized for ovicidal activity through exposure of the eggs of bean weevil to the leafy extracts of petroleum ether and acetone. Acetone extract has shown 85% lessening in the egg hatchability, while the petroleum ether extract exhibited the dropping up to 71% (Semwal et al., 2014).

7.7 Antiparasitic Activities

The anthelmintic activity of the aqueous extract obtained from the leaves of henna has been shown in adults of tropical flatworm, *Fasciola gigantica*. An effective increase in the mortalities and decrease in the motilities were documented at a significant concentration of 5% (w/v) against the oxyclozanide 5% (w/v) (Jeyathilakan et al., 2012). Dichloromethane and methanol leafy extracts exhibited strong antiparasitic activity against *Trypanosoma brucei* (1.5 μg/mL), Nematoda (410 μg/mL), *Plasmodium falciparum* 420 μg/mL), and *Leishmania donovani* (4100 μg/mL) (Okpekon et al., 2004).

7.8 Antidiabetic Activity

Polar leafy extracts have been explored by different researchers for determining their antidiabetic activities. Both in the basic and neutral media, the methanol extract obtained from the leaves was observed to exhibit hypoglycemic activity by a glucose oxidase assay. An ethanol extract obtained from the henna leaves administered to euglycemic and diabetic rats was seen to reduce the blood glucose level at a concentration of 200 mg/kg, compared to those treated with glibenclamide. It was further demonstrated that the ethanol leafy extract triggered a reduction in blood glucose, triglyceride, and total cholesterol concentrations ranging from 194 mg/dL to normal, 225.7–76.9 mg/dL and 148.9–55.3 mg/dL, respectively, in mice with an oral dose of 800 mg/kg (Semwal et al., 2014).

7.9 Anticataleptic Activity

Significant success was achieved through treating haloperidol-induced catalepsy in the mice from an aqueous henna extract. An enhancement in the superoxide dismutase activity was found, and a decrease in the cataleptic grooves was measured with a concentration of 400 mg/kg (Semwal et al., 2014).

8. SIDE EFFECTS AND TOXICITY

Henna is considered unsafe for use in children, especially in infants. Extremely large doses and extended use of henna may be a health risk due to the presence of potentially carcinogenic compounds. For this reason, GRAS recommends the limited use of henna, and it is advised not to take large amounts of henna fatty oil internally.

References

Agarwal, P., Alok, S., Verma, A., 2014. An update on Ayurvedic herb Henna (*Lawsonia inermis* L.): a review. International Journal of Pharmaceutical Sciences and Research 5, 330–339.

Al-Suwaidi, A., Ahmed, H., 2010. Determination of para-phenylenediamine (PPD) in henna in the United Arab Emirates. International Journal of Environmental Research and Public Health 7, 1681–1693.

Al Saif, F., 2016. Henna beyond skin arts: Literature review. Journal of Pakistan Association of Dermatologists 26, 58–65.
Alia, B., Bashir, A., Tanira, M., 1995. Anti-inflammatory, antipyretic, and analgesic effects of *Lawsonia inermis* L.(henna) in rats. Pharmacology 51, 356–363.
Babu, P.D., Subhasree, R., 2009. Antimicrobial activities of Lawsonia inermis-a review. Academic Journal of Plant Sciences 2, 231–232.
Bailey, L.H., Hortorium, L.H.B., Bailey, E., 1976. Hortus Third; a Concise Dictionary of Plants Cultivated in the United States and Canada.
Bartuska, W.R., Silverman, P., 1980. Henna Hair Coloring and/or Conditioning Compositions. Google Patents.
Basipogu, D., Syed, N.B., 2015. Gastro protective activity of *Lawsonia inermis* (Henna). A well-known traditional medicinal plant. IJAR 1, 833–837.
Cartwright-Jones, C., 2004. The Henna Page. Encyclopedia of Henna. Document Électronique. http://www.hennapage.com.
Cartwright-Jones, C., 2006. Developing Guidelines on Henna: A Geographical Approach. Kent State University.
Chaudhary, G., Goyal, S., Poonia, P., 2010. *Lawsonia inermis* Linnaeus: a phytopharmacological review. International Journal of Pharmaceutical Sciences and Drug Research 2, 91–98.
Chen, W., Nkosi, T.A., Combrinck, S., Viljoen, A.M., Cartwright-Jones, C., 2016. Rapid analysis of the skin irritant p-phenylenediamine (PPD) in henna products using atmospheric solids analysis probe mass spectrometry. Journal of Pharmaceutical and Biomedical Analysis 128, 119–125.
Chowdhury, M.S.H., Rahman, M.M., Koike, M., Muhammed, N., Salahuddin, K.M., Halim, M.A., Saha, N., Rana, M.P., Islam, M.J., 2010. Small-scale mehedi (*Lawsonia inermis* L.) farming in the Central Bangladesh: a promising NTFP-based rural livelihood outside the forests. Small-scale Forestry 9, 93–105.
Danzarami, D., Umar, M., Akafyi, D., Oko, J., Yusuf, I., Okeh, Q., Abdulkarim, M., Adamu, R., Phytochemical Screening and Chromatographic Analysis of Henna (*Lowsonia Inermis*) Plant Obtained from Zaria, Kaduna.
Dasgupta, T., Rao, A., Yadava, P., 2003. Modulatory effect of henna leaf (*Lawsonia inermis*) on drug metabolising phase I and phase II enzymes, antioxidant enzymes, lipid peroxidation and chemically induced skin and forestomach papillomagenesis in mice. Molecular and Cellular Biochemistry 245, 11–22.
El-Hag, A., Al-Jabri, A., Habbal, O., 2007. Antimicrobial properties of *Lawsonia inermis* (henna): a review. Australian Journal of Medical Herbalism 19, 114.
Endrini, S., Rahmat, A., Ismail, P., Hin, T.Y., 2002. Anticarcinogenic properties and antioxidant activity of henna (*Lawsonia inermis*). Journal of Medical Science 2, 194–197.
Goswami, M., Kulshreshtha, M., Rao, C.V., Yadav, S., Yadav, S., 2011. Anti-ulcer potential of *Lawsonia inermis* L. leaves against gastric ulcers in rats. Journal of Applied Pharmaceutical Science 1, 69.
Hasan, M.E., Al-Gehani, S.A.H., Abu-Harbah, A.A.O., The Importance of the Chemical Composition of Henna Tree Leaves (*Lawsonia Inermis*) and its Ability to Eliminate Tinea Pedis, With Reference to the Extent of Usage and Storage in the Saudi Society, Taif, KSA.
Hosein, H.K.M., Zinab, D., 2007. Phenolic compounds and antioxidant activity of henna leaves extracts (*Lawsonia inermis*). World Journal of Dairy & Food Sciences 2, 38–41.
Hsouna, A.B., Trigui, M., Culioli, G., Blache, Y., Jaoua, S., 2011. Antioxidant constituents from *Lawsonia inermis* leaves: Isolation, structure elucidation and antioxidative capacity. Food Chemistry 125, 193–200.
Jain, V., Shah, D., Sonani, N., Dhakara, S., Patel, N., 2010. Pharmacognostical and preliminary phytochemical investigation of *Lawsonia inermis* L. leaf. Romanian Journal of Biology-Plant Biology 55, 127–133.

Jeyathilakan, N., Murali, K., Anandaraj, A., Basith, S.A., 2012. In vitro evaluation of anthelmintic property of ethno-veterinary plant extracts against the liver fluke *Fasciola gigantica*. Journal of Parasitic Diseases 36, 26–30.

Kidanemariam, T.K., Tesema, T.K., Asressu, K.H., Boru, A.D., 2013. Chemical investigation of *Lawsonia inermis* L. leaves from Afar region, Ethiopia. Oriental Journal of Chemistry 29, 1129–1134.

Kirkland, D., Marzin, D., 2003. An assessment of the genotoxicity of 2-hydroxy-1, 4-naphthoquinone, the natural dye ingredient of Henna. Mutation Research: Genetic Toxicology and Environmental Mutagenesis 537, 183–199.

Kumar, S., Singh, Y., Singh, M., 2005. Agro-history, uses, ecology and distribution of henna (*Lawsonia inermis* L. syn. Alba Lam). In: Henna: Cultivation, Improvement, and Trade. Central Arid Zone Research Institute, Jodhpur, pp. 11–12.

Laouer, H., Zerroug, M.M., Sahli, F., Chaker, A.N., Valentini, G., Ferretti, G., Grande, M., Anaya, J., 2003. Composition and antimicrobial activity of Ammoides pusilla (Brot.) Breistr. essential oil. Journal of Essential Oil Research 15, 135–138.

Mikhaeil, B.R., Badria, F.A., Maatooq, G.T., Amer, M.M., 2004. Antioxidant and immunomodulatory constituents of henna leaves. Zeitschrift für Naturforschung C 59, 468–476.

Muhammad, A., 2014. Therapeutic flora in Holy Quran. African Journal of History and Culture 6, 141–148.

Mutluoğlu, M., Uzun, G., 2009. Can henna prevent ulceration in diabetic feet at high risk? Experimental Diabetes Research 2009.

Ogunbinu, A.O., Ogunwande, I.A., Walker, T.M., Setzer, W.N., 2007. Study on the essential oil of *Lawsonia inermis* (L) Lythraceae. Journal of Essential Oil Bearing Plants 10, 184–188.

Okpekon, T., Yolou, S., Gleye, C., Roblot, F., Loiseau, P., Bories, C., Grellier, P., Frappier, F., Laurens, A., Hocquemiller, R., 2004. Antiparasitic activities of medicinal plants used in Ivory Coast. Journal of Ethnopharmacology 90, 91–97.

Ozaslan, M., Zümrütdal, M., Daglıoglu, K., Kıılııc, I., Karagoz, I.D., Kalender, M., Tuzcu, M., Colak, O., Cengiz, B., 2009. Antitumoral effect of *L. inermis* in mice with EAC. IJP-International Journal of Pharmacology 5, 263–267.

Pasricha, J.S., Gupta, R., Panjwani, S., 1980. Contact dermatitis to henna (Lawsonia). Contact Dermatitis 6, 288–289.

Phirke, S.S., Saha, M., 2013. *Lawsonia inermis* L.: a rainfed ratoon crop. In: National Conference on Biodiversity: Status and Challenges in Conservation–FAVEO, pp. 189–193.

Priya, R., Ilavenil, S., Kaleeswaran, B., Srigopalram, S., Ravikumar, S., 2011. Effect of *Lawsonia inermis* on tumor expression induced by Dalton's lymphoma ascites in Swiss albino mice. Saudi Journal of Biological Sciences 18, 353–359.

Rahmat, A., Edrini, S., Ismail, P., Yap, T., Hin, Y., Bakar, M.A., 2006. Chemical constituents, antioxidant activity and cytotoxic effects of essential oil from *Strobilanthes crispus* and *Lawsonia inermis*. Journal of Biological Sciences 6, 1005–1010.

Rahmoun, N.M., Boucherit-Atmani, Z., Benabdallah, M., Boucherit, K., Villemin, D., Choukchou-Braham, N., 2013. Antimicrobial activities of the henna extract and some synthetic naphthoquinones derivatives. American Journal of Medical and Biological Research 1, 16–22.

Raja, W., Agrawal, R., Ovais, M., 2009. Chemopreventive action of *Lawsonia inermis* leaf extract on DMBA-induced skin papilloma and B16F10 melanoma tumour. Pharmacologyonline 2, 1243–1249.

Roy, P., 2004. Development of Agro-Techniques for Henna (*Lawsonia Inermis* L.) Production: Final Report ICAR Ad-Hoc Scheme.

Saadabi, M.A., 2007. Evaluation of *Lawsonia inermis* Linn.(Sudanese henna) leaf extracts as an antimicrobial agent. Research Journal of Biological Sciences 2, 419–423.

Semwal, R.B., Semwal, D.K., Combrinck, S., Cartwright-Jones, C., Viljoen, A., 2014. *Lawsonia inermis* L.(henna): Ethnobotanical, phytochemical and pharmacological aspects. Journal of Ethnopharmacology 155, 80–103.

Sghaier, K., Bagane, M., Peczalski, R., 2017. Estimation of the diffusion coefficient of water in henna leaves by adjustment of drying curves. In: Green Energy Conversion Systems (GECS), 2017 International Conference on. IEEE, pp. 1–4.

Sharaby, R., 2006. The bride's henna ritual: symbols, meanings and changes. Nashim: A Journal of Jewish Women's Studies & Gender Issues 11–42.

Sharma, V., 1990. Tuberculostatic activity of henna (*Lawsonia inermis* Linn.). Tubercle 71, 293–295.

Sherrow, V., 2006. Encyclopedia of Hair: A Cultural History. Greenwood Publishing Group.

Singh, D.K., Luqman, S., Mathur, A.K., 2015. *Lawsonia inermis* L.–A commercially important primaeval dying and medicinal plant with diverse pharmacological activity: a review. Industrial Crops and Products 65, 269–286.

Tadesse, M., Mesfin, B., 2010. A review of selected plants used in the maintenance of health and wellness in Ethiopia. Ethiopian e-Journal for Research and Innovation Foresight 2, 85–102.

Tripathi, R., Srivastava, H., Dixit, S., 1978. A fungitoxic principle from the leaves of *Lawsonia inermis* Lam. Experientia 34, 51–52.

Trivedi, P.C., 2006. Medicinal Plants: Traditional Knowledge. IK International Pvt Ltd.

Uddin, N., Siddiqui, B.S., Begum, S., 2013. Chemical constituents and bioactivities of *Lawsonia alba lam*.(henna). Journal of the Chemical Society of Pakistan 35, 476–485.

Wagini, N.H., Soliman, A.S., Badawy, E.-S.M., Abbas, M.S., Hanafy, Y.A., Some of Phytochemical, Pharmacological and Toxicological Properties of Henna (*Lawsonia Inermis* L.): A Review of Recent Researches.

Wurstbauer, K., Sedlmayer, F., Kogelnik, H.D., 2001. Skin markings in external radiotherapy by temporary tattooing with henna: improvement of accuracy and increased patient comfort. International Journal of Radiation Oncology, Biology, Physics 50, 179–181.

Zumrutdal, E., Ozaslan, M., 2012. A miracle plant for the herbal pharmacy; henna (*Lawsonia inermis*). International Journal of Pharmacology 8, 483–489.

CHAPTER

28

Himalayan Birch

Muhammad Raffi Shehzad[1], *Muhammad Asif Hanif*[1], *Rafia Rehman*[1], *Ijaz Ahmad Bhatti*[1], *Khem Raj Bhatta*[2]

[1] Department of Chemistry, University of Agriculture, Faisalabad, Pakistan; [2] Department of Chemistry, St. Xavier's College Maitighar, Kathmandu, Nepal

OUTLINE

Medicinal Plants of South Asia
https://doi.org/10.1016/B978-0-08-102659-5.00028-8

Copyright © 2020 Elsevier Ltd. All rights reserved.

1. BOTANY

1.1 Introduction

Himalayan birch (*Betula utilis*) (Fig. 28.1) is a perennial tree that belongs to Betulaceae family. In ancient times, the bark of Himalayan birch was used as a paper for writing lengthy scriptures. The genus *Betula* contains a range of 30–60 species. It contains shrubs and trees from different habitats in temperate climates and regions of the Northern hemisphere from the Subtropics to the Arctic, populating various habitats, including highlands, bogs, forest, and tundra (Furlow, 1990). The uncertainty in the exact number of species within the genus is largely attributed to variability among constituent specie. Variability is present in the morphologic traits such as in leaves. The flowers are monoecious. Male and female flowers are separate, but both sexes are present on same plant. Pollination is carried out by wind (Singh et al., 2012).

B. utilis is known by different names depending upon where you are in the world. In English, it is called Himalayan birch tree. In Hindi, as well as in Nepali, it is called bhoj patra; the other Hindi name is bhurja. In Tamil, it is called bhurjjamaram and purchcham. In Malayalam, it is called bhujapatram and bhurjjamaram. In Telugu, it is called bhujpatri. In Kannada, it is called bhuyapathra. In Sanskrit, it is called bahulavalkalah, bahupata, and bhurjpatraka. In Pakistan, it is called bhoj pattar (Bhattacharyya et al., 2006).

In the Himalayan region, Himalayan birch specie with significant broad leaves are usually found. Many different varieties and cultivars of Himalayan birch are used in landscaping throughout the world. These varieties include *Betula utilis* var. *jacquemontii* "Snow Queen," *Betula utilis* var. *jacquemontii* "Grayswood Ghost," *Betula utilis* var. *jacquemontii* "Silver Shadow," *Betula utilis* var. *jacquemontii* "Jermyns," *Betula utilis* var. *jacquemontii* "Doorenbos," *Betula utilis* "Bhutan Sienna," *Betula utilis* "Park Wood," *Betula utilis* "Forest Blush," *Betula utilis* "Nepalese Orange," *Betula utilis* "Mount Luoji," and *Betula utilis* var. *jacquemontii* "Inverleith."

Variation in plant height does not exist, but leaf size varies from 10 to 12 cm. Variation in leaf color also exists. The plant has glossy, rich green

FIGURE 28.1 Himalayan birch tree, stem, and bark.

leaves with a bright yellow to yellowish autumnal tints. The tree has lightly spreading to a large, dense crown of leaves (canopy). Himalayan birch's several forms have copper-, orange-, or white-colored bark (Meurer et al., 1988).

Various studies are available on the composition of essential oil of different *Betula* species from various parts of the world. Terpenes and methyl salicylate are the main components of oil of *Betula* species. Stem and bark of Himalayan birch mainly contain essential oil (Pal et al., 2015).

The statistics of production of Himalayan birch have not yet been calculated. The variability of Himalayan birch is reflected in the wide range of uses, which will be explored in more detail later in the chapter.

1.2 History/Origin

Himalayan birch is native to Himalaya and grows at elevations up to 4500 m. The genus name *Betula* comes from Latin "betumin" meaning mineral pitch or asphalt, because it is a diminutive borrowed from Gaulish (betu meaning resin or glue). The specific nickname "utilis" shows various uses of different parts of this tree. The Hindi name for this tree is bhoj patra or bhurja, which shares a similarity with the other Indo-European words that give the origin of the common name (birch). The common name birch derived from an ancient word meaning "to shine," probably relating to the white bark. In 1825, David Don in his *Prodromus Florae Nepalensis* gave the name *Betula utilis* to this tree. In the 19th century, Kashmiri pandits reported that all of their books were written on the bark of Himalayan birch. The bark of *B. utilis* was used till now for writing blessed mantras that are positioned in an amulet and worn around the neck for the protection or for a blessing (Müller, 1881). According to the legend, the followers of Lord Shiva use the bark of this tree as clothing (Chauhan, 1999). In the Indian epics, it is said that yantras written on the Himalayan birch are best to overcome problems and troubles.

1.3 Demography/Location

Himalayan birch is grown in a variety of climatic and environmental conditions. Himalayan birch best grows in temperate regions. It grows in a variety of soil types and pH ranges, but it prefers well-drained and heavy clay soil with a pH range of 5–7.5. The distribution range of Himalayan birch is from north of China to the south and the Himalayan region (Zobel and Singh, 1997). Himalayan birch is grown widely in the following countries: China (Hebei, Ningxia, Hubei, Shaanxi, Sichuan, Yunnan), Bhutan, Afghanistan, Tajikistan, India (Arunachal Pradesh, Sikkim, Uttaranchal Himachal Pradesh), Uzbekistan, Kyrgyzstan, Nepal, Kazakhstan, and in Pakistan (Northern Pakistani west and along the Himalayan region, Kashmir) (Shaw et al., 2014). In Uttarakhand, Himalayan birch grows along moraines around Bhojbasa, close to the snout of the Gangotri glacier in India (Bhattacharyya et al., 2006).

The statistics of essential oil production from Himalayan birch have not been calculated yet. Different countries use different parts of Himalayan birch for various purposes. These include Russia, the United States, and Germany. In Russia, tincture of Himalayan birch buds is used. In the

United States, betulinic acid derived from Himalayan birch is used, and Germans uses Himalayan birch leaves (Khare, 2004).

1.4 Botany, Morphology, Ecology

Himalayan birch is a medium size tree that grows up to 20 m in height. The bark of Himalayan birch is white or reddish white, shining, with white, horizontal, smooth lenticels. Layers are present on the outer bark of this tree. The leaves of this tree are elliptic, irregular serrate, and ovate acuminate. Flowers bloom in May and June, and flowers are monoecious. Male and female flowers are mostly separate, but both types of flowers are also found on the same plant. Seeds are thin and winged. The plant prefers a variety of soils including light sandy, medium loamy, and heavy clay soils, but the soil should be well drained. The plant can grow in acid, neutral, and alkaline soils. It can show good growth in semishade (light woodland) or no shade (Bean, 1981). This tree can tolerate winds but cannot tolerate maritime exposure. Himalayan birch is often cultivated as a decorative plant in temperate regions and thrives best in moist, sandy soils. It requires a mean annual temperature of 6.2°C. The mean annual rainfall requirement of Himalayan birch is 44.4 cm. Propagation is by seeds, which are sown as soon as mature or sown after stratification in sandy soil, which is kept moist and shady. Seedlings should be transplanted when they are about 1 year old. It can also be propagated by layering green wood cuttings under glass and grafting or budding on seedling stock (Chauhan, 1999).

2. CHEMISTRY

Himalayan birch is an aromatic plant (Batta and Rangaswami, 1973). The leaves of the plant have a peculiar, aromatic odor and a bitter taste (Handa, 2006). It contains flavonoids, tannins, and essential oils. The bark of Himalayan birch contains lupeol, oleanolic acid, betulin, betulitc acid, acetyloheanolic acid, lupenone, methyle betulonate, sitosterol, methyl betulonate, methyl betultriterpenoid, and methyl betulonate (Mishra et al., 2016; Jäger et al., 2009; Shukla et al., 2017). The sap of Himalayan birch contains many minerals, sugars, and vitamins. Himalayan birch also contains monoterpenes, phenylpropenes, terpene alcohol, sesquiterpenes, alkenals, esters, acid, and sesquiterpenoid alcohols (Pal et al., 2015).

Exact information about the chemical composition of Himalayan birch is not available yet. However, the literature showed that the sap of this tree contains minerals, sugars, and vital vitamins. The sugar is mainly fructose and glucose. Minerals present in this tree are calcium, manganese, potassium, magnesium, iron, zinc, sodium, and phosphorous. The main vitamins present in this tree are vitamin B and vitamin C. It is also source of several amino acids. The sap obtained from the Himalayan birch is used for making syrup that can be consumed directly or is used as a part of candies, soups, salad dressings, and even in drinks. The buds of the Himalayan birch tree contain antibiotic and diuretic properties, while its bark contains diuretic, digestive, and antipyretic properties.

Himalayan birch contains a number of phytochemicals that are responsible for its therapeutic properties like contraceptive, carminative, and antiseptic effects (Selvam, 2008). Its therapeutic constituents have antiseptic, aromatic, carminative, and contraceptive effects. Himalayan birch contains betulin up to 12% of its own weight. It has antiseptic effects (Selvam, 2008). Other important constituents include flavonoids, glanonoids, hyproside, tannin, saponins, and quercetin. Flavonoids present in the leaves of Himalayan birch include quercetin-3-*O*-glucuronide, quercetin-3-*O*-galactoside, quercetin-3-*O*-rhamnoside, and myricetin-3-*O*-galactoside. The major essential oil components of Himalayan birch include geranic acid, β-seleneol, β-linalool, and terragon. Minor essential oil components include palmitic acid, cadinene, and 2,4-decadienal. Other essential oil components of Himalayan birch are β-sesquiphellandrene, champacol, and 1,8-cineol. The essential oil components of Himalayan birch are monoterpenes, sesquiterpenoid alcohols, sesquiterpenes, phenylpropenes, alkenals, ester, acid, and terpene alcohols (Pal et al., 2015). The structures of some active components of Himalayan birch are shown in Fig. 28.2.

3. POSTHARVEST TECHNOLOGY

Himalayan birch has spreading branches and a low trunk. The best quality Himalayan birch must be collected during March to April when the plant is at the flowering stage. The large flakes of bark are collected by peeling off horizontally, which then dried, packed, and stored (Chauhan, 1999). The general method of harvesting the wood is a kind of lopping. Some branches of this tree are cut in first year, while other parts are cut until the tree contains only trunk. The trunk is cut at the end. The advantage of this method is that tree can live for many years and provides wood (Elvin, 1998).

4. PROCESSING

The major part of Himalayan birch that is used for various purposes is the bark. The bark of the plant is peeled horizontally then dried properly and used directly as a substitute for writing paper (Bhardwaj et al., 2014). Fresh bark is chopped and stored under refrigeration in glass till it is hydro-distilled for extraction of essential oil (Pal et al., 2015).

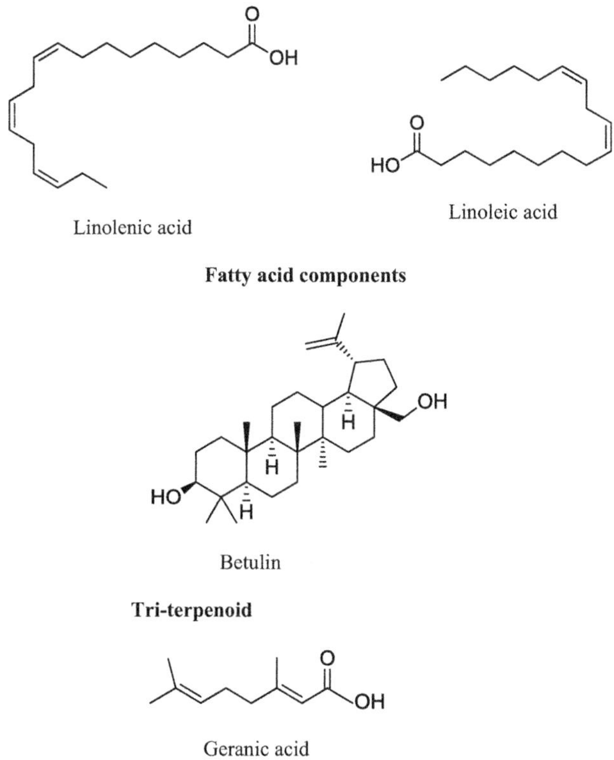

FIGURE 28.2 Structure of active component of Himalayan birch.

White bark of Himalayan birch is used for betulin extraction. For this purpose, bark is peeled from fresh trees. Then it is dried by air in the shade to protect it from direct sunlight. The process of drying takes 1 month until a steady weight is obtained. The dried bark is cut into very small pieces and crushed in a simple blender. Different solvents are then used for obtaining betulin from its bark (Šiman et al., 2016). The sap of Himalayan birch is used for making syrup. Sap can be collected at end of the winter. A large amount of sap is needed for making syrup. In the whole season, many trees are tapped for this purpose. After collection of sap, it is used instantly or stored below 5°C for a few days. Sap is made concentrated by evaporating the moisture and then utilizing for preparation of syrup. For future use, sap can also be frozen (Trummer and Malone, 2009).

5. VALUE ADDITION

The sap of Himalayan birch is used for making syrup and several dressings. The sap is obtained from the inner bark and is also used in cake making due to its nutrition values. Due to its ability of satisfying hunger, it is known as a famine food around the world. It is also used in making different perfumes, which include Iceland wintergreen and a very famous Russian leather fragrance. The buds of the Himalayan birch contain diuretic and antibiotic properties, while its bark contains diuretic, digestive, and antipyretic properties. Young leaves of Himalayan birch are used for making tea. This tea is used for gall bladder stone, dropsy, rheumatism, and for dissolving kidney stones.

6. USES

Himalayan birch is a multipurpose tree. The leaves of the tree, which contain lots of vitamin C, are used to make medicine. The leaves, roots, bark, sap, twigs, and wood of Himalayan birch are used for food purposes, for the construction of buildings, in pharmaceuticals, in paints, as well as several other purposes (Sher and Hussain, 2009). The bark is widely used for a packaging material. Its outer bark is also used as a waterproofing material and for the roofing houses. It is also used in cosmetics. It prevents hair dandruff and promotes the natural growth of hair. Extracts of this plant are used in cosmetics products such as soaps and shampoo. It is also use as a fire wood. Branches of Himalayan birch are used during the marriage ceremony. The bark is good for earache. The decoction of this bark is used as a wash in poisoned wounds and otorrhea. The extract of the bark is used as a carminative. Different parts of this plant inhibit growth of fungal strains and are used in traditional medicines. The birch species are used in folk remedies for abdominal and mammary cancers. Its leaves are used to make a diuretic tea for colds, dysentery, milky urine, and stomach ailments. In higher latitudes, pollens of birch tree are, however, considered to be the most allergic tree pollens (Udgirkar et al., 2012).

7. PHARMACOLOGICAL USES

7.1 Anticancer Activity

Himalayan birch contains botulin that can be converted into the betulinic acid very easily. Studies showed that betulinic acid decreases the growth of malignant melanoma and cancer of the lungs and liver. Betulinic acid is a famous growth inhibitor of neuroectodermal, malignant tumor cells and human melanoma. In these cells, betulinic acid also induces apoptosis. In chemoselective cells, anticancer agents with different activity also trigger apoptosis (Chadha, 1972). It alters the functions of mitochondria that play a role in apoptosis leading to cell death (Hirsch et al., 1997). A xenograft mouse model in a panel of cancer cells and primary tumor sample have been used for the antitumor cytotoxicity of the betulinic acid. The betulinic acid is reported to be very cytotoxic against melanoma cells. The cytotoxicity of betulinic acid has been studied against neuroectodermal tumor cells including Ewing sarcoma cells, medulloblastoma, glioblastoma, and neuroblastoma (Singh et al., 2012).

7.2 Anti-HIV Activity

Many derivatives of betulinic acid have been reported to inhibit HIV at the initial stage of the life cycle of virus. These compounds have the tendency to become a useful addition to anti-HIV therapy, which is based primarily on the enzymes protease and transcriptase (Singh et al., 2012).

7.3 Antimicrobial Activity

Betulinic acid obtained from bark of *B. utilis* has antibacterial activity against some important human pathogenic bacteria like *Citrobacter* sp., *Klebsiella pneumonia, Escherichia coli, Pseudomonas aeruginosa, Salmonella typhi, Proteus mirabilis, Salmonella paratyphi, Shigella boydii, Shigella sonnei, Shigella flexneri, Streptococcus faecalis,* and *Staphylococcus aureus,* and it also affects gram-positive bacteria (Kumaraswamy et al., 2008). A dried, stored sample of bark of Himalayan birch was found to be active against *Aspergillus niger* and *Aspergillus flavus.*

7.4 Antioxidant Activity

Betulinic acid obtained from the bark of Himalayan birch is found to possess potent antioxidant activity (Singh et al., 2012).

7.5 Antiinflammatory Activity

Inflammation is a response of the body to the cell damage by an external source. It has been reported that in the propagation of oxidation the methanolic and water extract of Himalayan birch stopped the initiation of free radicles or stopped the chain reaction. This shows that Himalayan birch is very useful in inflammation treatment. During inhibition, the activity of Himalayan birch was found to be less than lipoxygenase enzyme. By acting on free radicals, it may decrease the inflammation. Lipoxygenases are very responsive to the antioxidants, and mostly, they play a role in inhibition of lipid hydroperoxide formation due to the scavenging of lipidoxy or lipidperoxy radicals formed in the course of enzymic peroxidation. This can limit the availability of lipid hydroperoxide substrate necessary for the catalytic cycle of lipoxygenases (Rackova et al., 2007).

8. SIDE EFFECTS AND TOXICITY

Himalayan birch is possibly safe in medicinal amounts for most adults even for short periods of time. There is not enough reliable information about the use of Himalayan birch during breastfeeding or pregnancy. Stay on the safe side and avoid use. Himalayan birch pollen causes allergies. There is some concern about use of Himalayan birch leaf extract for people suffering from blood pressure, as it might increase the amount of sodium (salt) that the body retains, and this can make high blood pressure worse.

References

Batta, A., Rangaswami, S., 1973. Angiospermae dicotyledonae: crystalline chemical components of some vegetable drugs. Phytochemistry 12, 214–216.

Bean, W., 1981. Trees and shrubs hardy in Great Britain. In: Brickell, C. (Ed.), The AZ Encyclopedia of Garden Plants. London Dorling Kindersley Ltd.

Bhardwaj, A., Rani, S., Rana, J., 2014. Traditionally used common fibre plants in outer siraj area, Himachal Pradesh. Indian Journal of Products and Resources 190–194.

Bhattacharyya, A., Shah, S.K., Chaudhary, V., 2006. Would tree ring data of *Betula utilis* be potential for the analysis of Himalayan glacial fluctuations? Current science Bangalore 91, 754.

Chadha, Y., 1972. The Wealth of India: A Dictionary of Indian Raw Materials and Industrial Products. Council for Scientific and Industrial Research, New Delhi.

Chauhan, N.S., 1999. Medicinal and aromatic plants of Himachal Pradesh. In: Medicinal and Aromatic Plants of Himachal Pradesh. Indus Publishing, p. 632.

Elvin, M., 1998. Sediments of time: Environment and society in Chinese history. In: Sediments of Time: Environment and Society in Chinese History. Cambridge University Press, p. 820.

Furlow, J.J., 1990. The genera of Betulaceae in the Southeastern United States. Journal of the Arnold Arboretum 71, 1–67.

Handa, P., 2006. Ayurveda for health and Beauty. In: Ayurveda for Health and Beauty. Amazon.com, p. 144.

Hirsch, T., Marchetti, P., Susin, S.A., Dallaporta, B., Zamzami, N., Marzo, I., Geuskens, M., Kroemer, G., 1997. The apoptosis-necrosis paradox. Apoptogenic proteases activated after mitochondrial permeability transition determine the mode of cell death. Oncogene 15, 1573–1581.

Jäger, S., Trojan, H., Kopp, T., Laszczyk, M.N., Scheffler, A., 2009. Pentacyclic triterpene distribution in various plants—rich sources for a new group of multi-potent plant extracts. Molecules 14, 2016–2031.

Khare, C.P., 2004. In: Indian Herbal Remedies: Rational Western Therapy, Ayurvedic, and Other Traditional Usage, Botany. Springer science & business media, p. 532.

Kumaraswamy, M., Kavitha, H., Satish, S., 2008. Antibacterial evaluation and phytochemical analysis of *Betula utilis D. Don* against some human pathogenic bacteria. Advances in Biological Research 2, 21–25.

Meurer, B., Wiermann, R., Strack, D., 1988. Phenylpropanoid patterns in fagales pollen and their phylogenetic relevance. Phytochemistry 27, 823–828.

Mishra, T., Arya, R.K., Meena, S., Joshi, P., Pal, M., Meena, B., Upreti, D., Rana, T., Datta, D., 2016. Isolation, characterization and anticancer potential of cytotoxic triterpenes from *Betula utilis* bark. PLoS One 11, e0159430.

Müller, F.M., 1881. Selected Essays on Language, Mythology and Religion. Longmans, Green, and Company.

Pal, M., Mishra, T., Kumar, A., Baleshwar, Upreti, D., Rana, T., 2015. Chemical constituents and antimicrobial potential of essential oil from *Betula utilis* growing in high altitude of Himalaya (India). Journal of Essential Oil Bearing Plants 18, 1078–1082.

Rackova, L., Oblozinsky, M., Kostalova, D., Kettmann, V., Bezakova, L., 2007. Free radical scavenging activity and lipoxygenase inhibition of Mahonia aquifolium extract and isoquinoline alkaloids. Journal of Inflammation 4, 15.

Selvam, A., 2008. Inventory of vegetable crude drug samples housed in botanical survey of India. Phcog Rev.: Review Article Howrah, Pharmacognosy Reviews 2, 61–94.

Shaw, K., Roy, S., Wilson, B., 2014. *Betula utilis,* himalayan birch. In: Shaw, K., Roy, S., Wilson, B. (Eds.), The IUCN Red List of Threatened Species, International Union for Conservation of Nature and Natural Resources., United Kingdom, pp. 1–9.

Sher, H., Hussain, F., 2009. Ethnobotanical evaluation of some plant resources in Northern part of Pakistan. African Journal of Biotechnology 8, 1–11.

Shukla, S., Mishra, T., Pal, M., Meena, B., Rana, T.S., Upreti, D.K., 2017. Comparative analysis of fatty acids and antioxidant activity of *Betula utilis* bark collected from different geographical region of India. Free Radicals and Antioxidants 7, 80.

Šiman, P., Filipová, A., Tichá, A., Niang, M., Bezrouk, A., Havelek, R., 2016. Effective method of purification of betulin from birch bark: the importance of its purity for scientific and medicinal use. PLoS One 11, e0154933.

Singh, S., Yadav, S., Sharma, P., Thapliyal, A., 2012. Betula utilis: a potential herbal medicine. International Journal of Pharmaceutical & Biological Archives 3, 493–498.

Trummer, L., Malone, T., 2009. Some Impacts to Paper Birch Trees Tapped for Sap Harvesting in Alaska.

Udgirkar, R.F., Kadam, P., Kale, N., 2012. Antibacterial activity of some Indian medicinal plant: a review. International Journal of Universal Pharmacy and Bio Sciences 1, 1–8.

Zobel, D.B., Singh, S.P., 1997. Himalayan forests and ecological generalizations. BioScience 47, 735–745.

Further Reading

Cao, D., Zhao, G., Yan, W., 2007. Solubilities of betulin in fourteen organic solvents at different temperatures. Journal of Chemical & Engineering Data 52, 1366–1368.

Cascio, J., Barber, V., 2014. Backyard Birch Tapping & Syrup Basics 2014, 1–4.

Joshi, H., Saxena, G.K., Singh, V., Arya, E., Singh, R.P., 2013. Phytochemical investigation, isolation and characterization of betulin from bark of *Betula utilis*. Journal of Pharmacognosy and Phytochemistry 2.

Kumaraswamy, M., Satish, S., 2008. Free radical scavenging activity and lipoxygenase inhibition of Woodfordia fructicosa Kurz and *Betula utilis* Wall. African Journal of Biotechnology 7, 2013.

CHAPTER

29

Hollyhock

Muhammad Raffi Shehzad, Muhammad Asif Hanif, Rafia Rehman, Ijaz Ahmad Bhatti, Asma Hanif
Department of Chemistry, University of Agriculture, Faisalabad, Pakistan

OUTLINE

Medicinal Plants of South Asia
https://doi.org/10.1016/B978-0-08-102659-5.00029-X

Copyright © 2020 Elsevier Ltd. All rights reserved.

1. BOTANY

1.1 Introduction

Hollyhock *(Althaea rosea)* (Fig. 29.1) is an annual herb and belongs to the mallow family (Malvaceae). It has been used for many years and has become a vital ingredient in some traditional food preparations. In most of the literature, the synonym of *A.* is *Alcea*. In the past, *A.* and *Alcea* were

FIGURE 29.1 Hollyhock plants and flowers.

fused together because they are closely related. *A.* is a small genus having dozens of species distributed in Asia, Africa, and Europe (Shaheen et al., 2010; Munir et al., 2012). The great variability among the constituent species is prevalent in morphology, growth habit, flower color, leaves, stem, and chemical composition. The pollination is done by bees. Cross-pollination is favored by protandry, but true forms and colors do not appear in all hybrids. In the absence or shortage of pollinators, self-pollination is accomplished by stigmas coming in contact with anthers via style curvature (Vaidya, 2000; Abid et al., 2010).

A. rosea is known by different names depending where you are in the world. In English, it is typically called hollyhock. In Arabic, it is called khatmae; in Italian, it is called malva rosa, malvone, and rosoni; in Chinese, it is called shi kui and zhu kui; in French, it is called passe rose and rose papale; in Pakistan, especially in Punjab, it is called gulekhyra, gulkhaira, khatmi, and rishak hatmi; in Japanese, it is called han-ao; in Korean, it is called jeop-si-kkot; in Russian, it is called stockrosa; and in Spanish, it is called A. (Lim, 2012). In German, it is called augenpappel and rosenpappel; in Greek, it is called altaia; in Malta, it is called hollyhock and malvarose; in Romanian, it is called nalba de gardina; and in Tamil, it is called simaithuthi (Fahamiya et al., 2016).

1.2 History/Origin

A. rosea is native to Pakistan, India, and China, where it has been cultivated for years. The generic name comes from a Greek word *altho* meaning to heal in indication to its medicinal importance (Murray, 1989). The derivation of the common name hollyhock is obscure. It may come from holy plus the Anglo-Saxon word hoc for mallon (Singh and Panda, 2005). In the past, hollyhock had been used for the coloring of drinks. During the 15th century, hollyhock was imported in Europe from Southwestern China. William Turner, an herbalist of that time, gave it the name hollyhock, from which the English name was derived (Ammar et al., 2013). In Pakistan, India, China, and Japan, *A. rosea* is displayed in artwork and on the walls of historical buildings. Traders used to bring *A. rosea* from Asia to the Middle East, and it appeared in different parts of the Middle East in the 11th century. European visitors used to bring flowers and seeds of *A. rosea* to America. *A. rosea* can also be found in Neanderthal sites.

1.3 Demography/Location

A. rosea is grown in a variety of climatic and environmental conditions. Although hollyhock is a temperate specie, it also grows in the tropics.

Hollyhock prefers full sun and fertile, loamy, and well-drained soil. Wet soil is not tolerated, but light shade is (Lim, 2012; Munir et al., 2012). Hollyhock is an ornamental plant richly cultivated in gardens. Hollyhock has been growing from the Mediterranean region to south Asia and is widely dispersed in temperate regions of the world. Hollyhock was imported in Europe from Southwestern China during the 15th century. By 1870, Turkey was the major source of the commercial fabric dyes from these plants. Hollyhock is grown in the following countries: Taiwan, Pakistan, China (Sichuan), India, southern Europe, and in the tropical areas of Southeast Asia (Ammar et al., 2013).

1.4 Botany, Morphology, Ecology

Hollyhock is a straight, cylindrical, densely branched, herbaceous annually or short-lived perennial herb growing to 2 m high. The stem of hollyhock is pubescent, light green, and terete. The leaves are wide, serrated, like the heart. Outlines of leaves are orbicular, 6–16 cm across with five to eight shallow lobes, margins are crenate, and bases are cordate, and 3–6 cm long petioles. It has large flowers (Ghasemi and Atakishiyeva, 2016). The flowers of hollyhock have a range of colors from white to dark red including yellow, pink, orange, and maroon (Ammar et al., 2013). Along rachis, flowers present independently. Sometimes flowers also present in a small cluster. Flowers are 7–12 cm, showy and large, and when opened fully look like funnels in different colors. The flowers of hollyhock have five to nine bracts, ovate are six, sepals are green, and five petals that form a funnel that contains numerous threads like stigma below and stamens cover the tip. Fruit contains 15–20 oval seeds (Lim, 2012). Leaves are large, long petioled, cordate-ovate, and acutely lobed (Fahamiya et al., 2016). The brownish-black-colored, kidney-shaped seeds are about 6 mm with rugose and hair at margin. The seeds turn mucilaginous when soaked in water (Fahamiya et al., 2016).

A. rosea has potential to accumulate and tolerate higher concentrations of cadmium (Liu et al., 2009). The flowers of *A. rosea* have different colors. The plant is easily grown from the seeds (Ammar et al., 2013). Some of the varieties of *A. rosea* are *A. rosea* "powder puffs," *A. rosea* "nigra," *A. rosea* "chaters double" and *A. rosea* "summer carnival." The wide range of forms, colors, and sizes has elevated the ornamental importance of this plant in the recent years, and increasing the economic value of the plant globally. It is common to see *A. rosea* L. as an ornamental plant in public parks and home gardens. *A. rosea* grows best in winter in plain areas, but it also grows in rains where the monsoon is not heavy. *A. rosea* grows round the year in the plains under moderate climate. The flowering seasons vary in the plains and on the hills. Normally, the plant flowers in 4–5 months, but the early ones may take less time (Fahamiya et al., 2016; Pullaiah, 2006). *A. rosea* grows on both clayey and sandy soils, fortified with organic manure (Fahamiya et al., 2016). The pH of soil required for *A. rosea* ranges from 6 to 8.

2. CHEMISTRY

Hollyhock contains minerals, the highest amount of tannins, carbohydrates, cyanides, and *Althaea* mucilage. The roots of *A. rosea* contain pectin, flavonoids, mucilage, starch, pectin, sucrose, phenolic acids, asparagines, and tannins. Mucilage consists of several polysaccharides. The presence of kaempferol is reported from all the varieties of flowers. Besides kaempferol, pink- and orange-colored flowers contain herbacetin; mauve and red ones contain quercetin; white and yellow contain herbacetin and an unidentified pigment. The yellow variety is rich in anthoxanthins and contains herbacetin as the major pigment (Fahamiya et al., 2016).

A. rosea is significant because of different dyes obtained from its petals. Purple-red pigment is obtained directly with the help of macroporous resin from its petals. This plant is reported as an emmenagogue expectorant used in Unani medicine. Flavonoids extracted from hollyhock were used in pharmaceutical preparations as a raw material. Pectinic substances isolated from hollyhock stem gave rhamnose, arabinose, xylose, glucose, and uronic acid. Roots gave the same pectinic substances but in different percentages. Anthocyanin is a drug obtained from this plant, and this drug has antimicrobial effect, influence on the gastrointestinal tract, and antiinflammatory activity (Munir et al., 2012).

A. rosea contains minerals including calcium, magnesium, chromium, potassium, nickel, sodium, copper, and lead. The composition of amino acids from dry extract of stem is threonine, serine, glutamine, proline, glycine, cysteine, asparagines, and alanine (Lim, 2012).

Hollyhock flowers contain mucilaginous acidic polysaccharides having high molecular weight and are superior to mucilaginous substances of molecular weight 30,000 and 40,000 from red and black flowers, respectively. Hollyhock plant also contains doubly petalled flowers that contain pectin, food dye, and mucilage polysaccharide. The polysaccharides consist of glucose, galactose, arabinose, mannose, rhamnose, and xylose in the ratio of 1:2:1:2:7:7, respectively. Rhamnose and arabinose are the main polysaccharides. It also contains 0.7% nitrogen and 3.5% protein, which includes valine 0.1%, lysine 0.3%, histidine 0.2%, scrine 0.1%, glycine 0.45%, aspartic acid 0.5%, glutamic acid 0.6%, and threonine 0.1%. Pigment obtained from its flower consists of peonidin, malvidin, and delphinidin (Lim, 2012). *A. rosea* is bad for the stomach as a regular article of the diet. Rembert Dodoens (was a Flemish physician and botanist, also

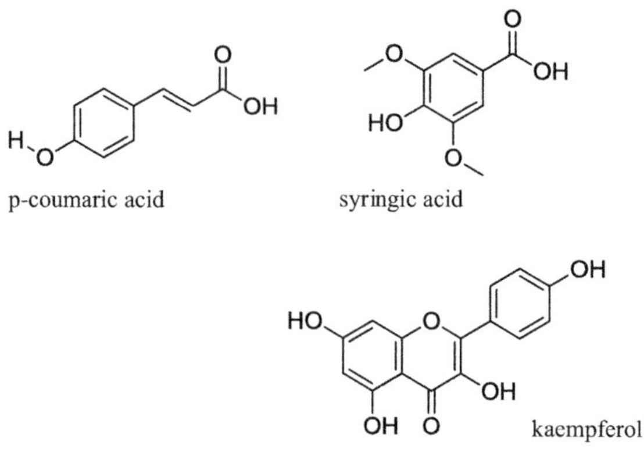

FIGURE 29.2 Hollyhock active components.

known under his Latinized name Rembertus Dodonaeus) found that hollyhock soften and loosen stomach and results in reduction in the organ's ability to retain food long enough to concoct or digest it (Tobyn et al., 2016).

A. rosea has aromatic odor. The lipid extract of seeds contains hydrocarbons, acids, esters, glycerides, and alcohols. The main constituent of seed oil is linoleic acid. The plant also contains alkaloids, glycosides, saponins, tannic acids, anthraquinones, flavonoids, steroids, and triterpenoids (Fahamiya et al., 2016). Five flavonoid compounds obtained from aerial parts of *A. rosea* are recognized as quercetin 3-*O*-β-D-glucuronopyranoside-8-C-β-D-glucopyranoside, kaempherol-3-*O*-β-D-rutinoside, kaempherol-4′-*O*-β-d-glucoside, kaempherol-3-*O*-β-d-glucoside, and kaempferol (Fahamiya et al., 2016). From the flowers of *A. rosea*, phenolic acids have been identified, which include p-coumaric acid, caffeic acid, ferulic acid, p-hydroxybenzoic acid, syringic acid, p-hydroxyphenylacetic, and vanillic acid. Among the phenolic acids analyzed, the prevailing acids include p-hydroxybenzoic acid, p-coumaric acid, and syringic acid (Dudek et al., 2006). Isobutyl alcohol, limonene, phellandrene, *p*-tolualdehyde, citral, terpineol, and *p*-sitosterol are also present (Fahamiya et al., 2016). Some active components are shown in Fig. 29.2.

3. POSTHARVEST TECHNOLOGY

Hollyhock plant is used as a garden flower, and after simple treatment, it can also be used as cut flowers. The stem secretes a sticky substance that make them unable to draw in water. Flowers will rapidly droop by sealing the ends with boiling water. Seeds of *A. rosea* ripen from

August to October. Seeds mature within 1 month after flowering. Mature seedpods may be harvested individually or by cutting whole plants. The pods are dried, and seeds are extracted manually by hand or by threshing. After cleaning, seeds are stored in a dry, cool place (Desai, 2004).

4. PROCESSING

A. rosea is a good source of herbal dyes. The major class is flavones having yellow and brown colors. Most of the flavonoids present in it are yellow-colored dyes. Mordant affects the fastness of the yellow dye. The main components responsible for the dye are althaein, anthocyanin, and altheanin. It is used in alkalimetry and acidimetry as a red dye (Gokhale et al., 2004). Dry flowers of hollyhock can be extracted with methyl alcohol/hydrochloric acid. For the precipitation of the pigment althaein chloride, ether is used. The crude product obtained is first converted into picrate and then reconverted into chloride by using methyl alcohol/hydrochloric acid. Althaein chloride gives one molecule of the glucose and also one molecule of myrtillidin on hydrolysis (Onslow, 2014).

After harvesting, seeds are separated from the pods. Seeds of hollyhock are flat. Seeds are attached to each other and needed to be separated from each other without any damage. Debris and chaff are removed by tweezers. Seeds that have shriveled and have a brownish appearance still contain some humidity, so it is necessary to dry them further before storing. Excess humidity can be removed by placing the seeds of *A. rosea* on wax paper or paper towel and drying them for a week. Do not dry the seeds of *A. rosea* for more than a week because after some time, they do not lose moisture, but they will absorb the moisture. Hollyhock seeds can last for longer periods of times if the dry seeds are properly stored in a sealed glass container.

5. VALUE ADDITION

A. rosea is important for making dyes. The flowers of all colors except black will make a yellow dye on fibers mordant with alum and cream of tartar. The black flowers produce lavender to purple dye depending upon the pH of the bath water. Green dye is obtained from fresh leaves and brown dye from the petals. These dyes are used for silk, cotton, and leather. It is also used in papermaking due to fibers present in the stem. Leaves, roots, stem, and flowers of *A. rosea* have edible properties. For the coloring of confectionery products, fruit jam, jelly, vines, sausages, nonalcoholic beverages, and some other food products, a red pigment is used that is obtained from the petals of hollyhock plant. In sausage manufacturing, pigments are used instead of sodium nitrate. It is thermally stable and does not change the organoleptic property of food. Petals and buds of flowers are used in salads. Herbal tea made from the petals of flowers is refreshing. Leaves of hollyhock can be consumed for cooking as a pot herb. From the roots of hollyhock, starch is obtained (Lim, 2012). *A. rosea* is commonly used for making papers that are used for packing and for making paper bags (Munir et al., 2012).

6. USES

Many herbs and spices contribute significantly to health despite the low amount of consumption, as they are full of antioxidants and certain mineral compounds. A warm, comforting tea made of hollyhock flowers will ease cystitis and cause frequent urination. Hollyhock will soothe skin inflammation, rashes, boils, and even abscesses when used as a lotion. The young branches and leaf powder when mixed with flour of wheat are used as a deworming and carminative agent in the cattle (Sher and Alyemeni, 2011). Root extract with water is applied to hair to remove dandruff. *A. rosea* is also used for cooking purposes. Young leaves are preferred in cooking. Hollyhock is also used as a pot herb. Flower buds, flower petals, and leaves of hollyhock are also used in salads. Flower petals are used for making tea.

Hollyhock have been used in traditional medicine as an antiinflammatory, demulcent, astringent, diuretic, febrifuge, and emollient. Hollyhock has been used as a mouthwash, and it also controls bedwetting and inflammation. The flowers of Hollyhock have been regarded as diuretic, emollient, and demulcent. Flowers are also used for treatment of chest pain, for the improvement of blood circulation, and for hemorrhage and constipation (Lim, 2012). It is also used in a decoction with milk given to pregnant ladies to ease delivery. The hairy leaves, stem, and pollen are irritants (Fahamiya et al., 2016). The herb, roots, and seeds are used to treat cough and lung diseases in Iranian traditional medicine. It is also used in treatment of cough and respiratory problems. Other uses include external application in skin inflammations and ulcers (Ammar et al., 2013). The flowers and roots are also used in Tibetan medicine where they are known to have a neutral potency and acrid, sweet taste. Loss of appetite, inflammation of womb, kidneys, and seminal discharge are also treated by hollyhock (Yashaswini et al., 2011). Hollyhock is a good source of calcium. Due to this reason, it can be used for the treatment of diseases like blood clotting, muscle contraction, and teeth and bone weakness.

7. PHARMACOLOGICAL USES

7.1 Antiurolithiatic Activity

A study showed that concentration of calcium oxalate deposit on kidney of rats can be decreased by the hydroalcoholic extract obtained from hollyhock roots. This effect was recognized most likely to the antiinflammatory and diuretic effects or presence of polysaccharides (mucilaginous) in the plant. The result showed that hollyhock plays an important role in the elimination of the calcium oxalate deposited on rat kidney (Ahmadi et al., 2012).

7.2 Immunomodulatory Activity

Immunomodulation means the change of immune response, which may decrease or increase. Immunosuppression is the enrichment in the immune response (Mukherjee et al., 2014). Polysaccharides obtained from hollyhock have immunomodulatory and antiulcer properties. Research conducted in 2012 showed that aqueous extract of polysaccharide obtained from Hollyhock exhibited immunomodulatory activity. Antibody response of albumen egg was increased by polysaccharide extract, and it acted as a lymphocyte polyclonal activator but has no effect on interferon and intericukin gene transcription (Lim, 2012).

7.3 Antimicrobial Activity

Antimicrobial agents reduce the growth of microorganisms or kill the microorganisms. For a specific microorganism, a specific antimicrobial medicine is used. For example, antifungal medicine is used against the fungi and antibiotics against the bacteria. Antimicrobial activity against *Listeria monocytogenes, Staphylococcus epidermis, Staphylococcus aureus, Escherichia coli, Streptococcus pyogenes, Salmonella typhi, Bacillus cereus,* and *Bacillus anthracis* has been observed for the leaf and flower extracts of hollyhock. The most resistant strain was *E. coli,* as indicated in the studies. In another study, the different extracts of *A. rosea* flowers were found to have antibacterial activity against different bacterial strains by disc diffusion method. The solvents used for the preparation of extracts were water, ethanol, n-hexane, methanol, and ethyl acetate. These extracts were found to possess an activity against *E. coli*, *S. aureus*, *S. epidermidis*, and *Salmonella typhimurium* (Lim, 2012).

7.4 Anticancer/Cytotoxicity Activity

Hollyhock has been observed to have anticancer activity. Phenolic compounds obtained from the ethanolic extract of hollyhock roots were suberic, scopolin, and sebacic acid. These compounds are slightly cytotoxic to human cancerous cells, lung carcinoma, ovary malignant ascites, skin melanoma, and colon adenocarcinoma. Cytotoxicity against brine shrimp was also observed for ethyl acetate extract of flowers. According to the literature, methanolic extracts obtained from hollyhock suppress the transformation of neoplastic cells by inhibiting the activity of kinase of the epidermal growth factor receptor (Lim, 2012).

7.5 Analgesic and Antiinflammatory Activity

Hollyhock flower ethanolic extract has antiinflammatory and analgesic properties (Lim, 2012).

7.6 Antiulcer Activity

Antiulcer activity is used to treat ulcer in the upper part of the small intestine and in the stomach. Polysaccharides obtained from stem of this plant have coating and resorptive properties and are used for lowering the forestomach lesion by following the ligature of pylorus. These activities are observed when these polysaccharides are present inside the intestine intravenously or intraperitoneally. Hollyhock plant is a good source of antiulcer polysaccharides (Lim, 2012).

8. SIDE EFFECTS AND TOXICITY

Hollyhock might be safe for most people, but the possible side effects are not known. However, hollyhock's hairy leaves, stem, and pollen are irritants.

References

Abid, R., Alam, J., Qaiser, M., 2010. Pollination mechanism and role of insects in *Abutilon indicum* (L.) sweet. Pakistan Journal of Botany 42, 1395–1399.

Ahmadi, M., Rad, A.K., Rajaei, Z., Mohammadian, N., Tabasi, N.S., 2012. Alcea rosea root extract as a preventive and curative agent in ethylene glycol-induced urolithiasis in rats. Indian Journal of Pharmacology 44, 304.

Ammar, N.M., El-Kashoury, E.-S.A., El-Kassem, L.T.A., El-Hakeem, R.E.A., 2013. Evaluation of the phenolic content and antioxidant potential of *A. rosea* cultivated in Egypt. Journal of The Arab Society for Medical Research 8, 48.

Desai, B.B., 2004. Seeds Handbook: Processing and Storage. CRC Press.

Dudek, M., Matławska, I., Szkudlarek, M., 2006. Phenolic acids in the flowers of *A. rosea* var. nigra. Acta Poloniae Pharmaceutica 63, 207–211.

Fahamiya, N., Shiffa, M., Aslam, M., 2016. A comprehensive review on *A. rosea* Linn. Journal of Pharmaceutical Research 6, 6888–6894.

Ghasemi, M., Atakishiyeva, Y., 2016. Investigation of the antibacterial effect of native *Peganum harmala, Mentha pulegium* and *Alcea rosea* hydro-alcoholic extracts on antibiotic resistant *Streptococcus pneumoniae* and *Klebsiella pneumonia* isolated from Baku, Azerbaijan. Infection, Epidemiology and Medicine 2, 12–14.

Gokhale, S., Tatiya, A., Bakliwal, S., Fursule, R., 2004. Natural dye yielding plants in India. Natural Product Radiance 3, 228–234.

Lim, T.K., 2012. Edible Medicinal and Non-medicinal Plants. Springer.

Liu, J.N., Zhou, Q.X., Wang, S., Sun, T., 2009. Cadmium tolerance and accumulation of *A. rosea* Cav. and its potential as a hyperaccumulator under chemical enhancement. Environmental Monitoring and Assessment 149, 419–427.

Mukherjee, P.K., Nema, N.K., Bhadra, S., Mukherjee, D., Braga, F.C., Matsabisa, M.G., 2014. Immunomodulatory Leads From Medicinal Plants, pp. 235–256.

Munir, M., Hussain, A., Ul-Haq, I., Qureshi, R., Munazir, M., Rshad, M., Khan, M., 2012. Callogenesis potential of cotyledonary explants of *A. rosea*. from Pakistan. Pakistan Journal of Botany 44, 271–275.

Murray, E., 1989. Monet's Passion: Ideas, Inspiration and Insights From the Painter's Garden. Pomegranate.

Onslow, M.W., 2014. The Anthocyanin Pigments of Plants. Cambridge University Press.

Pullaiah, T., 2006. Encyclopaedia of World Medicinal Plants. Daya books.

Shaheen, N., Khan, M.A., Yasmin, G., Hayat, M.Q., Munsif, S., Ahmad, K., 2010. Foliar epidermal anatomy and pollen morphology of the genera *Alcea* and *A.* (Malvaceae) from Pakistan. International Journal of Agriculture and Biology 12, 329–334.

Sher, H., Alyemeni, M., 2011. Pharmaceutically important plants used in traditional system of Arab medicine for the treatment of livestock ailments in the kingdom of Saudi Arabia. African Journal of Biotechnology 10, 9153–9159.

Singh, M.P., Panda, H., 2005. Medicinal Herbs With Their Formulations. Daya Books.

Tobyn, G., Denham, A., Whitelegg, M., 2016. The Western Herbal Tradition: 2000 Years of Medicinal Plant Knowledge. Singing Dragon.

Vaidya, K., 2000. Natural cross-pollination in roselle, *Hibiscus sabdariffa* L.(Malvaceae). Genetics and Molecular Biology 23, 667–669.

Yashaswini, S., Hegde, R., Venugopal, C., 2011. Health and nutrition from ornamentals. International Journal of Research in Ayurveda and Pharmacy 2, 375–382.

CHAPTER

30

Horseweed

Haq Nawaz[1], *Muhammad Asif Hanif*[1], *Rafia Rehman*[1], *Radosław Kowalski*[2]

[1] Department of Chemistry, University of Agriculture, Faisalabad, Pakistan; [2] Department of Analysis and Evaluation of Food Quality, University of Life Sciences in Lublin, Lublin, Poland

OUTLINE

https://doi.org/10.1016/B978-0-08-102659-5.00030-6

Copyright © 2020 Elsevier Ltd. All rights reserved.

1. BOTANY

1.1 Introduction

Horseweed (*Conyza canadensis* L.) is known by different names in the different parts of the world, out of which two are the most common, including jarayupriya in India and butterweed and horseweed in the United States; it is called fireweed and hogweed in Canada. The name of this species, "*canadensis*," means "of or from Canada and North America" and refers to its distribution (Nesom, 1990). All the associated species of this family are differentiated based on variability in growth habit, flower color morphology, leaves, stems, and chemical composition. Genus *Conyza* is found to be very rich in terpenoids like clerodanes, sesquiterpenes, and diterpenes (Zdero et al., 1990).

The genus name, "*Conyza*," came from Greek word "konops," which means "flea." It was first used by a natural philosopher and author Pliny for a fleabane. It is found in every part of the world but is infrequent in the tropical rainforest. Aquatic or semiaquatic species are also uncommon.

C. canadensis is now a common weed in temperate to tropical regions. It originated in North America. It is widely dispersed throughout the world including different areas of Pakistan like Punjab, the western Himalayas, and Kashmir. It is a mainly annual herbaceous weed, producing a high number of lightweight seeds that are dispersed by wind. It usually grows in undisturbed sites and is a problem in low-tillage systems such as plantations and orchards. But it can also be seen in some agricultural crops. It reduces crop yields by direct competition for resources and by producing harmful chemical substances that have deleterious and toxic effects on seed's germination and plant growth. It may be controlled by tillage at a suitable growth stage. In past few decades, horseweed gained resistance against herbicides and was found to be very first weed that showed resistance in 2001 (Kruger et al., 2010). The resistant species can be found in many countries of the world, as these are introduced internationally as a contaminant of forage seeds, cotton, and cereals. Horseweed had been reported to reduce soybean yield up to 85% (Bruce and Kells, 1990) Fig. 30.1.

FIGURE 30.2 Horseweed

1.2 History

In 1753, *C. Canadensis* was described by Linnaeus by using the term *Erigeron canadensis,* which was later, in 1943, transferred to *Conyza* genus by Cronquist. It is, however, still widely referred by its older name, *Erigeron canadensis. C. canadensis* has distinctive characters in the genus, while many other species (e.g., *Conyza bonariensis* and *Conyza sumatrensis*) can be confused with each other due their similarities. In the middle of the 17th century, it was introduced into Europe, along with Canadian furs through shipping in France. *C. canadensis* was introduced in London during World War II. And it was the among six other species that were distributed by bombing there.

1.3 Demography/Location

C. canadensis is found in broad climatic amplitude and widely located in temperate, mediterranean zones along with taiga, tundra, and subtropics. Usually the most suitable condition for its growth is higher altitudes. Although *Conyza canadensis* specie is native to North America (i.e., Canada and United States), it was also spread to Europe in the 1600s (Michael, 1977) and later was distributed to Australia and Asia. It can be seen in south and north subtropical areas of Africa (Holm, 1997). In Bhutan, higher elevations (above 2000 m) are the habitat of this weed. Although it is grown in other temperate climates, it is seldom commercially produced outside of the United States. The harvesting for the production of essential oil is done sporadically, and no large-scale production of oil is known.

1.4 Botany, Morphology, Ecology

The plant height is about 5–100 cm (2–40 in.). Leaves are alternate, stalkless, short-stalked, dense, withering early with blades (narrowly). Flowering time is July–September. Horseweed is an annual herb and grows both in summer and winter. A basal rosette of dark green hairy leaves is formed from seedlings. Leaves are thin and width is less than 1 cm, having toothed margins and distinct petioles. When stem starts to elongate, basal rosettes deteriorate. Horseweed plant stem is usually unbranched; however, flowering stems are present near the top. Central stem is rigid and erect. It has small flowering branches at the top portions and is covered with long white hair. The leaves are numerous and are around the stem in alternate arrangement and appear as almost whorled. Their length may differ somewhat but beneath the inflorescence they are of same length and in column shape. Their length is around 3–4 in. and 0.5 in. from corner to corner, narrowly pointed at the base, and having toothed outer margins with white fine hair. New leaves are 5 mm wide near the inflorescence and are delicate, more linear, not having any toothed margins. Flowering stems appear at the top after maturity of the plant. Branched flowering stems extend outward and upward and end in a pack of smaller fused flowers. Flower heads are smaller, about 3–5 mm in diameter (Frankton and Mulligan, 1987). These composite flowers have larger bracts, but they are about 1/8 in. across. Each composite flower comprises several yellow smaller flowers in the center with outer tiny erect white ray florets having no scent, which bloom in mid-summer to fall season but can grow throughout the year and last for 2–3 weeks. The seeds are light in weight and are distributed by wind. The root system of this plant is a branching taproot.

The plants preferentially grow in full sun in dry weather and fertile soil. The plant can also grow in soil that has high content of clay or gravel. Moisture conditions and soil fertility affect growth of plant. It usually develops rapidly in summer. Some of the lower leaves wilt or turn yellow in summer drought, but overall it can resist summer drought, although some of the lower leaves may turn yellow and fall away. It can reseed excessively in sunny locations with exposed top soil.

The whole year is its germinating season, but best growth occurs in early fall or spring (Holm, 1997; Buhler and Owen, 1997). Seed germination is influenced by different environmental factors including pH, soil moisture, temperature, and light. Temperature and light both significantly affect horseweed germination. Optimum temperature for seed germination is 24–30°C. Germination is negatively affected outside this temperature range. Horseweed seed germination is more favorable in light periods (Nandula et al., 2009).

Germination of horseweed decreases with increasing salt concentration. Germination decreases as osmotic potential increases. Although at lower incidence, horseweed can still germinate under moderate water-stress conditions. Seeds shows maximum emergence when planted on the surface, while seeds sown 0.5 cm or below show no emergence.

2. CHEMISTRY

Horseweeds contains essential oil, triterpenoids, phenolic acids, sphingolipids, acetylenes, and steroids. The chemical composition and yield of *C. canadensis* essential oil depends upon the kind of plant's organ, origin, and ontogenesis phase (Hrutfiord et al., 1988; Lis et al., 2003; Jirovetz et al., 1999). Structures of some active compounds are provided in Fig. 30.2.

3. POSTHARVESTING TECHNOLOGY

It is harvested during its flowering time and should be preferably used. Shelf life for dried herb is 1 year, and it should not be stored for more than 1 year.

(i) Spathulenol

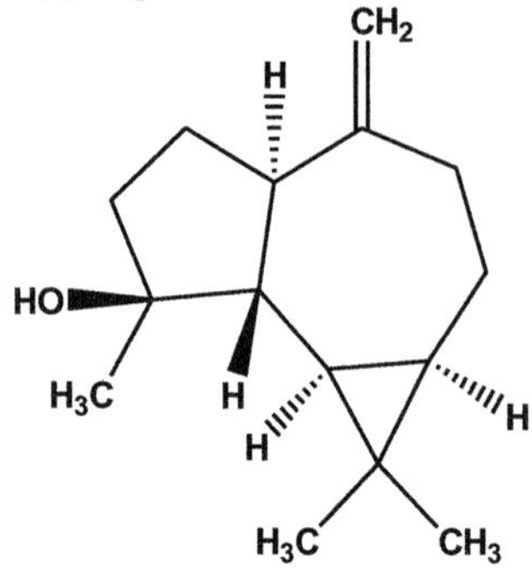

(ii) Limonene

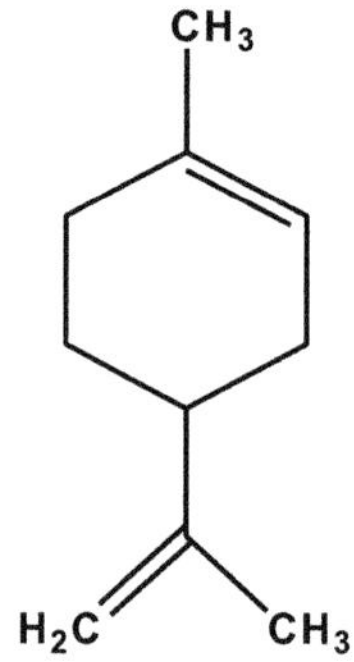

(iii) β pinene

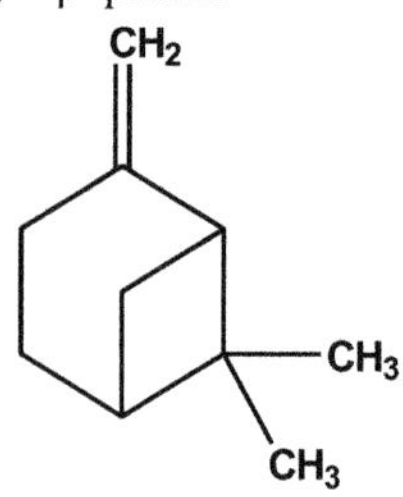

(iv) Lachnophyllum Ester

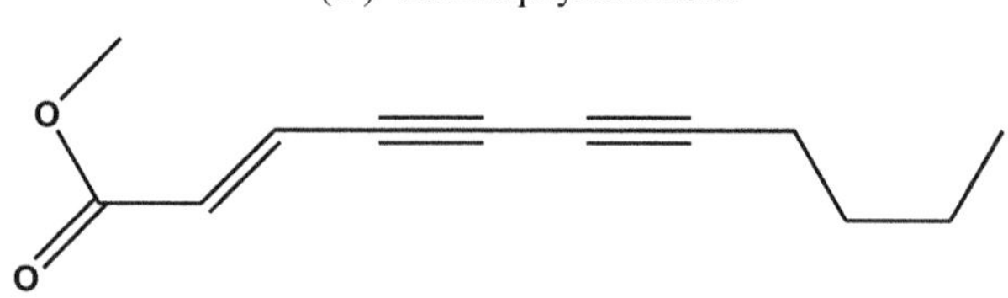

FIGURE 30.2 Structures of bioactive compounds.

4. PROCESSING

Fresh plant yields 0.3%–0.6% oil, while dry plants give 0.2% oil by distillation. Oil obtained by steam distillation is a colorless to pale yellow and has a slightly pungent, herbaceous odor. The oil tends to polymerize on exposure to air. A high yield of oil is obtained from aboveground parts. The oil yield is highest from flowering part. Oil content varies during the vegetation period. However, budding phase yield was highest. Stems and roots yield only a small amount of oil. The content of limonene in herb oil increases at the budding phase. Lower limonene contents are present in the beginning of vegetation and at the end of vegetation. In conclusion, early flowering phase is the most favorable phase for harvesting. In this phase, oil yield is high and is also of constant composition (Lis et al., 2003). Several secondary metabolites including triterpenoids, sphingolipids, acetylenes, steroids, and phenolic acids are present in the horseweed and isolated from the plant extract (Mukhtar et al., 2002; Xie et al., 2007).

5. VALUE ADDITION

It is an edible plant, especially the young leaves. The leaves are best dried and stored for later use to help flavor meals. The young seedlings are also edible. Native people used to grind young leaves and ate them raw with meals (the same as onion is used). The leaves are a good source of calcium and potassium and of proteins. Because of its powerful medicinal properties, it should be eaten sparingly, not in large amounts. Dried herb can be used as a spice in food, flavoring food with aromatic fragrance. Horseweed is used traditionally in northern areas of Pakistan as sweetening agent.

6. USES

The whole plant including aerial parts and the roots were used for medicinal purposes in ancient times. It was used traditionally or officially as herbal medicine for the treatment of gastrointestinal disorders, most commonly in dysentery and diarrhea, and as a diuretic agent. It was applied for the treatment of wounds, swellings, pain caused by arthritis, ulcers, diuretic, anthelmintic, insecticide, antirheumatic, antidiarrheal, and antiulcer. Leaves are used as a poultice externally as a wound disinfectant, against hemorrhoids, and as a vermifuge. In Chinese folk medicines, bronchitis and cystitis were also being treated by the volatile oil of horseweed. The weed is used in making unique fragrances and scents (Asolkar et al., 1992; Chiej, 1984).

The infusion of horseweed and flowers to 1% is a liver decongestant. The infusion at 3%–4% is used as a diuretic in diseases of the genitourinary and urethral washings. The decoction of the whole plant to 10% is used as a purifying antirheumatic agent, primarily to eliminate uric acid. It has also been recommended for bladder disorders. It cures colon trouble, summer complaint, and cholera. It is also beneficial in kidney gravel, diabetes, hemorrhages, fevers, cystitis, nosebleeds, tuberculosis, bronchitis, coughs, and dropsy.

7. PHARMACOLOGICAL USES

7.1 Prophylactic Effects

Its aqueous extract shows antioxidant as well as antiplatelet activity (Olas et al., 2006; Saluk-Juszczak et al., 2007). Peroxynitrite is an oxidant having toxic effects on blood platelets and causes the oxidation of thiols, carbonylation, and nitration of platelet proteins and lipids, resulting in peroxidation. Platelet protein may be protected against radical damage of peroxides, oxide, and nitrite by natural extract of *C. canadensis*. The natural extract from horseweed contains polysaccharides that proved useful as antiaggregator and antioxidative agents.

7.2 Antioxidant Activity

Both aqueous and ethanolic extracts are of equal importance and show medicinal properties. It is reported that extracts from *C. canadensis* may inhibit formation of the oxygen radical (O_2) and decrease generation of $ONOO^-$, which can lead to aggregation of the platelets (Olas et al., 2006).

7.3 Anticoagulant Activity

The commercial name of the ethanolic extract of the horseweed is "Hemorigen." It is used in excessive bleeding in hemorrhoidal and gynecological issues. These properties are related to tannins. The substances isolated from C. *canadensis* are found to show anticoagulant property (Pawlaczyk et al., 2011). The aqueous extract of *C. canadensis* from old or young plants, polysaccharide part, aglycon part, and glycoconjugate part at the concentrations >0.75 mg/mL strongly inhibited platelet aggregation induced by collagen. The polysaccharide part was observed to have the strongest inhibitory effect on platelet aggregation and has antiaggregatory properties.

7.4 Antimicrobial Activity

C. canadensis was extensively used in folk medicines for urinary infections. Urinary infections are usually caused by *E. coli*.

7.5 Anticoagulant and Antiplatelet Effects

The effect of the oil extracted from different parts of the plant is studied, *in vitro*, on the aggregation of the platelets. The strongly inhibited platelet aggregation was found at the concentrations of aglycon part, above 0.75 mg/mL. The polysaccharide part, glycoconjugate part, and aqueous extract of young or old plants were evaluated in a dose-dependent manner in presence of collagen (2 μg/mL). The polysaccharides showed strongest inhibitory effect induced by collagen. The *in vivo* anticoagulant activity was shown by phenolic polysaccharide extracts of this plant, and it was found that protamine sulfate neutralized its effects. Moreover, it also showed antiplatelet activity that is induced by arachidonic acid and limited to the pathway of cyclooxygenase. The anticoagulant activity of different fractions of the plant preparation was determined to know the fraction having highest activity. The effect of plant preparations and most active fraction on thrombin and factor Xa was analyzed. Moreover, the inactivation of thrombin and factor Xa by the antithrombin and inhibition of the thrombin by heparin cofactor II were analyzed. It was found that thrombin and amidolytic activities of the factor Xa were inhibited by both, but in the presence of antithrombin, for unfractionated heparin, higher concentrations were required to get the same effects. The reason of this anticoagulation property was found to be their interaction with heparin cofactor II and inactivation of the thrombin (Pawlaczyk et al., 2011). The analysis of the protective effects of the polysaccharides on the proteins of the platelets extracted from this plant against nitrative and oxidative damage caused by $ONOO^-$ against nitrative and oxidative damage were studied. The peroxynitrite induced the oxidative damage of platelet proteins, and estimation of the levels of carbonyl groups and nitrotyrosine (a marker of platelet protein nitration) helped to evaluate the protecting effects of this extract. These extracts have been evaluated for the generation of O_2 in the platelets by using cytochrome-C method, and its induction by ADP was also described. The extracts were found to decrease the oxidation and reduction of the proteins in the platelets treated with $ONOO^-$, leading to reduce the production of the O_2 in these cells, which supports the significance of the free radicals for the functioning of the platelets, particularly their aggregation (Olas et al., 2006).

7.6 Antiinflammatory Effects

In another study, extracts from the epigean part of this plant were reported to show antiinflammatory effect on the rats suffering from carrageenin and formalin edema. Notably, the best antiinflammatory activity was shown by the fractions extracted by petroleum ether, which were found to have active compounds including beta-santalene, alpha-curcumene, beta-gamma-cadinene, himachalene, and cuparene (Lenfeld et al., 1986). To study the mode of action the extracts of methanol from erigeron canadensis were evaluated by their exposure to the *Lipopolysaccharide* (LPS)-stimulated macrophage cells. It is shown that the extracts caused inhibition of the nitric oxide synthase–derived NO and cyclooxygenase2-derived *Prostaglandin* E2 (PGE2) production along with reduction of the LPS-induced nuclear translocations and *trans* activities of NFκB (Sung et al., 2014).

7.7 Anticancer Effects

For the evaluation of anticancer activity of the extracts of different species of this plant, aqueous and organic phases were studied in vitro by employing HeLa (cervix epithelial adenocarcinoma), MCF7 (breast epithelial adenocarcinoma), and A431 (skin epidermoid carcinoma) cells, with the help of MTT assay. It is reported that the extract from the roots by using n-hexane showed greater anticancer activity than those extracted from the other organs. Moreover, the MCF7 cells showed more sensitivity compared to other cell lines, as shown by the IC_{50} values. The IC_{50} values for HeLa cells were found to be 17.4–18.72, 6.47–12.94 μg/mL for herb and root extracts, respectively. Moreover, for the MCF7 cell line, the IC_{50} values, including 7.93–15.8 and 3.32–9.17 μg/mL, were observed for herb and root extracts, respectively. For A431 cells, these values were 11.6–21.46 and 9.47–20.12 μg/mL for herb and root extracts, respectively (Csupor-Löffler et al., 2009; Réthy, 2008; Réthy et al., 2007). The active components consisting of conyzapyranone B; 4E, 8 Z-matricaria-γ-lactone, and spinasterol appeared to be more potent against these cell lines compared to healthy fetal fibroblasts (MRC-5) of human (Csupor-Löffler et al., 2011).

7.8 Mutagenic Effects

The ability of the flavonoids present in this plant to cause mutation was evaluated by employing Ames test with the help of different strains of *S. typhimurium* including TA100, TA102, TA1535, TA97, TA98, and TA1538 with or without metabolic activation. The quercetin caused point mutations in the strains including TA97, TA98, TA100, and TA102. The presence of liver microsome fraction S9 of rats significantly increased the ability of the quercetin to cause mutagenic activity in these strains, while rhamnetin was found to be a much weaker mutagen. These compounds induced mutations in the strains *S. typhimurium* including TA97, TA98, and TA100 only induced by the metabolic activation. The comparison of the structures of these flavonoids with their mutagenicity led to the conclusion that the methoxy groups and hydroxyl groups present in the B ring, at 3′ and 4′ positions, can be an important factor for this activity (Czeczot et al., 1990).

The mice induced with gastric ulcer, by using HCl/ethanol, were protected by the ethanolic extracts of the aerial parts of this plant in a dose-dependent manner as indicated by reduction of the ulcer lesions from 74.4% to 14.4%. It is reported that for the animals pretreated with the extract of 100 mg/kg, the antiproliferative effect was higher compared to that caused by the sucralfate employed as a reference drug. The pretreatment with the extracts led to the reversal of the changes including inflammation, hemorrhage, edema, and loss of the epithelium cells (Park et al., 2013).

7.9 Depigmentation Effects

The effects of the extracts of this plant were evaluated on the melanogenesis and toxicity of the mouse melanoma cells. The extracts led to the downregulation of the melanin significantly at a nontoxic dose. The extract was divided into five fractions, and one of the fractions showed inhibition of the melanin by 48% at a dose of 100 mg/mL. Notably, this effect is 2.5 times more pronounced than the commercial arbutin, which was 17.5% (Hong et al., 2008).

8. SIDE EFFECTS AND TOXICITY

There is not enough information available to know if horseweed is safe. It may cause an allergic reaction in people who are sensitive to the Asteraceae/Compositae plant family.

References

Asolkar, L., Kakkar, K., Chakre, O., Chopra, R.N., Nayar, S., Chopra, I.C., 1992. Glossary of Indian Medicinal Plants. Publications & Information Directorate.

Bruce, J.A., Kells, J.J., 1990. Horseweed (*Conyza canadensis*) Control in No-Tillage Soybeans (Glycine Max) with Preplant and Preemergence Herbicides. Weed Technology, pp. 642–647.

Buhler, D.D., Owen, M.D., 1997. Emergence and survival of horseweed (*Conyza canadensis*). Weed Science 98–101.

Chiej, R., 1984. Encyclopaedia of Medicinal Plants. MacDonald. ISBNO-356-10541-10545.

Csupor-Löffler, B., Hajdú, Z., Zupkó, I., Molnár, J., Forgo, P., Vasas, A., Kele, Z., Hohmann, J., 2011. Antiproliferative constituents of the roots of *Conyza canadensis*. Planta Medica 77, 1183–1188.

Csupor-Löffler, B., Hajdú, Z., Réthy, B., Zupkó, I., Máthé, I., Rédei, T., Falkay, G., Hohmann, J., 2009. Antiproliferative activity of Hungarian Asteraceae species against human cancer cell lines. Part II. Phytotherapy Research: An International Journal Devoted to Pharmacological and Toxicological Evaluation of Natural Product Derivatives 23, 1109–1115.

Czeczot, H., Tudek, B., Kusztelak, J., Szymczyk, T., Dobrowolska, B., Glinkowska, G., Malinowski, J., Strzelecka, H., 1990. Isolation and studies of the mutagenic activity in the Ames test of flavonoids naturally occurring in medical herbs. Mutation Research: Genetic Toxicology 240, 209–216.

Frankton, C., Mulligan, G., 1987. Weeds of Canada (revised). Publication 948. In: Ministry of Supply and Services Canada. NC Press Limited, Toronto, Ontario.

Holm, L., 1997. World Weeds: Natural Histories and Distribution. John Wiley & Sons.

Hong, E.-S., Nguyen, D.T.M., Nguyen, D.H., Kim, E.-K., 2008. Inhibition of melanogenesis by Erigeron canadensis via down-regulating melanogenic enzymes in B16F10 melanoma cells. Korean Journal of Chemical Engineering 25, 1463–1466.

Hrutfiord, B.F., Hatheway, W.H., Smith, D.B., 1988. Essential oil of *Conyza canadensis*. Phytochemistry 27, 1858–1860.

Jirovetz, L., Puschmann, C., Buchbauer, G., Fleischhacker, W., Kaul, V., 1999. Essential oil analysis of *Erigeron canadensis* flowers from India using GC-FID, GC-MS and olfactometry. Scientia Pharmaceutica 67, 89–95.

Kruger, G.R., Davis, V.M., Weller, S.C., Johnson, W.G., 2010. Growth and seed production of horseweed (*Conyza canadensis*) populations after exposure to postemergence 2, 4-D. Weed Science 58, 413–419.

Lenfeld, J., Motl, O., Trka, A., 1986. Anti-inflammatory activity of extracts from *Conyza canadensis*. Die Pharmazie 41, 268–269.

Lis, A., Piggott, J.R., Góra, J., 2003. Chemical composition variability of the essential oil of *Conyza canadensis* Cronq. Flavour and Fragrance Journal 18, 364–367.

Michael, P., 1977. Some weedy species of Amaranthus (amaranths) and Conyza-Erigeron (fleabanes) naturalized in the Asian-Pacific region. In: Proceedings of the Sixth Asian-Pacific Weed Science Society Conference, Jakarta, Indonesia, pp. 87–95.

Mukhtar, N., Iqbal, K., Malik, A., 2002. Novel sphingolipids from *Conyza canadensis*. Chemical and Pharmaceutical Bulletin 50, 1558–1560.

Nandula, V.K., Eubank, T.W., Poston, D.H., Koger, C.H., Reddy, K.N., 2009. Factors Affecting Germination of Horseweed (*Conyza Canadensis*).

Nesom, G., 1990. Further definition of Conyza (Asteraceae: Astereae). Phytologia 68, 229–233.

Olas, B., Saluk-Juszczak, J., Pawlaczyk, I., Nowak, P., Kolodziejczyk, J., Gancarz, R., Wachowicz, B., 2006. Antioxidant and antiaggregatory effects of an extract from *Conyza canadensis* on blood platelets in vitro. Platelets 17, 354–360.

Park, W., Bae, J.-Y., Chun, M., Chung, H., Han, S., Ahn, M.-J., 2013. Suppression of gastric ulcer in mice by administration of *Erigeron canadensis* extract. Proceedings of the Nutrition Society 72.

Pawlaczyk, I., Czerchawski, L., Kuliczkowski, W., Karolko, B., Pilecki, W., Witkiewicz, W., Gancarz, R., 2011. Anticoagulant and anti-platelet activity of polyphenolic-polysaccharide preparation isolated from the medicinal plant *Erigeron canadensis* L. Thrombosis Research 127, 328–340.

Réthy, B., 2008. Antitumour Effect of Plant Extracts and Their Constituents on Cancer Cell Lines. Szte.

Réthy, B., Csupor-Löffler, B., Zupkó, I., Hajdú, Z., Máthé, I., Hohmann, J., Rédei, T., Falkay, G., 2007. Antiproliferative activity of Hungarian Asteraceae species against human cancer cell lines. Part I. Phytotherapy Research: An International Journal Devoted to Pharmacological and Toxicological Evaluation of Natural Product Derivatives 21, 1200–1208.

Saluk-Juszczak, J., Olas, B., Pawlaczyk, I., Gancarz, R., Wachowicz, B., 2007. Effects of the extract from Conyza canadensis on human blood platelet aggregation. General Physiology and Biophysics 26, 150–152.

Sung, J., Sung, M., Kim, Y., Ham, H., Jeong, H.-S., Lee, J., 2014. Anti-inflammatory effect of methanol extract from *Erigeron canadensis* L. may be involved with upregulation of heme oxygenase-1 expression and suppression of NFκB and MAPKs activation in macrophages. Nutrition Research and Practice 8, 352–359.

Xie, W.D., Gao, X., Jia, Z.J., 2007. A new c-10 acetylene and a new triterpenoid from *Conyza canadensis*. Archives of Pharmacal Research 30, 547–551.

Zdero, C., Ahmed, A., Bohlmann, F., Mungai, G., 1990. Diterpenes and sesquiterpene xylosides from east africanConyza species. Phytochemistry 29, 3167–3172.

CHAPTER

31

Indian Globe Thistle

Naima Tariq[1], *Anam Waheed*[1], *Muhammad Irfan Majeed*[1], *Muhammad Asif Hanif*[1], *Rafia Rehman*[1], *Mohamed Eddouks*[2]

[1] Department of Chemistry, University of Agriculture, Faisalabad, Pakistan; [2] Faculty of Sciences and Techniques Errachidia, Moulay Ismail University, Errachidia, Morocco

OUTLINE

https://doi.org/10.1016/B978-0-08-102659-5.00031-8

Copyright © 2020 Elsevier Ltd. All rights reserved.

1. BOTANY

1.1 Introduction

Indian globe thistle is a useful traditional medicinal plant belonging to Asteraceae family (Erenler et al., 2014). The genus name derives from the Greek words "ekhinos" meaning "hedgehog" It is also called *layboti* (Wichtl, 2004). The genus *Echinops* is a member of Asteraceae family. *Echinops* species have been used as traditional medicine for treatment of migraines, diuretic, heart diseases, urinary infections, as well as worms and hemorrhoids in Ethiopia (Erenler et al., 2014). The family Asteraceae or family Compositae known as aster, daisy, or sun flower family is a taxon of dicotyledonous flowering plants. The family name is derived from genus Asteraceae and refers to the star-shaped flower head members. The Asteraceae is the second largest family in the division Magnoliophyta with 1100 genera and over 20,000 recognized species. The Asteraceae are cosmopolitan in distribution, but mostly found in open or semiopen habitats rather than deep woods. Many genera and species are cultivated for ornamental purposes. *Echinops* and some other genera are one-flowered with individually involucrate heads aggregated into a secondary head with a secondary involucre.

These plants are hardy and are often considered to be highly ornate. This genus receives its common name from its globe-like flower that grows in shades of purple and white. The leaves of these plants are spiky from the edges and greenish gray, while its fruits are cylindric achene. The blossoms of these plants are round heads that grow in groups. These flower heads are on top of the ribbed stems of the plant, making the total height of the plant nearly 5 ft. (1.5 m). The plants attract swarms of bees and butterflies and are usually planted behind the borders in gardens. These plants are often utilized as cut flowers, as they remain fresh for weeks when placed in vases indoors. They are also used in dried floral arrangements and also for ornamental purposes (Rahman et al., 2008; Barreda et al., 2015).

Echinops echinatus Roxb is recognized by specific names in different parts of the world. In English, it is called Indian globe thistle and camel's thistle. In Hindi, it is called as *gokhru, uthkanta,* or *utakatira.* In Gujarat, it is known as *shuliyo, utkanto, or utkato.* In Sanskrit, it is recognized as *kantalu, kantaphala,* or *utati.* In Pakistan, it is called *utkantaka* in Urdu (Pranav et al., 2013). In Sindhi, it is known as *uthkattar, unt katara, bhattar,* or *luth.* In different areas, it is locally known as *catsori* (Khan et al., 2013), *oont kateli* (Sharma, 2001), *layboti,* globe thistle (Hayat et al., 2008), *shulio* (Kumar and Aggarwal, 2014) *kanderi bhattar,* and *ont katara* (Eram et al., 2013).

1.2 History/Origin

Asteraceae may represent as much as 10% of autochthon flora in many regions of the world. Most members of Asteraceae are herbaceous, but a significant number are also shrubs, vines, and trees. This family has a worldwide distribution and is most common in arid and semiarid regions of subtropical and lower temperate latitudes, and its species are also found in Eastern and Southern Europe, Tropical and North Africa, and Asia (Dhayalan et al., 2015).

1.3 Demography/Location

Indian globe thistle is native to India, Afghanistan, Pakistan, and Myanmar (Kadhim, 2013). It is located at Asia-temperate Western Asia, and in Asia-tropical Indian Subcontinent in Andhra Pradesh, Bihar, Himachal Pradesh, Jammu and Kashmir, Karnataka, Madhya Pradesh, Maharashtra, Rajasthan, Tamil Nadu, Uttar Pradesh, and West Bengal (Pranav et al., 2013). It also survives in Africa and the Mediterranean (Erenler et al., 2014).

1.4 Botany, Morphology, Ecology

Indian globe thistle is a thistle-like, young, stiff, and yearly herb with 1–3 ft height, and it contains widely spread branches from the base. The leaves are oblong, lobes triangular, spiny and sessile, pinnatifid, alternate, capped with cottony wool beneath; its spines are 2.5 cm long. Deeply pinnatifid leaves are 7–12 cm long. Flower heads are purple or white, approximately 1 in. in diameter and rigid, globosely folded at the ends of branches; involucres surrounded by tough white bristles approaching short popups, popup hairs yellowish, making a short cylinder-shaped brush above the achene domes that exist in solitary 3–5 cm white sphere-shaped balls (Qudsia et al., 2015). Petals of its flowers are 5 mm long (Pranav et al., 2013). The roots are tapering and of whitish-brown color (Dymock, 1893). Indian globe thistle is a miraculous plant. Its flowering period is from October to January and April to May. It is a herb and commonly found in fields (Hayat et al., 2008), silts among rocks, on hillsides, and sandy places (Akbar and Fatima, 2012).

2. CHEMISTRY

Indian globe thistle seeds possess a sweet taste, whereas a plant itself is bitter and increases appetite. Different alkaloids are present in different parts of this plant. Flavonoids, carbohydrates, and tannins have also been reported (Patel et al., 2011a,b). Plants that belong to this genus possess different compounds including flavonoids, alkaloids, lipids, polyacetylenes, steroids, and terpenoids (Kadhim, 2013). Studies revealed the presence of various bioactive diterpenoids, flavones, thiophene, and flavone glycosides (Erenler et al., 2014). The phytochemical tests on this genus displayed the existence of phenolics, tannins, and carbohydrates in root extracts. Extracts of aerial part of Indian globe thistle contain flavonoids, alkaloids, and carbohydrates (Patel et al., 2011a,b).

Various alkaloids have been isolated from different parts of Indian globe thistle. Seeds of this plant contain 2′,5,7-trihydroxy-3.6-dimethoxy flavone-7-O-b-D-galactopyranosyl–O-a-L-rhamnopyranoside. The aerial parts possess echinopsine, alkaloids, echinozolinone, echinacin, echinaticin, echinopsidine, taraxasterol acetate, and apigenin and derivatives. Apigenin, echitin (I), a new acylflavoneglucoside, apigenin 7-O-glucoside, and 7-hydroxyechinozolinone (I) are reported from the flowers of *Echinops echinatus*. Compounds reported from the flowers of *E. echinatus* are shown in Fig. 31.2.

Leaves of Indian globe thistle contain an antiinflammatory active flavanone glycoside 5,7-dihydroxy-8,4′-dimethoxyflavanone-5-O-a-L-rhamnopyranosyl-7-O-b-D-arabinopyranosyl-(1,4)-O-b-D-glucopyranoside A and dihydroquercetin-4′-methyl ether. Four phenolic compounds, echinacin (I), apigenin, echinaticin (II), and apigenin 7-O-glucoside, are reported from Indian globe thistle. Isomeric acyl flavone glycosides

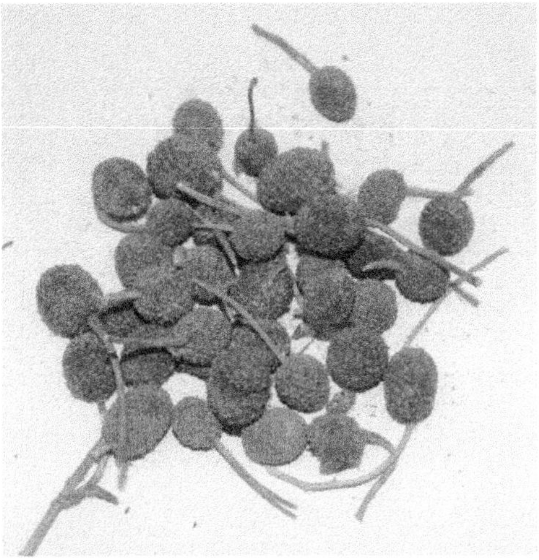

FIGURE 31.1 Dried Indian globe thistle.

echinacin (I) and echinaticin (II) were also found in Indian globe thistle. An alkaloid, echinozolinone, has been also obtained from Indian globe thistle. In addition to echinopsine and echinopsidine, a new alkaloid, echinozolinone, has been distinguished in Indian globe thistle as 3(2-hydroxyethyl)-4(3H)-quinazolinone. The extracted flavonoids from Indian globe thistle include kaempferol, kaempferol 7-methylether, kaempferol 4′ methylether, 5,7–8,4-dimethoxyflavanone-5-O-Lrhamnopyranosyl-7-O-D-arabinopyranosyl-(1,4)-O-D-glucopyranoside, and kaempferol 3-O-alpha-L-rhamnoside, myrecetin-3-O-alpha-L-rhamnoside dihydroquercetin-4′-methyl ether (Maurya et al., 2015).

3. POSTHARVEST TECHNOLOGY

Indian globe thistle is collected at flowering stage. At flowering stage, this plant is rich with aromatic compounds. The collected plant material is shade dried to preserve its bioactive compounds.

4. PROCESSING

Freshly harvested Indian globe thistle is processed to extract essential oil. For this purpose, Indian globe thistle is harvested at flowering stage before sun rise and immediately processed using hydro or steam distillation to extract essential oil. Indian globe thistle essential oil is enriched with highly bioactive compounds.

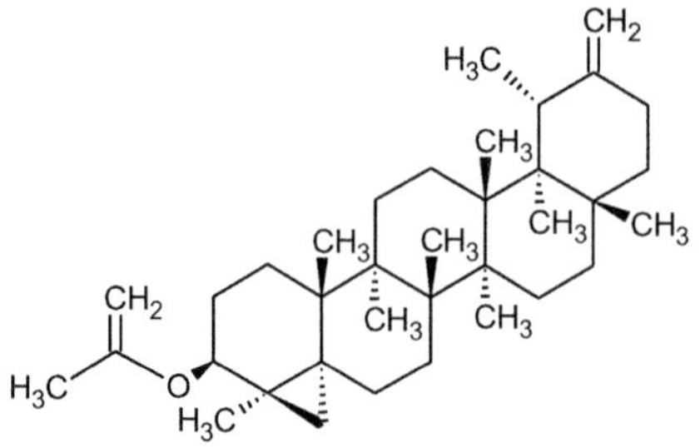

Taraxasterol acetate

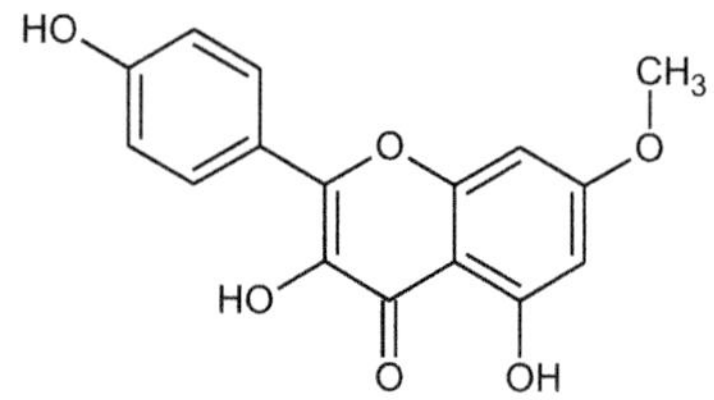

Echinoside

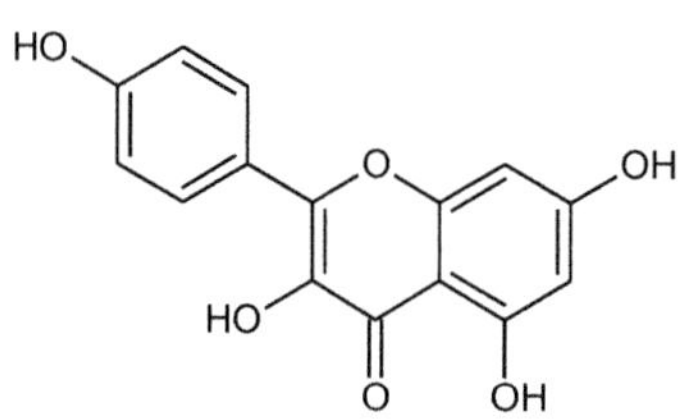

Kaempferol

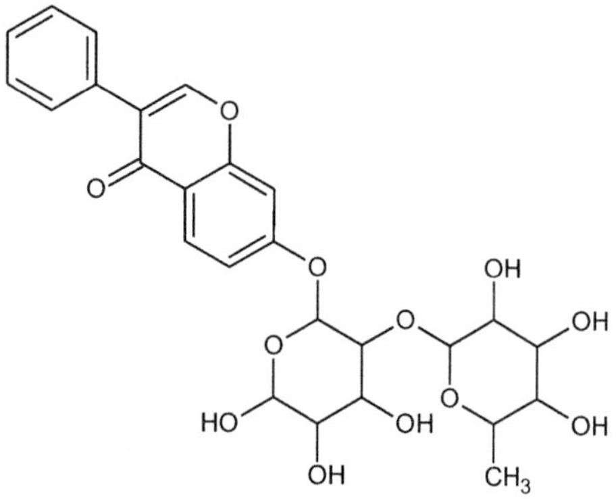

Kaempferol-3-O-alpha-I-rhamnoside echinoside

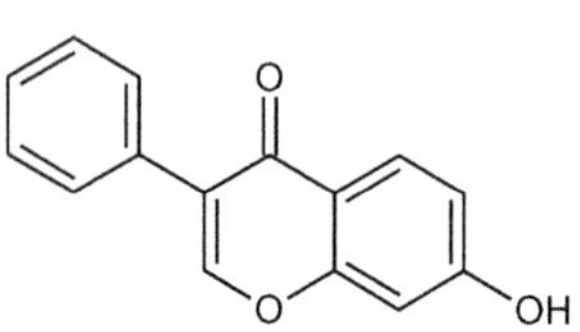

Aglycon

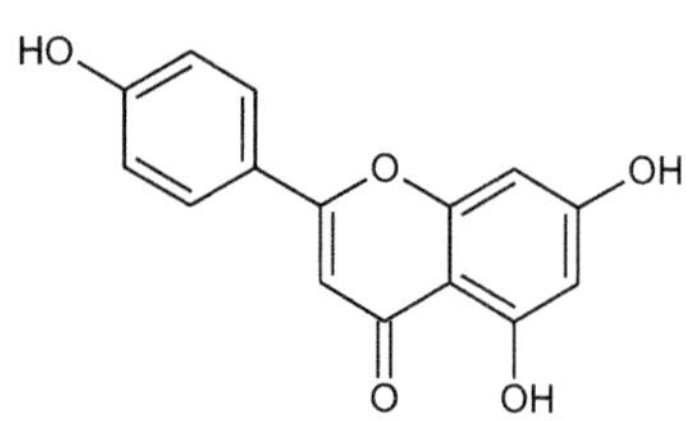

Apigenin

FIGURE 31.2 Compounds reported from the flowers of *E. echinatus*.

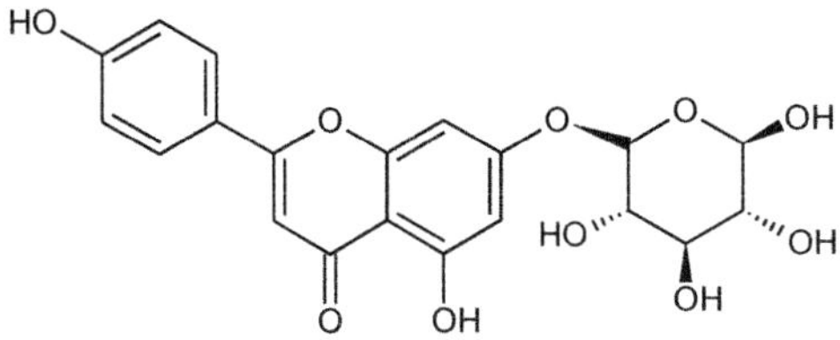

Apigenin-7-O-glucoside

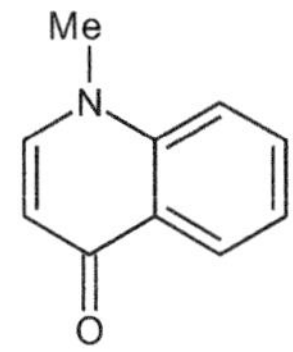

Echinacin

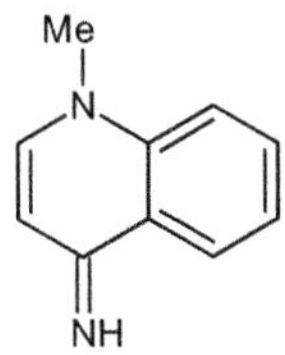

Echinopsidine

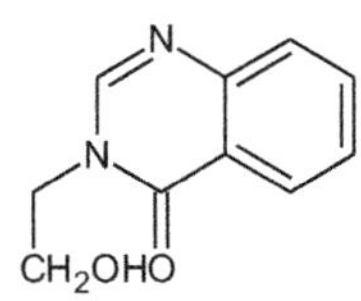

Echinozolinone

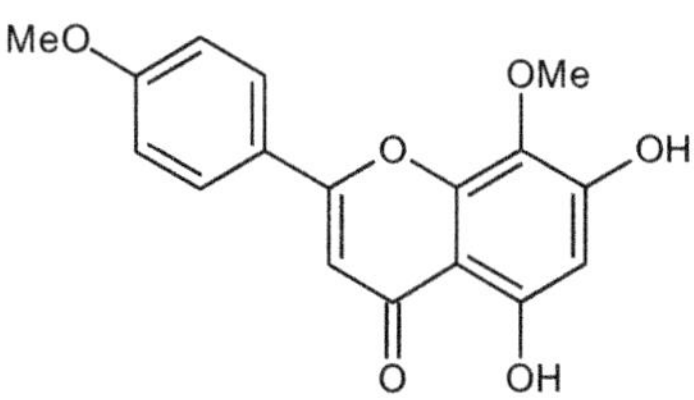

Dihydroquercetin-4 '-methyl ether

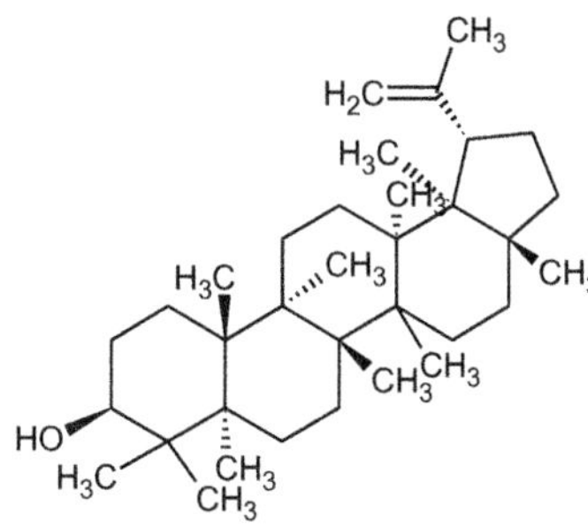

Lupeol

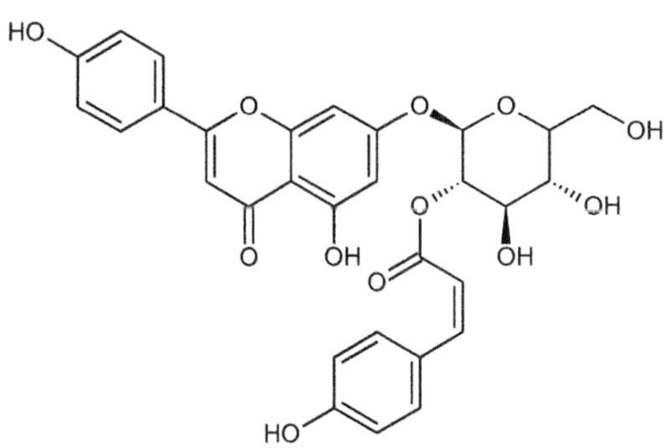

Echitin

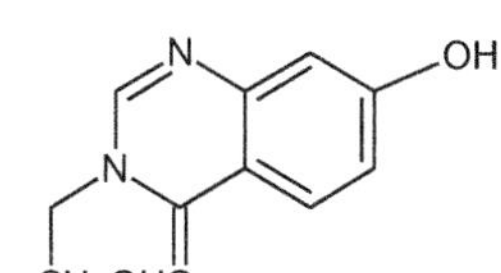

7-hydroxy-echinozolinone

FIGURE 31.2 Con'd

5. VALUE ADDITION

Natural food pigments are extracted from Indian globe thistle. It is used in chemical-free toothpaste and cosmetics.

6. USES

Most diseases have been cured by utilizing different plants and their products traditionally. Indian globe thistle is a common medicinal plant in Pakistan, India, and Sri Lanka, as each part is medicinally important. During the last 5 decades, apart from the chemical structure of the Indian globe thistle compounds, considerable progress has been made regarding the pharmaceutical applications and biologic activity of this plant. It is now considered a valuable source for the development of medicines against different diseases and also for the development of industrial products (Pranav et al., 2013). Indian globe thistle has an extended list of traditional medical applications. The plant is used in ophthalmia, as diuretic and nerve tonic, and to treat hysterica. This plant is also used against skin itching (Maurya et al., 2015). Root decoction is utilized for toothache (Jagtap et al., 2013). Indian globe thistle is used in the diseases of the brain, microbial infections (Desta, 1993), inflammations and pains in the joints. Root bark and roots of the plant are used in different aboriginal systems of medicine for treating various ailments. The root is used as aphrodisiac and abortifacient; infusion of the root is given in hysteria, impotence, seminal debility; and its decoction is given in scrofula, dyspepsia, fevers, and syphilis. Ethnomedicinal analysis revealed that the rural population of Kutch region in Gujarat state, India, utilized the suspension of powdered root bark in milk (100 g/250 mL) to cure diabetes. The traditional healers of Chhattisgarh in India apply this herb in distinctive ways both externally and internally and for treating sexual disorders. During delivery the paste of the root is applied to the lower abdominal region to accelerate the process. The paste is also used orally for safe and quick delivery. Aqueous paste of the root bark of Indian globe thistle is externally applied on the male genitals before intercourse in patients having poor sexual exuberance; in place of water, pure honey can also be utilized for making paste. The paste of Indian globe thistle is applied to the hair for 15 minutes to treat lice infestation (Maurya et al., 2015). Whole plant is used to treat jaundice (Imtiaz et al., 2013). During child birth, dried roots are tied directly on the hair of the lady for ease in delivery (Sharma, 2001). Indian globe thistle gained importance for its wound-healing activity (Sravanthi et al., 2010). The root paste is applied on wounds in cattle, two times a day for 3 days (Sikarwar & Kumar, 2005). Root is mixed with vinegar to make a tea for oral use. Juice of flowers is poured into eyes (Adnan et al., 2014). A mixture of root extract or leaf powder with honey is taken in the morning to expel round worms. For the treatment of leucorrhea the ash of the whole plant with ghee and butter is used (Ghatapanadi et al., 2011). The ash from spines of inflorescence is mixed with cow ghee and used locally for the treatment of eczema (Patil Sunil and Patil, 2012). Leaf paste is used externally for skin papules. People of Orissa used it to cure diarrhea, and in Maharashtra the decoction of the whole plant is used as a febrifuge. Indian globe thistle paste is smeared on the soles and palms for the treatment of heatstroke in Rajasthan. Asthma patients get symptomatic relief if the fumes of burned leaves of Indian globe thistle are inhaled. The root extract is effective in treating whooping cough and renal colic and also for the treatment of malarial fever (Maurya et al., 2015). This plant is used to treat liver disorders in Cholistan desert (Pakistan) and Gond tribes of Maharashtra and Bhandara (India). The root of Indian globe thistle is also used to relieve the pain of scorpion sting (Gayake et al., 2013). The spectral data of Indian globe thistle revealed the presence of a new alkaloid, echinozolinone: 3(2-hydroxyethyl)-4(3H)-quinazolinone. Quinazolinones are a class of drugs that act as a sedative or hypnotics that possess a 4-quiazolinone core. Their use has been proposed to treat cancer (Kadhim, 2013). The 50% ethanol extract of its root has been demonstrated to cause sperm antimotility (Reena et al., 2015), and it reduced sperm density in cauda epididymis (Banerjee et al., 2015). Whole plant and roots are used to treat inappropriate eating habits, indigestion, chronic liver diseases, and sexual weakness (Malik et al., 2015). The juice of seeds and roots is used to cure piles (Sharma et al., 2012). The ethanolic solution of the roots of Indian globe thistle is used during incision or excision (Arun et al., 2013). The plant is bitter, pungent, hot, and diuretic. It is used in hysteria, dyspepsia, and scrofula (Akbar and Fatima, 2012). Traditionally, Indian globe thistle root powder has been used as a general tonic. Root decoction is given in polyurea. Spermatorrhoea is cured by using root powder with milk (Hayat et al., 2008). The powder of its roots is also given to treat typhoid in India (Porte, 2014). Native people of Cholistan utilize this plant in hepatobiliary disorders. The herb is also reported to possess strong molluscidal, anthelmintic, vermicidal, antifertility, and antiinflammatory activities (Eram et al., 2013). About 10 pieces of roots are crushed and mixed in warm water and used for bathing to cure leprosy. It is also tied at the waist of a lady to assure fertilization (obviate abortion). The paste of its root is also applied on snakebite simultaneous to the juice being given orally. Its juice is also beneficial to cure bronchitis (Patel et al., 2012). Roots are used in skin diseases, acute mastitis, and hemorrhoids (Patil, 2008).

A paste prepared by mixing the root bark powder with the juice of *Blumea lacera* and *Datura stramonium* leaves is employed to avoid premature interjection syndrome. The fumes acquired after burning the roots and leaves of Indian globe thistle are very useful to treat patients suffering from respiratory troubles, specifically asthma for permanent and quick relief. This plant has also been narrated for different biologic activities including antiinflammatory, diuretic, hypoglycemic, antibacterial, antispasmodic, and antifungal properties (Murthy and Madhav, 2014). Researches also evaluated the analgesic activity from the methanolic extract of aerial parts and roots of Indian globe thistle (Patel et al., 2011a,b).

The seeds of this plant are aphrodisiac and sweet. The plant is stomachic, bitter, antipyretic, increases the appetite, stimulates the liver, analgesic, and is used in ophthalmia, pains in the joints, chronic fever, and inflammations (Pranav et al., 2013). It is useful as a cough suppressant, as well as being used in urinary disorder and ophthalmia. Fever and hoarse cough in children can be cured by using the roots of Indian globe thistle (Shaheen et al., 2014).

7. PHARMACOLOGICAL PROPERTIES

7.1 Antiulcer Activity

One of the most important capabilities of Indian globe thistle is its antiulcer activity. Two flavonoids, kaempferol and myricetin, are utilized to heal gastric ulcers. These polyphenolic compounds also possess anti-*Helicobacter pylori* activity and used as an additive and alternative agent in therapy. Thus researchers revealed that these flavonoids are less toxic and more efficient to cure gastrointestinal diseases, specifically peptic ulcers (Kadhim, 2013).

7.2 Antimicrobial Activity

Four phenolics have been separated from Indian globe thistle including apigenin, apigenin-7-O-glucoside, echinacin, and echinaticin. The methylation of echinacin and apigenin-7-O-glucoside permethylate gave the two derivatives, echinacin-permethyl ether and apigenin-5,4′-dimethyl ether, respectively. All these compounds were examined against conidia's germination of *Alternaria tenuissima* Wiltshire, which provokes leaf blight disease in pigeon pea (Cajanuscajan). Echinacin is highly active at 150^{-1} μg/mL and was considered the most effective in these compounds, and it was utilized as a control against *Alternaria* blight of pigeon pea (Qudsia et al., 2015).

7.3 Antiinflammatory Effects

Researchers conducted antiinflammatory studies on the ethanol extract of whole parts of Indian globe thistle. It hinders the acute inflammation by formaldehyde, carrageenan, and adjuvant and the chronic arthritis caused by adjuvant and formaldehyde. Indian globe thistle extract proved to be more advantageous to take parenterally than orally (Singh et al., 2008). Two antiinflammatory active compounds have been isolated from the leaves of Indian globe thistle, one is dihydroquercetin-4′-methyl ether and the other is a new flavanone glycoside known as 5,7–8,4-dimethoxyflavanone-5-O-α-L-rhamnopyranosyl-7-O-β-D-arabinopyranosyl-(14)-O-β-D glucopyranoside (Patel, 2012).

7.4 Cardiovascular Disease

Flavonoids are the main constituent of Indian globe thistle that play an important role in reducing the risk of cardiovascular diseases. A different mechanism takes place to decrease the incidence of stroke, including a decrease in low-density lipoproteins (LDL) oxidation by lipoxygenase suppression and devaluation of oxidative stress, inhibition of leucocyte–leucocyte adhesion, myeloperoxidase, reduced expression of inducible cyclooxygenase-2 and nitric oxide synthase, and by inhibiting platelet aggregation. The other factors included a reduction in oxidative stress, inhibition of nicotinamide adenine dinucleotide phosphate-oxidase, repression of metalloproteinase and vasodilatory properties, and recovery of nitric oxide owing to the inhibition of superoxide production. It was found previously that quercetin protected LDL against oxidative modifications effect, and in oxidative stress, it proved to be more protective (Kadhim, 2013).

7.5 Analgesic Activity

Methanolic extract of aerial parts and dried roots of Indian globe thistle contains alkaloids, flavonoids, carbohydrates, phenolic compounds, and tannins. It was found that high doses of methanolic extracts of the root and aerial parts of Indian globe thistle showed more analgesic activity than low dose of the root and aerial part methanolic extracts of Indian globe thistle. A high dose (500 mg/kg) of methanolic extract of aerial parts possesses significant analgesic activity compared to the standard analgesic drug, pentazocine. Indian globe thistle roots exhibited analgesic activity due to presence of phenolics, carbohydrates, and tannins, whereas presence of carbohydrates, alkaloids, and flavonoids are the main cause of analgesic activity of the aerial part of Indian globe thistle (Patel et al., 2011a,b).

7.6 Diuretic Activity

The methanolic extract of roots and aerial parts of Indian globe thistle diuretic capability was examined by Lipschitz test model. The extract exhibited significant diuretic activity. The low and high doses of the extracts of root and aerial parts of Indian globe thistle showed potent diuretic action by increasing the excretion of potassium and sodium salts. The high dose of the extracts of Indian globe thistle produced more diuretic potential than the low dose. Roots of Indian globe thistle contain phenolics, tannins, and carbohydrates. Due to these constituents, the high dose of the root extract depicts better diuretic activity compared to furosemide, a known diuretic drug. Diuretic activity of the aerial part of Indian globe thistle seems to be due to flavonoid, carbohydrate, and alkaloid contents (Patel et al., 2011a,b).

7.7 Antiandrogenic Activity

Many reports on Indian globe thistle indicate an antiandrogenic action for the plant and its utilization as a clinically effective medicine to cure benign prostatic hyperplasia. The root extract of Indian globe thistle with petroleum ether was investigated on male reproductive organs. The studies were carried out on two dose levels of 30 and 60 mg/kg body weight. Terpenoids present in the root extract depicted a reduction in a relative weight of the organs without affecting the final body weight and showed a decrease ($P < .01$) in the level of serum and the concentration of cauda epididymal sperm. The effect of this specie on prostate production plays a vital role in the development of new contraceptive modalities for males (Kadhim, 2013). The extract of Indian globe thistle has been demonstrated to decrease the rise of the prostatic per body weight ratio in which testosterone plays an important role. The butanol fraction of extract showed good activity. The levels of testosterone and prostate-specific antigen were monitored (Qudsia et al., 2015).

7.8 Hepaprotective Activity

The hepatoprotective activity of Indian globe thistle was tested and cited by many researchers. Flavonoids present in this plant bind to the subunit of DNA-dependent RNA polymerase I, and then trigger the enzyme. These phytocompounds increased the protein synthesis that leads to regeneration and produces hepatocytes. Apigenin, silymarin, kaempferol, and quercetin were revealed as potential therapeutic agents against microcrystin LR-induced hepatotoxicity. Myricetin and rutin were reported to exhibit hepatoprotective effects in experimental cirrhosis (Kadhim, 2013). The extract of aerial parts of Indian globe thistle showed the hepatoprotective effect on CCl_4-induced liver injuries through prevention of process of lipid per oxidation, downregulation of *CYP2E1* gene expression, and free radical scavengers (Eram et al., 2013).

7.9 Antidiabetic Activity

Steroidal glycoalkaloids and isobutylamide isolated from the roots of Indian globe thistle was found effective for treating diabetes mellitus (Pranav et al., 2013).

7.10 Antiirritant Activity

Indian globe thistle chloroform extract was reported to have anti-irritant activity. These fractions were tried on irritated and abraded skin of rabbits, and effective results were obtained (Maurya et al., 2015).

7.11 Antipyretic Activity

The ethanolic extract of Indian globe thistle was found to have anti-pyretic activity. This plant was used to reduce *Esherichia coli* lysate-induced pyrexia in rabbits (Maurya et al., 2015). This activity was observed at 500 mg/kg and 750 mg/kg of extract of Indian globe thistle, but was less than the positive control (Maurya et al., 2015).

7.12 Antibacterial Activity

A mother tincture of Indian globe thistle plant was tested for anti-bacterial activity against many microorganisms previously (Maurya et al., 2015; Ahmad et al., 2012).

7.13 Wound-Healing Activity

In incision and dead space models, extract of Indian globe thistle in chloroform, petroleum ether, ethanol, and distilled water was reported to have wound-healing activity (Maurya et al., 2015).

7.14 Antioxidant Activity

Indian globe thistle has a free radical-scavenging property described in many in vitro models, which includes scavenging of nitric oxide radical, superoxide anion, and radical of 2,2 diphenyl-1-picrylhydrazyl (Maurya et al., 2015; Rudrappa and Mohmoud, 2010).

8. SIDE EFFECTS AND TOXICITY

For most adults, Indian globe thistle is likely safe when taken by mouth. In some people, taking milk thistle extract can cause nausea, diarrhea, intestinal gas, loss of appetite, fullness or pain, and possibly headache. Indian globe thistle application to the skin is possibly safe for short periods of time. Reliable information regarding injection of Indian globe thistle into the body does not exist. Not much is known about use of Indian globe thistle during pregnancy and breastfeeding. Indian globe thistle may cause an allergic reaction in patients who are sensitive to the Asteraceae/Compositae plant family. Members of this family include chrysanthemums, ragweed, daisies, marigolds, and many others. Indian globe thistle should be employed in well-monitored doses during diabetes, as it contains certain chemicals that might lower blood sugar. Indian globe thistle might act like estrogen and could create hormone-sensitive conditions such as uterine cancer, breast cancer, endometriosis, ovarian cancer, or uterine fibroids worse.

References

Adnan, M., Ullah, I., Tariq, A., Murad, W., Azizullah, A., Khan, A.L., Ali, N., 2014. Ethnomedicine use in the war affected region of Northwest Pakistan. Journal of Ethnobiology and Ethnomedicine 10, 1.

Ahmad, M., Ghafoor, N., Aamir, M., 2012. Antibacterial activity of mother tinctures of cholistan desert plants in Pakistan. Indian Journal of Pharmaceutical Sciences 74, 465.

Akbar, G., Fatima, S., 2012. Floral Guide for Indus Ecoregion. WWF, Karachi, Pakistan.

Arun, M., Satish, S., Anima, P., 2013. Herbal boon for wounds. International Journal of Pharmacy and Pharmaceutical Sciences 6, 1–12.

Banerjee, M., Hazra, A., Bharitkar, Y.P., Mondal, N.B., 2015. Insights of spermicidal research: an update. Journal of Fertilization: In Vitro-IVF-Worldwide 2015.

Barreda, V.D., Palazzesi, L., Tellería, M.C., Olivero, E.B., Raine, J.I., Forest, F., 2015. Early evolution of the angiosperm clade Asteraceae in the Cretaceous of Antarctica. Proceedings of the National Academy of Sciences 112, 10989–10994.

Desta, B., 1993. Ethiopian traditional herbal drugs. Part II: antimicrobial activity of 63 medicinal plants. Journal of Ethnopharmacology 39, 129–139.

Dhayalan, M., Jegadeeshwari, A., Gandhi, N., 2015. Biological activity sources from traditionally used tribe and herbal plants material. Asian Journal of Pharmaceutical and Clinical Research 8, 11–23.

Dymock, W., 1893. Pharmacographia Indica.

Eram, S., Ahmad, M., Arshad, S., 2013. Experimental evaluation of *Echinops echinatus* as an effective hepatoprotective. Scientific Research and Essays 8, 1919–1923.

Erenler, R., Yilmaz, S., Aksit, H., Sen, O., Genc, N., Elmastas, M., Demirtas, I., 2014. Antioxidant activities of chemical constituents isolated from *Echinops orientalis* Trauv. Records of Natural Products 8, 32–36.

Gayake, D., Awasarkar, U., Sharma, P., 2013. Indigenous traditional medicinal plant resources from Ahmednagar district, Maharastra, India. Asian Journal of Biomedical and Pharmaceutical Sciences 3, 1.

Ghatapanadi, S., Johnson, N., Rajasab, A., 2011. Documentation of folk knowledge on medicinal plants of Gulbarga district, Karnataka. Indian Journal of Traditional Knowledge 10, 349–353.

Hayat, M.Q., Khan, M.A., Ahmad, M., Shaheen, N., Yasmin, G., Akhter, S., 2008. Ethnotaxonomical approach in the identification of useful medicinal flora of tehsil Pindigheb (District Attock) Pakistan. Ethnobotany Research and Applications 6, 035–062.

Imtiaz, S., Abdullah, S., Afzal, S., Rehman, G., Waheed, M., 2013. Medicinal plants used by traditional healers of Punjab, Pakistan. Canadian Journal of Applied Sciences 3, 496–515.

Jagtap, D., Patil, H., Jakhi, P., 2013. Ethno-medicinal survey of some plants from villages of Khatav Tahashil (MS) India. International Journal of Life Sciences 1, 264–269.

Kadhim, E.J., 2013. In: Phytochemical Investigation and Testing the Effect of Iraqi Echinops Heterophyllus Family Compositae on Wound Healing. Department of Pharmacognosy, University of Baghdad, Baghdad, p. 120.

Khan, R.U., Mehmood, S., Khan, S.U., Subhan, M., 2013. Ethnobotanical study of common weed flora of sugarcane in district Bannu, Khyber Pakhtunkhawa, Pakistan. Journal of Medicinal Plants Studies 1, 49–78.

Kumar, A., Aggarwal, S.G., 2014. Ecology and biodiversity status of Sachin Gidc and its surroundings with special reference to conservation measures for Indian Peafowl (*Pavo cristatus*) schedule-I bird species. Octa Journal of Environmental Research 2.

Malik, S., Ahmad, S., Sadiq, A., Alam, K., Wariss, H.M., Ahmad, I., Hayat, M.Q., Anjum, S., Mukhtar, M., 2015. A comparative ethno-botanical study of Cholistan (an arid area) and Pothwar (a semi-arid area) of Pakistan for traditional medicines. Journal of Ethnobiology and Ethnomedicine 11, 1.

Maurya, S.K., Kushwaha, A.K., Seth, A., 2015. Ethnomedicinal review of usnakantaka (*Echinops echinatus* Roxb.). Pharmacognosy Reviews 9, 149.

Murthy, E.N., Madhav, N.V., 2014. Ramagiri, Enumeration of Medicinal Plants of Forests. Hill, Department of Botany, p. 120.

Patel, A.J., 2012. Pharmacognostical Phytochemical and Pharmacological Investigation of Echinops Echinatus Roxb.

Patel, A.J., Patel, N.M., Patel, A.A., Patel, J., Patel, S., 2011a. Comparative diuretic activity of root and aerial part methanolic extracts of *Echinops echinatus* Roxb. Der Pharmacia Lettre 3, 168–172.

Patel, A.J., Patel, N.M., Patel, A.A., Patel, J., Patel, S., 2011b. Comparative analgesic activity of root and aerial part methanolic extracts of *Echinops echinatus* Roxb. International Journal of Pharmaceutical Innovation 1, 23–29.

Patel, K.C., Patel, R.S., Reddy, A.S., 2012. Ethnobotanical practices of some plant species of Danta and ambaji forest areas of North Gujarat (India). International Journal of Life Sciences Leaflets 192–199.

Patil, D.A., 2008. Herbal Cures Traditional Approach. Aavishkar Publishers, India, Jaipur. Distributors.

Patil, S.J., Patil, H.M., 2012. Ethnomedicinal Herbal Recipes from Satpura Hill Ranges of Shirpur Tahsil, Dhule, Maharashtra, India. Res J Recent Sci 1, 333–336.

Porte, S.M., 2014. Overview of folk medicine used for typhoid in India. International Journal of Research in Ayurveda and Pharmacy 5, 219–224.

Pranav, V., Vinay, J., Priya, M., Sudhir, B., Neha, A., Love, C., Indu, S., Rajkumar, S., Harsha, M., 2013. *Echinops echinatus* roxb. -A Nature's Drugstore: an overview. American Journal of PharmTech Research 3.

Qudsia, B., Wahid, M., Irfan, M., Chaurasiya, V., Iqbal, I., 2015. Ethno-pharmacological and phytochemical constituents review of *Echinops echinatus Roxb.* Journal of Applied Pharmacy 7, 69–74.

Rahman, A., Alam, M., Khan, S., Ahmed, F., Islam, A., Rahman, M.M., 2008. Taxonomic studies on the family Asteraceae (Compositae) of the rajshahi division. Research Journal of Agriculture and Biological Sciences 4, 134–140.

Reena, B., Sushma, K., Ashok, K., 2015. Screening of 166 antifertility medicinal plants. Review International Journal of Institutional Pharmacy and Life Sciences 162–196.

Rudrappa, J., Mohmoud, R., 2010. Free Radical Scavenging Activity of *Echinops echinatus* Roxb. Root.

Shaheen, H., Qureshi, R., Akram, A., Gulfraz, M., 2014. Inventory of medicinal flora from Thal desert, Punjab, Pakistan. African Journal of Traditional, Complementary and Alternative Medicines 11, 282–290.

Sharma, N.K., 2001. Ethanomedicine of Gadia Lohars of Rajhastan. Zoo's Print Journal 16, 593–594.

Sharma, J., Gairola, S., Gaur, R., Painuli, R., 2012. The treatment of jaundice with medicinal plants in indigenous communities of the Sub-Himalayan region of Uttarakhand, India. Journal of Ethnopharmacology 143, 262–291.

Sikarwar, R., Kumar, V., 2005. Ethnoveterinary knowledge and practices prevalent among the tribals of Central India. Journal of Natural Remedies 5, 147–152.

Singh, A., Malhotra, S., Subban, R., 2008. Anti-inflammatory and analgesic agents from Indian medicinal plants. International Journal of Integrative Biology 3, 57–72.

Sravanthi, K.C., Sarvani, M., Srilakshmi, S., Ashajyothi, V., 2010. Wound healing herbs-a review. International Journal of Pharmacy and Technology 2, 603–624.

Wichtl, M., 2004. Herbal Drugs and Phytopharmaceuticals: A Handbook for Practice on a Scientific Basis. Medpharm GmbH Scientific Publishers.

Further Reading

Afzal, S., Afzal, N., Awan, M.R., Khan, T.S., Gilani, A., Khanum, R., Tariq, S., 2009. Ethnobotanical studies from Northern Pakistan. Journal of Ayub Medical College, Abbottabad 21, 52–57.

Bisht, V., Purohit, V., 2010. Medicinal and aromatic plants diversity of asteraceae in Uttarakhand. Natural Science 8, 121–128.

Odelu, G., 2015. Preliminary studies on medicinal plants of huzurabad division, Karimnagar district, Telangana, India. International Journal of Innovative Research in Science, Engineering and Technology 4, 4483–4492.

CHAPTER

32

Indian Pennywort

Shahzia Yousaf[1], *Muhammad Asif Hanif*[1], *Rafia Rehman*[1], *Muhammad Waqar Azeem*[1], *Anca Racoti*[2]

[1] Department of Chemistry, University of Agriculture, Faisalabad, Pakistan; [2] The National Institute for Research & Development in Chemistry and Petrochemistry – ICECHIM, Bucharest, Romania

OUTLINE

Medicinal Plants of South Asia
https://doi.org/10.1016/B978-0-08-102659-5.00032-X

Copyright © 2020 Elsevier Ltd. All rights reserved.

1. BOTANY

1.1 Introduction

Indian pennywort *(Centella asiatica)* (Fig. 32.1) is a perennial, clonal, herbaceous plant belonging to *Umbelliferae* (*Apiaceae*) family that is found all over India, growing in damp areas up to an altitude of 1800 m. It occurs commonly in subtropical and tropical countries, including parts of Pakistan, India, Madagascar, Sri Lanka, Eastern Europe, South Africa, and

FIGURE 32.1 Dried Indian pennywort.

the South Pacific. Twenty species that are closely resembled to *C. asiatica* propagate in most parts of the tropic or wet pantropical parts like rice paddies, also in rocky places and higher altitudes (Bown, 1995). It is a tasteless, weakly scented plant that more frequently flourishes in water-rich places. Its leaves are green, small, fan-shaped with white or light pink to purple flowers, and it has small, elliptical fruit. For medicinal purposes the entire plant is used. It is mostly used for blood purification as well as treating memory improvement (intellect), high blood pressure, and other similar problems. In Ayurveda, *C. asiatica* is one of the main herbs for activating the nerves and brain cells. To treat emotional disorders, such as depression, hakims use *C. asiatica*. In Western medicine, in the mid-20th century, *C. asiatica* and its alcohol extracts were used in the treatment of leprosy (Duggina et al., 2015). *C. asiatica* was confused earlier with *Bacopa monnieri* in the Indian market because both were sold by the "Brahmi" name. Later on, the problem was solved by giving a separate name: brahmi for *B. monnieri* and mandookaparni for *C. Asiatic*. Due to extensive use at a large scale along with little cultivation, wild species of the plant are disappearing from sight, and attempts have not been made for its revival (Larsen and Olsen, 2007). Due to the same reason, International Union for Conservation of Nature and Natural Resources enlisted it as a rare and threatened plant species. Region-wise common names are gotu kola in Urdu and thankuni in Bengali, mandookparni in Hindi, pegaga in Malay, bekaparanamu in Telgu, vallarai in Tamil, and kodagam in Malayalam.

1.2 History/Origin

C. asiatica (Linn), native to Sri Lanka, South Africa, Madagascar, and Malaysia, has been used since ancient times by tribal groups and cultures as a medicinal herb. For about 2000 years, *C. asiatic* has been a part of Chinese herbal medicines and 3000 years in Indian Ayurvedic system of medicines (Gnanapragasam et al., 2004). In Ayurvedic medicines, mandukaparni is used for the treatment of skin diseases, leprosy, gastric catarrh, elephantiasis, kidney troubles, asthma, bronchitis, and leucorrhoea, while in Chinese herbal medicine, it is used for curing toxic fever and leucorrhoea. In Malaysia, *C. Asiatic* is used to cure anxiety, eczema, and mental fatigue. It is also eaten as salad when fresh (Goh et al., 1995). Fresh plant extracts have been used for many years by the people of Malay Peninsula and Java for healing wounds, internally and externally. The plant extracts are also an active part of brain tonics for mental impairment. The presence of pentacyclic triterpenes, madecassoside, madecassic acid, asiaticoside, and asiatic acid is responsible for these properties (Jamil et al., 2007).

1.3 Location/Demography

C. asiatic grows all over the tropical and subtropical areas of India up to 600 m altitude. It is also found at an altitude of 1200 m in Mount Abu (Rajasthan, India) and at even more altitude of 1550 m in Sikkim (north-eastern Indian state). Sandy loam (60%) type of soil proved to be the utmost choice for regeneration of it, rather than clayey soil (Devkota and Jha, 2009). In the wild, *Centella* plant is found in wet or moist soils of marshes, swamp, bogs, and along the margins of ponds, lakes, irrigation, streams, and drainage canals of irrigated paddy fields. It is also found in wet pine savannas, flat woods, and palmetto flats, often forming meadows. It grows in water or on land (Hamid et al., 2002). The species is pantropic, found in the United States from Delaware to southern Florida, in West Indies, Mexico, central and southern America, Australia, Sri Lanka, and the Philippines.

1.4 Botany, Morphology, Ecology

C. asiatic (L.) is a flat, creeping, stoloniferous, perennial herb, faintly aromatic that grows up to 15 cm height. Stem is glabrous, striated, rooting at nodes. Each node has one to three leaves with a width of 1.5–5 cm, petiole length 2–6 cm, leaves have crenate margins and both sides glabrous covering base leaf (George and Joseph, 2009). Flower coloring is white to purple or pink with three to four flowers at each umbel. April to June are the flowering months, and plant fruits throughout the growth. Fruits attain the length of 2 in. in spherical or rectangular shape with intensely thickened pericarp (Shukla et al., 1999). Seed embryo is horizontally compressed and pendulous. Flowers are in fascicled umbels, each umbel consisting of three to four white to purple or pink flowers. *C. asiatica* is found abundantly in secondary succession communities, but it can grow wild under a varied range of climatic conditions. There is a variety of factors that affect germination of plants, including the condition of the plant, size, the temperature, and duration of exposure to light. The smaller and more succulent the plant, the greater the vulnerability is to death or damage from temperatures that are too low or too high.

2. CHEMISTRY

C. asiatica is an aromatic plant that has a sweet taste and major pentacyclic triterpenoids including asiatic acid, brahmic acid, or madecassic acid, asiaticoside, and brahmoside. Other products include centelloside, centellose, and madecassoside (Schaneberg et al., 2003). The main chemical constituents of Indian pennywort are shown in Fig. 32.2.

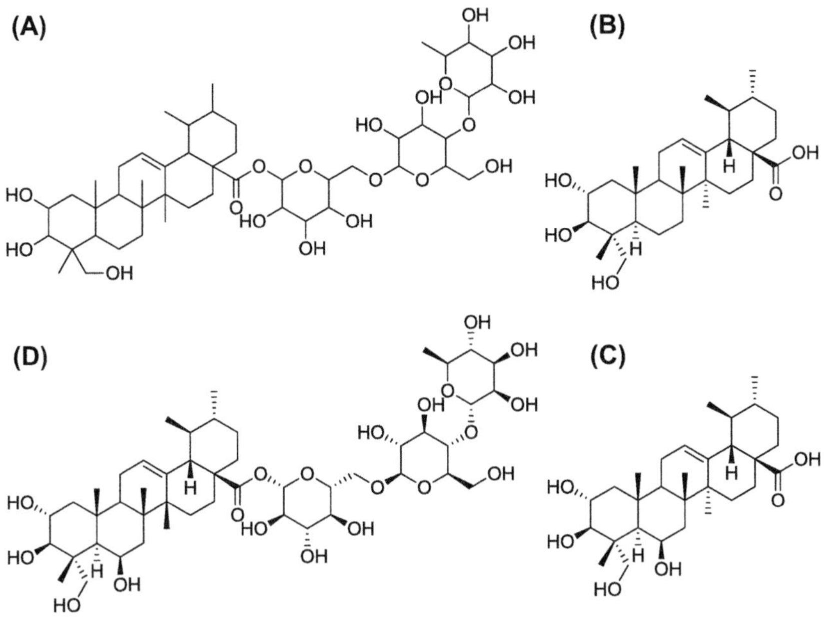

FIGURE 32.2 Main chemical constituents of Indian pennywort. (A) Asiaticoside. (B) Asiatic Acid. (C) Madecassoside. (D) Madecassic Acid.

A diverse range of biochemical compounds or secondary metabolites are present in *C. asiatic*, as witnessed by the scientific investigations. Due to the presence of biologically active components of triterpenes saponin, *Centella* plant plays a vital role in nutraceuticals and medicines (Plohmann et al., 1997). The triterpenes of *Centella* are made up of various compounds including asiaticoside (Fig. 32.2A), asiatic acid (Fig. 32.2B), brahmocide (Fig. 32.2C), madecassic acid (Fig. 32.2D), brahminoside, brahmic acid, centic acid, centelloside, centellic acid, thankiniside, and isothankunisode (Brinkhaus et al., 2000). Madecassic acid, madecassoside, asiatic acid, and asiaticoside are most important active triterpenes among them, biologically (Jayathirtha and Mishra, 2004). Moreover, it contains total phenolic contents (TPC) due to the presence of catechin, naringin, apigenin, rutin, quercetin, kaempferol, and volatile oils like farnesol and caryophyllene. *Centella* is also a rich source of vitamins A, B_1, B_2, and C, and niacin and carotene. The total ash contains iron, sodium, magnesium, calcium, phosphate, sulfate, and chloride (Zheng, 1989). A hundred grams of *Centella* contained 37 Kcal energy, 391 mg of potassium, 171 mg of calcium, and 2 g of protein (Thomas et al., 2010).

Based on its large medicinal uses, many phytochemical studies have been carried out on *C. asiatica*. Terpenoids, asiaticoside, and glucosides separated from plant were more active in treatment of leprosy. The primary chemical description of terpenoids fraction comprises the prime biologic active components (Tholon et al., 2002). The chemical composition of *Centella* plant plays a very important role in medicinal and nutraceutical applications, due to biologically active components of triterpenes saponins (Biradar and Rachetti, 2013). The presence of high concentrations of phytochemicals is found in leaves, compared to the petiole and roots. The most important biologically active compounds are madecassic acid, madecassosides, asiatic acid, and asiaticoside among these phytochemicals. These phytochemicals are the biomarkers for the quality of *C. asiatica*. Phytochemicals are found in higher concentrations in the leaves relative to the roots and petioles. Among these phytochemicals, triterpenes, the most significant biologically active compounds are the asiatic acid, madecassic acid, asiaticoside, and madecassoside. They have been used as the biomarker components for quality assessment of *Centella* (James et al., 2008). Moreover, it contains TPC due to the presence of catechin, naringin, apigenin, rutin, quercetin, kaempferol, and volatile oils like farnesol and caryophyllene (Thomas et al., 2010). Out of 15 different variants of *Centella,* found in Malaysia, there are only three recognized triterpenes (madecassoside, asiaticoside, and asiatic acid). The concentration of these triterpenes varies with conditions and environment in which it grows (James and Dubery, 2009).

3. POSTHARVEST

Leaves of gotu kola can be picked during the summer months. The best harvesting time is the morning, because at higher temperatures, the essential oil content decreases due to evaporation, as essential oils have some very volatile compounds in their composition. Flowers are harvested in months of April to June. The fresh leaves have flavor complexity, and aroma is lost in dried leaves due to evaporation of volatiles. On drying, flowers and leaves should not be broken; otherwise, flavor will be reduced due to loss of essential oil.

4. PROCESSING

Gotu kola (*C. asiatica*), like other herbs, is used in different ways for a variety of purposes. Its fresh leaves are used as a salad. In addition, dry leaves, powdered leaves, and essential oil are processed. Chopped leaves and the entire plant can be stored frozen in the dark with or without oil. It is dried in a dry medium or shade, to prevent oxidation or evaporation of essential oil. To avoid the loss of volatile compounds, the drying temperature should not exceed 40°C. The essential oil extracted from flowers is of good quality, and oil extracted from leaves is of poor quality. Hydro distillation process is used for extraction of essential oil from freshly harvested herb plant leaves (Kumar and Gupta, 2002).

5. VALUE ADDITION

The plant's young leaves can be used both as a salad and sandwiches for its spicy flavor and as a cooked food. It is also used in curries, soups, stir fries, and as a taste enhancer with fish and vegetables. Leaves are used in a decoction of milk and infusions in dried or fresh form. Gotu kola is utilized in Sri Lankan cuisine (Sastravaha et al., 2005). *Centella* is supposed to be a nutritious plant. Malluma contains coconut and gotu kola along with chili powder, turmeric powder, lemon juice, and crushed green chilis (Babu et al., 1995). In Sri Lanka, a nutritious rich porridge called kola kenda is used by the people. This porridge is prepared with boiled red rice fluid, crushed gotu kola, and coconut milk. Pennywort, a sweet drink, uses *Centella* leaves. Leaves are also used in Thailand and Vietnam as a salad and in drinks (Wang et al., 2003).

6. USES

C. asiatica is known as a common medicinal plant in different medicinal systems. In the Indian system of medicine called Ayurveda, it is used in compound formulations for curing gastrointestinal disorders and the diseases of the central nervous system and skin (Sharma and Kumar, 1998). It is one of the ingredients of the Indian summer drink thandaayyee. In Bangladesh, mashed *Centella* is eaten with rice. Dried leaves of *Centella* are used in tea and are recommended at a 0.33–0.68 g dosage three times a day. There is no evidence found for its side effects when staying within recommended limits. In oral application, asiaticoside 1 g/kg body weight has good tolerance and is nontoxic. It has a sweet aroma, so it is used in cosmetics (Pragada et al., 2004). Many investigators have described different biologic actions of *C. asiatica*. These activities are wound healing, antidepressant, antiaging, anticancer, etc. Madecassol, an extract of the plant having the constituents of asiatic acid, madecassic acid, and asiaticoside accelerates healing of grafted wounds by promoting fibroblast proliferation and extracellular matrix synthesis (Pandey et al., 1993).

7. PHARMACOLOGICAL USES

7.1 Antitumor Activity

Crude extract of *C. asiatica*, orally administered in mice, and its partly purified fractions produced apoptosis in solid tumors and improved the life span of these tumor-bearing mice. Asiatic acid has anticancer activity against skin cancer (Gupta et al., 2003).

7.2 Memory Enhancing Effects

An aqueous extract of this herb had major effects on enhancing memory and learning and reduced the levels of 5-hydroxytryptamine, dopamine, and norepinephrine. It also decreased the metabolites of these chemicals of neurotransmitter in brain (Nalini et al., 1992). *C. asiatica* is composed of brahmoside, isobrahmic acid, brahmic acid, and brahminoside, which has anticonvulsant, psychotropic, and sedative properties. It is beneficial in anxiety, dementia, and mental disorders too. Thus, in mental construction, the entire herb in a synergistic way produces the enhancement of attention, memory, and awareness in children who have a learning disability (Saha et al., 2002).

7.3 Cardioprotective Effects

The alcoholic extract of the entire plant exhibited strong cardioprotective action in controlling ischemia reperfusion–induced myocardial infraction in rats (Pragada et al., 2004).

7.4 Immunostimulating Effects

Immunomodulating pectin extracted from *C. asiatica* possesses triterpenoids saponins and methanol extracts that have initial immunomodulatory influence. Alcoholic extract of the whole plant exhibited antiprotozoal lactenin contradiction of *Entamoeba histolytica* (Dhar et al., 1968).

7.5 Mental Retardation Effects

Mentally retarded children who were orally given tablets of *C. asiatica* showed a substantial growth in behavior patterns and general ability (Dhanasekaran et al., 2009).

7.6 Antitubercular and Antileprotic Activity

In the treatment of different kinds of tuberculosis and leprosy, asiaticoside is very useful (Rao et al., 2005). Clinical trials performed on normal adult mice exposed that the drug improved the level of serum cholesterol, red blood cells, total protein, blood sugar, and blood urea. It maintains the central nervous system and has a soothing effect on the body (Appa Rao et al., 1967).

7.7 Wound Healing Effects

Asiaticoside in *C. asiatica* has wound curative ability by increasing angiogenesis and collagen formation. The asiaticoside enhances the stretching strength of the newly formed skin and promotes the wound's healing. It also prevents the inflammatory process, which may increase the capillary permeability and provoke hypertrophy in abrasions (Incandela et al., 2001). In the laboratory, antioxidant levels were observed on one animal to study the effects of asiaticoside, and the antioxidants played a key role in the healing of injuries (Shukla et al., 1999). The investigator studied the effects of asiaticoside on delayed as well as normal wound healing. In studies, it was observed that topical uses of 0.2% solution of asiaticoside resulted in an increase in collagen content, 57% in tensile strength, 56% in hydroxyproline, and well epithelization in guinea pig punch wounds. But healing is delayed in streptozotocin diabetic rats, so topical application of 0.4% solution of asiaticoside on punch injuries increased tensile strength, epithelization, hydroxyproline, and collagen contents there by smoothing the healing. Oral and topical administration of an alcoholic extract of *C. asiatica* was studied in the laboratory. Asiaticoside was more active orally at a 1 mg/kg dose in the guinea pig punch and 40 μg/disk in chick wound. The wounds treated by extract were observed to epithelialize quicker and the speed of wound contraction was greater, as compared to the control. These results showed that *C. asiatica* produced different actions on the different stages of scar repair in normal as well as delayed healing models (Shukla et al., 1999).

7.8 Effects on Venous Insufficiency

It was postulated that *C. asiatica* helps in connective tissue maintenance by strengthening the weakened veins. In the cure of scleroderma, it can also assist in soothing connective tissue development, decreasing its formation by stimulating the formation of chondroitin sulfate and hyaluronidase. *C. asiatica* acts on the vascular wall of the connective tissues, being active in venous insufficiency and hypertensive microangiopathy. It also reduces the rate of capillary filtration by refining microcirculatory parameters (Cesarone et al., 1992).

7.9 Sedative and Anxiolytic Properties

The effect of *C. asiatica* on the central nervous system was studied in the Indian literature for sedative, rejuvenant, stimulatory-nervine tonic, and intelligence- and tranquilizer-enhancing properties (Kumar and Gupta, 2002). It has been used conventionally as a sedative mediator in many Eastern cultures; the effect is primarily due to the brahminoside and brahmoside components. The anxiolytic activity is due to binding to cholecystokinin receptors, a group of gastrin protein joined receptors that fix the gastrin or the peptide hormones cholecystokinin. These receptors play a vital role in modulation of hunger, nociception, anxiety, and memory in humans and animals (Sairam et al., 2001).

7.10 Antidepressant Properties

The antidepressant activities of entire triterpenes from *C. asiatica* on the immobility periods in forced swimming mice and amount of amino acid in brain tissue of mice was detected (Chen et al., 2005; Gohil et al., 2010; Young and Jewell, 1996).

7.11 Antiepileptic Properties

Asian *C. asiatica* raises the cerebral levels of gamma aminobutyric acid, which is traditionally used as an anticonvulsant and anxiolytic. The steroids extracted from the plant are used for the treatment of leprosy. It also reduces the formation of spontaneous motor activity, lipid per oxidation products, hypothermia, potentiation in diazepam withdrawal-induced hyperactivity, and potentiation of pentobarbitone resting instant. The extract (200 mg/kg body weight) totally blocks pentylenetetrazol-induced convulsions. These conclusions suggest its probability as a central nervous system depressant, as well as antioxidant and anticonvulsant actions (Gnanapragasam et al., 2004).

7.12 Cognitive and Antioxidant Properties

C. asiatica is recognized to increase the nervous system, brain, attention span, concentration, and to combat aging (Brinkhaus et al., 2000). An investigation showed *Centella* has cognitive-enhancing and antioxidant properties in normal rats. Aqueous extracts of *C. asiatica* (300, 200, and 100 mg/kg for 21 days) were analyzed for their effect in intra-cerebroventricular, oxidative stress and streptozotocin-induced cognitive injury in rats (Kumar and Gupta, 2002). The rats cured with *C. asiatica* had a dose-dependent rise in plus-maze paradigms, cognitive behavior, and passive avoidance. To give more information about the mechanism of this neuroprotection by *C. asiatica*, one study described that the phosphorylation of cyclic adenosine mono phosphate (AMP) response constituent binding protein was increased in both a neuroblastoma cell line showing amyloid beta and embryonic cortical prime cell culture in rats (Xu et al., 2008). Furthermore, the involvement of two major constituents to the enhanced phosphorylation was studied. In one more study, oral treatment with 50 mg/kg/day of rough methanol extract of *C. asiatica* for 14 days meaningfully improved the antioxidant enzymes, like catalase, superoxide dismutase, and glutathione peroxidase in lymphoma-bearing mice. The antioxidants like ascorbic acid and glutathione were reduced in the animals. Derivatives of asiatic acid exert the main neuroprotective effects on cultured cortical cells by defense mechanism. Therefore, these components were proved to be effective in defending neurons from the oxidative harm produced by exposure to excess glutamate (Lee et al., 1999).

7.13 Effects on Gastric Ulcer

C. asiatica aqueous extract was found effective in curing gastric injuries caused by ethanol intake (Cheng and Koo, 2000). It also strengthened the gastric mucosal barrier and reduced free radical harmful effects. Animal studies in rats showed antiulcer prevention induced by cold and resistant stress. This property was associated with famotidine and sodium valproate. The reduction in gastric ulceration depends on the dose of herb extract or drug (Chatterjee et al., 1992). It was suggested that *C. asiatica* extract increased gamma-aminobutyric acid levels in the brain and made rats resistant to cold restraint ulceration (Zivkovic et al., 1982). In another study, *C. asiatic* fresh juice was tested in rats for its antiulcer activity against ethanol, cold restraint stress, aspirin, and pyrrolic ligation–induced gastric ulcer. The oral dose of 200 and 600 mg/kg twice a day for 5 days showed substantial protection against said experimental ulcer models, and results were comparable with those elicited by sucralfate. *C. asiatica* extracts showed little or no effect on offensive acid pepsin secretion. However, at 600 mg/kg, it significantly increased gastric juice mucin secretion and increased the mucosal cell glycoproteins, signifying an increase in cellular mucus (Sairam et al., 2001). One study presented that *C. asiatica* and its elements, asiaticosides, have an antiinflammatory property that carried out inhibition of nitric oxide and consequently assisted ulcer healing (Guo et al., 2004). Some other investigators also displayed the efficiency of *C. asiatica* by clinical and preclinical studies for curing gastric ulcers (Chen et al., 2005). *C. asiatica* has also been examined to prove its part in periodontal therapy (Sastravaha et al., 2005).

7.14 Antinociceptive and Antiinflammatory Properties

The effects of *C. asiatica* upon inflammation and pain in rodent models were described. The antinociceptive action of the aqueous *C. asiatica* extract (300, 100, 30, and 10 mg/kg) was determined by means of acetic acid–induced hot plate and writhing technique in mice, while the antiinflammatory activity of *C. asiatica* was calculated through prostaglandin E2-induced foot edema in rats (Somchit et al., 2004). The aqueous *C. asiatica* extract exposed important antinociceptive activity with both the models analogous to aspirin but is less effective than morphine and has more antiinflammatory activity similar to mefenamic acid (Bateman et al., 1998). Currently, antirheumatic arthritis influence of madecassoside in type II collagen-induced arthritis (CIA) in mice was conducted to examine the healing potential and mechanisms of madecassoside on CIA. Madecassoside dose (40, 20, and 10 mg/kg), orally directed from the day of the antigen test for 20 consecutive days, reduced strictness of the disease on mice (Liu et al., 2008).

7.15 Radioprotection Effects

It was proposed that *C. asiatica* might be valuable in inhibiting radiation-induced behavioral variations during clinical radiotherapy. The plant extracts were also used as a radioprotective at a sublethal dose of Co 60 gamma radiation (Sharma and Sharma, 2002). At 100 mg/kg dose, the survival period of the mice was significantly enlarged. Reduction in body weight of drug-treated group animals was significantly smaller in comparison with the animals that were only given radiation (Shobi and Goel, 2001).

7.16 Antidiabetic Activity

To check the antidiabetic effects of *C. asiatic*, methanolic and ethanolic extracts were prepared, and their antidiabetic action on the alloxan-induced diabetic rats was checked. Alcoholic extracts significantly reduced the blood glucose level (Emran et al., 2015).

8. SIDE EFFECTS AND TOXICITY

Indian pennywort is safe for pregnant women when applied on the skin. People suffering from liver diseases should not use Indian pennywort, as it enhances liver damage. Also, stop using this plant 2 weeks before surgery.

References

Appa Rao, M., Usha, S., Rajagopalan, S., Sarangan, R., 1967. Six Months Results of a Double Blind Trial to Study the Effect of Mandookaparni and Punarnava on Normal Adults.

Babu, T., Kuttan, G., Padikkala, J., 1995. Cytotoxic and anti-tumour properties of certain taxa of Umbelliferae with special reference to *Centella asiatica* (L.) urban. Journal of Ethnopharmacology 48, 53–57.

Bateman, J., Chapman, R., Simpson, D., 1998. Possible toxicity of herbal remedies. Scottish Medical Journal 43, 7–15.

Biradar, S.R., Rachetti, B.D., 2013. Extraction of some secondary metabolites & thin layer chromatography from different parts of *Centella asiatica* L.(URB). American Journal of Life Sciences 1, 243–247.

Bown, D., 1995. Encyclopedia of Herbs and Their Uses, vol. 165. Dorling Kindersley Limited, New York, p. 289.

Brinkhaus, B., Lindner, M., Schuppan, D., Hahn, E., 2000. Chemical, pharmacological and clinical profile of the East Asian medical plant *Centella aslatica*. Phytomedicine 7, 427–448.

Cesarone, M., Laurora, G., De Sanctis, M., Belcaro, G., 1992. Activity of *Centella asiatica* in venous insufficiency. Minerva Cardioangiologica 40, 137–143.

Chatterjee, T., Chakraborty, A., Pathak, M., Sengupta, G., 1992. Effects of plant extract *Centella asiatica* (Linn.) on cold restraint stress ulcer in rats. Indian Journal of Experimental Biology 30, 889–891.

Chen, Y., Han, T., Rui, Y., Yin, M., Qin, L., Zheng, H., 2005. Effects of total triterpenes of *Centella asiatica* on the corticosterone levels in serum and contents of monoamine in depression rat brain. Zhong Yao Cai= Zhongyaocai= Journal of Chinese Medicinal Materials 28, 492–496.

Cheng, C., Koo, M., 2000. Effects of *Centella asiatica* on ethanol induced gastric mucosal lesions in rats. Life Sciences 67, 2647–2653.

Devkota, A., Jha, P.K., 2009. Variation in growth of *Centella asiatica* along different soil composition. Botany Research International 2, 55–60.

Dhanasekaran, M., Holcomb, L.A., Hitt, A.R., Tharakan, B., Porter, J.W., Young, K.A., Manyam, B.V., 2009. *Centella asiatica* extract selectively decreases amyloid β levels in hippocampus of Alzheimer's disease animal model. Phytotherapy Research 23, 14–19.

Dhar, M., Dhar, M., Dhawan, B., Mehrotra, B., Ray, C., 1968. Screening of Indian plants for biological activity: Part I. Indian Journal of Experimental Biology 6, 232–247.

Duggina, P., Kalla, C.M., Varikasuvu, S.R., Bukke, S., Tartte, V., 2015. Protective effect of centella triterpene saponins against cyclophosphamide-induced immune and hepatic system dysfunction in rats: its possible mechanisms of action. Journal of Physiology and Biochemistry 71, 435–454.

Emran, T.B., Dutta, M., Uddin, M.M.N., Nath, A.K., Uddin, M.Z., 2015. Antidiabetic potential of the leaf extract of *Centella asiatica* in alloxaninduced diabetic rats. Jahangirnagar University Journal of Biological Sciences 4 (1), 51–59.

George, M., Joseph, L., 2009. Anti-allergic, anti-pruritic, and anti-inflammatory activities of *Centella asiatica* extracts. African Journal of Traditional, Complementary and Alternative Medicines 6.

Gnanapragasam, A., Ebenezar, K.K., Sathish, V., Govindaraju, P., Devaki, T., 2004. Protective effect of *Centella asiatica* on antioxidant tissue defense system against adriamycin induced cardiomyopathy in rats. Life Sciences 76, 585–597.

Goh, S.H., Chuah, C., Mok, J., Soepadmo, E., 1995. Malaysian Medicinal Plants for the Treatment of Cardiovascular Diseases. Petaling Jaya: Pelanduk Publications, ISBN 1089850565, 162pp.

Gohil, K.J., Patel, J.A., Gajjar, A.K., 2010. Pharmacological review on *Centella asiatica*: a potential herbal cure-all. Indian Journal of Pharmaceutical Sciences 72, 546.

Guo, J.S., Cheng, C.L., Koo, M.W., 2004. Inhibitory effects of *Centella asiatica* water extract and asiaticoside on inducible nitric oxide synthase during gastric ulcer healing in rats. Planta Medica 70, 1150–1154.

Gupta, Y., Kumar, M.V., Srivastava, A., 2003. Effect of *Centella asiatica* on pentylenetetrazole-induced kindling, cognition and oxidative stress in rats. Pharmacology Biochemistry and Behavior 74, 579–585.

Hamid, A.A., Shah, Z.M., Muse, R., Mohamed, S., 2002. Characterisation of antioxidative activities of various extracts of *Centella asiatica* (L) urban. Food Chemistry 77, 465–469.

Incandela, L., Cesarone, M., Cacchio, M., De Sanctis, M., 2001. Total triterpenic fraction of *Centella asiatica* in chronic venous insufficiency and in high-perfusion microangiopathy. Angiology 52, S9.

James, J.T., Dubery, I.A., 2009. Pentacyclic triterpenoids from the medicinal herb, *Centella asiatica* (L.) urban. Molecules 14, 3922–3941.

James, J.T., Meyer, R., Dubery, I.A., 2008. Characterisation of two phenotypes of *Centella asiatica* in Southern Africa through the composition of four triterpenoids in callus, cell suspensions and leaves. Plant Cell Tissue and Organ Culture 94, 91–99.

Jamil, S.S., Nizami, Q., Salam, M., 2007. *Centella asiatica* (Linn.) urban: a review. Natural Product Radiance 6, 158–170.

Jayathirtha, M., Mishra, S., 2004. Preliminary immunomodulatory activities of methanol extracts of *Eclipta alba* and *Centella asiatica*. Phytomedicine 11, 361–365.

Kumar, M.V., Gupta, Y., 2002. Effect of different extracts of *Centella asiatica* on cognition and markers of oxidative stress in rats. Journal of Ethnopharmacology 79, 253–260.

Larsen, H.O., Olsen, C.S., 2007. Unsustainable collection and unfair trade? Uncovering and assessing assumptions regarding Central Himalayan medicinal plant conservation. In: Plant Conservation and Biodiversity. Springer, pp. 105–123.

Lee, M.K., Kim, S.R., Sung, S.H., Lim, D., Kim, H., Choi, H., Park, H.K., Je, S., Ki, Y., 1999. Asiatic acid derivatives protect cultured cortical neurons from glutamate-induced excitotoxicity. Research Communications in Molecular Pathology and Pharmacology 108, 75–86.

Liu, M., Dai, Y., Yao, X., Li, Y., Luo, Y., Xia, Y., Gong, Z., 2008. Anti-rheumatoid arthritic effect of madecassoside on type II collagen-induced arthritis in mice. International Immunopharmacology 8, 1561–1566.

Nalini, K., Aroor, A., Rao, A., Karanth, K., 1992. Effect of *Centella asiatica* fresh leaf aqueous extract on learning and memory and biogenic amine turnover in albino rats. Fitoterapia 63, 231–238.

Pandey, N., Tewari, K., Tewari, R., Joshi, G., Pande, V., Pandey, G., 1993. Medicinal plants of Kumaon Himalaya: strategies for conservation. Himalayan Biodiversity Conservation Strategies 3, 293–302.

Plohmann, B., Bader, G., Hiller, K., Franz, G., 1997. Immunomodulatory and antitumoral effects of triterpenoid saponins. Die Pharmazie 52, 953–957.

Pragada, R., Veeravalli, K., Chowdary, K., Routhu, K., 2004. Cardioprotective activity of *Hydrocotyle asiatica* L. in ischemia-reperfusion induced myocardial infarction in rats. Journal of Ethnopharmacology 93, 105–108.

Rao, S.B., Chetana, M., Devi, P.U., 2005. *Centella asiatica* treatment during postnatal period enhances learning and memory in mice. Physiology & Behavior 86, 449–457.

Saha, A., Bhatia, B., Kulkarni, K.S., 2002. Evaluation of the Efficacy of Mentat in Children with Learning Disability: A Placebo-Controlled Double-Blind Clinical Trial.

Sairam, K., Rao, C.V., Goel, R., 2001. Effect of *Centella asiatica* Linn on physical and chemical factors induced gastric ulceration and secretion in rats. Indian Journal of Experimental Biology 39, 137–142.

Sastravaha, G., Gassmann, G., Sangtherapitikul, P., Grimm, W.-D., 2005. Adjunctive periodontal treatment with *Centella asiatica* and *Punica granatum* extracts in supportive periodontal therapy. Journal of the International Academy of Periodontology 7, 70–79.

Schaneberg, B., Mikell, J., Bedir, E., Khan, I., Nachname, V., 2003. An improved HPLC method for quantitative determination of six triterpenes in *Centella asiatica* extracts and commercial products. Die Pharmazie–An International Journal of Pharmaceutical Sciences 58, 381–384.

Sharma, B., Kumar, A., 1998. Biodiversity of medicinal plants of Triyugi Narain (Garhwal Himalaya) and their conservation. In: National Conference on Recent Trends in Spices and Medicinal Plant Research, Calcutta, WB, India, p. 78.

Sharma, J., Sharma, R., 2002. Radioprotection of Swiss albino mouse by *Centella asiatica* extract. Phytotherapy Research 16, 785–786.

Shobi, V., Goel, H., 2001. Protection against radiation-induced conditioned taste aversion by *Centella asiatica*. Physiology & Behavior 73, 19–23.

Shukla, A., Rasik, A., Jain, G., Shankar, R., Kulshrestha, D., Dhawan, B., 1999. In vitro and in vivo wound healing activity of asiaticoside isolated from *Centella asiatica*. Journal of Ethnopharmacology 65, 1–11.

Somchit, M., Sulaiman, M., Zuraini, A., Samsuddin, L., Somchit, N., Israf, D., Moin, S., 2004. Antinociceptive and antiinflammatory effects of *Centella asiatica*. Indian Journal of Pharmacology 36, 377.

Tholon, L., Neliat, G., Chesne, C., Saboureau, D., Perrier, E., Branka, J.-E., 2002. An in vitro, ex vivo, and in vivo demonstration of the lipolytic effect of slimming liposomes: An unexpected α_2 -adrenergic antagonism. Journal of Cosmetic Science 53, 209–218.

Thomas, M.T., Kurup, R., Johnson, A.J., Chandrika, S.P., Mathew, P.J., Dan, M., Baby, S., 2010. Elite genotypes/chemotypes, with high contents of madecassoside and asiaticoside, from sixty accessions of *Centella asiatica* of south India and the Andaman Islands: for cultivation and utility in cosmetic and herbal drug applications. Industrial Crops and Products 32, 545–550.

Wang, X.-S., Dong, Q., Zuo, J.-P., Fang, J.-N., 2003. Structure and potential immunological activity of a pectin from *Centella asiatica* (L.) urban. Carbohydrate Research 338, 2393–2402.

Xu, Y., Cao, Z., Khan, I., Luo, Y., 2008. Gotu Kola (*Centella asiatica*) extract enhances phosphorylation of cyclic AMP response element binding protein in neuroblastoma cells expressing amyloid beta peptide. Journal of Alzheimer's Disease: JAD 13, 341–349.

Young, G., Jewell, D., 1996. Creams for Preventing Stretch Marks in Pregnancy. The Cochrane Library.

Zheng, M., 1989. An experimental study of the anti-HSV-II action of 500 herbal drugs. Journal of Traditional Chinese Medicine 9, 113.

Zivkovic, B., Scatton, B., Dedek, J., Bartholini, G., 1982. GABA influence on noradrenergic and serotonergic transmissions: implications in mood regulation. In: New Vistas in Depression, vol. 40. Pergamon Press, Oxford, pp. 195–201.

CHAPTER

33

Indian Senna

Huma Naz[1], Haq Nawaz[1], Muhammad Asif Hanif[1], Muhammad Adnan Ayub[2], Selima Khatun[3]

[1] Department of Chemistry, University of Agriculture, Faisalabad, Pakistan;
[2] Department of Chemistry, University of Okara, Okara, Pakistan;
[3] Department of Botany, Government General Degree College, Simgur, Hooghly, India

OUTLINE

Medicinal Plants of South Asia
https://doi.org/10.1016/B978-0-08-102659-5.00033-1

Copyright © 2020 Elsevier Ltd. All rights reserved.

1. BOTANY

1.1 Introduction

Indian senna (*Cassia angustifolia*) is perennial shrub belonging to family Fabaceae. It has been used traditionally by various civilizations across the world as a laxative. The genus *Cassia* contains a range of 250–350 species that are native to tropical regions of Africa, Somalia, upper Egypt, Sudan, Pakistan, and India (Reddy et al., 2015). The uncertainty within the genus is largely attributed to the variability among the species. There is a great variability in morphology, growth habit, flower color, height, leaves, and chemical constituents (Tinworth et al., 2010). Indian senna is cross-pollinated by a variety of bees. There are a number of plants outside of this genus with common name similar to Indian senna including Alexandrian senna, Cassia senna, and East senna. Indian senna is known by different names depending on its origin. It is commonly known as Indian senna in English, sanaya and hindisana in Hindi, swarn patri in Sanskrit, nilavaka and chinnukki in Malayalam, and nelavrika and sonamukhi in Kannada. It is known as sanā in Arabic and Sana Makkahi in Urdu. Probably the most familiar senna is Indian senna (*C. angustifolia*).

1.2 History/Origin

C. angustifolia Vahi is native to Pakistan, India, North Africa, and Sudan. The plant derives its name from the Arabic sena and from the Hebrew word cassia, which means "peeled back," a reference to its peelable bark. Senna was given the name of purging cassia in Europe during the middle ages because it was used at that time in an Italian medical school as a purgative. There are several suggested origins for the word Indian senna. It came from Greek word *kasia*, according to Biblical Hebrew kiddah', i.e., "split," one of the principal spices of the holy anointing oil. It came from qatsa meaning "to cut off, strip off bark" of a cinnamon-like plant in late Old English. In China, they named senna Fan–Hsieh–Yeh, which means foreign country laxative herb. *C. angustifolia* was first discovered in the ancient city of Makkah, which was the heart of the old province Hijaz. The plant was first used as a

FIGURE 33.1 Indian senna dried leaves.

herbal medicine for Muslims by Holy Prophet Muhammad (Peace Be Upon Him) and was grown in abundance around Makkah. Holy Prophet Muhammad (Peace Be Upon Him) said, "Seena and the Sanoot (cumin) are indeed two plants for you which have cure of every disease except saam (death)" (Sultana et al., 2012).

It is traded under the name of senna or sana makkahi in the herbal shops of India, Pakistan, and Arabian countries (Revathi et al., 2013). The Arabian physicians first brought this drug into use, and later on, the Greeks noticed it. It is used as a laxative according to Ayurveda, useful in alleviating constipation and acting on the lower bowel (Larkin et al., 2008). In China, the *C. angustifolia* pods are taken during a day fast (Fig. 33.1).

1.3 Demography/Location

C. angustifolia is grown on all kinds of soil with adequate drainage and basically in an arid environment. Warm winters free from frost and mild tropical climate are the ecologic requirements of senna cultivation. It also requires bright sunshine and scanty rain for its cultivation (Hussain et al., 2007). Absolute figures for senna production are difficult to acquire. The total world production of *C. angustifolia* is estimated to be 4000 ha of cultivation of this crop that produces about 500 tons of pods, 7500 tons of fruit, and 2500 tons of leaves annually (Akerele and Heywood, 1991). The major country for Indian senna export is Germany followed by the United States, China, France, Spain, and Japan (Revathi et al., 2013).

1.4 Botany, Morphology, Ecology

Indian senna is an upright shrub, with ascending branches, 2–3 ft in height, and it has a pale to green color stem. The leaves have four to five pairs that spread on the branches and have pointed apex and spike shape. They are 1–2 in. long and 0.2–0.3 in. in breadth. The leaflets are lanceolate, margins are entire, ovate with unequal base, and apex is acute mucronate. The petioles are short, stout, and sometimes broken. The leaflets are 0.5–1.5 cm wide and 1.5–6.0 cm in length. The leaflets are young, and there are hairs on the upper and lower surfaces of the leaflets. The leaves have light yellow to pale olive color and have slightly bitter taste and mucilage odor (Patil et al., 2013). Flowers are usually yellow, small, subterminal, erect, and racemes. Flower possess pods that are compressed, smooth, ovate, and have a dark brown seed (Sreeramu, 2010). The calyx is enantiomorphic and corolla is monosymmetric. The arrangements of sepals are either clockwise or anticlockwise. The petals have a single main vein that extends up to tip and lower petals are concave. Indian senna requires warm weather and irrigated conditions. The plant cannot survive at low temperatures (Trease and Evans, 1972). It cannot bear poor drainage, humidity, and high rainfall (Tripathi, 1999). It grows well in soil with a pH ranging from 7.0 to 8.5. Indian senna is a sun-loving plant. Water logging and heavy irrigation, even temporary stagnation, can cause loss to the crop (Krishnaiah et al., 2009).

2. CHEMISTRY

Indian senna is an aromatic plant and is used as an herb and love sachet. Indian senna has a peculiar odor, faint and mucilaginous sweetish taste. The fragrance and aroma of Indian senna is due to the presence of essential oil contents in different parts of plant and leaves. Indian senna also contain flavonoids, anthraquinones, α –pinene, β-pinene, and phenolic contents. The chemical constituents vary depending on the type of species/cultivars of *C. angustifolia*. For flower coloration, flavonoids are the most important plant pigment that is responsible for the production of red and blue pigmentation in petals (Annapareddy, 2012). Fistulic acid and anthraquinone acid also impart color to the pods of the plant. Aroma in *C. angustifolia* is due to the limonene (*cis*-limonene oxide), β-caryophyllene, and estragole. Active components of Indian senna are shown Fig. 33.2.

It is also known for being a good source of minerals and carbohydrates. The fruit tissues are a rich source of potassium, calcium, iron, and manganese. Flowers also have trace amounts of alkaloids. Indian senna contains 29%–31% protein, 10%–12% minerals, and 8%–10% mucilage.

(i) Rhein

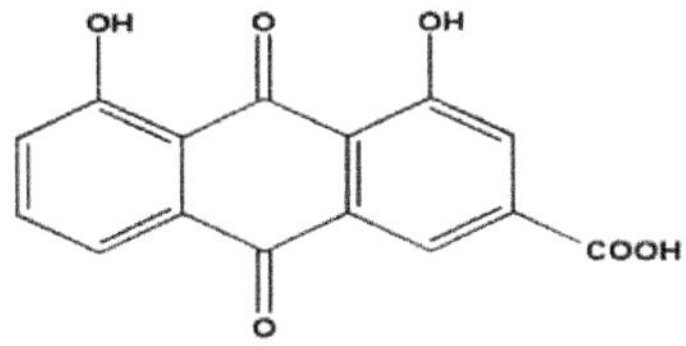

(ii) Aloe-emodin

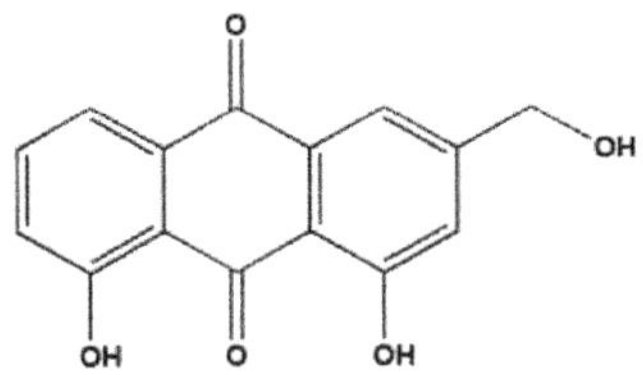

(iii) Tannin

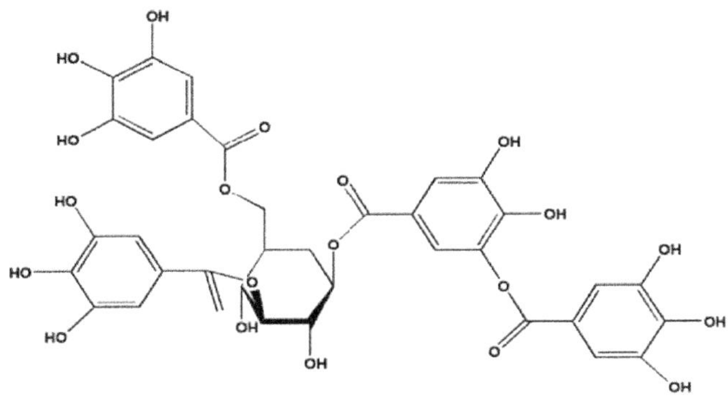

(iv) Chrysophanol

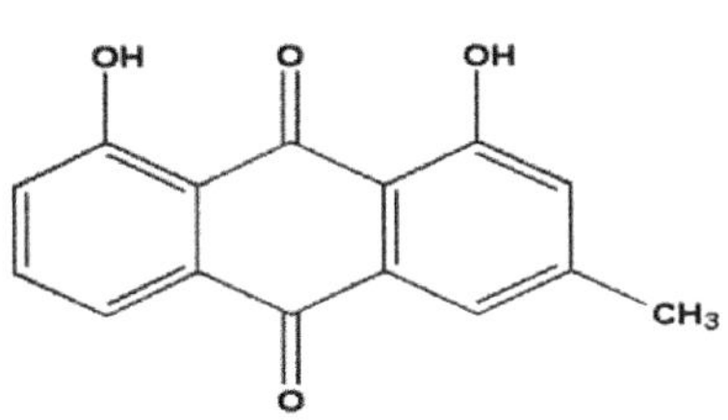

FIGURE 33.2 Active components of Indian senna.

Carbohydrates in the plant are 2%–3% polysaccharides, and others are mannose, glucose, and fructose. It contains two types of glucosides, myricyl and sennoside, and a rich source of anthraquinone derivatives. It is also known for flavonoids and antioxidant activities. However, it is realized that it should not be overused because of laxative action. It is advised not to take internally a large amount of the essential oils of *C. angustifolia*.

Anthraquinone derivatives are the main active constituents of senna, which are responsible for its laxative properties. Indian senna has an aromatic odor due to the presence of essential oils in the leaves. The volatile oil of the leaves of *C. angustifolia* contains terpenes, aldehydes, and phenols. Besides the essential oil, the plant also contains salicylic acid, mannitol, saponin, resin, and sodium potassium tartrate (Laghari et al., 2011). Isorhamnetin and kaempferol are also present in this plant (Majid et al., 2013). Other main chemical constitutions α-sitosterol, sennoside A, B (rhein and dianthrone), sennoside C, D (rhein, aloe-emodin, and heterodianthrone) and anthraquinones are present in free form. Small amounts of aloe-amine, rhein, chrysophanol, and their glycosides are detected in this plant. The seeds are a source of gums that are present in the form of galactomannan composed of mannose and galactose and have a basic structure with a main chain of (1 → 4)–linked β-D-mannopyranosyl units attached to α-(1 → 6)-D-galactopyranosyl (Chaubey and Kapoor, 2001).

C. angustifolia essential oil contains α-pinene, β-pinene, β-caryophyllene, octanol, γ-terpinolene, estragole, *cis*-limonene oxide, *trans*-anethole, caryophyllene oxide, and geranyl. It contains the highest percentage of *cis*-limonene oxide and lowest amount of α-pinene.

3. POSTHARVESTING TECHNOLOGY

Conventionally, harvesting of Indian senna is done when the leaves are thick, bluish, and fully grown. Maximum yield of leaves is obtained under irrigated conditions of crops. The first harvesting is done after 90 days of sowing, and second and third harvesting are usually done after 150 and 210 days of sowing, a little before the maturity of the pods. For seed production, pods are collected during the months of February through March, when getting brown in color. Dried leaves of senna can also be stored in black polyethylene bags, aluminum foil bags, and transparent polyethylene bags.

In full sun, the harvested leaves and pods are spread under shade on the floor for 6–10 h for the reduction of moisture contents up to 10%.

Drying is done in a well-ventilated room for maintenance of light green to yellowish green color of leaves with constant stirring. Within 10–20 days, leaves are completely dried in a well-ventilated room. After drying, seeds are separated from pods by beating pods with sticks. The leaves and pods are separated from each other by mechanical air blowing. For highest grade, bold pods and large leaves of yellowish green in color are separated for obtaining best price in the market. The next grade includes leaves and pods of brownish color. The lowest grade contains small/broken pods and leaves. At an industrial level, mature leaves have 2.0% –2.5% and pods have 2.5%–3.0% of sennosides contents. The leaves are stored in a cool and dry place after drying, grading, and packing. To reduce transportation cost, leaves are pressed by hydraulic press.

4. PROCESSING

Indian senna is consumed in a variety of ways and for various purposes. In addition, to its fresh leaves, other common processed forms of the plant include dried and graded leaves/pods and essential oil.

Indian senna is traditionally dried in a forced flow type machinal dryer at different air temperatures. The optimum temperature for drying is 45°C to avoid losses of sennosides. The dried leaves are stored in light- and moisture-proof packages.

Various extraction methods are available for the extraction of sennosides from leaves. Soxhlet, reflux, and maceration are the most common extraction methods in which alcoholic and aqueous solutions are used. The sennosides are extracted in the form of calcium sennosides using organic solvents like benzene and methanol. The extraction solutions are acidified with hydrochloric acid (or organic acids), and anhydrous calcium chloride is added in the presence of ammonia solution to precipitate out calcium sennoside. A microwave-assisted extraction technique has been developed for the extraction of calcium sennoside from leaflets of Indian senna using lesser amounts of organic solvents in a shorter time period.

5. VALUE ADDITION

Value-added products of Indian senna are flavored senna leaves, tea, sennosides, tablets, and other drugs. The products of Indian senna are available in various kinds of herbal formulations that are commonly used in pharmaceutical dosage in the form of tablets, capsules, powders, granules, liquids, pastes, and suppositories. Coated and uncoated tablets containing

40–60 mg of calcium sennoside contain 7.5–18 mg of hydroxyanthracene glycoside. Senna leaf powder is used as such and combined with some other powders such as cascara (bark from *Rhamnus purshiana*) and bgol (seed husks of psyllium, plantago ovate). Senna paste is used for skin ailments. Granules and liquids are available in the form of syrup or fluid that are with or without alcohol. In the market, senna herbal teas and chocolates are also found (Ambrose et al., 2016). It can be used in a combination of herbs such as ginger, cloves, fennel, cinnamon, and coriander, although addition of other aromatics is designed for antinauseous effect.

6. USES

Senna is full of antioxidants and is also a good source of carbohydrates and minerals. The fruits and leaves of Indian senna are laxative and diuretic. Senna product with cascara product is used for cathartics.

Indian senna has been considered a tonic for the entire body and a digestive system cleanser. Its seeds are used for abdominal trouble, skin diseases, anthelmintic digestive and to treat piles, and in several other diseases (Srivastava et al., 2010).

In cosmetic preparation, it is used as a hair dye with henna leaves to make hair black. Due to its high polysaccharide content, it is used as skin-conditioning agent. It can also retain water on the surface of skin and hair, and exhibit film-forming capacities to repair dry tissue. It can also be used for acne. In skin diseases, vinegar is mixed with senna leaf powder and applied. Sennoside contents of senna are used for the treatment of ringworm. Dried root powder was used for snakebite.

The leaves are also the important ingredient of Nilaavarai Churman, which is used for biliousness, vomiting, stomach distention, antianemic activity, dysentery, and fever. It is also used for breathing problems, bilious colic, hemorrhoids, and flatulence. The senna herb is an effective medication for the treatment of bloating, abdominal cramping, and diarrhea. It is used for skin diseases like scabies and itching (Aggarwal et al., 2011). It is used for the treatment of typhoid fever, jaundice, leukoderma, febrifuge, cholagogue, and as an anthelmintic.

Indian senna is known to have strong antioxidant activity due to the presence of flavonoids. Essential oils have many biological and pharmaceutical properties such as antidiabetic, anticancer, antimutagenic, antiviral, anticancer, and antiinflammatory ones (Raut and Karuppayil, 2014). Antioxidants are the most important substances for the protection of bodily damage that is induced due to free radicals. The flavonoids can also have antimicrobial, antiinflammatory, antitumor, and other protective activity on human health (Ayo, 2010).

7. PHARMACOLOGICAL USE

7.1 Antioxidant Activity

The purple methanol solution of flower of Indian senna was assayed for DPPH activity (Ghosh and Gaba, 2013). Leaves and flowers have the highest DPPH scavenging ability (Laghari et al., 2011).

7.2 Antimicrobial Activity

Different extracts of Indian senna exhibited strong antimicrobial activity against several microbes including *Staphylococcus aureus* (gram-negative bacteria), *Escherichia coli* (gram-positive bacteria), *Streptococcus mutan, Lactobacillus casei, Lactobacillus acidophilus, Bacillus megaterium, Pseudomonas aeruginosa, Aspergillus niger,* and *Aspergillus flavus.* It was also found that methanol extract exhibited the highest antimicrobial activities compared to other solvents (VijayaSekhar et al., 2016).

7.3 Antiinflammatory Activity

Indian senna is used to treat skin disorders, and its extracts exhibited significant antiinflammatory activity (Aggarwal et al., 2011). The extract of the plant inhibited edema that was caused due to the presence of 12-O-tetradecanoylphorbol-13-acetate, oxazolone, and arachidonic acid in single and multiple applications (Cuellar et al., 2001).

7.4 Antitumor Activity

It is used for the treatment of various types of tumors (Upadhyay et al., 2011). Water-soluble polysaccharides isolated from Indian senna leaves were further fractionated and on methylation showed the presence of 1,4-linked galacturonic acid, 1,2,4-rhamnose, 1,3,6-galactose, 1,3-arabinose, terminal galactose, and arabinose residues. The polysaccharide fraction was tested against sarcoma-180 (S-180) in tumor-bearing mice and showed encouraging results.

7.5 Antihepatoma Activity

Indian senna crude extract is found as one of the most effective drugs against hepatoma (Lin et al., 2002).

7.6 Hemorrhoids Repair Effects

Indian senna is the most important ingredient in colon-cleansing products that are used to repair the hemorrhoids. Senna contains sennosides that act on the bowel lining and cause the laxative affect. Hemorrhoids are caused by constipation (Raju et al., 2011).

8. SIDE EFFECTS AND TOXICITY

Indian senna is likely safe for short-term use for children over age 2 and most adults when taken by mouth. Indian senna is a Food and Drug Administration–approved nonprescription medicine. Indian senna can cause some side effects including cramps, diarrhea, and stomach discomfort. Indian senna is possibly unsafe for long-term use by mouth or in high doses. Long-term use can also change the amount or balance of some chemicals in the blood (electrolytes) that can cause heart function disorders, muscle weakness, liver damage, and other harmful effects.

References

Aggarwal, B.B., Prasad, S., Reuter, S., Kannappan, R., R Yadav, V., Park, B., Hye Kim, J., C Gupta, S., Phromnoi, K., Sundaram, C., 2011. Identification of novel anti-inflammatory agents from Ayurvedic medicine for prevention of chronic diseases:"reverse pharmacology" and "bedside to bench" approach. Current Drug Targets 12, 1595–1653.

Akerele, O., Heywood, V., 1991. Conservation of Medicinal Plants. Cambridge University Press.

Ambrose, D.C., Manickavasagan, A., Naik, R., 2016. Leafy Medicinal Herbs: Botany, Chemistry, Postharvest Technology and Uses. CABI.

Annapareddy, P., 2012. Identifying and quantifying flavonoids in three medicinal plants by HPLC. International Journal of Innovative Research and Development 1, 344–362. ISSN 2278–0211.

Ayo, R., 2010. Phytochemical constituents and bioactivities of the extracts of *Cassia nigricans* Vahl: a review. Journal of Medicinal Plants Research 4, 1339–1348.

Chaubey, M., Kapoor, V.P., 2001. Structure of a galactomannan from the seeds of *Cassia angustifolia* Vahl. Carbohydrate Research 332, 439–444.

Cuellar, M., Giner, R., Recio, M., Manez, S., Rıos, J., 2001. Topical anti-inflammatory activity of some Asian medicinal plants used in dermatological disorders. Fitoterapia 72, 221–229.

Ghosh, P.K., Gaba, A., 2013. Phyto-extracts in wound healing. Journal of Pharmacy & Pharmaceutical Sciences 16, 760–820.

Hussain, S., Siddiqui, S.U., Khalid, S., Jamal, A., Qayyum, A., Ahmad, Z., 2007. Allelopathic potential of senna (*Cassia angustifolia* Vahl.) on germination and seedling characters of some major cereal crops and their associated grassy weeds. Pakistan Journal of Botany 39, 1145.

Krishnaiah, D., Devi, T., Bono, A., Sarbatly, R., 2009. Studies on phytochemical constituents of six Malaysian medicinal plants. Journal of Medicinal Plants Research 3, 067–072.

Laghari, A.Q., Memon, S., Nelofar, A., Laghari, A.H., 2011. Extraction, identification and antioxidative properties of the flavonoid-rich fractions from leaves and flowers of *Cassia angustifolia*. American Journal of Analytical Chemistry 2, 871.

Larkin, P., Sykes, N., Centeno, C., Ellershaw, J., Elsner, F., Eugene, B., Gootjes, J., Nabal, M., Noguera, A., Ripamonti, C., 2008. The management of constipation in palliative care: clinical practice recommendations. Palliative Medicine 22, 796–807.

Lin, L.T., Liu, L.T., Chiang, L.C., Lin, C.C., 2002. In vitro anti-hepatoma activity of fifteen natural medicines from Canada. Phytotherapy Research 16, 440–444.

Majid, U., Siddiqi, T.O., Aref, I.M., Iqbal, M., 2013. Quantitative changes in proteins, pigments and sennosides of *Cassia angustifolia* vahl treated with mancozeb. Pakistan Journal de Botanique 45, 1509–1514.
Patil, S.G., Wagh, A.S., Pawara, R.C., Ambore, S.M., 2013. Standard tools for evaluation of herbal drugs: an overview. The Pharma Innovation 2.
Raju, B.S., Balakrishna, G., Kumar, D.S., Kumar, M.R., Prakash, Y.E., Aneela, K., 2011. A herbal plant *Cassia angustifolia*. Journal of Atoms and Molecules 1, 1.
Raut, J.S., Karuppayil, S.M., 2014. A status review on the medicinal properties of essential oils. Industrial Crops and Products 62, 250–264.
Reddy, N.R.R., Mehta, R.H., Soni, P.H., Makasana, J., Gajbhiye, N.A., Ponnuchamy, M., Kumar, J., 2015. Next generation sequencing and transcriptome analysis predicts biosynthetic pathway of sennosides from Senna (*Cassia angustifolia* Vahl.), a non-model plant with potent laxative properties. PLoS One 10, e0129422.
Revathi, P., Parimelazhagan, T., Manian, S., 2013. Ethnomedicinal plants and novel formulations used by Hooralis tribe in Sathyamangalam forests, Western Ghats of Tamil Nadu, India. Journal of Medicinal Plants Research 7, 2083–2097.
Sreeramu, B., 2010. Cultivation of Medicinal and Aromatic Crops. Universities Press.
Srivastava, M., Srivastava, S., Rawat, A., 2010. Chemical standardization of *Cassia angustifolia* Vahl seed. Pharmacognosy Journal 2, 554–560.
Sultana, S., Ahmad, M., Zafar, M., Khan, M.A., Arshad, M., 2012. Authentication of herbal drug Senna (*Cassia angustifolia* Vahl.): a village pharmacy for Indo-Pak subcontinent. African Journal of Pharmacy and Pharmacology 6, 2299–2308.
Tinworth, K.D., Harris, P.A., Sillence, M.N., Noble, G.K., 2010. Potential treatments for insulin resistance in the horse: a comparative multi-species review. The Veterinary Journal 186, 282–291.
Trease, G.E., Evans, W.C., 1972. Pharmacognosy, London: Bailliere Tindall 795pp. Caphaelis, Ipomoea, Datura, Hyoscyamus, Atropa. Digitalis, Valeriana.
Tripathi, Y., 1999. *Cassia angustifolia*, a versatile medicinal crop. International Tree Crops Journal 10, 121–129.
Upadhyay, A., Nayak, P.S., Khan, N.A., 2011. Sennoside contents in Senna (*Cassia angustifolia* Vahl.) as influenced by date of leaf picking, packaging material and storage period. Journal of Stored Products and Postharvest Research 2, 97–103.
VijayaSekhar, V., Prasad, M.S., Joshi, D.S.D.S., Narendra, K., Satya, A.K., Rao, K.R.S.S., 2016. Assessment of phytochemical evaluation and in-vitro antimicrobial activity of *Cassia angustifolia*. International Journey of Pharmacognosy and Phytochemical Research 8 (2), 305–312.

CHAPTER

34

Jujube

Zunaira Irshad[1], *Muhammad Asif Hanif*[1], *Muhammad Adnan Ayub*[2], *Asma Hanif*[1], *Hassan Imran Afridi*[3]

[1] Department of Chemistry, University of Agriculture, Faisalabad, Pakistan; [2] Department of Chemistry, University of Okara, Okara, Pakistan; [3] National Centre of Excellence in Analytical Chemistry, University of Sindh, Jamshoro, Pakistan

OUTLINE

Medicinal Plants of South Asia
https://doi.org/10.1016/B978-0-08-102659-5.00034-3

Copyright © 2020 Elsevier Ltd. All rights reserved.

1. BOTANY

1.1 Introduction

Ziziphus jujuba is a common plant typically known as jujube (occasionally jujuba) (Fig. 34.1). It is also known as Indian, red, Chinese, and Korean date. It belongs to the genus *Ziziphus* that is a member of the flowering plant family Rhamnaceae. It contains about 50–55 genera and about 870–900 species. The genus *Ziziphus* contains 135 to 170 species (Islam and Simmons, 2006). This family (Rhamnaceae) is found in hot climates and subtropical areas all around the world. Jujube is native to Asia and requires high temperatures with a sufficient amount of water for its optimum growth. Jujube fruit gets wrinkled on ripening, and it contains a single seed (Tripathi, 2014). There are two major types of jujube: the Indian jujube (*Ziziphus mauritiana* Lam) and the common jujube (*Ziziphus jujuba* Mill). These species are grown in many countries around the world. The Ziziphus is an Arabic term, and Greeks called the jujube ziziphon. It is known by the different names in different languages, in Sanskrit (rajabadari), in Punjabi (beri), in Bengali (kul), in Assamese (bogori), in Gujrati (bodori bordi), Hindi (ber), Marathi (bor), Malayalam (badaram), Tamil (vadari), Telugu (renu), Persian (Anab), Urdu (ber) and Sindhi (jangri).

FIGURE 34.1 Jujube trees and fruits.

1.2 History/Origin

The old books in India, Japan, and Korea narrate the use of jujube fruit and seeds; some scholars believe even as early as 9000 BC. This plant is found throughout Asia, and now it is cultivated even in Russia, Northern Africa, and the United States. This plant entered the United States in Christ's era. The jujube is considered a wild plant exclusively in the area of the Deccan plateau (Pareek, 2001). Jujube is considered an exclusive medication for the cure of diseases of whole-body organs, specifically kidneys and lungs. It is considered a holy plant in many of the religions around the world. Some superstitions are also associated with the jujube tree. It is considered a haunted tree in many regions of Pakistan. Though, the Sikhs do not consider it a sacred tree. The golden temple located in Amritsar contains only one jujube tree that is believed to eradicate the feeling of distress among the people. The myths linked to Ram and Shabari make it a beneficial fruit to be eaten. Prophet Adam (AS) ate jujube fruit, the first on earth among other fruits. Muslims use jujube leaves boiled in water to give a bath to dead bodies. Muslims also believe jujube is a paradise tree and that Prophet Muhammad (PBUH) saw this tree at the utmost boundary (Sidrat Al-Muntha) on Mi 'raj night.

1.3 Demography/Location

Jujube is mostly found in China, Pakistan, Iran, Lebanon, the Korean peninsula, and Northern India. It is believed to be a native plant in China, and it is highly cultivated in hot climatic areas of Europe, southern Asia, Africa, and Australia. This plant was grown in Madagascar and in Western Island. In the French Islands of the Caribbean, it is commonly called pomme suerte, also known as Indian jujube; it is different from the jujube found in southern California. The jujube plants found in Iran, Australia, Burma, and Syria possess different physical qualities than those from France (Munier, 1973), the United States, and Russia. Jujube grows mostly in the southern half of North America. Ripening of fruits occurs in the summer season and needs an excessive quantity of water. The optimum temperature for jujube cultivation is 5.5–22°C, optimum required annual rain fall is 87–2000 mm, and the required soil pH is 4.55–8.4. Actually, it can be considered that the arid and semiarid areas of the world are the best growth sites for jujube.

1.4 Morphology, Botany, Ecology

Jujube is a 5- to 12-m-high shrub with thorny branches. The leaves are green, have shiny surfaces, and are 2–7 cm wide and 1–3 cm broad, having three visible basal veins and toothed edges. The flowers are 3–7 mm wide, with five inconspicuous yellowish-green petals. The fruit is oval, 1–3 cm long, drupe, and when immature, it is smooth green. There is a single hard stone similar to an olive stone. The morphology is different in different regions of the world: it ranges from very small to medium-sized trees that are erect with a branching system. Height varies from 3 to 20 m on average. The bark of the plant is dark brown to reddish in color. Branches spread erect, becoming flexuous and dull brown gray. Fruiting branches are not deciduous. Leaf laminae are elliptic to ovate or nearly orbicular. The apex is rounded, obtuse or subacute to emarginated, the base rounded, sometimes cuneate, mostly symmetrical or nearly so. Margins are minutely seriate. There are three marked nerves almost to the apex, the nerves being depressed in the upper part, light or dark green, with glabrous surface. The lower surface is whitish due to persistent dense hairs but may be buff colored. Occasionally the lower surface is glabrous. Leaves are petiolate, 1.1–5.8 mm long, and stipules are mostly spines, in each pair one hooked and one straight, or both hooked, or more rarely not developed into a spine. Flowers have sepals that are dorsally tomentose, a disk about 3 mm in diameter, and a two-celled ovary, immersed in the disk. Styles are two, 1 mm long, and connate for half their length. Flowers tend to have an acrid smell. Flowers are born in cymes or small axillary clusters. Cymes can be sessile or shortly pedunculate, peduncles 1–4 mm tomentose. Pedicels are also tomentose and are 2–4 mm at flowering and 3–6 mm at fruiting (Majumdar).

2. CHEMISTRY

Jujube fruit is sometimes considered the equivalent to the date due to its high nutrition. There is not any authentic writing that has reported the presence of essential oil in the jujube leaves or any other parts. But there are a lot of other compounds present in all parts of the plant. The composition of the plant depends upon the geography and techniques used for its processing. Seventy-seven percent of the fruit consists of carbohydrates including fructose and glucose exclusively. Vitamins A, B, and C and minerals like Ca and K and many others are also found in jujube plant. Jujube fruit is used as a food supplement since a lot of glucoside saponins, including jujubiside A and B, have been identified (Buxton). The recorded amount of pulp is 81%–97% (Pareek, 1983), and in another report the considered range of the pulp percentage is 91.6%–92.9% (Jawanda et al., 1980).

The jujube fruit contains ascorbic acid, minerals, and carbohydrates and mostly is taken as fresh fruit (Pareek, 2001). The jujube fruit contains water 17.38 to 22.52%, ash 2.26 to 3.01%, reducing sugar 57.61 to 77.93%, soluble fibers 0.57 to 2.79%, insoluble fibers 5.24 to 7.18%, and lipids 0.37 to 1.02%. It contains a large amount of various minerals including Zn, Cu, Na, K, P, Mn, and Ca and an excessive amount of vitamins B and C and P (bioflavonoid). World Health Organization recommended that one jujube fruit can fulfill the daily requirements for vitamins C and B.

A large number of phytochemicals (flavonoids, terpenoids, alkaloids) are present in various parts of the jujube plant (Pareek, 2001). Seeds and fruits contain isoprinosin, apigenin-6-C-b D-glucopyranoside, isovitexin-2″-O-b-D-glucopyranoside, 6‴-feruloylspinosin (Cheng et al., 2000). Almost 10 flavonoids, which include quercetin 3-O-rutinoside, quercetin 3-O-β-D¬galactoside, quercetin 3-O-α-L-arabinosyl-(1 → 2)-α-L-rhamnoside, quercetin 3-O-β-D-glucoside, kaempferol 3-O¬robinobioside, 3′,5′-Di-C-β-D-glucosylphloretin, and kaempferol 3-orutinoside, are present in different parts of the plant (Cheng et al., 2000; Pawlowska et al., 2009). The jujube fruit contains triterpenoic acid including alphitolic acid, colubrinic acid, 3-O-transpcoumaroylalphitolic acid, 3-O-cis-p-coumaroylalphitolic acid 3-O-cis-p-coumaroylmaslinic acid, betulonic acid, oleanolic acid, and

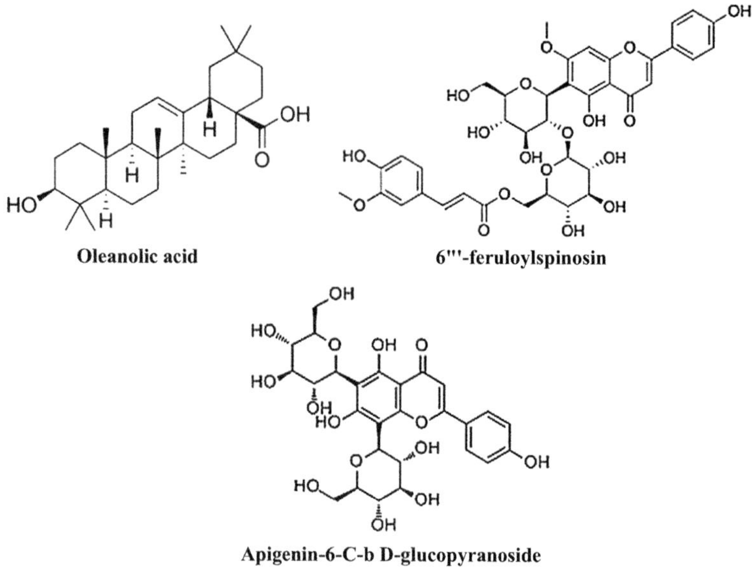

FIGURE 34.2 Important chemical components of jujube.

zizyberanalic acid (Lee et al., 2003). The triterpene esters like dimethyl ester and 2-oprotocatechuoyl alphitolic acid, ceanothic acid, and affeoyl alphitolic acid are present in different parts (Eiznhamer and Xu, 2004). Jujube also contains many phenolic compounds (Pawlowska et al., 2009). Betulinic acid is found in all parts of the plant; it is naturally occurring and may cause antitumor activity. It is found that it only kills the cancerous cells, while it does not damage the living and normal cells. It also has antiinflammatory effects. In some studies, it is reported that it can also kill *Staphylococcus aureus* and *Escherichia coli* (Eiznhamer and Xu, 2004). Some important components isolated from jujube are shown in Fig. 34.2.

3. POSTHARVEST TECHNOLOGY

Fruits picked off from the tree when they are green (still not ripened). These fruits gradually increase in size and the color changes to reddish yellow or brown. The rate of color change is dependent upon the temperature. Jujube fruits produce very little ethane but respond to the ethane treatment actively. The treatment of 100-ppm ethylene gas increases the rate of ripening. Similar results can be obtained from the treatment of ethephon. After harvesting, the jujube fruit cannot remain fresh for a long time; for this purpose the fruits are frozen to preserve their freshness, aroma, and taste The temperature required for freezing the fruits for storage is −32 to −22°C. It has been proven that the fruits frozen in liquid remain fresh for a longer time compared to the fruits frozen in the cold air chamber. This has been reported that for long-term storage, jujube fruits have to be frozen properly in the cold liquid media under −22°C (Wang et al., 2008).

4. PROCESSING

Jujubes are used in many ways and for the various purposes. The ripened fruits are stored after drying in open air or sunlight. The heat drying is the best method for preserving as it does not affect the vitamin C. The red jujube can be used as the snack. The stored fruits can be consumed several months later (Guo and Shan, 2010). Full red fruits contain about 65%–75% vitamin C. Jujube is also used for the preparation of pastries in the baking industry in China (Yao, 2013). Jams, juices, and vinegar are also made from the jujube. Asian, American, and Chinese people use jujube as a raw or cooked food.

5. VALUE ADDITION

The jujube contains a high amount of sugar and vitamin C, which makes it fit for various value-added products. Jujube extracts are added in bread to enhance the taste, and addition of jujube does not affect the bread texture. Addition of 50% jujube showed good effects compared with control bread in color, flavor, taste, and general palatability. There are many other similar products available (Ashton, 2006). Commonly the fruits are dried in the sun and then crushed along with tamarind, red chilies and salt. In Tamil Nadu (state of India) fresh fruit is crushed with tamarind, salt, and chilies and dried in sunlight afterward to make cakes. The jujube is also eaten as snacks; usually, they are smoked to enhance their taste. In China and Korea, sweetened syrup is used, and canned jujube tea and tea bags are also found in the market. Jujube vinegar is used in Bengal and Bangladesh; in Vietnam, fruits are consumed fresh; and jujube honey is used in Morocco.

6. USES

The jujube is considered an important medicinal plant to treat many diseases. About all parts of this plant are used, and many studies have reported their medicinal uses. China is one of the biggest markets of jujube fruit. The fruit is prized as a health food as well as a tasty treat. It is believed the jujube fruit's sweet smell can make teenagers fall in love. Jujube is used as a charm for fertility and placed in the newly married couple's room. Because of its fresh smell, jujube leaves are used as a potpourri in Bhutan. The leaves are used as an insect repellent. The wood of jujube is used to make the body of a Korean traditional instrument, the taepyeongso, and the Go bowls. There are many diseases that are believed to be cured by using jujube; these include liver disorders, muscular conditions, ulcers, diarrhea, and muscle fatigue.

The fruits and seeds are employed as antistress agents (Goetz, 2009) and also have the potential to be used as antifungal, antiinflammatory, antibacterial (Jiang et al., 2007), anticontraceptive, hypotensive, immunostimulant, antioxidant, and wound healing agents (Mahajan and Chopda, 2009). Acute constipation can be controlled by the use of jujube fruits (Naftali et al., 2009), and it is also very active against jaundice (Ebrahimi et al., 2013). Syrups can also be prepared from the fruits for the treatment of cough, flu, and cold. A decoction of the leaves of jujube is made for the treatment of throat. The leaves can also be used to treat the obesity. The fresh juice of jujube is used for reducing small pox intensity. Urinary disorders can be prevented by using the fresh leaves of jujube along with cumin (Oudhia, 2003). The roots of this plant are also used to cure coughs, biliousness, and headache. The bark is used to treat digestive disorders. Leucorrhoea and the eye problems are cured by the use of jujube fruits (Thompson, 2015).

7. PHARMACOLOGICAL USES

7.1 Antioxidant Activity

Jujube peel is a potential source of the antioxidant compounds. The antioxidant activity has been found by the DPPH and reducing power assay (Esteki and Urooj, 2012). The methanolic extracts of jujube have showed analgesic effect (Adzu et al., 2001). The high levels of the antioxidant present in the jujube fruits prove them to be a potential source of the compounds that may decrease or prevent the progression of the cancer cells.

7.2 Immunostimulant Effects

The chemotactic and intracellular potential to kill human neutrophil potency has been shown by the leaf extracts of the jujube (Ganachari et al., 2004). The wound healing potential of jujube root has been reported recently (Ansari et al., 2013).

7.3 Wound Healing Activity

Jujube extract in methanol possesses the ability to heal wounds in albino rats (Arutla et al., 2012).

7.4 Cardiovascular Activity

A compound called neolignan is found in jujube leaves that enhances the release of endogenous prostaglandin I_2 from the aorta, so it is a potential vasodilator (Ganachari et al., 2004).

7.5 Antifertility/Contraceptive Property

Jujube bark is considered to possess components that are antisteroidogenic, block the estrus cycle, and decrease the size of the ovaries. It has great impact on the blood and serum composition of the rats. The crude extracts are considered to have the opposite effects (Rekha and Chandrasekhara, 2014).

7.6 Antiinflammatory Activity

The leaves possess significant antiinflammatory potential against carrageenan-induced paw edema in rats (Kumar et al., 2004). The inflammation, toe swelling, and granulation tissue proliferation in rats can be inhibited using the jujube extracts. The antiinflammatory activity is shown by the betulinic acid found in jujube that decreases the action of enzymes that cause leukotriene biosynthesis, especially 5-lipoxygenase (Ramadoss et al., 2000).

7.7 Antiulcer Activity

Jujube also possesses potential to inhibit ulcer formation in rats that is considered to be due to antisecretory and cytoprotective action (Ganachari and Kumar, 2004). Sarcoma-180 can be avoided up to 61% by using jujube for 14 days consecutively.

7.8 Sweetness Inhibition Effects

The sweetness inhibiting compound triterpenoid is found in jujube. It has been noticed that the ability to taste sweet things in flies, rats, and hamsters is reduced by the use of jujube leaf extracts. There are a lot of compounds that act as sweetness inhibitors, including saponins II, III, IV, V, and VI. Jujubiside is present in leaves and seeds, while fruits contain *Ziziphus* saponins. It has been proved that ziziphin and jujube saponins II and III help to reduce obesity and diabetes in overweight persons due to their ability to suppress the sweet taste of sucrose. These compounds are four times more active compared to any other sweetness inhibitor (Suttisri et al., 1995).

7.9 Antiallergic Activity

The antiallergic activity shown by jujube extracts has been studied by the inhibitory effect on hyaluronidase (bovine test) (Su et al., 2002). Alcoholic extracts are more active than water extracts against *Trichophyton*. Jujube extracts are used for diarrhea that is caused by castor oil (Rao and Lakshmi, 2012). The excretion of leukotriene D_4 in the human peripheral circulatory system from white blood cells (basophilic cells) is suppressed by jujube fruit extracts. The antiallergic activity has been shown in alcoholic extracts of jujube in asthma and allergy in animals. It has proven to be effective in preventing eosinophilia and degranulation of mesenteric cells. Jujube extracts show the presence of triterpenoids, alkaloids, proteins, and flavonoids.

7.10 Hypoglycemic Effects

The hypoglycemic effect has been said to be possessed by the jujube fruit and seeds. A study was performed on two groups, and ANOVA was applied on data that confirmed that the glucose–triglyceride–cholesterol and very-low-density lipoprotein (VLDL) levels significantly decreased in people that were treated with the jujube fruits extracts (Rao and Lakshmi, 2012).

7.11 Antiobesity Activity

The alcoholic extracts of jujube leaves decrease body weight and food uptake and reduced the glucose level in the blood of rats. Sibutramine is an antiobesity medicine that showed the same effect as that of the jujube extracts (Ganachari et al., 2007).

7.12 Hepatoprotective Effects

Water extract of jujube reduce the liver injury that is caused by ischemia. This effect might be due to the antioxidant potential of the jujube (Chen et al., 2010). The jujube extracts showed protective activity against CCl_4-induced hepatic injury. Jujube extracts have been proven as a good source of hepatoprotective phytochemicals because the extracts are a rich source of sugars, vitamins, and triterpenoids.

7.13 Sedative-Hypnotic and Anxiolytic Effects

The anxiolytic activity has been shown by the seed extracts of jujube. It is considered that seed extracts can suppress nervous system activity by stimulating sleep but cannot cause the muscle relaxation or anti-convulsion behavior. The anxiolytic and sedative-hypnotic activity is said to be possessed by the seeds and the leaves of the jujube (Peng et al., 2000).

7.14 Anticancer Activity

The triterpenes and triterpenoic acids showed cytotoxicities against tumor cell lines. The action of the 3-O-p-coumaroyl alphitolic acids is much greater than the noncoumaroic triterpenoids, which proves that the C-3 position of the coumaroyl group in the lupene-type triterpenes enhances the cytotoxic activity (Lee et al., 2003). The cytotoxicity has been verified against the tumor cells using jujube extract. The lupene is a terpenoid that shows the cytotoxicities. The cytotoxicity of 3-O-p-coumaroylalphitolic acids is superior to noncoumaroic triterpenoids (Lee et al., 2003). Triterpenoic acid and betulinic acid showed activity against selective melanoma cells. Betulinic acid is developing in the medicinal industry. It is thought that betulinic acid is a major causative agent that can act against cancerous cells.

7.15 Antimicrobial Activity

The antifungal effects of jujube extracts have been studied (Mahajan and Chopda, 2009). Jujube extracts show strong action against various fungal strains including *Aspergillus niger, Aspergillus flavus, Malassezia furfur, Candida albicans,* and *C. tropicalis*. Root extracts are found to be active against 20 strains of bacteria (Mahajan and Chopda, 2009). The leaf extract possesses the ability to act against *Pseudomonas* spp., *Bacillus subtilis, Klebsiella* spp., and *Proteus vulgaris* (Chowdary and Padashetty, 2000).

7.16 Permeability Enhancement Activity

Water extracts of jujube displayed permeability enhancement activity (Eley and Dovlatabadi, 2002).

7.17 Cognitive Activities

Jujube extracts are effective against Alzheimer disease due to the presence of oleamide (Heo et al., 2003). It has been proven that in the brain of Alzheimer patients, choline acetyltransferase enzyme (responsible for controlling the release of acetyl choline) becomes diminished. Oleamide is the major compound that acts against the disease.

8. SIDE EFFECTS AND TOXICITY

Jujube prevents pregnancy. It can interfere with blood sugar control during or after an operation as it lowers the blood sugar level.

References

Adzu, B., Amos, S., Wambebe, C., Gamaniel, K., 2001. Antinociceptive activity of *Zizyphus spina-christi* root bark extract. Fitoterapia 72, 344–350.

Ansari, M.J., Al-Ghamdi, A., Usmani, S., Al-Waili, N.S., Sharma, D., Nuru, A., Al-Attal, Y., 2013. Effect of jujube honey on *Candida albicans* growth and biofilm formation. Archives of Medical Research 44, 352–360.

Arutla, R., Swaroopa, D., Rao, K.S., 2012. Wound healing potential of *Ziziphus jujuba* bark extract on Albino rats. International Journal of Research in Ayurveda and Pharmacy 3.

Ashton, R.W., 2006. Jujube: The Chinese Date.

Buxton, D.R., US Department of Agriculture Agricultural Research Service.

Chen, C., Lee, J., Wang, D., Shen, C., Shen, K., Lin, M., 2010. Water Extract of *Zizyphus jujube* Attenuates Ischemia/Reperfusion–Induced Liver Injury in Rats (PP106), Transplantation Proceedings. Elsevier, pp. 741–743.

Cheng, G., Bai, Y., Zhao, Y., Tao, J., Liu, Y., Tu, G., Ma, L., Liao, N., Xu, X., 2000. Flavonoids from *Ziziphus jujuba* mill var. spinosa. Tetrahedron 56, 8915–8920.

Chowdary, N., Padashetty, N., 2000. In Vitro Screening of Antibacterial Activity of Leaves of Ber, vol. 29. Current Research-University of Agricultural Sciences (Bangalore), pp. 78–79.

Ebrahimi, S., Ashkani-Esfahani, S., Poormahmudi, A., 2013. Investigating the efficacy of *Zizyphus jujuba* on Neonatal Jaundice.

Eiznhamer, D., Xu, Z., 2004. Betulinic acid: a promising anticancer candidate. Idrugs: The Investigational Drugs Journal 7, 359–373.

Eley, J.G., Dovlatabadi, H., 2002. Permeability enhancement activity from *Ziziphus jujuba.* Pharmaceutical Biology 40, 149–153.

Esteki, T., Urooj, A., 2012. Antioxidant components and activity in the peel of *Ziziphus jujuba* mill. Journal of Pharmacy Research 5, 2705–2709.

Ganachari, M., Kumar, S., 2004. Anti-ulcer properties of *Ziziphus jujuba* Lam leaves extract in rats. Journal of Natural Remedies 4, 103–108.

Ganachari, M., Kumar, S., Alagawadi, K., 2007. Anti-obese activity of *Ziziphus jujuba* Lam leaves extract in dietary obese rats. Journal of Natural Remedies 7, 102–108.

Ganachari, M., Kumar, S., Bhat, K., 2004. Effect of *Ziziphus jujuba* Leaves Extract on Phagocytosis by Human Neutrophils.

Goetz, P., 2009. Demonstration of the psychotropic effect of mother tincture of *Zizyphus jujuba* mill. Phytothérapie 7, 31–36.

Guo, Y.-x., Shan, G.-h., 2010. The Chinese Jujube. Shanghai Scientific and Technical Publishers, Shanghai, China.

Heo, H.-J., Park, Y.-J., Suh, Y.-M., Choi, S.-J., Kim, M.-J., Cho, H.-Y., Chang, Y.-J., Hong, B., Kim, H.-K., Kim, E., 2003. Effects of oleamide on choline acetyltransferase and cognitive activities. Bioscience, Biotechnology and Biochemistry 67, 1284–1291.

Islam, M.B., Simmons, M.P., 2006. A thorny dilemma: testing alternative intrageneric classifications within *Ziziphus* (Rhamnaceae). Systematic Botany 31, 826–842.

Jawanda, J., Bal, J., Josan, J., Mann, S., 1980. Studies on the storage of ber fruits. I. Room temperature. Punjab Horticultural Journal.
Jiang, J.-G., Huang, X.-J., Chen, J., Lin, Q.-S., 2007. Comparison of the sedative and hypnotic effects of flavonoids, saponins, and polysaccharides extracted from semen *Ziziphus jujube*. Natural Product Research 21, 310–320.
Kumar, S., Ganachari, M., Nagoor, V., 2004. Anti-inflammatory activity of *Ziziphus jujuba* Lam leaves extract in rats. Journal of Natural Remedies 4, 183–185.
Lee, S.M., Min, B.S., Lee, C.-G., Kim, K.-S., Kho, Y.H., 2003. Cytotoxic triterpenoids from the fruits of *Zizyphus jujuba*. Planta Medica 69, 1051–1054.
Mahajan, R.T., Chopda, M., 2009. Phyto-Pharmacology of *Ziziphus jujuba* mill-a plant review. Pharmacognosy Reviews 3, 320.
Majumdar, G., Vedic Plants in BC. Law Commemoration 1.
Munier, P., 1973. Le palmier-dattier. Techniques agricoles et productions tropicales, vol. 24.
Naftali, T., Feingelernt, H., Lesin, Y., Rauchwarger, A., Konikoff, F.M., 2009. *Ziziphus jujuba* extract for the treatment of chronic idiopathic constipation: a controlled clinical trial. Digestion 78, 224–228.
Oudhia, P., 2003. Research Note on Medicinal Herb of Chhattisgarh, India Having Less Known Traditional Uses, vol. IX.
Pareek, O., 1983. The Ber (The ber).
Pareek, O., 2001. Fruits for the Future 2: Ber, International Centre for Underutilized Crop, vol. 38. Redwood Books, Wiltshire, pp. 15–20.
Pawlowska, A.M., Camangi, F., Bader, A., Braca, A., 2009. Flavonoids of *Zizyphus jujuba* L. and *Zizyphus spina-christi* (L.) Willd (Rhamnaceae) fruits. Food Chemistry 112, 858–862.
Peng, W.-H., Hsieh, M.-T., Lee, Y.-S., Lin, Y.-C., Liao, J., 2000. Anxiolytic effect of seed of *Ziziphus jujuba* in mouse models of anxiety. Journal of Ethnopharmacology 72, 435–441.
Ramadoss, S., Jaggi, M., Siddiqui, M.J.A., 2000. Use of Betulinic Acid and its Derivatives for Inhibiting Cancer Growth and a Method of Monitoring This. Google Patents.
Rao, G.H.J., Lakshmi, P., 2012. Anti diarrhoeal activity of *Ziziphus jujuba* leaf extract in rats. International Journal of Pharma Bio Sciences 3, 532–538.
Rekha, S., Chandrasekhara, S., 2014. Antifertility effect of Ziziphus jujuba mill. World Journal of Pharmacy and Pharmaceutical Sciences 3 (3), 1363–1370.
Su, B.-N., Cuendet, M., Farnsworth, N.R., Fong, H.H., Pezzuto, J.M., Kinghorn, A.D., 2002. Activity-guided fractionation of the seeds of *Ziziphus jujuba* using a cyclooxygenase-2 inhibitory assay. Planta Medica 68, 1125–1128.
Suttisri, R., Lee, I.-S., Kinghorn, A.D., 1995. Plant-derived triterpenoid sweetness inhibitors. Journal of Ethnopharmacology 47, 9–26.
Thompson, A.K., 2015. Fruit and Vegetable Storage: Hypobaric, Hyperbaric and Controlled Atmosphere. Springer.
Tripathi, S., 2014. *Ziziphus jujuba*: A Phytopharmacological Review.
Wang, Z., Xue, J., Liu, L., Deng, X., Wei, T., 2008. Effects of freezing methods and storage temperatures on the flesh firmness of jujube fruits. I International Jujube Symposium 840, 505–512.
Yao, S., 2013. Past, present, and future of Jujubes—Chinese dates in the United States. HortScience 48, 672–680.

CHAPTER

35

Lemon Balm

Kinza Waheed[1], *Haq Nawaz*[1], *Muhammad Asif Hanif*[1], *Rafia Rehman*[1], *Isiaka A. Ogunwande*[2]

[1] Department of Chemistry, University of Agriculture, Faisalabad, Pakistan; [2] Natural Products Research Unit, Department of Chemistry, Faculty of Science, Lagos State University, Ojo, Lagos, Nigeria

OUTLINE

Medicinal Plants of South Asia
https://doi.org/10.1016/B978-0-08-102659-5.00035-5

Copyright © 2020 Elsevier Ltd. All rights reserved.

1. BOTANY

1.1 Introduction

Lemon balm is an important medicinal plant in the world trade, and it has huge applications in drugs and flavoring industries (Aharizad et al., 2012). Genus *Melissa* includes a small number of species, and it is native to Europe and Asia. Lemon balm undergoes cross-pollination (Moradkhani et al., 2010). The bees are primary pollinators of the lemon balm (Chwil, 2009). The name lemon balm must not be confused with "bee balm," which belongs to another genus named *Monarda.*

Melissa officinalis was assigned different names according to different regions in the world. In English, it is known by lemon balm, mountain balm, or sweet balm (Khare, 2008). It is known as badranjbooye in Iran (Joharchi and Amiri, 2012). In Urdu, it is known by baranjiboya and in Hindi, it is called billilotan.

M. officinalis has a wide range of varieties and cultivars varying in their scent, flavor, and uses. Popular examples of cultivars are *M. officinalis* citronella, *M. officinalis* lemonella, *M. officinalis* quedlinburger, *M. officinalis* lime, and *M. officinalis* Aurea (Moradkhani et al., 2010).

M. officinalis is morphologically and chemically highly variable, and these variations appear to be strongly influenced by the environment. The origin, source, and growing conditions of *M. officinalis* have an impact on the uses, flavors, aromas, and medicinal applications of the plant. The variability of *M. officinalis* is reflected in the wide ranges of uses, and these uses will be discussed in this chapter.

1.2 History/Origin

M. officinalis is native to Northern Africa and Southern Europe, northern Iran, Afghanistan, in tropical regions, and in eastern and western Mediterranean regions where it has been cultivated for 2000 years (Pereira et al., 2009). Its name is originated from the Greek term for "honey bees," as lemon balm draws honey bees, and the term balm also originated from the Greek term "balsamons," which means "balsam," which stands for oily and sweet-smelling resins. This herb is known by different titles such as Melissa, sweet balm, plant labiates, honey plant, bee's leaf, bee balm, and apiatrum. It has a great history of its spread during the 1500s to 1700s, when a wave of European settlers came to the America. They also brought their livestock and most prized plants with them. Lemon balm seeds and lemon balm came with the first American settlers.

In medieval times, lemon balm was used as a spreading herb. Lemon balm was used to convey messages between lovers and was known as a consolation due to its application in relaxing medicines (Kowalchik and Hylton, 1987). Many cultures held the belief that mystical soothing powers are associated with lemon balm. In the 11th century, Avicenna, an ancient Arab physician, also approved that lemon balm "caused the mind and heart to become merry." This unique property of the herb launches it in medicine to cure feelings of anxiety and hopelessness.

Similarly in 1530, the German Braunschweig stated in the *Book of Distillation* that sweet balm contributed to a "sharp wytte" and "good memory" and restores those who were angry to be "mery and refressht again". The ancient Greeks believed that a strong relationship exists between bees and lemon balm; for example, there was a concept that honey bees will not leave their hive if this herb is present adjacent to their hive. The bees considered this herb as a sign of finding the way of their hive after moving away from their hive. Because of their view, Greeks used this herb to rub the hives of bees to create the feelings of welcome for the bees. The Prince of Wales, Prince Llewellyn, said that he drunk the tea of *Melissa* every day, which was the secret of his long age of 108 years.

John Evelyn, a famous English writer, described the lemon balm as a "ruler of brain." It gives mental strengthening and reduces melancholia. Essential oil of lemon balm was termed "bal-smin," which means king of all the essential oils. Avicenna suggested that lemon balm can nourish the heart (BAĞDAT and COŞGE, 2006).

1.3 Demography/Location

Lemon balm can grow in a variety of environments and climatic conditions but grows well in a moderate temperature climate. The herb requires sunny days for better growth and development (Balm, 2015). It is widely grown in Italy, Romania, North America, Bulgaria, Pakistan, Iran, Brazil, Turkey, Egypt, Kurdistan, and Poland (Norouzi and Pasha Zanousi, 2012; Silva et al., 2005; Cosge et al., 2009; Aziz and El-Ashry, 2009; Taherpour et al., 2012; Patora et al., 2003).

1.4 Botany, Morphology, Ecology

Lemon balm is a perennial plant having opposite pairs of toothed leaves of ovate shape, growing in the form of square. Lemon balm is a bushy plant, variation has been found in its height and width, its height ranges from 8 in. to 5 ft, and its width ranges from 12 to 24 in. (Small, 1997). The plant's fruit is very small sized and leaves may be smooth or hairy (Moradkhani et al., 2010). Flowers of lemon balm are two-lipped and grow in the form of whorled clusters. Bilateral symmetry has been found in flowers of lemon balm. Different colors have been exhibited by the flowers such as pale yellow, purplish, bluish, pinkish, and white, and it has nonglandular hair (Brickell and Zuk, 1997). Almost five sepals or petals are present in each flower. Sepals and petals are fused and form a tube or cup. Leaves of the lemon balm are 30–50 mm in length, with shiny upper surface, deeply veined, and wrinkled. The surface of the leaf is deeply vined and coarse and edge of the leaf is toothed or scalloped (Meftahizade et al., 2013). It gives the scent of lemon with a slight hint of mint. Due to the presence of the very strong and agreeable scent of lemon, it is known as lemon balm (Bakhru, 1992).

Temperature required for rapid growth of lemon balm is 15–35°C, and precipitation of 500–600 mm is required, which must be well distributed throughout the growing season. In case of lack of proper precipitation, it should be irrigated. Full sunny days are good for its rapid growth, but this herb can also germinate in the regions of partial shade (Janina, 2003). Sandy and loamy fertile soil is required for the cultivation of the lemon balm, which must be well drained. Lemon balm grows well in soil having a pH range of 5–7 (Moradkhani et al., 2010; Janina, 2003). It is vulnerable to shortage of water in the year of establishment.

2. CHEMISTRY

Lemon balm has a large number of ecotypes based on its flavor, taste, and many other phenotypic properties. The distinctiveness of fragrance and aroma in many lemon balm species/cultivars is due to the presence of essential oils in leaves and other parts of the plant. The components that are responsible for its strong aroma are β-caryophyllene, citronellal, geranial, geraniol, and neral (Franke, 1977; Mihajlov et al., 2013). Lemon balm has a citrus-like aroma due to the presence of citral isomers such neral and geranial, as well as a minute quantity of geranyl acetate and citronellal (Shakeri et al., 2016).

Lemon balm contains both thiamine (vitamin B) and vitamin C. It has been observed that lemon balm contains about 254 mg vitamin C per 100 mL of solution (Franke, 1977). Lemon balm contains many elements such as iron, zinc, potassium, calcium, and magnesium. Lemon balm was found to contain 14,602 ± 1.20 ppm potassium, 17522 ± 0.65 ppm calcium, 166.5 ± 0.56 ppm iron, 793 ± 0.91 ppm magnesium, 4,21 ± 0.76 ppm zinc, and small amounts of vitamins and proteins in a previous study (Tomescu et al., 2015). Lemon balm exhibits a great aroma of lemon, but its lemony aroma and flavor are due to presence of citronellal and citral (neral and geranial) compounds (Patora et al., 2003). Its oil also contains flavonglychoside acid, rosmarinic acid, and terpinene in very small amounts (Tagashira and Ohtake, 1998; Hohmann et al., 1999). The chemical structures of active components of lemon balm are shown in Fig. 35.1.

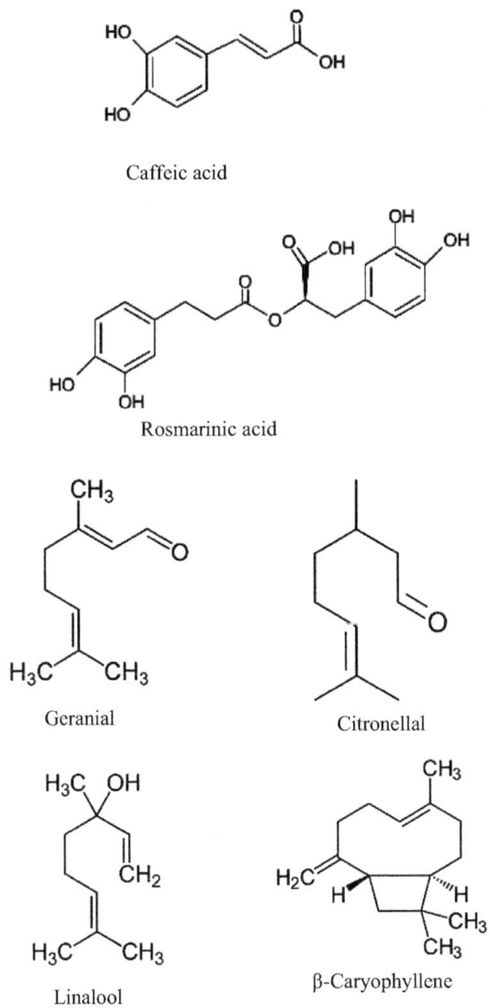

FIGURE 35.1 Chemical structures of active components of lemon balm.

3. POSTHARVESTING TECHNOLOGY

The best time to harvest lemon balm is 10:00 a.m., after the evaporation of morning dew and before the sun shines brightly. Lemon balm must be harvest when it is in robust health and looking alive and full of energy (Wynn and Fougere, 2006). The time of harvest influences oil contents. In a previous study, the best yield of essential oil was obtained by harvesting in morning after water sprayed was on the crop to prevent the loss of volatile terpene (Sorensen, 2000). Fresh material contains a greater amount of essential oil compared to dry. Lemon balm can be utilized in dried or fresh form. While using the fresh leaves of lemon balm, great care must be taken to avoid injury or bruising. Leaves of lemon balm dry very quickly; they will not be as fragrant as fresh. The leaves of lemon balm are much tastier when fresh. The dry leaves must be stored in airtight containers.

Lemon balm is dried in the shade to protect its chemical composition. Very intense sunlight will cause its volatile components to disappear. Dry leaves are stored in bags that allow air flow. Moisture must be avoided. Plastic bags can cause fungous growth if moisture is present. Essential oil can be packaged in smaller or bulk quantities. Leaves of lemon balm can also be dried by placing them on a drying rack or oven on a straw tray with a paper towel on it (Wynn and Fougere, 2006).

4. PROCESSING

Generally, lemon balm herb is dried in the shade to preserve the chemical composition of the plant. Direct exposure of intense sunlight will cause the disappearance of the volatile component of the essential oils. Epoxy-coated aluminum containers, ceramics, dark glass, and fluorinated plastic are used for its packaging. Great care is required to handle its essential oil because it is volatile in nature. It must be kept in a dark or cool place to prevent oxidation. Keep it airtight and do not expose it to heavy metals or heat.

Different methods are used to extract the chemical components of lemon balm. It can be steam distilled right after harvesting. Steam distillation is used to obtain the volatile components from the dried herb. The quality of the leaves will be reduced if their color turns to brown. So, the crop must be steam distilled immediately after harvest.

In a previous study, maximum essential oil obtained at flowering stage was 0.073%, and minimum oil yield 0.064% was obtained after the flowering stage (Ayanoglu et al., 2005). Lemon balm can be harvested two times (in the first year) to three times (in second and third years) per vegetation year. In India, maximum dry and fresh herbage and oil yield was obtained by harvesting after 160 days of planting, and the lowest yield was obtained by harvesting after 120 days of planting (Singh et al., 2014).

5. VALUE ADDITION

Lemon balm is commonly used as a flavoring in herbal tea and ice cream, sometimes in combination with other herbs like spearmint. Lemon balm can be paired with candies or dishes. It has been used in different dishes like fish and is an important constituent in pesto of lemon balm (Balm, 2007). Essential oil is released by crushing and squeezing its fresh leaves, and then this oil is used to add a flavor in fruits dishes, candies, pastries, and summer drinks. Lemon balm combines well with other spices like thyme, parsley, pepper, and chervil. Lemon balm is added to sweet foods, fish, and fowl (just before serving). Its leaves are used for garnishing purpose such as garnishing in salad, cake, and beverages. Twigs of lemon balm are used to decorate pitchers and water jars. Fresh lemon balm gives a very strong lemon fragrance and flavor, making it delightful for custards, tea, and fruit dishes (Small, 1997; Rinzler, 1990).

6. USES

Many spices and herbs contribute significantly to health despite the low amounts of consumption, as they are full of antioxidants and certain mineral compounds. Lemon balm is a good source of certain minerals and fibers. Lemon balm imparts a refreshing flavor to food, and lemon balm tea is available at many health food stores. Whatever is your taste and preference, lemon balm can be a great addition to your kitchen, and it adds flavor to food/dishes with an added health benefit.

Many cultures held the belief that mystical soothing powers are associated with lemon balm, and Dioscorides used this plant in drinks to relieve patients and for scorpion and dog bites. The ancient Arabs used it for the treatment of heart diseases. In earlier times, a spray of this herb was used to stop the blood of wound and to help to cure crooked necks, headache, pregnancy sickness, toothache, and baldness. Lemon balm was also used against catarrh, flatulence problems, and fever.

This herb is used as a bathing herb, where most of the leaves of the charming fragrance sprinkled on the water, or its mixture is added in the water. Lemon balm has great applications to regulate the mental processes.

Lemon balm has many culinary applications; leaves of lemon balm can be used in salads, desserts, dressings, confections, soups, and sauces. In the case of beverages, its leaves in dried form are utilized in drinks, vinegars, and teas. It imparts a lemon or mint-like flavor in soft drinks. Essential oil of lemon balm is also used to make furniture polish that has the great scent of lemon. It is utilized in the perfume industry. The leaves, essential oil, and flowers of this herb are used to make potpourri. It must be harvested in late summer for its best yield and effects. The herb has been valued as a medicinal herb, cosmetic, and culinary herb. Fresh leaves of lemon balm are used to top drinks and main dishes or to decorate salads, while its dried leaves are most commonly used for tea. It is a medicinal herb and has memory-enhancing properties and is used as a mood-elevating drug, as a mild depressant, sleep aid, and as a digestive aid to relieve gas, nausea, stomach pain, hypertension, and migraine headache, along with being used with fever to increase perspiration. Lemon balm is a good toner for skin since it smooths wrinkle and closes pores. Lemon balm works as an insect repellent, and it can be rubbed on hands to avoid bee stings. Bee keepers grow lemon balm to attract new swarms of bees.

7. PHARMACOLOGICAL USES

Traditionally, *M. officinalis* has had many applications in the field of medicine such as antispasmodic, surgical dressing, diaphoretic, carminative, tonic, diaphoretic, relief of stress-induced headache, sedative, hypnotic, and strengthening the memory (Blumenthal et al., 2000). Essential oil of *M. officinalis* is used in aromatherapy, which is useful for treatment of mild depression. And this is native to the northern Mediterranean (Tavares et al., 1996). Essential oil of lemon balm also shows sedative properties, and this activity is dose dependent, where doses in fewer amounts are more active. The researchers proposed that its sedative activity is due to presence of caryophyllene (Werker, 1993). Extract of lemon balm is beneficial to lower the lipids of liver tissues and reduce total cholesterol and also to raise the level of glutathione in the tissues (Bolkent et al., 2005). Lemon balm extract has been used as soothing hypnotic, to relief stress-induced headache, and as antiviral to boost healing of herpes simplex cold sores (Blumenthal, 2000). Researchers reported that the sedative properties are shown by essential oil of *M. officinalis* in mice and seemed to be dose dependent. It was demonstrated that lemon balm can be used as a tranquilizer, and methanolic extract of lemon balm and its components showed GABA-T inhibitory activity during an in vitro study on the brains of rats (Awad et al., 2009).

7.1 Antioxidant Activity

Lemon balm showed strong antioxidant effects, which were 10 times stronger than the antioxidant effects of vitamin C and vitamin B (Popova et al., 2016). The herb has been investigated to determine antioxidant activity and total contents of carotenoids, L-ascorbic acid, and phenolics. In an animal study, the efficacy of aqueous extract of *M. officinalis* was investigated in decreasing magnesium-induced brain oxidative stress in mice. Aqueous extract of lemon balm has potent neuroprotective properties and antioxidative properties, validating its efficacy in attenuating Mn-induced oxidative stress in the mouse brain (Martins et al., 2012).

7.2 Antiviral Activity

Hot water extract of lemon balm was injected into embryonated eggs for protection against the fatal effect of Newcastle, herpes simplex viruses, Semliki Forest, and vaccinia. The plaques produced by these viruses in the embryotic cells of chicks were suppressed by applying lemon balm extract–impregnated antibiotic sensitivity discs to the agar overlay surface. Injecting 10% gelatin into the eggs before applying the lemon balm extract largely eliminate the antiviral effect. It has been found that this activity is due to presence of tannin and tannin-like polyphenol that act on the surface of cells (Cohen et al., 1964). Tannin free polyphenol fraction of an aqueous extract inhibited vaccine viruses and herpes simplex in egg and cell culture systems. Aqueous extracts of the leaves have also been reported to have activity against Semliki Forest virus (Chevallier, 2016). Ten percent aqueous extract of lemon balm restricted the replication in vitro of influenza virus A2, herpes simplex virus type 2, mycoviruses, and influenza viruses in vitro and vaccine virus. Hemagglutination induced by mumps virus or Newcastle disease virus was inhibited by tannin, which is isolated from the aqueous extract of lemon balm (Parameswari et al., 2009).

7.3 Antispasmodic Activity

Those agents that are used to suppress muscle spasms are called antispasmodic, and this activity has been shown by ethanolic extract of lemon balm. Guinea pigs are a rodent specie belonging to Caviidae family and *Cavia* genus. An ethanol extract of the leaves and essential oil inhibited histamine- and barium-induced contractions of guinea pig ileum in vitro and exhibited smooth muscle relaxant activity in tracheal muscle, while an aqueous extract was inactive (Forster et al., 1980; Reiter and Brandt, 1984).

7.4 Gastrointestinal Tract Effects

Lemon balm has been used to promote digestion, to treat gastrointestinal disorders, and for gripping pain of the abdomen.

7.5 Antimicrobial Activity

Antimicrobial activity of lemon balm was tested against six fungi and 13 bacterial strains (Mimica-Dukic et al., 2004). The essential oil of lemon balm showed great antimicrobial activity with small minimum inhibitory concentration values and large inhibition zones against all microorganisms tested. Essential oil of lemon balm can be used in natural therapies of infectious diseases in plants, animals, and humans as well as in preservation of food (Abdellatif et al., 2014).

7.6 Antidepression and Antianxiety Activity

Antidepressant activity of aqueous extract of lemon balm was investigated by evaluating its influence on the behavior and relevant neurotransmitters of rats (Lin et al., 2015). The behavioral effects of an acute or subacute orally administered ethanol extract of lemon balm were evaluated in male and female Wistar rats in elevated plus-maze, forced swimming, and open field tests. The potential psychoactive properties of lemon balm may provide a unique pharmacological alternative for certain psychiatric disorders; however, the efficacy seems to be dependent on both gender and administration length (Taiwo et al., 2012). Leaves of lemon balm reduced the number of anxious patients in comparison to placebo and reduced the frequency of palpitation episodes. No side effect has been shown by lemon balm. It can be concluded that lyophilized aqueous extract of lemon balm leaves may be a proper and safe herbal drug for the treatment of benign palpitation and can be used as an antianxiety and antidepression agent (Alijaniha et al., 2015).

7.7 Antiagitation Activity

A state of nervous excitement and anxiety is called agitation, and lemon balm has antiagitation activity. Lemon balm inhibited the binding of t-butyl bicyclophosphorothionate to the rat forebrain gamma-aminobutyric acid receptor channel but had no effect on N-methyl-D-aspartate, alpha-amino-3-hydroxy-5-methyl-4-isoxazolepropianate, or nicotinic acetylcholine receptors. Lemon balm caused a significant dose-dependent reduction in both excitatory and inhibitory transmission, with a net depressant effect on neurotransmission that is in contrast to the classical GABA (gamma-aminobutyric acid) antagonist picrotoxinin, which evoked profound epileptiform burst firing in these cells. It was reported that neural membranes are unlikely to reflect a sedative interaction with any of the ionotropic receptors (Huang et al., 2008). It is recommended that leaves of lemon balm be used in differentiation of neuroblast, to increase cell proliferation and integration into granule cells by lowering serum corticosterone level and raising GABA levels in the mice (Yoo et al., 2011).

7.8 Anticancer Activity

Lemon balm contains agents that are used to inhibit and control the growth of cancer cells. Proliferation of colon carcinoma cell is inhibited by extract of lemon balm. Lemon balm extract also induces apoptosis through formation of reactive oxygen species. Extract of lemon balm also has anticancer effect, which was investigated in different human cancerous cells. Result shows that hydroalcoholic extract of lemon balm possesses a strong potential to inhibit the proliferation of various cancerous tissues in a dose-dependent manner. It has been investigated that the antiproliferative effect of lemon balm is specific to tumor type. And hormone-dependent cancer was more responsive to the antitumor effects of essential oil of lemon balm (Weidner et al., 2015).

8. SIDE EFFECTS AND TOXICITY

When taken by mouth, lemon balm can cause some side effects, including increased appetite, nausea, vomiting, abdominal pain, dizziness, and wheezing. If applied to the skin, the lemon balm may cause skin irritation and increased cold sore symptoms.

References

Abdellatif, F., Boudjella, H., Zitouni, A., Hassani, A., 2014. Chemical composition and antimicrobial activity of the essential oil from leaves of Algerian *Melissa officinalis* L. EXCLI Journal 13, 772.

Aharizad, S., Rahimi, M.H., Moghadam, M., Mohebalipour, N., 2012. Study of genetic diversity in lemon balm (*Melissa officinalis* l.) populations based on morphological traits and essential oils content. Annals of Biological Research 3, 5748–5753.

Alijaniha, F., Naseri, M., Afsharypuor, S., Fallahi, F., Noorbala, A., Mosaddegh, M., Faghihzadeh, S., Sadrai, S., 2015. Heart palpitation relief with *Melissa officinalis* leaf extract: double blind, randomized, placebo controlled trial of efficacy and safety. Journal of Ethnopharmacology 164, 378–384.

Awad, R., Muhammad, A., Durst, T., Trudeau, V.L., Arnason, J.T., 2009. Bioassay-guided fractionation of lemon balm (*Melissa officinalis* L.) using an in vitro measure of GABA transaminase activity. Phytotherapy Research 23, 1075–1081.

Ayanoglu, F., Arslan, M., Hatay, A., 2005. Effects of harvesting stages, harvesting hours and drying methods on essential oil content of lemon balm grown in Eastern Mediterranean. International Journal of Botany 1, 138–142.
Aziz, E., El-Ashry, S., 2009. Efficiency of slow release urea fertilizer on yield and essential oil production of lemon balm (*Melissa officinalis* L.) plant. American-Eurasian Journal of Agricultural & Environmental Sciences 5, 141–147.
Bakhru, H., 1992. Herbs that Heal: Natural Remedies for Good Health. Orient Paperbacks.
Balm, L., 2007. A Herb Society of America Guide, vol. 7. The Herb Society of America.
Balm, L., 2015. Lemon Balm (*Melissa officinalis* L.) an Herbal Medicinal Plant with Broad Therapeutic Uses and Cultivation Practices: A Review.
BAĞDAT, R.B., COŞGE, B., 2006. The essential oil of lemon balm (*Melissa officinalis* L.), its components and using fields. Anadolu Journal of Agricultural Sciences 21, 116–121.
Blumenthal, M., German Federal Institute for Drugs and Medical Devices, 2000. Commission E. Herbal Medicine: Expanded Commission E Monographs. Integrative Medicine Communications, Newton, MA, pp. 160–169.
Blumenthal, M., Goldberg, A., Brinckmann, J., 2000. Herbal Medicine. Expanded Commission E Monographs. Integrative Medicine Communications.
Bolkent, S., Yanardag, R., Karabulut-Bulan, O., Yesilyaprak, B., 2005. Protective role of *Melissa officinalis* L. extract on liver of hyperlipidemic rats: a morphological and biochemical study. Journal of Ethnopharmacology 99, 391–398.
Brickell, C., Zuk, J., 1997. The American Horticultural Society: AZ Encyclopedia of, Garden Plants.
Chevallier, A., 2016. Encyclopedia of Herbal Medicine. Penguin.
Chwil, M., 2009. Flowering biology and nectary structure of *Melissa officinalis* L. Acta Agrobotanica 62.
Cohen, R.A., Kucera, L.S., Herrmann, E.C., 1964. Antiviral activity of *Melissa officinalis* (lemon balm) extract. Experimental Biology and Medicine 117, 431–434.
Cosge, B., Ipek, A., Gurbuz, B., 2009. GC/MS analysis of herbage essential oil from lemon balms (*Melissa officinalis* L.) grown in Turkey. The Journal of Applied Biological Sciences 3.
Forster, H., Niklas, H., Lutz, S., 1980. Antispasmodic effects of some medicinal plants. Planta Medica 40, 309–319.
Franke, W., 1977. On the contents of vitamin C and thiamin during the vegetation period in leaves of three spice plants (*Allium schoenoprasum* L., *Melissa officinalis* L. and *Petroselinum crispum* (mill.) nym. ssp. crispum). In: I International Symposium on Spices and Medicinal Plants, vol. 73, pp. 205–212.
Hohmann, J., Zupkó, I., Rédei, D., Csányi, M., Falkay, G., Máthé, I., Janicsák, G., 1999. Protective effects of the aerial parts of *Salvia officinalis*, *Melissa officinalis* and *Lavandula angustifolia* and their constituents against enzyme-dependent and enzyme-independent lipid peroxidation. Planta Medica 65, 576–578.
Huang, L., Abuhamdah, S., Howes, M.J.R., Dixon, C.L., Elliot, M.S., Ballard, C., Holmes, C., Burns, A., Perry, E.K., Francis, P.T., 2008. Pharmacological profile of essential oils derived from *Lavandula angustifolia* and *Melissa officinalis* with anti-agitation properties: focus on ligand-gated channels. Journal of Pharmacy and Pharmacology 60, 1515–1522.
Janina, M., 2003. Melissa officinalis. International Journal of Aromatherapy 10, 132–139.
Joharchi, M.R., Amiri, M.S., 2012. Taxonomic evaluation of misidentification of crude herbal drugs marketed in Iran. Avicenna Journal of Phytomedicine 2, 105.
Khare, C.P., 2008. Indian Medicinal Plants: An Illustrated Dictionary. Springer Science & Business Media.
Kowalchik, C., Hylton, W., 1987. Rodale's Illustrated Encyclopedia of Herbs, vi+. Emmaus, Pennsylvania, 536 p.

Lin, S.-H., Chou, M.-L., Chen, W.-C., Lai, Y.-S., Lu, K.-H., Hao, C.-W., Sheen, L.-Y., 2015. A medicinal herb, *Melissa officinalis* L. ameliorates depressive-like behavior of rats in the forced swimming test via regulating the serotonergic neurotransmitter. Journal of Ethnopharmacology 175, 266–272.

Martins, E.N., Pessano, N.T., Leal, L., Roos, D.H., Folmer, V., Puntel, G.O., Rocha, J.B.T., Aschner, M., Ávila, D.S., Puntel, R.L., 2012. Protective effect of *Melissa officinalis* aqueous extract against Mn-induced oxidative stress in chronically exposed mice. Brain Research Bulletin 87, 74–79.

Meftahizade, H., Sargsyan, E., Moradkhani, H., 2013. Investigation of antioxidant capacity of *Melissa officinalis* L. essential oils. Journal of Medicinal Plants Research 4, 1391–1395.

Mihajlov, L., Ilieva, V., Markova, N., Zlatkovski, V., 2013. Organic cultivation of lemon balm (*Melissa officinalis*) in Macedonia. Journal of Agricultural Science and Technology B 3, 769.

Mimica-Dukic, N., Bozin, B., Sokovic, M., Simin, N., 2004. Antimicrobial and antioxidant activities of *Melissa officinalis* L.(Lamiaceae) essential oil. Journal of Agricultural and Food Chemistry 52, 2485–2489.

Moradkhani, H., Sargsyan, E., Bibak, H., Naseri, B., Sadat-Hosseini, M., Fayazi-Barjin, A., Meftahizade, H., 2010. *Melissa officinalis* L., a valuable medicine plant: a review. Journal of Medicinal Plants Research 4, 2753–2759.

Norouzi, T.S., Pasha Zanousi, M.M., 2012. Essential oil component in leaf and flower of Lemon balm (*Melissa officinalis* L.). Research in Pharmaceutical Sciences 7, S749.

Parameswari, G., Meenatchisundaram, S., Subbraj, T., Suganya, T., Michael, A., 2009. Note on pharmacological activities of *Melissa officinalis* L. Ethnobotanical Leaflets 2009, 25.

Patora, J., Majda, T., Góra, J., Klimek, B., 2003. Variability in the content and composition of essential oil from lemon balm (*Melissa officinalis* L.) cultivated in Poland. Acta Poloniae Pharmaceutica 60, 395–400.

Pereira, R.P., Fachinetto, R., de Souza Prestes, A., Puntel, R.L., da Silva, G.N.S., Heinzmann, B.M., Boschetti, T.K., Athayde, M.L., Bürger, M.E., Morel, A.F., 2009. Antioxidant effects of different extracts from *Melissa officinalis*, *Matricaria recutita* and *Cymbopogon citratus*. Neurochemical Research 34, 973–983.

Popova, A., Dalemska, Z., Mihaylova, D., Hristova, I., Alexieva, I., 2016. *Melissa officinalis* L.-GC profile and antioxidant activity. International Journal of Pharmacognosy and Phytochemical Research 8 (4), 634–638.

Reiter, M., Brandt, W., 1984. Relaxant effects on tracheal and ileal smooth muscles of the Guinea pig. Arzneimittel Forschung 35, 408–414.

Rinzler, C.A., 1990. The Complete Book of Herbs, Spices, and Condiments: From Garden to Kitchen to Medicine Chest. Facts on File.

Shakeri, A., Sahebkar, A., Javadi, B., 2016. *Melissa officinalis* L.—A review of its traditional uses, phytochemistry and pharmacology. Journal of Ethnopharmacology 188, 204–228.

Silva, S.d., Sato, A., Lage, C.L.S., Gil, S., da Silva, R.A., Azevedo, D.d.A., Esquibel, M.A., 2005. Essential oil composition of *Melissa officinalis* L. in vitro produced under the influence of growth regulators. Journal of the Brazilian Chemical Society 16, 1387–1390.

Singh, S., Haider, S., Chauhan, N., Lohani, H., Sah, S., Yadav, R., 2014. Effect of time of harvesting on yield and quality of *Melissa Officinalis* l. In doon Valley, India. Indian Journal of Pharmaceutical Sciences 76, 449.

Small, E., N.R.C. Canada, 1997. Culinary Herbs. NRC Research Press.

Sorensen, J.M., 2000. Melissa officinalis. International Journal of Aromatherapy 10, 7–15.

Tagashira, M., Ohtake, Y., 1998. A new antioxidative 1, 3-benzodioxole from *Melissa officinalis*. Planta Medica 64, 555–558.

Taherpour, A., Maroofi, H., Rafie, Z., Larijani, K., 2012. Chemical composition analysis of the essential oil of *Melissa officinalis* L. from Kurdistan, Iran by HS/SPME method and calculation of the biophysicochemical coefficients of the components. Natural Product Research 26, 152–160.

Taiwo, A.E., Leite, F.B., Lucena, G.M., Barros, M., Silveira, D., Silva, M.V., Ferreira, V.M., 2012. Anxiolytic and antidepressant-like effects of *Melissa officinalis* (lemon balm) extract in rats: influence of administration and gender. Indian Journal of Pharmacology 44, 189.

Tavares, A., Pimenta, M., Goncalves, M., 1996. Micropropagation of *Melissa officinalis* L. through proliferation of axillary shoots. Plant Cell Reports 15, 441–444.

Tomescu, A., Rus, C., Pop, G., Alexa, E., Radulov, I., Imbrea, I.M., Negrea, M., 2015. Researches regarding proximate and selected elements composition of some medicinal plants belonging to the lamiaceae family. Agronomy Series of Scientific Research/Lucrari Stiintifice Seria Agronomie 58.

Weidner, C., Rousseau, M., Plauth, A., Wowro, S., Fischer, C., Abdel-Aziz, H., Sauer, S., 2015. *Melissa officinalis* extract induces apoptosis and inhibits proliferation in colon cancer cells through formation of reactive oxygen species. Phytomedicine 22, 262–270.

Werker, E., 1993. Function of essential oil-secreting glandular hairs in aromatic plans of Lamiacea—a review. Flavour and Fragrance Journal 8, 249–255.

Wynn, S.G., Fougere, B., 2006. Veterinary Herbal Medicine. Elsevier Health Sciences.

Yoo, D.Y., Choi, J.H., Kim, W., Yoo, K.-Y., Lee, C.H., Yoon, Y.S., Won, M.-H., Hwang, I.K., 2011. Effects of *Melissa officinalis* L.(lemon balm) extract on neurogenesis associated with serum corticosterone and GABA in the mouse dentate gyrus. Neurochemical Research 36, 250–257.

CHAPTER

36

Ma-Huang

Adan Iqbal[1], Rasheed Ahmad Khera[1], Muhammad Asif Hanif[1], Muhammad Adnan Ayub[2], Muhammad Nadeem Zafar[3]

[1] Department of Chemistry, University of Agriculture, Faisalabad, Pakistan; [2] Department of Chemistry, University of Okara, Okara, Pakistan; [3] Department of Chemistry, University of Gujrat, Gujrat, Pakistan

OUTLINE

Medicinal Plants of South Asia
https://doi.org/10.1016/B978-0-08-102659-5.00036-7

Copyright © 2020 Elsevier Ltd. All rights reserved.

1. BOTANY

1.1 Introduction

Ma-huang (Fig. 36.1) (*Ephedra gerardiana)* belongs to the primitive plant family Ephedraceae that relates to Gymnosperms (meaning naked seeds). Genus *Ephedra* comprises more than 50 species (Schaneberg et al., 2003). Some of the primary species of *Ephedra* genus are represented by *Ephedra major* Host, *Ephedra sinica, Ephedra gerardiana* Wall, *Ephedra intermedia* Schrenk & Meyer, and *Ephedra equisetina* Bunge (Morton, 1977). Species of genus Ephedra grow well in temperate and subtropical regions of China, Afghanistan, Bhutan, Pakistan, India, and North and Central America. It uncovers its use as an antipyretic, circulatory stimulant, diaphoretic, and as a cough medicine.

There are around nine species of genus *Ephedra* present in Pakistan. The members of genus *Ephedra* are evergreen and short perennial shrubs

FIGURE 36.1 Dried Ma-huang.

that are frost and drought resistant. Taxonomy of *Ephedra* species is controversial due to its morphologic features that are overlapping, making them difficult to distinguish and differentiate.

E. gerardiana falls in the category of endangered plants due to its extensive use for medicinal purposes, export, overgrazing, and overharvesting that has resulted in loss of biologic diversity. *E. gerardiana* has several common names depending on its location in the world. In Tibetan, it is known as amsania, ma houng, budagur, khanda, and chefrat, and in China as ma-huang. In Pakistan, it is known as narom, oman, raci, sang kaba, sikkim ephedra, som kalapna, and soma kalpa. In India, it is known as tootagantha and somalata, names in the Sanskritn language. The Arabic names for this herb are alend and zanbul khail.

1.2 History/Origin

In China, it is known as ma-huang, which literally means yellow horsetail or hemp yellow, typically due to the bitter taste associated with it. It is one of the oldest known medicinal plants. The use of *E. gerardiana* that dates to BCE was in sacred vessels at the temples for worship of fire. Among other old uses was in ritual vessels, which has been confirmed though its traces found mixed along with poppy pollen. The use of this plant along with other entheogens is also known in psychoactive applications.

In India, it has been known since the Vedic period, where it is used as a substitute for the psychoactive plant whose identity has been lost. Those psychoactive plant somas were used for the preparation of a sacred beverage until losing their identity and being substituted with other plants. Among them, one substitute is *E. gerardiana*, whose stimulating effects were even more spectacular. That is why it was named as plant of the moon or somalata. The oldest sacred book *Sanskrit* also describes its use for long life, for which its juice extract was used. During the period of the Roman Empire, its benefits were also confirmed by the literature.

At the end of the 16th century, Ephedra dried stems were transported to Japan, where they gained much attention from pharmacists due to its medicinal benefits. Nagayoshi Nagai, a Japanese pharmacist working in collaboration with a chemist named August Wilhelm von Hofmann, isolated ephedrine in 1885. Pharmacological activities of ephedrine proved to be toxic, which made people stop working on it and created a gap of about 30 years in the research related to ephedrine. But reinvestigations resulted in the discovery of its role as a cardiac stimulant, relaxant of smooth muscles of bronchi, and in elevating blood pressure, much like the hormone adrenaline. The stimulating role of ephedrine raised its sales in 1940, but those declined during the 1980s due to its street drug reputation. Overdose of this alkaloid can result in failure of the heart and hypothermia, as well as possibly causing death. About a 1-2 g dose of ephedrine alkaloid is considered lethal.

1.3 Demography/Location

Ephedra species require a sandy and dry environment for growth and are native in subtropical and temperate regions of Pakistan, China, India, Afghanistan, Mongolia, and regions of North and Central America (Blumenthal and King, 1995). *E. gerardiana* grows in the region of the Himalayan Mountains from Afghanistan to Bhutan because its growth requires high and dry mountain deserts. In Nepal, it is found most often in association with *Rhododendron* and *Juniperus recurva* species. *E. gerardiana* is located in Pakistan in the regions of Urak, Chitral, Swat, South Waziristan, Razmak, Tirch Mir, Swat, Kalam, Gilgit Baltistan, Char, above Skardu, Shaksgan, and Kashmir.

As mentioned earlier the concentration of alkaloid content varies depending on the demographic distribution. The American and Chilean species do not contain any of the active compounds and only the common species in Europe (*Ephedra distachya*) is reported to contain little alkaloid content depending on the habitat (that is the Atlantic coast). However, the Pakistan, Indian, and Chinese species of the genus *Ephedra* that contain a significant amount of alkaloids include *E. intermedia, E. equisetina, E. sinica,* and *E. gerardiana*. The production of commercial medicines and other uses of these species depend on the alkaloid content present in them. Almost all commercial applications of *Ephedra* extracts derive from the ephedrine alkaloids found in the stems in many Eurasian species. Ma-huang is the best reported drug made from *Ephedra*. This traditional Chinese medicine has been used for 5000 years for the treatment of fever, asthma, and nasal congestion. In Pakistan, an annual production of about 1500 tons is estimated, of which about 1200 tons is exported to Japan and Singapore. Other popular names of ma-huang are desert tea, Mormon tea, American ephedra, Chinese ephedra, European ephedra, Pakistani ephedra, and ephedra.

1.4 Botany, Morphology, Ecology

Rigid plant of *E. gerardiana* reaches approximately a height of 60–90 cm. It is surrounded by numerous branches that are densely packed and arise from a woody base with scales present at the joints of branches. Rhizomes growing through underground buds are responsible for the wide distribution of this plant. Decussate leaves that are in opposite direction are found in three or four whorls. The mode of growth of branches is basitonic and surrounded by lateral branches that are responsible for maintaining the shrubby habitat of this medicinal plant. Small-sized yellow flowers arise from stalks that are primarily surrounded by small leaves. Edible red fruit of round shape ripens in autumn season (Morton, 1977). Bitter, astringent taste and greenish-yellow color and aromatic odors are the key features of stem of *E. gerardiana* (Khandelwal, 2008). Nodes are cylindrical in appearance with furrows and stipulate leaves and internodes also grow on nodes.

Dioecious bearing both male and female strobili are present in same plants; however, monoecious species are also reported in genus *Ephedra*. In monoecious plants, ovules are present above the base, while stamens are present at the base. Male strobili are surrounded at the node or the axial of the leaf and comprised of a short axis on which thick bracts are arranged. Microsporangiophores or male strobili have about 11 sporangia at their tips, which are ovate in appearance. Female strobili or megasporangia are composed of a central axis and are present two to four per node. About one to three female strobili grow in the female flowers that bear an ovule or megasporangium at their end. Seeds result from double fertilization and are one to two in number. These are triangular or ovoid in shape and are without any dormant period. Seeds grow in to a cotyledon and a new plant grows from it. Seed dispersal mode is responsible for making dry or fleshy bract. Polyplicate type of pollen is characteristic of the species of the genus *Ephedra*. Branching pattern, size of grains, along with shape of ridges are key characteristics for the species of this genus.

E. gerardiana growing at 2700–4500 m altitude is found in subtropical and temperate regions. Frost and drought resistance are the requirements for the growth of this plant. Alkaloid content is badly affected by the rainfall, due to which it dramatically decreases. May through August and October through November are the months during which alkaloid content reaches the maximum value. Ability to survive in rocky/sandy soil, little water, and highly salty soil are promising features of this plant (Christian, 1998).

2. CHEMISTRY

Its bitter and astringent taste is responsible for giving the tongue a sensation of numbness. The major active constituents of *E. gerardiana* include alkaloids (Tai, 2003). The concentration of alkaloids of this plant ranges from 0.5% to 2.5% (Soni et al., 2004), while in the aerial parts, it ranges from 0.02% to 3.4%, including six other alkaloids that are optically active, and these are mostly present in the internodes. Of the optically active alkaloids isolated from *E. gerardiana*, (−) ephedrine is the major isomer among them, comprising 30%–90% of total alkaloids. Moreover, (+) pseudoephedrine and other trace amounts of ephedrine are also extracted along with ephedrine. Species of *Ephedra*, time and year of harvest, weather conditions, and altitude all affect the alkaloid content of

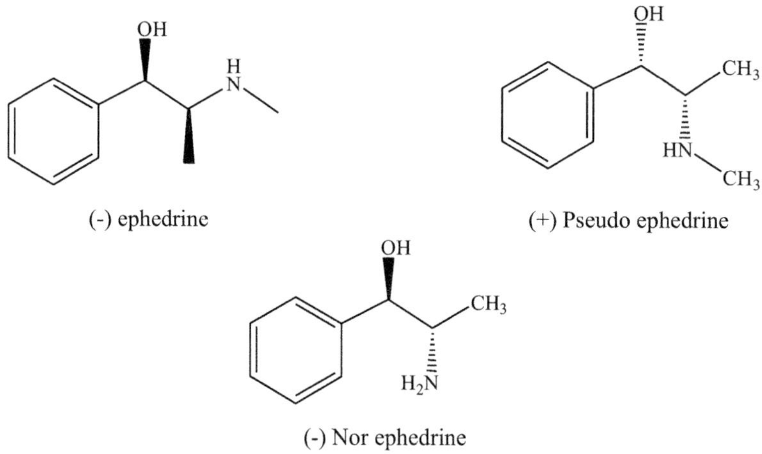

FIGURE 36.2 Structures of major active constituent of Ma-huang.

the plant. It is also a good source of catechins and (−) epicatechin, which are oxidizing components and help in weight loss. Beside this, it also contains polyphenols, herbacetin, and quercetin. Quercetin is well known for its antioxidant abilities. Gallic acid, which is known for its anticancer and antiinflammatory activity, is also present in it. Other chemical constituents that have been isolated up till now are found to be macrocyclic spermine alkaloids, cyclo polyglycine, kynurenic acid derivatives, methanoproline amino acid, flavones, flavonols, tannins, ephedradinnes A−D, carboxylic acid, and volatile terpenes. Out of these, ephedrannine A as well as maokonine have been found to possess hypotensive activity. Structures of major active components of ma-huang are shown in Fig. 36.2.

The (−) ephederine was first reported in 1887 by the Japanese pharmacist Nagai as the major constituent. The second major constituent is the (+) pseudoephiderine that is also the diastereoisomer of (−) ephederine, and it was isolated in 1889 by Ladenburg and Oelschlägel. The other four optically active isomers include (−) *N*-methylephedrine, (−) norephedrine, (+) *N*-methylpseudoephedrine, and (−) nor pseudoephiderine that were reported in the late 1920s. Usually (−) ephedrine isomer comprises about 60%−90% of total alkaloids content, (+) pseudoephederine is the second major isomer, while other alkaloids are present in the trace amounts. Other bioactive compounds reported include ephedroxane, (−) ephedrine oxazolidone derivative that is responsible for attributing antiinflammatory activity. The Food and Drug administration (FDA) reported that beyond the optically active compounds that have been reported, the plant also contains other phytoconstituents that are responsible for enhancing pharmacological properties and toxicity of products, making it difficult to know the toxicity level just by looking at the contents of an ephedra product.

Secondary metabolites of different types such as flavones, tannins, flavanols, carboxylic acids, and other volatile terpenes have also been reported in *E. gerardiana* (Abourashed et al., 2003). The astringent taste of ephedra products is attributed to the presence of tannins (ellagic and gallic tannins). Volatile oil as a constituent has also been found. (−) EPH, oxazolidone analogue, and ephedroxane, mostly obtained from the aerial parts of *Ephedra* species have been found in at least six more species that contain ephedrine alkaloids (Konno et al., 1985).

3. POSTHARVEST TECHNOLOGY

Collection of *E. gerardiana* plant at the proper time is essential as any delay before or after proper time collection may affect the quality and ultimately results in lowering the quality of the drug. The best time for collection is the months of August and October. The ideal/proper time for plant collection is estimated when it contains the maximum amount of therapeutically active agents. Beside this, collected plant parts suffer from a drastic decrease that results in lowering the ephedrine-type alkaloid content. After the collection of plant parts, they are dried, and the most commonly used method is sun drying due to its simplicity and ease. However, factors like consumption of time, labor, and its dependence on weather also affect it. It also suffers from the drawback that long exposure to sun rays may bleach the active constituents and degrade them, affecting the quality of chemical constituents. To preserve the quality for a longer period, effective drying is required. Humidity as well as dew that falls at night also affects the active constituents. *E. gerardiana* dried parts serve as the raw material for pharmaceutical preparations, and usually, they are mostly dried by means of sunlight to preserve the quality of medicinal plants (Gaur and Sharma, 2011). Sun drying is also found to be effective in reducing the growth of microorganisms. Drying via means of mechanical methods may also be used as it offers the advantage of drying at the fastest rate. Followed by drying, packing of the dried plant parts is done. Unsuitable methods of packaging and storage also make it difficult to maintain the proper quality of alkaloid contents (Fukushima, 2004). Packaging of the dried plant parts is mostly done through storing in ventilated polythene-lined gunny bags. By applying the suitable conditions, it has been found that polythene-lined gunny bags result in less destruction of alkaloid content in comparison to other materials that are used for traditional drying such as cotton cloths, sacks, and bags. From all these findings, it appears that packing material nature also results in affecting the quality of alkaloid contents. In the same way, proper storage is also required to sustain the quality of alkaloid contents.

4. PROCESSING

Ephedra content in its pure form is most commonly used rather than as extracted from its herb. The ephedra content is extracted by means of acid base extraction procedure. By following this extraction procedure, dried stems powdered material is made alkaline by means of base, and after this, it is removed by means of $CHCl_3$. After distilling the solvent, a dilute acid is used for dissolving the residue. Carbon is decolorized, and after this, filtration is carried out. Alkalization of the filtrate is followed by reextraction by means of some solvent such as diethyl ether, which is then evaporated to obtain crystals that are washed through hot water and dried. Crystals obtained are of (−) ephedrine. Hydrochloride of L-ephedrine, pseudoephedrine in the form of its base, oxalate of l-methylephedrine, D-bitartrate of d-methylpseudoephedrine, and sulfate of nor-d-pseudoephedrine all have been isolated.

5. VALUE ADDITION

E. gerardiana fruit pulp is a rich source of many amino acids and used for the preparation of jams (Hussain et al., 2006). This fruit pulp is found to be good in taste and can also be used as a source of food for birds and rodents. It also finds its use in the preparation of tea for the treatment of cough, as an antipyretic, circulatory stimulant, and diaphoretic. The chemical constituents of *E. gerardiana,* catechin and epicatechin are used in the preparation of weight loss drinks and pills. A drink from this plant is used as a rich source of energy. As a source of exerting eye soothing and fresh looks, it is used on lawns and in home decoration. Yaks and goats living on mountainous regions feed on it, so these animals also enjoy the stimulating effect of this perennial herb.

6. USES

E. gerardiana is used as a diet supplement for the reduction of weight, to enhance the physical performance of athletes, as a cardiac stimulator, and in nasal decongestants that are used for the treatment of asthma. Traditional uses of *E. gerardiana* include the use of its tea for the treatment of diseases associated with respiratory tracts. *E. gerardiana* along with other *Ephedra* species such as *Ephedra vidris* are used for making Mormon tea, which was used by the native people of America to cure venereal diseases and to purify the blood and help in cleansing the kidneys. Mormon tea was introduced by the people of the West, who used it as a source of refreshing and awakening. Due to its diuretic properties, it is used for the treatment of gonorrhea and syphilis. Hay fever and cold are cured by the dried stems of this plant. *E. gerardiana* dried bundles are used for burning incense during ceremonies. Fruit pulp of *E. gerardiana* is used for making jams, and it is also used as a food for goats and hawks and serves as source of food in mountainous regions.

For medication the dried stems of *E. gerardiana* are used in traditional Chinese ways to relieve influenza, bronchitis, asthma, hay fever, and nasal congestion. To completely eradicate the symptoms of allergy and hay fever, ephedrine as well as pseudoephedrine are mostly used, making them popular constituents of medicines against allergy (Chaturvedi and Dass, 2011). These also find their use for treating fever, lack of perspiration, low blood pressure, hives, and headache (Musselman, 1996). Folk use of medicine involves its crushing into a fine powder that is then boiled in water and used in the evening as well as morning for treating cold and hay fever. For treating diaphoretic indications, stem parts of *E. gerardiana* are used, while its rhizomes and root parts are used to treat night sweating due to its antiperspirant activity (Leung, 1990). Hindus and Parsees during their ceremonies use ephedra to produce a feeling of exhilaration. *E. gerardiana* dried bundles in ceremonies are burned to produce incense. Due to the spicy as well as pleasant smoke, its smell is comparable to that of a forest fire. Its ashes are used as a snuff. The smoke is inhaled by Lamas and Shaman while performing their ritualistic ceremonies.

In Ayurvedic medicine, tea made from dried parts of the *E. gerardiana* plant is used for the elevation of the blood pressure, for diuretic use, and to constrict the vessels of blood. This stimulatory effect appears after 6–8 hours of consumption. To treat hay fever, respiratory disorders, and asthma, people of Nepal use the incense or a tea made from it. For the problems caused by rejuvenation, a preparation made from *E. gerardiana* is used by the people of Tibet (Manandhar, 1980). However, when consumed at high dosage, it can cause nausea, heart palpitations, and sweating, and it is not recommended for heart patients. The extracts from the plant are found to enhance the psychotic effect of other medicines when used in combination, such as with mushrooms that are used for the preparation of psychotic medicines (Baker, 2010). The salts of ephedra are used in the treatment of swelling and congestion in the form of nasal sprays. To prevent hypotension, for anesthesia duration, ephedrine in the form of subcutaneous injection is used. However, it is also used in oral form for the treatment of nocturnal enuresis, angioneurotic edema, myasthenia gravis, urticaria, and epilepsy. If taken through oral pathways, pseudoephedrine could also be used as a nasal decongestant effectively (Morton, 1977).

7. PHARMACOLOGICAL USES

7.1 Antiasthmatic Effects

The antiinflammatory and bronchial smooth muscle relaxation effects of *E. gerardiana* are observed previously, making it antiasthmatic. Its ethanolic extracts are found to significantly reduce the albumin-induced eosinophilic inflammation of rats at a concentration of 100–200 mg/kg (Chaitanya et al., 2013). Another therapeutic effect against asthma is the antagonism of specific chemicals (mediators) that are released in response to foreign stimuli by mast cells. These chemicals include histamines, leukotrienes, and platelet activation factors that cause immediate bronchial inflammation. Degranulation of mast cells is important for release of these chemicals, and this has been characterized as a requirement for positive anaphylaxis. The chemicals are responsible for the crucial inflammatory responses including bronchial contraction, airway eosinophilia, and airway hyperresponsiveness. A mast cell stabilizer reduces synthesis of leukotrienes, histamines, inflammatory mediators, and serotonin and consequently causes blockage of the mediator receptors. These mast cell stabilizers have been extracted from several plant extracts and utilized as constituents of drugs for curing asthma ranging from mild to severe. The antiinflammatory activity of ethanolic extract of *E. gerardiana* is due to inhibition of these mediators, i.e., H_1 receptor agonist activity.

The *E. gerardiana* ethanolic extracts are also found to show activity against inflammation, and it can be compared to that of other reference drugs used such as dexamethasone. This inhibitory activity is attributed to inhibition of histamine, prostaglandins, and serotonin mediators that are released during inflammation and bronchial asthma, thus blockage of these mediators results in antiinflammatory activity. Further the ethanolic ma-huang extract also results in inhibiting the cyclooxygenase enzyme that causes inhibition of the synthesis of prostaglandin mediators. It appears from these observations that antagonistic effect to certain mediators exerts a protective mechanism against inflammatory activity, and response of an antigen could protect from it.

7.2 Weight Loss Effects

During the past few years, the use of ephedrine obtained from *E. gerardiana* extract and other members of the family Ephedraceae has gained much attention and has been extensively used in preparation of supplements for weight loss. However, the overuse may be lethal, as it badly affects the heart and may lead to cardiac attack. Ephedrine alkaloids have been reported to stimulate weight loss in both animal and plant studies, thus solving the problem of obesity. In a review conducted for 8 weeks, it was found that ephedrine, which is the major isomer of *E. gerardiana*, is responsible for weight loss in the shortest time and about 0.9 kg weight loss/month has been found. The possible mechanism of this shortest term response proceeds through the release of norepinephrine. The mechanism possesses effect on the hypothalamus (part of brain), which through release of norepinephrine exerts an anorexic effect (Andraws et al., 2005). The major accepted mechanism to promote weight loss appears through the breakdown of adipose tissues through release of catecholamines in response to stimulation of β receptors of fat. This stimulation leads to enhancement of lipogenesis and results in weight loss; this process is more prominent in persons with a slow rate of metabolism. This activity can be increased through synergistic effect with other botanical preparations such as caffeine and aspirin, which on administration in a cup of hot tea promote thermogenesis. These substances promote weight loss by acting as antagonists of adenosine receptors that result in inhibiting cellular phosphodiesterase activity. These synergistic compounds are said to promote the effect of ephedra and other alkaloids (Ang-Lee et al., 2001). In another study, it was found that ephedra alone results in 14% decrease in body weight along with a decrease in body fat of about 42%. However, when consumed with caffeine and theophylline, it resulted in 25% and 75% decrease in body weight. Theophylline and caffeine alone resulted in less observable decrease in body weight because the increased weight loss can be attributed to increased fat break down resulting from ephedrine and promoted by caffeine and theophylline. One of the better designed studies on the use of ephedrine/caffeine combination determined the safe use and efficiency of ma-huang herb–containing supplements with ephedrine (90 mg/day) along with koala nut (192 mg/day) and caffeine given to subjects for 6 months in a study conducted on weight loss. Randomly taken, 167 subjects received ephedrine/caffeine combination. The primary changes of these subjects were monitored such as blood pressure, weight loss, and heart rate. Moreover, it also results in decreasing blood lipid and body fat in men and women suffering from obesity. Compared with other treatments the herbal supplement results in minimum side effects (Tutin et al., 1993).

7.3 Antidiabetic Effects

The antidiabetic effect is attributed to the presence of five active glycans including ephedrans A, B, C, D, and E, which are found to lower the level of glucose in blood both in normal and alloxan-induced diabetic mice. Two compounds belonging to the family of epinephrine are found potent inhibitors of dipeptidyl peptidase-4. The names of these two compounds include (−) ephedrine and (+) pseudoephedrine. Both these compounds are reported to possess hypoglycemic activity (Celine et al., 2016). This antidiabetic activity is thought to proceed by stimulating the release of epinephrine that results in lowering the level of sugar.

7.4 CNS Stimulation Effects

The active constituent of *E. gerardiana* that exerts a stimulating effect on the central nervous system (CNS) is ephedrine, which is a strong CNS stimulant (Chen et al., 2004). In the species that are not found in America, norpseudoephedrine is present, which is more a stimulant of the CNS than ephedrine. Ephedrine and norpseudoephedrine pass through the blood−brain barrier and exert their stimulating effect on the hypothalamus and limbic system neurons. This, in turn, causes the dopamine release (Maglione et al., 2005), blood pressure regulation, adrenaline production, and regulates the heartbeat. Moreover, the release of dopamine is also responsible for decreasing fatigue, increasing physical activity, and a feeling of joy. Due to the chemical similarity between ephedrine and amphetamine, there exists strong concern for its misuse. The reinforcing effects of amphetamine in humans are like those of ephedrine, although it is not found to be as strong as that of amphetamines. Norepinephrine release is associated with a higher dosage level that is responsible for feelings of restlessness, anxiety, and insomnia.

7.5 Muscle Endurance Effects

Ephedrine when used in combination with other stimulants or either alone results in increasing the performance of anaerobic exercise. Ephedrine is responsible for the release of catecholamine, and it also stimulates the central nervous system, which results in increasing the performance. *E. gerardiana* finds its use in manufacturing the products that are used for the development of muscles. However, it has been found that chemical constituents of this plant are not involved in affecting the cells that are thought to be associated with hypertrophy of muscles. But it results in heart palpitations and increasing the risk of gastrointestinal, autonomic, and psychiatric symptoms (Peters et al., 2005).

7.6 Antimicrobial Activity

E. gerardiana is found to possess strong antimicrobial activities against *Escherichia coli*, *Pseudomonas aeruginosa*, *Staphylococcus aureus*, *Bacillus anthracis*,*Staphylococcus aureus*, and *Pseudomonas aeruginosa* (Baltch and

Smith, 1994; Taylor et al., 2002). The volatile oil of Ephedra demonstrated inhibitory activity against influenza virus (Soltan and Zaki, 2009).

7.7 Cardiovascular Action

E. gerardiana contains alkaloids that cause systolic as well as diastolic blood pressure elevation, increase performance of cardiac indices, and stimulate heartbeat. Ephedrine, epinephrine, and norepinephrine causes cardiac stimulation and vasoconstriction via exciting the sympathetic nervous system (Mack, 1996; Maglione et al., 2005; WHO, 1999).

7.8 Nasal Decongestant Effects

Ephedrine, a major component of *E. gerardiana*, acts on the sympathomimetics neurons and stimulates the α-receptor activity that results in vasoconstriction as well as blanching if it is applied on the mucosal surface of nose and pharynx. Pseudoephedrine and ephedrine both are used as nasal decongestants for the treatment of allergic rhinitis, although it is not as effective for the nasal decongestant treatment resulting from colds (Abourashed et al., 2003).

7.9 Diaphoretic Effects

E. gerardiana provides ventilation to the lungs, and it is commonly used to relieve algor venti syndrome, the symptoms of which are fever, chills, headache, lack of perspiration, nasal obstruction, body aches, and aversion to cold as well as exterior-excess. It induces sweating that releases to the exterior, so it is used to halt asthmatic attacks by dispersing stagnated qi of the lungs and separates the function, opening the peripheral collaterals and channels.

7.10 Diuretic Effects

Pseudoephedrine has a diuretic function greater than ephedrine, which exerts its diuretic effect possibly through dilating renal vessels, increasing renal blood flow, or inhibiting sodium ion reabsorption of renal tubes (Hong et al., 2011; Kwon et al., 2001).

7.11 Mydriatic Effects

Ephedrine produces mydriasis (dilation of pupil), and it occurs without pupillary light reflex blocking (Chen and Poth, 1929; Howard and Lee, 1927).

7.12 Antioxidant Activity

E. gerardiana was found to possess strong antioxidant activities in several studies (Abdelhady et al., 2011; Bell, 1980; Rice-Evans et al., 1997).

7.13 Radioprotective Role

E. gerardiana was widely investigated for its role as a radioprotectant to gain better results from natural sources.

8. SIDE EFFECTS AND TOXICITY

The FDA during the year 2000 had received reports that described about 1000 injuries resulting from the undiscriminating use of ephedrine-containing products. As a result, the FDA has banned all over-the-counter drugs that contain ephedrine. Due to its adverse effects, the World Anti-Doping Agency banned ephedrine, methylephedrine, and pseudoephedrine for their use in the form of stimulants. Moreover, the chemical synthesis of these products is also monitored. The bad effects resulting from the use of ephedra products include palpitations, nervousness, headaches, myocardial infarction, anxiety, seizures, nausea, hypertension, strokes, hyperthermia, and death. To add to this, uncontrolled hypertension, myocardial ischemia/infarction, and dysrhythmias have also been reported. *E. gerardiana* has been reported in the list of products whose consumption results in hepatic injury. Other indications reported until now include seizures, vascular ischemia, hemorrhage, and vasculitis (Caveney et al., 2001). It is the most commonly used herb by the parturient (Caveney and Starratt, 1994). During the past 2 decades, preparation of crystal meth (methamphetamine) has been synthesized by one-step simple reduction. Ephedrine is illegally converted to crystal meth, which has led to epidemic of addiction. This addiction causes acute psychosis and severe damage to the nervous system. Therefore, strict regulations of laws are required to stop this epidemic of addiction.

References

Abdelhady, M.I., Motaal, A.A., Beerhues, L., 2011. Total phenolic content and antioxidant activity of standardized extracts from leaves and cell cultures of three Callistemon species. American Journal of Plant Sciences 2, 847.

Abourashed, E.A., El-Alfy, A.T., Khan, I.A., Walker, L., 2003. Ephedra in perspective—a current review. Phytotherapy Research 17, 703–712.

Andraws, R., Chawla, P., Brown, D.L., 2005. Cardiovascular effects of ephedra alkaloids: a comprehensive review. Progress in Cardiovascular Diseases 47, 217–225.

Ang-Lee, M.K., Moss, J., Yuan, C.-S., 2001. Herbal medicines and perioperative care. JAMA 286, 208–216.

Baker, G., 2010. Garden of Eden: the Shamanic use of psychoactive flora and fauna and the study of consciousness. Australian Journal of Medical Herbalism 22, 107–108.
Baltch, A.L., Smith, R.P., 1994. Pseudomonas aeruginosa: Infections and Treatment. Infectious Disease and Therapy Series.
Bell, E., 1980. Possible significance of secondary compounds in plants. Encyclopedia of Plant Physiology. New Series.
Blumenthal, M., King, P., 1995. Ma Huang: Ancient Herb, Modern Medicine, Regulatory Dilemma. HerbalGram, USA.
Caveney, S., Charlet, D.A., Freitag, H., Maier-Stolte, M., Starratt, A.N., 2001. New observations on the secondary chemistry of world Ephedra (Ephedraceae). American Journal of Botany 88, 1199–1208.
Caveney, S., Starratt, A., 1994. Glutamatergic signals in ephedra. Nature 372, 509.
Celine, S., Tomy, S., Ujwala, T., Chander, S.J.U., 2016. A detailed overview of medicinal plants having hypoglycemic activity. International Journal of Phytomedicine 8.
Chaitanya, B., Sagi, S.R., Shashikanth, P., Karunakar, K., 2013. Evaluation of anti-asthmatic activity of ethanolic extract of Ephedra gerardiana wall in mice by ovalbumin induced method. Asian Journal of Pharmaceutical and Clinical Research 1, 166–169.
Chaturvedi, S., Dass, S., 2011. Traditional medicinal and economic uses of gymnosperms. Bulletin of Environment, Pharmacology and Life Sciences 1, 70–72.
Chen, C., Biller, J., Willing, S.J., Lopez, A.M., 2004. Ischemic stroke after using over the counter products containing ephedra. Journal of the Neurological Sciences 217, 55–60.
Chen, K., Poth, E.J., 1929. Racial differences as illustrated by the mydriatic action of cocaine, euphthalmine, and ephedrine. Journal of Pharmacology and Experimental Therapeutics 36, 429–445.
Christian, R., 1998. The Encyclopedia of Psychoactive Plants: Ethnopharmacology and Its Applications. Park Street Press, Rochester.
Fukushima, K., 2004. Bioactivity of Ephedra: Integrating Cytotoxicity Assessment with Real-Time Biosensing.
Gaur, R., Sharma, J., 2011. Indigenous knowledge on the utilization of medicinal plant diversity in the Siwalik region of Garhwal Himalaya, Uttarakhand. Journal of Forest and Environmental Science 27, 23–31.
Hegazi, G.A.E.-M., El-Lamey, T.M., 2011. In vitro production of some phenolic compounds from ephedra alata decne. Journal of Applied Environmental and Biological Sciences 1, 158–163.
Hong, H., Chen, H.-B., Yang, D.-H., Shang, M.-Y., Wang, X., Cai, S.-Q., Mikage, M., 2011. Comparison of contents of five ephedrine alkaloids in three official origins of Ephedra Herb in China by high-performance liquid chromatography. Journal of Natural Medicines 65, 623–628.
Howard, H., Lee, T., 1927. The effect of instillations of ephedrine solution upon the eye. Experimental Biology and Medicine 24, 700–702.
Hussain, M., Shah, G.M., Khan, M.A., 2006. Traditional medicinal and economic uses of Gymnosperms of Kaghan valley, Pakistan. Ethnobotanical Leaflets 2006, 7.
Khandelwal, K.R., 2008. Practical Pharmacognosy. Pragati Books Pvt. Ltd.
Konno, C., Mizuno, T., Hikino, H., 1985. Isolation and hypoglycemic activity of ephedrans A, B, C, D and E, glycans of Ephedra distachya Herbs1. Planta Medica 51, 162–163.
Kwon, Y.-B., Lee, J.-D., Lee, H.-J., Han, H.-J., Mar, W.-C., Kang, S.-K., Beitz, A.J., Lee, J.-H., 2001. Bee venom injection into an acupuncture point reduces arthritis associated edema and nociceptive responses. Pain 90, 271–280.
Leung, A.Y., 1990. Chinese medicinals. In: Advances in New Crops. Timber Press, Portland, OR, pp. 499–510.
Mack, R., 1996. "All but death, can be adjusted". Ma Huang (ephedrine) adversities. North Carolina Medical Journal 58, 68–70.

Maglione, M., Miotto, K., Iguchi, M., Jungvig, L., Morton, S.C., Shekelle, P.G., 2005. Psychiatric effects of ephedra use: an analysis of Food and Drug Administration reports of adverse events. American Journal of Psychiatry 162, 189–191.
Manandhar, N., 1980. Medicinal Plants of Nepal Himalayas. Ratna Pustak Bhandar, Bhotahity, Kathmandu.
Morton, J.F., 1977. Major Medicinal Plants: Botany, Culture and Uses.
Musselman, L.J., 1996. Encyclopedia of common natural ingredients used in food, drugs, and cosmetics, ed. 2. Albert T. Leung, and steven foster. Economic Botany 50, 422.
World Health Organization (WHO)., 1999. WHO Monographs on Selected Medicinal Plants (vol. 2). World Health Organization.
Peters, C.M., O'neill, J.O., Young, J.B., Bott-Silverman, C., 2005. Is there an association between ephedra and heart failure? a case series. Journal of Cardiac Failure 11, 9–11.
Rice-Evans, C., Miller, N., Paganga, G., 1997. Antioxidant properties of phenolic compounds. Trends in Plant Science 2, 152–159.
Schaneberg, B.T., Crockett, S., Bedir, E., Khan, I.A., 2003. The role of chemical fingerprinting: application to Ephedra. Phytochemistry 62, 911–918.
Soltan, M.M., Zaki, A.K., 2009. Antiviral screening of forty-two Egyptian medicinal plants. Journal of Ethnopharmacology 126, 102–107.
Soni, M.G., Carabin, I.G., Griffiths, J.C., Burdock, G.A., 2004. Safety of ephedra: lessons learned. Toxicology Letters 150, 97–110.
Tai, W., 2003. The Essentials of Traditional Chinese Herbal Medicine. Foreign Languages Press.
Taylor, G., Gravel, D., Johnston, L., Embil, J., Holton, D., Paton, S., 2002. Canadian Hospital Epidemiology Committee. Canadian Nosocomial Infection Surveillance Program. Prospective surveillance for primary bloodstream infections occurring in Canadian hemodialysis units.
Tutin, T., Burges, N., Chater, A., Edmondson, J., Heywood, V., Moore, D., Valentine, D., Walters, S., Webb, D., 1993. Flora Europaea, vol. 1. Cambridge University, p. 187.

CHAPTER

37

Mango

Farooq Khalid[1], *Haq Nawaz*[1], *Muhammad Asif Hanif*[1], *Rafia Rehman*[1], *Abdullah Mohammed Al-Sadi*[2]

[1] Department of Chemistry, University of Agriculture, Faisalabad, Pakistan;
[2] Department of Crop Sciences, College of Agricultural and Marine Sciences, Sultan Qaboos University, Muscat, Oman

OUTLINE

Medicinal Plants of South Asia
https://doi.org/10.1016/B978-0-08-102659-5.00037-9

Copyright © 2020 Elsevier Ltd. All rights reserved.

1. BOTANY

1.1 Introduction

Mango (*Mangifera indica*) (Fig. 37.1), a commonly known evergreen tree, belongs to the Anacardiaceae family (Bally, 2006). A great variety exists in family Anacardiaceae, as it contains 70 genera with 600 species. In different regions, it is designated by different names, such as in Sanskrit, it is called madhuula, madhulualaka, and ambrah; in English,

FIGURE 37.1 Mango tree and fruits.

mango; French, mangot; Urdu: aam; Punjabi, amb; and in Portuguese, manga and mangueir. The name used for mango around the world reflects the language and the culture of the people who grow it (Shah et al., 2010). Mango is the national fruit of the Philippines, Pakistan, and India. Mango is also known as the national tree of Bangladesh (Soujanya et al., 2017). Mangos (metric tons) are widely produced in India (13,557,100), China (4,140,290), Thailand (2,469,810), Indonesia (2,150,000), Pakistan (1,728,000), Mexico (1,509,270), Brazil (1,197,690), Nigeria (831,489), Bangladesh (828,168), and the Philippines (771,441) (Saave, 2011). Mango is one of the most consumed fresh fruits in the world, with worldwide production exceeding 45 million metric tons in 2014 (FAO, 2017). To obtain an exact figure for the total dried mango production worldwide is a hard nut to crack. A large amount of dried mango is imported by the countries possessing full processing equipment. However, statistics may reveal that the United States is the major importer of fresh and dried mango followed by China, Netherland, Germany, and the United Kingdom. The world import of mango was 1,206,768 t/year. Among the largest dried mango importing countries are the United States (332,108 t/year), the Netherlands (142,035 t/year), China (115,140 t/year), Germany (48,451 t/year), the United Kingdom (47,578 t/year), Canada (46,648 t/year), France (32,130 t/year), Japan (10,543 t/year), and Spain (32,233 t/year) (Saave, 2011).

Mango is called the "King of Fruits," and Chaunsa is known as "King of Mangoes." Chaunsa mango has a golden yellow color. It is soft, almost fibreless, and has an aromatic, pleasant, sweet flavor (Rajwana et al., 2011). The world most famous aromatic and delicious taste Chaunsa only grows well in Pakistan in the region of South Punjab.

1.2 History of *Mangifera indica*

Many researchers consider that *Mangifera indica* originated mainly from India, owing to the wide cluster of varieties native to this country. Taxonomical and molecular analyses have also ascertained an evolution of *M. indica* within a region that includes northeastern India, Bangladesh, and Myanmar (Shamili et al., 2012). Trading routes have been the source for *M. indica*'s spread from its origin. Arabs are considered responsible for *Mangifera's* spread in east Africa. In this way, *M. indica*'s cultivation spread widely not only in tropical but also in subtropical regions (Dillon et al., 2013). The origin of the English word mango is a Malayalam word "manna," also known as manga in Portuguese. It was first used in 1498 in Kerala during the spice trade (Mandavilli, 2016; Shah et al., 2010). Mango has been widely cultivated in Southeast Asia since the 15th century. Mango name comes from Malayalam. In 1658, *Mangifera* was given as a name for the first time by botanists. *Mangifera* arbor was referred by Linnaeus in 1747. Later on, in 1753, it finally got its name as *M. indica* (Yadav and Singh, 2017). Around 100,000 mango trees were reported to be planted by former Mughal King Akbar (1556–1605) in Darbhanga, a city in eastern India (Chandra and Jain, 2017). The traders from Spain and Portugal are considered responsible for mango's spread in tropical and subtropical parts of the world. At the start of the 20th century, various cultivars of mango from Asia and India were brought to a mango development center in Florida, where many cultivars were chosen and distributed widely. Selection criteria of these cultivars was based on aroma, milder taste, colorful skin, and larger fruit size (Kuhn et al., 2017).

1.3 Demography/Location of *Mangifera indica*

Mangifera indica prefers a warm, wild, frost-free climate with suitable winter and a dry season. Reduction in fruit yields have been observed in *M. indica* owing to rain plus high humidity during growth of fruit and flowering. Tree of *M. indica* normally flowers from middle to late winter, and fruits mature from early to mid-summer months (Bally, 2006). Mango grows in wide environmental conditions. Generally, mango is cultivated in very hot, arid and humid to cool environments. Mango tree grows well in areas with temperature of 45°C. However, the optimum range of temperature for proper growth and production is 27–36°C. Mango plant prefers sunny conditions. Although mangoes are drought tolerant, they do not grow properly in case of deficiency of water. Mango requires fertile soil along with essential nutrients. Excellent growth of mango tree has been observed when pH of soil ranges from 6 to 7.2. Mature mango tree requires 11,000 cubic meters per hectare per year (De Villiers, 1998).

1.4 Botany, Morphology, Ecology

M. indica is considered a long-lived evergreen herbal tree having a height of 15–30 m. Most of the cultivated trees are between 3 and 10 m height when they are fully mature. Moreover, height depends on the amount and variety of pruning. Noncultivated and seedling trees have often reached a height of 15 m. Besides, under favorable conditions in forest conditions, they can reach up to 30 m height (Bally, 2006). *M. indica* has a dome shape with dense foliage, and it carries more branches coming from a stout trunk. Leaves of *M. indica* appear reddish in color and produce a sweet odor on crushing. When ripe, fruit of *M. indica* has a single seed with yellowish pulp and thick yellowish to reddish or green skin (Shah et al., 2010). *M. indica* has a long tap root that branches just below the ground level and has two or four major anchoring taproots that can reach down 6 m to the water table. The finer roots are present from the surface, down to 1 m. Distribution of finer root changes seasonally according to moisture distribution in soil (Latha et al., 2012). The flowering season of *M. indica* comes in the last months of every year. Four to six weeks are required for the flowering period, with a night temperature that can vary from 8 to 15°C and day temperature up to 20°C. During December to April, flowering has been observed to occur in some parts of world (Bally, 2006).

2. CHEMISTRY

Mango is a fascinating aromatic plant cultivated mainly for its juicy taste and edible fruit. Many mango cultivars are differentiated due to their characteristic flavor, which depends upon the ratio of palmitic to palmitoleic acid as fruit ripens. A ratio less than 1 of palmitic to palmitoleic acid attributes to the strong flavor and characteristic fragrance or aroma of a mango cultivar (Maróstica and Pastore, 2007). It has been found experimentally that the difference in the concentration among classes of four compounds, which include esters, sesquiterpenes, lactones, and monoterpenes, decide the characteristic flavor of a mango cultivar. A large number of volatile compounds were found in Indian mango cultivars. Mango cultivars from Sri Lanka possess different aroma along with flavor due to high concentration of sesquiterpenes and terpenes hydrocarbons (Maróstica and Pastore, 2007).

Mango contains a minimal amount of fat. It has been investigated that 100 g of fresh mango provides 60 kcal. Mango is also a healthy source of vitamin A, vitamin C, and other nutrients. Vitamins A, B, and C along with other nutrients are the existing constituents of mango (Sultana et al., 2012). Due to the presence of phenols and flavonoids, mango leaves show considerable antioxidant activities (Fernández-Ponce et al., 2012). 100 g of mango leaves contain 1490 I.U. of vitamin A, 72 mg of phosphorous, 29 mg of calcium, 1.6% of fiber, 16.5% of carbohydrates, 0.4% of fat, and 3.0% of protein, along with some other nutrients. Mango seed is also a rich source of protein, fat, and carbohydrates (Nagarajan, 2012).

Essential oil from leaves and peels of mango can be extracted by hydro distillation method. The volatile oil was found to contain monoterpenes, sesquiterpenes, oxygenated sesquiterpenes, and oxygenated monoterpenes (Džamić et al., 2010). Processing of mango yields peel and kernel as by-products. Oil from mango seed kernel contains 44%–48% of saturated fatty acids and 52%–56% of unsaturated fatty acids. Structures of some bioactive compounds present in mango are shown in Fig. 37.2.

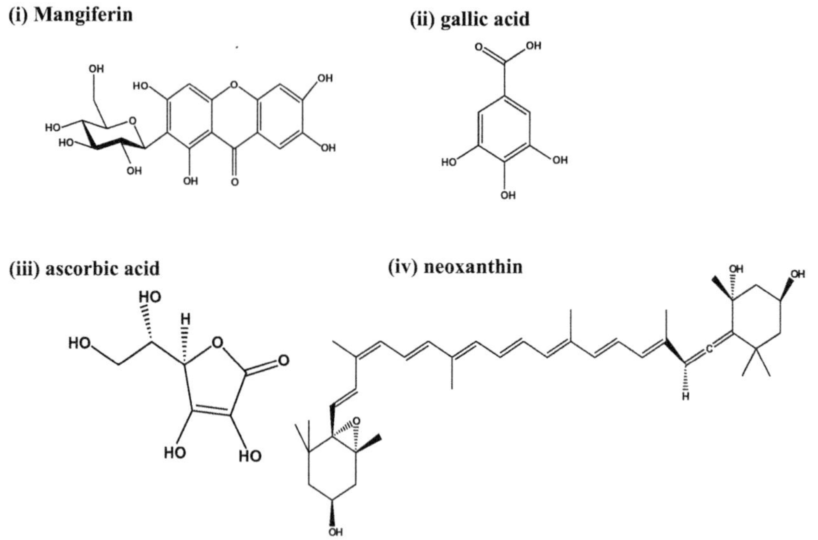

FIGURE 37.2 Structures of some bioactive compounds present in mango.

3. POSTHARVEST TECHNOLOGY

M. indica's fruits are best harvested by employing clippers. Large trees and fruits have also been harvested using picking poles. But the method involves some limitations as it is not favorable for time beyond the completion of harvesting. A color change has been considered another indication for *mango* maturity. Owing to the variety of mangoes, mature green mango can be stored at room temperature for 4–10 days. By employing chemical treatment, precooling, and low temperature, shelf life of fruit can be extended. Low temperature can aid for only 3–4 weeks. Fruit is harvested early in the season for mainly capturing the market. For the preparation of unripe fruit, mango must be dipped in ether solution at 52°C for 5 minutes.

4. PROCESSING

Almost all parts of mango fruit are consumed in numerous ways. Mostly, it is consumed as fresh, ripe, and unprocessed fruit. Mango also finds its applications in various forms such as slices, flakes, pickles, puree, and juice. However, mango processing is largely related with industrial environments and favorable technology development (Saave, 2011). *M. indica* has been processed into chutneys, pulp, and jams (Ganeshan et al., 2016). The final quality of mango depends not only on the physiologic processes occurring during ripening, but also on processes during fruit development and maturation. Commercially, the drying process of mango involves washing of mango slices with 15 ppm of chlorine solution to remove any microbial activity. Then, a temperature of 75°C is provided to the internal parts of fruit followed by 10 minutes boiling with water to inactivate enzymes. Mango pieces placed in plastic containers are further processed in a pulp machine. Mango pieces are then passed through a mesh to achieve a desirable size of 0.5 mm with continuous stirring, and temperature of 95°C is provided for 10 minutes to prevent any sort of microbial spoilage. Furthermore, additives are added to increase the shelf life of pulp and to reduce microbial growth, followed by packing. Finally, cooling of hot pulp is performed with fresh water at lower temperature. Essential oil from leaves and peels of mango can be extracted by hydro distillation method. In a previous study, fresh mango leaves and peels yielded 0.02% and 0.05%, respectively, of essential oil by hydro distillation method (Baloch et al., 2017).

5. VALUE ADDITION

Aqueous extract of *M. indica*'s leaves have been marketed as an antioxidant under brand name VIMANG (Garrido et al., 2004). Mango kernel oil finds its applications as an adequate and rational constituent of various cosmetics, mainly to get rid of dry skin and frost bite, rashes, blemishes and wrinkles, itching skin, for a sun screen, wounds and cracks, dermatitis, stretch marks during pregnancy, muscle fatigue, aches, and tension. Mango kernel oil is regarded as an alternative to cocoa butter, an essential part of confectionaries and chocolates. Currently, mango kernel oil is being utilized for the production of vanaspati, milk powder, health boosters, cookies, baking tray coating, nutraceutical products, dairy whitener, cakes, muffins, cake molds, frozen desserts, cream products, milk shakes, and ice creams. Mango has potential applications in the cosmetic industry as sunscreen, base, and shaving cream. Due to the presence of natural minerals and antioxidants, mango is extensively preferred over other lipids. Copper, zinc, and selenium are the existing minerals within mango. Mango kernel oil is effectively efficient for the storage of oil and fat due to its longer induction period and higher total phenolic contents compared to other vegetable oils. Functional properties involving mango kernel oil can be improved commercially by using methods such as interesterification, transesterification, and fractionation. Mango and various value-added products relevant to mango such as pickles, chutneys, slices, and nectar have acquired worldwide popularity (Nadeem et al., 2016).

6. USES

Historically, the Jain goddess Ambika was shown sitting beneath a mango tree (Chandra and Jain, 2017). In India, Lord Ganesha has been found holding a fully ripe mango fruit, paying respect to the devotees for their potential perfection. Mango flowers are also used to worship Saraswati, a goddess. There is no concept of celebrating new year's day in Telugu, known as Ugadpasses, without eating a delicious dish Ugadi pachadi made with mango pieces (Chandra and Jain, 2017). Seeds and skin of dried mango have been utilized in ancient Ayurvedic medicine (Parvez, 2016). For decorating doors and during wedding celebrations, mango leaves are used as an ornament as Ganesh Chaturthi (Yadav and Singh, 2017). Many embroidery styles in India such as Kanchipuram silk sarees and Kashmiri shawls contain mango paisleys along with motifs (Gopalakrishnan, 2013). Due to its pre-Islamic Zoroastrian background, paisleys are very common in Iranian art (Chandra and Jain, 2017). As a sign of fortune, doors are decorated with mango leaves in India during festivals (Yadav and Singh, 2017). With respect to flavor and juicy taste, mango is regarded as one of the three royal fruits in Tamil Nadu along with banana and jackfruit (Subrahmanian et al., 1997, 2010). The expression in the West Indies "to go mango walk" means to steal someone's mango fruit. Praises of mango were sung by Kalidasa, a Sanskrit poet. In China, mango became popular during the cultural revolution and was considered a symbol of love for the people from Mao Zedong (Zedong, 2011).

Mangifera indica contains high concentrations of vitamins, minerals, polyphenolic flavonoids, and prebiotic dietary fiber as antioxidant compounds (Ara et al., 2014). The latest research revealed that *M. indica* plays a very valuable role in protection against breast and colon cancer, owing to existence of polyphenolic antioxidants in *M. indica* (Abbasi et al., 2015). The fruit of *M. indica* is a vital source of vitamin A and flavonoids. Carotene-rich fruits are employed for protection against oral cavity plus lungs cancers (Shah et al., 2010). A large quantity of potassium is extracted from fresh mangoes. Potassium has been considered a most valuable constituent of cells that aids in controlling blood pressure and heart rate (Yatnatti et al., 2014). *M. indica* is also an active source of vitamin B_6 that aids in mainly brain production (Fowomola, 2010). Homocysteine level has been found to be controlled by *M. indica* in blood; otherwise, it may cause strokes and coronary artery diseases. *M. indica* also possesses a minor quantity of copper, which mainly is regarded as a cofactor in many valuable enzymes. To produce erythrocytes, copper has been utilized (Fowomola, 2010). *M. indica* is also highly rich in phytonutrients as pigments and antioxidants. Mango products, particularly leaves and bark, possess high quantities of phenolic antioxidant compounds, mainly phenolic acids, flavonoids, benzophenones, xanthones, and gallotannins. Mango polyphenols have shown several types of valuable activities like antioxidant, antifungal, antidiabetic, antimicrobial, antipyretic, immunomodulatory, antiinflammatory, and analgesic activities (Fernández-Ponce et al., 2015).

Along with fruit the *M. indica* leaves are also healthy. Its leaves are used for the cure of patients suffering from diabetes by regulating insulin level (Shah et al., 2010). Due to being rich in vitamin A, it helps to aggrandize eye sight and prevents night blindness. *M. indica* has also been suggested by beauty experts to be used for body scrubs to make skin smoother and softer. Several activities have been attributed to various parts of *M. indica* (Nadeem et al., 2016). For instance, bark of *M. indica* is considered a cure for diarrhea, fully ripe fruit of *M. indica* is used for treatment of consistent constipation, seeds have been utilized as an astringent to bowls, and piles are also treated by *Mangifera* to enhance the quality of a patient's life (Shah et al., 2010). A lot of medicinal applications have been associated with *M. indica* including antidiabetic, antioxidant, anti–lipid per oxidant, glucosyl xanthone, wound healing, and cardiotonic activities. *M. indica* is the most famous among other tropical fruits. Several parts of *M. indica* have been exercised as antiseptic, dentifrice, vermifuge, astringent, stomachic, diaphoretic, laxative, tonic, asthma, cough, hypertension, as diuretic, also for the treatment involving dysentery, piles, diarrhea, and anemia. Smoke of burning *Mangifera* leaves is used for the treatment of throat infections (Shah et al., 2010; Wauthoz et al., 2007).

7. PHARMACOLOGICAL USES

7.1 Antioxidant Activity

Seeds and peels of *M. indica* have been found to possess flavanol, gallotannins, benzophenone derivatives, ascorbic acid, phenolic compounds, and xanthone glycosides.

7.2 In Vitro Antioxidant Activity

M. indica's leaf extract was subjected to in vitro antioxidant analysis by employing a suitable testing method. The desired study exhibited a very strong activity of hydroxyl radicals, hypochlorous acid, and proved that *M. indica* is an iron chelator. An obvious inhibitory action on rat brain was observed due to peroxidation. Moreover, copper-phenanthroline system was responsible for damage of phospholipids and DNA (Martınez et al., 2000).

7.3 Antiinflammatory Effects

M. indica's aqueous extract has been utilized to combat stress to improve people's standard of living. Owing to the results of performed analysis, *M. indica*'s seed kernel has shown considerable antiinflammatory activity (Kabuki et al., 2000). Uba mango juice has shown considerable potential effects against metabolic risk of obesity caused by adiposity and inflammation. Mango juices have reversed drastic effects of high-fat diets (Natal et al., 2016).

7.4 Antimicrobial Effects

Owing to investigations performed, antimicrobial activity has also been attributed to *M. indica*'s ethanolic extract (El-Gied et al., 2012).

7.5 In Vitro and In Vivo Antiinflammatory Activity

The aqueous extract of *M. indica*'s leaf has been marketed as an antioxidant under the brand name VIMANG (Garrido et al., 2004).

7.6 Immunomodulatory Activity

M. indica alcoholic extract has shown considerable effects on hormonal and cell-mediated components of immune systems in mice (Makare et al., 2001).

7.7 Anticarcinogenic Activity

Anticarcinogenic potential has been attributed to the existence of phenolic and flavonoid constituents within *M. indica*. Owing to anticancer activities, *M. indica* has been the center of gravity for research in recent years. Researchers mainly focus on mangiferin (xanthonoid), a major component in *M. indica*. Mangiferin has been found to block the inflammatory NF-KB pathway. Hence, growth and formation of cancer is retarded. Mangiferin also inhibits the NF-KB pathway to counter skin cancer. Mangiferin originates from the bark extract of *M. indica* tree. Another compound that exhibits anticancer activities is gallic acid. Gallic acid has also been recovered from mango tree. A sufficient quantity of mangiferin also exists in leaves of *M. indica*. Mangiferin also finds its applications as a cure for heart diseases and shows considerable effects against breast cancer (Gold-Smith et al., 2016).

Carcinogenesis can be prevented by using chemoprevention. Cancer creative effects have been observed in various bioactive compounds. ATPases along with glycoprotein play an important role in carcinogenesis. The effect of mangiferin on glycoprotein and ATPases was seen in controls and cancer-bearing mice. Mangiferin has been found to revert the higher concentration of glycoprotein and ATPases to a normal level. Hence, mangiferin does possess anticancer effects (Rajendran et al., 2008). Selective cytotoxic activity in triple negative breast cancer cells and regulation of some cancer drug target genes by chloromangiferamide indicate that it can be used to develop a potential chemotherapeutic agent for triple negative breast cancer cells (Ediriweera et al., 2016).

8. SIDE EFFECTS AND TOXICITY

Mangoes can raise blood sugar levels. Overeating mangoes causes weight gain and diarrhea. Ripening agents like calcium carbide, banned in many countries, can cause allergic reaction.

References

Abbasi, A.M., Guo, X., Fu, X., Zhou, L., Chen, Y., Zhu, Y., Yan, H., Liu, R.H., 2015. Comparative assessment of phenolic content and in vitro antioxidant capacity in the pulp and peel of mango cultivars. International Journal of Molecular Sciences 16, 13507–13527.

Ara, R., Motalab, M., Uddin, M., Fakhruddin, A., Saha, B., 2014. Nutritional evaluation of different mango varieties available in Bangladesh. International Food Research Journal 21.

Bally, I.S., 2006. Mangifera Indica (Mango): Traditional Trees of Pacific Islands. Their Culture, Environment, and Use, pp. 441–464.

Baloch, F.S., Tahir, S.S., Sherazi, S.T.H., Jilani, N.S., Khokhar, A.L., Rajput, M.T., 2017. Evaluation of essential oil components from the fruit peelings of sindhri and langra varieties of mango (*Mangifera indica* L.). Pakistan Journal of Botany 49, 1479–1484.

Chandra, S., Jain, A., 2017. Foundations of Ethnobotany (21st Century Perspective). Scientific Publishers.

Džamić, A.M., Marin, P.D., Gbolade, A.A., Ristić, M.S., 2010. Chemical composition of *Mangifera indica* essential oil from Nigeria. Journal of Essential Oil Research 22, 123–125.

De Villiers, E., 1998. The Cultivation of Mangoes. ARC.

Dillon, N.L., Bally, I.S., Wright, C.L., Hucks, L., Innes, D.J., Dietzgen, R.G., 2013. Genetic diversity of the Australian national mango genebank. Scientia Horticulturae 150, 213–226.

Ediriweera, M.K., Tennekoon, K.H., Adhikari, A., Samarakoon, S.R., Thabrew, I., De Silva, E.D., 2016. New halogenated constituents from *Mangifera zeylanica* Hook. f. and their potential anti-cancer effects in breast and ovarian cancer cells. Journal of Ethnopharmacology 189, 165–174.

El-Gied, A.A.A., Joseph, M.R., Mahmoud, I.M., Abdelkareem, A.M., Al Hakami, A.M., Hamid, M.E., 2012. Antimicrobial activities of seed extracts of mango (*Mangifera indica* L.). Advances in Microbiology 2, 571–576.

Fernández-Ponce, M.T., Casas, L., Mantell, C., de la Ossa, E.M., 2015. Use of high pressure techniques to produce *Mangifera indica* L. leaf extracts enriched in potent antioxidant phenolic compounds. Innovative Food Science & Emerging Technologies 29, 94–106.

Fernández-Ponce, M.T., Casas, L., Mantell, C., Rodríguez, M., de la Ossa, E.M., 2012. Extraction of antioxidant compounds from different varieties of *Mangifera indica* leaves using green technologies. The Journal of Supercritical Fluids 72, 168–175.

Fowomola, M., 2010. Some nutrients and antinutrients contents of mango (*Magnifera indica*) seed. African Journal of Food Science 4, 472–476.

Ganeshan, G., Shadangi, K.P., Mohanty, K., 2016. Thermo-chemical conversion of mango seed kernel and shell to value added products. Journal of Analytical and Applied Pyrolysis 121, 403–408.

Garrido, G., González, D., Lemus, Y., Garcıa, D., Lodeiro, L., Quintero, G., Delporte, C., Núñez-Sellés, A.J., Delgado, R., 2004. In vivo and in vitro anti-inflammatory activity of *Mangifera indica* L. extract (VIMANG®). Pharmacological Research 50, 143–149.

Gold-Smith, F., Fernandez, A., Bishop, K., 2016. Mangiferin and cancer: mechanisms of action. Nutrients 8, 396.

Gopalakrishnan, S., 2013. Marketing system of mangoes in India. World Applied Sciences Journal 21, 1000–1007.

Kabuki, T., Nakajima, H., Arai, M., Ueda, S., Kuwabara, Y., Dosako, S.i., 2000. Characterization of novel antimicrobial compounds from mango (*Mangifera indica* L.) kernel seeds. Food Chemistry 71, 61–66.

Kuhn, D.N., Bally, I.S., Dillon, N.L., Innes, D., Groh, A.M., Rahaman, J., Ophir, R., Cohen, Y., Sherman, A., 2017. Genetic map of mango: a tool for mango breeding. Frontiers of Plant Science 8.

Latha, K., Latha, M., Vagdevi, H., 2012. Comparative studies on anthelmintic activity of *Mangifera indica* L. var. *thotapuri* and *Mangifera indica* L. var. *neelam* root crude extracts. International Journal of Phytopharmacy 2, 21–24.

Makare, N., Bodhankar, S., Rangari, V., 2001. Immunomodulatory activity of alcoholic extract of *Mangifera indica* L. in mice. Journal of Ethnopharmacology 78, 133–137.

Mandavilli, S.R., 2016. On the origin and spread of languages: propositioning Twenty-first century axioms on the evolution and spread of languages with concomitant views on language dynamics. ELK Asia Pacific Journal of Social Science 3.

Maróstica Jr., M.R., Pastore, G.M., 2007. Tropical fruit flavour. In: Berger, R.G. (Ed.), Flavours and Fragrances Chemistry, Bioprocessing and Sustainability. Springer-Verlag, Berlin, Heidelberg, pp. 189–200.

Martınez, G., Delgado, R., Pérez, G., Garrido, G., Núñez-Sellés, A., León, O.S., 2000. Evaluation of the in vitro antioxidant activity of *Mangifera indica* L. extract (Vimang). Phytotherapy Research 14, 424–427.

Nadeem, M., Imran, M., Khalique, A., 2016. Promising features of mango (*Mangifera indica* L.) kernel oil: a review. Journal of Food Science and Technology 53, 2185–2195.

Nagarajan, G., 2012. A Study on Production and Marketing of Mango in Dindigul District.

Natal, D.I.G., de Castro Moreira, M.E., Milião, M.S., dos Anjos Benjamin, L., de Souza Dantas, M.I., Ribeiro, S.M.R., Martino, H.S.D., 2016. Ubá mango juices intake decreases adiposity and inflammation in high-fat diet-induced obese Wistar rats. Nutrition 32, 1011–1018.

Parvez, G.M., 2016. Pharmacological activities of Mango (Mangifera indica): a review. Journal of Pharmacognosy and Phytochemistry 5, 1.

Production of mangoes, mangosteens, and guavas in 2017, Crops/Regions/World list/Production Quantity (pick lists). UN Food and Agriculture Organization, Corporate Statistical Database (FAOSTAT). 2017. Retrieved 16 April 2019.

Rajendran, P., Ekambaram, G., Magesh, V., Sakthisekaran, D., 2008. Chemopreventive efficacy of mangiferin against benzo (a) pyrene induced lung carcinogenesis in experimental animals. Environmental Toxicology and Pharmacology 26, 278–282.

Rajwana, I.A., Khan, I.A., Malik, A.U., Saleem, B.A., Khan, A.S., Ziaf, K., Anwar, R., Amin, M., 2011. Morphological and biochemical markers for varietal characterization and quality assessment of potential indigenous mango (*Mangifera indica*) germplasm. International Journal of Agriculture and Biology 13.

Saave, N., 2011. Report. In: Marty, O., Conte, K., Vonnahme, C. (Eds.), Export Factsheet Ecowas: Mangoes, vol. 34. International Trade Center, Switzerland.

Shah, K., Patel, M., Patel, R., Parmar, P., 2010. *Mangifera indica* (mango). Pharmacognosy Reviews 4, 42.

Shamili, M., Fatahi, R., Hormaza, J., 2012. Characterization and evaluation of genetic diversity of Iranian mango (*Mangifera indica* L., Anacardiaceae) genotypes using microsatellites. Scientia Horticulturae 148, 230–234.

Soujanya, B., Kumar, A.K., Sreedhar, M., Aparna, K., Reddy, K.R., 2017. Quantification of lupeol in selected commercial coloured cultivars of mango (*Mangifera indica* L.) cultivated in Telangana region. International Journal of Pure & Applied Bioscience 5, 2141–2146.

Subrahmanian, N., Hikosaka, S., Samuel, G.J., Thiagarajan, P., 1997. Tamil Social History. Institute of Asian Studies.

Subrahmanian, N., Hikosaka, S., Samuel, G.J., 2010. Mango. Scribd.

Sultana, B., Hussain, Z., Asif, M., Munir, A., 2012. Investigation on the antioxidant activity of leaves, peels, stems bark, and kernel of mango (*Mangifera indica* L.). Journal of Food Science 77, C849–C852.

Wauthoz, N., Balde, A., Balde, E.S., Van Damme, M., Duez, P., 2007. Ethnopharmacology of *Mangifera indica L.* bark and pharmacological studies of its main C-glucosylxanthone, mangiferin. International Journal of Biomedical and Pharmaceutical Sciences 1, 112–119.

Yadav, D., Singh, S., 2017. Mango: history origin and distribution. Journal of Pharmacognosy and Phytochemistry 6, 1257–1262.

Yatnatti, S., Vijayalakshmi, D., Chandru, R., 2014. Processing and nutritive value of mango seed kernel flour. Current Research in Nutrition and Food Science Journal 2, 170–175.

Zedong, M., 2011. Mao Zedong.

Further Reading

Bal, D.D., 2013. Mango. SCRIBD.

DaMatta, F.M., 2007. Ecophysiology of tropical tree crops: an introduction. Brazilian Journal of Plant Physiology 19, 239–244.

Deliya, M., Thakor, C., Parmar, B., 2012. A study on "differentiator in marketing of fresh fruits and vegetables from supply chain management perspective". Abhinav: National Monthly Referred Journal of Research in Commerce and Management 1, 40–57.

Flowerree, D., 2010. Commercial Processing of Mangoes, Nutrition/Health. https://www.themangofactory.com/nutrition/commercial-processing-of-mangoes/. Archived on 10/01/2019.

Garrido, G., González, D., Delporte, C., Backhouse, N., Quintero, G., Núnez-Sellés, A.J., Morales, M.A., 2001. Analgesic and anti-inflammatory effects of *Mangifera indica* L. extract (Vimang). Phytotherapy Research 15, 18–21.

Jain, S., Singh, H., 2014. (4) India's notable presence in Linnaeus' Botanical Classification.

Lizada, C., 1993. Mango, Biochemistry of Fruit Ripening. Springer, pp. 255–271.

Marcos-Filho, J., 2014. Physiology of Recalcitrant Seeds. Revolvy.

Medina, J., García, H., 2002. In: Mejia, D., Lewis, B. (Eds.), MANGO: Post-Harvest Operations.

Mishra, H.K., 2006. Goddesses and the Centres of Goddess Worship in Early Rajasthan (7th to 15th Century).

Moore, M., 2015. Mango. Revolvy.

Morgan, C., 2013. Mango. Scribd.

National Agricultural Library, U., 2016. Mangoes. Revolvy.
Nelson, S.C., 2008. Mango Powdery Mildew.
Noratto, G.D., Bertoldi, M.C., Krenek, K., Talcott, S.T., Stringheta, P.C., Mertens-Talcott, S.U., 2010. Anticarcinogenic effects of polyphenolics from mango (*Mangifera indica*) varieties. Journal of Agricultural and Food Chemistry 58, 4104.
Prasad, R., Shivay, Y., Nene, Y., 2016. Asia's contribution to the evolution of agriculture: creativity, history, and mythology. Asian Agri-History 20.
Radhakrishnan, M., 2017. Benefits and Uses of Mango Leaves. StyleCraze.
Saiarvind, 2014. Benefits and Superstitues of Mango. Scribd.

CHAPTER

38

Moringa

Farwa Nadeem[1], *Muhammad Asif Hanif*[1], *Ijaz Ahmad Bhatti*[1], *Shahzad Maqsood Ahmed Basra*[2]

[1] Department of Chemistry, University of Agriculture, Faisalabad, Pakistan;
[2] Department of Agronomy, University of Agriculture, Faisalabad, Pakistan

OUTLINE

Medicinal Plants of South Asia
https://doi.org/10.1016/B978-0-08-102659-5.00038-0

Copyright © 2020 Elsevier Ltd. All rights reserved.

1. BOTANY

1.1 Introduction

Moringa (*Moringa oleifera* Lam) (Fig. 38.1) is naturally growing, widely cultivated, and a highly valued medicinal plant belonging to genus "*Moringa*" and family "Moringaceae," commonly found in tropical and subtropical areas having approximately 13 different species (Anwar et al., 2007). *Moringa* has different names in different languages like "horse-radish tree," "Ben tree," and "drumstick tree" in English, "la ken" in Chinese, "moringa" and "moringueiro" in Portuguese, "moringa," "Ben," and "Ángela" in Spanish, "morungue" and "moringe à graine ailée" in French, "rawag" in Arabian, "tikshnagandhaa," "haritashaaka," "akshiva," and "raktaka" in Ayurvedic, "sahajan" in Unani, "soanjna" and "sohanjna" in Punjabi, "sigru" and "murinna" in Malayalam, "munaga" and "mulaga" in Telugu, "morigkai" in Tamil, "suragavo" in Gujarati, "sainjna" and "saguna" in Hindi, "subhanjana" in Sanskrit, and "Moringa oleifera" in Latin (Mishra et al., 2011). This fast-growing, soft, woody plant is indigenous to Northern India and Pakistan along with some areas of Nepal and is known to have tremendous therapeutic potential, as it can treat more than 300 diseases (Ganguly, 2013). The word "Moringa" is derived from the Tamil word "murungai" meaning "twisted pods."

1.2 History/Origin

The history of Moringa initiated on the Indian subcontinent approximately 2000 BCE. Moringa has extensively been used by Indians,

FIGURE 38.1 Moringa plant and seeds.

Egyptians, Greeks, and Romans for thousands of years, with writings dating as far as 150 AD, while the history of moringa dates back to 150 BC. Some historical evidences revealed that ancient queens and kings used leaves and fruit of moringa in their diets to ensure healthy skin and to maintain mental alertness. Some ancient Indian Mauryan warriors were fed intentionally with extracts of moringa leaves on the war front. Furthermore, drink of the elixir was believed to add some extra energy that relieves pains and stresses incurred during warfare. These courageous soldiers were the ones who defeated Alexander the Great (Manzoor et al., 2007; Zaku et al., 2015). Moringa is mostly found in southern hills of Himalayas and was introduced in numerous tropical and subtropical areas mainly by the Asian migrants.

1.3 Demography and Location

Moringa grows well in tropical to subtropical countries having specific environmental features characterized by dry to relatively moist climate with mean annual rainfall ranging from 760 to 2500 mm with a temperature range of 18–28°C. The best soil for growth of moringa is waterlogged heavy clay with an approximate pH between 4.5 and 8.0 at an altitude of about 2000 m (Nouman et al., 2014).

1.4 Botany, Morphology, and Ecology

Moringa oleifera is a deciduous, evergreen, small to medium plant that can grow up to a height ranging from 10 to 12 m, with a spreading open crown of typical umbrella shape. The root system is deep, and bole is crooked, usually containing one stem, however, sometimes forked from the base. The grayish bark is corky with drooping and fragile branches having feathery foliage. Young shoots and twigs are covered by dense short hairs that are greenish white to purplish in color. Leaves of Moringa are alternate in arrangement with approximate length ranging from 7 to 60 m. They are the tripinnate compounds in which each pinnate bears four to six pairs of leaflets with dark green coloration and elliptical to obovate shape, with length of 1–2 cm. The inflorescences of moringa are 10–20 cm long, constituting spreading panicles bearing a large number of fragrant flowers. The flower of moringa is zygomorphic and pentamerous with an average length in the range of 7–14 mm and light cream in color. Fruit is typically a three-valved capsule with length in the range of 10–60 cm, often known as "pods" that look like a drumstick, thus are named accordingly. Younger fruit is green in color that turns brown when the fruit matures. Fruit at the stage of maturation splits or opens along each angle to expose seeds completely. Capsule constitutes 15–20 rounded seeds containing a high amount of oil with the approximate diameter of 1 cm surrounded by three papery wings up to the length of 2.5 cm. Moringa seeds are known to have a significant amount of oil (Orwa et al., 2009).

2. CHEMISTRY

Some recent investigations have revealed that oil of moringa contains 15 major components, out of which the three most commonly occurring fatty acids are α-monoolein or glycerylmonooleate (5.27%), palmitic acid (26.16%), and oleic acid (58.88%), which accounts for 94.39% of total seed oil, whereas nonfatty acid components constitute only 4.34%. The chemical structures of some important fatty acids present in moringa are shown in Fig. 38.2.

Different parts of moringa are known to have variety of chemical compounds depending on the changing climatic conditions and plant

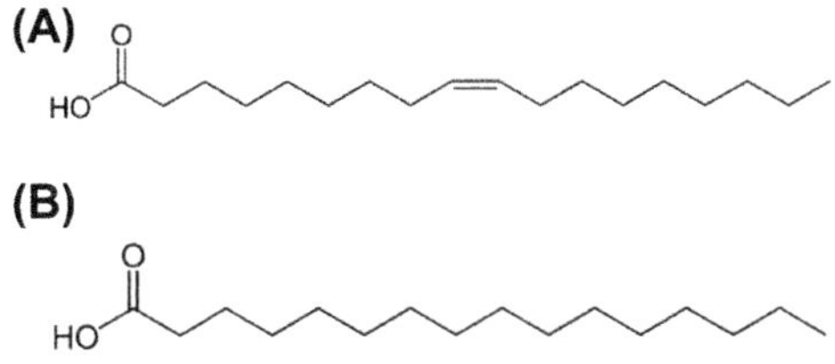

FIGURE 38.2 Important fatty acids present in Moringa. (A) Oleic acid. (B) Palmitic acid.

species. Ethanolic extract prepared by pods of moringa constitutes *O*-ethyl-4-[(α-L-Rhamnosyloxy)-benzyl] carbamate, β-sitosterol, *O*-[2′-hydroxy-3′-(2″-heptenyloxy)]-propyl undecanoate, and methyl hydroxybenzoate (Faizi et al., 1998), whereas aqueous extract of pods is characterized by L-rhamnose, L-arabinose, D-galacturonic acid, 6-*O*-Me-D-galactose, and D-galactose with a molar ratio of 1:1:1:1 (Roy et al., 2007). Moreover, phenolic compounds and glucosinolates are also isolated from different parts of moringa (Bennett et al., 2003). Roots of this plant constitute a high concentration of benzyl glucosinolates and 4-(α-l-rhamnopyranosyloxy)-benzylglucosinolate (Mishra et al., 2017). Aqueous extract of seeds, fruit, and leaves of moringa was reported to have ferulic acid, vanillin, quercetin, kaempferol, ellagic acid, chlorogenic acid, and gallic acid as characterized by MS/MS and HPLC techniques (Singh et al., 2009). Essential oil or volatile components are not only confined to leaves of moringa, but flowers are also known to have a significant amount of scented oil as analyzed by gas chromatography, mass spectrometry, and simultaneous distillation extraction. Approximately 74 different chemical compounds have been identified, constituting 99.8% of total essential oil yield. Volatile fractions were characterized by oxygenated sesquiterpenes (13.3%), monoterpene hydrocarbons (15.7%), oxygenated monoterpenes (16%), nitrogenous compounds (16.6%), and aliphatic compounds (34.2%). Some major chemical components were benzyl isothiocyanate (6.4%), α-terpineol (7.8%), and (E)-nerolidol (13.3%) (Pino, 2013).

3. POSTHARVEST TECHNOLOGY

Leaves of *Moringa oleifera* can be harvested from densely grown fields when plants attain a height ranging from 1.5 to 2.0 m, which normally requires 60–90 days in properly hydrated, well-drained, and fertile soil. Harvesting of leaves is usually practiced by cutting with a sharp knife along leaf stems at an approximate height of 20–45 cm above the surface of the ground. This type of harvesting practice will help to promote new shoot development. Further, subsequent harvesting can be done after every 35–40 days, and even shoots of moringa, when to be used as fodder, can be harvested after an interval of 75 days. In case of intercropping, plants can even be harvested after 2–4 months of complete growth. Initial cutting can be done manually at a height in the range of 20 cm to 1.5 m. Harvesting of this plant should be done at a height where they are long enough not to be shaded by any other companion crop. Heaping of freshly harvested moringa leaves should be avoided as it tremendously increases the deterioration process. Freshly harvested leaves readily lose moisture, so they should be used or preserved immediately after cutting (Amaglo, 2006). The best harvesting time for seeds of moringa is immediately after maturity when the seed coat dries and seeds turns brown. Seeds are collected after extraction and bagged to be stored in a cool dry place.

4. PROCESSING

Processing of the leaves initiates with stripping of leaflets from the leaf petiole, which can directly be picked from branches, and damaged or diseased leaves are discarded. Leaves are then washed by using portable water to eliminate dirt and then soaked in 1% solution of saltwater for about 3–5 minutes to avoid all types of microbial infections. Finally, the leaves are removed by saline water and again washed with clean water followed by draining and drying. Draining is the process involving straining water from leaves through a perforated strainer. These leaflets are then spread on food-grade mesh before going into a dryer. Drying of moringa can be achieved by three basic processes: (1) room drying, (2) solar drying, and (3) mechanical drying. Leaves should be dried until its moisture contents are below 10%. Nevertheless, this mechanical method can only be made applicable for large-scale production. After drying, the next step is milling of dried leaves using stainless-steel hammer mills, pestle mortar, or burr mills to form fine powders. This finely ground powder is then sieved using different mesh sizes according to the requirement depending on application. This powder still contains high moisture content, so it is dried at about 50°C for more than 30 minutes to maintain humidity less than 7.5%. Packing and storage are the last but very important steps that require intensive care regarding personal hygiene. These packed materials are then labeled and marketed to be used for various purposes (Mishra et al., 2012).

5. VALUE ADDITION

Leaf powder of moringa is extensively used as a food supplement or medicine in animal feed. Moringa seeds and seed oil have potential applications in the production of biofuel, industrial oil, lubricants, perfumes, and cosmetic products. Cake of moringa seeds, also known as "sapal," is used as a biofertilizer and has many applications in purification or treatment of wastewater. Extract or tea prepared by leaves of moringa helps to promote quality of sleep and also acts as a detoxifying agent. Different products prepared by different parts of moringa help to fight against a number of bacterial, viral, and fungal infections, healing gastric ulceration and all skin problems. Seeds of the plant when mixed with coconut oil help in treatment of several health problems like epilepsy, painful boils, urinary problems, cramps, gout, rheumatism, and arthritis. Some major and highly valuable marketable products of moringa are (1) tea of dried moringa leaves, (2) moringa polvoron, (3) moringa powder, (4) moringa cookies, (5) moringa noodles, (6) moringa sauces for soups, (7) moringa hotdogs, (8) moringa biscuits, (9) burgers, and (10) bread of moringa. However, further investigations are required to increase the quality of foodstuffs and shelf-life of all useful products (Mani et al., 2007).

6. USES

Moringa oleifera is multipurpose tree of tropical regions mainly used as food and known to have numerous agricultural, medicinal, and industrial applications including animal feed and food supplements (Denton et al., 2004). Every part of moringa such as flower, roots, pods, and leaves is used for edible purposes, and the leaves are specifically considered to be an excellent source of green vegetable. Roots are used as an alternative or substitute for horseradish, but they are slightly toxic in nature. Leaves are known to be the most nutritious part of the plant as they contain appreciable quantities of minerals, vitamin A, vitamin B, and vitamin C, so they are recommended for nursing mothers, pregnant women, and young children. Leaves of *Moringa oleifera* can be boiled or pan fried to be used as salads or mixed with sauces and soups. Leaf powder helps infants restore from malnutrition. The flowers are used to make tea or added in sauces to make pasta. Younger pods taste like asparagus and also are used for edible purposes, while older pods are added in curries and sauces to enhance bitterness. Immature seeds are cooked in different ways, whereas mature seeds are roasted and eaten like peanuts. Seeds constitute almost 30%–40% edible "ben oil" that can be used for dressing of salads and in cooking in place of olive oil, as it is highly resistant to rancidity and provides substantial quantities of tocopherol, sterols, and oleic acid (Yu et al., 2005). Oil of moringa has several industrial applications such as (1) manufacturing of perfumes and fragrances, (2) lubricants, and (3) preparation of paints (Denton et al., 2004). Bark of moringa is soaked in alcohol or boiled with water to obtain infusions that can be used to treat severe stomach ailments, ulceration, stomach pains, hypertension, anemia, diabetes, joint pains, and poor vision (Abe and Ohtani, 2013), along with uterine disorders, hemorrhoids, and toothache (Yabesh et al., 2014). Moringa seeds are used to settle down the impurities in wastewater (Popoola and Obembe, 2013).

Infusion of roots can be an effective remedy for treatment of number of diseases, as it acts as a potential antiparalytic and anthelmintic component (Anwar et al., 2007). Flowers are used to produce useful aphrodisiac substances for the treatment of enlarged spleen, tumors, hysteria, muscular diseases, and severe inflammations (Yabesh et al., 2014). Some other nonmedicinal and nonfood applications of moringa are of extreme importance, as they can be used as a natural plant growth enhancer because leaves are known to be a rich source of zeatin that is an important plant hormone belonging to cytokinins. Some recent investigations have shown that leaf extract also stimulates plant growth and increases crop yield (Ashfaq et al., 2012). Seed powder of *Moringa oleifera* is used in water purification through replacement of dangerous and explosive chemicals like aluminum sulfate (Popoola and Obembe, 2013). Furthermore, seed and leaf extract exhibits tremendous biopesticidal activities and was found effective against adults and larvae of *Trogoderma granarium*, resulting in significant reductions in fungal infections (Ashfaq et al., 2012). Seeds of moringa are an excellent source of fatty acids, thus also used for production of biodiesel to cope with increasing energy demands and numerous environmental problems.

7. PHARMACOLOGICAL USES

7.1 Antiinflammatory Activity

Aqueous extract of moringa roots were used in rats to check their antiinflammatory potentials using indomethacin as a standard drug, and intentionally edema was induced through subcutaneous carrageenin injections. At dose of 750 mg per kilogram, moringa was able to inhibit development of edema at 1, 3, and 5 hours with an approximate reduction of 53.5%, 44.6%, and 51.1%, respectively (Ndiaye et al., 2002).

7.2 Antioxidant Activity

Aqueous and hydroalcoholic extracts of roots and leaves of *Moringa oleifera* were tested for antioxidant activities using two basic extraction techniques, reflux and shaking (Sultana et al., 2009), and results indicated that they have strong antioxidant potentials. Another recent research investigated antioxidant potential of aqueous extract of seeds, fruit, and leaves of *Moringa oleifera* (Singh et al., 2009).

7.3 Anticancer Activity

Extract of moringa was used for cytotoxicity by using MTT (3-(4,5-dimethylthiazol-2-yl)-2,5-diphenyltetrazoliumbromide) assay using tumor cell lines, hemolysis assay, sea urchin eggs assay, and brine shrimp lethality assay (Suresh et al., 2011). Another research was conducted to investigate potential cytotoxic effects of leaf extract of *Moringa oleifera* using human multiple myeloma cell lines for evaluation of efficiency. In organic extract, methanolic extract of leaves of moringa showed the least viability at maximum concentration of dose (Parvathy and Umamaheshwari, 2007).

7.4 Antifertility Activity

Aqueous extract of moringa was tested for antiimplantation activity inside reproductive organs of female rats, and antifertility effects were studied through aqueous extract of plant roots. Oral administration of extract significantly increased uterine wet weight of bilaterally ovariectomized rats. This entire estrogenic activity was fully supported by uterine stimulation histoarchitecture. When extract was conjointly given along with estradiol dipropionate (EDP), successive reduction in uterine wet weight was observed when compared to the gain with EDP alone, and histological structures of uterine were also strongly inhibited (Shukla et al., 1988).

7.5 Hepatoprotective Activity

Seed extract of moringa was studied for hepatoprotective effects on liver fibrosis that was intentionally induced through oral administration of 20% carbon tetrachloride. With this, seed extract of *Moringa oleifera* was also administered orally on a daily basis. Incorporation of this seed extract tremendously decreased carbon tetrachloride–induced elevation of serum aminotransferases activity and level of globulin. The elevation of myeloperoxidase activity and hepatic hydroxyproline content were also reduced by treatment with moringa extract. Further, immunohistochemical investigations revealed that moringa significantly reduced amount of smooth muscle α-actin–positive cells and rich accumulation of collagens I and III inside liver (Hamza, 2010).

7.6 Cardiovascular Activity

Lyophilized hydroalcoholic extracts of moringa were tested for cardioprotective effects in an isoproterenol-induced myocardial infarction model inside the body of male Wistar albino rats. Chronic treatments by using *Moringa oleifera* resulted in marked modulations of several biochemical enzymes, such as creatine kinase-MB, catalase, lactate dehydrogenase, glutathione peroxidase, and superoxide dismutase, which failed to exhibit significant effects on reduction of glutathione as compare to Isoproterenol (ISP) control group. Treatment with moringa significantly prevented increase in peroxidation of lipids in myocardial tissues (Nandave et al., 2009).

7.7 Antiulcer Activity

Aqueous extract of leaves of *Moringa oleifera* was employed for antiulcer activities on bodies of adult Holtzman albino rats of either sex (male or female) using ondansetron as a standard drug (Debnath and Guha, 2007).

7.8 Analgesic and Antipyretic Activity

Aqueous leaf extracts of *Moringa oleifera* were investigated to evaluate wound healing potentials in bodies of male Swiss albino mice. Decrease in scar area and significant increase in granuloma dry weight, hydroxy proline content, granuloma breaking strength, skin breaking strength, and wound closure rate were observed (Rathi et al., 2006). Similarly, wound healing capacity and antipyretic effects were also studied using ethyl acetate and ethanolic extracts of the leaves of *Moringa oleifera*. Ethyl acetate and ethanolic extracts of seeds exhibited marked antipyretic effects in bodies of rats, while ethyl acetate extract formed by dried leaves showed remarkable wound healing potentials on granulomas, dead spaces, incisions, and excisions in the bodies of rats. That is why almost 10% extract is generally used in different skin ointments (Hukkeri et al., 2006).

7.9 Antidiabetic Activity

Leaves of *Moringa oleifera* were tested for antidiabetic potentials on glucose tolerance in Wistar and Goto-Kakizaki rats. Moringa was known to significantly decrease blood glucose level in bodies of Wistar rats. However, changes in blood glucose were remarkably higher in the Goto-Kakizaki rats. The action of *Moringa oleifera* was greater in Goto-Kakizaki rats compared to Wistar rats (Ndong et al., 2007). Furthermore, antidiabetic activities were also tested with aqueous extracts of leaves of Moringa on body weight, urine protein, urine sugar, hemoglobin, glycemic control, and total protein contents (Jaiswal et al., 2009).

7.10 Diuretic and Antiurolithiatic Activity

Alcoholic and aqueous extracts of root wood of *Moringa oleifera* was investigated for antiurolithiatic activity on the calcium oxalate urolithiasis in bodies of male Wistar albino rats. Oral administration of these alcoholic and aqueous extracts significantly reduced elevated levels of urinary oxalate, showing a proper regulatory action on synthesis of endogenous oxalate. The elevated level of stone-forming constituents in kidneys of calculogenic rats was significantly reduced by preventive and curative treatments using alcoholic and aqueous extracts (Karadi et al., 2006). Furthermore, antiurolithiatic potentials from alcoholic and aqueous extracts of root bark of *Moringa oleifera* were also investigated. Both these extracts significantly decreased kidney retention level of phosphates, calcium, and oxalates and urinary excretion. Moreover, elevated serum levels of uric acid, creatinine, and urea nitrogen were tremendously reduced by using these extracts (Karadi et al., 2008).

7.11 Central Nervous System Activity

Aqueous extract of root of *Moringa oleifera* was tested for anticonvulsant potentials, and these effects were studied on norepinephrine, dopamine, and brain serotonin levels, locomotor behavior, and penicillin-induced convulsion in Holtzman-strain adult albino rats (Ray et al., 2003). Similarly, ethanolic extract of leaves of *Moringa oleifera* was tested for alteration of monoamines of brain such as serotonin, dopamine, and norepinephrine along with EEG wave pattern in Alzheimer disease in affected rats. Treatment with leaf extract of *Moringa oleifera* restored monoamine levels of brain regions to near control levels (Prabsattroo et al., 2015).

7.12 Local Anesthetic Activity

Methanolic extract of root bark of *Moringa oleifera* was incorporated in bodies of guinea pigs and frogs to check local anesthetic effects. Both these model animals exhibited the same effects after treatment, which indicated that Moringa possesses strong local anesthetic effects (Mishra et al., 2011).

7.13 Antiallergic Activity

Ethanolic extracts of the seeds of *Moringa oleifera* were investigated to check inhibitory actions on local and systemic anaphylaxis. The anti-anaphylactic effects of these extracts were studied in a model mouse of Compound 48/80-induced systemic anaphylactic shock. Passive cutaneous anaphylaxis activated by antigen-antibody was also used to assess effect of extract (Mahajan and Mehta, 2007).

7.14 Anthelmintic Activity

Ethanolic extracts of *Moringa oleifera* were tested for anthelmintic effects against Indian earthworm named *Pheritima posthuma*. Several different concentrations of these extracts were applied, and experimental results were shown in terms of time of death and paralysis of worms. Piperazine citrate (10 mg/mL) was used as a reference standard and distilled water was used as control group (Trapti et al., 2009).

7.15 Antitubercular Activity

Recently, it was investigated that fatty oil of *Moringa oleifera* showed strong antitubercular effects when applied on locally isolated strains of *Mycobacterium tuberculosis* at 25% (volume/volume) while using rifampicin as control drug at a concentration of 0.09 micrograms per milliliter (Santhosh and Suriyanarayanan, 2014).

7.16 Antimicrobial Activity

Aqueous, ethyl acetate, chloroform, and methanolic extracts of *Moringa oleifera* were investigated to check antibacterial potentials against four major bacterial strains *Pseudomonas fluorescens, Bacillus megaterium, Citrobacter freundii,* and *Staphylococcus aureus* by using erythromycin as a positive control (Zaffer et al., 2014). Moringa extracts were found active against *Saccharomyces cerevisiae, Candida tropicalis,* and *Candida albicans*. Aqueous and ethanolic extracts showed maximum activity against *S. cerevisiae* (Patel et al., 2014).

7.17 Immunomodulatory Activity

Ethanolic and methanolic extracts of leaves of *Moringa oleifera* were evaluated to check immunomodulatory effects (Sudha et al., 2010) inside cyclophosphamide-induced immunodeficient mice (Gupta et al., 2010). Results of these investigations revealed that leaf extract of *Moringa oleifera* showed a significant increase in serum immunoglobulins, percent neutrophils, and white blood cells, suggesting that it tremendously stimulated humoral and cellular immune responses. The same results were also confirmed by many other investigations done by the various researchers in different regions of the world (Asiedu-Gyekye et al., 2014).

8. SIDE EFFECTS AND TOXICITY

Chemicals in the root, bark, and flowers can make the uterus contract, and this might cause a miscarriage.

References

Abe, R., Ohtani, K., 2013. An ethnobotanical study of medicinal plants and traditional therapies on Batan Island, the Philippines. Journal of Ethnopharmacology 145, 554–565.

Amaglo, N., 2006. How to Produce Moringa Leaves Efficiently? Workshop.

Anwar, F., Latif, S., Ashraf, M., Gilani, A.H., 2007. *Moringa oleifera*: a food plant with multiple medicinal uses. Phytotherapy Research 21, 17–25.

Ashfaq, M., Basra, S.M., Ashfaq, U., 2012. Moringa: a miracle plant for agro-forestry. Journal of Agriculture and Social Sciences 8.

Asiedu-Gyekye, I.J., Frimpong-Manso, S., Awortwe, C., Antwi, D., Nyarko, A., 2014. Micro- and macroelemental composition and safety evaluation of the nutraceutical Moringa oleifera leaves. Journal of Toxicology 2014.

Bennett, R.N., Mellon, F.A., Foidl, N., Pratt, J.H., Dupont, M.S., Perkins, L., Kroon, P.A., 2003. Profiling glucosinolates and phenolics in vegetative and reproductive tissues of the multi-purpose trees *Moringa oleifera* L.(horseradish tree) and *Moringa stenopetala* L. Journal of Agricultural and Food Chemistry 51, 3546–3553.

Debnath, S., Guha, D., 2007. Role of Moringa oleifera on enterochromaffin cell count and serotonin content of experimental ulcer model. Indian Journal of Experimental Biology 45 (8), 726–731.

Denton, O., Schippers, R., Oyen, L., Siemonsma, J., 2004. Ressources végétales de l'Afrique tropicale 2 Légumes.

Faizi, S., Siddiqui, B.S., Saleem, R., Aftab, K., Shaheen, F., 1998. Hypotensive constituents from the pods of *Moringa oleifera*. Planta Medica 64, 225–228.

Ganguly, S., 2013. Indian ayurvedic and traditional medicinal implications of indigenously available plants, herbs and fruits: a review. International Journal of Research in Ayurveda and Pharmacy 4 (4), 623–625.

Gupta, A., Gautam, M.K., Singh, R.K., Kumar, M.V., Rao, C.V., Goel, R., Anupurba, S., 2010. Immunomodulatory effect of *Moringa oleifera* Lam. extract on cyclophosphamide induced toxicity in mice. Indian Journal of Experimental Biology 48 (11), 1157–1160.

Hamza, A.A., 2010. Ameliorative effects of *Moringa oleifera* Lam seed extract on liver fibrosis in rats. Food and Chemical Toxicology 48, 345–355.

Hukkeri, V., Nagathan, C., Karadi, R., Patil, B., 2006. Antipyretic and wound healing activities of *Moringa oleifera* Lam. in rats. Indian Journal of Pharmaceutical Sciences 68, 124.

Jaiswal, D., Rai, P.K., Kumar, A., Mehta, S., Watal, G., 2009. Effect of *Moringa oleifera* Lam. leaves aqueous extract therapy on hyperglycemic rats. Journal of Ethnopharmacology 123, 392–396.

Karadi, R., Palkar, M., Gaviraj, E., Gadge, N., Mannur, V., Alagawadi, K., 2008. Antiurolithiatic property of *Moringa oleifera* root bark. Pharmaceutical Biology 46, 861–865.

Karadi, R.V., Gadge, N.B., Alagawadi, K., Savadi, R.V., 2006. Effect of *Moringa oleifera* Lam. root-wood on ethylene glycol induced urolithiasis in rats. Journal of Ethnopharmacology 105, 306–311.

Mahajan, S.G., Mehta, A.A., 2007. Inhibitory action of ethanolic extract of seeds of *Moringa oleifera* Lam. on systemic and local anaphylaxis. Journal of Immunotoxicology 4, 287–294.

Mani, S., Jaya, S., Vadivambal, R., 2007. Optimization of solvent extraction of Moringa (Moringa oleifera) seed kernel oil using response surface methodology. Food and Bioproducts Processing 85, 328–335.

Manzoor, M., Anwar, F., Iqbal, T., Bhanger, M., 2007. Physico-chemical characterization of Moringa concanensis seeds and seed oil. Journal of the American Oil Chemists' Society 84, 413–419.

Mishra, G., Singh, P., Srivastav, S., Verma, R.K., 2017. Moringa Olifera-An Important Medicinal Plant: A Review of its Traditional Uses, Phytochemistry and Pharmacological Properties. February 2011.

Mishra, G., Singh, P., Verma, R., Kumar, S., Srivastav, S., Jha, K., Khosa, R., 2011. Traditional uses, phytochemistry and pharmacological properties of *Moringa oleifera* plant: an overview. Der Pharmacia Lettre 3, 141–164.

Mishra, S.P., Singh, P., Singh, S., 2012. Processing of Moringa oleifera leaves for human consumption. Bulletin of Environment, Pharmacology and Life Sciences 2, 28–32.

Nandave, M., Ojha, S.K., Joshi, S., Kumari, S., Arya, D.S., 2009. Moringa oleifera leaf extract prevents isoproterenol-induced myocardial damage in rats: evidence for an antioxidant, antiperoxidative, and cardioprotective intervention. Journal of Medicinal Food 12, 47–55.

Ndiaye, M., Dieye, A., Mariko, F., Tall, A., Sall, A.D., Faye, B., 2002. Contribution to the study of the anti-inflammatory activity of *Moringa oleifera* (Moringaceae). Dakar Medical 47, 210–212.

Ndong, M., Uehara, M., Katsumata, S.-i., Suzuki, K., 2007. Effects of oral administration of *Moringa oleifera* Lam on glucose tolerance in Goto-Kakizaki and Wistar rats. Journal of Clinical Biochemistry & Nutrition 40, 229–233.

Nouman, W., Basra, S.M.A., Siddiqui, M.T., Yasmeen, A., Gull, T., Alcayde, M.A.C., 2014. Potential of *Moringa oleifera* L. as livestock fodder crop: a review. Turkish Journal of Agriculture and Forestry 38, 1–14.

Orwa, C., Mutua, A., Kindt, R., Jamnadass, R., Anthony, S., 2009. Agroforestry Database: A Tree Reference and Selection Guide Version 4 0. World Agroforestry Centre, Kenya.

Parvathy, M., Umamaheshwari, A., 2007. Cytotoxic effect of *Moringa oleifera* leaf extracts on human multiple myeloma cell lines. Trends in Medical Research 2, 44–50.

Patel, P., Patel, N., Patel, D., Desai, S., Meshram, D., 2014. Phytochemical analysis and antifungal activity of *Moringa oleifera*. International Journal of Pharmacy and Pharmaceutical Sciences 6, 144–147.

Pino, J.A., 2013. Floral scent composition of moringa oleifera Lam. Journal of Essential Oil Bearing Plants 16, 315–317.

Popoola, J.O., Obembe, O.O., 2013. Local knowledge, use pattern and geographical distribution of *Moringa oleifera* Lam.(Moringaceae) in Nigeria. Journal of Ethnopharmacology 150, 682–691.

Prabsattroo, T., Wattanathorn, J., Iamsaard, S., Somsapt, P., Sritragool, O., Thukhummee, W., Muchimapura, S., 2015. Moringa oleifera extract enhances sexual performance in stressed rats. Journal of Zhejiang University – Science B 16, 179.

Rathi, B., Bodhankar, S., Baheti, A., 2006. Evaluation of aqueous leaves extract of *Moringa oleifera* Linn for wound healing in albino rats. Indian Journal of Experimental Biology 44 (11), 898–901.

Ray, K., Hazra, R., Guha, D., 2003. Central inhibitory effect of *Moringa oleifera* root extract: possible role of neurotransmitters. Indian Journal of Experimental Biology 41, 1279–1284.

Roy, S.K., Chandra, K., Ghosh, K., Mondal, S., Maiti, D., Ojha, A.K., Das, D., Mondal, S., Chakraborty, I., Islam, S.S., 2007. Structural investigation of a heteropolysaccharide isolated from the pods (fruits) of Moringa oleifera (Sajina). Carbohydrate Research 342, 2380–2389.

Santhosh, R.S., Suriyanarayanan, B., 2014. Plants: a source for new antimycobacterial drugs. Planta Medica 80, 9–21.

Shukla, S., Mathur, R., Prakash, A.O., 1988. Antifertility profile of the aqueous extract of *Moringa oleifera* roots. Journal of Ethnopharmacology 22, 51–62.

Singh, B.N., Singh, B., Singh, R., Prakash, D., Dhakarey, R., Upadhyay, G., Singh, H., 2009. Oxidative DNA damage protective activity, antioxidant and anti-quorum sensing potentials of *Moringa oleifera*. Food and Chemical Toxicology 47, 1109–1116.
Sudha, P., Asdaq, S.M.B., Dhamingi, S.S., Chandrakala, G.K., 2010. Immunomodulatory activity of methanolic leaf extract of moringa oleifera in animals. Indian Journal of Physiology and Pharmacology 54 (2), 133–140.
Sultana, B., Anwar, F., Ashraf, M., 2009. Effect of extraction solvent/technique on the antioxidant activity of selected medicinal plant extracts. Molecules 14, 2167–2180.
Suresh, V., Sruthi, V., Padmaja, B., Asha, V., 2011. In vitro anti-inflammatory and anti-cancer activities of Cuscuta reflexa Roxb. Journal of Ethnopharmacology 134, 872–877.
Trapti, R., Vijay, B., Komal, M., Aswar, P., Khadabadi, S., 2009. Comparative studies on anthelmintic activity of *Moringa oleifera* and *Vitex negundo*. Asian Journal of Research in Chemistry 2, 181–182.
Yabesh, J.M., Prabhu, S., Vijayakumar, S., 2014. An ethnobotanical study of medicinal plants used by traditional healers in silent valley of Kerala, India. Journal of Ethnopharmacology 154, 774–789.
Yu, L.L., Parry, J.W., Zhou, K., 2005. Oils from herbs, spices, and fruit seeds. Bailey's Industrial Oil and Fat Products.
Zaffer, M., Ahmad, S., Sharma, R., Mahajan, S., Gupta, A., Agnihotri, R.K., 2014. Antibacterial activity of bark extracts of *Moringa oleifera* Lam. against some selected bacteria. Pakistan Journal of Pharmaceutical Sciences 27, 1857–1862.
Zaku, S., Emmanuel, S., Tukur, A., Kabir, A., 2015. Moringa oleifera: an underutilized tree in Nigeria with amazing versatility: a review. African Journal of Food Science 9, 456–461.

CHAPTER

39

Oleander

Asma Seher[1], Muhammad Asif Hanif[1], Muhammad Adnan Ayub[2], Muhammad Idrees Jilani[3], Mohamad Fawzi Mahomoodally[4]

[1] Department of Chemistry, University of Agriculture, Faisalabad, Pakistan;
[2] Department of Chemistry, University of Okara, Okara, Pakistan;
[3] Department of Chemistry, University of Sahiwal, Sahiwal, Pakistan;
[4] Department of Health Sciences, Faculty of Science, University of Mauritius, Mauritius

OUTLINE

Medicinal Plants of South Asia
https://doi.org/10.1016/B978-0-08-102659-5.00039-2

Copyright © 2020 Elsevier Ltd. All rights reserved.

1. BOTANY

1.1 Introduction

Nerium oleander L., also known as oleander (Fig. 39.1), is a perennial evergreen flowering shrub that belongs to the dogbane family (Apocynaceae). The common oleanders are of two main types, *Thevetia peruviana* (Pers.) K. Schum and *Nerium oleander* L., along with various cultivators and varieties. Globally, oleander grows throughout the tropical and subtropical regions. Oleander is broadly cultivated as a casual hedge in dry subtropical and hot temperate areas (Langford and Boor, 1996).

New cultivars of oleander are red, orange, and purple. The name of ordinary oleander that has a pink color is *Nerium oleander* (Wong et al., 2013). Its genus name is derived from the Greek word used for moist, "neros," referring to the favorable atmosphere for the growth of this species (Pagen, 1987). The species of *N. oleander* are *N. indicum* and *N. odorum* (Wong et al., 2013). For the plants such as *N. oleander, N. odorum,* and *N. indicum,* oleander is an idiom (Khan et al., 2010). The common names of oleander are kaner in Pakistan, laurier-rose in France, oleander in German, oleandrio in Italy, kyochiku-to in Japan, adelfa in Spain, and rosebay in the United Kingdom (Pagen, 1987).

The plant is used for the cure of dermatological diseases. The decoction of leaves is consumed for the treatment of inflammation. Due to its toxicity, this plant is used externally for the cure of different diseases. The plant consists of antinociceptive, antiinflammatory, antimicrobial, locomotor, anticancer, diuretic, antileukemic, and central nervous system (CNS) depressant activities (Zibbu and Batra, 2010).

1.2 History and Origin

Oleander is native to Mediterranean areas of Asia, Europe, and Africa. As an ornamental plant, this species has been grown in the subtropical and tropical regions of the world; though, it is regarded as a poisonous wild plant in some areas of the world. This plant was identified by the Greeks under three names: *rhododendron*, *rhododaphne*, and *nerion*. It is well illustrated by Pliny the Elder, who cited its noxious character and rose-like flowers. In greenhouses, the ordinary oleander has long been cultured, and several species have been introduced. Sweet oleander is a species of the same family that is smaller in size with flowers having a vanilla odor. Oleander is generally grown in the open air in temperate

FIGURE 39.1 Oleander plant at various stages.

FIGURE 39.1 (Continued)

FIGURE 39.1 (Continued)

countries such as Japan, Canada, Africa, and New Zealand (Jashemski and Meyer, 2002). Oleander spreads through seeds in different regions of the world from its native areas. However, they can also be propagated by cutting a branch of plant without flowers and then placing it into water until the growth of roots is about 2–3 cm. Then, it is planted in a pot, and as the plant becomes well rooted, it is transplanted to a definitive place, whereby growth of the new plant takes place quickly.

1.3 Demography/Location

Oleander is an attractive fluorescent plant having white or pink blossoms particularly suitable to arid and bright locations. Oleander is commonly found in a damp environment on places having good exposure to the sunlight. Particularly, it appears in the beds and along the banks of streams (Adome et al., 2004). This plant was introduced only in the areas where they are openly grown for decorative projects. The plant is also grown in different countries of Asia. Oleanders are dispersed in Himalaya from Nepal to Kashmir up to 6510 ft, in central Waziristan and south Salt Range, as well as in Japan and China. Several species of oleander are found in Punjab, Pakistan.

The different cultivars of oleander plant grow at different heights. Many of the cultivars grow from 8.0 to 12.0 ft in height. The adult plants might reach about 20.0 ft in height in the protected areas. Some of the dwarf cultivars grow up to 3.0–5.0 ft. Oleander mostly grows normally as well as quickly, for a height of 1.0–2.0 ft per year. The plants grow again very quickly from the base area if they are damaged from cold. Due to the rapid rate of growth and thick multistemmed habit of oleander, they are ideal for their usage as an informal or screening hedge (Schuch, 2009).

1.4 Botany, Morphology, and Ecology

Oleander is an evergreen shrub. Leaves are 10–22 cm long, narrow, untoothed, and short-stalked, and dark or gray-green in color. Some cultivars have variegated leaves with white or yellow. All leaves have a prominent midrib, are "leathery" in texture, and usually arise in groups of three from the stem (Zibbu and Batra, 2010). Oleander leaves are in short stalks, in threes, wrinkled, 11–14 cm extended, linear-lanceolate, tapering into short, acuminate, shining above and dark green, nerves abundant dispersing parallel, and midrib solid. The width of leaves of oleander is commonly 1.0 inches. Flowers of oleander are rose colored and odorous (Hussain and Gorsi, 2004). Flowers of different colors are produced by oleanders during the hot season because of reproduction of seed cases. Terminal flowery heads are produced by the plant, normally white or pink (Zibbu and Batra, 2010). Though, about 400 varieties have been bred. These species present a broad range of various colors of flowers: pale pink to deep pink, carmine, lilac, salmon, purple, apricot, copper, white, and orange (Soundararajan and Karrunakaran, 2010). All flowers are five petalled, and their diameter is approximately 5.0 cm (Zibbu and Batra, 2010). Every plant has long projections that are petal-like in appearance. The throat of every flower is fringed. Intermittently, binary flowers are encountered during cultivars. A long, thin capsule is composed by fruit (i.e., 6.0–8.0 mm in diameter and 10.0–12.0 cm in length). The opened, fluffy seeds are dispersed. In cultivated plants, fruiting is not common. When a branch or twig is cut or broken, a thick sap, white in color, is excreted. The corolla is 3.7 cm in diameter, aromatic, lobes are curved. Follicles are 14–22 cm in length, firm, and isolated length-wise. Seeds are about 12 cm in length. The tip of seeds consists of a coma of hairs. These hairs are of light brown color (Hussain and Gorsi, 2004). The optimum temperature for the germination is 20°C, along with the growing temperature of 13–55°C. The best development of the plant occurs in full sunlight. But in partially shaded areas, they still produce flowers. They can grow in any type of soil like well-drained sand, heavy clay, marshy plots, and locations that have a higher level of chloride, sodium, and other salts. The plant grows best in the pH of 6.5–7.5.

2. CHEMISTRY

Due to the presence of essential oil in flowers, oleanders have a characteristic odor. Glycosides are the main toxic components present in different parts of the plant. One of the widely studied components is oleandrin (Fig. 39.2) (Lahoz et al., 2009). More than 10 other glycosides whose chemical structures are well known include five α-cardenolides (such as uzangenin-type) and five β-cardenolides. The names of some of the common glycosides are: glucosyloleandrin, oleandrin, gentiobiosyl-oleandrin, digitalose/diginose, nerigoside, diginose, and gentiobiose-diginose. The glucosides (rosaginoside, corteneroside, and nerioside) are also present in the bark. The steroids are present in the roots. The seeds of oleander consist of glucosides (odorosides, adigoside, and oleandrin). The compound having no cardiac effect is adyregenin (Kumar et al., 2013). Furthermore, the lymph of oleander is also full of mineral deposits such as ursolic acid and α-tocopherol, which is an essential antioxidant. Digitoxigenin is extensively scattered in plants. It either acts as a toxin or insect deterrent (Hussain and Gorsi, 2004; Zibbu and Batra, 2010).

Oleander is fine source of saponin, alkaloids, glycosides, flavonoids, and tannic acid. Triterpenoids and steroids are found in oleander leaves. Oleander has a considerably larger quantity of flavonoids (14.43%),

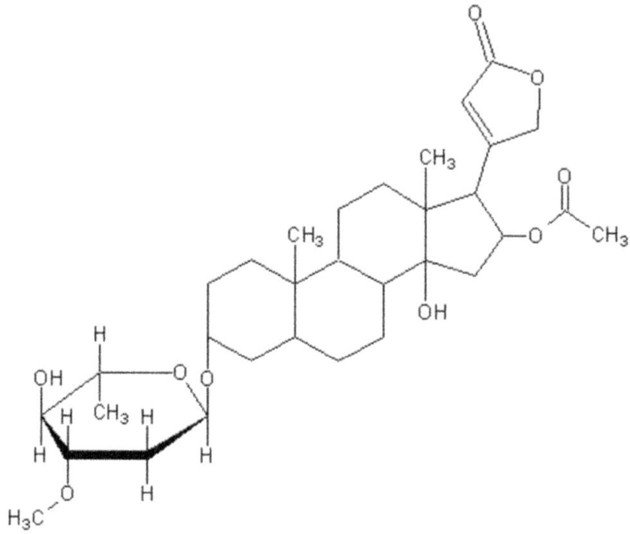

FIGURE 39.2 Oleandrin is an active component of oleander.

alkaloids (0.30%), glycosides (0.30%), steroids, tannic acid (14.5%), saponins (71.0%), and triterpenoids (7.0%) in leaves. The total amount of tannins, vitamin C, and phenols is also considerably larger in leaves (600, 430, and 600 mg/g respectively). In the gemmomodified extract of oleander, the quantity of K, Cr, and Zn is greater. However, the amount of Mn and Fe is higher in leaves, and there is no particular difference in the amount of Cu, Co, Ca, Na, and Mg in the leaves and gemmomodified extract of oleander (Zafar et al., 2014).

A vast variety of compounds obtained from plants are recognized as secondary metabolites. About 34 volatile components, presenting 93.22% of the whole composition, were identified in the oleander oil from the flowers. The mainly rich constituents found in the oil of flower were neriine (22.55%), digitoxigenin (11.24%), α-pinene (5.53%), amorphane (8.12%), calarene (5.11%), β-phellandrene (4.83%), limonene (5.02%), terpinene 4-ol (3.97%), isoledene (2.93%), sabinene (3.21%), 3-carene (2.55%), β-pinene (2.02%) cymen-8-ol (1.66%), and humulene (2.28%) (Derwich et al., 2010). The popular values of oleander are because of two glycosides (oleandrin and neriin), an alkaloid (having cardio-stimulatory effect) (Zibbu and Batra, 2010), and due to three glycosides (gentiobiosyl-nerigoside, gentiobiosyl-beaumontoside, and gentiobiosyl-oleandrin) obtained from leaves (Zibbu and Batra, 2010).

3. VALUE ADDITION

Due to the presence of toxic components such as oleandrin and neriin (cardiac glycosides), oleander cannot be used in dietary products (Khan et al., 2010). Oleander is used as a preservative for wood (Goktas et al., 2009). However, Aloe vera–based *Nerium oleander* extract (NAE-8) is used as an age-defying active ingredient in antiaging skin care products (Benson et al., 2014).

4. POSTHARVESTING

Oleander is harvested in May through June. The sample is harvested when the leaves, stems, roots, and flowers become mature. The sample is picked up by cutting at the height of 3–4 m. After cutting, the sample is cleaned and dried in oven at the temperature of 40°C. Afterward, the sample is stored at the temperature of −20°C. The cutting should not be done without gloves at any stage of harvest and processing. During handling of plant, it is very important to protect yourself because this plant contains toxic components that can be absorbed through the skin (Tayoub et al., 2014).

5. PROCESSING

For the preparation of extracts, the flowers are dried under the shade. The dried leaves are powdered roughly with the help of a grinding mill. To defat the powder, the extraction is done by using petroleum ether, and afterward, it is macerated with an equal ratio of water and ethanol along with the continuous stirring. The filtration of extractive incorporated solvent is done, and the remaining extractives are squeezed out by marc pressing (Singhal and Gupta, 2011).

6. USES

Oleander has a plethora of traditional uses such as an astringent, acrid, anthelmintic, stomachic, aphrodisiac, febrifuge, emetic, diuretic, expectorant, anticancer, and cardio tonic, among others. The plant is consumed traditionally for management of renal and vesicle calculi, cardiac asthma, skin-related problems, chronic stomach, snake bites, joint pains, cancer, leprosy, and ulcers. In the treatment of scabies, the decoction of the leaves is used externally to lessen inflammations. Flowers and leaves are also utilized for managing malaria. Traditionally, leaves and flowers are used as an abortifacient to induce the embryo termination in unwanted pregnancies. For ulcer around genitals and hemorrhoids, the powder of roots is used externally. As a parasitic, rat poison, and insecticide, the bark and leaves are used. The whole plant is poisonous if eaten, and skin irritation may be caused by contact with it.

The flowers and leaves of oleander are diaphoretic, cardiotonic, anticancer, diuretic, antifungal, antibacterial, and expectorant (Chopra et al., 1986; Zibbu and Batra, 2010). A decoction of the leaves has been used topically in the cure of scabies and to decrease inflammation. This is an extremely toxic plant, consisting of a dominant cardiac toxin and must only be used with much precaution. The root is strongly resilient and consumed in the shape of adhesives and used in cancer management due to its toxic nature. It is compressed into a paste along with water and applied to wounds (Zibbu and Batra, 2010). The husk is sour and is consumed as a therapeutic, febrifuge, and against infrequent fever. Essential oil made from the husk of the root is consumed in the cure of leprosy and skin infections of a scaly type. Oleander is consumed as a laxative in dropsy and rheumatism. All parts of the plant have anticancer activities (Abe and Yamauchi, 1992).

In the drug store, the dried or green leaves, a mixture, plaster, decoction, or salve is prepared from the foliage; the decoction is from bark or a powdered bark; a gum is prepared from the roots, and the dehydrated flowers are utilized. It has been consumed to aggravate menstrual cycle, as an abortive, and as an antispasmodic in the cure of angina pectoris. It is consumed for the cure of all types of skin disorders as an exterior medication for itchiness, ringworm, scabies, lice, boils, and leprosy, to cure skin breakouts or pain in herpes, and to wipe out worms in cuts. Oleander has also been consumed for the cure of cancer (Wang et al., 2000): the foliage, flowers, latex, or leaf extract, roots, and bark have been utilized to combat warts, corns, carcinoma, cancerous ulcers, and ulcerating or hard tumors. Oleander is an antitumor, antiinflammatory, moisturizing, and potentializes apoptosis. The aqueous and hydroalcoholic juices obtained from the flowers are an inhibitor of sensation of pain and hae a tonic effect on the heart. The seeds and leaves raise toxicity for uneasiness of stomach, emesis, delirium, bradyarrhythmia, and ventricular hyperkalemia that can rapidly finish to passing away.

Though it was traditionally used to manage several diseases, there is still a dearth of validation of such claims. Currently, there is no high-quality data in vitro, in vivo, and clinical data to validate the use of oleander plant as used traditionally. In recent times, the anticancer effects of oleander and its components have received much attention and advanced research. Oleandrin is the main component that inhibits certain transcription factors, kinases, and provocative negotiators, along with the tumor necrosis issue. To overcome tumor genesis and inflammation, a molecular base for the capability of oleander can be granted. For numerous diseases in addition to arthritis, oleander may have many functions, but all of them have the requirement of more information (Manna et al., 2000). The utilization of herbal medications as healing remedies is as old as humankind. In Iran, the extract from leaves in dried form has been consumed as a diuretic and cardiotonic in edema (Zargari, 1995). It is recognized as a traditional medication in Cuba (Carbajal et al., 1991). In Malaysia, the plant has been utilized for its antitumor activity (Ilham et al., 1995). In Pakistan, India, and Bangladesh, it has also been consumed for its antibacterial activity (Srinivasan et al., 2001). On the other hand, this plant has been recognized as very poisonous and can be lethal to animals and man (De Pinto et al., 1981; Langford and Boor, 1996; Mazumder et al., 1994).

7. PHARMACOLOGICAL USES

7.1 Antinociceptive Activity

The ethanolic and aqueous juices obtained from the leaves of oleander have a particular property of inhibiting the sensation of pain. The ethanolic extract was more marked for this purpose. Both extracts were exposed to persuade ulcerogenicity in mice. Flowers of oleander either dehydrated or fresh also have strong effects in the inhibition of the sensation of pain (Erdemoglu et al., 2003).

7.2 Antiinflammatory Activity

The fresh and dried flowers of *Nerium oleander* and their ethanolic extracts displayed powerful antiinflammatory action against carrageenan-induced hind paw edema model in rats devoid of inducing any gastric injure (Erdemoglu et al., 2003).

7.3 Antimicrobial Activity

In the vascular plant the availability of the components that have antibacterial and antifungal effects is well recognized. The antimicrobial activity of leaves and roots of oleander is prominent against *Bacillus pumilus, Staphylococcus aureus, Aspergillus niger, Escherichia coli,* and *Bacillus subtilis* (Zibbu and Batra, 2010).

7.4 Locomotor Activity

It has been described that decontaminated portions derived from the methanol extract of green leaves of oleander have a depressant activity for the CNS (i.e., decline in locomotor activity). Leaf extract also exhibited considerable painkilling activity (as pointed out by inhibitory properties on acetic acid-persuaded and elevated time of reaction to thermal test) (Zia et al., 1995).

7.5 Anticancer Activity

The juicy extract of oleander has gotten attention due to its cancer inhibiting effect. Oleandrin, irigenin, and aglyconeole are the bioactive components that possess anticancer properties. Anvirzel, an extract of oleander, has been reported to possess cytotoxicity in the cell linings of tumors in humans (apoptosis was a chief method of cancerous cell death) (Smith et al., 2001). The antitumor response of oleander plant extract, and the corresponding capabilities of oleandrigenin and oleandrin to stop fibroblast growth factor-2 exported from two human prostate cancer cell lines, PC3 and DU145, were also studied previously. Oleandrin and Anvirzel are the extracts of oleander plant that persuade cell fatality in cancer cells (Smith et al., 2001). Toxicity of oleander and *Calotropis procera* have an action in the antitumor human cell line test with ED50 varied in the range of 0.007–2.12 μg/mL, depending upon the cell line (Zibbu and Batra, 2010). Oleander extracts are also used in ulcers and cancers of the penis (Pathak et al., 2000).

7.6 Diuretic Effects

Oleandrin was obtained from oleander to trigger the function of heart and to have a diuretic effect (Kumar et al., 2013). The consequence of odorin on the heart of dogs and rabbits was indistinguishable along with that of digitalis group. Neriodin was two times as vigorous as digitoxin in digitalis and showed a response resembling that of oleandrin (Pathak et al., 2000).

7.7 CNS Depressant Activity

Numerous oleandrins have been segregated from oleander, and their pharmacological properties have also been assessed (Covacevich et al., 1987). Experimentations have established that the raw alcoholic extract from the foliage has CNS-depressant activity. Oleander consists of at least 2% cardiac glycosides. From the oleander husk, a compound named rosagenin can be obtained and has an effect like strychnine (Zibbu and Batra, 2010).

8. SIDE EFFECTS AND TOXICITY

The ordinary oleander plants are highly toxic plants, due to the nondigitalis cardiac glycosides present in different parts of this plant. Oleandrin and neriine are two cardiac glycosides present in oleander. Both of these toxic constituents of cardiac glycosides are closely similar to foxglove, which is also toxic in nature. Both components consist of negative chronotropic, positive ionotropic, and cross-reactivity. This comprises the venoming of the Na–K pump of the heart and expanded vagotonia by the glycosides. The majorities of the manifestations, from the toxicity of oleander, are gastrointestinal and cardiac in nature and come into view 4 hours after its intake. Many of the plants, along with oleander and foxglove, have been recognized as consisting of cardiac glycosides including oleandroside, oleandrin, digitoxigenin, thevetoxin, nerioside, and thevetin. In oleander, the cardiac glycosides cause more gastrointestinal disorders than those present in digoxin, and the warning signs range from vomiting and nausea to bloody diarrhea and cramping. Also, irritability is caused by it to the mucosal membranes that results in gleaming in the area of the mouth and expanded salivation. Confusion, drowsiness, dizziness, mydriasis, visual disturbances, and weakness are the symptoms associated with toxicity in the CNS (Khan et al., 2010). Cardiac irregularities, including a variety of ventricular dysrhythmias, bradycardia heart block, and tachyarrhythmia, are the most severe toxicologic effects of oleander plant. ECG generally exposed an amplified and a decreased QRS-T interval, T wave flattening or inversion, and PR interval. It is considered that these medical demonstrations are the consequence of both direct cardiac glycoside poisoning and expanded vagotonia. The cure of toxicity due to oleander relied on the cure of digitalis-glycoside poisoning and comprised of sustaining the patient hemodynamically. This may take account for managing atropine for serious bradyarrhythmia; consuming lidocaine hydrochloride or phenytoin to overcome cardiac dysrhythmia; putting a momentary transvenous cardiac pacing; or electric countershock and controlling digoxin-immune Fab immunoglobulin fragments. Other techniques for the cure are intended at eliminating the lethal stuff from the abdomen by vomiting. Particular interest has to be devoted to a patient with bradyarrhythmia earlier than vomiting is persuaded due to the probability of a vasovagal response and deterioration of the bradyarrhythmia. Glycosides that are not absorbed properly might be constrained slightly, according to the particular glycoside, by a variety of joining components in the gut (Khan et al., 2010). These constituents are expected to be efficient in captivating nonpolar glycosides, like digitoxin, in contrast to the highly polar glycosides, such as digoxin (e.g., colestipol and cholestyramine resin). In the avoidance of more intakes of the cardiac glycosides, activated charcoal was consumed by disruption of the dissemination of the glycosides from liver to the intestine (Khan et al., 2010).

References

Abe, F., Yamauchi, T., 1992. Cardenolide triosides of oleander leaves. Phytochemistry 31, 2459–2463.

Adome, R., Gachihi, J., Onegi, B., Tamale, J., Apio, S., 2004. The cardiotonic effect of the crude ethanolic extract of *Nerium oleander* in the isolated Guinea pig hearts. African Health Sciences 3, 77–82.

Benson, K.F., Newman, R.A., Jensen, G.S., 2014. Antioxidant, anti-inflammatory, anti-apoptotic, and skin regenerative properties of an Aloe vera-based extract of *Nerium oleander* leaves (nae-8 (®)). Clinical, Cosmetic and Investigational Dermatology 8, 239–248.

Carbajal, D., Casaco, A., Arruzazabala, L., Gonzalez, R., Fuentes, V., 1991. Pharmacological screening of plant decoctions commonly used in Cuban folk medicine. Journal of Ethnopharmacology 33, 21–24.

Chopra, R., Varma, B., Chopra, I., 1986. Supplement to Glossary of Indian Medicinal Plants. Publications & Information Directorate-CSIR.

Covacevich, J., Davie, P., Pearn, J., 1987. Toxic Plants & Animals. Queensland Museum.

De Pinto, F., Parlermo, D., Milillo, M.A., Iaffaldano, D., 1981. Oleander (*Nerium oleander*) poisoning in cattle. Clinica Veterinaria, La 104, 15–18.

Derwich, E., Benziane, Z., Boukir, A., 2010. Antibacterial activity and chemical composition of the essential oil from flowers of *Nerium oleander*. Electronic Journal of Environmental, Agricultural and Food Chemistry 9, 1074–1084.

Erdemoglu, N., Küpeli, E., Yeşilada, E., 2003. Anti-inflammatory and antinociceptive activity assessment of plants used as remedy in Turkish folk medicine. Journal of Ethnopharmacology 89, 123–129.

Goktas, O., Ozen, E., Duru, M.E., Mammadov, R., 2009. Determination of the color stability of an environmentally-friendly wood stain derived from oleander (*Nerium oleander* L.) leaf extracts under UV exposure. Wood Research 54, 63–72.

Hussain, M., Gorsi, M., 2004. Antimicrobial activity of *Nerium oleander* Linn. Asian Journal of Plant Sciences 3, 177–180.

Ilham, M., Yaday, M., Norhanom, A., 1995. Tumour promoting activity of plants used in Malaysian traditional medicine. Natural Product Sciences 1, 31–42.

Jashemski, W.F., Meyer, F.G., 2002. The Natural History of Pompeii. Cambridge University Press.

Khan, I., Kant, C., Sanwaria, A., Meena, L., 2010. Acute cardiac toxicity of *Nerium oleander*/indicum poisoning (kaner) poisoning. Heart Views: The Official Journal of the Gulf Heart Association 11, 115.

Kumar, A., De, T., Mishra, A., Mishra, A.K., 2013. Oleandrin: a cardiac glycosides with potent cytotoxicity. Pharmacognosy Reviews 7, 131.

Lahoz, E., Caiazzo, R., Carella, A., Porrone, F., Porrone, F., 2009. Colletotrichum acutatum Simmonds as agent of anthracnose and stem blight on *Nerium oleander* in Italy. Floriculture & Ornamental Biotechnology 3, 62–66.

Langford, S.D., Boor, P.J., 1996. Oleander toxicity: an examination of human and animal toxic exposures. Toxicology 109, 1–13.

Manna, S.K., Sah, N.K., Newman, R.A., Cisneros, A., Aggarwal, B.B., 2000. Oleandrin suppresses activation of nuclear transcription factor-κB, activator protein-1, and c-Jun NH2-terminal kinase. Cancer Research 60, 3838–3847.

Mazumder, P., Rao, P., Kumar, D., Dube, S., Gupta, S.D., 1994. Toxicological evaluation of *Nerium oleander* on isolated preparations. Phytotherapy Research 8, 297–300.

Pagen, F., 1987. Oleanders; Nerium L. And the Oleander Cultivars. Series of Revisions of Apocynaceae XX. Agricultural University Wageningen.

Pathak, S., Multani, A.S., Narayan, S., Kumar, V., Newman, R.A., 2000. AnvirzelTM, an extract of *Nerium oleander*, induces cell death in human but not murine cancer cells. Anti-Cancer Drugs 11, 455–463.

Schuch, U., 2009. Pruning Shrubs in the Low and Mid-elevation Desert in Arizona.

Singhal, K., Gupta, G., 2011. Some central nervous system activities of *Nerium oleander* Linn (Kaner) flower extract. Tropical Journal of Pharmaceutical Research 10, 455–461.

Smith, J.A., Madden, T., Vijjeswarapu, M., Newman, R.A., 2001. Inhibition of export of fibroblast growth factor-2 (FGF-2) from the prostate cancer cell lines PC3 and DU145 by Anvirzel and its cardiac glycoside component, oleandrin. Biochemical Pharmacology 62, 469–472.

Soundararajan, T., Karrunakaran, C., 2010. Micropropagation of *Nerium oleander* through the immature pods. Journal of Agricultural Science 2, p181.

Srinivasan, D., Nathan, S., Suresh, T., Perumalsamy, P.L., 2001. Antimicrobial activity of certain Indian medicinal plants used in folkloric medicine. Journal of Ethnopharmacology 74, 217–220.

Tayoub, G., Sulaiman, H., Alorfi, M., 2014. Analysis of oleandrin in oleander extract (*Nerium oleander*) by HPLC. Journal of Natural Products 7, 73–78.

Wang, X., Plomley, J.B., Newman, R.A., Cisneros, A., 2000. Analyses of an oleander extract for cancer treatment. Analytical Chemistry 72, 3547–3552.

Wong, S.K., Lim, Y.Y., Chan, E.W., 2013. Botany, uses, phytochemistry and pharmacology of selected Apocynaceae species: a review. Pharmacognosy Communications 3, 2.

Zafar, F., Jahan, N., Zafar, W.-U.-I., Aslam, S., 2014. Comparative evaluation of phytochemical, mineral and vitamin contents of gemmomodified extracts and leaves of two indigenous medicinal plants. International Journal of Agriculture and Biology 16.

Zargari, A., 1995. Medicinal Plants. Tehrari University Publications.

Zia, A., Siddiqui, B.S., Begum, S., Siddiqui, S., Suria, A., 1995. Studies on the constituents of the leaves of *Nerium oleander* on behavior pattern in mice. Journal of Ethnopharmacology 49, 33–39.

Zibbu, G., Batra, A., 2010. A review on chemistry and pharmacological activity of *Nerium oleander* L. Journal of Chemical and Pharmaceutical Research 2, 351–358.

CHAPTER

40

Olive

Ayesha Mushtaq[1], *Muhammad Asif Hanif*[1], *Muhammad Adnan Ayub*[2], *Ijaz Ahmad Bhatti*[1], *Mehrez Romdhane*[3]

[1] Department of Chemistry, University of Agriculture, Faisalabad, Pakistan; [2] Department of Chemistry, University of Okara, Okara, Pakistan; [3] Laboratoire de Recherche: Energie, Eau, Environnement et Procédés, ENIG, University of Gabès- Tunisia

OUTLINE

https://doi.org/10.1016/B978-0-08-102659-5.00040-9

Copyright © 2020 Elsevier Ltd. All rights reserved.

1. BOTANY

1.1 Introduction

Olive (*Olea europaea* L.) (Fig. 40.1) is the main cultivated specie belonging to the monophyletic Oleaceae family, which contains 30 genera and 600 species (Cronquist, 1981). Temperate and tropical regions of Malaysia and Asia are best areas for olive growth (Pérez et al., 2005). The genus name *Olea* comes from the Greek word "elaia," but it is known by 80 different names (Medail et al., 2001). The genus *Olea* contains approximately 30–35 species and is distributed in Asia, Africa, Europe, and Oceania (Bracci et al., 2011). *O. europaea* L. is the only edible specie of *Olea genus* (Hashmi et al., 2015). The Mediterranean basin is a traditional region for olive cultivation, containing 95% of total olive orchards present in the world. Olive tree leaves (*O. europaea* L.) have been extensively used in traditional medicines as extracts, herbal teas, and powder in Mediterranean and European countries. Olive is known by different names depending where you are in the world. It is commonly known as olivo (Spain), elia (Greece), jaitun (India), olive (England), oulivie (France), and zaitun (Pakistan) (Hashmi et al., 2015). Different cultivars can be distinguished from each other by drupe color and shape, oil composition, leaf morphology, and phenology. The main descriptive points correspond to 42 characteristics of fruit, leaf, and stone morphology for their identification (Breton et al., 2008). Olive cultivars can be used for eating purpose or for oil production. Olive is a slow-growing and extremely long-lived species, with a life expectancy up to 1000 years (Rhizopoulou, 2007).

1.2 History/Origin

Olives were first planted in the East Mediterranean region and then expanded toward Western regions in the next era. Cultivation of olives spread from Crete to Israel, Syria, Cyprus, Palestine, Egypt, and Turkey. Olive culture was expanded from Greece to North Africa and Southern Italy, in the 8th century BC and then reached into France. In Palestine and

FIGURE 40.1 Olives.

Israel, olive culture was expanded by King David and King Solomon. Around 1400 years ago, Prophet Muhammad (P.B.U.H.) suggested to his believers to apply olive oil on their bodies. Oil of this plant is commonly used in many cultures and religions. Holy olive oil was used as an ointment during baptism in churches of Christians. Christian preachers carried olive plants to North America with them for food and ceremonial uses. Oil was also used by the ancient Greek kings and Jews for massage. Fossil evidences indicated that olive was planted in Palestine and Syria from the 4th century BC. Moreover, mummies discovered from Egypt were wearing wreaths of olives. However, recent archaeological study has provided a strong evidence of presence of olives in Palestine in ancient times. In early times, olives were cultivated at various sites in Western Australia, New South Wales, Victoria, and South Australia (Smyth, 2002). In the late 20th century, interest in olive planting coincided with the migration of people from European countries who were conversant with olive planting.

1.3 Demography/Location

Olive is a typical component of the Mediterranean climate, characterized by warm, dry summers and rainy, cool winters. The olive tree is a thermophile species and is adapted to tolerate drought and salinity stress (Muzzalupo, 2012; Rhizopoulou, 2007). It grows on a wide range of soils, but it prefers sandy loam soils of moderate depth (Spennemann and Allen, 2000). Olives are planted in the following countries: Asia (Pakistan, India, Syria, Jordan, China, Iraq, Palestine, Turkey, Lebanon, and Iran), Oceania (New Zealand and Australia), America (Uruguay, the United States, Chile, Mexico, Argentina, and Peru), Europe (Italy, Cyprus, Montenegro, Spain, France, Albania, Portugal, and Greece) and Africa (Morocco, South Africa, Tunisia, Egypt, and Algeria) (Therios, 2009). This plant also grows in the subtropical hilly areas of Baluchistan and Khyber Pakhtoonkhwa in Pakistan (Afzal et al., 2017).

1.4 Botany, Morphology, Ecology

The olive tree generally grows up to 10 m in height. Its stem possesses a large diameter, usually twisted or bent. Leaves are short, ovate, narrow, leathery, lanceolate, attenuate, oblong, and glabrous. Leaves are silvery whitish in color. Its petiole is 1–3 cm wide and 5–10 cm long. Numerous flowers are present that are small, hermaphrodite, subsessile, and creamy white in color. Calyx is short with four small teeth. Corolla is 1–2 mm long. Fruit is small, ovoid, blackish-violet after maturation, usually 1–2.5 cm in length, larger in cultivated varieties compared to wild plants (Hashmi et al., 2015). Color of bark is usually pale grey.

Olive tree can be grown in a variety of climates. Optimum temperature required for olive growth is 40°C, with the growing temperatures of 15–20°C. Cultivation of olives is not suitable in areas of high altitude because of frost danger. This plant can be easily grown even in dry and calcareous soils. However, sandy loam soils having suitable amounts of nitrogen, potassium, and phosphorous are best for the growth of olive plants. They are considered drought-tolerant plants. Soils having pH less than 8.5 are suitable for olive growth. Moreover, these plants can grow in soils having high content of boron without toxicity issues (Kiritsakis and Shahidi, 2017).

2. CHEMISTRY

Olive oil is highly beneficial for health due to the presence of large quantities of monounsaturated and other bioactive compounds such as carotenoids, tocopherols, and phospholipids (Covas, 2008). The distinctive taste of olive oil is also due to the aforementioned compounds. Various factors including temperature, soil conditions, pH, harvesting regime, and processing technologies greatly influence the composition of oil constituents (Covas, 2008). Olive varieties in which oil content is less than 12%, like Manzanillo, Ascolano, and Kalamata, are extensively used to produce table olives. However, the varieties having high oil content like Verdial, Gemlik, Hojiblanca, Arauco, and Nychati are preferably used for the production of oil (Ryan and Robards, 1998). Water is responsible for half of the weight of olive leaves and fruits. Cellulose and sugar are the major carbohydrates found in olives. Reduction of carbohydrate contents during maturation is due to increase in oil content. A variety of minerals including copper, manganese, magnesium, potassium, and iron are also present in both olive leaf and fruits (Boudhrioua et al., 2009). Various polyphenols including oleuropein (60–90 mg/g weight of dry leaves), vanillic acid, elenolic acid derivatives, hydroxytyrosol, *p*-coumaric acid, tocopherol, and tyrosol are also present in olive leaves (Ryan et al., 2003).

Olive seeds contain 2–4 g oil. Lignin (20.63%–25.11%), hemicellulose (21.45%–27.64%), and cellulose (29.79%–34.35%) are the main constituents present in olive stone (Ghanbari et al., 2012). Considerable quantities of proteins, fats, and phenolics are also present. Flesh components of olives comprise two main components, saponifiable and unsaponifiable fats, which are transformed to oil.

The former, consisting of free fatty acids, partial glycerides, phosphatides, and triglycerides contribute around 98% of oil composition, but the latter, composed of minor constituents including phenolics, tocopherols, coloring pigments, and phytosterols, represent nearly 1%–2% total oil composition. Olive oil is simply a mixture of volatile components, fats, vitamins, soluble compounds, and small olive bits. The free fatty acids concentration should be <0.8%, while concentration of polyphenols must be in the range of 200–500 mg/kg. Olive oil contains large quantities of nutritive fats and supplies 120 calories per tablespoon (9 calories/g). Many scientists suggested that olive oil is cholesterol free, having no adverse effects on human health (Gulfraz, 2006).

O. europaea has aromatic odor due to presence of various compounds including alcohols, ketones, esters, and hydrocarbons (Vichi et al., 2003). About 280 different compounds are identified in the volatile fractions of olive oils (Boskou et al., 2006). Like other edible oils, olive oil also contains significant quantities of pigments (carotenoids and chlorophylls). Stearic acid (C18:0), palmitic acid (C16:0), palmitoleic acid (C16:1), oleic acid (C18:1), and linolenic acid (C18:3) are the major fatty acids present in olives. Gadoleic (C20:1), myristic (C14:0), and margaric (C17:0) acids are present in small amounts. Traces of eicosenoic and 11-*cis*-vaccenic acids have been detected with the help of 13C-NMR spectroscopic approach (Ghanbari et al., 2012).

Hydroxytyrosol and tyrosol, ligustroides and flavonoids are main phenolic compounds found in olive oil. Oleuropein is the most abundant minor component of olive oil (Therios, 2009) (Fig. 40.2).

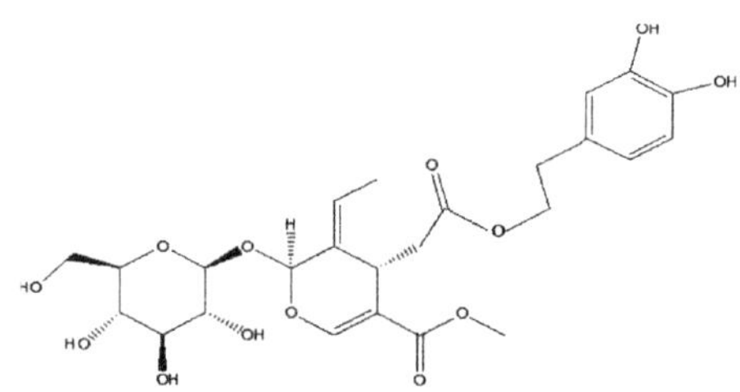

Oleuropein

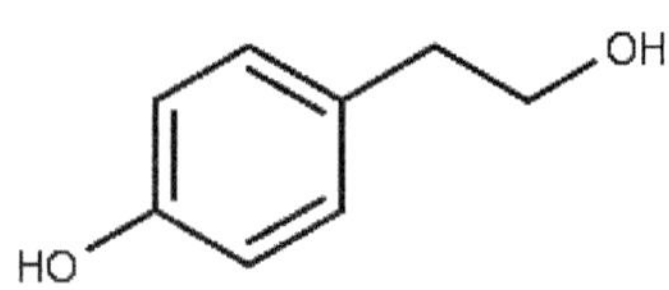

Tyrosol

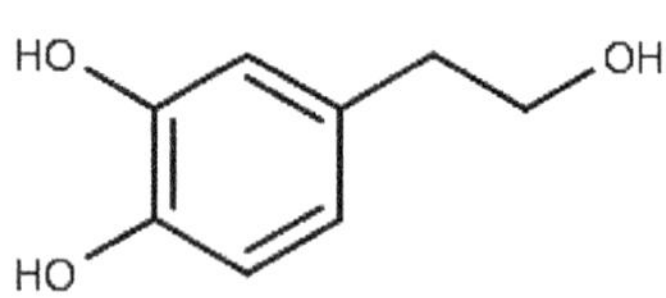

Hydroxytyrosol

FIGURE 40.2 Structures of some important phytochemicals present in olive oil.

3. POSTHARVEST TECHNOLOGY

Olive harvesting is done during a relatively short period of time, when fruit offers a better yield for oil production. Thus, green, immature olives resist postharvest handling better than the mature ones (García and Yousfi, 2006). In Spain, the harvesting is normally carried out during the months of December, January, or February, when fruit reaches a red or black skin color. After harvesting, which is normally done with the cheapest process called "vareo" (shaking the olive branches with rods), the olives, laying on the soil from the beaten branches, are picked up and put in dump truck boxes, sometimes even using digging machines. Storage of these olives is performed by simple heaping in fruit piles on the courtyard of the olive mills, awaiting their processing. During these conditions, various degenerative processes and pathogenic contaminations develop in the fruit in a very short time. Normally, oils obtained from these fruits are characterized by presenting hydrolytic and oxidative deteriorations measured by their high acidity values, peroxide value, or ultraviolet absorbance at 232 and 270 nm, produced respectively by lipases, lipoxygenases, and lyases of both olive and parasitic origin (Garcia and Streif, 1991).

To avoid this physical and biological deterioration of the fruit, which determines the production of a low-quality virgin oil, the industry is currently following a main strategy of reducing the interval between harvesting and processing, as much as possible, and recommends steps for the olive during this period: First, the separation of harvested olives from the ones previously falling onto the soil is strongly recommended. Second, the use of perforated boxes is recommended for keeping the olives.

The use of refrigeration in fruit storage is based on the reduction of the speed of the metabolic processes induced by low temperatures. The main metabolic effect of cool temperature is the decrease in the effectiveness of fruit respiration. A high level of respiration determines two undesirable effects: a shorter commercial life for the product stored and a loss in weight and nutritional value. Furthermore, refrigeration delays the deterioration activity of the pathogens, which are mainly responsible for postharvest loss of fruits. The optimum storage temperature range is 5–7°C. Some fruits are sensitive to low temperature and during cold storage develop a physiologic disorder commonly named "chilling injury." Chilling injury symptoms in olives begin as an internal browning of the flesh around the pit; subsequently, external pitting appears dull in color on the skin of this area, and browning becomes darker and progressively occupies all the internal flesh and the skin surface. Black olives can be stored at 2–5°C because they are less susceptible to injury than green olives (García and Yousfi, 2006). A relative humidity of 90–95% is recommended to minimize water loss from fresh olives.

Postharvest storage of olives under a CO_2 atmosphere for a period of 12 days resulted in color and flavor development and reduced bitterness. The storage of olives under CO_2 atmosphere resulted in pronounced increases in total phenyl and total flavonoid contents, mainly within the first 3–5 days, but total anthocyanin exhibited a different pattern of evolution (Dourtoglou et al., 2006).

4. PROCESSING

About 90% of olives are used for oil production and 10% for table olives. Edible table olives are prepared in different ways. Mainly three commercial types of table olives are processed in olive-producing countries. They are Kalamata olives, green olives, and black olives. It is estimated that 39.7% of world olive production is processed as green olives, 40% is produced as black, and the rest (20.3%) is used for preparation of all the other commercial types (Ünal and Nergiz, 2003). Green olive processing includes initial treatment of olives with lye, which causes the hydrolysis of oleuropein (glycoside responsible for the taste of olives) (Brenes and de Castro, 1998). This treatment ruptures the olive skin to increase permeability of the plasma membrane for the movement of some biochemical compounds into the neighboring liquid such as free fatty acids and alpha-tocopherol (Marsilio et al., 1996). Then, olives are washed with excess water to remove alkalis. After washing, olives are stored for 2–6 months in brine solution (5%–10% NaCl). Development of acidity and pH drop are determining factors for the completion of fermentation (Sakouhi et al., 2008).

Olive oil processing encompasses four technological steps, each affecting the final quality and quantity of olive oils obtained: leaf removal and washing, crushing, malaxation, and oil separation (Di Vincenzo, 2002). Currently, olive oil is produced commercially by both batch (traditional press) and continuous (centrifugation) methods, though the latter has been far more widely used (Servili et al., 2012). Extra virgin olive oil of high quality is extracted first. The pomace can then be processed again to yield the lower quality refined virgin olive oil. Further extraction with organic solvents can be undertaken to produce low-quality refined husk oil (Owen et al., 2000).

5. VALUE ADDITION

Typical value-added products include jams, jellies, preserves, fruit sauces and spreads, pickles, preserved vegetables, hot chili sauces, extra virgin olive oil, and vinegar.

6. USES

The olive wood is heavy, very tough, and is used for high-end furniture, inlays, turned objects, and handcrafts (Maldonado et al., 2016). It is also appreciated as firewood because it burns even when wet (Breton et al., 2009). Scientists are trying to understand how a balanced diet promotes human health and how it prevents chronic diseases like cardiovascular disorders, neurodegenerative diseases, and tumors.

Olive has many uses ranging from culinary to religious. It is generally regarded as a sign of honor, purity, success, and euphoria. The shoots of this plant are used in many countries to honor peace, wisdom, and success (Therios, 2009). Now, olive oil is utilized in religious functions. Leafy branches of this tree were found in Tutankhamen's tomb. In ancient times, olive oil was used in Greece for body fitness (Gooch, 2005). Oil of this plant was previously used to massage athletes and kings in Greece. It was burnt in the holy lamps of temples. Victors of Olympic games were crowned with olive leaves. Olive wood was used to fashion the Greek cult figures (*xoana*). Athenians claimed that the olive was first cultivated in Athens. According to Greek literature, an olive tree was appeared when Athena (the goddess of wisdom) struck her spear onto the surface of earth. The gift of Athena (olive), which was commonly used as perfume, oil, medicine, and food, was thought to be a more pacific gift than Poseidon's horse (Psilakis, 1996).

O. europaea has been extensively used as a traditional medicine for curing variety of diseases in many countries. Leaves, wood, fruits, seeds, oil, and bark of this tree are used in various forms, either alone or in combination with other herbs. Seed oil is used as a laxative, and it can also be used externally as a balm to cure swelling (Al-Khalil, 1995). Decoctions of dried leaves and fruit can be taken orally to cure infections of urinary and respiratory tracts, diarrhea, stomach and intestinal ailments, and as a mouth cleanser (Bellakhdar et al., 1991). Hair loss can be prevented by applying olive oil continuously (Khan et al., 2007). Boiled extract of olive leaves is used for the treatment of hypertension (Darias et al., 1986, 1996). Hot water extract of olive leaf can be used for the treatment of high blood pressure (Lawrendiadis, 1961). Essential oil of olive fruits is taken orally for the treatment of renal lithiasis. It is also applied externally for the treatment of rheumatism, sores, burns, and to enhance circulation (De Feo et al., 1992). A drink prepared from fresh olive leaves is used to cure inflammation (Pieroni et al., 1996). A tincture prepared from olive leaf is used as a febrifuge and is also applied externally as an ointment to enhance the growth of nails (De Feo and Senatore, 1993). Leaves of this plant are used to cure intestinal and stomach problems, and their essential oil is taken orally to treat liver diseases and constipation (Bellakhdar et al., 1991).

Olive oil is taken orally for the treatment of agitation and hypertension. Infusion of dried or fresh leaves is used to cure asthma and is also used for induction of diuresis. The oil is applied externally over fractured limbs, and fruit is also known to be a skin cleanser (Khan et al., 2007a). Boiled extract of leaves is used as a tonic for the treatment of inflammation (Pieroni et al., 1996). Ointments prepared from boiled extracts of leaves are used to cure eye infections (Guerin and Reveillere, 1985). Hot water extracts of olive leaves are used to induce diuresis (Ribeiro et al., 1988). A mixture of olive oil and lemon juice is used for the treatment of gallstones (Hashmi et al., 2015).

7. PHARMACOLOGICAL USES

7.1 Antimicrobial Activity

Antimicrobial effect of aqueous extract of olive leaf was examined against *Bacillus subtilis, Cryptococcus neoformans, Bacillus cereus, Staphylococcus aureus, Klebsiella pneumoniae, Candida albicans, Escherichia coli,* and *Pseudomonas aeruginosa*. Stronger antimicrobial action was observed against *S. aureus and E. coli* than other disease-causing organisms (Pereira et al., 2007).

7.2 Antioxidant Activity

The best deliberate antioxidant phytonutrient found in olives is oleuropein, which has been shown to function as an antioxidant nutrient to decrease oxidation of LDL cholesterol and to help guard nerve cells from oxygen-related injury. Oxidized LDL is the most damaging form of cholesterol and can initiate damage to arterial tissues, thereby promoting atherosclerosis. Also, caffeic acid was reported to have antioxidant activity through the scavenging of superoxide anion (Chimi et al., 1995).

7.3 Anticancer Activity

Components of *O. europaea* exhibited strong anticancer effects on several types of cancers (Casaburi et al., 2013). Antiproliferative and apoptotic activities of erythrodiol were investigated in human colorectal carcinoma HT-29 cells (Juan et al., 2008). It retarded the cell growth without causing any toxicity at a dose of 100 μM in colon adenocarcinoma cells. Similar studies were conducted on methanolic and water extracts of olive leaf against endothelial and cancer cells. These extracts were helpful to prevent cell division of human breast cancer (MCF-7), human urinary bladder carcinoma (T-24), and bovine brain capillary endothelial. Hydroxytyrosol is a strong antitumor agent as it protects cells from harmful effects of hydrogen peroxide, DNA from dangerous effects of peroxy nitrite, and may block cell division at G1-phase and induce apoptosis (Fabiani et al., 2002). Maslinic acid (triterpenoid isolated from olive fruits and leaf) showed strong antitumor activities against the HT29 colon cancer cells (Bianchi et al., 1994).

7.4 Antinociceptive Activity

Studies conducted on antinociceptive activities of olive extracts on Wistar rats indicated that extracts doses of (50–200 mg/kg) produce strong analgesic effects, and administration of this drug (200 mg/kg) in peritoneum resulted in significant reduction in pain responses during formalin test (Esmaeili-Mahani et al., 2010).

7.5 Antidiabetic Activities

In a previous study, it was found that antidiabetic effect of olive leaf is directly related with the presence of both oleanolic acid and oleuropein (Sato et al., 2007). In another study, antidiabetic activity of hydroxytyrosol and oleuropein separated from olive leaves was found due to their abilities of oxidative stress reduction that are widely related with pathologic impediments of diabetes (Jemai et al., 2009).

7.6 Cardiovascular Disorders

Compounds such as oleuropein present in olives decrease the risks of cardiovascular diseases by preventing LDL oxidation, which results in reduction of atherosclerotic plaque formation (Visioli and Galli, 1994).

7.7 Hypolipidemic Activity

Hypolipidemic and hypoglycemic activities of olive leaf were tested on laboratory animals. Oleuropein was reported to be an active component, with a proposed mechanism of action of potentiation of glucose-induced insulin release and an increase in peripheral blood glucose uptake (Khan et al., 2007).

7.8 Neuroprotective Activities

The effect of maslinic acid (a terpenoid present in olive leaf) on diabetic rats was studied by different scientists. Rats were injected with streptozotocin for induction of neuronal death. A significant neuroprotective activity was exhibited by maslinic acid in dose-dependent mode

(Teng et al., 2010). A previous study show that oleuropein and olive leaves extract are useful for the treatment of Parkinson disease (Pasban-Aliabadi et al., 2013). Oleuropein aglycone reduces the production of poisonous amyloid aggregates in brain and, hence, reduces the chances of Alzheimer disease (Diomede et al., 2013).

8. SIDE EFFECTS AND TOXICITY

There is not enough reliable information about the safety of taking olives in pregnancy or surgery available.

References

Afzal, F., Adnan, M., Rahman, I.U., Noor, M., Khan, A., Iqbal, S., Shah, J.A., Khan, M.A., Roman, M., Wahid, F., 2017. Growth response of olive cultivars to air layering. Pure and Applied Biology (PAB) 6, 1403–1409.

Al-Khalil, S., 1995. A survey of plants used in Jordanian traditional medicine. International Journal of Pharmacognosy 33, 317–323.

Bellakhdar, J., Claisse, R., Fleurentin, J., Younos, C., 1991. Repertory of standard herbal drugs in the Moroccan pharmacopoea. Journal of Ethnopharmacology 35, 123–143.

Bianchi, G., Pozzi, N., Vlahov, G., 1994. Pentacyclic triterpene acids in olives. Phytochemistry 37, 205–207.

Boskou, D., Tsimidou, M., Blekas, D., 2006. Polar phenolic compounds. In: Boskou, D. (Ed.), Olive Oil, Chemistry and Technology. AOCS Press, Champaign, IL, 73–92 p.

Boudhrioua, N., Bahloul, N., Slimen, I.B., Kechaou, N., 2009. Comparison on the total phenol contents and the color of fresh and infrared dried olive leaves. Industrial Crops and Products 29, 412–419.

Bracci, T., Busconi, M., Fogher, C., Sebastiani, L., 2011. Molecular studies in olive (*Olea europaea* L.): overview on DNA markers applications and recent advances in genome analysis. Plant Cell Reports 30, 449–462.

Brenes, M., de Castro, A., 1998. Transformation of oleuropein and its hydrolysis products during Spanish-style green olive processing. Journal of the Science of Food and Agriculture 77, 353–358.

Breton, C., Pinatel, C., Medail, F., Bonhomme, F., Berville, A., 2008. Comparison between classical and Bayesian methods to investigate the history of olive cultivars using SSR-polymorphisms. Plant Science 175, 524–532.

Breton, C., Terral, J.-F., Pinatel, C., Médail, F., Bonhomme, F., Bervillé, A., 2009. The origins of the domestication of the olive tree. Comptes Rendus Biologies 332, 1059–1064.

Casaburi, I., Puoci, F., Chimento, A., Sirianni, R., Ruggiero, C., Avena, P., Pezzi, V., 2013. Potential of olive oil phenols as chemopreventive and therapeutic agents against cancer: a review of in vitro studies. Molecular Nutrition & Food Research 57, 71–83.

Chimi, H., Morel, I., Lescoat, G., Pasdeloup, N., Cillard, P., Cillard, J., 1995. Inhibition of iron toxicity in rat hepatocyte culture by natural phenolic compounds. Toxicology in Vitro 9, 695–702.

Covas, M.-I., 2008. Bioactive effects of olive oil phenolic compounds in humans: reduction of heart disease factors and oxidative damage. Inflammopharmacology 16, 216–218.

Cronquist, A., 1981. An Integrated System of Classification of Flowering Plants. Columbia University Press.

Daniyal, M., Determination of Some Macro-Minerals of *Olea europaea* L. Growing in Pakistan.

Darias, V., Abdala, S., Martin, D., Ramos, F., 1996. Hypoglycaemic Plants from the Canary Islands. Phytotherapy Research, United Kingdom.

Darias, V., Bravo, L., Barquin, E., Herrera, D.M., Fraile, C., 1986. Contribution to the ethnopharmacological study of the Canary Islands. Journal of Ethnopharmacology 15, 169–193.

De Feo, V., Aquino, R., Menghini, A., Ramundo, E., Senatore, F., 1992. Traditional phytotherapy in the Peninsula Sorrentina, Campania, Southern Italy. Journal of Ethnopharmacology 36, 113–125.

De Feo, V., Senatore, F., 1993. Medicinal plants and phytotherapy in the Amalfitan coast, Salerno province, Campania, Southern Italy. Journal of Ethnopharmacology 39, 39–51.

Di Vincenzo, D., 2002. Influence of olive processing on virgin olive oil quality. European Journal of Lipid Science and Technology 104, 587–601.

Diomede, L., Rigacci, S., Romeo, M., Stefani, M., Salmona, M., 2013. Oleuropein aglycone protects transgenic *C. elegans* strains expressing Aβ42 by reducing plaque load and motor deficit. PLoS One 8, e58893.

Dourtoglou, V.G., Mamalos, A., Makris, D.P., 2006. Storage of olives (*Olea europaea*) under CO_2 atmosphere: effect on anthocyanins, phenolics, sensory attributes and in vitro antioxidant properties. Food Chemistry 99, 342–349.

Esmaeili-Mahani, S., Rezaeezadeh-Roukerd, M., Esmaeilpour, K., Abbasnejad, M., Rasoulian, B., Sheibani, V., Kaeidi, A., Hajializadeh, Z., 2010. Olive (*Olea europaea* L.) leaf extract elicits antinociceptive activity, potentiates morphine analgesia and suppresses morphine hyperalgesia in rats. Journal of Ethnopharmacology 132, 200–205.

Fabiani, R., De Bartolomeo, A., Rosignoli, P., Servili, M., Montedoro, G., Morozzi, G., 2002. Cancer chemoprevention by hydroxytyrosol isolated from virgin olive oil through G1 cell cycle arrest and apoptosis. European Journal of Cancer Prevention 11, 351–358.

Garcia, J., Streif, J., 1991. The Effect of Controlled Atmosphere Storage on Fruit and Oil Quality of'Gordal'Olives/Die Wirkung einer Lagerung in kontrollierter Atmosphäre auf die Frucht-und Ölqualität von'Gordal'Oliven. Gartenbauwissenschaft 233–238.

García, J.M., Yousfi, K., 2006. The postharvest of mill olives. Grasas Y Aceites 57, 16–24.

Ghanbari, R., Anwar, F., Alkharfy, K.M., Gilani, A.-H., Saari, N., 2012. Valuable nutrients and functional bioactives in different parts of olive (*Olea europaea* L.)—a review. International Journal of Molecular Sciences 13, 3291–3340.

Gooch, E., 2005. Ten plus one things you may not know about olive. Epikouria Magazine.

Guerin, J., Reveillere, H., 1985. Antifungal activity of plant extracts used in therapy. 2. Study of 40 plants extracts against 9 fungi species [*Saccharomyces pastorianus, Candida albicans, Rhizopus nigricans, Aspergillus niger, Aspergillus fumigatus, Penicillium digitatum, Botrytis cinerea, Fusarium oxysporum, Trychophyton mentagrophytes*]. Annales Pharmaceutiques Françaises.

Gulfraz, M., 2006. Determination of essential oil content of wild olive and its comparison with olive oil. Ethnobotanical Leaflets 2006, 1.

Hashmi, M.A., Khan, A., Hanif, M., Farooq, U., Perveen, S., 2015. Traditional uses, phytochemistry, and pharmacology of *Olea europaea* (olive). Evidence-based Complementary and Alternative Medicine 2015.

Jemai, H., El Feki, A., Sayadi, S., 2009. Antidiabetic and antioxidant effects of hydroxytyrosol and oleuropein from olive leaves in alloxan-diabetic rats. Journal of Agricultural and Food Chemistry 57, 8798–8804.

Juan, M.E., Wenzel, U., Daniel, H., Planas, J.M., 2008. Erythrodiol, a natural triterpenoid from olives, has antiproliferative and apoptotic activity in HT-29 human adenocarcinoma cells. Molecular Nutrition & Food Research 52, 595–599.

Khan, Y., Panchal, S., Vyas, N., Butani, A., Kumar, V., 2007. *Olea europaea*: a phytopharmacological review. Pharmacognosy Reviews 1, 114–118.

Kiritsakis, A., Shahidi, F., 2017. Olives and Olive Oil as Functional Foods: Bioactivity, Chemistry and Processing. John Wiley & Sons.
Lawrendiadis, G., 1961. Contribution to the knowledge of the medicinal plants of Greece. Planta Medica 9, 164–169.
Maldonado, N.G., López, M., Caudullo, G., de Rigo, D., 2016. *Olea europaea* in Europe: Distribution, Habitat, Usage and Threats. European Atlas of Forest Tree Species, Publ. Off. EU, Luxembourg, p. e01534b.
Marsilio, V., Lanza, B., Angelis, M.D., 1996. Olive cell wall components: physical and biochemical changes during processing. Journal of the Science of Food and Agriculture 70, 35–43.
Medail, F., Quezel, P., Besnard, G., Khadari, B., 2001. Systematics, ecology and phylogeographic significance of *Olea europaea* L. ssp. maroccana (Greuter & Burdet) P. Vargas et al., a relictual olive tree in south-west Morocco. Botanical Journal of the Linnean Society 137, 249–266.
Muzzalupo, I., 2012. Olive Germplasm–Italian Catalogue of Olive Varieties. InTech, Rijeka.
Owen, R.W., Giacosa, A., Hull, W.E., Haubner, R., Würtele, G., Spiegelhalder, B., Bartsch, H., 2000. Olive-oil consumption and health: the possible role of antioxidants. The Lancet Oncology 1, 107–112.
Pasban-Aliabadi, H., Esmaeili-Mahani, S., Sheibani, V., Abbasnejad, M., Mehdizadeh, A., Yaghoobi, M.M., 2013. Inhibition of 6-hydroxydopamine-induced PC12 cell apoptosis by olive (*Olea europaea* L.) leaf extract is performed by its main component oleuropein. Rejuvenation Research 16, 134–142.
Pereira, A.P., Ferreira, I.C., Marcelino, F., Valentão, P., Andrade, P.B., Seabra, R., Estevinho, L., Bento, A., Pereira, J.A., 2007. Phenolic compounds and antimicrobial activity of olive (*Olea europaea* L. Cv. Cobrançosa) leaves. Molecules 12, 1153–1162.
Pérez, J.A., Hernández, J.M., Trujillo, J.M., López, H., 2005. Iridoids and secoiridoids from Oleaceae. Studies in Natural Products Chemistry 32, 303–363.
Pieroni, A., Heimler, D., Pieters, L., Van Poel, B., Vlietinck, A., 1996. In vitro anticomplementary activity of flavonoids from oliva (*Olea europaea* L.) leaves. Die Pharmazie 51, 765–767.
Psilakis, N., 1996. Mythology of Crete Island. Karmanor, Heraclion, Greece.
Rhizopoulou, S., 2007. *Olea europaea* L. A botanical contribution to culture. American-Eurasian Journal of Agricultural & Environmental Sciences 2, 382–387.
Ribeiro, R.d.A., de Barros, F., de Melo, M.M.R.F., Muniz, C., Chieia, S., das Graças Wanderley, M., Gomes, C., Trolin, G., 1988. Acute diuretic effects in conscious rats produced by some medicinal plants used in the state of Sao Paulo, Brasil. Journal of Ethnopharmacology 24, 19–29.
Ryan, D., Prenzler, P.D., Lavee, S., Antolovich, M., Robards, K., 2003. Quantitative changes in phenolic content during physiological development of the olive (*Olea europaea*) cultivar Hardy's Mammoth. Journal of Agricultural and Food Chemistry 51, 2532–2538.
Ryan, D., Robards, K., 1998. Critical review. Phenolic compounds in olives. Analyst 123, 31R–44R.
Sakouhi, F., Harrabi, S., Absalon, C., Sbei, K., Boukhchina, S., Kallel, H., 2008. α-Tocopherol and fatty acids contents of some Tunisian table olives (*Olea europaea* L.): changes in their composition during ripening and processing. Food Chemistry 108, 833–839.
Sato, H., Genet, C., Strehle, A., Thomas, C., Lobstein, A., Wagner, A., Mioskowski, C., Auwerx, J., Saladin, R., 2007. Anti-hyperglycemic activity of a TGR5 agonist isolated from *Olea europaea*. Biochemical and Biophysical Research Communications 362, 793–798.
Servili, M., Taticchi, A., Sordini, B., Esposto, S., Urbani, S., 2012. Technological Aspects of Olive Oil Production. INTECH Open Access Publisher.

Smyth, J., 2002. Perspectives of the Australian olive industry. Advances in Horticultural Science 280–288.

Spennemann, D., Allen, L., 2000. Feral olives (*Olea europaea*) as future woody weeds in Australia: a review. Animal Production Science 40, 889–901.

Teng, G., Yi-Song, Q., Huang, M.-H., Huang, L.-F., Xu-Zhen, T., Yun-Man, L., Hong-Bin, S., 2010. Neuroprotection of maslinic acid, a novel glycogen phosphorylase inhibitor, in type 2 diabetic rats. Chinese Journal of Natural Medicines 8, 293–297.

Therios, I.N., 2009. Olives. CABI.

Ünal, K., Nergiz, C., 2003. The effect of table olive preparing methods and storage on the composition and nutritive value of olives. Grasas Y Aceites 54, 71–76.

Vichi, S., Pizzale, L., Conte, L.S., Buxaderas, S., López-Tamames, E., 2003. Solid-phase micro-extraction in the analysis of virgin olive oil volatile fraction: characterization of virgin olive oils from two distinct geographical areas of northern Italy. Journal of Agricultural and Food Chemistry 51, 6572–6577.

Visioli, F., Galli, C., 1994. Oleuropein protects low density lipoprotein from oxidation. Life Sciences 55, 1965–1971.

CHAPTER

41

Pholi (Wild Safflower)

Naima Tariq, Muhammad Irfan Majeed, Muhammad Asif Hanif, Rafia Rehman

Department of Chemistry, University of Agriculture, Faisalabad, Pakistan

OUTLINE

Medicinal Plants of South Asia
https://doi.org/10.1016/B978-0-08-102659-5.00041-0

Copyright © 2020 Elsevier Ltd. All rights reserved.

1. BOTANY

1.1 Introduction

Carthamus oxycantha M. Bieb (Fig. 41.1) is provincially known as pholi and kandyari. It is a versatile wild-growing weed. The genus *Carthamus* belongs to the Asteraceae/Compositae family and includes 16 identified species. *Carthamus tinctorius* (safflower) is the only cultivated specie belonging to this genus; the remaining are either weeds or wild (Bibalani and Shadkami-Til, 2011). *C. oxyacantha* is one of the wild species that is widespread in Turkey, Iran, subtropical regions of western Iraq, Pakistan, northwest India, throughout Kazakhstan, Uzbekistan, and Turkmenistan. This species is more important than other species in this genus because it is claimed to be the ancient male parent of *C. tinctorius* (cultivated safflower) (Sheykhlou et al., 2012). Pholi is commonly recognized as wild safflower. It is an annual spiny-leaved herb that is up to 1.5 m tall. It is a xerophytic and rugged noxious weed that grows in winter crops. It lessens the yield of a cereal crop. It shows similarity to the other spiny-leaved plants of genus *Carthamus* that are not eaten by livestock, which allows it to disseminate on grazing lands. It is a premium source of drying and nonedible oil (28%–29% oil content) from waste lands (Biodiesel - Feedstocks, 2012). *C. oxycantha* Bieb is less utilized wild safflower that is an annual, bushy, and thorny medium-sized herb. During the months of April to July, it is found to be a troublesome weed in the harvested fields

FIGURE 41.1 Pholi plant and seeds.

FIGURE 41.1 continued

of Rabi crops (Dwivedi et al., 2012). Based on taxonomy, the 16 species of genus *Carthamus* have been divided into three different groups (*Odonthagnathis*, *Carthamus*, and *Atractylis*), this classification is based on the chromosomal numbers and the relationship between different species. Species like *C. tinctorius*, *C. palaestinus*, and *C. oxyacanthus* that contain 12 sets of chromosomes are included in section *Carthamus*. The wild species *C. palaestinus* and *C. oxyacanthus* can be hybrid with the only cultivated species *C. tictorius* (Talebi et al., 2012).

C. oxycantha M. Bieb is recognized by different names in different parts of world. In English, it is called wild safflower or thistle. In Pakistan and India, it is commonly known as pholi (Ahmad et al., 2006). In Persian, it is called golerang-e-zard (Ghasemi et al., 2013). The other names include kantiari, kasumbi, azghai, and jeweled distaff thistle (Turgumbayeva et al., 2014).

C. oxyacantha is a self-pollinating plant, but it has the capability to outcross (10%) through pollen transfer by insects. Wild safflower closely resembles cultivated safflower and is a spiny-leaved herb of 1.5 m length. The orange-yellow flowers grow in flower heads approximately 2–3 cm in diameter, and the leaves are covered with spines. Wild safflower can be differentiated from cultivated safflower by its pointed, large spines (Al Fadal and Al-Fredan, 2015). Its flowering period is from March until June (Shah, 2013). It is spinose branched herb and usually found in grain fields after harvesting season. The leaves are alternate, whorled, or opposite and often deeply lobed with dentate margins. It possesses a nonwoody stem. It has a deep tap root. The fruit of *C. oxyacantha* is achene, and the seeds are elongated and flat (BiBi et al., 2014). It is resistant to hard environmental conditions and dry climate. *C. oxycantha,* the closely related wild species of *C. tinctorius,* is enormously used in developing drought-resistant varieties and also varieties resistant to various diseases and pests like *Ramularia* leaf spot, *Alternaria* leaf spot, rust, safflower fly, and powdery mildew (Zadeh et al., 2011).

1.2 History

The wild safflower is native to Ethiopia. It has two primary centers of origin, the mountainous regions of Afghanistan and Ethiopia, and it also belongs to the plains of Myanmar (Burma), Pakistan, and India. The maximum production of wild safflower is in Mexico, whereas India is the second largest producer of *C. oxyacantha* (Ahmad et al., 2015).

1.3 Demography/Location

C. oxyacantha grows adequately in the harvested fields in India (Himachal Pradesh, Uttar Pradesh, Haryana, Rajasthan, and Punjab) and Pakistan (Baluchistan and Northwest Frontier Province and extending westward) (Dwivedi et al., 2012). It is generally distributed in Azerbaijan, Afghanistan, Kyrgyzstan, Tajikistan, Turkmenistan, Pakistan, India, Iran, and Iraq (Biodiesel - Feedstocks, 2012).

This specie has appreciable genetic diversity in Iran, and its quality of oil is equivalent to the cultivated safflower. It is believed that the two species *C. tinctorius* and *C. oxycantha* are the two precursors that arose from different regions of the world and provide worthwhile information regarding the utilization and conservation of safflower germplasm (Talebi et al., 2012). It is a common weed along the boundaries of wheat fields and mostly found in open and dry places, plains, and mountains (Ahmad et al., 2007).

1.4 Botany, Morphology, Ecology

C. oxycantha is an annual, erect, bushy, spiny-leaved herb up to 1.5 m tall. It has a long tap root system, with a nonwoody stem. Leaves possess secretary canals with latex or resin. Morphologic studies reveal that leaves are simple, alternate, and deeply lobbed. Seeds are elongated and flat. Flowers are orange-yellow colored, and flower heads are about 1–1.5 in. In recent studies, pholi has spines and lobed leaves. Its spiny nature confirms it to be a xerophyte. Spines reduce the loss of water and prevent desiccation (BiBi et al., 2014). The fruit is elliptic, 2–3.5 mm wide, and 3–5.5 mm long (Turgumbayeva et al., 2014). Seed is asymmetric and the color ranges from whitish-gray, grayish-brown, or black. The dispersion of seeds is due to the rotation of dry bushes (Khalid and Shad, 1990).

The regional climate and the soil characteristics of the site are important for the distribution of *C. oxycantha* (weed) species in the fields. Composition of weed species, population density, and their distribution changes by different geographic regions. Other factors also affect the population of weeds. Many researches confirmed that cultivation, choice of crop rotation, harvest time, soil management, sowing techniques and time, chemical and mechanical weed control methods, and fertilizing are the main factors that affect the weed flora (Hassannejad and Ghafarbi, 2013).

Transverse sections of leaf stem and roots of *C. oxycantha* showed that the epidermis in leaves surround a thick cortex. Vascular bundles are present in the shape of rings surrounded by a bundle of sheath cells. A cuticle layer is present in the stem around the epidermis, and a parenchymatous cortex is beneath the epidermis. Vascular bundles are present in the endodermis (inner layer). In roots, epidermis surrounds the cortex. Pith is made up of ground tissues. Around the pith, vascular bundles are located in a circle, showing a high capacity for water conduction (BiBi et al., 2014).

C. oxyacantha ($2n = 24$) consists of three pairs of submetacentric (sm) and nine pairs of metacentric (m) chromosomes, and the longest two pairs possess satellites. The total length of one set of chromosomes is 10.03 μm (Shanmugasundaram and Heidelberg, 2011). The actual length of chromosomes ranges from 0.54 to 1.10 μm, and the average length is approximately 0.83 μm. Chromosomes are arranged in descending order on the basis of their relative length (Carapetian and Zarei, 2015).

2. CHEMISTRY

C. oxyacantha is a wild-growing weed in wheat fields. It is used as a vegetable in cooking, and its seeds are also utilized in different food products as well as to produce essential oil. Different metal ions are also present in *C. oxyacantha* (Souri et al., 2010). The color of its seed changes from white to black. Its seeds possess oleic and linoleic oil, but the concentration of oil content varies according to the location and demography of the specie (Raza et al., 2015).

C. oxycantha contains crude fat in its seed, but leaves are calorie-free. It is also known as a good source of protein, fiber, minerals, and carbohydrates. This plant is also known for its flavonoids and antioxidant properties. Different trace elements are also present in *C. oxycantha* including calcium, cadmium, chromium, copper, iron, potassium, magnesium, manganese, molybdenum, sodium, nickel, phosphorus, and zinc (Bukhsh et al., 2007).

C. oxycantha seeds yield two types of oils: linoleic oil and oleic oil. Fatty acid oil composition is stearic acid (1.5%–2%), palmitic acid (5%–6%), linoleic acid (13%–18%), and oleic acid (74%–80%) (Fig. 41.2) (ul Hassan et al., 2013). Different varieties of pholi with lower linoleic acid and higher oleic acid would be preferred to achieve better stability. Cultivars with very high levels of oleic or linoleic acid would increase the value of oils (Muralidharudu et al., 1993).

The fruit of *C. oxycantha* comprises 20%–25% proteins, 2%–15% residual fat, and 60% hull. Flowers of *C. oxycantha* contain two main pigments: the dye carthamin, a flavanone, which is orange-red, and the water-soluble yellow carthamidin. They possess 0.3%–0.6% carthamin. Flavonoids, glycosides, sterols, and serotonin derivatives have been identified from seeds and flowers. Two new glycosides, beta-D-fructofuranosyl carthamoside and 2-O-methylglucopyranosyl-carthamoside, and a known compound 3′,4′,5,7-tetrahydroxyflavanone have been separated from *C. oxyacantha* using preparative HPLC (Ahmad et al., 2007). The leaves contain proteins, potassium, carbohydrates, calcium, magnesium, phosphorus, and sodium. Seeds possess crude fats, fiber, calcium, potassium, magnesium, and phosphorus. The two glycosides,

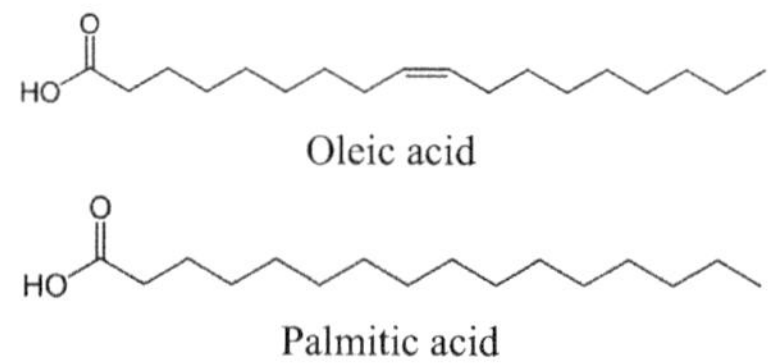

FIGURE 41.2 Important fatty acids present in pholi.

2-O-methylglucopyranosyl-carthamoside and beta-D-fructofuranosyl carthamoside, along with 3′,4′,5,7-tetrahydroxyflavanone have been isolated from pholi (Turgumbayeva et al., 2014). The structure of bioactive compounds in *C. oxycantha* is shown in Fig. 41.2. Medicinal plants contain essential and trace elements derived from the herbs and their extracts. The seeds of *C. oxyacantha* are a good source of crude fiber (35.5 ± 7.2%), crude fats (35.6 ± 5.9%), K (2200 ± 80%), Ca (2600 ± 20%), Fe (49.8 ± 7.9%), Na (1000 ± 10%), and Cu (27.4 ± 8.7%) as compared to the leaves, because the leaves are a better source of total carbohydrates (18.9 ± 4.2%), crude proteins (21.87 ± 2.8%), Mg (726 ± 12.6%), and P (799 ± 17.9%) compared to the seeds (Bukhsh et al., 2007).

3. POSTHARVEST TECHNOLOGY

Pholi flowering time is from April to September. It can be harvested from May to December.

4. PROCESSING

Safflower seed oil extracted by pressing is used as cooking oil, while the low-quality oil extracted through solvent extraction is used as a paint solvent. Safflower flower is used to color cosmetics and dye fabrics.

5. VALUE ADDITION

C. oxyacantha is utilized as a natural food colorant alternative of saffron. Seeds are used for cooking purpose. In winter, roasted seeds are used as dry fruit. The young leaves of this plant are also used as a vegetable. Oleic and linoleic oils have been extracted from this specie that can act as an ointment and are used for lightning. The flower contains two pigments, water-insoluble carthamin dye and water-soluble carthamidin dye, that impart red, brown, yellow, and purple color to cloth during the dyeing process (Souri et al., 2010; Hassan et al., 2010). These dyes are also used for religious and traditional purpose. Flowers and seeds of pholi possess flavonoids, serotonin, sterols, and glycosides, which shows antioxidant potential that can help in controlling aging and cancer (El-Haci et al., 2013; Drummen et al., 2004). Flowers of pholi are mixed with bread, pickles, and rice to give them orange color (ul Hassan et al., 2013).

C. oxyacantha is traditionally used as a medicinal plant. Various bioactive compounds are present in pholi that can be utilized to reduce antifungal and antibacterial infections, e.g., gonorrhea, eye infections, pneumonia, and mycotic. *C. oxycantha* can also be used as a cost-effective biodiesel feedstock to replace petroleum materials (Raza et al., 2015).

6. USES

Plants provide numerous applications in the fields of medicinal and health sciences. Due to the ecologic issues and variability, it is very difficult to precisely study the medicinal nature of weeds. The constituents of *C. oxycantha* are a biosynthetic precursor or utilized in valuable medicinal products (Ahmad et al., 2007). The screening of *C. oxycantha* proves its antihyperglycemic, antihyperlipidemic, hepatoprotective, and antinephrolithiatic activities (Meshram et al., 2011).

Many herbs and weeds that grow without any cost and care possess significant applications in the field of medicine. The presence of valuable proteins in the Asteraceae species make it suitable for the synthesis of chemical drugs and shows its application in medical and nutritional sciences (Leila et al., 2011).

C. oxycantha possesses wide pharmaceutical, industrial, and biologic significance. The seeds of pholi are utilized in cooking, whereas the young leaves are used as a vegetable, and the fruits are used as a bird feed. In many countries, pholi and its straw are used as a premium green fodder.

The flowers are utilized to induce labor and are beneficial for blood circulation, heart diseases, male infertility, cerebral thrombosis, bronchitis, and rheumatism. Pholi-based medicines are also effective during pain and swelling associated with trauma (Ahmad et al., 2007). Flowers of *C. oxycantha* are used to treat jaundice (Mahmood et al., 2011), while the seeds are beneficial as a laxative. The sap is used to reduce salivation. The oil is important to treat scabies (Parvaiz, 2014).

The oil extracted from seeds is edible, utilized to make salad dressings, margarine, and used in cooking. In batik industry, it has been used traditionally to make roghan wax. The flowers are used to extract dye. It imparts a yellow, brown, purple, and red color to cloth based on addition of mordants, colorants, and procedure of dyeing. At a small scale, this dye is still manufactured for religious and traditional purposes (Ahmad et al., 2007).

C. oxycantha is a true substitute for saffron as a natural food colorant. Flowers of pholi are mixed with bread, rice, pickles, and other food to give them an attractive orange color. The seed cake is used as an animal feed. The cake obtained from undecorticated seeds contain matairesinol-glucoside that is suitable for ruminants. *C. oxycantha* meal and flour from decorticate seed are identified for the preparation of high-protein human diet supplements (Ahmad et al., 2007). It is used as an antidiabetic therapy by local healers. Additionally, its seeds are important to make biofuel. It is also beneficial to cure tinnitus, nocturnal enuresis, and earache (Shah, 2013).

In guinea pig ileum, crude extract of pholi and its fractions are a major cause of atropine spasmogenic effect *C. oxycantha* shows significant antihyperlipidemic activity (Ahmad et al., 2006). It is effective for local people who suffer from irritant problems by different kinds of weeds (ul Hassan et al., 2013). Oil extracted from this plant is used as a brain tonic (Khan et al., 2013). The oil of this species is used for overcoming ulcer, in the production of varnishes and soft soaps, and as an important remedy for itching. The wild safflower is a good source of genes for disease resistance (Dwivedi et al., 2012). A decoction of its leaves is given to remove worms (Leila et al., 2011).

C. oxyacantha grows naturally without any cost and care. *C. oxyacantha* seed yield is 833–939 (kg/ha), and the estimated quantity of its oil is 24%–32%. *C. oxycantha* seed is also utilized for the production of biodiesel (Zadeh et al., 2011; Khanahmadzadeh et al., 2011).

Vegetable oils are a renewable source of energy, biodegradable, readily available, and cost effective. Numerous methods are derived for the enzymatic and chemical modification of oil by improving its properties. These modified oils are used as a feedstock that can replace petroleum-derived materials in number of applications including lubricants and plasticizers. *C. oxycantha* with a good iodine value containing oleic acid and linoleic acid was epoxidized using peroxy acid by the reaction of carboxylic acid and hydrogen peroxide in the presence of acidic cation exchange resin. These plant-based epoxides proved to be a sustainable, biodegradable and renewable source of energy that can replace petroleum-based materials (Meshram et al., 2011).

Pholi contains essential trace elements such as copper, magnesium, chromium, zinc, manganese, potassium, and phosphorus that play an important role in the biologic system. The biological effects of the trace elements in living systems depend upon their concentration, so care should be taken, especially when drugs and herbs are investigated in humans.

In the drier parts of Haryana (India), local people use the powder of dried pholi petals with honey to cure acute cough. The plant at the fruiting and flowering stages is cut into small pieces along with *Ocimum americanum* to cure sunstroke and headache. In Rajasthan (India), the decoction of the roots is given to control arthritis. In different areas of the world, poor people use the seed oil of pholi for illumination and edible purposes. The paste of the crushed seeds is applied for ripening of a tumor that is developed because of hair uprooting. A massage ointment of the seed oil mixed with garlic is beneficial for the treatment of rheumatism and arthritis. Seed cake is given to animals to control gastric problems. In

Rajasthan, its oil is mixed with mustard oil to cure skin diseases and relieve body pain. Seeds are parched and eaten with wheat or gram to control piles pain. During famine, seeds are fed to goats and camels (Dwivedi et al., 2012). In some areas of Pakistan, the seeds of C. *oxyacantha* are utilized for the treatment of diabetes (Bukhsh et al., 2014).

7. PHARMACOLOGICAL USES

7.1 Anticancer Activity

C. *oxycantha* is used in different geographic regions as an antiulcer remedy (Khalid and Shad, 1990). Different bioactive compounds are present in the extract of pholi (glycosides, tannins, terpenoids, flavonoids, and saponins) that are responsible for antiulcer activity. Many mechanisms and factors are involved in ulcerogenesis. The studies revealed that methanolic extract of C. *oxycantha* seeds in a dose-dependent manner caused an inhibition of induced ulcer. C. *oxycantha*'s capability of ulcer protection is by increasing glycoprotein levels and mucin contents, thereby maintaining the mucosal epithelium, and by decreasing gastric acid secretion and ulcers. The protection of the gastric mucosa against ulceration is due to antisecretory and cytoprotective properties of C. *oxycantha* (Hussain et al., 2015).

7.2 Antistress Activity

Oxidative stress is the major cause of neurodegenerative disorders like Alzheimer and Parkinson diseases. Methanolic and dichloromethane extracts of C. *oxyacantha* showed the greatest protective effect against stress (Firuzi et al., 2012).

7.3 Hepatoprotective Activity

The hepatoprotective studies of C. *oxyacantha* were conducted by different researchers to investigate the preventive effects of the pholi extracts against liver damage. The protective effects of C. *oxyacantha* seed extracts in alcohol against CCl_4-induced hepatic damage was studied previously. C. *oxycantha* showed significant hepatocurative activity. This activity in alcoholic extract of C. *oxyacantha* seeds is because of the existence of various antioxidants that behave as scavengers and remove the formation of free radicals. These antioxidants also have potential to prevent the peroxidation process and also improve the health of hepatocytes (Bukhsh et al., 2014).

7.4 Antiinflammatory Activity

Different members of *Carthamus* including *C. oxycantha* have been used to manage inflammatory conditions and reduce pain traditionally. *C. oxycantha* shows a marked analgesic effect like morphine by inhibiting prostaglandins action. *C. tinctoriuos* (safflower) also possess similar analgesic effect. The plant extract also contains some constituents including flavonoid glycosides and flavonoids that contain inflammatory effect and are used in folk medicines. Moreover, the studies have revealed that the scavenging properties of the free radicals of glycosides and flavonoid components of the plant are the major cause of antiinflammatory activity. It is reported that the *C. oxycantha* also produces profound antiphlogistic effects similar to standard drug diclofenac (Bukhari, 2013).

8. SIDE EFFECTS AND TOXICITY

The seed oil of *C. oxyacantha* is considered safe for most people when taken by mouth. It is possibly to safely inject *C. oxyacantha* emulsion intravenously under administration of a healthcare professional. *C. oxyacantha* is likely unsafe during pregnancy, as it can bring on menstrual periods, make the uterus contract, and cause miscarriages. Side effects of *C. oxyacantha* for breastfeeding mothers are still not known. *C. oxyacantha* is dangerous to use during surgery or bleeding disorders, as it can slow blood clotting and increase the risk of bleeding in people with bleeding disorders. Persons suffering from allergies should not use *C. oxyacantha*. *C. oxyacantha* oil might increase blood sugar and interfere with diabetes.

References

Ahmad, M., et al., 2006. Ethnomedicinal demography and ecological diversification of some important weeds from District Attock-Pakistan. Pakistan Journal of Weed Science Research. Weed Science Society of Pakistan, Department of Weed Science.

Ahmad, M., et al., 2007. A review on *Carthamus oxycantha*. Pakistan Journal of Pharmaceutical Sciences 20 (23), 1.

Ahmad, W., et al., 2015. Evaluation of steroidal and metabolic effect of Tukhm-e-qurtum (*Carthamus tinctorius*, Linn seed). International Journal of Advances in Pharmacy Medicine and Bioallied Sciences 3 (1), 32–36.

Al Fadal, S.M., Al-Fredan, M.A.-E., 2015. Ecophysiological studies on Saudi wild safflower (*Carthamus oxyacantha* Bieb) seed ecotypes. American Journal of Environmental Sciences 11 (3), 125.

Bibalani, G.H., Shadkami-Til, H., 2011. Investigation on stem biomass of *Carthamus oxycantha* in Shanejan Rangelands, East Azerbaijan, Iran. Reasearch Journal of Biological Sciences 6 (3), 96–99.

BiBi, H., Afzal, M., Muhammad, A., 2014. Morphological and anatomical studies on selected dicot xerophytes of district Karak, Pakistan. American-Eurasian Journal of Agricultural & Environmental Sciences 14 (11), 1201–1212.

Biodiesel - Feedstocks, 2012. Production and Applications. In Tech.

Bukhari, I., 2013. The central analgesic and anti-inflammatory activities of the methanolic extract of *Carthamus oxycantha*. Journal of Physiology & Pharmacology 64, 369–375.

Bukhsh, E., Malik, S.A., Ahmad, S.S., 2007. Estimation of nutritional value and trace elements content of *Carthamus oxyacantha, Eruca sativa* and *Plantago ovata*. Pakistan Journal of Botany 39 (4), 1181.

Bukhsh, E., et al., 2014. Hepatoprotective and hepatocurative properties of alcoholic extract of *Carthamus oxyacantha* seeds. African Journal of Plant Science 8 (1), 34–41.

Carapetian, J., Zarei, G., 2015. Karyotype analysis of three wild species of safflower from west Azerbaijan, Iran. Advances in Environmental Biology 117–123.

Drummen, G.P., et al., 2004. Attenuation of lipid peroxidation by antioxidants in rat-1 fibroblasts: comparison of the lipid peroxidation reporter molecules cis-parinaric acid and C11-BODIPY 581/591 in a biological setting. Biochimica et Biophysica Acta (BBA) - Molecular and Cell Biology of Lipids 1636 (2), 136–150.

Dwivedi, N.K., et al., 2012. Management of wild safflower (*Carthamus oxyacantha* Bieb.) germplasm from arid and semi-arid regions of India. Journal of Oilseeds Research 20, 52–54.

El-Haci, I.A., et al., 2013. Phenolics content and antioxidant activity of some organic extracts of endemic medicinal plant *Anabasis aretioides* Coss. & Moq. from Algerian Sahara. Pharmacognosy Journal 5 (3), 108–112.

Firuzi, O., et al., 2012. Evaluation of the protective effects of some Iranian medicinal plants against neurotoxin-induced apoptosis in neuronal cells. Research in Pharmaceutical Sciences 7 (5).

Ghasemi, P.A., Momeni, M., Bahmani, M., 2013. Ethnobotanical study of medicinal plants used by Kurd tribe in Dehloran and Abdanan districts, Ilam province, Iran. African Journal of Traditional, Complementary and Alternative Medicines 10 (2), 368–385.

Hassan, Z., et al., 2010. Two new carthamosides from *Carthamus oxyacantha*. Natural Product Communications 5 (3), 419–422.

Hassannejad, S., Ghafarbi, S.P., 2013. Effect of environmental factors on weed species distribution. Technical Journal of Engineering and Applied Sciences 3 (4), 355–360.

Hussain, L., et al., 2015. Antiulcerogenic effect of *Carthamus oxyacantha* M. Bieb (Asteraceae) in mice and rat models. Pakistan Journal of Zoology 47 (2), 529–534.

Khalid, S., Shad, R., 1990. Importance of structure and shape in the spread and adaptability of weed seed. Pakistan Journal of Agricultural Research 11 (2), 111–132.

Khan, A., et al., 2013. Study of wild medicinal xerophytes of district Bannu, Kpk. Journal of Medicinal Plants Studies 1 (4), 79–86.

Khanahmadzadeh, A., et al., 2011. Assessment of wild safflower oil methyl ester as potential alternative fuel. Journal of Applied Environmental Biological Sciences 1 (9), 325–328.

Leila, J., Ghassem, H., Hamide, S., 2011. Introduction the medicinal species of Asteraceae family in Ilkhji and Sharafaldin regions of Esat Azarbaijan in Iran. Journal of American Science 7 (5), 455–458.

Mahmood, A., et al., 2011. Ethno medicinal survey of plants from district Bhimber Azad Jammu and Kashmir, Pakistan. Journal of Medicinal Plants Research 5 (11), 2348–2360.

Meshram, P.D., Puri, R.G., Patil, H.V., 2011. Epoxidation of wild safflower (*Carthamus oxyacantha*) Oil with peroxy acid in presence of strongly acidic cation exchange resin IR-122 as catalyst. International Journal of ChemTech Research 3 (3), 1152–1163.

Muralidharudu, Y., Sujatha, M., Singh, M., 1993. Extent of variation in seed oil content and fatty acid composition of cultivated and two closely related wild species of safflower. In: International Safflower Conference.

Parvaiz, M., 2014. Ethnobotanical studies on plant resources of Mangowal, district Gujarat, Punjab, Pakistan. Avicenna Journal of Phytomedicine 4 (5), 364.

Raza, M.A., Mukhtar, F., Danish, M., 2015. Cuscuta reflexa and *Carthamus oxyacantha*: potent sources of alternative and complimentary drug. SpringerPlus 4 (1), 76.

Shah, R.A., 2013. Studies on Floral Biodiversity of District Mianwali and Allied Areas with Emphasis on Ethnobotanical Uses. Quaid-i-Azam University, Islamabad.

Shanmugasundaram, S., 2011. Wild crop relatives: genomic and breeding resources. In: Kole, C. (Ed.), Oilseeds. Springer, Heidelberg, Dordrecht, London, ISBN 978-3-642-14870-5, p. 295. Experimental Agriculture 48(01) (2012), 156.

Sheykhlou, N., et al., 2012. Study of new dry land safflower cultivars yield under different planting seasons in Zanjan area. The International Journal of Agriculture and Crop Sciences 4 (20), 1546–1550.

Souri, E., et al., 2010. Antioxidative activity of sixty plants from Iran. Iranian Journal of Pharmaceutical Research 55–59.

Talebi, M., et al., 2012. Molecular characterization of *Carthamus tinctorius* and *C. oxyacanthus* germplasm using sequence related amplified polymorphism (SRAP) markers. Plant Omics 5 (2), 136.

Turgumbayeva, A., et al., 2014. Phytochemical screening and biological activities of the camel thorn (*Alhagi kirghisorum*) and safflower flowers (*Carthamus tinctorius* L.) grown in Kazakhstan. American-Eurasian Journal of Agricultural & Environmental Sciences 14 (12), 1487–1491.

ul Hassan, S.S., et al., 2013. Counter irritant activity of *Carthamus oxyacantha*. Pakistan Journal of Pharmaceutical Sciences 26 (4), 665–672.

Zadeh, A.K.A., et al., 2011. Suitability of *Carthamus oxyacantha* plant as biodiesel feedstock. Australian Journal of Crop Science 5 (12), 1639.

CHAPTER

42

Puncture Vine

Muhammad Asif Hanif, Shahzia Yousaf, Rafia Rehman, Asma Hanif, Raziya Nadeem
Department of Chemistry, University of Agriculture, Faisalabad, Pakistan

OUTLINE

Medicinal Plants of South Asia
https://doi.org/10.1016/B978-0-08-102659-5.00042-2

Copyright © 2020 Elsevier Ltd. All rights reserved.

1. BOTANY

1.1 Introduction

The puncture vine (*Tribulus terrestris*) (Fig. 42.1) is a herbal plant belonging to the genus *Tribulus* and the family Zygophyllaceae. The genus *Tribulus* has nearly 20 species distributed worldwide. Three species, *Tribulus terrestris, Tribulus cistoides,* and *Tribulus alatus,* are found in Pakistan and India (Chhatre et al., 2014). It is a tap-rooted, perennial

FIGURE 42.1 Puncture vine.

flowering plant, but in colder environments, it grows annually. It is commonly found all over the world, mostly in warm temperate and tropical areas, in southern Europe, southern Asia, Africa, and in northern Australia. It can also grow in sterile land and poor climates that are not habitable by other plants (Hashim et al., 2014). The herb *T. terrestris* is used in Indian and Chinese medicine for the treatment of various diseases (Bourke et al., 1992).

The common names of *T. terrestris* are puncture vine, cat's head, bull-head, caltrop, burra gokharu, yellow vine, goat head, bindii, and devil's horn. Its fruits are prickly, which makes it hazardous for animals to graze on it. The fruit of *T. terrestris* contains beta-carboline alkaloids that cause staggers in sheep. It has a fine network of rootlets that arise from the taproot to give a large surface area. This enables the plant to absorb the maximum amount of water from the soil and allows it to live in very arid environments (Cowan, 1999). The dried fruits contain semidrying oils, peroxides, diastase, glucosides, resins, protein, and a large amount of inorganic matter (Lamba et al., 2011).

1.2 History/Origin

T. terrestris was first found in the Pacific Northwest in the sand alongside line tracks at Whitman County, Wawawai, Washington and southern Europe. It was commented by the California's Department of Food and Agriculture that *T. terrestris* perhaps first originated from Saharan regions and then spread into Mediterranean regions (Wu et al., 1996). This plant was supposed to be accidently imported into United States from Mediterranean regions. However, it was first reported in 1903 in California. Various species of this plant were introduced in India, Turkey, Tajikistan, Syria, Saudi Arabia, Mongolia, Lebanon, Kyrgyzstan, Kazakhstan, Jordan, Japan, Israel, Iraq, Iran, Georgia, Cyprus Sinai, China, Azerbaijan, Armenia, Afghanistan, and Pakistan, and some more important countries of Asia. This plant is also found in abundance in north Australia. The plant has spread and been introduced into the Pacific Northwest of the United States by humans. The fruits of *T. terrestris* look like a caltrop, a metal device positioned on the ground in medieval warfare, with one spike up to slow proceeding armies (Hajmohamadi, 2013). This wild plant belongs to a small genus comprised of about 12 species in the Caltrop family (Zygophyllaceae). The name of this genus originates from the Latin word tribo, "to tear," which denotes the spiky fruits. This wild plant harms the eyes, mouths, feet, digestive tracts, and hides of livestock. *T. terrestris* flowers might have high levels of nitrates, which are enough to poison livestock. Photosensitization that causes bighead in sheep, as reported by some researchers, is due to a poison produced in *T. terrestris*. This toxin damages the liver of sheep, which impairs its ability to remove phylloerythrin, a derivative of chlorophyll (Khushdil et al., 2012). This plant has abundantly been used for a number of medicinal purposes since time immemorial. In China and India, these medicinal plants have been used for 5000 years. In Ayurvedic system of medicines, this herb is known for its strong antiurolithiatic, aphrodisiac, and diuretic potentials (Deepak et al., 2002).

1.3 Location/Demography

T. terrestris is a plant that can grow in both temperate and warm regions that have loose, dry, and sandy soil, near loose blown soil and sand dunes at the margins of fields. Light, moisture, and appropriate temperature are the three basic conditions for the proper germination and growth of this plant. However, it can also grow in heavier soils, more specifically when the soil is compacted, and moist ones such as found in playgrounds or roadsides. This plant is rarely a serious problematic weed among intensively cropped fields. Nevertheless, burs are very difficult to remove from seeds when found in crops like peas and lima beans (Khan, 2009). The seeds of this plant germinate in accordance with their direction of bur when conditions are ideal for proper growth and development. Although, it is not considered a prolific seed producer because it grows by itself and needs to face no competition over millions of seeds. According to some recent researches, it was found that a single plant can produce 576,000 fruits in a season, and by assuming that a single fruit produces two seeds, the average number of seeds is about 1,152,000. Similarly, spread of seeds is facilitated by spiny fruit segments. Smaller and larger seed spines tend to arrange themselves at different angles no matter where the seeds are falling. Nevertheless, at least one spine must be pointed upward to meet the unwary hoof, foot, or vehicle tire. The sharp spines of this plant can readily penetrate skin, tires, rubbery materials, and leathery substances, thereby transporting them at larger distances before being picked off or knocked out. It is expected that seeds would probably have crossed the oceans in the wool of European sheep. This plant is one of the few weeds that have spread via aircraft. First infestation in some areas occurred in cracks of airport runways where seeds were brushed from tires of the aircraft. In fields, germination starts during warm spring weather and continues till frost (Duke, 2002).

1.4 Botany, Morphology, Ecology

T. terrestris is a small horizontal herb, 10–60 cm long, having yellow flowers and pinnate leaves, a silky, hairy shrub. It bears opposite leaves, unequal paripinnate, five to eight pairs of pinnae and elliptical or oblong lanceolate. Carpel fruits are of distinguishing, stellate shape, round-shaped to some extent, flattened, five cornered, and covered with prickles of very light yellow color (Jamil et al., 2012). Several seeds are present in each crocus with diagonal partitions between them. The seeds are oily in nature. The fresh root is light brown, slender, fibrous, cylindrical, often branched, containing a number of small rootlets and 7–18 cm long and 0.3–0.7 cm in diameter (Duke, 2002). The roots have an aromatic odor and sweetish taste. The fruits and roots are mostly used as a traditional medicine for the cure of various diseases. The fruits of this herb are known as goat head in the United States and "chih-hsing" in China. The spiky fruit of this plant appears like a cloven hoof of a cow, so it is termed "go-ksura," meaning "cow hoof" These spiky fruits are greenish yellow in coloration with faint appearance and spiny surface. Fruit of this plant is globose constituting five, muriculate, glabrous, wedge-shaped, woody cocci having two pairs of sharp spines with one pair longer than the other. Tips of spines meet in pairs, thereby forming a pentagonal framework all around the fruit. The outermost surface of the schizocarp is quite rough in appearance, while several seeds occur in between transverse partitions. The smell of this fruit is faintly aromatic, whereas its taste is slightly acrid (Chhatre et al., 2014).

Seedlings of this plant tend to emerge in the form of flushes following the proper irrigation and appreciable rainfall. New plant seeds are usually dormant, with less than 10% germination rate throughout the year after production. Flowers appear 3 weeks after first germination, and fruits produce with viable seeds 1 or 2 weeks later. Puncture vine does not emerge from soil having depth greater than 2 in.

T. terrestris plants usually have many fruits, averaging 200–5000 per plant. Each fruit/burr usually contains five nutlets (cocci), and each nutlet (coccus) can contain two to five seeds. Seed dispersal of each fruit section (coccus) has two sharp divergent spines and several other spines and warty protuberances, enabling the *T. terrestris* fruits to easily attach to anything and be dispersed (Al-Bayati and Al-Mola, 2008).

The flowers of *T. terrestris* are small and yellow, with five petals. It grows with a multiple stem system that emerges from a single crown. Its leaves are pinnate that are made of two rows of little leaflets having approximate length less than a quarter inch. Leaves of this plant are oppositely arranged all around the stem. The fruits appear in a week after blooming, and its seeds are stiff and firm having two sharp spines with average length of about 10 mm. *T. terrestris* is responsible for severe diseases in perennial crops like vine and asparagus. Puncture vine's rigid spines are a strong irritant on bare grounds like barn lots, school grounds, railway roads, road shoulders, and various other areas normally used for storage and parking. However, these spines can cause severe injuries to the pets and people living in close vicinity and can puncture the inflated balls and tires of bicycles. The seed closer to the pointed end of bur is the largest and generally sprouts first (Khushdil et al., 2012). Mostly fresh ripened *T. terrestris* seeds are dormant and need a seasoning period of almost 6–12 months. The dormancy of seeds has been observed to be greater in both fresh and 3- to 6-year-old seeds. The germination rate of 6-month-old seeds is 84%, while freshly harvested seeds have a germination rate of 10%. Water imbibition of dormant and nondormant seeds differs: dormant seeds have a water imbibition capacity that is half of that of a nondormant seed. The largest seed within a coccus germinates first; the residual seeds may grow or stay dormant subject to moisture availability. This biggest seed is typically positioned near the base of the coccus (Ernst and Tolsma, 1988).

This weed is very common in all South Asian countries. In Australia, the daily maximum temperatures in summer are more than 84°F (29°C), where *T. terrestris* is a major weed. In accordance with records collected from the northwestern United States, various species of this plant are found in the regions characterized by average minimum temperature of July at around 19.4°C (73°F) and average minimum temperature of January at around −8.3°C (17°F). However, the maximum average temperature of January reaches up to −4.4°C (24°F). Natural occurrence of this plant is in the areas having maximum and minimum mean annual rate of precipitation 38 cm (15 in.) and 28 cm (11 in.) respectively. The seedlings of this plant were apparently very poor when observed in the shaded areas.

2. CHEMISTRY

Puncture vine is an important aromatic medicinal herb (Sun et al., 2002). It contains essential oils, vitamins, saponins (Fig. 42.2), and various alkaloids. Plant extracts of *T. terrestris* can play a role in supporting both performance and health status of poultry (Wang et al., 1997). This extract is harmless for humans and animals and contains biologically active substances such as flavonoids, saponins, alkaloids, phytosterols, glycosides, and other constituents (Wu et al., 1996). *T. terrestris* has steroids,

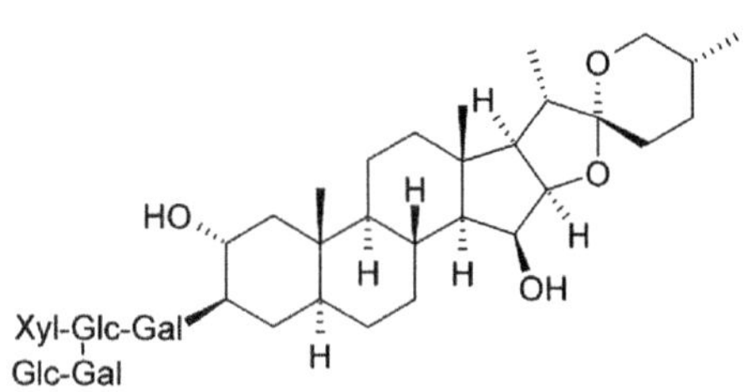

FIGURE 42.2 Saponins are important chemical components of puncture vine.

saponins, glycoside, flavonoids, alkaloids, some amide, unsaturated fatty acids, and vitamins (Charan, 2000).

Furostanol and spirostanol saponins of sarsasapogenin, chlorogenin, diosgenin, hecogenin, gitogenin, neotigogenin, and tigogenin types are abundantly found in this type of plant. In addition to this, four sulfated saponins of tigogenin and diosgenin were also isolated.

Furostanol glycosides containing protogracillin and protodioscin were found, of which protodioscin was the most dominant, and saponin and spirostanol glycosides were present in small quantities (Kostova and Dinchev, 2005). This showed that the flavonoids in *T. terrestris* should be developed, studied, and more used for medicinal purposes (Wu et al., 1999). Kaempferol-3-glucoside, kaempferol, tribuloside (kaempferol-3-β-d-(6″-p-coumaroyl) glucoside, and kaempferol-3-rutinoside, as well as 18 flavonoids including quercetin, caffeoyl derivatives, and glycosides were isolated from the fruits and leaves (Xu et al., 2009).

3. POSTHARVEST TECHNOLOGY

Generally, *T. terrestris* can be harvested any time, but the best timing is early in the morning because essential oil content is strongest in morning at lower temperatures. The fresh leaves and fruits have more alkaloids, complexity, and intensity than old dried leaves. Leaves and fruits should not be shredded when drying, as essential oils will be lost. The specific yield reduces with the plant density rise (Han et al., 2008).

4. PROCESSING

T. terrestris contains variety of chemical constituents that are used for various purposes. An alcoholic or an ethereal extract of the ground fruits gives water a crystalline residue precipitated by ammonia and possesses the characteristics of an alkaloid associated with hydrochloric acid or alkaline chlorides (Lozanov et al., 2013). The sugar of the saponins in plant samples is extracted from different topographic areas by chopping leaves and fruits. The freshly extracted juice of the whole plant contains inorganic nitrates, typically potassium nitrate, which have a diuretic property (Hashim et al., 2014).

The leaves and the fruits are washed, then dried in the shade to avoid the oxidation of essential oil. *T. terrestris* should be dried soon after harvest. The temperature should not be raised to minimize the evaporation of volatile compounds. The essential oil extracted from the plant can be separated into two types: (1) the oil from the leaves and (2) the oil from fruits, which is of higher quality. The fruits also contain ascorbic acid, alkaloids, essential oils, and fixed oils (14%). The amount of ascorbic acid increases from roots to the fruits in *T. terrestris*. It is also rich in iron. Distillation is used to extract the essential oil from the herb (Kostova and Dinchev, 2005).

5. VALUE ADDITION

This plant is of great industrial importance, as it is a major source of a number of medicinal preparations and pharmaceutical mixtures based on saponins. Food supplements prepared for general stimulating actions are known to have appreciable concentration of these biochemical materials that are currently marketed in the United States of America and all over Europe. *T. terrestris* is a well-known plant and has traditionally been used in folk systems of medicines in many countries all over the world for treatment of different types of diseases. Relatively higher saponin contents of this plant make it unique and valuable among all other medicinal plants. Similarly, very little work has been done on this medicinal plant, and still, its database is not available. Hence, development of novel analytical techniques for determination of different saponins in a mixture is of great importance, and its molecular exploration is the emerging area for a number of future scientific researches.

Different kinds of sugars such as glucose, rhamnose, galactose, and xylose have been found in this plant and vary depending upon on the origin of the plant. *T. terrestris* still remains an object of further study (Hashim et al., 2014).

6. USES

Lots of herbs and spices significantly improve health regardless of low consumption because they are full of mineral compounds and certain antioxidants. A number of derivatives of sarsasapogenin, chlorogenin, diosgenin, tigogenin, neogitogenin, ruscogenin, neotigogenin, hecogenin, neohecogenin, and gitogenin are abundantly found in different parts of this herb. However, the concentration of each component depends on the nature of the plant and part of its extract. The sugar moieties of the isolated furostanol and spirostanol saponins are oligosaccharides, which give energy to the body. Further studies on the biological actions of individual compounds and extracts are of extraordinary importance.

T. terrestris is used in traditional medicines as an aphrodisiac, astringent, palliative, and stomachic tonic. For genitourinary tract ailments, the dried fruit of the herb is very effective. It is an important part of Gokshuradi Guggulu, a potent Ayurvedic medicine used for eliminating urinary and genitourinary tract stones. This plant is in extensive usage as a folk medicine for treatment of impotence in Bulgaria. In some traditional Chinese formulations, the fruit of this plant is used to treat abdominal distension, eye trouble, emission, edema, morbid leucorrhea, sexual dysfunctioning, and to strengthen the muscles (Dey and De, 2012). The essential oil of *T. terrestris* has an important commercial worth. It is used in many commercial products including soap and detergents due to its aroma, also making it suitable for use in perfumes (Chhatre et al., 2014).

7. PHARMACOLOGICAL USES

7.1 Diuretic Effects

A decoction of *T. terrestris* is used to raise the electrolyte and urine volume over a specific period. It considerably lowers the sodium and serum potassium levels throughout the body. Hence, it is recommended as a diuretic agent, which is helpful in hypertensive and renal disease patients (Mukherjee et al., 2008). In albino rats, the diuretic potential of *T. terrestris* has been assessed. The diuretic effect was attributed due to the existence of potassium salts, essential oil, and nitrates, which are present mainly in the fruits and seeds, so *T. terrestris* has a slight side effect in albino rats (Mazaro-Costa et al., 2010). The aqueous extract of *T. terrestris*, when given orally at a dose of 5 g/kg, produced a positive diuresis. Chloride and sodium concentrations in the urine were higher. The smooth muscles increased tonicity and, together with its diuretic action, assisted in the movement of stones in the urinary tract (Al-Ali et al., 2003). Diuretic effects of this plant were comparable with a reference furosemide that is known to have potassium-sparing activity. Diuretic action of this plant makes it highly useful as an effective antihypertensive medicinal agent (Al-Ali et al., 2003).

7.2 Antihypertensive Property

T. terrestris possesses antihypertensive activity (Karimi Jashni et al., 2012). The compounds of *Tribulus* extracts promote the release of nitric oxide from nerve endings and the endothelium. It reduces hypertension by relaxing smooth muscles and increasing angiotensin production that alters enzyme inhibition (Lubna Fatima et al., 2014). The extract of *T. terrestris* electrically enhanced nitroglycerine and produced a prompt relaxation in a rabbit with a pro-erectile function. The mechanism responsible for the antihypertensive action is yet not fully known. Important antihypertensive effects of an aqueous extract of *T. terrestris* were described, and it was proposed that they might be accredited to its inhibitory effect on angiotensin converting enzyme activity (Sharifi et al., 2003). Treatment of rats with an aqueous extract of *T. terrestris* meaningfully decreases substantial acetylcholinesterase (ACE) action in all tissues of the rat. The role of arterial smooth muscle vasodilation in the antihypertensive effect of *T. terrestris* looked improbable, particularly due to the failure to detect any relaxant effect of *T. terrestris* in the rabbit aorta given the analogous amount of the extract repressed peristaltic movement in rabbit jejunum and sheep ureter (Karimi Jashni et al., 2012). Improvement and dilatation of coronary arteries has been observed with prolonged use of *T. terrestris,* without any side effects (Gauthaman and Adaikan, 2005).

7.3 Analgesic Activity

The extract of the entire dried plant with chloroform, when injected intraperitoneally into mice, exhibited an analgesic effect (Lubna Fatima et al., 2014).

7.4 Anticholinergic Activity

The dried fruit, when directed through gastric intubation to mice in a preparation containing *Aconitum sinese, Mentha arvensis, Sophora flavescens,* and *Alpinia* species exhibited anticholinergic effect (Phillips et al., 2006).

7.5 CNS Activity

T. terrestris extract demonstrated pharmacological activity in a prior screening, presented marked CNS-stimulating action (Taur and Patil, 2011).

7.6 Antiarthritic Activity

A methanolic extract of *T. terrestris* fruit was tested for antiarthritic activity. The herbal extract was administered at a dose of 300 mg/kg and 200 mg/kg for 21 days and showed antiarthritic effect in rat paws (Sellandi et al., 2012). It was found that the modulation of membrane stability was an important mechanism of antiarthritic activity. This activity is possibly due to the presence of flavonoids. These flavonoids possess surface charge counterbalancing effects. It was established that the injection of *T. terrestris* (200 and 300 mg/kg bodyweight) results in prevention of leukocyte movement, which has a valuable effect in joint protection. This effect may be due to the occurrence of a steroidal glycoside (Phillips et al., 2006). The investigators have revealed that calcium levels in the blood serum increase due to *T. terrestris* in broiler parents (Ross, 2007).

7.7 Skeletal Muscle Relaxant Activity

The ethanol (95%) extract of the dried entire plant, administered intraperitoneally to mice, demonstrated skeletal muscle relaxant activity (Taur and Patil, 2011).

7.8 Smooth Muscle Relaxant Activity

The ethanol (95%) extract of the dried aerial parts of the plant showed a smooth muscle relaxant activity on rabbit duodenum (Taur and Patil, 2011).

7.9 Anticholesteremic Effects

The saponin fraction of the dried root, administered by gastric intubation to rabbits, decreased the development of protein, carbohydrate, and lipid dystrophy of the liver in cholesterol-loaded animals (Ross, 2001).

7.10 Hypoglycemic and Hypolipidemic Effects

The extract of *T. terrestris* could reduce the blood glucose levels in diabetic rats (Sellandi et al., 2012).

7.11 Antioxidant Property

T. terrestris has shown prominent antioxidant properties. In an experimental study, the investigators showed that the extract from *T. terrestris* decreased the formation of hydro peroxide significantly, suggesting that this species has potent natural antioxidants (Majumdar et al., 2008).

7.12 Antibacterial Effects

The saponin extract of *T. terrestris* has antimicrobial activity against a number of microorganisms, and rather than simply changing the surface tension of the extracellular medium, the extract exhibited an inhibitory effect on gram-positive bacteria, inducing membranolytic properties, so it was influenced by microbial population thickness (Mohammed, 2008). Flavonoids are present in gram-negative and gram-positive bacteria, representing the existence of a wide spectrum antibiotic compounds or general metabolic poisons in the plant (Sisto et al., 2012). Antimicrobial and antifungal activity of aqueous and organic extracts of the leaves, roots, and fruits of *T. terrestris* was studied. It was found that the antibacterial activity of the extracts is possibly due to their ability to form a complex with bacterial cell walls and extracellular soluble proteins

(Al-Bayati and Al-Mola, 2008). Additionally, lipophilic flavonoids could also interrupt microbial membranes. The antifungal activity of *T. terrestris* may be due to various chemicals noticeable in its extracts such as saponins (Sisto et al., 2012).

7.13 Antitumor Activity

The selective antitumor activity in vitro of *T. terrestris* on human cancer cells was studied (Costa-Lotufo et al., 2005). Saponins in *T. terrestris* extracts, structurally related to diosgenin, may suppress propagation, block the cell cycle, and promote apoptosis in human tumor cell lines. It was found that *T. terrestris* has a protective ability against UVB-induced carcinogenesis (Ukani et al., 1997). The mechanism of action by which *T. terrestris* saponins control cell death demonstrates its potential to be developed into a new medicine for cancer patients (Wang et al., 1990).

7.14 Cardiotonic Activity

Saponins of *T. terrestris* improve coronary circulation. Coronary heart disease patients were cured in a clinical trial. The total effective rate of remission from angina pectoris was 82.3%, and improved ECG (52.7%) was greater than that of control group 35.8% (Costa-Lotufo et al., 2005; Xu et al., 2009). *T. terrestris* plays a major role in the cure of various heart diseases including myocardial infarction, the sequelae of cerebral thrombosis, and coronary and cerebral arteriosclerosis (Zhang et al., 2010). The basic mechanism for the defensive influence of tribulosin from *T. terrestris* against cardiac reperfusion damage was studied in rats. Tribulosin protects the myocardium from reperfusion injury by protein kinase C epsilon stimulation (Lubna Fatima et al., 2014). The treatment of rats by tribulosin caused a major reduction of aspartate transaminases, lactate dehydrogenase activity, malondialdehyde, myocardial apoptosis, and creatine kinase. Aqueous extract of fruit of this plant has substantial in vitro ACE inhibitory properties. Methanolic and aqueous extracts possesses excellent antihypertensive properties through direct relaxation of arterial smooth muscles along with membrane hyperpolarization in acutely hypertensive rats (Phillips et al., 2006). It also improves heart function and even protects heart cells following a heart attack (Zhang et al., 2010).

7.15 Antidiabetic Activity

The saponins of *T. terrestris* possess hypoglycemic activities (Li et al., 2002). It considerably decreases the level of serum triglyceride, serum cholesterol, and serum glucose but increases serum superoxide dismutase activity in alloxan-induced diabetic mice. Gluconeogenesis is inhibited in mice by a decoction of *T. terrestris* (Li et al., 2001). The ethanolic extract had a protective influence at 2 g/kg of body weight by preventing oxidative tension in streptozotocin-induced diabetic rats. This extract of *T. terrestris* showed 100% inhibition of aldose reductase at a dose of 30 μg/mL using dl-glyceraldehyde as a substrate and 70% inhibition of α-glucosidase at 500 μg/mL when maltose was used as a substrate (Amin et al., 2006). Postprandial blood glucose level was decreased substantially in rats after injection of saponins from *T. terrestris*. The coronary artery was dilated, which enhanced coronary circulation. It is recommended, in Ayurveda, for the cure of angina pectoris and other cardiac complications of diabetes by dropping lipid and blood glucose levels through the action of antioxidants (Lamba et al., 2011).

8. SIDE EFFECTS AND TOXICITY

It is likely unsafe to eat the spine-covered fruit of *T. terrestris*, as it can cause a serious lung problem.

References

Al-Ali, M., Wahbi, S., Twaij, H., Al-Badr, A., 2003. *Tribulus terrestris*: preliminary study of its diuretic and contractile effects and comparison with Zea mays. Journal of Ethnopharmacology 85, 257–260.

Al-Bayati, F.A., Al-Mola, H.F., 2008. Antibacterial and antifungal activities of different parts of *Tribulus terrestris* L. growing in Iraq. Journal of Zhejiang University - Science B 9, 154–159.

Amin, A., Lotfy, M., Shafiullah, M., Adeghate, E., 2006. The protective effect of *Tribulus terrestris* in diabetes. Annals of the New York Academy of Sciences 1084, 391–401.

Bourke, C., Stevens, G., CARRIGAN, M.J., 1992. Locomotor effects in sheep of alkaloids identified in Australian *Tribulus terrestris*. Australian Veterinary Journal 69, 163–165.

Charan, A.A.S.D.S., 2000. Herbal Cure. Pustak Mahal.

Chhatre, S., Nesari, T., Somani, G., Kanchan, D., Sathaye, S., 2014. Phytopharmacological overview of *Tribulus terrestris*. Pharmacognosy Reviews 8, 45.

Costa-Lotufo, L.V., Khan, M.T.H., Ather, A., Wilke, D.V., Jimenez, P.C., Pessoa, C., de Moraes, M.E.A., de Moraes, M.O., 2005. Studies of the anticancer potential of plants used in Bangladeshi folk medicine. Journal of Ethnopharmacology 99, 21–30.

Cowan, M.M., 1999. Plant products as antimicrobial agents. Clinical Microbiology Reviews 12, 564–582.

Deepak, M., Dipankar, G., Prashanth, D., Asha, M., Amit, A., Venkataraman, B., 2002. Tribulosin and β-sitosterol-D-glucoside, the anthelmintic principles of *Tribulus terrestris*. Phytomedicine 9, 753–756.

Dey, A., De, J.N., 2012. Ethnobotanical survey of Purulia district, West Bengal, India for medicinal plants used against gastrointestinal disorders. Journal of Ethnopharmacology 143, 68–80.

Duke, J.A., 2002. Handbook of Medicinal Herbs. CRC press.

Ernst, W., Tolsma, D., 1988. Dormancy and germination of semi-arid annual plant species, Tragus berteronianus and *Tribulus terrestris*. Flora 181, 243–251.

Gauthaman, K., Adaikan, P., 2005. Effect of Tribulusterrestris on nicotinamide adenine dinucleotide phosphate-diaphorase activity and androgen receptors in rat brain. Journal of Ethnopharmacology 96, 127–132.

Hajmohamadi, S., 2013. Physiological and pharmaceutical effects of *Tribulus terrestris* as a multipurpose and valuable medicinal plant. International Journal of Advanced Biological and Biomedical Research 1, 1289–1295.

Han, M., Yang, L., Han, D., Yang, L., 2008. Study on fitting growth density and best harvest time of *Tribulus terrestris*. Zhongguo Zhong yao za zhi = Zhongguo zhongyao zazhi = China Journal of Chinese Materia Medica 33, 750–753.

Hashim, S., Bakht, T., Bahadar, K., Jan, S., 2014. Medicinal properties, phytochemistry and pharmacology of *Tribulus terrestris* L.(Zygophyllaceae). Pakistan Journal of Botany 46, 399–404.

Jamil, M., Ansari, J., Ali, A., Ahamad, J., Ali, M., Tamboli, E., 2012. Pharmacological scientific evidence for the promise of *Tribulus terrestris*. International Research Journal of Pharmacy 3, 403–406.

Karimi Jashni, H., Malekzadeh Shiravani, S., Hoshmand, F., 2012. The effect of the *Tribulus terrestris* extract on spermatogenesis in the rat. Journal of Jahrom University of Medical Sciences 9, 9.

Khan, F.M., 2009. Ethno-veterinary medicinal usage of flora of greater Cholistan desert (Pakistan). Pakistan Veterinary Journal 29, 75–80.

Khushdil, M., Chand, N., Khan, S., Qureshi, M., Tanweer, A., 2012. Comparative effect of different schedules of administration of medicinal plants (Allium sativum, berberislycium, Eclipta alba and Mangiferaindica) infusion on the immunity and overall performance of broiler chicks. Sarhad Journal of Agriculture 28, 319–326.

Kostova, I., Dinchev, D., 2005. Saponins in *Tribulus terrestris*–chemistry and bioactivity. Phytochemistry Reviews 4, 111–137.

Lamba, H.S., Bhargava, C.S., Thakur, M., Bhargava, S., 2011. α-glucosidase and Aldose reductase inhibitory activity in vitro and anti-diabetic activity in vivo of *Tribulus terrestris* L.(Dunal). International Journal of Pharmacy and Pharmaceutical Sciences 3, 270–272.

Li, M., Qu, W., Chu, S., Wang, H., Tian, C., Tu, M., 2001. Effect of the decoction of *Tribulus terrestris* on mice gluconeogenesis. Zhong yao cai = Zhongyaocai = Journal of Chinese Medicinal Materials 24, 586–588.

Li, M., Qu, W., Wang, Y., Wan, H., Tian, C., 2002. Hypoglycemic effect of saponin from *Tribulus terrestris*. Zhong yao cai = Zhongyaocai = Journal of Chinese Medicinal Materials 25, 420–422.

Lozanov, V., Markov, T., Bozhanov, S., Georgieva, E., Mitev, V., 2013. Antitumor activity of Bulgarian herb *Tribulus terrestris* L. on human breast cancer cells. Journal of Bioscience and Biotechnology 2, 25–32.

Lubna Fatima, M., Arshiya Sultana, M., Saad Ahmed, M., Shabiya Sultana, M., 2014. Pharmacological activities of *Tribulus terrestris* Linn: a systemic review. World Journal of Pharmacy and Pharmaceutical Sciences 4.

Majumdar, A., Saraf, M., Andrades, N., Kamble, R., 2008. Preliminary studies on the antioxidant activity of *Tribulus terrestris* and Eclipta alba. Pharmacognosy Magazine 4, 102.

Mazaro-Costa, R., Andersen, M.L., Hachul, H., Tufik, S., 2010. Medicinal plants as alternative treatments for female sexual dysfunction: utopian vision or possible treatment in climacteric women? The Journal of Sexual Medicine 7, 3695–3714.

Mohammed, M.J., 2008. Biological activity of saponins isolated from *Tribulus terrestris* (fruit) on growth of some bacteria. Tikrit Journal of Pure Science 13.

Mukherjee, P.K., Kumar, V., Kumar, N.S., Heinrich, M., 2008. The Ayurvedic medicine Clitoria ternatea—from traditional use to scientific assessment. Journal of Ethnopharmacology 120, 291–301.

Phillips, O.A., Mathew, K.T., Oriowo, M.A., 2006. Antihypertensive and vasodilator effects of methanolic and aqueous extracts of *Tribulus terrestris* in rats. Journal of Ethnopharmacology 104, 351–355.

Ross, I.A., 2001. *Tribulus terrestris*, Medicinal Plants of the World. Springer, pp. 411–426.

Ross, I.A., 2007. Medicinal Plants of the World. In: Chemical Constituents, Traditional and Modern Medicinal Uses, vol. 3. Springer Science & Business Media.

Sellandi, T.M., Thakar, A.B., Baghel, M.S., 2012. Clinical study of *Tribulus terrestris* Linn. in Oligozoospermia: a double blind study. Ayu 33, 356.

Sharifi, A.M., Darabi, R., Akbarloo, N., 2003. Study of antihypertensive mechanism of *Tribulus terrestris* in 2K1C hypertensive rats: role of tissue ACE activity. Life Sciences 73, 2963–2971.

Sisto, M., Lisi, S., D'Amore, M., De Lucro, R., Carati, D., Castellana, D., La Pesa, V., Zuccarello, V., Lofrumento, D.D., 2012. Saponins from *Tribulus terrestris* L. protect human keratinocytes from UVB-induced damage. Journal of Photochemistry and Photobiology B: Biology 117, 193–201.

Sun, W., Gao, J., Tu, G., Guo, Z., Zhang, Y., 2002. A new steroidal saponin from *Tribulus terrestris* Linn. Natural Product Letters 16, 243–247.

Taur, D., Patil, R., 2011. Mast cell stabilizing and antiallergic activity of *Abrus precatorius* in the management of asthma. Asian Pacific Journal of Tropical Medicine 4, 46–49.

Ukani, M., Nanavati, D., Mehta, N., 1997. A review on the ayurvedic herb *Tribulus terrestris* L. Ancient Science of Life 17, 144.

Wang, B., Ma, L., Liu, T., 1990. 406 cases of angina pectoris in coronary heart disease treated with saponin of *Tribulus terrestris*. Zhong xi yi jie he za zhi = Chinese Journal of Modern Developments in Traditional Medicine/Zhongguo Zhong xi yi jie he yan jiu hui (chou), Zhong yi yan jiu yuan, zhu ban 10, 85–87, 68.

Wang, Y., Ohtani, K., Kasai, R., Yamasaki, K., 1997. Steroidal saponins from fruits of *Tribulus terrestris*. Phytochemistry 45, 811–817.

Wu, G., Jiang, S., Jiang, F., Zhu, D., Wu, H., Jiang, S., 1996. Steroidal glycosides from *Tribulus terrestris*. Phytochemistry 42, 1677–1681.

Wu, T.-S., Shi, L.-S., Kuo, S.-C., 1999. Alkaloids and other constituents from *Tribulus terrestris*. Phytochemistry 50, 1411–1415.

Xu, T., Xu, Y., Liu, Y., Xie, S., Si, Y., Xu, D., 2009. Two new furostanol saponins from *Tribulus terrestris* L. Fitoterapia 80, 354–357.

Zhang, S., Li, H., Xu, H., Yang, S., 2010. [Effect of gross saponins of *Tribulus terrestris* on cardiocytes impaired by adriamycin]. Yao xue xue bao = Acta pharmaceutica Sinica 45, 31–36.

CHAPTER

43

Saffron

Maryam Khan[1], *Muhammad Asif Hanif*[1], *Muhammad Adnan Ayub*[2], *Muhammad Idrees Jilani*[3], *Shahzad Ali Shahid Chatha*[4]

[1] Department of Chemistry, University of Agriculture, Faisalabad, Pakistan; [2] Department of Chemistry, University of Okara, Okara, Pakistan; [3] Department of Chemistry, University of Sahiwal, Sahiwal, Pakistan; [4] Department of Chemistry, Government College University, Faisalabad, Pakistan

OUTLINE

Medicinal Plants of South Asia
https://doi.org/10.1016/B978-0-08-102659-5.00043-4

Copyright © 2020 Elsevier Ltd. All rights reserved.

1. BOTANY

1.1 Introduction

Crocus sativus (Saffron) (Fig. 43.1) belongs to family Iridaceae that contains 78 genera and 1750 species. The plants of this family have been used for ornamental and medicinal uses as well as for condiments. The genus *Crocus* contains 85 species, and some of them are ornamental plants having beautiful flowers (Kafi, 2006). Saffron is also famous by its common names in different regions and languages like zafran (in Urdu), kesar (in Hindi), zafferano (in Italian), safran (in French), croci (in Latin), and krokos (in Greek).

FIGURE 43.1 Saffron.

1.2 History/Origin

Saffron derived its name from the Arabic word "zafaran," meaning yellow. Saffron has a long history dating back almost 3000 years, traversing many civilizations, continents, and cultures. It is native to Southwest Asia and was first time cultivated in or near Persia (Deo, 2003; Melnyk et al., 2010). It has also been reported that saffron was first harvested from the wild *Crocus cartwrightianus*, a mutant of which is *C. sativus* (Negbi, 2003). Different theories have been presented about the origin of saffron. However, the documented evidences reveal that it originated in Iran, in Zagross, and Alvand mountains, most probably. The oldest evidence regarding the utilization of saffron is in "Achaemenian," an ancient Persian dynasty.

1.3 Demography/Location

Saffron shows best growth in a Mediterranean climate but can tolerate wide ranges of climate. Iran, Morocco, Greece, Azerbaijan, Pakistan, Italy, and Spain are the largest producers of saffron (Kafi, 2006). In Spain, saffron is classified by its quality in three grades: Mancha, Rio, and Sierra (Lozano et al., 1999). The annual production of saffron in 1991 was estimated to be about 300 tons (300,000 kg) in terms of saffron stigmas. Iran is the leading producer of saffron. In 2005, the annual production of saffron in Iran was 230 tons (230,000 kg) of dry stigmas, which was 93.7% of the total world production of that year. Greece was the second-ranked producer in 2005, producing 5.7 tons (5700.0 kg). Pakistan and Morocco were in the third position with 2.3 tons (2300.0 kg) of production. Smaller scale cultivation of saffron is carried out in Tasmania, Egypt, China, Israel, France, Mexico, New Zealand, Turkey, Central Africa, and California. The yield of saffron greatly depends upon the environmental conditions and cultivation methods.

1.4 Botany, Morphology, Ecology

Saffron is herbaceous plant attaining a height of 20–30 cm. It consists of a solid underground corm terminating at adventitious roots. A number of membranous sheaths called cataphylls enclose and protects leaves, bracts, bracteole, and flowering stalk, which originates from the corm. Leaves are navicular, sword shaped, and 10–18 cm long. Saffron flowers appear before any vegetative growth. It bears usually four flowers that are elongated funnel shaped or cup shape. The flowers are usually purple, white, golden, or blue. Each flower contains three vivid crimson stigmata at the distal end of a carpel. The stigmata are collected, dried, and used as seasoning and coloring agent in foods (Kafi, 2006).

2. CHEMISTRY

Saffron contains a surplus amount of some valuable compounds like vitamin A, vitamin C, and folic acid in addition to proteins, fats, and carbohydrates. Saffron is a rich source of minerals like iron, calcium, magnesium, copper manganese, potassium, phosphorus, zinc, and sodium (Hamidpour et al., 2014). Saffron also contains some vital secondary metabolites, monoterpenoids, flavonoids, carotenoids, and anthocyanins (Rios et al., 1996). Saffron contains different chemical constituents whose concentration fluctuate greatly depending upon the growing conditions (Melnyk et al., 2010). Analyses of chemical composition showed that saffron contains approximately 50%–63% sugars (dextrin, starch, pectin, gums, and pentosans), 12%–13% protein, 10%–12% moisture, 5%–8% fat, 5%–7% minerals, and 5%–10% crude fiber (Rios et al., 1996). The essential oil of saffron stigmas and tepals differs in composition. Stigma essential oil contains crocetin, crocin, picrocrocin, and safranal as the major components (Rubio-Moraga et al., 2010). Tepals essential oil contains kaempferol, quercetin, rutin, hesperidin, and luteolin. It also has been reported that tepals essential oil also contain tannins and anthocyanins (Srivastava et al., 2010). Some bioactive components of saffron are shown in Fig. 43.2.

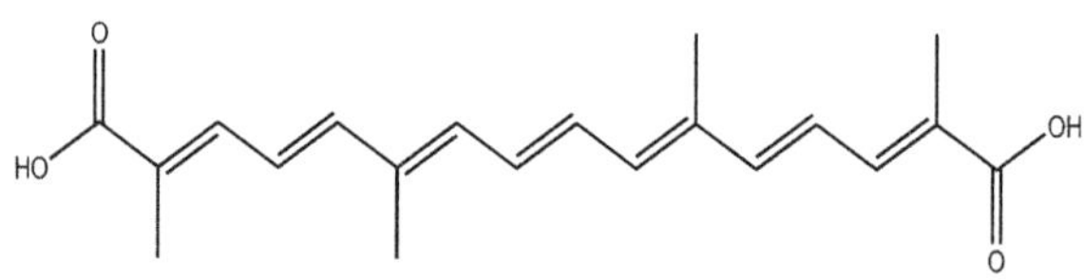

Crocetin

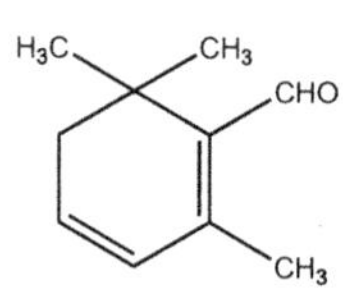

Safranal

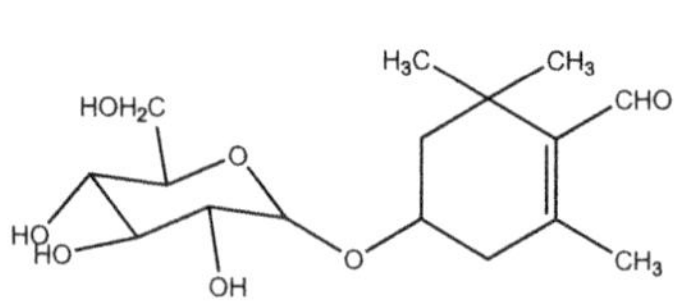

Picrocrocin

FIGURE 43.2 Bioactive components of saffron.

3. POSTHARVEST TECHNOLOGY

Different countries have their own sowing times based on their agro-climatic conditions and weather. In Pakistan and India, those are between the middle of July to end of August; in central Italy, corms are sown in the second half of August, in Spain, after the middle of June, in Greece, between May and September, and in Morocco, from the last week of August to first week of September (Negbi, 1999). However, higher yield with better quality of saffron is obtained with earlier sowing (end of July) than with a later one (end of August) (Gresta et al., 2008). Stigmas are hand-picked from saffron flowers before daylight to preserve the desirable volatile components (Melnyk et al., 2010). Drying of stigmas is a very careful job as incomplete drying may cause total loss of the product due to decomposition and molding. Dried, uncontaminated stigmas are ready to store and market. Safranal is produced from picrocrocin during the drying process of saffron. Safranal is not found in the fresh stigma, and its concentration depends on both the storage conditions and drying processes. Dried saffron can be used for a long period of time (Lozano et al., 2000).

4. PROCESSING

For long term storage of saffron, it is essential to dry it. A maximum of 110°C temperature can be used for drying saffron; however, air heated up to 70°C can also be used for 6 min for this purpose. A dehydrator can also be used for drying of saffron at 48°C for 3 h. A quick and high-temperature drying process produces a brightly colored product, while a poor-quality product is obtained through slow drying. There must be 10% moisture in the final dried product for long-term storage (Kafi, 2006). Steam distillation, supercritical fluid extraction, microsimultaneous steam distillation solvent extraction (MSDE) have also been reported for extraction of essential oil from saffron. Each method gives a different aromatic composition (Tarantilis and Polissiou, 1997). The supercritical fluid extraction has been reported as a nondestructive method for extraction of saffron (Lozano et al., 2000). It is reasonable to use the appropriate blend of solvents for the extraction of essential oil of desired composition.

5. VALUE ADDITION

Saffron along with hematoxylin and phloxine is used to improve animal and human histological staining methods (Negbi, 2003). Saffron has been used from prehistoric times as a spice in many dishes due to its specific aroma, color, and taste. It is the most expensive spice in the world, with a price of nearly $40–50 per gram. 75,000 flowers or 225,000 stigmas of saffron just give one pound of spice. Saffron is widely used in confectionary as well as in alcoholic and nonalcoholic beverages (Hill, 2004; Melnyk et al., 2010). Addition of saffron in foods not only helps to develop aroma and taste but also enhances their medicinal and health attributes.

6. USES

The dried stigmas of saffron are used in industry, with multiple applications such as textile dye, drug, and culinary adjunct. It is used as a food additive, flavoring, and coloring agent and to develop taste (Lozano et al., 1999; Madan et al., 1966). Saffron is used mainly as a spice in foods as coloring and flavoring agent due to the analgesic attributes. Herbal practitioners have been using saffron to cure a number of health disorders (Culleré et al., 2011). It is also used in a number of herbal products. In the recent years, public interest toward natural food additives instead of synthetic chemicals has been developing. The use of saffron as a coloring agent for sausages and oleomargarines is rising, and it is also used in dairy products such as cheese, butter, and ice cream to improve their color and flavor (Kafi, 2006). Crocin, a dark red color component of stigmas of saffron, forms an orange color in water, so it is used as a natural food colorant. Saffron stigmas are also used to dye cloth a yellow color (Melnyk et al., 2010). After separation of styles the remaining flowers of saffron are considered a waste that contain a significant amount of red plant pigment. Anthocyanin, which with other flavonoids, gives a violet color to saffron petioles. It is converted to red color in acidic conditions after extraction, so it is a rich source of red color to be used for various applications. Similarly, a blue or green dye can also be made from the petals of the flowers (Kafi, 2006).

Saffron has been used as a traditional medicine for expectorant exhilarating agent, relaxant, spasm calmative, and digestion stimulant. Saffron has also been used for the treatment of bloody diarrhea, measles, fever, hepatitis, urine infections, diabetes, cholera, and skin diseases. Essential oil of saffron is a relaxant and useful in insomnia of nervous origin. Saffron is also found to have useful effects on bronchitis and lung diseases. In India, saffron is used to cure kidney diseases, liver vesical diseases, and cholera. External application of saffron tincture is famous with herbal practitioners for dermal diseases (Kafi, 2006).

7. PHARMACOLOGICAL USES

7.1 Antioxidant Activity

The antioxidant activity of stigma, corm, and leaf of saffron has been assessed by many certified assays such as DPPH radical scavenging assay, nitroblue tetrazolium reduction, and ferric reducing power assay. It has also been presented by researchers that ethanolic extracts of saffron showed higher antioxidant activity than aqueous extracts. Comparative investigation also reports that the highest antioxidant activity was presented by ethanolic extract of stigma followed by corm and leaf. It has also been reported that antioxidant activity of various parts of saffron is concentration dependent. The antioxidant activity increases with increasing concentration of the extract and vice versa. It was found that saffron extract made plants more tolerant to stress by reducing methyl viologen–induced lipid peroxidation, chlorophyll damage, and protein oxidation. While in bacteria and yeast, tolerance against H_2O_2-mediated oxidative stress has been observed by application of saffron extract. The antioxidant activity of saffron extracts is due to the presence of flavonoids, phenolics, and carotenoids in saffron. Phenolic compounds are well recognized to scavenge the free radicals by their hydroxyl groups, while flavonoids suppress reactive oxygen formation, scavenge reactive species, and chelate trace elements responsible for free-radical production, thereby playing role in antioxidant potential (Baba et al., 2015). Antioxidant potential of aqueous extract of saffron in human vascular endothelial cells has also been reported. In addition to acting as a direct antioxidant agent, saffron extract also behaves as a modulator of micro-RNAs and RNAs profile on human vascular endothelial cells under stress conditions, so it could be used in the treatment of oxidative stress (Rodriguez-Ruiz et al., 2016). Safranal and crocin in saffron are reported to have antioxidant activity by trapping free radicals (Chen et al., 2008).

7.2 Antiulcer Potential

Saffron plays a significant role in gastric disorders. Some scientific investigations claim that saffron showed significant antiulcer potential in mice. Ethanolic saffron extract, commercial safranal, and commercial crocin were used and reported to exhibit significant antiulcer potential. All three treatments reduced ulcer formation (Kianbakht and Mozaffari, 2009). Xu et al. (2009) also reported the antiulcer effect of crocin, a carotenoid in the saffron extract. Indomethacin (an antiinflammatory drug) causes peptic ulcers. Crocin stopped damage to the stomach mucosa; however, damage occurred in the control rats treated with indomethacin (Xu et al., 2009). Saffron was also reported to improve digestion.

Aqueous extract of saffron was reported to increase pepsin and gastric acid secretions in mice, which in turns helps in protein digestion by lowering the pH. The possible mechanism was thought to be the activation of nitric oxide synthetase by saffron extract, which increases release of histamine from cells, thereby increasing gastric acid and pepsin secretions (Nabavizadeh et al., 2009).

7.3 Anticancer Activity

Antitumor activity of ethanol extract of saffron stigma has been investigated on different cancer cell lines (Tavakkol-Afshari et al., 2008). Ethanol saffron extract showed selective cytotoxic effect against epithelial-like human hepatocellular carcinoma cells (HepG2) as well as human cervical carcinoma cells (HeLa) but remain nontoxic toward normal mouse fibroblast cells (L929). Saffron extract has also been reported to show antitumor effect by stimulating apoptosis in malignant cells (Tavakkol-Afshari et al., 2008). The effect of saffron extract against three colorectal cancer cell lines (HCT-116, SW-480, and HT-29) was studied, and the results indicated its anticancer properties. Crocin, an important component of saffron, significantly reduced the proliferation of colorectal cancer cells, showing no toxicity to normal cells (Aung et al., 2007). A significant antitumor effect of saffron extract against breast cancer and pancreatic cancer has also been reported. Crocin and crocetin were found to be the bioactive compounds responsible for breast and pancreatic antitumor potential (Chryssanthi et al., 2007; Dhar et al., 2009). It is found that carotenoids of saffron (crocetin, crocin, and dimethylcrocetin) bind themselves to DNA grooves, resulting in conformational fluctuations in DNA. Crocin caused change in gene expression inducing apoptotic death of cancerous cells (Bathaie et al., 2007). Crocetin show antitumor effect by reducing synthesis of DNA, RNA, and protein in neoplastic cells, RNA polymerase II inhibition, and interaction with histone H1 and H1-DNA structures. Safranal also was found to show mild antitumor activity (Milajerdi et al., 2016).

7.4 Cardioprotective Activity

Many bioactive components of saffron like crocetin, crocin, and safranal have reported to show cardioprotective activity. Crocetin, the natural carotenoid antioxidant in saffron, was administrated in white rabbits to determine its effects on atherosclerosis. Three different diets, namely a standard, a high lipid (HLD), or a high lipid plus crocetin diet, were given to rabbits for 8 weeks. The rabbits in the HLD group were found to develop hypercholesterolemia and atherosclerosis. However, the high lipid plus crocetin diet group showed fewer harmful effects than the high lipid diet. Although crocetin did not reduce the plasma lipid levels, it was found to decrease aorta cholesterol deposits, atherosclerotic lesions, atheroma, and foam cells in the crocetin-fed group. It was reported that crocetin suppresses the activation of nuclear factor kappa B in aortas, resulting in decreases in vascular cell adhesion molecule-1 (VCAM-1) expression. VCAM-1 is responsible for increased lipids within arterial walls, so its reduction helps in reducing lipid deposition in arterial walls, which in turns reduces atherosclerosis and aortic lesions (Zheng et al., 2005).

Sheng et al. (2006) reported the crocin antiatherosclerotic properties. Crocin was reported to inhibit the serum triglycerides and low-density lipids cholesterol and also promoted the fecal excretion of fat and cholesterol. This resulted in decrease cholesterol level and atherosclerosis in the crocin-treated group. Crocin was also found to inhibit pancreatic and gastric lipase activity. Pancreatic lipase has a role in fat absorption, and inhibition in its activity resulted in low lipid absorption (Sheng et al., 2006). Safranalor, the aqueous extract of saffron, was reported to reduce level of malondialdehyde content lipid peroxidation and level of malondialdehyde in heart tissue and also suppressed histopathologic changes in heart tissue, resulting in minimal chances of cardiac diseases (Mehdizadeh et al., 2013).

7.5 Role in Insulin Resistance

To evaluate the effect of crocetin against insulin resistance, dexamethasone (that causes insulin resistance) and dexamethasone plus crocetin were administered in rats for 6 weeks. The dexamethasone plus crocetin—treated group showed significantly lower levels of serum insulin, triglycerides, tumor necrosis factor-α (TNF-α), and free fatty acids than the pure dexamethasone group. In case of insulin resistance, TNF-α is overexpressed, resulting in reduced glucose transporter 4 activity, causing decreased insulin-stimulated glucose absorption. In the crocetin-treated group, a reduction in TNF-α and serum insulin levels is evidence of the role of crocetin against insulin resistance (Xi et al., 2005). In another study, Xi et al. (2007) reported the effect of crocetin extracted from saffron against insulin resistance. Rats that were fed a high fructose diet were administrated different concentration doses of crocetin. Results showed a decrease in epididymal adipose tissue, blood pressure, serum triglycerides, low density and high-density lipid cholesterol, serum insulin, and free fatty acids in the high-dose crocetin group. However the fructose-fed control group showed opposite results compared to the crocetin group, showing potential of crocetin against insulin resistance (Xi et al., 2007).

7.6 Antidepressant Activity

To check the antidepressant effect of ethanolic extract of saffron stigma, saffron stigma extract and an antidepressant drug, imipramine, were compared against depression. The antidepressant activity was checked by Hamilton Depression Rating Scale (HAM-D). According to the results, both saffron extract and the drug imipramine showed significant antidepressant activity, with no significant difference. Hence, saffron can be used as a potential antidepressant drug (Akhondzadeh et al., 2004). In another study, two groups of people were provided 30 mg/day of saffron and placebo for 6 weeks. The antidepressant activity was evaluated using the HAM-D. It was reported that saffron extract showed significant antidepressant activity compared to the placebo group. Crocin was mainly consider to show antidepressant effect in saffron (Akhondzadeh et al., 2005). Antidepressant effects of saffron petal extracts were also evaluated in humans. Two groups of people were administered ethanolic petal extract (30 mg/day) and placebo capsules for 8 weeks, and the antidepressant activity was evaluated by HAM-D. The results showed lower ratings for HAM-D in the saffron group, showing its antidepressant effect (Moshiri et al., 2006). In another study, the effect of petal extract (30 mg/day) was compared to fluoxetine (20 mg/day) (an antidepressant drug), and results showed a significant decrease in HAM-D score, with no significant results in both groups (Basti et al., 2007). Although crocin is not present in petal extract, kaempferol, a natural flavonoid present in petals extract of saffron, was found to show antidepressant effects (Hadizadeh et al., 2010).

7.7 Effects on Premenstrual Syndrome

Premenstrual syndrome is characterized as a group of emotional disorders occurring in the second half of the menstrual cycle. A test was conducted in two groups of women treated with either 30 mg/day saffron or placebo. The placebo-treated group was considered the control. Premenstrual Daily Symptoms questionnaire and HAM-D were used to evaluate the results, which show improved results for both tests for saffron-treated women compared to pretreatment symptoms. Results were also quite significantly improved in comparison to the control group (Agha-Hosseini et al., 2008).

7.8 Antianxiety and Hypnotic Activity

Saffron has a potential to cure anxiety. The antianxiety activity of crocin compared to diazepam (antianxiety drug) was reported by using a light/dark test on rats. The rats were placed in a two-chambered box, having one light and one dark chamber, and time in each box spent by rats was measured as a gauge of anxiety. There was idea that rat experiencing high anxiety would choose the dark area. Results showed that intraperitoneal-administered crocin reduced anxiety comparable to diazepam. It is revealed by increased inactivity in entering the dark chamber and enhanced time spent in the light chamber (Pitsikas et al., 2008).

Insomnia is difficulty in staying asleep, causing anxiety and tiredness (Morin, 2000). It was reported that there is an effect of saffron, safranal, and crocin on the hypnotic activity, locomotor activity, anxiolytic activity, and coordination and balance in mice. Different concentrations of aqueous extract of saffron, crocin, and safranal were administrated in mice intraperitoneally. Pentobarbital sleeping time test, an elevated plus maze test, an open field test, and a rotarod test were used to test hypnotic activity, anxiolytic activity, locomotor activity, and motor coordination and balance, respectively. Doses of 56 and 80 mg/kg saffron in mice showed a significant anxiolytic effect, which was found to exist in saffron due to safranal, while a dose of 560 mg/kg saffron showed high hypnotic activity. Though, crocin showed no noticeable hypnotic, anxiolytic, or myorelaxation effects (Hosseinzadeh and Noraei, 2009).

7.9 Hypolipidemic Effects

Crocin was found to decrease the cholesterol levels in hyperlipidemic rats that were fed for 2 months with excessive cholesterol. It was reported that crocin repressed the cholesterol and dietary fat absorption. Furthermore, crocin was also found to increase the fecal excretion of fat and cholesterol in rats, with no effect on the elimination of bile acids. Crocin also exhibited inhibitory action on pancreatic lipase enzyme, causing inhibition of dietary fat absorption (Sheng et al., 2006). Due to hypolipidemic effect, saffron has a potential as an antiobesity drug (Nam et al., 2010; Poma et al., 2012). It acts as an antiobesity drug by (1) inhibiting pancreatic lipase activity, which blocks the dietary fat digestion and hence decreases calorie intake; (2) acting as an antioxidant, thereby suppressing the adipocyte differentiation; (3) increasing satiety by raising the level of neurotransmitters, hence suppressing food intake; and (4) enhancing lipid and glucose metabolism (Mashmoul et al., 2013).

8. SIDE EFFECTS AND TOXICITY

Some possible side effects include dry mouth, anxiety, dizziness, drowsiness, nausea, change in appetite, and headache. Allergic reactions can occur in some people. High doses can cause poisoning, including yellow appearance of the skin, eyes, and mucous membranes; vomiting; dizziness; bloody diarrhea; bleeding from the nose, lips, and eyelids; numbness; and other serious side effects. Doses of 12–20 g can cause death.

References

Agha-Hosseini, M., Kashani, L., Aleyaseen, A., Ghoreishi, A., Rahmanpour, H., Zarrinara, A., Akhondzadeh, S., 2008. *Crocus sativus* L.(saffron) in the treatment of premenstrual syndrome: a double-blind, randomised and placebo-controlled trial. BJOG: An International Journal of Obstetrics and Gynaecology 115, 515–519.

Akhondzadeh, S., Fallah-Pour, H., Afkham, K., Jamshidi, A.-H., Khalighi-Cigaroudi, F., 2004. Comparison of *Crocus sativus* L. and imipramine in the treatment of mild to moderate depression: a pilot double-blind randomized trial [ISRCTN45683816]. BMC Complementary and Alternative Medicine 4, 1.

Akhondzadeh, S., Tahmacebi-Pour, N., Noorbala, A.A., Amini, H., Fallah-Pour, H., Jamshidi, A.H., Khani, M., 2005. *Crocus sativus* L. in the treatment of mild to moderate depression: a double-blind, randomized and placebo-controlled trial. Phytotherapy Research 19, 148–151.

Aung, H., Wang, C., Ni, M., Fishbein, A., Mehendale, S., Xie, J., Shoyama, A., Yuan, C., 2007. Crocin from *Crocus sativus* possesses significant anti-proliferation effects on human colorectal cancer cells. Experimental Oncology 29, 175.

Baba, S.A., Malik, A.H., Wani, Z.A., Mohiuddin, T., Shah, Z., Abbas, N., Ashraf, N., 2015. Phytochemical analysis and antioxidant activity of different tissue types of *Crocus sativus* and oxidative stress alleviating potential of saffron extract in plants, bacteria, and yeast. South African Journal of Botany 99, 80–87.

Basti, A.A., Moshiri, E., Noorbala, A.-A., Jamshidi, A.-H., Abbasi, S.H., Akhondzadeh, S., 2007. Comparison of petal of *Crocus sativus* L. and fluoxetine in the treatment of depressed outpatients: a pilot double-blind randomized trial. Progress in Neuro-Psychopharmacology and Biological Psychiatry 31, 439–442.

Bathaie, S.Z., Bolhasani, A., Hoshyar, R., Ranjbar, B., Sabouni, F., Moosavi-Movahedi, A.-A., 2007. Interaction of saffron carotenoids as anticancer compounds with ctDNA, Oligo (dG. dC) 15, and Oligo (dA. dT) 15. DNA and Cell Biology 26, 533–540.

Chen, Y., Zhang, H., Tian, X., Zhao, C., Cai, L., Liu, Y., Jia, L., Yin, H.-X., Chen, C., 2008. Antioxidant potential of crocins and ethanol extracts of Gardenia jasminoides ELLIS and *Crocus sativus* L.: a relationship investigation between antioxidant activity and crocin contents. Food Chemistry 109, 484–492.

Chryssanthi, D.G., Lamari, F.N., Iatrou, G., Pylara, A., Karamanos, N.K., Cordopatis, P., 2007. Inhibition of breast cancer cell proliferation by style constituents of different Crocus species. Anticancer Research 27, 357–362.

Culleré, L., San-Juan, F., Cacho, J., 2011. Characterisation of aroma active compounds of Spanish saffron by gas chromatography–olfactometry: quantitative evaluation of the most relevant aromatic compounds. Food Chemistry 127, 1866–1871.

Deo, B., 2003. Growing saffron—the world's most expensive spice. Crop & Food Research 20, 1–4.

Dhar, A., Mehta, S., Dhar, G., Dhar, K., Banerjee, S., Van Veldhuizen, P., Campbell, D.R., Banerjee, S.K., 2009. Crocetin inhibits pancreatic cancer cell proliferation and tumor progression in a xenograft mouse model. Molecular Cancer Therapeutics 8, 315–323.

Gresta, F., Lombardo, G., Siracusa, L., Ruberto, G., 2008. Saffron, an alternative crop for sustainable agricultural systems. A review. Agronomy for Sustainable Development 28, 95–112.

Hadizadeh, F., Khalili, N., Hosseinzadeh, H., Khair-Aldine, R., 2010. Kaempferol from saffron petals. Iranian Journal of Pharmaceutical Research 2, 251–252.

Hamidpour, R., Hamidpour, S., Hamidpour, H., Shahlari, M., 2014. Effect of *Crocus sativus* and its active compounds for the treatment of several diseases: a review. International Journal of Case Reports and Images 4, 666–670.

Hill, T., 2004. Contemporary Encyclopedia of Herbs & Spices. J. Wiley.

Hosseinzadeh, H., Noraei, N.B., 2009. Anxiolytic and hypnotic effect of *Crocus sativus* aqueous extract and its constituents, crocin and safranal, in mice. Phytotherapy Research 23, 768–774.

Kafi, M., 2006. Saffron (*Crocus sativus*): Production and Processing. Science Publishers.

Kianbakht, S., Mozaffari, K., 2009. Effects of saffron and its active constituents, crocin and safranal, on prevention of indomethacin induced gastric ulcers in diabetic and nondiabetic rats. Journal of Medicinal Plants 1, 30–38.

Lozano, P., Castellar, M., Simancas, M., Iborra, J., 1999. A quantitative high-performance liquid chromatographic method to analyse commercial saffron (*Crocus sativus* L.) products. Journal of Chromatography A 830, 477–483.

Lozano, P., Delgado, D., Gomez, D., Rubio, M., Iborra, J., 2000. A non-destructive method to determine the safranal content of saffron (*Crocus sativus* L.) by supercritical carbon dioxide extraction combined with high-performance liquid chromatography and gas chromatography. Journal of Biochemical and Biophysical Methods 43, 367–378.

Madan, C., Kapur, B., Gupta, U., 1966. Saffron. Economic Botany 20, 377–385.

Mashmoul, M., Azlan, A., Khaza'ai, H., Yusof, B.N.M., Noor, S.M., 2013. Saffron: a natural potent antioxidant as a promising anti-obesity drug. Antioxidants 2, 293–308.

Mehdizadeh, R., Parizadeh, M.R., Khooei, A.-R., Mehri, S., Hosseinzadeh, H., 2013. Cardioprotective effect of saffron extract and safranal in isoproterenol-induced myocardial infarction in wistar rats. Iranian Journal of Basic Medical Sciences 16, 56–63.

Melnyk, J.P., Wang, S., Marcone, M.F., 2010. Chemical and biological properties of the world's most expensive spice: Saffron. Food Research International 43, 1981–1989.

Milajerdi, A., Djafarian, K., Hosseini, B., 2016. The toxicity of saffron (*Crocus sativus* L.) and its constituents against normal and cancer cells. Journal of Nutrition & Intermediary Metabolism 3, 23–32.

Morin, C.M., 2000. The nature of insomnia and the need to refine our diagnostic criteria. Psychosomatic Medicine 62, 483–485.

Moshiri, E., Basti, A.A., Noorbala, A.-A., Jamshidi, A.-H., Abbasi, S.H., Akhondzadeh, S., 2006. *Crocus sativus* L.(petal) in the treatment of mild-to-moderate depression: a double-blind, randomized and placebo-controlled trial. Phytomedicine 13, 607–611.

Nabavizadeh, F., Salimi, E., Sadroleslami, Z., Karimian, S.M., Vahedian, J., 2009. Saffron (*Crocus sativus*) increases gastric acid and pepsin secretions in rats: role of nitric oxide (NO). African Journal of Pharmacy and Pharmacology 3, 181–184.

Nam, K.N., Park, Y.-M., Jung, H.-J., Lee, J.Y., Min, B.D., Park, S.-U., Jung, W.-S., Cho, K.-H., Park, J.-H., Kang, I., 2010. Anti-inflammatory effects of crocin and crocetin in rat brain microglial cells. European Journal of Pharmacology 648, 110–116.

Negbi, M., 1999. Saffron Cultivation, Saffron: *Crocus sativus* L. CRC Press, pp. 1–17.

Negbi, M., 2003. Saffron: *Crocus sativus* L. CRC Press.

Pitsikas, N., Boultadakis, A., Georgiadou, G., Tarantilis, P., Sakellaridis, N., 2008. Effects of the active constituents of *Crocus sativus* L., crocins, in an animal model of anxiety. Phytomedicine 15, 1135–1139.

Poma, A., Fontecchio, G., Carlucci, G., Chichiricco, G., 2012. Anti-inflammatory properties of drugs from saffron crocus. Anti-Inflammatory & Anti-allergy Agents in Medicinal Chemistry 11, 37–51.

Rios, J., Recio, M., Giner, R., Manez, S., 1996. An update review of saffron and its active constituents. Phytotherapy Research 10, 189–193.

Rodriguez-Ruiz, V., Barzegari, A., Zuluaga, M., Zunooni-Vahed, S., Rahbar-Saadat, Y., Letourneur, D., Gueguen, V., Pavon-Djavid, G., 2016. Potential of aqueous extract of saffron (*Crocus sativus* L.) in blocking the oxidative stress by modulation of signal transduction in human vascular endothelial cells. Journal of Functional Foods 26, 123–134.

Rubio-Moraga, A., Trapero, A., Ahrazem, O., Gómez-Gómez, L., 2010. Crocins transport in *Crocus sativus*: the long road from a senescent stigma to a newborn corm. Phytochemistry 71, 1506–1513.

Sheng, L., Qian, Z., Zheng, S., Xi, L., 2006. Mechanism of hypolipidemic effect of crocin in rats: crocin inhibits pancreatic lipase. European Journal of Pharmacology 543, 116–122.

Srivastava, R., Ahmed, H., Dixit, R., Saraf, S., 2010. *Crocus sativus* L.: a comprehensive review. Pharmacognosy Reviews 4, 200.

Tarantilis, P.A., Polissiou, M.G., 1997. Isolation and identification of the aroma components from saffron (*Crocus sativus*). Journal of Agricultural and Food Chemistry 45, 459–462.

Tavakkol-Afshari, J., Brook, A., Mousavi, S.H., 2008. Study of cytotoxic and apoptogenic properties of saffron extract in human cancer cell lines. Food and Chemical Toxicology 46, 3443–3447.

Xi, L., Qian, Z., Shen, X., Wen, N., Zhang, Y., 2005. Crocetin prevents dexamethasone-induced insulin resistance in rats. Planta Medica 71, 917–922.

Xi, L., Qian, Z., Xu, G., Zheng, S., Sun, S., Wen, N., Sheng, L., Shi, Y., Zhang, Y., 2007. Beneficial impact of crocetin, a carotenoid from saffron, on insulin sensitivity in fructose-fed rats. The Journal of Nutritional Biochemistry 18, 64–72.

Xu, G.-L., Li, G., Ma, H.-P., Zhong, H., Liu, F., Ao, G.-Z., 2009. Preventive effect of crocin in inflamed animals and in LPS-challenged RAW 264.7 cells. Journal of Agricultural and Food Chemistry 57, 8325–8330.

Zheng, S., Qian, Z., Tang, F., Sheng, L., 2005. Suppression of vascular cell adhesion molecule-1 expression by crocetin contributes to attenuation of atherosclerosis in hypercholesterolemic rabbits. Biochemical Pharmacology 70, 1192–1199.

CHAPTER

44

Sesame

Ayesha Mushtaq[1], Muhammad Asif Hanif[1], Muhammad Adnan Ayub[2], Ijaz Ahmad Bhatti[1], Muhammad Idrees Jilani[3]

[1] Department of Chemistry, University of Agriculture, Faisalabad, Pakistan; [2] Department of Chemistry, University of Okara, Okara, Pakistan; [3] Department of Chemistry, University of Sahiwal, Sahiwal, Pakistan

OUTLINE

Medicinal Plants of South Asia
https://doi.org/10.1016/B978-0-08-102659-5.00044-6

Copyright © 2020 Elsevier Ltd. All rights reserved.

1. BOTANY

1.1 Introduction

Sesame seed (*Sesamum indicum* L.) (Fig. 44.1) is one of the world's most essential and traditional oil seed–producing crops (Shyu and Hwang, 2002). It has been widely used for many years as a seed of global significance for edible oil, cake, flour, paste, and confectionary purposes due to its highly stable oil contents, nutritious proteins, and savory, nutty roasted flavor (Hama, 2017). Sesame belongs to genus *Sesamum* (Hassan, 2012).

FIGURE 44.1 Sesame plants and seeds.

The genus *Sesamum* (family Pedaliaceae) is comprised of about 37 species. *S. indicum* is a well-known member of this genus and is widely cultivated. Other species including *Sesamum angustifolium*, *Sesamum auriculatum*, *Sesamum prostratum*, and *Sesamum radiatum* are cultivated at a smaller scale in Pakistan and other south Asian countries (Hwang, 2005). Sesame seeds contain a huge amount of oil (50%–60%), due to which it is regarded as the queen of oil seeds (Pham et al., 2010). Sesame is grown over 50 countries in the world (Akbar et al., 2011). China, India, Myanmar (Burma), Uganda, Turkey, Nigeria, Thailand, Bangladesh, Ethiopia, Mexico, and Pakistan are the major producers of sesame. Sesame is known by different names depending where you are in the world. It is usually known as til (Urdu, Hindi), hu-ma (Chinese), sesame (French), goma (Japanese), gergelim (Portuguese), and ajonjoli (Spanish) (Anilakumar et al., 2010).

1.2 History

S. indicum L. (sesame) is an ancient crop and has been cultivated in Asia for more than 5000 years (Akbar et al., 2011). Archeologic evidences showed that sesame seeds were cultivated in the Indus Valley between 1750 and 2250 BC, and more recent charred sesame seeds have been found in Shahi-Tump and Qalat in Pakistan (Nadeem et al., 2015).

1.3 Demography

Sesame grows well in tropical and subtropical regions about 40°N to 40°S latitude (Stamatov and Deshev, 2014). This annual crop is cultivated mainly in areas having well-drained soils and moderate rainfall with long growing seasons (Anilakumar et al., 2010). It requires long periods of sunshine and generally it is a short-day plant. Different varieties are observed that are unaffected by the length of the day. It is propagated by seeds sown in spring and takes about 4 months for the seeds to ripen fully. Sesame is frost susceptible, so its cultivation is restricted to frost-free regions of the world. This crop is well suited for cultivation in crop rotation (Anilakumar et al., 2010). India and China are largest producers, both with an annual harvest of about 750,000 tons, followed by Sudan (300,000 tons) and Myanmar (425,000 tons). The average yield of sesame seed is around 400 kg/ha worldwide. Sesame seed oil, with an annual production of 760,000 MT in 2003, is the 12th largest edible oil produced in the world, higher in quantity than safflower and olive oil. The production of sesame oil increased 20% in the last 10 years. China has almost doubled the production of sesame oil (from 142,000 to 210,000 MT), whereas India has decreased the production by 44% (from 236,000 to 131,000 MT) in that period.

1.4 Botany, Morphology, Ecology

Sesame is an annual, erect herb that may grow between 0.5 and 2.5 m in length, depending upon cultivar and growth environment. It has 26 somatic chromosomes (2n = 26). The stems may have branches and are obtusely quadrangular, longitudinally furrowed, and densely hairy. The degree and type of branching of the stem are also important varietal characters. Sesame leaves are hairy on both sides and are highly variable in shape and size, not only among different varieties but also on the same plant. The lower leaves are opposite, ovate, dull green in color, 3–17.5 cm in length and 1–7 cm in width, and the petiole is 5 cm long. The upper leaves are alternate or subopposite, lanceolate, with a few coarse teeth, and petiole is (1–2) cm long. Large, white, and bell-shaped flowers are present, which are bilaterally symmetrical (zygomorphic), geniculate, and short-pedicelled. They may born singly or two to three together, in the axils of upper leaves (Hwang, 2005). The calyx is small and five parted, and the segments are ovate-lanceolate and 0.5–0.6 cm long. The corolla is 3–4 cm long, five-lobbed, two-lipped, widened upward, purple, white, or pinkish in color, with spots, stripes, and blotches (yellow or purple) on the inner surface. The stamens are four in number, didynamous, and inserted on the base of the corolla; the anthers are sagittate. The ovary is superior and two celled. The fruit of sesame is a capsule (2–5 cm long and 0.5–2 cm in diameter), and it is erect, oblong, brown or purple in color, quadrilateral in section, intensely grooved with a short, triangular beak. The capsules with a wider diameter will usually have higher rows of seeds, and the total number of seeds per capsule can be as high as 100–200.

Sesame seeds are small (1.5–2 mm in width and 3–4 mm in length), smooth or reticulate, flat and ovate. Seed color varies from red, white, yellow, gray, brown, to black. The weight of 1000 seeds is around 2.5–3.5 g. Sesame seeds consists of testa (exo and endo), endosperm, and cotyledon. The oil drops are located in the cotyledon. It is generally believed that the light-colored seeds with thin coats are higher in quality and oil content than the dark-colored seeds (Hwang, 2005).

Sesame is considered a tropical and subtropical crop, and it is usually present below 1250-m altitudes, while some cultivars are locally adapted up to 1500 m above sea level (Adam, 2015). Generally, sesame is a short-day plant that may grow in long-day areas. Sesame is a drought-tolerant crop, but it is very intolerant to waterlogging and excessive rainfall (Adam, 2015; Nadeem et al., 2015). The crop is usually grown in areas having annual rainfall of 625–1100 mm. The best soil temperature is 20°C, and air temperature should not be lower than 30°C. Sesame can be grown in a variety of soils, but it prefers well-drained, fertile, and neutral to slightly alkaline (pH 5–8) soils (Langham et al., 2010).

2. CHEMISTRY

Sesame seeds are generally included in healthy foods. They contain many vitamins, nutrients, antioxidants, and minerals that are essential for human health (Anilakumar et al., 2010). Sesame seed is comprised of fats (45%–50%), proteins (20%–25%), carbohydrates (3%–14%), and ash (3%–5%) contents. Sesame seed has about 17% seed weight as hull, which is rich in oxiric acid (2%–3%), crude fiber, and calcium. Oxiric acid may form complexes with calcium. Therefore, it is recommended to remove sesame seed hulls before using sesame seed meal as human food. After proper dehulling of sesame seeds, the concentration of oxiric acid can be reduced to 0.25% of the seed weight. Dehulling of seeds decreases the concentrations of carbohydrate, fiber, and ash contents. However, fat and protein contents are raised during this process (Hwang, 2005). *S. indicum* has a characteristic aromatic odor because of the presence of essential oil, which is mainly confined in the green leaves. This volatile oil is chiefly comprised of the most potent antioxidants such as sesamol, sesamolin, and sesamin, glycerol esters of different fatty acids, and lignans and myristic acid. Oil extracted from the seeds is called fixed oil and is mainly composed of triterpenes, triterpene alcohols, tocopherols, and sesame lignans. Campesterol, stigmasterol, cholesterol, Δ^5-avenasterol, Δ^7-ave-nasterol, β-sitosterol, and Δ^7-stigmasterol are major sterols found in sesame seed oil. Among them, β-sitosterol is present in abundance. Besides the essential or fixed oils, sesame seeds also contain α, γ, and δ-tocopherols. Hence, sesame is considered an important source of phytonutrients such as omega-6 fatty acids, flavonoid phenolic antioxidants, vitamins, and dietary fiber. The information available about the chemical properties of sesame seeds is centered on the numerous studies that have been conducted in various parts of the world. Therefore, it is expected that chemical ingredients of sesame may differ due to geographic influences. Sesamolin is a characteristic constituent of sesame seeds and is not found in other species; however, sesamin is present in all species. Other lignans including sesamol, sesaminol and sesamolinol are present in trace amount. Lignan glycosides (glycosylated forms of lignans) are found in sesame meal. Sesaminol glycoside, sesamolinol glycoside, and pinoresinol glycoside are the major lignan glycosides in sesame (Hwang, 2005). Major fatty acids present in sesame seed oil are oleic (C18:1), linoleic (C18:2), stearic (C18:0), and palmitic acids (C16:0) (Dawodu et al., 2014). Structures of some important chemical compounds found in sesame seed oil are shown in Fig. 44.2.

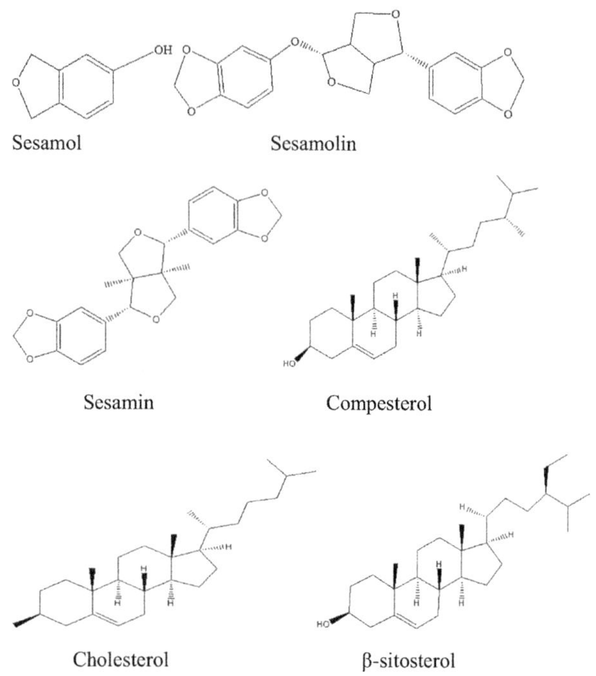

FIGURE 44.2 Structures of some important chemical compounds found in sesame seed oil.

3. POSTHARVEST TECHNOLOGY

Sesame seeds are harvested when the color of 50% of the capsules changes from green to yellow. Other signs of optimum harvesting time include lowest capsule turns brown, stem and leaves turn yellow and begin to fall off, and end of blossoming (Terefe et al., 2012). Artificial drying methods must be used if the critical 6% moisture contents cannot be obtained through sun drying, because sesame seeds become docked above 5.1% for moisture weight, and above 6.9% moisture, there are quality discounts. After the seeds have been harvested and dried, the storage bags must be tested and free of insects. Packed seeds must be stored in the absence of light. Storage temperature should be less than 18°C.

4. PROCESSING

Three major types of oils processed from sesame are roasted sesame oil (obtained after treatment of roasted seeds), small mill oil (obtained from dehulled seeds), and refined sesame oil (obtained from unroasted sesame seeds). Refined sesame oil (food-grade oil) is produced after alkali refining, deodorization, and decolorization (bleaching) of crude sesame oil. Sesame seed cakes are obtained after oil extraction still contains residual oil (18%–20%), which can be extracted by solvent extraction technique (Yoshida et al., 2000).

After oil removal, sesame cake can be processed into food-grade sesame flour. Prior to oil extraction, sesame seeds are roasted at a temperature range of 140–200°C to obtain roasted sesame oil. Roasted sesame oil having strong characteristic essence of roasted sesame seeds is believed to be beneficial to health, and it is most commonly consumed in Japan, Korea, and China.

5. VALUE ADDITION

Dehulled seeds are utilized to enhance aesthetic value, texture, and taste of various bakery products such as cookies, bread, sesame bars, and bread sticks (Tunde-Akintunde et al., 2012). Flavor becomes more pronounced when seeds are roasted gently for a short time period over low flame. Seeds are also used for preparing tahini (fine paste of ground seeds), which is used as an ingredient in halva (in Greece), sprinkled on the bread surface (Netherland and Germany), and for making pastries, cakes, crackers, and rolls in commercial bakeries. In the Middle East and South and East Asia, sesame seeds are mixed with syrup or honey for making candies. In Japan, starch and sesame paste are mixed and used for the preparation of *goma-dofu*. Sesame seeds are also used for the preparation of margarine and sprinkled over desserts and salads. It can be used in other dishes such as East Asian and Mexican cuisines. In industries, sesame seed oil can be used as a solvent in intravenous drip solutions or in injected drugs, as a cosmetics carrier oil, and to coat stored grains to avoid weevil attacks. Seed oil also shows synergic effect with some insect repellents. Low-grade sesame oil can be used in illuminants, soaps, and lubricants. Seed oil is also used in the preparation of paints, pharmaceuticals, and inks. The oil also has additional use in the manufacture of perfumes, moisturizers, cosmetics (skin conditioning agents, hair preparations, bath oils, hand products, and make-up), insecticides, and paints and varnishes. Sesame oil can also be used as alternative diesel fuel by mixing with sodium hydroxide and methanol. However, all of these uses are rather insignificant in terms of the quantities used.

6. USES

Sesame seeds have been widely employed in culinary as well as traditional medicines for their nutritive, preventive, and curative properties. Sesame seed oil has been used as a laxative due to its emollient, nutritive, and demulcent properties (Anilakumar et al., 2010). The leaves are rich in a gummy matter; when it is mixed with water, a rich bland mucilage is formed that is used for the treatment of infant cholera, diarrhea, dysentery, cataract, boils, carbuncle, menstrual irregularities, polyurea, stomach trouble, serious burns, skin diseases, alopecia, and as a tonic (Morton, 1990).

7. PHARMACOLOGICAL USES

7.1 Antioxidant Activity

Sesame increases vitamin E recycling, improves functions of the liver, and provides defense against oxidative stress induced by alcohol. Sesamin increases the levels of high-density lipoproteins while decreasing the cholesterol levels (Anilakumar et al., 2010). Sesame oil enhances the detoxification of chemicals in the liver, minimizes the chances of mammary tumors induced by chemicals, and provide defense against oxidative stress, which is involved in the pathogenesis of endotoxin intoxication (Anilakumar et al., 2010). Oxidative stress is mainly caused by reactive oxygen intermediates. Reactive oxygen intermediates such as H_2O_2, nitric oxide, free radicals, and singlet oxygen all are mediators of cell damage and play a putative role in oxidative stress in endotoxin intoxication (Takashi et al., 2003). The effects of ethanolic extract of sesame coat were investigated against nitric oxide production and oxidation of low-density lipoprotein (LDL). The results revealed that extract at the doses of 0.01–0.8 mg/mL was effective against cellular injury induced by H_2O_2 and LDL oxidation induced by copper; it shows that ethanolic extract has a protective effect on life molecules and generation of inflammatory mediators in vitro (Wang et al., 2007). Clinically, it was found that consumption of sesame oil increases glutathione peroxidase (GPx), superoxide dismutase, and catalytic activities as well as reducing oxidative stress in the patients of hypertension (Sankar et al., 2006).

7.2 Antihyperlipidemic Activity

One of the most important capabilities of sesame seeds in recent times has been their antihyperlipidemic activity. Dietary supplementation with sesame seeds significantly reduces total cholesterol and LDL concentrations in rabbits under a high-fat diet (Asgary et al., 2013). The effect of diet with sesame seed powder at doses of 5% and 10% was examined for a period of 4 weeks. Administration of sesame seed powder resulted in a significant decrease in total lipid and cholesterol level, and LDL level, but the concentration of high-density lipoprotein was increased (Visavadiya and Narasimhacharya, 2008). In another study, the rats were fed with 18% sesame protein isolate for 28 days. The results showed that protein isolate of sesame decreases plasma cholesterol, triacylglycerols, and LDL levels and mitigates lipid peroxidation in hypercholesterolemic as well as normal diet groups and increases HDL-C levels (Biswas et al., 2010). In a recent study, the ethanolic extracts of sesame seeds (400 mg/kg and 700 mg/kg, orally) were examined against hepatotoxicity induced by CCl_4 in rats. The results revealed that elevated levels of serum glutamic-pyruvic transaminase and serum glutamic oxaloacetic transaminase were restored to normal levels. These results proved that sesame seed extract is a strong hepatoprotective agent against hepatic damage induced by CCl_4 in rats(Azab, 2014). These anti-hyperlipidemic activities of sesame could be credited to the induction of cholesterol turnover by increasing fecal excretion of sterol and production of hepatic bile acid (Biswas et al., 2010).

7.3 Antimicrobial Activity

Sesame is a natural antibacterial agent against common skin disease–causing microbes including *Streptococcus*, *Staphylococcus*, and athlete's foot fungus. It destroys *Streptococcus* and other common cold bacteria. It is also used for the treatment of psoriasis and dry skin disorders. It is a natural protector against ultraviolet radiations. In a study, the results revealed that minimum inhibitory concentration (MIC) of sesame oil against *Salmonella typhi* is 10 μL/mL. However, for other organisms the MIC values were in the range of 350–500 μL/mL. The sesame oil shows the best antimicrobial activity, equal with standard Kanamycin, and it also shows highest zone of inhibition against *Salmonella typhi* (Saleem, 2011). In another study, antimicrobial activities of aqueous and ethanolic extracts of sesame were examined against some microbes including *S. typhi*, *Staphylococcus aureus*, *Klebsiella pneumonia*, and *Escherichia coli* using various concentrations (100 mg/mL, 200 mg/mL, and 400 mg/mL). The ethanolic extract (400 mg/mL) showed a strong inhibitory effect against *E. coli*, the growth of *K. pneumonia* was mildly inhibited, while growth of *S. aureus* was not affected by ethanolic extract. On the other hand, aqueous extract had no inhibitory effect against tested microbes (Ogunsola and Fasola, 2014).

7.4 Anticancer Properties

Sesame oil has been found to inhibit the growth of malignant melanoma in vitro and the proliferation of human colon cancer cells (Smith and Salerno, 1992). Sesame seed consumption increases the levels of γ-tocopherols in plasma and improves the activities of vitamin E, so it reduces the risks of cancer and heart disorders (Cooney et al., 2001). Cephalin (a compound isolated from sesame seed) shows a strong hemostatic activity (Anilakumar et al., 2010). Historically, fiber has been used as an antiulcer, antidiabetic, laxative, cardioprotective, and antitumor agent. Myristic acid has cancer preventive capability, and it is found in sesame seed ranging from 328 to 1728 ppm (Shasmitha, 2015).

7.5 Antihypertensive Activities

In a study, it was revealed that sesamin and its active metabolites can induce antihypertensive effects in experimental animal models (Nakano et al., 2006). Another study indicated that sesame seed oil consumption reduced oxidative stress and increased catalase, superoxide dismutase, and GPx activities in hypertensive patients (Shasmitha, 2015). These results support the hypothesis that consumption of sesame oil enhances the antioxidant defense system of humans. Sesamin was suggested as a useful prophylactic treatment in cardiovascular hypertrophy and hypertension (Nakano et al., 2002). In another study, among the hypertensive patients using nifedipine (calcium channel blocker), sesame oil was compared along with other edible oils. Among the groups, sesame appeared to be the most promising against blood pressure rise (Sankar et al., 2005).

7.6 Antipyretic and Antiinflammatory Activity

A significant antipyretic activity, almost comparable to paracetamol, was produced by sesame oil. In a study, sesame oil was administered as a dietary supplement, and it produced antipyretic, analgesic, and antiinflammatory effects in animal models (Saleem et al., 2011). The antiinflammatory activity was assessed on the basis of paw edema inhibition induced by the injection of carrageenan (an edematogenic agent) into the subplantar region of the right hind paw of the rat. The results showed that the sesame oil and sesamin inhibited the formation of pleural exudate and the leukocyte migration, which confirms the antiinflammatory activity (Monteiro et al., 2014; Patel et al., 2012).

7.7 Wound Healing Activity

Free radicals are produced at the place of a wound, which reduce the healing process by damaging lipids, nucleotides, cell membranes, and proteins. Various antioxidants, including vitamin E and curcumin, are reported to give protection to tissues against oxidative damage. The process of wound healing is promoted by the use of antioxidants. Sesame seeds and their oil extract have potential antioxidant activity, which helps to prevent oxidative damage and promotes the healing process (Fukuda et al., 1986). Recently, a study was conducted to examine the effects of sesame seeds and its oil on experimentally induced incision, excision, dead space, and burn wounds in mice. A formulation of sesame seeds and oil prepared in Carbopol at 2.5% and 5% concentrations was applied to wounds. There was significant reduction (approximately 50%) in the period of wound healing in burn and excision wound models. However, a significant increase in breaking strength was observed in the case of incision wounds. Treatment of seeds and oil (250 and 500 mg/kg) in models of dead space wounds resulted in a substantial increase in hydroxyproline content, dry weight, and breaking strength. These results suggest the use of sesame seeds and oil in wound healing (Kiran and Asad, 2008).

7.8 Sesame in Lipid Metabolism

Considering the chemical composition, the dietary intake of sesame oil is expected to improve the condition, preventing any postprandial lipemia or lipid oxidation. Although many reports are available concerning the effect of sesamin on lipid metabolism, only a few studies using intact sesame oil as a diet are available (Chandra and Kuvibidila, 2012). It is apparent that sesame rich in lignans more profoundly affects hepatic fatty acid oxidation and serum triacylglycerol levels. Therefore, consumption of sesame rich in lignans results in physiologic activity to alter lipid metabolism in a potentially beneficial manner (Sirato-Yasumoto et al., 2001). Sesamol has been shown to reduce lipopolysaccharide-induced oxidative stress and upregulate phosphatidylinositol 3-kinase/Akt/endothelial nitric oxide synthase pathways (Ying et al., 2011).

7.9 Analgesic Activity

Analgesic effect of the ethanolic extract of sesame seeds was tested by writhing model induced by acetic acid in mice by different scientists. Liberation of endogenous substances by acetic acid causes algesia in rats (Trongsakul et al., 2003). The results revealed that a significant writhing inhibition was produced by the extract at the dose of 500 mg/kg, which is analogous to the standard drug ibuprofen at a dose of 50 mg/kg.

7.10 As a Prophylactic Agent

The liver, being a primary site for alcohol metabolism, is highly susceptible to alcohol-related injuries. However, dietary supplements may relieve some deleterious effects of alcohols. Recently, the prophylactic effect of *S. indicum* was examined against toxicity induced by ethanol in rats. Aqueous extract of *S. indicum* (200 mg/kg body wt.) was administrated orally in rats for 28 days. The results revealed that aqueous extract of sesame was more effective compared to hexane and methanolic extracts. The effectiveness of aqueous extract was due to the presence of tannins, phenols, flavonoids, and saponins. It was also noticed that ethanol administration causes significant elevation in serum triglyceride levels, which is dangerous for the liver and other organs. These adverse effects were significantly reduced after the treatment of *S. indicum* aqueous extracts, which indicated the prophylactic effect of *S. indicum* against ethanol toxicity (Oyinloye et al., 2014).

8. SIDE EFFECTS AND TOXICITY

In theory, sesame might affect blood sugar control in people with diabetes.

References

Adam, M., 2015. Effect of White & Red Sesame Cake Varity on Broiler Performance. UOFK.

Akbar, F., Rabbani, M.A., Masood, M.S., Shinwari, Z.K., 2011. Genetic diversity of sesame (*Sesamum indicum* L.) germplasm from Pakistan using RAPD markers. Pakistan Journal of Botany 43, 2153–2160.

Anilakumar, K.R., Pal, A., Khanum, F., Bawa, A.S., 2010. Nutritional, medicinal and industrial uses of sesame (*Sesamum indicum* L.) seeds-an overview. Agriculturae Conspectus Scientificus 75, 159–168.

Asgary, S., Rafieian-Kopaei, M., Najafi, S., Heidarian, E., Sahebkar, A., 2013. Antihyperlipidemic effects of *Sesamum indicum* L. in rabbits fed a high-fat diet. Science World Journal 2013.

Azab, A., 2014. Hepatoprotective effect of sesame oil against lead induced liver damage in albino mice: histological and biochemical studies. American Journal of BioScience 2, 1–11.

Biswas, A., Dhar, P., Ghosh, S., 2010. Antihyperlipidemic effect of sesame (*Sesamum indicum* L.) protein isolate in rats fed a normal and high cholesterol diet. Journal of Food Science 75, H274–H279.

Chandra, L.C., Kuvibidila, S., 2012. Health benefits of sesame oil on hypertension and atherosclerosis. American Journal of Bio-Pharmacology Biochemistry and Life 1, 36–42.

Cooney, R.V., Custer, L.J., Okinaka, L., Franke, A.A., 2001. Effects of dietary sesame seeds on plasma tocopherol levels. Nutrition and Cancer 39, 66–71.

Dawodu, F., Ayodele, O., Bolanle-Ojo, T., 2014. Biodiesel production from *Sesamum indicum* L. seed oil: an optimization study. Egyptian Journal of Petroleum 23, 191–199.

Fukuda, Y., Nagata, M., Osawa, T., Namiki, M., 1986. Contribution of lignan analogues to antioxidative activity of refined unroasted sesame seed oil. Journal of the American Oil Chemists' Society 63, 1027–1031.

Hama, J.R., 2017. Comparison of fatty acid profile changes between unroasted and roasted brown sesame (*Sesamum indicum* L.) seeds oil. International Journal of Food Properties 20, 957–967.

Hassan, M.A., 2012. Studies on Egyptian sesame seeds (*Sesamum indicum* L.) and its products 1-physicochemical analysis and phenolic acids of roasted Egyptian sesame seeds (*Sesamum indicum* L.). World Journal of Dairy & Food Sciences 7, 195–201.

Hwang, L.S., 2005. Sesame Oil. Bailey's Industrial Oil and Fat Products.

Kiran, K., Asad, M., 2008. Wound Healing Activity of *Sesamum indicum* L Seed and Oil in Rats.

Langham, D., Riney, J., Smith, G., Wiemers, T., Pepper, D., Speed, T., 2010. Sesame Producers Guide. Sesaco Corp.

Monteiro, É.M.H., Chibli, L.A., Yamamoto, C.H., Pereira, M.C.S., Vilela, F.M.P., Rodarte, M.P., de Oliveira Pinto, M.A., da Penha Henriques do Amaral, M., Silvério, M.S., de Matos Araújo, A.L.S., 2014. Antinociceptive and anti-inflammatory activities of the sesame oil and sesamin. Nutrients 6, 1931–1944.

Morton, J., 1990. Mucilaginous plants and their uses in medicine. Journal of Ethnopharmacology 29, 245–266.

Nadeem, A., Kashani, S., Ahmed, N., Buriro, M., Saeed, Z., Mohammad, F., Ahmed, S., 2015. Growth and yield of sesame (*Sesamum indicum* L.) under the influence of planting geometry and irrigation regimes. American Journal of Plant Sciences 6, 980.

Nakano, D., Itoh, C., Takaoka, M., Kiso, Y., Tanaka, T., Matsumura, Y., 2002. Antihypertensive effect of sesamin. IV. Inhibition of vascular superoxide production by sesamin. Biological and Pharmaceutical Bulletin 25, 1247–1249.

Nakano, D., Kwak, C.-J., Fujii, K., Ikemura, K., Satake, A., Ohkita, M., Takaoka, M., Ono, Y., Nakai, M., Tomimori, N., 2006. Sesamin metabolites induce an endothelial nitric oxide-dependent vasorelaxation through their antioxidative property-independent mechanisms: possible involvement of the metabolites in the antihypertensive effect of sesamin. Journal of Pharmacology and Experimental Therapeutics 318, 328–335.

Ogunsola, O., Fasola, T., 2014. The antibacterial activities of *Sesamum indicum* Linn. leaf extracts. Methods 18.

Oyinloye, B., Nwozo, S., Amah, G., Awoyinka, A., Ojo, O., Ajiboye, B., Tijani, H., 2014. Prophylactic effect of aqueous extract of Sesamum indicum seeds on ethanol-induced toxicity in male rats. Nutrition Research and Practice 8, 54–58.

Patel, M., Murugananthan, G., Gowda, K., 2012. In vivo animal models in preclinical evaluation of anti-inflammatory activity—a review. International Journal of Pharmaceutical Research and Allied Sciences 1, 1–5.

Pham, T.D., Thi Nguyen, T.-D., Carlsson, A.S., Bui, T.M., 2010. Morphological evaluation of sesame (*Sesamum indicum* L.) varieties from different origins. Australian Journal of Crop Science 4, 498.

Saleem, T.M., 2011. Anti-microbial activity of sesame oil. International of Journal of Research in Phytochemistry and Pharmacology 1, 21–23.

Saleem, T.M., Basha, S.D., Mahesh, G., Rani, P.S., Kumar, N.S., Chetty, C., 2011. Analgesic, anti-pyretic and anti-inflammatory activity of dietary sesame oil in experimental animal models. Pharmacologia 2, 172–177.

Sankar, D., Sambandam, G., Rao, M.R., Pugalendi, K., 2005. Modulation of blood pressure, lipid profiles and redox status in hypertensive patients taking different edible oils. Clinica Chimica Acta 355, 97–104.

Sankar, D., Rao, M.R., Sambandam, G., Pugalendi, K., 2006. Effect of sesame oil on diuretics or ß-blockers in the modulation of blood pressure, anthropometry, lipid profile, and redox status. Yale Journal of Biology & Medicine 79, 19.
Shasmitha, R., 2015. Health benefits of Sesamum indicum: a short review. Asian Journal of Pharmaceutical and Clinical Research 8, 1–3.
Shyu, Y.-S., Hwang, L.S., 2002. Antioxidative activity of the crude extract of lignan glycosides from unroasted Burma black sesame meal. Food Research International 35, 357–365.
Sirato-Yasumoto, S., Katsuta, M., Okuyama, Y., Takahashi, Y., Ide, T., 2001. Effect of sesame seeds rich in sesamin and sesamolin on fatty acid oxidation in rat liver. Journal of Agricultural and Food Chemistry 49, 2647–2651.
Smith, D.E., Salerno, J., 1992. Selective growth inhibition of a human malignant melanoma cell line by sesame oil in vitro. Prostaglandins, Leukotrienes and Essential Fatty Acids 46, 145–150.
Stamatov, S., Deshev, M., 2014. Selection approaches for the sesame forms suitable for mechanized harvesting. Bulgarian Journal of Agricultural Science 20, 1435–1438.
Takashi, I., Kushiro, M., Takahashi, Y., Shinohara, K., Fukuda, N., Sirato-Yasumoto, S., 2003. Sesamin, a sesame lignan, as a potent serum lipid-lowering food component. Japan Agricultural Research Quarterly: Japan Agricultural Research Quarterly 37, 151–158.
Terefe, G., Wakjira, A., Berhe, M., Tadesse, H., 2012. Sesame Production Manual. EIAR and Embassy of the Kingdom of the Netherlands, pp. 1–34.
Trongsakul, S., Panthong, A., Kanjanapothi, D., Taesotikul, T., 2003. The analgesic, antipyretic and anti-inflammatory activity of Diospyros variegata Kruz. Journal of Ethnopharmacology 85, 221–225.
Tunde-Akintunde, T., Oke, M., Akintunde, B., 2012. Sesame Seed, Oilseeds. InTech.
Visavadiya, N.P., Narasimhacharya, A., 2008. Sesame as a hypocholesteraemic and antioxidant dietary component. Food and Chemical Toxicology 46, 1889–1895.
Wang, B.-S., Chang, L.-W., Yen, W.-J., Duh, P.-D., 2007. Antioxidative effect of sesame coat on LDL oxidation and oxidative stress in macrophages. Food Chemistry 102, 351–360.
Ying, Z., Kherada, N., Kampfrath, T., Mihai, G., Simonetti, O., Desikan, R., Selvendiran, K., Sun, Q., Ziouzenkova, O., Parthasarathy, S., 2011. A modified sesamol derivative inhibits progression of atherosclerosis. Arteriosclerosis, Thrombosis, and Vascular Biology 31, 536–542.
Yoshida, H., Hirakawa, Y., Takagi, S., 2000. Roasting influences on molecular species of triacylglycerols in sesame seeds (*Sesamum indicum*). Journal of the Science of Food and Agriculture 80, 1495–1502.

Further Reading

Abou-Gharbia, H.A., Shehata, A.A.Y., Shahidi, F., 2000. Effect of processing on oxidative stability and lipid classes of sesame oil. Food Research International 33, 331–340.
Anilakumar, K., Nagaraj, N., Santhanam, K., 2001. Effect of coriander seeds on hexachlorocyclohexane induced lipid peroxidation in rat liver. Nutrition Research 21, 1455–1462.
Chakraborthy, G., Sharma, G., Kaushik, K., 2008. Sesamum indicum: a review. Journal of Herbal Medicine and Toxicology 2, 15–19.
Chavali, S., Zhong, W., Forse, R., 1998. Dietary α-linolenic acid increases TNF-α, and decreases IL-6, IL-10 in response to LPS: effects of sesamin on the Δ-5 desaturation of ω6 and ω3 fatty acids in mice. Prostaglandins, Leukotrienes and Essential Fatty Acids 58, 185–191.
Moazzami, A.A., 2006. Sesame Seed Lignans Diversity, Human Metabolism and Bioactivities.

Nzikou, J., Matos, L., Bouanga-Kalou, G., Ndangui, C., Pambou-Tobi, N., Kimbonguila, A., Silou, T., Linder, M., Desobry, S., 2009. Chemical composition on the seeds and oil of sesame (*Sesamum indicum* L.) grown in Congo-Brazzaville. Advance Journal of Food Science and Technology 1, 6–11.

Revathi, S., John Joel, A., Manivannan, N., 2012. Research note genetic variability in sesame (*Sesamum indicum* L.). Electronic Journal of Plant Breeding 3, 692–694.

Saeed, F., Qamar, A., Nadeem, M.T., Ahmed, R.S., Arshad, M.S., Afzaal, M., 2015. Nutritional composition and fatty acid profile of some promising sesame cultivars. Pakistan Journal of Food Sciences 25, 98–103.

Saleem, T., Chetty, C.M., Kavimani, S., 2012. Sesame oil enhances endogenous antioxidants in ischemic myocardium of rat. Revista Brasileira de Farmacognosia 22, 669–675.

Vishwanath, H., Anilakumar, K., Harsha, S., Khanum, F., Bawa, A., 2012. In vitro antioxidant activity of Sesamum indicum seeds. Asian Journal of Pharmaceutical and Clinical Research 5, 56–60.

CHAPTER

45

Sweet Lemon

Adan Iqbal[1], *Rasheed Ahmad Khera*[1], *Muhammad Asif Hanif*[1], *Muhammad Adnan Ayub*[2], *Abdullah Mohammed Al-Sadi*[3]

[1] Department of Chemistry, University of Agriculture, Faisalabad, Pakistan;
[2] Department of Chemistry, University of Okara, Okara, Pakistan;
[3] Department of Crop Sciences, College of Agricultural and Marine Sciences, Sultan Qaboos University, Oman

OUTLINE

Medicinal Plants of South Asia
https://doi.org/10.1016/B978-0-08-102659-5.00045-8

Copyright © 2020 Elsevier Ltd. All rights reserved.

1. BOTANY

1.1 Introduction

Sweet lemon (*Citrus limetta* Risso) (Fig. 45.1) is a small evergreen tree with irregular branches. It is a member of the family Rutaceae. This family is found in temperate and tropical regions of the world and is economically very important due to its fruit usage. The Rutaceae family comprises 150 genera and 1600 species that are distributed worldwide. All these genera are closely related to one another due to interhybridization. Interhybridization seems to be possible between different citrus plants and those plants that are not categorized as citrus, results in generation of new varieties. Four taxa of the citrus fruits that include orange, lemon, grapefruit, and tangerine are in many cases propagated asexually, and this results in loss of their characteristic traits when bred. However, these hybrids are frequently produced, which makes the citrus family tree a complicated one. *C. limetta* is known by different common names in the world. It is known as sweet lemon, Mediterranean sweet lemon in English, and mittha in Pakistan. Limu Shirin is its name in Iran, kannada in India, and malta in Bangladesh. Although the most common citrus is sweet lemon (*C. limetta*), many other varieties and cultivars of this genus are available that vary in flavors, scents, and uses. The most common examples include lemon, grapefruit, pomelo, tangerine, and tangelo (Gulsen and Roose, 2001). *C. limetta* flavors can range from sweet to acidic, while all varieties possess a pleasant odor.

1.2 History/Origin

The generic name citrus was derived from Latin, in which it was traditionally used to describe the species "*Citron medica.*" It originated, however, from the Greek word cedar. The exact origin of citrus fruits has not been clearly identified. Several hypotheses by different researchers

Sweet lemon

FIGURE 45.1 *Citrus limetta* tree and fruit (Mousambi/Mittaha).

have been proposed. It was identified that the important centers of origin of citrus and its genera are subtropical and tropical regions of India, Southeast Asia and China. The evidence reporting the presence of citrus in China was presented in the book "Tribute of Yu," during the kingdom of Ta Yu in 2205–2197 BC. Literature of the Chou dynasty 1027–256 BC and Chin and Han dynasties 202 BC–220 AD provides evidence regarding the presence of citrus fruits. In Chinese culture, citron was considered to be the hand of Buddha, and it was exchanged for good luck. Some historians consider that it was brought to Palestine by Jews upon their return from Babylonian exile in about 500 BC, while others think that it was brought by the return of Alexander from India in about 300 AD. With the conquest of Spain by the Moors, oranges were introduced there around 900 AD. The Spanish introduced citrus fruits to the modern world around the 16th century, and they propagated it to the Europe and North and South America, specifically Florida and Mexico, which are today considered the largest citrus orchards in the world. World citrus production started around the 19th century with the advent of railroads and steam ships.

1.3 Demography/Location

Citrus fruits are found in a variety of environmental conditions and geographic regions. The most suitable conditions for their growth are considered to be found in tropical and subtropical regions. Sand, water, heat, and drought tolerance are major ecologic factors affecting their growth. They exhibit drought tolerance but no heat or water tolerance. *C. limetta's* main production is seen in Pakistan, India, China, Philippines, Burma, Thailand, Indonesia, Guinea, Australia, Vietnam, Malaysia, Palestine, and Egypt. It is rare in Europe. FAO (2016) statistics showed that, worldwide, citrus species are grown in more than 140 countries, accounting for 124246.0 thousand tons of fruit that were produced in the year 2016. Among them, the largest producers of the citrus fruits (in thousand tons) are the China (32,705.9), Brazil (16555.1), the United States (7829.0), India (9755.8), Spain (6682.0), and Mexico (6634.0). Pakistan is among top 10 leading citrus-producing countries of world with a yearly production of 1907.4 thousand tons. FAO (2016) statistics showed that major lemon- and lime-producing countries (in thousand tons) are India (2613.8), China (2405.9), and Mexico (2270.0) (FAO, 2016).

1.4 Botany, Morphology, Ecology

C. limetta is a small tree with irregular branches and numerous thorns that may reach up to 5–8 m in height at maturity. The leaf is of simple type with pinnate venation and alternate leaf arrangement. The leaf margins are crenate with glossy leaf texture, and the leaf possesses a pleasant odor. Leaves are oval and rounded. Flowers appear in the spring season, possess a pleasant smell, and are white in color. Flower size ranges from 1.5 to 3 cm. Petals are reflexed. The fruit is oval to pear shape and consists of a pericarp of green color initially, which changes to yellow at the end of the ripening season. The pulp is sweetish to slightly acidic in taste. An extensive root system of palisade form is found in which there is a central one around which develops many secondary roots (Cossmann, 1940). Production of fruit starts when the tree is 5–7 years old, and maximum production is achieved when it reaches the age of 10–20 years.

Sweet limes are picked when they are ripened and achieve the size of a tennis ball, with a luster of greenish yellow, indicating the right picking season (Sharma et al., 2004).

2. CHEMISTRY

C. limetta essential oil is extracted from the peel. The stages that contain the maximum amount of essential oil of *C. limetta* are stages one and two, in which a large number of glands are present for the production of essential oils. The essential oil contains several components, the most abundant of which is D-limonene, renowned for its antibacterial properties (Khan et al., 2016). Limonene is the most abundant and is present in concentrations of 45%–90% of the total essential oil, and it is responsible for the antimicrobial properties (Murdock and Allen, 1960). *C. limetta* contains a wide range of phytonutrients that include acetophenones, low molecular phenolic acids (hydroxy benzoic acid and hydroxyl cinnamic acids), flavonoids, terpenoids, and limonoids. The essential oil contains a low amount of fatty acids that typically include linoleic acid, palmitic acid, oleic acid, and stearic acid.

C. limetta is a rich source of calories and other essential components that include minerals, proteins, and vitamins. It is considered to have a lower fat content, and this quality makes it an ideal food for weight-reducing diets and for people suffering from heart, coronary, and cancer diseases (Khan et al., 2016). In general, one cup of fruit or 100% fruit juice, or ½ cup of dried fruit, can be considered one cup from the fruit group. One serving of sweet lemon fruit contains 43 calories, 0–0.3 g of fat, 9–9.3 mg of carbohydrates, 0–0.8 g of proteins, and 40–50 mg of vitamin C. Additionally, it also provides 480–490 mg of potassium, 0–0.7 mg iron, 30–40 mg calcium, and 25–30 mg phosphorus. It also serves as a source of vitamin A, flavonoids, limonoids, dietary fibers, and carotenoids (Okwu and Emenike, 2006). Chemical constituents of the fruits are affected by changes in climate, growing conditions, type of cultivar or rootstock used, and difference in maturity stage. The high nutritional content of *C. limetta* makes it beneficial for health and an ideal food. The most abundant terpenes found in *C. limetta* essential oil are monoterpenes such as D-limonene, known for its antibacterial activity and citral for antifungal applications. Linalool has fungistatic properties (Colecio-Juárez et al., 2012). Sweet lime is also a good source of flavonoids. Citrus essential oil can be utilized for edible applications, as pharmaceutical components, in nutritional supplements, in aromatherapy, and in the cosmetic industry. Safe, natural sources and wide acceptance by users are responsible for the use of the essential oils on a large scale. Citral, D-limonene, and linalool are mainly present in lime essential oil (Fig. 45.2).

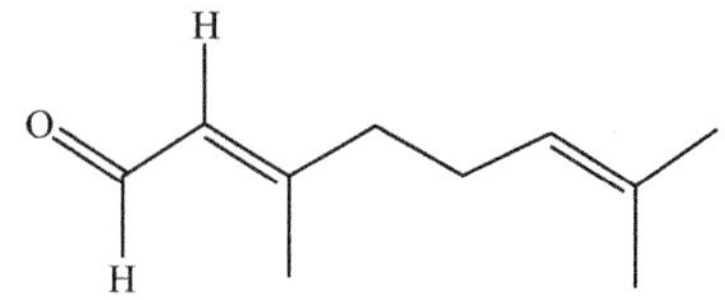

Citral

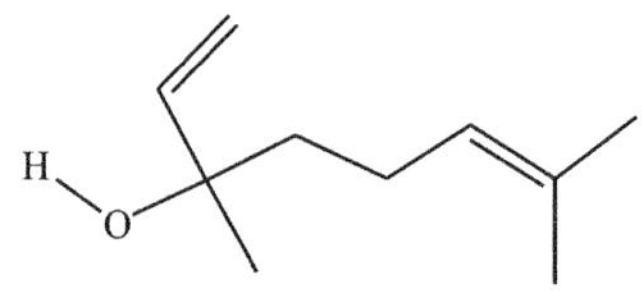

Linalool

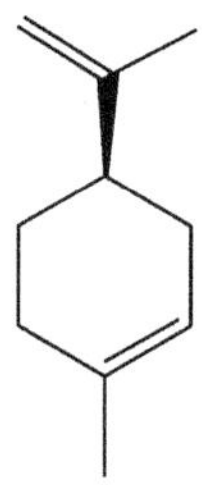

d-limonene

FIGURE 45.2 Structures of major chemical constituents of sweet lemon essential oil.

3. POSTHARVEST TECHNOLOGY

The most appropriate season for the vegetative growth of sweet limes is the rainy season, which includes the months of July, August, and November. March is also considered a peak season for growth of other citrus varieties. Losses during fresh fruit handling range from 5% to 10% in most developed countries, but from 25% to 30% or more in developing countries. However, citrus fruits are nonclimacteric and have a relatively long shelf life compared to mangos, bananas, and other tropical fruits. The dry and hot climate in the most parts of the tropical world render fruit unsuitable for marketing if not properly handled and stored. Ethylene has a role in the ripening and abscission of citrus and needs to be used wisely to benefit consumers and handlers. Removing ethylene from the fruit storage environment is challenging; although, its removal extends shelf life. Other plant growth-regulating substances, such as cytokinins, auxins, gibberellins, and polyamines, also profoundly affect fruit quality. Understanding the mechanism of action and key processes regulated by the plant regulators and bioregulators may lead to generation of strategies for modifying fruit characteristics to improve quality and delay senescence.

An ethylene inhibitor, 1-methyl cyclopropane (1-MCP), that apparently binds to the cellular ethylene receptor has been found to reduce chilling injury, stem-end rot, and volatile off flavor. This nontoxic, odorless compound also delays color development, so it has been recommended for use in green citrus fruits (Porat et al., 1999). During grading, immature, unripened, and decayed fruits are discarded by visual inspection, and fruits are packed, according to their size, in sacks or wooden or cardboard boxes before being sold as fresh fruits in local markets. To increase the shelf life to 25–26 days and to avoid postharvest damage even at room temperature, they can be treated with wax after harvesting the fruits. The shelf life can also be extended by wrapping the fruit with polythene sheets that are heat shrinkable. The coating process, using different types of waxes such as aqueous waxes, solvent-based waxes and wax solutions, reduces water loss and has been practiced for centuries to extend the storage life of citrus fruits. Nowadays, edible coatings, such as esters, lipids, sucrose, and cellulose enable the process of packaging without the use of hazardous chemicals (Krochta et al., 1994). The ideal storage temperature is 10–12°C with 90%–95% relative humidity. Local market transportation involves the use of cargo without refrigeration, and it often results in severe damage due to decay and fungal infections. For transportation of export quality, it is essential to regulate the temperature during storage and transport. This cool chain involves the presence of a cold store on the farm, in transportation vehicles, and at airports, etc. Citrus fruits are considered to be nonclimacteric, and they have a long shelf life. However, if they are not properly stored, postharvest damage can occur, such as the development of physiologic disorders, loss of taste, chilling injury, and various fungal diseases.

4. PROCESSING

Citrus fruits are consumed either as fresh fruits or in the forms of their by-products that include juices, jams, marmalades, etc. Almost one-third of the total citrus fruit production includes its use in the form of processed products. Citrus juices are consumed either as fresh juice or ready-to-serve juices. Fresh juice involves freshly squeezed juice, traditionally used at home (Okwu and Emenike, 2007). The shelf life of frozen juice may be up to 6 months. For the preparation of ready-to-serve juices, juice is packed in glass or plastic containers or cardboard cartons. This juice has a shelf life depending on the pasteurization procedures used and is available in the market as a ready-to-serve source. Sweet lime itself remains fresh for about 2 weeks at room temperature and 4–8 weeks when refrigerated. Slices of fruit can also be stored when frozen by submerging the slices in sweet syrup within an airtight glass container. Without submerging, the limonoid content can cause the pulp to be bitter in taste. The essential oil is composed of volatile compounds and is typically extracted from citrus peel using various techniques such as hydro distillation, solvent extraction, peel pressing, cold pressing, steam distillation, and microwave-assisted hydro distillation, depending on yield and quality required. Other processing procedures involve the preparation of jams and marmalades. Jam is an intermediate-moisture food prepared by boiling fruit pulp with sugar (sucrose), pectin, acid, and other ingredients (preservative, coloring, and flavoring materials) to a reasonably thick consistency, firm enough to hold the fruit tissues in position (Baker, 1994; Barrett et al., 2004). As a result of processing, peel is discarded as waste material.

5. VALUE ADDITION

It is used either as fresh fruit, in the form of pickles, eaten as a salad, as a flavoring in beverages, and foods, etc. (Ferguson, 1990). In cooking, lime is valued both for the acidity of its juice and the floral aroma of its zest. It is very commonly used in authentic Mexican, Southwestern US, Vietnamese, and Thai dishes. It is used as an ingredient in many cocktails because of its refreshing aroma. Traditionally, it is also used as a pickle and used in many countries, especially in India and Pakistan. The use of dry lime and loomi as a flavoring is typical of Persian and Iraqi cuisine as well as in Gulf style Baharat (a spice mixture that is also called kabsa or kebsa). Lime is an essential ingredient of any cuisine from India, and many varieties of lime dishes are made, such as sweetened lime pickle, salted pickle, and lime chutney. Lime is also used as an herb in south and east Asia, where it is used to add flavor to cold and hot water, tonics, and beverages. The pulp and peel are used as natural food additives, as consumer awareness toward chemical additives results in demand for natural food additives, and research regarding the use of *C. limetta* peel extract as a natural additive also proves it to possess in vitro antifungal activities against food-borne pathogens. In this way, it has proved itself to be an excellent food additive. In the traditional indigenous medicinal system, sweet lime juice is valued for curing fever, malaria, and jaundice (Cowan, 1999).

6. USES

C. limetta contributes significantly in maintaining health and providing essential nutrients. *C. limetta* is a rich source of calories and has a low fat content. It contributes a lot to human health through its bioactive components, mineral contents, and dietary fibers. It is used a lot as a food additive, and its peel, a waste from processing, is also used for several purposes. The essential oil and extract of *C. limetta* have been screened for their potential uses as alternative remedies for the treatment of many infectious diseases, preservation of food, as a food flavoring agent, and as adsorbent in the perfume industry. In recent years, citrus processing waste has been used for the production of single cell protein, bio-hydrogen, bioethanol, and multiple enzymes. In India, sweet lime is used in tantara for removing evil spirits. It is also combined with Indian chilies to make a protective charm to protect from evil eye. Furthermore, it is believed that hanging lime over sick people cures them of all illness by removing all evils lurking inside the body.

7. PHARMACOLOGICAL USES

7.1 Antimicrobial Activity

C. limetta peel oil showed significant antifungal and antibacterial activity against different food-borne pathogens including bacteria such as *Staphylococcus aureus, Bacillus subtilis, Bacillus cereus, Lactobacillus acidophilus, Escherichia coli,* and fungi including *Aspergillus niger, Aspergillus flavus, Aspergillus fumigatus,* and *Aspergillus ficuum*. These food-borne pathogens cause illness such as dysentery, pneumonia, food poisoning, whooping cough, and cholera. Higher contents of monoterpenoid hydrocarbon as a component of essential oil are usually considered to be the main reason for the antimicrobial activity (Shahzad et al., 2009).

7.2 Antioxidant Activity

Methanolic extract of *C. limetta* fruit peel shows significant *in vitro* antioxidant activity against superoxide radicals, nitric oxide radicals, and 2,2-diphenyl-1-picrylhydrazyl (DPPH) scavenging radicals, and it also results in inhibition of lipid peroxidase. DPHH radicals pair off by accepting hydrogen from the methanolic extract of *C. limetta*. This pairing results in the formation of stable DPHH. Nitric oxide radicals are generated endogenously and influence the function of vasodilation, synaptic plasticity, neurotransmission, and memory in the central nervous system. Superoxide radical formation occurs by activated phagocytes such as macrophages, monocytes, and eosinophils, when they attack bacteria and viruses (Halliwell and Gutteridge, 2015). Methanolic extract has also shown inhibition against ferrous and ferric ion–induced lipid peroxidation.

7.3 Cardiovascular Diseases

Owing to the antihypertensive effect of flavonoids, hesperidin and naringin as components of peel, and inner parts of the *C. limetta* fruit, it antagonizes the effect of angiotensin II (Touyz and Schiffrin, 2000). Angiotensin II is a potent vasoconstrictor and is responsible for immediate elevation of blood pressure due to its peptide nature and is responsible for cardiovascular diseases such as heart attack and stroke by causing thrombosis (blood clotting). This antagonistic effect of *C. limetta* extract is considered to be a physiologic process, i.e., without antagonizing the effect of any specific binding receptor.

7.4 Antiinflammatory Activity

The leaves, fruits, and peel of *C. limetta* extract, owing to xanthine oxidase activity, serve as a natural source to treat gout by stopping biosynthesis of uric acid, and they have fewer side effects compared to synthetic drugs. Potent xanthine oxidase inhibitors include polyphenols, flavonoids, coumarins, ellagic acid, valoneic acid, and dilactones, which are present in this medicinal plant.

7.5 Anticancer Potential

C. limetta fruit peel significantly reduces the volume of tumor cells and results in a decrease of cancerous activity, i.e., abnormal cell division, as have been proven by the previous studies on rat and increase in hemoglobin content and number of red blood cells. Thus, the citrus peel is responsible for maintaining the hematologic profile. Citrus flavonoids can inhibit the invasion of chick heart fragments and syngeneic mice liver by malignant mouse tumor cells. Hesperidin and diosmin have been shown to exhibit anticarcinogenic activities in various in vivo studies. The polymethoxylated flavones have been shown in numerous in vitro studies to exert strong antiproliferative action against cancer cells (Yang et al., 1997), antigen-activated T lymphocytes, gastric cancer cells, prostate cancer cells (Hirano et al., 1995), and squamous cell carcinoma (Bracke et al., 1994), and antimetastatic actions against human breast cancer cells have also been observed with tangeretin. Naringin, hesperidin, nobiletin, and tangeretin inhibit the bacterial mutagenesis (Hirano et al., 1995). The medicinal value of *C. limetta* derives from the bioactive phytochemical constituents that are responsible for the significant physiologic effect on the human body. *C. limetta* plants are estimated to contain about 40 limonoids, of which limonin and nomilin are abundant. These limonoids are responsible for the bitter taste of citrus species. Limonoids are responsible for inhibiting tumor formation and show anticancer activity.

This anticancer activity is attributed to the ability to stimulate the GST (Glutathione S-Transferase) enzyme. GST is responsible for catalyzing the reaction of dangerous electrophiles with glutathione and converting them to less toxic and hydrophilic substances, which are ultimately excreted from the body. In this way, they show anticancer activity (Okwu and Emenike, 2006).

7.6 Antidiabetic Effects

Methanolic extract of *C. limetta* fruit peel showed antihyperglycemic activity in streptozotocin-induced diabetic rats (Arai et al., 2000). The fruit peel of *C. limetta* is rich in flavonoids such as hesperidin and naringin. Both of these flavonoids possess a potent hypoglycemic activity that is responsible for mediating glucose-regulating enzymes. Both hesperidin and naringin have significantly lowered the blood glucose level, partly by increasing glycolysis and glycogen concentration or by lowering the gluconeogenesis process in the liver. Thus, the presence of flavonoids in the peel extract of *C. limetta* might be responsible for the anti-hyperglycemic activity and prove it a safe antidiabetic agent. In another study, quercetin flavonoid was shown to stimulate the release of insulin and increase the uptake of calcium by the islet cell, suggesting a way for the treatment of noninsulin-dependent cells.

The inhibitory effect of the aqueous *C. limetta* peel extract on the metabolism of carbohydrates was studied. The extract inhibited primarily the enzyme α-amylase by 49.6% at a concentration of 20 mg/mL and to a lesser extent the enzyme α-glucosidase with an inhibition of 28.2% at the same concentration. This inhibition was likely due to the high polyphenol content in the *C. limetta* peel (19.1 mg GAE/g).

7.7 Repellent/Larvicidal Activity

C. limetta peel extract has been found to possess larvicidal activity against *Aedes aegypti* and *Anopheles stephensi*. *C. limetta* peel extract in hexane and petroleum ether solvent were found to be effective against malarial and dengue fever vectors. The extract of hexane possesses 1.9-fold larvicidal potential compared to the extract using petroleum ether as solvent (Kumar et al., 2012).

The repellent activity can be attributed to the active ingredients of alkaloids, flavonoids, saponins, and phenolics present in the extract of citrus peels that exert an inhibitory influence on the lactic acid receptor cells. These receptor cells are responsible for attracting the mosquito to bite and feed on the blood. Due to this inhibitory effect, the mosquitoes are no longer attracted to bite, and this blood feeding response is prevented.

7.8 Antiatherosclerotic Activity

C. limetta possesses antiatherosclerotic effects due to the flavonoids that exert influence on the arteries of the vascular system and can oxidize cholesterol of plaque due to its antioxidative property. In this way, it protects the body against coronary heart diseases. Research has shown that regular consumption of flavonoids can reduce the risk of death from coronary heart diseases (Okwu, 2005). There is a correlation between flavonoids and cholesterol concentration of plasma. High consumption of flavonoids results in a decrease in cholesterol concentration of plasma and ultimately reduced risk of mortality from coronary heart diseases. Oxidative damage has been suggested to be a contributory factor in the development and complication of thrombosis. Recently, the beneficial effects of antioxidants from citrus fruits against thrombosis have gained interest (Okwu, 2005). Cholesterol-lowering effects of *C. limetta* are produced by limonene. Furthermore, polymethoxylated flavones (PMFs) are present in citrus fruit peel, which can lower cholesterol more effectively than some prescription drugs, without showing any side effects. Among the variety of citrus fruits containing PMFs, the most common PMFs were tangeretin and nobiletin. PMFs work like statin drugs that inhibit the synthesis of cholesterol and triglycerides inside the liver.

7.9 Dissolution of Gallstone

In *C. limetta*, the most abundant monoterpenoid, D-limonene (appears to be 45%–90% of the total), results in the dissolution of gallstones. D-limonene is a clinically proven solvent of cholesterol-containing gall stones, and it can dissolve a human gall stone within 2 hours. In animals, D-limonene infusion in gall bladder results in dissolution and disintegration of gallstones that were extracted from the bile juice in the bile duct (Koyuncu et al., 1998). In postgallstone surgery, infusion of D-limonene in patients results in dissolutions of gallstones that were left during surgery. This dissolution occurs after three infusions of 20 mL of D-limonene. Citrus fruits are highly recommended for persons suffering from kidney stones, gout, and arthritis. Citrates (citric acid salts) with citrus juice, containing especially potassium citrate, prevent the formation of kidney stones and ease their dissolution.

8. SIDE EFFECTS AND TOXICITY

There is no doubt lime juice carries plentiful health benefits; however, it may have a number of side effects that might cause health hazards like the decay of tooth enamel and ulcers within the stomach. Gastroesophageal reflux disorder is a digestive disorder triggered by limes or acidic foods. The vitamin C in lime juice, also called ascorbic acid, affects the absorption of iron in the body. Lime juice contains sulfites that can cause reactions and allergies in some people. Though lime juice helps treat kidney stones and is good for healthy kidneys, people suffering from kidney ailments should not consume lime juice because it might cause renal damage. Citrus fruits generally contain high potassium levels. This might be bad for those suffering from kidney disease. The diseased kidneys would not be able to maintain the electrolyte balance. This leads to abnormal levels of potassium, sodium, and phosphorous.

References

Arai, Y., Watanabe, S., Kimira, M., Shimoi, K., Mochizuki, R., Kinae, N., 2000. Dietary intakes of flavonols, flavones and isoflavones by Japanese women and the inverse correlation between quercetin intake and plasma LDL cholesterol concentration. Journal of Nutrition 130, 2243–2250.

Baker, R.A., 1994. Potential Dietary Benefits of Citrus Pectin and Fiber. Food technology, USA).

Barrett, D.M., Somogyi, L., Ramaswamy, H.S., 2004. Processing Fruits: Science and Technology. CRC Press.

Bracke, M., Bruyneel, E., Vermeulen, S.J., Vennekens, K.l., Vanmarck, V., Mareel, M., 1994. Citrus flavonoid effect on tumor invasion and metastasis. Food Technology 48, 121–124.

Colecio-Juárez, M.C., Rubio-Núñez, R.E., Botello-Álvarez, J.E., Martínez-González, G.M., Navarrete-Bolaños, J.L., Jiménez-Islas, H., 2012. Characterization of volatile compounds in the essential oil of sweet lime (*Citrus limetta* Risso). Chilean Journal of Agricultural Research 72, 275.

Cossmann, K., 1940. Citrus roots: their anatomy, osmotic pressure and periodicity of growth. Palestine Journal of Botany 3, 65–104.

Cowan, M.M., 1999. Plant products as antimicrobial agents. Clinical Microbiology Reviews 12, 564–582.

FAO, 2016. Food and Agriculture Organization of the United Nations, Citrus Fruit – Fresh and Processed Statistical Bulletin. http://www.fao.org/economic/est/est-commodities/citrus-fruit/en/1-63.

Ferguson, U., 1990. Citrus Fruits Processing. Horticultural Science, Florida, pp. 117–118.

Gulsen, O., Roose, M., 2001. Lemons: diversity and relationships with selected citrus genotypes as measured with nuclear genome markers. Journal of the American Society for Horticultural Science 126, 309–317.

Halliwell, B., Gutteridge, J.M., 2015. Free Radicals in Biology and Medicine. Oxford University Press, USA.

Hirano, T., Abe, K., Gotoh, M., Oka, K., 1995. Citrus flavone tangeretin inhibits leukaemic HL-60 cell growth partially through induction of apoptosis with less cytotoxicity on normal lymphocytes. British Journal of Cancer 72, 1380.

Khan, A.A., Mahmood, T., Siddiqui, H.H., Akhtar, J., 2016. Journal of Chemical and Pharmaceutical Research 8 (3), 555–563.

Koyuncu, H., Berkarda, B., Baykut, F., Soybir, G., Alatli, C., Gül, H., Altun, M., 1998. Preventive effect of hesperidin against inflammation in CD-1 mouse skin caused by tumor promoter. Anticancer Research 19, 3237–3241.

Krochta, J.M., Baldwin, E.A., Nisperos-Carriedo, M.O., 1994. Edible Coatings and Films to Improve Food Quality. Technomic Publ. Co.

Kumar, S., Warikoo, R., Mishra, M., Seth, A., Wahab, N., 2012. Larvicidal efficacy of the *Citrus limetta* peel extracts against Indian strains of *Anopheles stephensi* Liston and *Aedes aegypti* L. Parasitology Research 111, 173–178.

Murdock, D., Allen, W., 1960. Germicidal effect of orange peel oil and D-limonene in water and orange juice. Food Technology 14, 441–445.

Okwu, D., 2005. Phytochemicals, vitamins and mineral contents of two Nigerian medicinal plants. International Journal of Molecular Medicine and Advance Sciences 1, 375–381.

Okwu, D., Emenike, I., 2006. Evaluation of the phytonutrients and vitamins content of citrus fruits. International Journal of Molecular Medicine and Advance Sciences 2, 1–6.

Okwu, D., Emenike, I., 2007. Nutritive value and mineral content of different varieties of citrus fruits. Journal of Food Technology 5, 105–108.

Porat, R., Weiss, B., Cohen, L., Daus, A., Goren, R., Droby, S., 1999. Effects of ethylene and 1-methylcyclopropene on the postharvest qualities of 'Shamouti'oranges. Postharvest Biology and Technology 15, 155–163.

Shahzad, K., Nawaz, S., Ahmad, R., Mahmud, S., Iqbal, Z., Saeed, K., 2009. Evaluation of antioxidant and antimicrobial activity of essential oil of Tangerine fruit peel. Pakistan Journal of Biochemistry and Molecular Biology 42, 4–7.

Sharma, B., Hore, D., Gupta, S., 2004. Genetic resources of citrus of North-Eastern India and their potential use. Genetic Resources and Crop Evolution 51, 411–418.

Touyz, R.M., Schiffrin, E.L., 2000. Signal transduction mechanisms mediating the physiological and pathophysiological actions of angiotensin II in vascular smooth muscle cells. Pharmacological Reviews 52, 639–672.

Yang, M., Tanaka, T., Hirose, Y., Deguchi, T., Mori, H., Kawada, Y., 1997. Chemopreventive effects of diosmin and hesperidin on N-butyl-N-(4-hydroxybutyl) nitrosamine-induced urinary-bladder carcinogenesis in male ICR mice. International Journal of Cancer 73, 719–724.

CHAPTER

46

Tomato

Kinza Waheed, Haq Nawaz, Muhammad Asif Hanif, Rafia Rehman

Department of Chemistry, University of Agriculture, Faisalabad, Pakistan

OUTLINE

Medicinal Plants of South Asia
https://doi.org/10.1016/B978-0-08-102659-5.00046-X

Copyright © 2020 Elsevier Ltd. All rights reserved.

1. BOTANY

1.1 Introduction

Tomato (*Solanum lycopersicum*) (Fig. 46.1) is an important shrub of family Solanaceae. It also acts as a model system for the development of fruit and plant defense (Aoki et al., 2010). The genus *Solanum* contain 1500–2000 species (Machida-Hirano, 2015). Varieties of *S. lycopersicum* are red-fruited, and great variability has been shown with respect to quality, shape, intensity of fruit color, habitat for growth of plant, and morphology of leaf (Schauer et al., 2005). *S. lycopersicum* undergoes self-pollination and does not require insects or birds for pollination. There are 13 well-known

FIGURE 46.1 Tomato (*Solanum lycopersicum)* grows on a sprawling vine. One tomato plant yield buckets of tomatoes.

species in *Solanum lycopersicum,* which show different phenotypes and can be crossed with cultivated tomatoes (Kimura and Sinha, 2008a).

Tomato is known by different names in the world, but commonly, it is known as *love apple* (Jenkins, 1948). In Spanish, it is known as *jitomate.* In German, it is known as *tomate.* And it is known as *garden tomato* or *tomato* in English. In Urdu, it is known as tamater.

The most familiar specie of *Solanum* is *S. lycopersicum,* but a wide range of varieties and cultivars exists. *S. lycopersicum* is cultivated throughout the world, and it can be crossed with other species including *Solanum neorickii, Solanum chmielewskii, Solanum pimppinelifolium, Solanum pennellii,* and *Solanum habrrochaites,* and these all are collectively called *S. lycopersicum.*

The major difference between *S. lycopersicum* and all these varieties is the difference in their phenotype; mostly these differ in fruits (Schauer et al., 2005). Tomato plant has a sympodial shoot, fleshy fruits, and compound leave, which make it prominent among other plants (Kimura and Sinha, 2008b). *S. lycopersicum* is grown as an annual plant, and it lives only for a short period of time and is a perennial plant (Bauer et al., 2009).

1.2 History/Origin

S. lycopersicum is native to South America, and it is domesticated in Central and South Asia. In 1753, Linnaeus placed the tomato in the genus *Solanum* as *S. lycopersicum* L. The word *lycopersicum* is derived from *lyco,* mean wolf, and *persicum,* meaning peach, i.e., wolf peach. In 1754, Phillip Miller separated a new genus, *lycopersicon,* from *Solanum* and named the tomato *Solanum esculentum* Mill to accommodate tomato and many other species, where *esculentum* means edible. At that time, Miller emphasized that tomato is edible, but there was a concept among people about the poisonous nature of tomato. The name *S. esculentum* is against the international code of botanical nomenclature, so the name *lycopersicum* is used technically (Kimura and Sinha, 2008a). In early 16th century, the tomato was known by the name "mala auraea," which means golden apple. A phrase was used in early times to describe the tomato, "poma amoris fructu rubri," which means apple of love with red fruit (Smith, 2001).

Tomato has a great history, and Christopher Columbus was the first who encountered tomatoes in Europe in 1493. Heman Cortes was the first man who observed its valuable potential in Tenochtitlan and took its seeds back to Europe. In Spanish country the tomato spread in the 1540s and became a regular food in the 17th century. It became popular in other European countries in 1548. Europe is now very popular for the manufacturing of ketchup and production of tomato varieties. Tomato arrived in England in 1597.

1.3 Demography/Location

Tomato can be grown in a variety of environmental conditions, air, temperatures, and climates, but the optimum conditions for the growth of tomato plants are found in countries that have a warm climate. Tomato is very sensitive to frost at any stage of growth (Fontana et al., 2008). Germination of seeds is very slow below the optimum temperature, and its growth rate also reduces. The tomato is grown widely in the following countries: Pakistan, China, India, Iran, Italy, the Middle East, Spain, Britain, Mesoamerica, North America, and in North Africa. Canada is the largest exporter, and it exports more than 40% of the tomato and tomato-based products to America.

1.4 Botany, Morphology, Ecology

S. lycopersicum is a small shrub that has a soapy-smelling green fruit that contains a very irritating flavor. The length of plant is 1–3 m and contains a very weak, woody stem (Shukla et al., 2013). The stem is glandular, hairy, and coarse, found within the soil. The leaves are compound and small, and its leaflets are very rounded and thick. Its leaves are spirally arranged, the width of leaves is 10–30 cm, and length is 15–50 cm. Leaflets are ovate to oblong, and these are surrounded by glandular hair. In the large leaflets, a small pinnate appears. Inflorescence produces 6 to 12 flowers, and size of petiole is 3–6 cm (van Dam et al., 2005). In the primary morphogenesis stage, its leaflets start from the marginal blast zone and follow the same developmental stages as leaves (Shukla et al., 2013). Tomato plant produce flowers of yellow color that turn into fruit of different colors such as red, yellow, orange, and brown. Flowers are regular, bisexual, and their diameter is 1.5–2 cm. Flowers grow between the leaves. There is a small and hairy calyx tube and continual sepals. Fruit is highly colored, and the size of the berry is 1–2 cm in diameter. Unripe tomato is hairy and green in color. Tomato is usually furrowed, smooth, and round. Botanically, it is thought that tomato is a berry, a subset of the fruit, and nutritionally, it is related to vegetables (van Dam et al., 2005).

Tomatoes require warm conditions, and it is sensitive to frost. The optimum soil temperature for the germination of tomato seed is 20°C and rate of germination below 16°C is very slow. Air temperature ranges from 25 to 35°C are best for the ideal growth of fruit, and development is best with appropriate soil moisture. The maximum temperature that a tomato plant can tolerate is 38°C. When temperature falls below 20°C or rises above 38°C, then fruit development is badly affected, and quality is severely reduced (van Dam et al., 2005). Biosynthesis of lycopene is strongly inhibited at temperatures below 12°C, and temperatures above

32°C stop this process altogether (Dumas et al., 2003). Arranging the plant in rows north to south will provide the plants maximum sunlight, as plant develops well in full sunshine, (direct or bright sunlight is required for at least 6–8 hours per day). The tomato plant grows well in the pH range of 5.5–6.8. Water is required for the tomato plant when the soil dries to 1–2 in. in depth (van Dam et al., 2005).

2. CHEMISTRY

Tomato plant produces an irritating smell, taste, and very unpleasant feel to protect the plant from birds and insects that might destroy by feeding. The irritating smell of tomato is only to protect from pests, but it is intoxicating for humans. The characteristic smell of the tomato plant is due to its oil components and due to the sticky yellow secretion produced during harvesting of tomatoes. Tomato plant produce flowers of yellow color that turns into fruit of different colors such as red, yellow, orange, and brown. The volatile compounds present in tomato are linalool, humulene, beta-phellandrene, caryophyllene, 1,8 cineole, eugenol, (E)-2-hexenal, hexenal, limonene, and (Z)-3-hexenal (Buttery et al., 1987). Carotenoid is the pigment that is present in the plant, and it is responsible for the color of tomato fruit, leaves, and flowers (Mendelová et al., 2013).

Tomato fruit contains about 94.5% water per 100 g of its fresh weight, 71.27% carbohydrate, 16% protein per 100 g of dry weight, dietary fibers about 21.82%, 15,145.45 IU of vitamin A, and about 230.91 mg ascorbic acid (Mutari and Debbie, 2011). Tomato is a good source of calcium, magnesium, phosphorus (Srivastava and Kulshreshtha, 2013), and potassium (Willcox et al., 2003). It also contains a great amount of trace elements such as copper, iron, zinc, and manganese (Ahmed et al., 2011). Tomato (100 g) contains 11 mg sulfur and 51 mg chlorine, and these have a great role in the process of detoxification. Folic acid is present in tomato. Carotenoids are present in seed oil. Lycopene, chlorogenic acid, and p-coumaric acid are present in tomato, which are helpful in reducing cancer risk. Vegetable oil of the tomatoes includes steroidal alkaloids, steroidal saponins, flavonoids, carotenoids, and glycoside and fatty acid derivatives (Bhowmik et al., 2012). Structures of some bioactive compounds present in tomato are shown in Fig. 46.2.

3. POSTHARVEST TECHNOLOGY

The suitable regions for the growth of tomato are areas having a hot climate. In early July and mid-October, tomatoes are ready for harvest.

(i) Lycopene

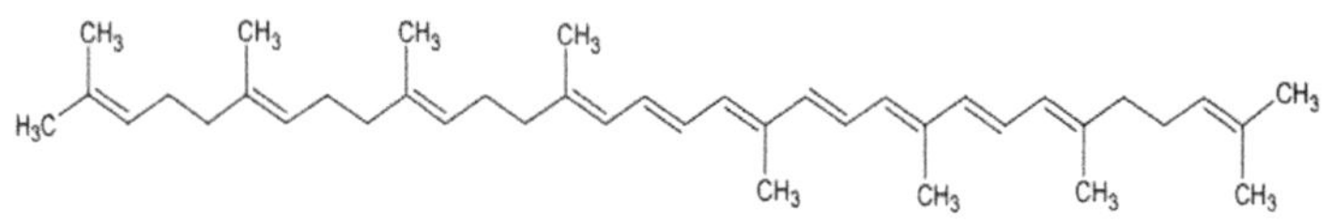

(ii) Carotenoids

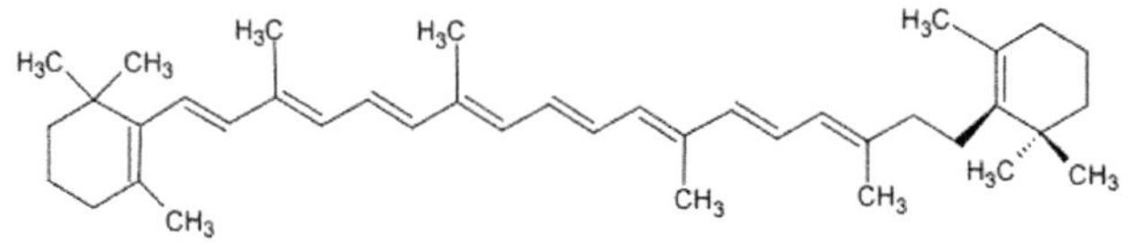

(iii) Vitamin C

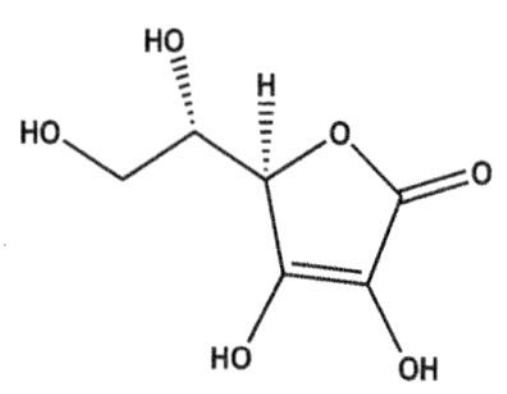

(iv) Folic acid

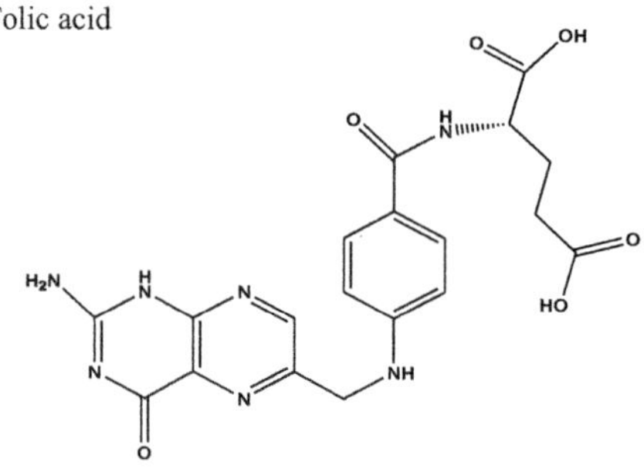

FIGURE 46.2 Structures of some bioactive compounds present in tomato.

Growers prefer to harvest the tomato crop after the sunset to avoid the daytime heat. In tropical regions, the main factor that is associated with the shelf life of tomatoes is increased respiration, which leads to faster ripening of tomato fruit and deterioration of fruit quality. Mechanical harvesters are used for picking the tomato in the field and shaking the tomatoes off the vine. Tomatoes are also sensitive to chilling. These can be stored in the refrigerator for 2 or 3 days. Sometimes, tomatoes are treated with calcium to increase shelf life. Calcium helps in sustaining the integrity of the cell wall and reduces the activity of cell wall degradation enzyme and also reduces softening of fruit. Storage is the main concern after harvesting to transport fruit to the market. Tomatoes are very sensitive to chilling, and storage temperatures that are recommended are different depending on the maturity of fruit. For slow ripening, 13°C temperature and 90%–95% humidity are recommended. At this temperature, tomatoes can be stored for 2–3 weeks, and color change occurs very slowly. In cold storage, 8 to 10°C temperature and humidity of 85%–90% are recommended for storing unripe tomatoes for 4 weeks. A temperature of 7°C and 90% humidity is required to store fully ripe tomatoes for 1 week (Gould, 2013).

4. PROCESSING

Tomatoes are consumed in a variety of ways such as in cooking, salads, and in making various products. Tomatoes are transported immediately from the field to a grading station, then to a processing plant where tomatoes are peeled, sliced, diced, and sauced into the familiar canned tomato products seen on store shelves. Bamboo baskets and plastic crates are used for packing of tomatoes for local markets. Plastic crates can be piled one on the other, and a contoured rim protects the tomatoes and keeps their freshness and allows circulation of air. Packing should be rigid to protect the tomatoes from being damaged.

Tomato ketchup is an important ingredient of our daily food. Tomato ketchup is prepared from fully ripe tomatoes. Pulp of tomato is extracted by boiling, then crushing and straining. Tomato fruit pulp is heated by adding 1/3 of sugar. Sauces of different flavors are added, mixed gently, and again heated to boil. Salt and sugar is added and heated until it becomes thick. Sodium benzoate dissolved in a small amount of water is added to the mixture, and it is kept covered for 5 minutes. Then it is poured into sterilized bottles (Gould, 2013).

Tomato volatile oil components can be extracted by supercritical CO_2 extraction method (Roy et al., 1996). Tomato seed oil is extracted by immersion in hexane and ultrasound-assisted solvent extraction (Arabani et al., 2015).

5. VALUE ADDITION

Tomato is an important food crop of the world. Ripe tomato is known as a fruit, and green tomato is considered a vegetable. Different types of minerals, fibers, and vitamins are present in tomato. Tomatoes can be used in the preparation of soups, pickles, purees, salads, and sauces and also used as a vegetable in many other ways (Shi and Maguer, 2000).

6. USES

Tomato is useful for human health and it decreases the chances of cancer, cardiovascular disease, and osteoporosis. Those people who eat tomato on daily basis have a low risk of contracting cancer diseases such as oral, colorectal, breast, cervical, stomach, prostate, lung, and many other kinds of cancer. Tomato in combination with garlic has a preventive effect against cancer. But how it prevents cancer, the mechanism of its anticancer activity, is still not known. Tomato juice restores the health of kidney patients during dialysis and is known as an energy drink. Fully ripened red tomato has a powerful antioxidant activity. Tomato is a valuable vegetable or fruit for rapid replacement of skin cells. Tomato juice is known as a great sport drink to restore from sleepiness and fatigue (Bhowmik et al., 2012).

For a long time, tomato has been appreciated due to its culinary properties and its organoleptic nature. In recent times, tomato has also been valued for to its health traits and dietetic and nutritional properties. Numerous medical researchers have shown that tomato is a good source of antioxidant compounds (Serio et al., 2006). For genomic and genetic studies, tomato is a model species. Certain enzymes and oils are present in the leaves of tomato that give them a very strong fresh tomato aroma. At the end of cooking, adding a few leaves of tomato to the batch of soup or in the pasta sauce boosts the fresh flavor and gives a very fresh aroma, just like adding basil.

Tomato is a rich source of carbohydrates, calcium, iron, chlorine, sulfur, copper, potassium, sodium, phosphorous, oxalic acid, magnesium, ascorbic acid, niacin, riboflavin, thiamine, vitamin A, fat, fiber, energy, moisture, and also vitamin C (Ahmed et al., 2011; Bhowmik et al., 2012; Mutari and Debbie, 2011). Due to presence of vitamin C, tomato juice can be used for the healing of sunburn (Bhowmik et al., 2012).

S. lycopersicum is the second most consumed and produced fruit or vegetable throughout the world. About 80% of the lycopene in American food comes from tomatoes and tomato-based products such as sauce, tomato ketchup, and tomato paste. Lycopene in the tomato is not always the same; it can vary from 50 mg/kg in the red tomato to 5 mg/kg in the yellow tomato. Foods such as papaya, grapefruit, and watermelon also contain lycopene but in a smaller amount compared to tomatoes (Paoli et al., 2008). Traditionally, tomatoes have been used in case of diarrhea and frequent urination. Half-ripe tomatoes are mostly used in dyspepsia, in homeopathic remedies. These are also used in case of body aches and hay fever. Tomatoes are useful in case of itching, ulcer, back pain, headache, sciatica, running sores, and to treat burns. Tomatoes are good for the kidneys, providing essential antioxidant activity, lowering the cholesterol level in body, and also curing vasodilatation and dysentery. Tomato is helpful in respiratory treatment such as husky voice and pain in chest. Tomato is very useful in the treatment of sharp pain of pectoral muscle (Shukla et al., 2013).

7. PHARMACOLOGICAL USES

7.1 Antiplatelet and Antiaggregation Activity and Endothelial Protection

Antiplatelet activity was found in ripe tomato fruit. Bioactive compounds of tomato were isolated, their structure identified, and their effect on platelet secretion, adhesion, and aggregation studied (Fuentes et al., 2012). It has been examined that tomato has the property of platelet antiaggregation *in vivo* and *in vitro* by inhibiting platelet aggregation induced by collagen and adenosine diphosphate (ADP). Aqueous and methanol extract of tomato showed similar platelet antiaggregation activity *in vivo*. Inhibition of platelet aggregation takes place by both types of extracts induced by ADP. Both aqueous and methanol extracts, under the influence of different temperature, maintained their platelet antiaggregation activity, showing that active compounds responsible for platelet antiaggregation activity were not affected by heat (Palomo et al., 2012).

Through bioassay-guided isolation, it was possible to isolate, identify, and detect a bioactive component with antiplatelet activity. The antiplatelet activity is specific for adenosine. Adenosine is also present in extract of tomato and in its processed byproducts, so it may be used as a functional ingredient to induce antiplatelet activities in processed foods, which may be helpful in primary prevention of cardiovascular diseases (Fuentes et al., 2012).

7.2 Oxidative Stress and Antioxidant Activity

The potent antioxidant lycopene is present in fresh tomatoes and in processed tomato products. Lycopene prevents damaging of DNA and also protects from mutation that leads to cancer (Rao and Agarwal, 1999).

7.3 Antiinflammatory Activity

Significant antiinflammatory activity has been shown by methanolic extract of the leaves of tomato. The antiinflammatory reaction of tomato extract on lipopolysaccharide stimulated the macrophages. By observing the production of prostaglandin E_2, its inflammatory and inhibitory activity was investigated. Tomato leaf extract reduced the production of Cyclooxygenase2 (PGE2) (Li et al., 2014).

7.4 Antifungal Activity

Extract of tomato leaves shows a great antifungal activity. The antifungal activity of tomato leaf volatiles was determined against three types of pathogenic fungi: *Fusarium oxysporum, Glomerella cingulata*, and *Botryotinis fuckeliana*. The growth of *G. cingulata* and *B. fuckeliana* was inhibited completely by tomato leave volatiles. These results showed that tomato leaves have a defense response against pathogenic fungi and tomato volatile leaves are biologic control agents (Kobayashi et al., 2012).

7.5 Anticarcinogenic Activity

Lycopene, chlorogenic acid, and p-coumaric acids are present in tomato, which is helpful in reducing the chances of cancer. Those people who eat tomato on daily basis have a low risk of oral, colorectal, breast, cervical, stomach, prostate, lung, and many other kinds of cancer. Anticarcinogenic activity of tomato was studied in mice in which carcinoma was induced by croton oil (carcinogen). The experimental group that was treated with *S. lycopersicum* had a prolonged time for the appearance of papillomas compared to the animals that were administrated with carcinogen only. The decrease in tumor cells may be due to effect in the promotional stage of tumorigenesis that stops the depletion of free radicals. Tomato and lycopene have a great importance in prevention of prostate and many other kinds of cancer (Agrawal et al., 2009).

7.6 Antimutagenicity

The clastogenic effect (those agents that break the chromosome and cause mutation) of tomato fruit extract has been observed against cyclophosphamide-induced chromosomal aberration in bone marrow cells of mice. Treatment with single but different test doses of *S. lycopersicum* fruit extract provided defense, when given 24 hour prior to single administration of cyclophosphamide (50 mg/kg body weight). A dose-dependent inhibition of chromosomal aberration was investigated. *S. lycopersicum* fruit extract seems to have preventive potential against cyclophosphamide-induced chromosomal aberration in the bone marrow cells of mice. These results shows that tomato fruit extract has a genotoxic potential (Raja et al., 2010).

7.7 Hypercholesterolemia Effects

A high level of cholesterol in blood is called hypercholesterolemia. Cholesterol is synthesized in the animal body, not in plants, and both lycopene and carotene present in tomatoes require cholesterol for their initial synthetic pathway (Fuhrman et al., 1997).

7.8 Effects on Immune Function

Tomato helps in the regulation of the immune system of the body. Nine adult women were subjected to intake of 24 g of tomato puree for 14 days. This tomato puree increased the carotenoid concentration in lymphocyte and plasma. This increase was related to an increase in lymphocyte resistance to oxidative stress. Before and after the ingestion of tomato puree, concentration of carotenoids was determined by High-performance liquid chromatography (HPLC) and resistance of lymphocyte to oxidative stress by comet assay (it detects DNA strand break down). By the intake of tomato puree, carotenoid concentration was increased in plasma and lymphocytes, so a 50% reduction occurred in lymphocyte DNA damage. It has observed that a small amount of tomato and tomato products including pasta sauces, soups, or juices added to food for a short period of time can increase oxidative resistance of lymphocytes and concentration of carotenoids (Boosalis et al., 1996).

7.9 Antiobesity Activity

Tomato is known to have antioxidant activities and lipid-lowering effects. Ethanolic extract of tomato has an antiobesity activity. Paste of *S. lycopersicum* reduced the plasma level of malondialdehyde and increased the activities of catalase, glutathione peroxide, and superoxide dismutase in hamsters. Tomatine is the major component of the green tomato. It forms the tomatine cholesterol complex, which is used to lower the serum LDL cholesterol (Friedman et al., 2000; Shukla et al., 2013). Use of tomato on a daily basis decreases the weight of adipose tissues, body mass index, food intake, and also body weight. Due to the antioxidant property of tomato, the ethanolic extract of *S. lycopersicum* produced an antiobesity effect (Shukla et al., 2013).

7.10 Anthelmintic Activity

Anthelmintic drugs are parasite repellents. Anthelmintic activity was significantly shown by the methanolic extract of tomato leaves. Even a small amount of methanolic extract of tomato is more effective in causing paralysis and death of earthworms than the standard drug (Manthri et al., 2011).

8. SIDE EFFECTS AND TOXICITY

Tomatoes are a safe food. Use of tomatoes leaves in large amounts is unsafe and can cause poisoning. The symptoms of poisoning may include severe throat and mouth irritation, headache, dizziness, diarrhea, mild spasms, and even death in severe cases. Not enough is known about the safety of the tomato vine.

References

Agrawal, R., Jain, R., Raja, W., Ovais, M., 2009. Anticarcinogenic effects of Solanum lycopersicum fruit extract on Swiss albino and C57 Bl mice. Asian Pacific Journal of Cancer Prevention 10, 379–381.

Ahmed, L., Martin-Diana, A.B., Rico, D., Barry-Ryan, C., 2011. The antioxidant properties of whey permeate treated fresh-cut tomatoes. Food Chemistry 124, 1451–1457.

Aoki, K., Yano, K., Suzuki, A., Kawamura, S., Sakurai, N., Suda, K., Kurabayashi, A., Suzuki, T., Tsugane, T., Watanabe, M., 2010. Large-scale analysis of full-length cDNAs from the tomato (*Solanum lycopersicum*) cultivar Micro-Tom, a reference system for the Solanaceae genomics. BMC Genomics 11, 1.

Arabani, A.A., Hosseini, F., Anarjan, N., 2015. Pretreatment and extraction of oil from seeds of tomato pomace using ultrasound. International Journal of Biosciences 6, 261–268.

Bauer, M., Barney, D.L., Robbins, J.A., 2009. Growing Tomatoes in Cool, Short-Season Locations. University of Idaho Extension.

Bhowmik, D., Kumar, K.S., Paswan, S., Srivastava, S., 2012. Tomato-a natural medicine and its health benefits. Journal of Pharmacognosy and Phytochemistry 1, 33–43.

Boosalis, M.G., Snowdon, D.A., Tully, C.L., Gross, M.D., 1996. Acute phase response and plasma carotenoid concentrations in older women: findings from the nun study. Nutrition 12, 475–478.

Buttery, R.G., Ling, L.C., Light, D.M., 1987. Tomato leaf volatile aroma components. Journal of Agricultural and Food Chemistry 35, 1039–1042.

Dumas, Y., Dadomo, M., Di Lucca, G., Grolier, P., 2003. Effects of environmental factors and agricultural techniques on antioxidantcontent of tomatoes. Journal of the Science of Food and Agriculture 83, 369–382.

Fontana, E., Nicola, S., Tibaldi, G., 2008. Tomato production systems and their application to the tropics. International Symposium on Tomato in the Tropics 821, 27–34.

Friedman, M., Fitch, T., Yokoyama, W., 2000. Lowering of plasma LDL cholesterol in hamsters by the tomato glycoalkaloid tomatine. Food and Chemical Toxicology 38, 549–553.

Fuentes, E., Castro, R., Astudillo, L., Carrasco, G., Alarcón, M., Gutiérrez, M., Palomo, I., 2012. Bioassay-guided isolation and HPLC determination of bioactive compound that relate to the antiplatelet activity (adhesion, secretion, and aggregation) from *Solanum lycopersicum*. Evidence-based Complementary and Alternative Medicine 2012.

Fuhrman, B., Elis, A., Aviram, M., 1997. Hypocholesterolemic effect of lycopene and β-carotene is related to suppression of cholesterol synthesis and augmentation of LDL receptor activity in macrophages. Biochemical and Biophysical Research Communications 233, 658–662.

Gould, W.A., 2013. Tomato Production, Processing and Technology. Elsevier.

Jenkins, J.A., 1948. The origin of the cultivated tomato. Economic Botany 2, 379–392.

Kimura, S., Sinha, N., 2008a. Tomato (*Solanum lycopersicum*): a model fruit-bearing crop. Cold Spring Harbour Protocols 2008 pdb. emo105.

Kimura, S., Sinha, N., 2008b. Tomato (*Solanum lycopersicum*): a model fruit-bearing crop. CSH Protocols 3, 1–9.

Kobayashi, F., Ishida, K., Ikeura, H., Odake, S., Hayata, Y., 2012. Application of tomato (*Solanum lycopersicum*) leaf volatiles as antifungal agents against plant pathogenic fungi. Journal of Agricultural Science 4, 231.

Li, H., Deng, Z., Liu, R., Loewen, S., Tsao, R., 2014. Bioaccessibility, in vitro antioxidant activities and in vivo anti-inflammatory activities of a purple tomato (*Solanum lycopersicum* L.). Food Chemistry 159, 353–360.

Machida-Hirano, R., 2015. Diversity of potato genetic resources. Breeding Science 65, 26–40.

Manthri, S., Sravanthi, K.C., Sidagonde, S., 2011. Anthelmintic activity of tomato leaf extract. Journal of Phytology 3.

Mendelová, A., Fikselová, M., Mendel, Ľ., 2013. Carotenoids and lycopene content in fresh and dried tomato fruits and tomato juice. Acta Universitatis Agriculturae et Silviculturae Mendelianae Brunensis 61, 1329–1337.

Mutari, A., Debbie, R., 2011. The effects of postharvest handling and storage temperature on the quality and shelf of tomato. African Journal of Food Science 5, 340–348.

Palomo, I., Fuentes, E., Padró, T., Badimon, L., 2012. Platelets and atherogenesis: platelet anti-aggregation activity and endothelial protection from tomatoes (*Solanum lycopersicum* L.). Experimental and therapeutic medicine 3, 577–584 (Review).

Paoli, S.d., Dias, A.P., Capriles, P.V., Costa, T.E., Fonseca, A.S., Bernardo-Filho, M., 2008. Effects of a tomato (*Solanum lycopersicum*) extract on the labeling of blood constituents with technetium-99m. Revista Brasileira de Farmacognosia 18, 190–196.

Raja, W., Agrawal, R., Ovais, M., 2010. Effects of *Solanum lycopersicum* fruit extract on Cyclophosphamide-induced chromosome aberrations in mouse bone marrow cells. Pharmacologyonline 1, 909–914.

Rao, A., Agarwal, S., 1999. Role of lycopene as antioxidant carotenoid in the prevention of chronic diseases: a review. Nutrition Research 19, 305–323.

Roy, B.C., Goto, M., Hirose, T., 1996. Temperature and pressure effects on supercritical CO_2 extraction of tomato seed oil. International Journal of Food Science and Technology 31, 137–141.

Schauer, N., Zamir, D., Fernie, A.R., 2005. Metabolic profiling of leaves and fruit of wild species tomato: a survey of the *Solanum lycopersicum* complex. Journal of Experimental Botany 56, 297–307.

Serio, F., Ayala, O., Bonasia, A., Santamaria, P., 2006. Antioxidant Properties and Health Benefits of Tomato. Studium Press, Houston, TX, USA, pp. 163–183.

Shi, J., Maguer, M.L., 2000. Lycopene in tomatoes: chemical and physical properties affected by food processing. Critical Reviews in Food Science and Nutrition 40, 1–42.

Shukla, P., Bajpai, K., Tripathi, S., Kumar, S., Gautam, G.K., 2013. A review on the taxonomy, ethnobotany, chemistry and pharmacology of (Solanum lycopersicum L.). International Journal of Chemistry and Pharmaceutical Sciences 1 (8), 521–527.

Smith, A.F., 2001. The Tomato in America: Early History, Culture, and Cookery. University of Illinois Press.

Srivastava, S., Kulshreshtha, K., 2013. Nutritional content and significance of tomato powder. Annals of Arid Zone 52, 121–124.
van Dam, B., Goffau, M.d., van Lidt de Jeude, J., Naika, S., 2005. Cultivation of Tomato: Production, Processing and Marketing.
Willcox, J.K., Catignani, G.L., Lazarus, S., 2003. Tomatoes and Cardiovascular Health.

Further Reading

Diop, N., Jaffee, S., 2005. Fruits and vegetables: global trade and competition in fresh and processed product markets. Global Agricultural Trade and Developing Countries 237–257.
Elson, C.E., Peffley, D.M., Hentosh, P., Mo, H., 1999. Isoprenoid-mediated inhibition of mevalonate synthesis: potential application to cancer. Experimental Biology and Medicine 221, 294–311.
Giuffrè, A.M., Capocasale, M., Zappia, C., Sicari, V., Pellicanò, T.M., Poiana, M., Panzera, G., 2016. Tomato seed oil for biodiesel production. European Journal of Lipid Science and Technology 118 (4), 640–650.
Ibrahim, H.S., Ahmed, L.A., El-din, M.M.E., 2008. The functional role of some tomato products on lipid profile and liver function in adult rats. Journal of Medicinal Food 11, 551–559.
Li, Z., Palmer, W.M., Martin, A.P., Wang, R., Rainsford, F., Jin, Y., Patrick, J.W., Yang, Y., Ruan, Y.L., 2011. High invertase activity in tomato reproductive organs correlates with enhanced sucrose import into, and heat tolerance of, young fruit. Journal of Experimental Botany 63 (3), 1155–1166.
Suárez, M.H., Rodríguez, E.R., Romero, C.D., 2008. Chemical composition of tomato (*Lycopersicon esculentum*) from Tenerife, the canary islands. Food Chemistry 106, 1046–1056.

CHAPTER

47

Tree Turmeric

Ammara Shamim[1], *Haq Nawaz*[1], *Muhammad Asif Hanif*[1], *Muhammad Idrees Jilani*[2], *Ubirajara Lanza Júnior*[3]

[1] Department of Chemistry, University of Agriculture, Faisalabad, Pakistan; [2] Department of Chemistry, University of Sahiwal, Sahiwal, Pakistan; [3] Federal University of Goiás – Special Academic Unit of Health Sciences-Regional Jataí, Course of Biomedicine, Campus Jatobá – University City-Brazil

OUTLINE

https://doi.org/10.1016/B978-0-08-102659-5.00047-1

Copyright © 2020 Elsevier Ltd. All rights reserved.

1. BOTANY

1.1 Introduction

Indian barberry or tree turmeric or zarishk are common names of *Berberis aristata*. It is a rhizomatous herbaceous perennial plant, its family is Berberidaceae, and the genus is *Berberis* (pronounced bẽr′ber-is). This plant is often found in deserts and is eaten by the people of the surrounding areas. The fruit is full of delicious juice and used as supplement to reduce thirst. It is a good source of natural sugar and is a complete diet source. Tree turmeric includes about 450–500 species around the world, and it is mostly found in the moderate and subtropical regions of Africa, Asia, and Europe. Tree turmeric is most commonly and widely found in Nepal and in Himalaya regions of Pakistan and India. It is present in nature in the damp regions of Sri Lanka (Potdar et al., 2012).

1.2 History/Origin

Tree turmeric is native to southern Asia. It needs a considerable amount of annual rainfall to thrive and a temperature range of 20–30°C. Turmeric is the most important part of Siddha and has been used for thousands of years in Asia. The name:"turmeric" originated from early English as turmeryte or tarmaret. It is said that it is of Latin origin from terra merita (merited earth).

1.3 Demography/Location

B. aristata is commonly found throughout Pakistan and India. This plant has been announced as an endangered species (Mazumder et al., 2011).

1.4 Botany, Morphology, Ecology

B. aristata, with a height of 2–3 m, is a little, straight, spiny shrub. It is a bark-containing plant thus is a woody plant; its bark seems to be yellow to brown from outside, but deep inside, it is yellow. Three leaves, which are modified like thorns, cover the bark, and in the longitudinal strips, they can be removed by hand. The leaves are approximately 4.9 cm long, 1.8 cm broad, and are arranged in tufts of 5–8. From the dorsal surface the leaves are deep green, and on the ventral surface, they are light green. With pinnate venation the leaves are simple. There are several to many sharp depressions along the margins of the leaf, and leaves are toothed and bear a leathery texture (Ali et al., 2008). The roots are yellowish-brown and cylindrical in appearance and are thick textured. Their roots are covered with thin, brittle bark, which is deeply furrowed, and its anatomy shows that it as rough and pale brown in color. The roots are odorless and bitter in taste and short textured. If the surface is cut down, it is bright yellow in color. The bark on the root is porous, diffused, rough, fibrous with fine ridges, and the growth rings are discernible and are fracture hard. The stem of this plant is 0.2–1.0 cm in diameter and nearly cylindrical, having longitudinal fine wrinkles running parallel to each other on the external surface. The scars of the lateral are branching with few lenticels seen on the external surface. Moreover, a transversely cut and smoothened surface shows a yellowish bark about half of the size of the radius of the stem and a yellow wood with a distinct central pith; the fracture is short and slightly fibrous, having a bitter taste (Rathi et al., 2013).

2. CHEMISTRY

A group of compounds that are curcuminoids consisting of curcumin (diferuloylmethane), demethoxycurcumin, and bisdemethoxycurcumin are the most crucial chemical components of *B. aristata*. A yellow matter, salts, ferric oxide, starch, albumen, volatile oil, resin, and fatty substances are abundantly present in *B. aristata*. Resins present in this plant consist of glucoside and turpethin, which are insoluble in ether but soluble in alcohol. The fraction of these resins present in the roots is brown in color and cannot be removed by animal charcoal. This plant contains curcumin, which is mixture of phenolic compounds. Turmerone, zingiberene, cineole, and other monoterpenes are present in the volatile oil of plant. The plant also contains proteins, a high quantity of vitamin A, starch, and many other vitamins.

The root bark of *B. aristata* contains berberine (a quaternary ammonium salt from the group of isoquinoline alkaloids) (2,3–methylenedioxy-9,10-dimethoxyprotoberberine chloride). About 2.3% protein, 2% ash, 12% sugars, 0.4% pectin, and 0.6% tannin are present in the fruit. Per 100 mL, juice of the fruit contains about 4.6 mg of vitamin C (Kamboj, 2000). Turmeric plant shows fluorescence due to the presence of curcumin. It makes its fluorescence if sprinkled into alcohol and illuminated by ultraviolet light (yellow-green fluorescence). It is not soluble in water. Turmeric can be used as an indicator since in acidic solutions it remains yellow and in alkaline solutions it gives a red color.

Taxilamine, palmatine, karachine, aromoline, berbamine, barberine, and oxyberberine are essential chemicals that are present in the plant. *B. aristata* consists of alkaloids like palmatine, dehydrocaroline, berbamine, berberine, epiberberine, oxycanthine, jatrorrhizine, aromaline, columbamine, taximaline, karachine, dihyrokarachine, and oxyberberine. Pseudopalmatine chloride, pseudoberberine chloride, pakistanine, and 1-O methyl pakistanine are the four alkaloids that are extracted from *B. aristate* due to medicinal properties. Another alkaloid called simple isoquinoline, or secobisbenzlisoquinoline alkaloid, is also separated from *B. aristata*. The alkaloids abundantly found in *B. aristata* are berberine and palmatine. The potassium and moisture levels in the soil influence contents of berberine present in this plant (Mazumder et al., 2011). A lot of vitamins are present in the plant; these vitamins are B1, B2, B3, C, E, and A. Mineral content of *B. aristata* is very high; it consists of many important minerals like zinc, selenium, potassium, manganese, chromium, iron, phosphorous, and calcium. The plant is enriched with many carotenoids, indicated by bright yellow color, xanthophylls indicated by orange color, and carotenes indicated by red color. The structures of some bioactive compounds of tree turmeric are shown in Fig. 47.1.

3. POSTHARVEST TECHNOLOGY

B. aristata plant parts are usually harvested at full bloom and stored in a dry place for further use or processing. Tree turmeric is traditionally dried by hanging washed bundles inverted in a dry and shaded place or placing whole spread leaves between sheets of papers to prevent oxidation and discoloration. Forced warm air drying is used for industrial production. Tree turmeric leaves should be dried immediately after harvesting because they darken if exposed to open air for an extended period. Drying should be done in temperatures not exceeding 40°C for the minimum loss of volatile compounds. The dried form of tree turmeric can be stored for 1 year when protected from heat, light, and moisture (Patel et al., 2012).

4. PROCESSING

B. aristata (tree turmeric), like other herbal plants, is consumed in a variety of ways and for various purposes. In addition to fresh leaves, other common processed forms of tree turmeric include whole dry

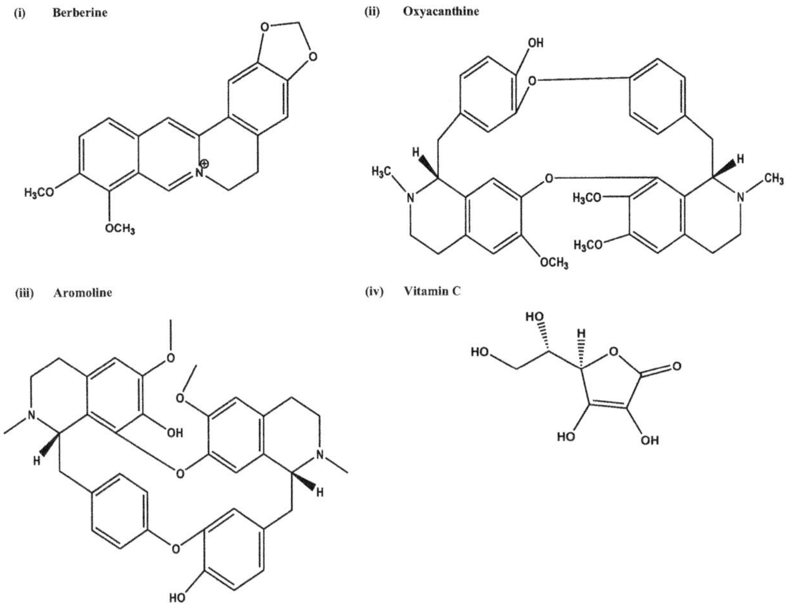

FIGURE 47.1 Structure of some bioactive compounds present in tree turmeric.

leaves, frozen, powdered leaves, and extracted essential oils. The whole plant or chopped leaves can be stored frozen, with and without oils, to be used for an extended time beyond its fresh shelf life. The leaves, bark, and roots of tree turmeric are dried in the shade and powdered.

5. VALUE ADDITION

Tree turmeric fresh leaves consist of the essential oil that is used by the people of India in their religious purposes. It is used in homeopathic medicines to cure many diseases. Drying of the leaves of tree turmeric loses its taste. Leaves are chilled to preserve their taste. Under low temperature, these leaves are preserved for long-term use in many ways. But the storage length increases by drying leaves. If preservation is required to last for a long time, leaves are dried under the shade at a high temperature and kept away from moisture to avoid fungal attack (Kader, 2002). In many countries like Malaysia, Thailand, and India, tree turmeric is used in food in many ways. Fresh leaves of the plant are used in food, and dried leaves are soaked in water to get the extract for its use in cooking food. Tree turmeric is used in curry powder, and it is a food coloring agent. Its leaves are used to degas, reduce vomiting sensation, and help to improve digestion. Turmeric helps prevent diabetes, atherosclerosis, and cures many cancers like ovarian cancer. In cosmetics, tree turmeric is used to cure eczema. From ancient times till now, the paste of turmeric has been used to cure dryness of skin, for fairer complexion, for smooth and soft skin, for glowing skin, and to help to remove blemishes and spots on the skin. In medicine, it is used in the preparation of antioxidants, so it helps to remove harmful free radicals from the body. It also poses an antiinflammatory property, which reduces the chances of osteoarthritis and rheumatoid arthritis (Chaturvedi, 2009).

6. USES

For more than 2500 years, *B. aristata* played a crucial role in herbal healing. It avoids plague. For the treatment of dysentery, it was used by India's Ayurvedic healers. For the cure of liver diseases and gallbladder sickness, European herbalists used it in the early Middle Ages. For treatment of abnormal uterine bleeding, inflammation, and high blood pressure, it is used by Russian healers. Barberry was distinguished by American Indians due to similarity with "Oregon grapes." Bitter tonic tinctures are made from *B. aristata*. These tinctures are used for the treatment of many specific diseases like intermittent and remittent fever and for cholagogue (agent to promote discharge of bile), stomachic, antiperiodic, and alterative. The tincture is also used in debility (state of mind weakness) and is also very popular in the treatment of periodic neuralgia and menorrhagia. *Berberis* when mixed with honey is called rausat, which is used in the effective treatment of sores, abrasions, aphthous (oral ulcers) and skin ulcerations. This plant is useful in the treatment of jaundice, enlarged spleen treatment, and as an emmenagogue (makes blood flow toward pelvis, starts menstruation). Rausat drug is also used as a diaphoretic, laxative, antipyretic, and antiseptic. The extract obtained by boiling (decoction) root bark of the plant is used for ulcers, hemorrhoids, and as an eyewash for eye infections. For the treatment of leprosy, the decoction is used by the Unani system of medication. Decoction of root bark is also used for curing skin infections and purification of blood (Mazumder et al., 2011).

Paste of this herb is used externally to reduce pain and inflammation, and the best use of this herb is in eye-related disorders. It is also used for quick healing of wounds, and paste of this herb is also useful in syphilis ulcers and anal fistula. It helps to cure spasmodic pain and is very beneficial for patients of menorrhagia and leucorrhoea. Roots of daruharidra are hepatoprotective in nature and fight against liver toxicity. It gives wonderful results in the sufferers of jaundice. It is also antidiabetic in nature. Root powder of this herb triggers the extra formation of insulin and protects the body from harmful effects of diabetes. It is a very effective herb with anticancer properties that fights against human colon cancer. This herb fights against various infections caused by microbes in both males and females. In females, it is used to cure uterine infections. It is also used in earaches and ear discharge. It is also used for constipation and piles. This herb helps to stimulate sweating, and it is used in intermittent fever. It is a much more effective herb in malaria than other herbs. Fruit of this herb is an appetizer, helps to stimulate digestive fire, and it is very effective in patients of dysentery. Paste of this herb is applied over eyelids to reduce conjunctivitis. Decoction of this herb is used in the form of gargles in throat infections and to improve the voice (Potdar et al., 2012). For diverse medicinal properties, the chemical components like bereberine and many similar alkaloids were given credit (Mokhber-Dezfuli et al., 2014)

In Ayurveda, its stem, roots, and fruits are taken for preparation of medicines. In an animal model, the aqueous roots extract of *Berberis* showed antimalarial activity. *Berberis* is used foremost for dyeing and as a tanning agent for leathers. Good dyeing properties are observed on cotton by root extract (Komal et al., 2011).

7. PHARMACOLOGICAL USES

7.1 Antimicrobial Activity

The antimicrobial action of hydroalcoholic concentrates of *B. aristata* were tried against 11 bacteria including the following: *Micrococcus luteus, Bacillus subtilis, Berberis cereus, Enterobacter aerogenes, Escherichia coli, Klebsiella pneumoniae, Proteus mirabilis, Pseudomonas aeruginosa, Staphylococcus aureus, Salmonella typhimurium, Streptococcus pneumoniae,* and eight contagious strains including *Aspergillus nodulins, Candida albicans, Aspergillus terreus, Trichophyton rubrum, Aspergillus spinulosus, Cryptococcus albidus, Aspergillus flavus,* and *Aspergillus niger*. The hydroalcoholic concentrates of base of *B. aristata* were successful against a large portion of the tried microscopic organisms (Mazumder et al., 2011).

7.2 Effects on Central Nervous System

Berberine, an alkaloid disengaged from *B. aristata* Linn, has been utilized as a part of the Indian arrangement of prescriptions as a stomachic, intense tonic, antiamebic, and furthermore in the treatment of Oriental wounds. In purely experimental models a large number of preclinical evidences support a possible role of berberine in the management of mental depression, schizophrenia, anxiety, Alzheimer disease, and cerebral ischemia.

It is shown that berberine has focal sensory system exercises, especially the capacity to restrain monoamine oxidase-A, a protein required in the dysfunction of norepinephrine and serotonin. Berberine (5, 10, 20 mg/kg) repressed the idleness time frame in mice in both constrained swim and tail suspension test; in any case, the impact was not dose dependent (Vuddanda et al., 2010). Berberine (5 and 10 mg/kg) additionally turned around the reserpine-induced behavioral mental depression. Berberine (5 mg/kg) upgraded the counterstability impact of subeffective dosages of different run of the mill yet not atypical stimulant medications in constrained swim test. Berberine (5 mg/kg), taking after its intense organization in mice, brought about expanded levels of norepinephrine (31%), serotonin (47%), and dopamine (31%) in the entire mind. Incessant organization of berberine (5 mg/kg) for 15 days significantly expanded the levels of norepinephrine (29%), serotonin (19%), and dopamine (52%). However, at higher measurements (10 mg/kg), there was no adjustment in the norepinephrine (12%) levels; yet a huge increment in the serotonin (53%) and dopamine (31%) levels was found. The energizer-like impact of berberine (5 mg/kg) in constrained swim test was avoided by pretreatment with L-arginine (750 mg/kg) or sildenafil (sold as the brand name Viagra) (5 mg/kg). Pretreatment of mice with 7-nitroindazole (7-NI) (25 mg/kg) or methylene blue (10 mg/kg) potentiated the impact of berberine (2 mg/kg) in the constrained swim test (Kulkarni and Dhir, 2010).

Pretreatment of mice with (+)- pentazocine (2.5 mg/kg), a high-affinity signal receptor agonist, delivered synergism with subeffective dosage of berberine (2 mg/kg). Pretreatment with different sigma receptor foes, viz., progesterone (10 mg/kg), rimcazole (5 mg/kg), and N-[2-(3,4-dichlorophenyl) ethyl]-N-methyl-2(dimethylamino) ethylamine (BD1047; 1 mg/kg, i.p.) turned around the counterfixed status impacts of berberine (5 mg/kg) (Kulkarni and Dhir, 2010).

Berberine at lower measurements did not influence the locomotor action and barbiturate-initiated rest time. It created mellow hypothermic activity in rats and showed pain relieving impact in mice. Taken together, proposals findings exhibit that berberine applied energizer-like impact in different behavioral ideal models of sadness potentially by balancing mind biogenic amines (norepinephrine, serotonin, and dopamine) (Vuddanda et al., 2010). Further, nitric oxide pathway or potentially sigma receptors are included in interceding its upper-like action in mouse constrained swim test.

7.3 Role in Diabetes Mellitus

An uncontrolled clinical trial explored the impact of berberine on 60 patients with sort 11 diabetes mellitus. The patients fluctuated in seriousness of this issue. Oral dosages (0.3–0.5 g three times each day) were recommended for 1–3 months, together with a helpful eating regimen endorsed for 1 month (Ni et al., 1995). Real side effects of diabetes vanished while enhancing the quality of life for patients, pulses wound up noticeably ordinary, and blood lipids diminished. Fasting glycogenic levels in 60% of patients were controlled. Additionally, testing in creature models demonstrated that treatment with berberine prompted more advantageous pancreatic tissue contrasted with controls. It is recommended that the component of activity of berberine might be related with advancing recovery and utilitarian recuperation of pancreatic β-cells (Ni et al., 1995).

7.4 Hepatoprotective Activity

Berberine, a known compound from *B. aristata* plant, was considered for its conceivable hostility to hepatotoxic action in rats. Pretreatment with berberine 4 mg/kg, orally twice a day for 2 days kept the acetaminophen- or CCl_4-prompted ascension in serum levels of basic phosphatase. Posttreatment with three progressive oral measurements of berberine 4 mg/kg each 6 hours lessened the hepatic harm prompted by acetaminophen, while CCl_4-incited hepatotoxicity was not adjusted, recommending a specific healing impact against acetaminophen (Janbaz and Gilani, 2000).

7.5 Inotropic Activity

B. aristata is an edible plant utilized in South Asian customary medication; specifically, its organic product is utilized as a tonic solution for liver and heart. In confined cardiovascular tissues, *B. aristata* naturally displays a positive inotropic activity (Gilani et al., 1999; Sabnis, 2006).

7.6 Cardiovascular Activity

The literature reports that berberine suppresses oxidative stress by inducing factor-2 nuclear factor-2 related to erythroid-2 (Nrf2) pathway (Hsu et al., 2013) leading to transcription of phase II and cytoprotective enzymes genes of proteins such as NAD (P) H quinone oxidoreductase-1 (NQO-1) and heme oxygenase-1 (HO-1) with a wide range of activities in the regulation of redox state. Currently, Nrf2 is recognized as an important mediator of berberine in reducing oxidative stress (Mo et al., 2014). In conclusion, berberine increases the bioavailability of endogenous nitric oxide in different systems. Perhaps this increase may improve cardiovascular function by potent vasodilatation determined by nitric oxide.

8. SIDE EFFECTS AND TOXICITY

There is not enough information to know if tree turmeric is safe for adults in medicinal amounts. It is unsafe to take tree turmeric during breastfeeding because it contains a chemical called berberine. Berberine can be transferred to the infant through breast milk, and it might cause harm.

References

Ali, M., Malik, A., Sharma, K.R., 2008. Vegetative propagation of *Berberis aristata* DC. An endangered Himalayan shrub. Journal of Medicinal Plants Research 2, 374–377.

Chaturvedi, T., 2009. Uses of turmeric in dentistry: an update. Indian Journal of Dental Research 20, 107.

Gilani, A.H., Janbaz, K.H., Aziz, N., Herzig, M.J., Kazmi, M.M., Choudhary, M.I., Herzig, J.W., 1999. Possible mechanism of selective inotropic activity of the n-butanolic fraction from *Berberis aristata* fruit. General Pharmacology: The Vascular System 33 (5), 407–414.

Hsu, Y.-Y., Tseng, Y.-T., Lo, Y.-C., 2013. Berberine, a natural antidiabetes drug, attenuates glucose neurotoxicity and promotes Nrf2-related neurite outgrowth. Toxicology and Applied Pharmacology 272, 787–796.

Janbaz, K.H., Gilani, A.H., 2000. Studies on preventive and curative effects of berberine on chemical-induced hepatotoxicity in rodents. Fitoterapia 71 (1), 25–33.

Kader, A.A., 2002. Postharvest Technology of Horticultural Crops. UCANR Publications.

Kamboj, V., 2000. Herbal medicine. Current Science-Bangalore 78, 35–38.

Komal, S., Ranjan, B., Neelam, C., Birendra, S., Kumar, S.N., 2011. Berberis aristata: a review. International Journal of Research in Ayurveda and Pharmacy 2, 383–388.

Kulkarni, S., Dhir, A., 2010. Berberine: a plant alkaloid with therapeutic potential for central nervous system disorders. Phytotherapy Research: An International Journal Devoted to Pharmacological and Toxicological Evaluation of Natural Product Derivatives 24, 317–324.

Mazumder, P.M., Das, S., Das, M.K., 2011. Phyto-pharmacology of *Berberis aristata* DC: a review. Journal of Drug Delivery and Therapeutics 1.

Mo, C., Wang, L., Zhang, J., Numazawa, S., Tang, H., Tang, X., Han, X., Li, J., Yang, M., Wang, Z., 2014. The crosstalk between Nrf2 and AMPK signal pathways is important for the anti-inflammatory effect of berberine in LPS-stimulated macrophages and endotoxin-shocked mice. Antioxidants and Redox Signaling 20, 574–588.

Mokhber-Dezfuli, N., Saeidnia, S., Gohari, A.R., Kurepaz-Mahmoodabadi, M., 2014. Phytochemistry and pharmacology of berberis species. Pharmacognosy Reviews 8, 8.

Ni, Y.-x., Liu, A.-q., Gao, Y.-f., Wang, W.-h., Song, Y.-g., Wang, L.-h., Zhang, Y.-h., 1995. Therapeutic effect of berberine on 60 patients with non-insulin dependent diabetes mellitus and experimental research. Chinese Journal of Integrated Traditional and Western Medicine 1, 91–95.

Patel, D.K., Patel, K., Dhanabal, S., 2012. Standardization of *Berberis aristata* extract through conventional and modern HPTLC techniques. Asian Pacific Journal of Tropical Disease 2, S136–S140.

Potdar, D., Hirwani, R., Dhulap, S., 2012. Phyto-chemical and pharmacological applications of *Berberis aristata*. Fitoterapia 83, 817–830.

Rathi, B., Sahu, J., Koul, S., Kosha, R., 2013. Detailed pharmacognostical studies on Berberis aristata DC plant. Ancient Science of Life 32, 234.

Sabnis, M., 2006. Chemistry and Pharmacology of Ayurvedic Medicinal Plants. Chaukhambha Amarabharati Prakashan.

Vuddanda, P.R., Chakraborty, S., Singh, S., 2010. Berberine: a potential phytochemical with multispectrum therapeutic activities. Expert Opinion on Investigational Drugs 19, 1297–1307.

CHAPTER

48

Vanilla

Hafsa Ahmad[1], *Rasheed Ahmad Khera*[1], *Muhammad Asif Hanif*[1], *Muhammad Adnan Ayub*[2], *Muhammad Idrees Jilani*[3]

[1] Department of Chemistry, University of Agriculture, Faisalabad, Pakistan;
[2] Department of Chemistry, University of Okara, Okara, Pakistan;
[3] Department of Chemistry, University of Sahiwal, Sahiwal, Pakistan

OUTLINE

Medicinal Plants of South Asia
https://doi.org/10.1016/B978-0-08-102659-5.00048-3

Copyright © 2020 Elsevier Ltd. All rights reserved.

1. BOTANY

1.1 Introduction

Vanilla (*Vanilla planifolia*) is perennial climbing herb that belongs to family (Orchidaceae). Vanilla has been used for many years and has become an important part in herbal medicine and in flavoring (Bory et al., 2008). The genus *Vanilla Swartz* contains nearly 110 species (Purseglove et al., 1981). The genus is very diverse in the Pacific islands (3 species), tropical America (52 species), Southeast Asia and New Guinea (31 species), Africa (14 species), and Indian Ocean islands (10 species) (Bouriquet, 1954).

Vanilla is an edible fruit that contains flavor and aroma compounds. It is known by different names; In English, it is typically called vanilla. In Urdu and Hindi, it is called vanila. In Bangali, it is called bhenila. In Indonesia, it is called paneli, panili, or vanili. It is known as fanilya in Arabic. Vanilla is one of the most famous scents, which is sweet, comforting, warm, and complex.

1.2 History/Origin

Vanilla beans mainly originated in Mexico and Central American countries such as Costa Rica and Honduras. The word "vanilla" is derived from the Spanish word vanilla, which means little black pods. The vanilla's family Orchidaceae comes from the Greek word orchis meaning testicles. In 1520, vanilla was introduced to Europe through Spanish Conquistadores, and nearly after 300 years, with the discovery of hand pollination of vanilla flowers, commercial production of vanilla started. In the 16th century, Spaniards imported vanilla beans to Spain for chocolate flavoring. In 1807, veins grown in Charles Graville's collection in Paddington were cut and sent to Paris and Antwerp's botanical gardens. In 1819, two plants were sent to Java and Buitenzorg from Antwerp, but only one plant survived the journey. This plant produced flowers but did not give fruit because a Mexican bee was not found in Indonesia and the hand pollination technique was not known. Then plants were sent from Paris to Reunion and in 1840 to Mauritius and from there to Madagascar in 1872. In these countries, the plants grew well, but they did not grow fruits due to the absence of natural pollination (Bruman, 1948). Prof. Charles Morren of Leige discovered the artificial pollination method in 1836. Edmond Albius practically performed this method in 1841. After that, this hand pollination method was used in the tropics, and vanilla beans were produced commercially in all these countries. In 1846, on a systemic basis, vanilla cultivation was introduced into Java (Lubinsky et al., 2008).

In the last century, vanilla has been produced in the tropics due to favorable climate conditions, but vanilla has not had great attention in Mexico, Indonesia, Tahiti, Madagascar, Reunion, Uganda, and Seychelles. It is grown in small amounts in Brazil, Dominica, Central America, Trinidad, Guadeloupe, Malaysia, Fiji, Sri Lanka, Tonga, India, China, and the Philippines. The cost of vanilla rises and falls according to the international market's demand. To pollinate vanilla, the practical artificial method was discovered in the mid-19th century, and this raised vanilla production in the Indian Ocean countries (Reunion Island, Madagascar, Comoros), which become the major vanilla-exporting countries in the world (Sharp et al., 2012).

1.3 Demography/Location

The soil that has good drainage and is rich in humus is suitable for the cultivation of vanilla. Vanilla grows well in a humid tropical climate with annual rainfall of 200–300 cm from sea level to 1500 m above sea level. The ideal temperature range for the growth of vanilla is 21–32°C. Northeastern regions of India (Tamil Nadu, Kerala, Karnataka, and Andaman and Nicobar Island) are the best places for the cultivation of vanilla plant. It is also cultivated in Indonesia, China, Tonga, Turkey, Mexico, Togo, Madagascar, Uganda, Comoros, Guadeloupe, Reunion, and French Polynesia (Bruman, 1948; Lubinsky et al., 2008; Rodolphe et al., 2011).

1.4 Botany, Morphology, Ecology

Vanilla is a creeping plant that can climb as long as 15 m. Leaves are 6–9 in. long, sessile, thick, and elliptical. It is attached to trees and climbs via roots. Roots are whitish, aerial, and long, having a diameter of 2 mm. Flowers are clusters of trumpets shaped with the vine and vary in color from cream to yellow and then to green. There are three sepals that are 4–7 cm long. The stems are 1–2 cm thick, flexible, green, and succulent, having many branches. The veins toss out roots that grow opposite the leaves and through which they cling to trees. Racemes (flowers stems) are not conspicuous and produce more than 20 flowers that are greenish-yellow in color. Rostellum (thin flap) separates the anther (male organ) from the stigma (female organ). The pollen cannot reach from anther to stigma if the rostellum is removed. Vanilla is planted in open land through providing shade for growth. It produces the flowers in the third year after planting. For maximum production of flowers, almost 7–8 years are required. Flowers produce fruits in the form of pods or capsules if the pollination is successful. These pods are commercially called beans (Correll, 1953). Flavor constituents are found inside the vanilla beans and secreted onto the placental region around the seeds. The release of flavor constituents to flavor compounds is catalyzed by the degradative or hydrolytic enzymes, which are located outside the fruit wall region. The epidermis is composed of isodiametric ground epidermal cells in which prominent chloroplasts are absent. Epidermal cells are bounded by thickened, pitted cell walls and have a rhomboidal crystal of calcium oxalate. Stomata are very spaced. Fruit wall consists of 15 unbranched vascular bundles, each having both phloem and xylem along sclerotic bundle sheath. The xylem contains helical and reticulate elements. Outside the vascular bundles, tissue is composed of parenchyma cells that are several times longer than wide. Parenchyma cells have chloroplast and rhomboidal calcium oxalate crystals. In the outer fruit wall, raphide vessels are present in large amounts and secrete a viscous fluid (Havkin-Frenkel et al., 2004).

The depth of soil and relative humidity for the growth of vanilla should be 40 cm or more and 80%, respectively. Vanilla grows well between 10 and 20 degrees latitude south and north, from the sea level up to almost 700 m. According to the climate, vanilla can be grown to a height of 800–1000 m. Vanilla grows in a moist, hot, and tropical climate with frequent but not great rain. Violent winds and rain situations are harmful to the vanilla plant. Optimum temperature range is 21–31°C, average temperature is 27°C, and annual rainfall 2000–2500 mm. The main properties of soil that are suitable for vanilla production are pH value range 6 to 7, good drainage, and abundant organic material (Bongers et al., 2013).

2. CHEMISTRY

Vanillin, vanillic acid, 4-hydroxybenzoic acid, p-hydroxybenzyl alcohol, and 4-hydroxy-benzaldehyde are the main chemical constituents obtained from the vanilla beans. Several variables including growth conditions, species, soil composition, fruit maturity, and processing affect chemical components of vanilla beans (Leong et al., 1989). The flavor famously associated with vanilla results from a complex and varied mixture of chemical compounds. Nearly 170 volatile components are present in vanilla that occur below 1 ppm. Vanillin is used as a main flavor backbone, which is 1.52%–2.42% of bean dry weight. Vanilla also has phenolics, glycosyl, galactose, mannose, and rhamnose. Nearly 2% vanillin is present in fully fermented fruit. The aroma constituents in vanilla are p-hydroxybenzyl methyl ether and p-hydroxybenzaldehyde, and based on the fermentation and storage, alcohol or carboxylic acid can also be present (Havkin-Frenkel and Dorn, 1997). Essential oil is extracted from the seeds, roots, beans, and pods of vanilla (Lucien and Romagnan, 1952). Vanillin and piperonal present in vanilla are responsible for delicious taste and have advantageous health effects (Ranadive, 2006). Glucovanillin is a flavor precursor in vanilla beans. Beta-glucosidase and glycosyl hydrolase are the hydrolytic and degradative enzymes that catalyze the flavor precursor to flavor constituent and are often present in the outer fruit wall.

The quantity of components in vanilla is as follows: 25% carbohydrates, 15%–30% cellulose, 25% sugars, 6% minerals, and 15% fat (Dignum et al., 2001). Vanilla's extract comprises a small quantity of B-complex groups of vitamins including pantothenic acid, niacin, riboflavin, thiamin, and vitamin B-6. The minerals that are found in small amount in vanilla include magnesium, zinc, calcium, iron, manganese, and potassium (Odoux and Brillouet, 2009). Some active components of vanilla are shown in Fig. 48.1.

3. POSTHARVESTING TECHNOLOGY

After pollination, vanilla beans require 7–9 months to mature. A green vanilla bean is like a large green bean that is filled with thousands of small seeds. When the blossom end tips become pale yellow, vanilla beans are plucked before beans ripen. After that, beans develop specific flavor character and also undergo a long, complex curing and drying process.

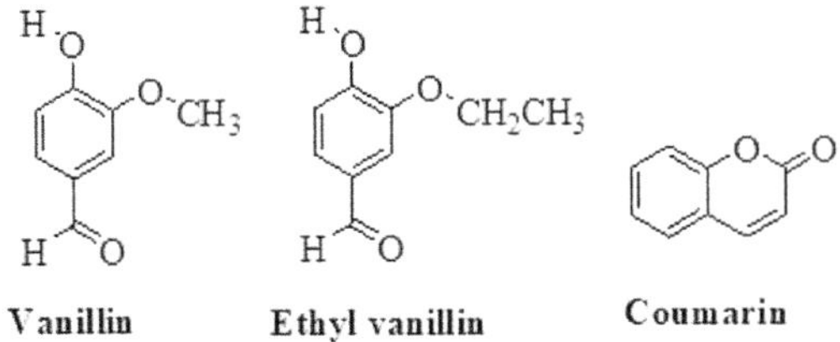

FIGURE 48.1 Some active components of vanilla.

Vanilla is harvested early in the morning. The harvested vanilla beans are green and flavorless, and afterward, these are subjected to 3–6 month's curing process, which is based on different curing procedures in various production regions. The main objective of the curing process is to enhance the vanilla flavor and to dry the beans for extraction that produces the specific vanilla flavor. Harvest includes cutting of beans and packing or saving in the bag. To avoid any physical damage to beans, cutting is done by hand instead of any tool (Choudhury et al., 2018).

4. PROCESSING

The main purpose of processing is to dry the vanilla beans and enhance the characteristic vanilla flavor. Killing, sweating, drying, and conditioning are the main steps in the curing process (Krishnakumar et al., 2007). Modern methods that are commonly used for killing are sun killing, oven killing, hot killing, killing via freezing, and killing through scratching. The major purpose of killing method is to bring the cessation of vanilla bean's vegetative life (Theodose, 1973). Vanilla beans are allowed to sweat after killing. During sweating, vanilla beans that are killed develop the specific vanilla aroma, flavor, and color. Vanilla beans are held at high temperature range 45–65°C for 7–10 days and at high humidity (Jones and Vicente, 1949). The main objective of the sweating stage is to retrain enough wetness, which is necessary to allow enzymes to catalyze different oxidative and hydrolytic processes. During the sweating stage, nonenzymatic reactions may also occur. At the same time, to prevent spoilage by microorganisms, some wetness is allowed to escape and also to reduce water content. Beans contain nearly 60%–70% moisture content at the end of this stage and are dried conventionally for refuge against microbial spoilage and to block further enzymatic activity. Moisture content reaches 25%–30% of bean weight at the end of drying stage. In the curing process, drying is a very tough stage to control. Sometimes, uneven drying may be possible due to the differences in bean moisture content, variable environmental situations, and varying bean size (Leong et al., 1989).

5. VALUE ADDITION

Vanilla is used in rubber tires, industrial chemicals, plastics, and odor-masking paints. Vanilla by-product beans residue is utilized to prepare a high-fiber formulation of functional foods. Vanilla is a natural flavor that is extensively utilized in many industries such as in beverages, pharmaceutics, sodas, food, tobacco, cosmetics, and traditional crafts. Vanilla is versatile and blends well along with various other scents such as spicy, sweet, and floral. Vanilla is also utilized in chocolate, dairy products, perfumes, mist sprays, inhalers, aroma lamps, light bulb rings, baths, massages, liquid candles, desserts, confectionery products, and pastries (Kanisawa et al., 1994).

6. USES

Small vanilla seeds, full fruit, powder, or fruit extract are specifically used in confectionery and sweet foods, as a flavoring agent in foods, and to reduce the quantity of sugar that is necessary for food sweetening. Vanilla plays a very important role in preparation of perfumes for fragrance (Gatfield et al., 2007). Medicinally, vanilla is used to relieve fever, gastric complaints, and used as a stimulant and aphrodisiac. Vanillin, which is the major flavoring component, shows antioxidant and antimicrobial activities. Vanilla oil is used in aromatherapy and found to be useful in improving one's physiologic and emotional health (Shankaracharya and Natarajan, 1973).

Vanilla extract is also useful in the menstruation regulation. It can be utilized to regulate the menstruation cycles of women who do not have regular periods. Slight problems can be easily treated by the extract of vanilla through taking in the diet. Vanilla extract is also useful to lose weight. As exercise and diet play an important role in loss of weight, for more benefits, vanilla extract can be taken. Traditionally, vanilla plant has been used for the treatment of different diseases such as hysteria, ulcers, preventing dental caries, dyspepsia, fever, toothache, and dysmenorrhea. It also has biologic activities like antiinflammatory, antispasmodic, and analgesic. Moreover, antioxidant, antisickle cell, antimicrobial, and antimutagenic effects of vanilla have also been studied. It is utilized for deep coughs, respiratory pain, stomach problems, and congestion. Vanillin, which is the major component in vanilla, has anticarcinogenic and anticancer activity. Essential oil of vanilla has been found to be an impressive mood lifter and is used to cure patients who are suffering from libido loss, erectile dysfunction, impotency, and sexual frigidity. Oil is used to stimulate secretion of human hormones that help to promote arousal.

7. PHARMACOLOGICAL USES

7.1 Antidepressant Activity

The main part of vanilla, vanillin hydroxybenzaldehyde, is a most effective antidepressant. The antidepressant component helps to relax the nerves and calm the brain. Antidepressant activity of *V. planifolia* was evaluated in mice (Dhingra and Sharma, 2005). It was found that vanillin at the dose of 100 mg/kg showed antidepressant activity that was comparable with fluoxetine. Due to the agonistic action at opioid or $\alpha2$ adrenergic receptors and antioxidant properties, vanillin exhibited antidepressant activity. Moreover, it has been seen that magnesium contents in vanilla help to maintain nerve metabolic balance in the body and in optimal neural transmission (Shoeb et al., 2013).

7.2 Antioxidant Activity

Vanilla is used as antioxidant agent that protects the body from wear and tear by neutralizing free radicals. The results showed that alcoholic vanilla extract contained higher antioxidant activity compared to the pure vanilla. Vanilla extract can used as an antioxidant agent in health supplements and for preservation of food (Karathanos et al., 2007).

Aqueous alcoholic extract of vanilla and pure standard components were screened for antioxidant activity by using 2,2-diphenyl-1-picrylhydrazyl (DPPH) and β-carotene-linoleate in vitro methods at the concentration of 200 ppm. It was been found that DPPH shows 43% while β-carotene-linoleate shows 26% antioxidant activity compared to values of 92% and 93% for butylated hydroxyanisole. More surprisingly, it was seen that 4-hydroxy-3-methoxybenzyl alcohol shows 65% antioxidant activity by using β-carotene-linoleate and 90% activity by using DPPH, while 4-hydroxybenzyl alcohol exhibits 45% antioxidant activity by β-carotene-linoleate and 50% by DPPH (Shyamala et al., 2007).

7.3 Antibacterial Activity

Vanilla is widely used to inhibit bacterial quorum sensing. The term quorum sensing is used to explain bacterial cell-to-cell signaling that is involved in bacterial physiology determination (Fuqua and Greenberg, 2002). Antibacterial activity of vanilla extract against pathogenic bacteria *Chromobacterium violaceum* was determined by the disc diffusion method. Vanilla extract was made by using 75% aqueous methanol. It was found that this vanilla extract decreased violacein production, due to which bacterial quorum sensing was inhibited (Klimes and Lamparsky, 1976). The results showed that methanol extract of vanilla has high antibacterial activity compared to the other extracts.

7.4 Antimutagenic Activity

Vanillin has the ability to inhibit chemical carcinogenesis. Antimutagenic activity of vanillin was evaluated in mice. During this evaluation, the effect of vanillin on growth and metastasis of cells was the center of interest. It has been found that vanillin inhibits the growth of cancer cells and has potential against metastasis by reducing the cancer cell's invasiveness (King et al., 2007).

7.5 Anticancer Activity

The essential oil of vanilla has ability to inhibit the growth of cancerous cells due to anticarcinogenic properties. It also acts as an antioxidant that kills free radicals in the body and inhibits the growth of cancerous cells. The bioactive compound called 3-hydroxy-4-methoxybenzaldehyde present in vanilla has great potent anticancer activity (Kundu and Mitra, 2016). It is also reported that vanilla helps to inhibit migration of human lung cancer cells, which are induced by the hepatocyte growth factor and inhibit angiogenesis. In HT-29 human colorectal cancer cell lines with a concentration of 400 μg/mL and NIH/3T3 normal cell lines with concentration 1000 μg/mL, vanillin induces apoptosis (Kumar et al., 2012).

7.6 Antiinflammatory Activity

Vanillin has antiinflammatory activity that provides protection against fibrosis and liver injury, tissue injury, and immune reactions. Vanilla extracts decrease inflammation by preventing the discharge of proinflammatory cytokines and mediators. Antiinflammatory activity of vanilla was evaluated against CCl_4-induced hepatotoxicity in rat. CCl_4 brings about a reduction in the antioxidant enzymes in the liver and thus causes the liver injury. Vanilla has the ability to stop CCl_4-generated free radicals (Makni et al., 2012).

7.7 Antibiofilm Activity

Vanillin is used as a potent inhibitor of biofilm. Antibiofilm activity of vanilla extract was determined by the broth microdilution method against the fungal pathogen *Candida albicans*. Minimum inhibitory concentration was developed at a concentration of 1 mg/mL of vanillin (Raut et al., 2013).

7.8 Antigenotoxic Effects

The antigenotoxic activity of vanillic acid in chromosome and DNA levels was determined by the alkaline comet attempt and cytokinesis block micronucleus test. Vanilla has shown protective effects against DNA damage against the mitomycin C (Erdem et al., 2012).

7.9 Aphrodisiac Activity

The aphrodisiac activity of vanillin was evaluated in rats. The results of a previous study show that vanillin has aphrodisiac properties at the dose of 200 mg/kg in male rats (Maskeri et al., 2012).

7.10 Hypolipidemic Activity

Hypolipidemic activity of vanillin on lipid profile was evaluated in mice. A cholesterol-rich, high-fat diet induced hypolipidemia in mice. Different doses of vanillin were given to mice, but a significant reduction in triglycerides and cholesterol took place at a dose of 200 mg/kg and 400 mg/kg of vanillin compared to atorvastatin group and hypolipidemic control, respectively. It has been found that other doses have no significant effect on lipid profile.

7.11 Antinociceptive Activity

Antinociceptive activity of vanillin was observed in mice. Vanillin was orally administered at doses of 1–10 mg/kg, and it was found that vanillin's antinociceptive action was dose dependent and maintained for 30 minutes. The results show that acetic acid extracts of vanilla develop the antinociceptive property. Antinociceptive activity may be brought about by opioid and α_2-adrenergic receptors (Park et al., 2009).

7.12 Antifungal Activity

Antifungal activity of vanillin was evaluated against medically important yeast. The minimal fungicidal concentration and minimal inhibitory concentration were observed. The result showed that vanillin inhibits the growth of the germ cell that causes the fungal property (Boonchird and Flegel, 1982).

7.13 Antisickle Cell

Antisickling effect for vanillin has been examined in vitro in mice and clinical studies (Zhang et al., 2004). There is an increase in hypoxic conditions and reduction in percentage of sickle cells. The results showed that 1 g vanillin has the most effective antisickling effects in mice and clinical studies (Abraham et al., 1991).

8. SIDE EFFECTS AND TOXICITY

Skin contact can cause irritation and swelling (inflammation). It might also cause headache and sleep problems (insomnia), especially for people who manufacture vanilla extract.

References

Abraham, D.J., Mehanna, A.S., Wireko, F.C., Whitney, J., Thomas, R.P., Orringer, E.P., 1991. Vanillin, a potential agent for the treatment of sickle cell anemia. Blood 77, 1334–1341.

Bongers, F., Charles-Dominique, P., Forget, P.-M., Théry, M., 2013. Nouragues: Dynamics and Plant-Animal Interactions in a Neotropical Rainforest. Springer Science & Business Media.

Boonchird, C., Flegel, T., 1982. In vitro antifungal activity of eugenol and vanillin against *Candida albicans* and *Cryptococcus neoformans*. Canadian Journal of Microbiology 28, 1235–1241.

Bory, S., Lubinsky, P., Risterucci, A.-M., Noyer, J.-L., Grisoni, M., Duval, M.-F., Besse, P., 2008. Patterns of introduction and diversification of vanilla planifolia (Orchidaceae) in reunion island (Indian ocean). American Journal of Botany 95, 805–815.

Bouriquet, G., 1954. Le vanillier et la vanille dans le monde. Paul Lechevalier, Paris.

Bruman, H., 1948. The culture history of Mexican vanilla. The Hispanic American Historical Review 28, 360–376.

Choudhury, D., Dobhal, P., Srivastava, S., Saha, S., Kundu, S., 2018. Role of botanical plant extracts to control plant pathogens - A review. Indian Journal of Agricultural Research 52, 341–346.

Correll, D.S., 1953. Vanilla-its botany, history, cultivation and economic import. Economic Botany 7, 291–358.

Dhingra, D., Sharma, A., 2005. Evaluation of antidepressant-like activity of glycyrrhizin in mice. Indian Journal of Pharmacology 37, 390.

Dignum, M.J., Kerler, J., Verpoorte, R., 2001. Vanilla production: technological, chemical, and biosynthetic aspects. Food Reviews International 17, 119–120.

Erdem, M.G., Cinkilic, N., Vatan, O., Yilmaz, D., Bagdas, D., Bilaloglu, R., 2012. Genotoxic and anti-genotoxic effects of vanillic acid against mitomycin C-induced genomic damage in human lymphocytes in vitro. Asian Pacific Journal of Cancer Prevention 13, 4993–4998.

Fuqua, C., Greenberg, E.P., 2002. Signalling: listening in on bacteria: acyl-homoserine lactone signalling. Nature Reviews Molecular Cell Biology 3, 685.

Gatfield, I., Hilmer, J.-M., Weber, B., Hammerschmidt, F., Reiss, I., Poutot, G., Bertram, H.-J., Meier, D., 2007. Raw materials: chemical and biochemical changes occurring during the traditional madagascan vanilla curing process effects on the sensory qualities of cured beans. Perfumer & Flavorist 32.

Havkin-Frenkel, D., Dorn, R., 1997. Vanilla.

Havkin-Frenkel, D., French, J.C., Graft, N.M., Pak, F., Frenkel, C., Joel, D., 2004. Interrelation of curing and botany in vanilla (*Vanilla planifolia*) bean. Acta Horticulturae 93–102.

Jones, M.A., Vicente, G.d.C., 1949. Criteria for testing vanilla in relation to killing and curing methods. Journal of Agricultural Research 78, 425–434.

Kanisawa, T., Tokoro, K., Kawahara, S., 1994. Flavor Development in the Beans of Vanilla Planifolia, Olfaction and Taste XI. Springer, pp. 268–270.

Karathanos, V.T., Mourtzinos, I., Yannakopoulou, K., Andrikopoulos, N.K., 2007. Study of the solubility, antioxidant activity and structure of inclusion complex of vanillin with β-cyclodextrin. Food Chemistry 101, 652–658.

King, A.A., Shaughnessy, D.T., Mure, K., Leszczynska, J., Ward, W.O., Umbach, D.M., Xu, Z., Ducharme, D., Taylor, J.A., DeMarini, D.M., 2007. Antimutagenicity of cinnamaldehyde and vanillin in human cells: global gene expression and possible role of DNA damage and repair. Mutation Research 616, 60–69.

Klimes, I., Lamparsky, D., 1976. Vanilla volatiles–a comprehensive analysis. International flavours and food additives 7 (6), 272–273.

Krishnakumar, V., Bindumol, G., Potty, S., Govindaraju, C., 2007. Processing of vanilla (*Vanilla planifolia* Andrews) beans-influence of storing fresh beans, killing temperature and duration of killing on quality parameters. Journal of Spices and Aromatic Crops 16, 31–37.

Kumar, R., Sharma, P., Mishra, P.S., 2012. A review on the vanillin derivatives showing various biological activities. International Journal of Pharm Tech Research 4, 266–279.

Kundu, A., Mitra, A., 2016. Methoxybenzaldehydes in plants: insight to the natural resources, isolation, application and biosynthesis. Current Science 111, 1325–1334.

Leong, G., Uzio, R., Derbesy, M., 1989. Synthesis, identification and determination of glucosides present in green vanilla beans (*Vanilla fragrans* Andrews). Flavour and Fragrance Journal 4, 163–167.

Lubinsky, P., Bory, S., Hernández, J.H., Kim, S.-C., Gómez-Pompa, A., 2008. Origins and dispersal of cultivated vanilla (*Vanilla planifolia* jacks.[Orchidaceae])[1]. Economic Botany 62, 127–138.

Lucien R, inventor, Romagnan A, assignee, 1952. Process for extracting aromatic oils from vanilla beans. United States patent US 2, 601–635.

Makni, M., Chtourou, Y., Fetoui, H., Garoui, E.M., Barkallah, M., Marouani, C., Kallel, C., Zeghal, N., 2012. Erythrocyte oxidative damage in rat treated with CCl4: protective role of vanillin. Toxicology and Industrial Health 28, 908–916.

Maskeri, R., Ullal, S.D., Belagali, Y., Shoeb, A., Bhagwat, V., 2012. Evaluation of aphrodisiac effect of vanillin in male wistar rats. Pharmacognosy Journal 4, 61–64.

Odoux, E., Brillouet, J.-M., 2009. Anatomy, histochemistry and biochemistry of glucovanillin, oleoresin and mucilage accumulation sites in green mature vanilla pod (*Vanilla planifolia*; Orchidaceae): a comprehensive and critical reexamination. Fruits 64, 221–241.

Park, S.-H., Sim, Y.-B., Choi, S.-M., Seo, Y.-J., Kwon, M.-S., Lee, J.-K., Suh, H.-W., 2009. Antinociceptive profiles and mechanisms of orally administered vanillin in the mice. Archives of Pharmacal Research 32, 1643.

Purseglove, J., Brown, E., Green, C., Robbins, S., 1981. Vanilla, Spices, vol. 2. Longman Inc, New York.

Ranadive, A., 2006. Vanilla-inside look: chemistry and biochemistry of vanilla flavor-A survey of the latest research. Perfumer & Flavorist 31, 38–45.

Raut, J.S., Rajput, S.B., Shinde, R.B., Surwase, B.S., Karuppayil, S.M., 2013. Vanillin inhibits growth, morphogenesis and biofilm formation by *Candida albicans*. Journal of Biologically Active Products from Nature 3, 130–138.

Rodolphe, G., Séverine, B., Michel, G., Pascale, B., 2011. Biodiversity and evolution in the Vanilla genus. The dynamical processes of biodiversity-case studies of evolution and spatial distribution 1–27.

Shankaracharya, N., Natarajan, C., 1973. Vanilla—chemistry, technology and uses. Indian Food Packer 27, 29–36.

Sharp, M.D., Kocaoglu-Vurma, N.A., Langford, V., Rodriguez-Saona, L.E., Harper, W.J., 2012. Rapid discrimination and characterization of vanilla bean extracts by attenuated total reflection infrared spectroscopy and selected ion flow tube mass spectrometry. Journal of Food Science 77, C284–C292.

Shoeb, A., Chowta, M., Pallempati, G., Rai, A., Singh, A., 2013. Evaluation of antidepressant activity of vanillin in mice. Indian Journal of Pharmacology 45, 141.

Shyamala, B., Naidu, M.M., Sulochanamma, G., Srinivas, P., 2007. Studies on the antioxidant activities of natural vanilla extract and its constituent compounds through in vitro models. Journal of Agricultural and Food Chemistry 55, 7738–7743.

Theodose, R., 1973. Traditional methods of vanilla preparation and their improvement. Tropical Science 15, 47–57.

Zhang, C., Li, X., Lian, L., Chen, Q., Abdulmalik, O., Vassilev, V., Lai, C.S., Asakura, T., 2004. Anti-sickling effect of MX-1520, a prodrug of vanillin: an in vivo study using rodents. British Journal of Haematology 125, 788–795.

CHAPTER

49

Walnut

Zarghouna Chudhary[1], *Rasheed Ahmad Khera*[1], *Muhammad Asif Hanif*[1], *Muhammad Adnan Ayub*[2], *Lamia Hamrouni*[3]

[1] Department of Chemistry, University of Agriculture, Faisalabad, Pakistan;
[2] Department of Chemistry, University of Okara, Okara, Pakistan;
[3] Laboratory of management and valorisation of forest resources, INRGREF-University of Carthage, Ariana, Tunisia

OUTLINE

Medicinal Plants of South Asia
https://doi.org/10.1016/B978-0-08-102659-5.00049-5

Copyright © 2020 Elsevier Ltd. All rights reserved.

1. BOTANY

1.1 Introduction

Juglans regia L. is commonly known as walnut tree, a deciduous plant of Juglans specie, that can attain a height up to 40 m. The bark of walnut tree is grayish black and intensely furrowed (Williams, 1990). English as well as Persian walnut belongs to Juglandaceae family and is highly valuable due to its good quality timber, nuts, and nut oil (Malvolti et al., 1995). There are more than 15 species of walnut trees, and the most common are *Juglans cinerea* (white or butter walnut), English or Persian walnut (*J. regia*), and *Juglans nigra* (black walnut) (LA, 1909). In English *J. regia* is typically called walnut or common walnut. In Urdu, it is commonly known as akhrot (Fig. 49.1). In German, it is called walnub, and in Spanish and French, known as nogal and noyer, respectively.

1.2 History/Origin

The word nut derives from the Latin word "nux" that means an edible kernel surrounded by hard shell. *J. regia* is the botanical name of walnut tree given by Romans, and Juglans means "acorn of Jupiter" and regia means royalty (Manchester, 1987). In Afghani language, it is pronounced as charmarghz, meaning "four brains," as it looks like a human brain (Menninger, 1977).

Walnut is a native plant of central Asia that spread throughout Europe, western China, Afghanistan, Persia, and Caucasus, where it is found in semicultivated as well as wild state (Dandekar et al., 2005). Walnut and its oil were used by Egyptians for the embalming of mummies and for making the strings of musical instruments (Facciola, 1990). In ancient

FIGURE 49.1 Leaves, stems, trees and fruits of Walnuts.

FIGURE 49.1 (Continued)

times, people used the bark of walnut tree to make dark brown or black dye (Galey et al., 1991).

1.3 Demography/Location

J. regia is found in different countries of Asia including Iran, Iraq, Turkey, Azerbaijan, Bhutan, Myanmar, Uzbekistan, Kazakhstan, Nepal, Armenia, Kyrgyzstan, China, Pakistan, Afghanistan, Bangladesh, Russia, Tajikistan, Georgia, and India (Kolov, 1998). Walnut is the second major nut-producing plant after almond. According to an international survey, the global walnut production in year 2010 was 1,500,000 tons, and China is one of the leading producers and exporters of walnuts, followed by America, which covers 34% of global walnut production (Frederick et al., 2014). Global demand and consumption of walnuts are increased day by day due to extensive health benefits. It has been reported that regular consumption of walnut fruit reduces the risk of heart diseases (Hemery, 2004).

1.4 Botany, Morphology, Ecology

J. regia is a large, deciduous tree attaining heights of 25–40 m, and a trunk up to 2 m diameter, commonly with a short trunk and broad crown, though taller and narrower in dense forest competition. It is a light-demanding species, requiring full sun to grow well. Walnut trees flourish in a well-drained and loamy soil that is free from alkali. Soil needs to be moist all the time. The pH of the soil should be in the range of 6–8 (Cogliastro et al., 1997). Walnuts need protection from wind (Polunin, 1977). The bark is smooth, olive-brown when young and silvery-grey on older branches, and it features scattered broad fissures with a rougher texture (Jackson and Looney, 1999). Like all walnuts, the pith of the twigs contains air spaces; this chambered pith is brownish in color. The leaves are alternately arranged, 25–40 cm long, odd-pinnate with five to nine leaflets, paired alternately with one terminal leaflet (Gleeson, 1982). The largest leaflets are the three at the apex, 10–18 cm long and 6–8 cm broad; the basal pair of leaflets is much smaller, 5–8 cm long, with the margins of the leaflets entire. The male flowers are in drooping catkins 5–10 cm long, and the female flowers are terminal, in clusters of two to five, ripening in the autumn into a fruit with a green, semifleshy husk and a brown, corrugated nut (Abrams, 1997). Walnuts grows best in a Mediterranean climate, with a growing temperature in the range of 0–10°C during winter, a frost-free period during flowering, and temperatures below 38°C during summer.

2. CHEMISTRY

Juglone is a major chemical component present in almost all of parts of walnut tree. It is a toxic isomer of lawsone that acts as a growth stunner. A small amount of alpha-hydrojuglone is also present in walnut (Kong et al., 2008). L-arginine, a precursor to nitric oxide, is also present in significant quantity in walnut trees (Crews et al., 2005). Walnut kernels have high nutritional value, containing 63%–65% kernel oil, 3%–4% water, around 7% dietary fiber, 14%–15% protein, and 12%–14% carbohydrates. The majority of nut oils contain highly unsaturated fatty acids that are not beneficial to human health, but in contrast, walnut oil contains a large amount of polyunsaturated fatty acids like 58% linoleic acid and 14% alpha-linolenic acid (Jackson and Looney, 1999). Walnut is a rich source of phytochemicals and contains a significant amount of phenols, flavonoids, sterols, tannins, phospholipids, terpenes, essential fatty acids, sphingolipids quinines, hydrocarbons, triterpenes, quinones, and oils (Crews et al., 2005). It is also contains essential minerals α-tocopherol, cellulose, albumin, and mucilage. Gallic acid, ethyl gallate, pyrogallol, vanillic acid,

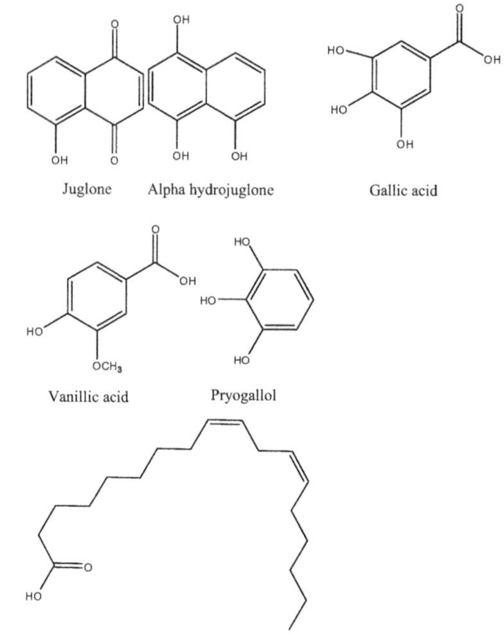

FIGURE 49.2 Important bioactive compounds of walnuts.

proto-catechuic acid, and p-hydroxybenzoic acid are the main phenolic acids present in walnut (Davis et al., 2007). Walnut seed oil contains sterols, acrid resin, myristin, free fatty acids, sterol esters, palmitin, and stearin (Calzoni et al., 1990). Important bioactive compounds of walnuts are shown in Fig. 49.2.

3. POSTHARVESTING TECHNOLOGY

Before the age of mechanization, the traditional September harvesting of walnuts consisted of shaking the trees by hand using long hooked poles to knock the nuts to the ground, from where they could be easily collected. This breaks off many of their points, thereby causing the production of new spurs, which will probably bear fruit-bearing flowers (Jackson and Looney, 1999). Today, the trees are shaken by a machine, while another machine uses vacuum suction to collect the fallen nuts. Commercial hot air dehydrators with blower fans circulate warm air to reduce the moisture of the walnuts between 12% and 20% to preserve their shelf life.

In past centuries, walnuts were simply left on drying racks away from the sun, until they were properly dried (Gandev, 2007). Postharvest treatment involves sieving the walnuts to remove sticks and leaves, hulling to remove hulls, washing, and then grading into different sizes. Hand sorting of nuts is used to remove damaged nuts. The nuts are then put into a drier to reduce the moisture content of the kernel from 10%–30% to around 5%–8%.

4. PROCESSING

Once the walnuts are harvested, precleaners are used to clean the walnuts so they are ready for hulling. A huller removes the outer green hull, and the nut is mechanically dehydrated (air dried) to the optimum 8% moisture level, preventing deterioration of the nut and protecting its quality during storage. Hulled walnuts are transported to nearby packing plants and are stored until needed for cracking (Mapelli et al., 1995). Drying may be done in the sun, in kilns, portable field bins, or in fruit bins. In all but the first method, heat applied from a gas or oil burner may be used up to 110°C. Above this temperature, kernel quality is decreased. Some growers prefer to use fan-forced air, without heating, to maintain nut quality at the expense of drying time. Packaging should be moisture-proof. Shelled products should be packaged in airtight, moisture-proof, opaque, or foil packages to maximize shelf life. Unroasted kernels are less likely to take up moisture than roasted kernels. Dried nuts last for about 4 months at room temperature before becoming rancid, and 1–2 years when stored in the freezer. The primary objectives of storage are to maintain the low water content attained after preliminary drying (to suppress enzyme activity, retain texture, and reduce microbial activity) and to limit exposure to O_2 (to minimize rancidity). The optimum temperature range for storage is 0–10°C (32–50°F), with the lower temperature being better.

5. VALUE ADDITION

Walnuts are delicious way to add extra flavor, nutrition, and crunch to a meal. Walnuts can 'be taken with onions, salt, and honey. Walnuts are a common food source and are used in cooking and baking (Rosengarten, 1984). Green walnuts are completely edible but are a bit sour. They form an ideal ingredient for pickles, jams, and marmalades (Payne, 1985).

6. USES

Walnuts have various benefits that are useful for humans. They are considered a major source of Ω-3fatty acid. They have been used as nuts from ancient time, and walnut oil is considered beneficial for our health, skin, and hair. In last few decades, walnut consumption has seen several changeable developments (Hocking, 1997). Several commercial industries use the finely ground walnut shells, in the aeronautical industry for polishing of metals, for removal of graffiti and paints, and for cleaning of electronic circuits, jet engines, and of automobiles. Walnut is used in face powder, exfoliating cleanser, and soaps. Shells of walnut are used in cleaning of stone, plastic, wood, fiberglass, and soft metals (Bown, 1995).

7. PHARMACOLOGICAL USES

7.1 Antimicrobial Activity

Walnut tree is a potent source of antimicrobial agent. It has been reported that different parts of this plant show strong antimicrobial activity against various fungal as well as bacterial strains. Aqueous extract of six different green husk varieties were found potentially active against *Bacillus subtilis, Cryptococcus neoformans, Escherichia coli, Bacillus cereus, Candida albicans, Pseudomonas aeruginosa*, and *Staphylococcus aureus* strains (Oliveira et al., 2008). Walnut ethyl acetate extract exhibited potential antifungal activity against the *Candida albicans* strains (Noumi et al., 2010). Powerful antimicrobial activity was shown from the pyroligneous acid, which is obtained from the branches of pyrolyzing walnut tree (Wei et al., 2010). Leaves and barks of walnut, especially its aqueous and other solvent extracts, showed antifungal potential against a vast range of fungal strains, determined by using agar dilution reddish method and disc diffusion method (Pereira et al., 2008). Acetone, methanol, chloroform, and ethyl acetate extract of bark exhibited antifungal activity against *Aspergillus niger, Alternaria alternata, Trichoderma virens, Fusarium solani, Pichia guiliermondii*, and *Pichia jadinii* (Ahmad et al., 1973).

7.2 Antioxidant Activity

Antioxidant potential of walnut seeds alcoholic extract assessed by *DPPH* (1,1-diphenyl-2-picrylhydrazyl) *radical* scavenging activity assays that showed positive results with high scavenging of free radicals (Carvalho et al., 2010). A similar pattern is mentioned in another literature report in which walnut stem and bark showed significant scavenging of free radicals (Ngoc et al., 2008). Moreover, the aqueous extract of walnut green husk exhibited in vivo antioxidant potential against cyclophosphamide-induced urotoxicity in the mice. Walnut extract restored all antioxidants significantly in the bladder (Bhatia et al., 2006).

7.3 Antimutagenic Activity

Walnuts extracts, i.e., aqueous and acetone, are quite effective and exhibited antimutagenic activity. These extracts were considered very efficient for inhibiting the 2-aminoflourene 2-AF–induced mutagenicity in strains of *Salmonella* (Kaur et al., 2003).

7.4 Anticancer Activity

Prostate cancer is one of the most common types of cancer reported in men. It has been reported that regular consumption of walnut seeds reduces the risk of prostate cancer in males (Spaccarotella et al., 2008). Juglone is one of the major components of walnut tree and is reported to have anticancer potential against melanoma cell lines (Aithal et al., 2009). In another study, it was reported that walnut extract and fractions with excess amount of juglone and ellagic acid impressively reduced the cancer cell proliferation. Walnut seed is a rich source of omega-3 that may decrease the risk of breast cancer (Hardman and Ion, 2008).

7.5 Cholesterol Lowering Activity

High cholesterol level in blood is one of the major factors of cardiovascular diseases. It has been reported that daily consumption of walnut seed significantly reduces the body cholesterol level (Sabate et al., 1993). Up to 70% risk of arteries clogging and stroke is reduced by consuming high amount of walnuts (De Lorgeril et al., 1994). Endothelial dysfunction coupled with the cardiovascular disease is restored by several food components. Walnut diet showed improved endothelial function in hypercholesterolemic patients. Walnut significantly minimized total cholesterol and low-density lipoprotein (LDL) cholesterol level (Ros et al., 2004). Similarly, it reduces the concentration of triglyceride and serum cholesterol (Rajaram et al., 2009).

7.6 Antidiabetic Activity

It has been reported that daily use of the diet rich in walnut seeds significantly reduced the sugar level in patients with diabetes mellitus type 2 (Gillen et al., 2005). Dietary fats can be influenced with the walnuts to produce the reductions in fasting insulin levels. The plant extract of walnuts inhibited efficiently alpha-glycosidase activity for both sucrose and maltase enzymes and showed no change in the insulin and glut-4 genes expression (Teimori et al., 2010).

7.7 Antiaging Effects

Walnut seed oil is a rich source of omega-3 natural antioxidants. Therefore, it has been used as a natural moisturizing and antiaging agent in beauty products (Espín et al., 2000). Moreover, the ethanolic extract of its leaves could be suggested as a good and new source of skin-lightning agents. Walnuts polyphenolic constituents showed significant skin lightning and were found more superior to the skin-lightning agents such as arbutin and ascorbic acid (Aitani and Shimoda, 2005).

7.8 Antiviral Activity

It has been illustrated that walnut leaf extracts prepared from 95% ethanol as well as ethyl acetate significantly reduced the production of tobacco mosaic virus (Zhai et al., 2006). Methanol extract of walnut at a minimum concentration of 1.5 μg/mL inhibited the Sindbis virus (Mouhajir et al., 2001).

7.9 Antianesthetic/Depressant Activity

The in vivo study results showed that walnut extracts significantly reduced depression. The walnut extracts open the closed coronary arteries of rabbit's heart (Auyong et al., 1963). The hexane extract of walnut fruit exhibited efficient antidepressant activity at 100 and 150 mg/kg doses of body weight of male Wistar rat compared with the standard drug fluoxetine. Tail suspension and forced swimming tests were used to evaluate the antidepressant activity (Rath and Pradhan, 2009).

7.10 Anti-Alzheimer Activity

In traditional medicinal systems, different parts of walnut have been used to reduce Alzheimer disease. It has been reported in different scientific reports that walnut extracts exhibited anti-Alzheimer activity. In one of the scientific reports, it was mentioned that walnut extract strongly defibrillated the beta-amyloid plaques (Chauhan et al., 2004). In another in vivo study, walnut significantly enhance the cognitive as well as motor performance in rats (Willis et al., 2009). Similarly, methanol extract of walnut has the ability to reduce the β-amyloid protein formation in Alzheimer patients. Moreover, it has been reported that the polyphenolic compounds present in walnut may attribute for its antiamyloidogenic activity (Chauhan et al., 2004).

7.11 Effects on Bone Health

Walnuts promote bone health. Walnut oil is a rich source of α-linoleic acid and Ω-3 fatty acids, which are essential for normal bone growth. It has been reported that the consumption of walnut in daily food stuff significantly reduced a marker of bone break down, i.e., N-telo-peptides serum concentration. Moreover, walnuts have osteoblastic potential that improves bone formation (Griel et al., 2007).

8. SIDE EFFECTS AND TOXICITY

Walnuts have been reported to have extensive toxic effects on different organisms, and it has been studied widely for its toxicity. The mechanism of walnut toxicity is well understood by either electrophilicity or by oxidative stress (Gant et al., 1986). Generally, juglones were more cytotoxic than the parent compound 1, 4-naphthoquinone (Doherty et al., 1987). Walnut promoted skin tumors in female Swiss mice that were pretreated with a subcarcinogenic dose of DMBA (di-methyl benzene anthracene), which was followed by juglone's dermal applications for a year. Though, the skin tumor was not observed, followed by the treatment of juglone without pretreatment with DBMA (Van Duuren et al., 1978). Juglone is considered an equivocal tumorigenic agent (thorax, lungs, respiration, appendages, and skin) by the Registry of Toxic Effects of Chemical Substances based on the 29-week skin painting study of mice.

References

Abrams, D., 1997. Sexual Symbolism and Merkavah Speculation in Medieval Germany: A Study of the Sod ha-egoz Texts. Mohr Siebeck, 13, 147.

Ahmad, S., Wahid, M.A., Bukhari, A., 1973. Fungistatic action of Juglans. Antimicrobial Agents and Chemotherapy 3, 436.

Aitani, M., Shimoda, H., 2005. The effect of ascorbic acid and arbutin on B16 melanoma cells. Japan Food Science 44, 58–63.

Aithal, B.K., Kumar, M.S., Rao, B.N., Udupa, N., Rao, B.S., 2009. Juglone, a naphthoquinone from walnut, exerts cytotoxic and genotoxic effects against cultured melanoma tumor cells. Cell Biology International 33, 1039–1049.

Auyong, T., Westfall, B., Russell, R., 1963. Pharmacological aspects of juglone. Toxicon 1, 235–239.

Bhatia, K., Rahman, S., Ali, M., Raisuddin, S., 2006. In vitro antioxidant activity of *Juglans regia* L. bark extract and its protective effect on cyclophosphamide-induced urotoxicity in mice. Redox Report 11, 273–279.

Bown, D., 1995. The Royal Horticultural Society Encyclopedia of Herbs & Their Uses. Dorling Kindersley Limited.

Calzoni, G., Speranza, A., Caramiello, R., Piccone, G., Zannini, P., 1990. Wall ultrastructure and biochemical features of the *Juglans regia* L. and *Juglans nigra* L. male gametophyte. Sexual Plant Reproduction 3, 139–146.

Carvalho, M., Ferreira, P.J., Mendes, V.S., Silva, R., Pereira, J.A., Jerónimo, C., Silva, B.M., 2010. Human cancer cell antiproliferative and antioxidant activities of *Juglans regia* L. Food and Chemical Toxicology 48, 441–447.

Chauhan, N., Wang, K., Wegiel, J., Malik, M.N., 2004. Walnut extract inhibits the fibrillization of amyloid beta-protein, and also defibrillizes its preformed fibrils. Current Alzheimer Research 1, 183–188.

Cogliastro, A., Gagnon, D., Bouchard, A., 1997. Experimental determination of soil characteristics optimal for the growth of ten hardwoods planted on abandoned farmland. Forest Ecology and Management 96, 49–63.

Crews, C., Hough, P., Godward, J., Brereton, P., Lees, M., Guiet, S., Winkelmann, W., 2005. Study of the main constituents of some authentic hazelnut oils. Journal of Agricultural and Food Chemistry 53, 4843–4852.

Dandekar, A., Leslie, C., McGranahan, G., Litz, R., 2005. *Juglans regia* walnut. Biotechnology of Fruit and Nut Crops 307–323.

Davis, L., Stonehouse, W., Mukuddem-Petersen, J., van der Westhuizen, F.H., Hanekom, S.M., Jerling, J.C., 2007. The effects of high walnut and cashew nut diets on the antioxidant status of subjects with metabolic syndrome. European Journal of Nutrition 46, 155–164.

De Lorgeril, M., Renaud, S., Salen, P., Monjaud, I., Mamelle, N., Martin, J., Guidollet, J., Touboul, P., Delaye, J., 1994. Mediterranean alpha-linolenic acid-rich diet in secondary prevention of coronary heart disease. The Lancet 343, 1454–1459.

Doherty, M.D.A., Rodgers, A., Cohen, G.M., 1987. Mechanisms of toxicity of 2-and 5-hydroxy-1, 4-naphthoquinone; absence of a role for redox cycling in the toxicity of 2-hydroxy-1, 4-naphthoquinone to isolated hepatocytes. Journal of Applied Toxicology 7, 123–129.

Espín, J.C., Soler-Rivas, C., Wichers, H.J., 2000. Characterization of the total free radical scavenger capacity of vegetable oils and oil fractions using 2, 2-diphenyl-1-picrylhydrazyl radical. Journal of Agricultural and Food Chemistry 48, 648–656.

Facciola, S., 1990. Cornucopia: A Source Book of Edible Plants.

Frederick, C., Tang, C., Wu, B., 2014. China Tree Nuts Annual. USDA, Foreign Agricultural Service. GAIN CH14051. http://gain.fas.usda.gov.

Galey, F., Whiteley, H., Goetz, T., Kuenstler, A., Davis, C., Beasley, V., 1991. Black walnut (*Juglans nigra*) toxicosis: a model for equine laminitis. Journal of Comparative Pathology 104, 313–326.

Gandev, S., 2007. Budding and grafting of the walnut (*Juglans regia* L.) and their effectiveness in Bulgaria (review). Bulgarian Journal of Agricultural Science 13, 683.

Gant, T.W., Doherty, M.d.A., Odowole, D., Sales, K.D., Cohen, G.M., 1986. Semiquinone anion radicals formed by the reaction of quinones with glutathione or amino acids. FEBS Letters 201, 296–300.

Gillen, L.J., Tapsell, L.C., Patch, C.S., Owen, A., Batterham, M., 2005. Structured dietary advice incorporating walnuts achieves optimal fat and energy balance in patients with type 2 diabetes mellitus. Journal of the American Dietetic Association 105, 1087–1096.

Gleeson, S.K., 1982. Heterodichogamy in walnuts: inheritance and stable ratios. Evolution 892–902.

Griel, A.E., Kris-Etherton, P.M., Hilpert, K.F., Zhao, G., West, S.G., Corwin, R.L., 2007. An increase in dietary n-3 fatty acids decreases a marker of bone resorption in humans. Nutrition Journal 6, 2.

Hardman, W.E., Ion, G., 2008. Suppression of implanted MDA-MB 231 human breast cancer growth in nude mice by dietary walnut. Nutrition and Cancer 60, 666–674.

Hemery, G.E., 2004. Genetic and Silvicultural Research Promoting Common Walnut (*Juglans regia*) for Timber Production in the United Kingdom.

Hocking, G.M., 1997. A dictionary of natural products. In: Terms in the Field of Pharmacognosy Relating to Natural Medicinal and Pharmaceutical Materials and the Plants, Animals and Minerals from Which They are Derived. Plexus Publishing, Inc.

Jackson, D., Looney, N.E., 1999. Temperate and Subtropical Fruit Production. CABI.

Kaur, K., Michael, H., Arora, S., Harkonen, P.L., Kumar, S., 2003. Studies on correlation of antimutagenic and antiproliferative activities of *Juglans regia* L. Journal of Environmental Pathology, Toxicology and Oncology 22.

Kolov, O., 1998. Ecological characteristics of the walnut-fruit forests of southern Kyrgyzstan. In: Blaser, J., Carter, J., Gilmour, D. (Eds.), Biodiversity and Sustainable Use of Kyrgyzstan's Walnut-Fruit Forests - Gland (IUCN), pp. 59–61.

Kong, Y.-h., Zhang, L., Yang, Z.-y., Han, C., Hu, L.-h., Jiang, H.-l., Shen, X., 2008. Natural product juglone targets three key enzymes from *Helicobacter pylori*: inhibition assay with crystal structure characterization. Acta Pharmacologica Sinica 29, 870.

LA, D., 1909. Contribution to the study of the genus Juglans (English translation by RE Cuendett). Bulletin de la Societe Dendrologique de France 11, 22–90.

Malvolti, M., Fineschi, S., Morgante, M., Vendramin, G., 1995. Mating system of a naturalized *Juglans regia* L. population Italy. Populations Genetics and Genetic Conservation of Forest Trees 305–308.

Manchester, S.R., 1987. The fossil history of the Juglandaceae. Monographs in Systematic Botany from the Missouri Botanical Garden 21, 1–137.

Mapelli, S., Lombardi, L., Brambilla, I., Iulini, A., Bertani, A., 1995. Walnut plant selection to hypoxic soil resistance. In: III International Walnut Congress, 442, pp. 129–136.

Menninger, E.A., 1977. Edible Nuts of the World. Horticultural Books.

Mouhajir, F., Hudson, J., Rejdali, M., Towers, G., 2001. Multiple antiviral activities of endemic medicinal plants used by Berber peoples of Morocco. Pharmaceutical Biology 39, 364–374.

Ngoc, T.M., Hung, T.M., Thuong, P.T., Kim, J.-C., Choi, J.S., Bae, K., Hattori, M., Choi, C.-S., Lee, J.S., Min, B.-S., 2008. Antioxidative activities of galloyl glucopyranosides from the stem-bark of *Juglans mandshurica*. Bioscience Biotechnology and Biochemistry 72, 2158–2163.

Noumi, E., Snoussi, M., Hajlaoui, H., Valentin, E., Bakhrouf, A., 2010. Antifungal properties of Salvadora persica and *Juglans regia* L. extracts against oral Candida strains. European Journal of Clinical Microbiology & Infectious Diseases 29, 81–88.

Oliveira, I., Sousa, A., Ferreira, I.C., Bento, A., Estevinho, L., Pereira, J.A., 2008. Total phenols, antioxidant potential and antimicrobial activity of walnut (*Juglans regia* L.) green husks. Food and Chemical Toxicology 46, 2326–2331.

Payne, T., 1985. California Walnuts and Light Foods. Cereal Foods World, USA.

Pereira, J.A., Oliveira, I., Sousa, A., Ferreira, I.C., Bento, A., Estevinho, L., 2008. Bioactive properties and chemical composition of six walnut (*Juglans regia* L.) cultivars. Food and Chemical Toxicology 46, 2103–2111.

Polunin, O., 1977. Trees and Bushes of Britain and Europe. Paladin.

Rajaram, S., Haddad, E.H., Mejia, A., Sabaté, J., 2009. Walnuts and fatty fish influence different serum lipid fractions in normal to mildly hyperlipidemic individuals: a randomized controlled study. American Journal of Clinical Nutrition 89, 1657S–1663S.

Rath, B., Pradhan, D., 2009. Antidepressant activity of *Juglans regia* L. fruit extract. International Journal of Toxicological and Pharmacological Research 1, 24–26.

Ros, E., Núñez, I., Pérez-Heras, A., Serra, M., Gilabert, R., Casals, E., Deulofeu, R., 2004. A walnut diet improves endothelial function in hypercholesterolemic subjects a randomized crossover trial. Circulation 109, 1609–1614.

Rosengarten Jr., F., 1984. The Book of Edible Nuts. Walker Publishing, New York, New York, USA, pp. 239–262.

Sabate, J., Fraser, G.E., Burke, K., Knutsen, S.F., Bennett, H., Lindsted, K.D., 1993. Effects of walnuts on serum lipid levels and blood pressure in normal men. New England Journal of Medicine 328, 603–607.

Spaccarotella, K.J., Kris-Etherton, P.M., Stone, W.L., Bagshaw, D.M., Fishell, V.K., West, S.G., Lawrence, F.R., Hartman, T.J., 2008. The effect of walnut intake on factors related to prostate and vascular health in older men. Nutrition Journal 7, 1.

Teimori, M., Montasser Kouhsari, S., Ghafarzadegan, R., Hajiaghaee, R., 2010. Study of hypoglycemic effect of *Juglans regia* leaves and its mechanism. Journal of Medicinal Plants 1, 57–65.

Van Duuren, B., Segal, A., Tseng, S., Rusch, G., Loewengart, G., Maté, U., Roth, D., Smith, A., Melchionne, S., Seidman, I., 1978. Structure and tumor-promoting activity of analogs of anthralin (1, 8-dihydroxy-9-anthrone). Journal of Medicinal Chemistry 21, 26–31.

Wei, Q., Ma, X., Dong, J., 2010. Preparation, chemical constituents and antimicrobial activity of pyroligneous acids from walnut tree branches. Journal of Analytical and Applied Pyrolysis 87, 24–28.

Williams, R.D., 1990. *Juglans nigra* L., black walnut. Silvics of North America 2, 391–399.

Willis, L.M., Shukitt-Hale, B., Cheng, V., Joseph, J.A., 2009. Dose-dependent effects of walnuts on motor and cognitive function in aged rats. British Journal of Nutrition 101, 1140–1144.

Zhai, M.-z., Zhang, F.-y., Wei, H., Wang, W., 2006. A study on the bioactivity of secondary metabolites from walnut green gull. Journal of Northwest Forestry University 21, 122.

CHAPTER

50

Yarrow

Fariha Ijaz[1], *Haq Nawaz*[1], *Muhammad Asif Hanif*[1], *Paulo Michel Pinheiro Ferreira*[2]

[1] Department of Chemistry, University of Agriculture, Faisalabad, Pakistan; [2] Department of Biophysics and Physiology, Laboratory of Experimental Cancerology, Federal University of Piauí, Teresina, Brazil

OUTLINE

https://doi.org/10.1016/B978-0-08-102659-5.00050-1

Copyright © 2020 Elsevier Ltd. All rights reserved.

1. BOTANY

1.1 Introduction

Yarrow (*Achillea millefolium*) (Fig. 50.1) is a perennial herb belonging to the family Asteraceae (Lakshmi et al., 2011). It is believed that yarrow is among those six medicinal plants whose pollens were found in a *Homo neanderthalensis* grave at Shanidar, dated to 65,000 BP (Leroi-Gourhan, 1975). *Achillea* contains around 130 species and varieties native to Europe and temperate areas of Asia and only some in North America. A great variability among the species due to morphology, growth, habit, flower, color, leaves, stems, and chemical composition is found (Saeidnia et al., 2011). *Achillea* species are insect pollinated, and cross-pollinating is common and it takes place between tetraploid individuals of *Achillea, Achillea collina, Achillea distans* ssp. *styriaca, and Achillea pratensis* (A. *millefolium* complex, Compositae) (Vetter et al., 1997). There is no plant

FIGURE 50.1 Dried yarrow.

outside the genus *Achillea* with the common name yarrow. *A. millefolium* is known by different folk names around the world. In English, it is commonly called yarrow or common yarrow. In Persian, *bumadaran* is a typical name used for many species of Achillea; *plumajillo* or *little feather* in New Mexico and Southern Colorado; *yarroway*, in eastern countries; *herbe des charpentiers* in French; *millefeuille schafgarbe* in German; *Achillea millefoglie* in Italy; *Biranjasif* or *Gomadar* in Urdu; *Gandrain* or *Puthkandain* Hindi *chiliofila* in Greek, and *yang shi cao* in Chinese (Saeidnia et al., 2011). Certainly, the most familiar yarrow is white yarrow (*A. millefolium*). Generally, the variety used has been superseded with enhanced trait cultivars varying in flavors, scents, and uses in conventional gardens. There are many special attractive cultivars available including *A. millefolium "Paprika"* and *A. millefolium*.

1.2 History, Origin, and Demography

The generic name, Achillea, comes from Achilles, an ancient hero. He frequently used yarrow for himself and for his soldiers. Millefolium means coming of 1000 leaves. Still, it is not possible to recognize whether usage since then has been continuous; it has certainly been persistent (Duke, 1986).

Pliny's *Natural History* showed a plant very similar to the yarrow, known by names *achilious, siderites,* and *millefolia*. The first two names were also used for various other species. Even some were different in the description. But these were considered very helpful in the wound recovery process. Three main surviving Old English medical texts from ca. 950–1000 CE were chiefly translations of prior Latin works, still including some influence from pre-Christian Northern European traditions, and all schools of thought suggest yarrow is a very helpful component in multiple recipes.

Although yarrow is grown in specific environmental conditions, the optimum conditions are found in countries with cooler highlands and in subtropical regions. Cool, little sun, and dry environment are key ecologic requirements for yarrow cultivation, though it is frost tolerant (Lakshmi et al., 2011).

1.3 Botany, Morphology, and Ecology

The plant can be shrubby or herbaceous, varying in size from 0.2 to 1 m (0.66–3.28 ft) in height, with dispersed rhizomatous expansion depending on the species. Leaves have hairiness and are feathery, cauline, and clasping. Its leaves are alternate and pinnately divided with a pointed whitish tip arranged at the base of stem. They are 5–20 cm long, with varying hairiness. Yarrow shows inflorescence as a flat top cluster.

Flowering heads have a diameter of 3–5 cm with varying colors and small pedunculate. In each capitulum, there are five outer florets with a corolla tube that is compressed. Fruit is without pappus, oblong, and compressed achene. Flower color ranges from white to pink. Most common varieties of yarrow are treated as perennial. A wide range of yarrow is grown in gardens as an ornamental plant, increasing the economic value globally. Yarrow is morphologically and chemically variable, and these variations appear to be influenced by environment. The origin, source, and growing conditions of yarrow have an impact on the plant's uses, flavors, aromas, and medical uses.

Yarrow requires temperate regions as it is a drought-tolerant specie. Optimum temperature for seed germination is 18–24°C, and it requires light for germination. Yarrow gives good response to poorly developed and well-developed soil (Lakshmi et al., 2011). It grows well in soil with pH ranging between 4.7 and 8.0. Yarrow has roots that can penetrate to 20 cm deep. It can withstand even on a poor, dry soil.

2. CHEMISTRY

There are many morphologic types of yarrow depending on different phenotype and genotype, which obviously change flavors. Color of yarrow leaves varies from grayish to green. White morphologic type of yarrow contains higher levels of essential oils and has a sweet smell like honey. Meanwhile, the pink morphologic type has higher content of flavonoids (Bimbiraitė et al., 2008). In general, yarrow contains flavonoids, sugars, energy value, proteins, saturated fatty acids, and other lipids, and the wild yarrow consists of more carbohydrates, unsaturated fatty acids, and organic and phenolic acids. Chemical compounds like isovaleric acid, asparagine, sterols, and salicylic acid have been isolated from yarrow, and aromas are associated with tannins and coumarins (Lakshmi et al., 2011).

Since yarrow contains larger quantities of fat and saturated fatty acids, it has higher caloric value (Dias et al., 2013). It is also a good source of vitamin C and minerals like aluminum, cobalt, potassium, copper, and calcium salts. Resins, gum, alkaloids, and tannins are also important constituents of yarrow. Ascorbic acid is also present in leaves of yarrow. Furthermore, essential oil composition is altered during maturation, and extraction methods significantly affect product composition (Costescu et al., 2014). *A. millefolium* has an attractive aroma because of the presence of essential and volatile oils, especially found in the flowers. Volatile oil contains flavonoids ranging from 0.05% to 0.07%, but yarrow does not present fixed oils (Saeidnia et al., 2011).

Yarrow is also well-known for flavonoids and alkaloids. In the United States, yarrow is unrestricted sold since it is labeled as GRAS (generally recognized as safe). Its sale is legalized as a supplement, and it is regulated by the Food and Drug administration. *A. millefolium* has good agronomic potentials as a natural dye in Iran, and recently, it was introduced as a new source of natural dye for wool dyeing due to the presence of the flavonoids luteolin V and apigenin VI (Kiumarsi et al., 2009).

Its chemical composition varies geographically. Essential oil content of yarrow is quite variable between species and cultivars and is thought to be related to growing conditions (Judzentiene and Mockute, 2010). When harvested at flowering, dried aerial parts show 0.10%–0.40%, and the inflorescence contains up to 0.50% of volatile oils (Costescu et al., 2014). The most important constituents of essential oils from yarrow are sabinene, β-pinene, 1,8-cineole, artemisia ketone, linalool, α-thujone, β-thujone, camphor, borneol, fenchyl acetate, bornyl acetate, (E)-β-caryophyllene, germacrene D, caryophyllene oxide, β-bisabolol, δ-cadinol, and chamazulene (Orav et al., 2006). Structures of yarrow bioactive compounds are shown in Fig. 50.2.

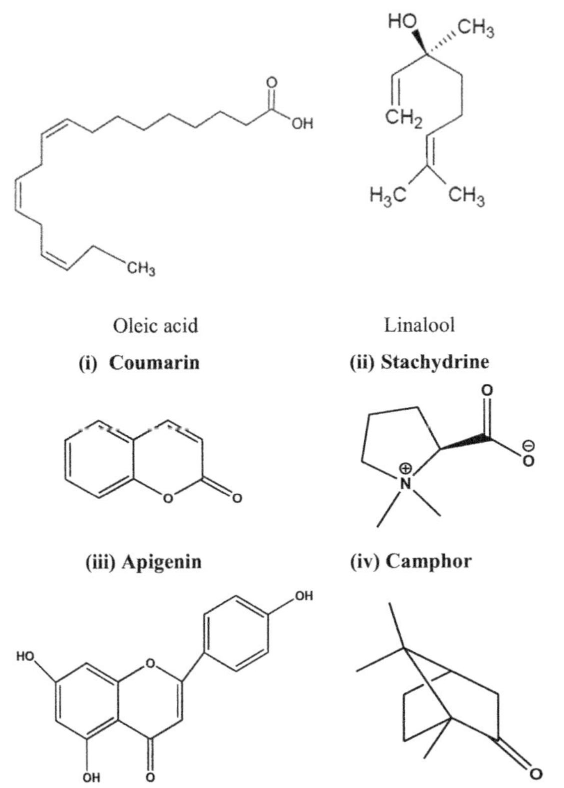

FIGURE 50.2 Structures of yarrow bioactive compounds.

3. POSTHARVEST TECHNOLOGY

The good harvesting times for yarrow collection are dry days when morning dew has evaporated and day temperature starts increasing because the plant exhibits more scent in the dawn and is full of essential oils. However, there is no data about fresh and dried yarrow in relation to flavor contents, but fresh leaves have more flavor and aroma. Yarrow leaves can be stored in a glass jar after drying in shade or freezing. Fresh flowers and leaves cannot be stored for long term. Therefore, leaves should be dried first for long storage. There is no data about broken leaves and reduced aroma while drying leaves. Yarrow leaves should be dried under shade. The proper timing for the harvesting of the younger leaves is in the start of spring season when they are soft. Later, they become very tough and dry so cannot be used for medicinal purposes. During the months of July to August, these leaves can be harvested because their specimens are in full flower, and the best time for the collection of flowers is peak of the summer season. Thus aerial parts that appear healthier can be broken off or snipped. To have full benefits from the flowers, they should be collected and used when they are fully opened, but not when these look yellow or brownish because at that time these are full of aromatic oil and tannins. When they are opened, they have vibrant look; however, there is no specific time for their harvesting, but they are more useful in early summer and spring. Even the roots of this magnificent species are applied as a painkiller. It is harvested in the fall.

For its safe storage, this plant should be dried carefully to avoid dew. Just like many herbs, the whole specimen other than roots can be bundled and cut before hanging upside-down to dry. It is the most easy and suitable way to obtain a large quantity of plant material. One can get plant leaves and flowers from the stem by this method, and these can be dried later. As dried mass is lighter in weight, drafts should be avoided. If this plant is placed in the sunlight, it fades, so it must be kept in a dark container or away from light sources.

4. PROCESSING

Most of medicinal properties of yarrow are commonly associated to oils, tinctures, vinegars, teas, and glycerites. For tincture, yarrow can be used fresh or dried, and for oils and salves, fresh specimens are used. Since it does not have much water, oil maintains quality for prolonged periods and exhibits low oxidation. Its leaves should be trimmed from the stems and spread out on a paper-lined tray to dry in a bright, airy space, out of direct light. Blossoms may be left on the stems and hung in small bunches upside-down in a very warm room. These dried flowers should be kept separately, and dried stems should be cut into small pieces before their storage in an airtight, dark glass container, labeled to show the contents and the date and place of harvest.

5. VALUE ADDITION

The unpleasant bitter taste of the yarrow can be avoided by cooking it with honey or any other sweet product. There is another useful product of yarrow that is formed by warm water by chewing the leaves, making a wonderful compress for wounds. It can also be applied to stings. It has a fragrant as well as preservative effect when some drops of the yarrow essential oil are added to yarrow glycerite. Its essential oil has best blending effect with oils of verbena, oak moss, and cedarwood. Although yarrow does not have a good taste, when its added with a mixture containing 70% glycerine and 30% water, its taste is improved. Its fresh leaves, obtained in early summer, add good taste and aroma to salads and homemade drinks.

For treatment of high blood pressure, yarrow is mixed with hawthorn, lime blossom, and mistletoe. Yarrow can be used instead of chamomile. It is found in many alcoholic preparations (Prajapati et al., 2003).

6. USES

Besides its pharmaceutical uses, it has many other beneficial effects in daily life: whenever it is subjected to heat, it emits a scent that spreads slowly in the room, so it is used as a house cleanser. It also used to prevent the hypertension because it is an important part of aromatherapy. Yarrow is also very important in religious aspects. There are many proofs that yarrow was used as a sacred plant in prehistorical times. Soup of the herbs were used as a sacred recipe that was presented on the Monday and Thursday before Easter. This soup contained nine holy healing herbs, including yarrow, and it was believed that this soup was the remedy of all the diseases and infections to be encountered in the coming year. In Italy, it is used as a perfect medicine for urinary, menstrual, stomach, and intestinal disorders. It is also found to lessen toothache. In some countries, it is used as a sedative. In Hungary, it has the trivial name cickafark and is applied in recovery of wounds, burns, and ailments. Some physicians report about yarrow that its tea is the best drug for kidney, bronchitis, and urinary problems and diarrhea. In Pakistan, India, Nepal, and China, yarrow has been also used to prevent bleeding and to cure sores, snakebites, inflammation, hemorrhoids, varicose veins, dysmenorrhea, and tuberculosis. It is a first vulnerary that staunches bleeding, and juices or dried powders can be applied in bleeding wounds. A strong tea can be used for internal bleeding, as its antiinflammatory action decreases swellings and heals inflamed cuts or wounds internally.

It has many effects on the nervous system, as it soothes and relaxes the nerves. It also makes the functioning of muscle, abdominal, and uterine systems more effective. Its action on the liver and gallbladder improves the activity of digestive system. Cystitis is an acute urinary infection, but yarrow acts as a marvelous drug due to its antiinflammatory and diuretic effects in this bad condition. It is considered a women's herb because it is used when menstruation is delayed, decreasing the pain in periods, lessening menstrual cramps, and reducing excessive bleeding. Thus its fresh juice is recommended as a tonic. Yarrow also performs a very vital role in the cardiac system. It improves the peripheral circulatory system by dilating the vessels of blood. So, it prevents cardiac patients from hypertension and severe angina. It is the best-known herb for producing a cooling sweat during high fever. As a result, fever is reduced. For the treatment of the rheumatism, its cleansing and toning activity along with antiinflammatory action is very useful. So yarrow is considered the best tonic to improve the functioning of the all main systems of the body with the passage of time. And it is used in the treatment of all acute problems. The essential oil of this magnificent herb is antiinflammatory in nature. It can easily handle inflammation of any kind, whether it is of nasal or respiratory tracts resulting from the cold, the digestive system caused by overeating of spicy food stuff, or acute inflammation of circulatory system resulted from a toxin (venom, narcotics, etc.) found in the bloodstream.

As it plays a vital role in better circulation of blood, it is very good for arthritic and rheumatic patients. Improved circulation prevents the storage of uric acid and other nitrogenous wastes in the muscles and joints, so it is good for the aforementioned patients. Diaphoretic properties of its oil help to eliminate these hazardous substances from the body by means of perspiration. This essential oil is also used as a relaxant for respiratory, nervous, and muscular systems.

When yarrow is applied to oily skin and hair, it dries these. Yarrow is also used in facial steam, washing, baths, or warm compresses. It beautifies our skin as it removes blackheads and ugly patches of acne. For porous skin, it is applied to tighten the skin. Along with the pharmacological uses, its aromatic effects create a pleasant atmosphere of the room. For restful sleep, many people use its dilute extract in tea before going to the bed. It is used in sweet dreams sleep tincture for times when falling asleep is difficult. Yarrow essential oil can reduce bleeding or hemorrhage by contracting the blood vessels. The essential oil of yarrow has astringent effects and increases digestion and secretion of digestive fluids (acids and bile). Moreover, it is involved in improving the working of the liver and digestive organs.

A tea bath of this herb is considered a reliable drug for rheumatic joints and varicose veins. Its roots can be used to cure toothaches and are placed in homemade drinks for chewing when they are needed. Yarrow is used against *Shigella* and other bacteria that are sources of infectious diarrhea. Yarrow has been used over the course of history to influence the blood circulatory system. It both stimulates blood platelets and breaks down blood clots. It dilutes the blood and allows the congestion to move in the portal vein, the pancreas, and the lungs.

It is one of the bitter herbs in Vermouth and is used as a hops substitute in beer brewing. Various physicians suggest yarrow for feeble situations of the digestive tract, including "precarious appetite, passive looseness of bowels, and the acute nervous prostration." Yarrow can also treat poor appetite due to very low digestive secretions and general inflammations that can make the tissues not function well. Yarrow brings blood to the surface of the body, thereby inducing sweating. It may lower blood pressure by moving blood to the skin and reducing the burden on the heart. It is very specific for all kinds of skins including hot, dry, constricted skin. From prehistoric times, herbalists have used yarrow for hemorrhage.

Yarrow lessens the flow of blood and removes painful clots. Yarrow is used to protect the skin from excessive sun and hot wind. It makes a wonderful facial steam that improves the complexion. The aromatics in yarrow open the lungs or sinuses and cool inflamed tissue. The aromatic constituent of this plant facilitates the respiratory system by opening the lungs and most of the inflamed tissues. Yarrow is used for physiologic and spiritual protection, especially to descriptive boundaries.

In Chinese medical literature, *Achillea* may have three main activities: clear external wind (diaphoretic), tonic deficiency, and clear heart phlegm (antihypertension). Most of its therapeutic practices have been confirmed by new experimental and clinical diagnosis.

The consumption of herbal teas from different species of *Achillea*, especially for treatment of the gastrointestinal tract, is common in folk medicine in South Asia. The bitter compounds that the tongue detects are due to flavonoids such as saponins and unpleasant tasting but powerful alkaloids like achilleine, trigonelline, and betonicine.

Use of different species of *Achillea* as tea increases the activity of the digestive system and makes the gastrointestinal tract more effective and is considered a native medicine in central Southern Asia. Yarrow oil has antioxidant and antimicrobial characteristics. The volatile oils work as antibacterial, antiinflammatory, and diuretic agents. The main components of tea tannins are aggressive astringents. The alkaloids possess two major properties: hypotensive and hypoglycemic. Yarrow even possesses coumarin in its cells, which has as an antithrombotic to lessen high blood pressure conditions. The bitterness of its extracts is due to the presence of the flavonoids like saponins and specific alkaloids like achilleine and betonicine. Actually, these are proven evidences for yarrow's actions in the digestive system, tissues, and the bloodstream (Lakshmi et al., 2011).

It also deals with ear infection of domestic pets (Lans et al., 2008). Moreover, an extract for treatment and diagnosis of hysteromyoma, synthesized from *A. millefolium* along with *Inula, Calami, Urtica, Arnica, Capsella* and some other useful medicinal plants, has been reported. The medicine is reported to be useful in treatment of hysteromyoma, especially hormone-dependent tumor. The pharmacological properties of *A. millefolium* are recognized worldwide, and the plant is included in the national pharmacopoeias of most countries (Saeidnia et al., 2011). The immunostimulant action of yarrow or its mechanism of action comprises three main types: phagocytosis activation, fibroblast stimulation, and the enhancement of respiratory activity that results in augmentation of leukocyte mobility. The root extract of the yarrow plant improves NK cell numbers by enhancing their production at the site of bone formation, resulting in the increase of numbers of spleen cells and also an increase in antitumor and cytolytic activity in aged mice.

7. PHARMACOLOGICAL USES

7.1 Antimicrobial Activity

Yarrow exhibits antibacterial activity against *S. typhimurium* and *S. aureus* with predicted MICs (minimum inhibitory concentrations) value on the order of tens of μg/mL or tens of mg/mL, respectively. The antibacterial properties of the extracts of yarrow are indicated by appearance of growth inhibition of *Escherichia coli*, whereas the aqueous extracts of flower, leaves, roots, shoots do not show any promising result. Furthermore study of essential oil and methanolic extracts did not show any results again (Lakshmi et al., 2011). A 50% ethanol extract of the flowers inhibits the growth of *Shigella dysenteriae*, but not that of *E. coli* or *Salmonella enteritidis*, in vitro at a concentration of 50 μL/agar plate (Caceres et al., 1990). It has been theorized that phenolic compounds such as flavonoids and phenol carbonic acids may underlie the experimental antimicrobial activity (Barel et al., 1991). The antimicrobial activity of essential oils in particular has formed the basis of many applications of different aromatic plants. This aspect assumes a particular relevance due to an increased resistance of some bacterial strains to the most common antibiotics and antimicrobial agents for food preservation.

7.2 Antioxidant Activity

Yarrow oil constituents strongly reduced the diphenyl picrylhydrazyl radical (IC(50) = 1.56 μg/mL) and thus showed hydroxyl radical scavenging effect in the Fe^{3+}-EDTA-H_2O_2 deoxyribose system (IC(50) = 2.7 μg/mL). It also prevents the nonenzymatic lipid-based peroxidation of rat liver homogenate (IC (50) = 13.5 μg/mL). Antioxidant activity of the extract is also shown due to the polar nature (Vardar-Ünlü et al., 2003).

7.3 Anticancer Activity

Yarrow exhibits good anticancer activity. Flavonoids and sesquiterpenoids present in yarrow show intense antiproliferative effects against mouse P-388 leukemia cells and cervix epithelial adenocarcinoma (HeLa), breast epithelial adenocarcinoma (MCF-7), and skin epidermoid carcinoma (A431) cells (Csupor-Löffler et al., 2009).

7.4 Antiinflammatory Activity

Oil of this magnificent plant has good antiinflammatory effects and is generally consumed to treat acute rheumatism, muscle cramps, and heal severe cuts or wounds. Its phytochemical components with flavonoids influence prostaglandin stimulation and are responsible for antiinflammatory activity. Azuline comprises half of the weightage of whole composition of the yarrow, and it is also known as a powerful antiinflammatory agent (Tunon et al., 1995).

7.5 Antihypertensive Activity

Yarrow shows a high angiotensin transforming enzyme inhibition and is practiced in the treatment of hypertension.

7.6 Cholagogue Activity

The cholagogue activity of yarrow is due to the presence of unsaturated fatty acids (Lakshmi et al., 2011).

7.7 Antinociceptive Activity

The hydroalcoholic extracts of *A. millefolium* L. (AM) and *Artemisia vulgaris* L. (AV) were analyzed by hot plate, writhing formalin, along with intestinal transit tests in an attempt to make sure of its prehistoric consumption as analgesic, antiinflammatory, and antispasmodic mediator.

Furthermore, none of the extracts produced even minor differences in the intestinal transit in mice, nor in the response time in the hot plate or in the immediate or late responses in the formalin-based test. In HPLC/DAD (high-performance liquid chromatography with a diode-array detector) analyses, results of both hydroalcoholic extracts showed the same flavonoid glycoside as a principal constituent, which was recognized as rutin. A high percentage of caffeic acid derivatives were also found in both extracts. The main differences were observed at 240 nm: AM had a higher content of rutin, while in AV the hydroxy benzoic acid derivative was the major part (Pires et al., 2009).

7.8 Antipyretic Activity

Oral intake of a hot diluted extract or the juice of the aerial parts of the magnificent plant was given to rabbits, at a dose of 30–55 g/kg, respectively, and was found to lessen body temperature, while the 95% of the ethanol extract was not effective.

7.9 Abortifacient Activity

Yarrow also has prehistorically been used as an abortifacient, emmenagogue, contraceptive, and for stimulating uterine wall contractions. For this reason, it is not recommended to consume in pregnancy (Boswell-Ruys et al., 2003).

7.10 Antiulcer Activity

The hot aqueous extract of *A. millefolium* exhibited effectiveness in inhibiting the gastric mucosa against acute gastric lesions stimulated by ethanol and indomethacin and in curing chronic gastric lesions induced by formic acid (Saeidnia et al., 2011).

8. SIDE EFFECTS AND TOXICITY

It is very safe to take the yarrow in a medicinal amount. In allergic patients, when yarrow comes in contact with the skin, it might cause irritation.

References

Barel, S., Segal, R., Yashphe, J., 1991. The antimicrobial activity of the essential oil from *Achillea fragrantissima*. Journal of Ethnopharmacology 33, 187–191.

Bimbiraitė, K., Ragažinskienė, O., Maruška, A., Kornyšova, O., 2008. Comparison of the chemical composition of four yarrow (*Achillea millefolium* L.) morphotypes. Biologija 54, 208–212.

Boswell-Ruys, C.L., Ritchie, H.E., Brown-Woodman, P.D., 2003. Preliminary screening study of reproductive outcomes after exposure to yarrow in the pregnant rat. Birth Defects Research Part B: Developmental and Reproductive Toxicology 68, 416–420.
Caceres, A., Cano, O., Samayoa, B., Aguilar, L., 1990. Plants used in Guatemala for the treatment of gastrointestinal disorders. 1. Screening of 84 plants against enterobacteria. Journal of Ethnopharmacology 30, 55–73.
Costescu, C.I., Rădoi, B.P., Hădărugă, N.G., Gruia, A.T., Riviș, A., Pârvu, D., David, I., Hădărugă, D.I., 2014. Obtaining and characterization of *Achillea millefolium* L. extracts. Journal of Agroalimentary Processes and Technologies 20, 142–149.
Csupor-Löffler, B., Hajdú, Z., Zupkó, I., Réthy, B., Falkay, G., Forgo, P., Hohmann, J., 2009. Antiproliferative effect of flavonoids and sesquiterpenoids from *Achillea millefolium* sl on cultured human tumour cell lines. Phytotherapy Research 23, 672–676.
Dias, M.I., Barros, L., Dueñas, M., Pereira, E., Carvalho, A.M., Alves, R.C., Oliveira, M.B.P., Santos-Buelga, C., Ferreira, I.C., 2013. Chemical composition of wild and commercial *Achillea millefolium* L. and bioactivity of the methanolic extract, infusion and decoction. Food Chemistry 141, 4152–4160.
Duke, J.A., 1986. Isthmian Ethnobotanical Dictionary.
Judzentiene, A., Mockute, D., 2010. Essential oil composition of two yarrow taxonomic forms. Open Life Sciences 5, 346–352.
Kiumarsi, A., Abomahboub, R., Rashedi, S.M., Parvinzadeh, M., 2009. *Achillea millefolium*, a new source of natural dye for wool dyeing. Prog. Color Colorants Coat 2, 87–93.
Lakshmi, T., Geetha, R., Roy, A., Kumar, S.A., 2011. Yarrow (*Achillea millefolium* linn). A herbal medicinal plant with broad therapeutic use-A review. International Journal of Pharmaceutical Sciences Review and Research 9, 136–141.
Lans, C., Turner, N., Khan, T., 2008. Medicinal plant treatments for fleas and ear problems of cats and dogs in British Columbia, Canada. Parasitology Research 103, 889–898.
Leroi-Gourhan, A., 1975. The flowers found with Shanidar IV, a Neanderthal burial in Iraq. Science 190, 562–564.
Orav, A., Arak, E., Raal, A., 2006. Phytochemical analysis of the essential oil of *Achillea millefolium* L. from various European Countries. Natural Product Research 20, 1082–1088.
Pires, J.M., Mendes, F.R., Negri, G., Duarte-Almeida, J.M., Carlini, E.A., 2009. Antinociceptive peripheral effect of *Achillea millefolium* L. and *Artemisia vulgaris* L.: both plants known popularly by brand names of analgesic drugs. Phytotherapy Research 23, 212–219.
Prajapati, N.D., Purohit, S., Sharma, A., Kumar, T., 2003. Medicinal Plants, third ed. Agrobios published company, India, p. 353.
Saeidnia, S., Gohari, A., Mokhber-Dezfuli, N., Kiuchi, F., 2011. A review on phytochemistry and medicinal properties of the genus *Achillea*. Daru Journal of Pharmaceutical Sciences 19.
Tunon, H., Olavsdotter, C., Bohlin, L., 1995. Evaluation of anti-inflammatory activity of some Swedish medicinal plants. Inhibition of prostaglandin biosynthesis and PAF-induced exocytosis. Journal of Ethnopharmacology 48, 61–76.
Vardar-Ünlü, G., Candan, F., Sökmen, A., Daferera, D., Polissiou, M., Sökmen, M., Dönmez, E., Tepe, B., 2003. Antimicrobial and antioxidant activity of the essential oil and methanol extracts of Thymus pectinatus Fisch. et Mey. Var. pectinatus (Lamiaceae). Journal of Agricultural and Food Chemistry 51, 63–67.
Vetter, S., Franz, C., Giasl, S., Kastner, U., Saukel, J., Jurenitsch, J., 1997. Inheritance of sesquiterpene lactone types within the *Achillea millefolium* complex (Compositae). Plant Breeding 116, 79–82.

Index

'*Note:* Page numbers followed by "f" indicate figures.'

C

D

F

H

M

Q

U

CPI Antony Rowe
Eastbourne, UK
December 16, 2019